Food Microbiology

Food Microbiology

Edited by **Dorothy Green**

R CALLISTO
REFERENCE

New York

Published by Callisto Reference,
106 Park Avenue, Suite 200,
New York, NY 10016, USA
www.callistoreference.com

Food Microbiology
Edited by Dorothy Green

International Standard Book Number: 978-1-63239-642-6 (Hardback)

The publisher's policy is to use permanent paper from mills that operate a sustainable forestry policy. Furthermore, the publisher ensures that the text paper and cover boards used have met acceptable environmental accreditation standards.

Printed in the United States of America.

Contents

Preface

Over the recent decade, advancements and applications have progressed exponentially. This has led to the increased interest in this field and projects are being conducted to enhance knowledge. The main objective of this book is to present some of the critical challenges and provide insights into possible solutions. This book will answer the varied questions that arise in the field and also provide an increased scope for furthering studies.

Food microbiology focuses on microbes that have both favourable and undesirable effects on the safety and quality of food. This book aims to understand the multiple branches that fall under the discipline of food microbiology such as food-borne microbes and their interactions with various foods, industrial and biotechnological exploitation of microbial diversity, microbiology of fermented foods, evolutionary dynamics of food-borne pathogenic microbes, genomics and functional genomics of pathogenic and value-adding technological microbes, development of probiotics as food supplements, etc. It also delves into its applications to food and process optimization, risk assessment, etc. This book covers the unexplored aspects of this area and serves as a reference text for the professionals and students engaged in this field.

I hope that this book, with its visionary approach, will be a valuable addition and will promote interest among readers. Each of the authors has provided their extraordinary competence in their specific fields by providing different perspectives as they come from diverse nations and regions. I thank them for their contributions.

Editor

Hha has a defined regulatory role that is not dependent upon H-NS or StpA

Carla Solórzano[1], Shabarinath Srikumar[2], Rocío Canals[2], Antonio Juárez[1,3], Sonia Paytubi[1]* and Cristina Madrid[1]*

[1] Departament de Microbiologia, Universitat de Barcelona, Barcelona, Spain, [2] Institute of Integrative Biology, University of Liverpool, Liverpool, UK, [3] Institut de Bioenginyeria de Catalunya, Parc Científic de Barcelona, Barcelona, Spain

Edited by:
Dongsheng Zhou,
Beijing Institute of Microbiology
and Epidemiology, China

Reviewed by:
Yanping Han,
Beijing Institute of Microbiology
and Epidemiology, China
Shane Dillon,
Trinity College Dublin, Ireland

*Correspondence:
Sonia Paytubi and Cristina Madrid,
Departament de Microbiologia,
Universitat de Barcelona,
Avenida Diagonal 643,
08028 Barcelona, Spain
s.paytubi@gmail.com;
cmadrid@ub.edu

The Hha family of proteins is involved in the regulation of gene expression in enterobacteria by forming complexes with H-NS-like proteins. Whereas several amino acid residues of both proteins participate in the interaction, some of them play a key role. Residue D48 of Hha protein is essential for the interaction with H-NS, thus the D48N substitution in Hha protein abrogates H-NS/Hha interaction. Despite being a paralog of H-NS protein, StpA interacts with HhaD48N with higher affinity than with the wild type Hha protein. To analyze whether Hha is capable of acting independently of H-NS and StpA, we conducted transcriptomic analysis on the *hha* and *stpA* deletion strains and the *hha*D48N substitution strain of *Salmonella* Typhimurium using a custom microarray. The results obtained allowed the identification of 120 genes regulated by Hha in an H-NS/StpA-independent manner, 38% of which are horizontally acquired genes. A significant number of the identified genes are involved in functions related to cell motility, iron uptake, and pathogenicity. Thus, motility assays, siderophore detection and intra-macrophage replication assays were performed to confirm the transcriptomic data. Our findings point out the importance of Hha protein as an independent regulator in *S.* Typhimurium, highlighting a regulatory role on virulence.

Keywords: Hha, H-NS, StpA, *Salmonella*, motility, pathogenicity island, gene regulation

Introduction

One of the relevant features of bacterial cells is their ability to sense and adapt to a usually rapidly changing environment. Bacteria have developed several mechanisms to detect and transduce external stimuli resulting in modifications of the gene expression pattern. Nucleoid-associated proteins play relevant roles in bacteria, both organizing the chromosome and influencing gene expression. A well-known example is the nucleoid-associated protein H-NS. The H-NS protein is widely distributed within Gram-negative bacteria and is one of the best characterized examples of a modulator that influences gene expression in response to environmental stimuli (Dorman, 2007). In *Escherichia coli*, up to 5% of the genes are subjected to H-NS regulation (Hommais et al., 2001). In *Salmonella* Typhimurium, approximately 9% of the genes show an H-NS-dependent regulation (Lucchini et al., 2006; Navarre et al., 2006). Moreover, a significant 77% of temperature-dependent genes described in *S.* Typhimurium are modulated by H-NS (Ono et al., 2005). The H-NS protein consists of an N-terminal dimerization domain separated from a C-terminal DNA-binding domain by a linker region (Tendeng and Bertin, 2003). H-NS binding sites typically show curvature given by A-T rich sequences, a common trait found at promoters (Fang and Rimsky, 2008). The H-NS protein is not only capable of interacting with DNA but also with itself and other proteins. One of the best known examples of H-NS interacting protein is its paralog, StpA, that can form homomeric or heteromeric complexes *in vivo* mediated by the N-terminal domains of the proteins (Williams et al., 1996; Cusick and Belfort, 1998; Free et al., 1998; Johansson and Uhlin, 1999). Oligomerization of H-NS, forming extended protein filaments along target sequences, is critical for the regulatory role of the protein (Spurio et al., 1997; Dame et al., 2000; Badaut et al., 2002; Stella et al., 2005).

The Hha family of nucleoid associated proteins includes a group of sequence related low-molecular mass proteins. They are uniquely encoded by members of the Enterobacteriaceae (Madrid et al., 2007b) and are involved in modulation of virulence gene expression in response to environmental cues (Madrid et al., 2007a). Good examples of that are the regulation of α-haemolysin and the *esp* operons in *E. coli* (Nieto et al., 1991; Sharma and Zuerner, 2004; Sharma et al., 2005), and the modulation of *hilA* and SPI-2 virulence genes in *S.* Typhimurium (Fahlen et al., 2001; Coombes et al., 2005; Silphaduang et al., 2007; Vivero et al., 2008; Queiroz et al., 2011; Fàbrega and Vila, 2013). Genes coding for such proteins are present in one or more copies per chromosome or in transmissible elements such as conjugative plasmids (Madrid et al., 2007a). In addition to *hha*, the genome of *Salmonella* contains the *ydgT* gene, which encodes an Hha paralog (Paytubi et al., 2004).

Hha interacts with H-NS to fine-tune its modulatory activity (Nieto et al., 2000; Madrid et al., 2002b, 2007a; Vivero et al., 2008). In addition to modulate housekeeping genes, H-NS plays a relevant role in silencing large stretches of DNA that may have been acquired by lateral gene transfer (Lucchini et al., 2006; Navarre et al., 2006; Oshima et al., 2006). Hence, H-NS appears as a regulatory element facilitating the incorporation of horizontally acquired genes (HGT). Previous data have shed some light on the field indicating that whereas H-NS homo-oligomers modulate expression within the core genome, the preferential target of H-NS/Hha complexes are HGT genes (Baños et al., 2009). In other words, Hha-like proteins interact with H-NS allowing H-NS to discriminate between HGT and core genome and thus silencing xenogeneic DNA (Baños et al., 2009).

H-NS amino acid residues responsible for the interaction with Hha are located mainly within helices H1 and H2 of the H-NS N-terminal domain (García et al., 2006), being the arginine residue at position 12 (R12) of H-NS critical for Hha binding. Indeed, mutagenesis of amino acid R12 was shown to dramatically reduce the interaction of H-NS with Hha (García et al., 2006).

The three dimensional structure of Hha consists of four α-helical segments separated by loops (Yee et al., 2002). On the subject of the Hha protein, amino acid residues interacting with H-NS are scattered along the full length molecule (Nieto et al., 2002; García et al., 2005). Recent studies describe a non-homogeneous charge distribution of the Hha-like proteins, i.e., its positively and negatively charged residues cluster on opposing surfaces of the molecule (Paytubi et al., 2011; Ali et al., 2013).

The predominantly basic surface of Hha points away from H-NS, indicating that these positively charged residues are essential for the regulatory control. This suggests that Hha could potentially provide an additional interaction surface for the nucleoprotein complex (Ali et al., 2013). On the other hand, site directed mutagenesis of conserved negatively charged residues on Hha allowed the identification of residues E25 and D48 as critical for Hha-H-NS interaction (de Alba et al., 2011). Removal of the negative charge at position 25 severely compromises the interaction with H-NS although it does not suppress it, whilst aspartic acid at position 48 is strictly required for the complex formation. The mutagenesis of aspartic acid at position 48 totally impairs the binding of Hha to the N-terminal domain of H-NS and in consequence its capability to silence *hlyABCD* expression (de Alba et al., 2011).

The lack of a clear DNA binding domain in Hha has suggested that the interactions with H-NS/StpA are required for Hha to modulate gene expression. However, it cannot be ruled out that Hha may modulate gene expression independently of H-NS/StpA. In this work we identify the set of genes of *Salmonella enterica* serovar Typhimurium SV5015 that are under the regulation of the Hha protein in an H-NS/StpA-independent manner.

Materials and Methods

Bacterial Strains, Plasmids, and Culture Media
Bacterial strains and plasmids used in this work are listed in **Table 1**. Cells were grown at 37°C in Luria–Bertani (LB) medium (10 g l^{-1} NaCl, 10 g l^{-1} tryptone, 5 g l^{-1} yeast extract) or LPM medium [5 mM KCl, 7.5 mM $(NH_4)_2SO_4$, 0.5 mM K_2SO_4, 38 mM glycerol (0.3% v/v), 0.1% casamino acids, 8 μM $MgCl_2$, 337 μM PO_4^{-3} and 80 mM MES (for titration to pH 5.8)] (Coombes et al., 2004). The antibiotics used were kanamycin (Km) 50 μg ml^{-1}, ampicillin (Ap) 100 μg ml^{-1} and chloramphenicol (Cm) 25 μg ml^{-1}.

Genetic Manipulations and Molecular Techniques
The *hha* gene from SV5015 was amplified by PCR using the oligonucleotides pairs HhaS_BamHI and HhaS_HindIII. The generated fragment, containing the putative promoter and 67 nucleotides of the non-coding sequence downstream *hha* was cloned into pACYC184, generating plasmid pACYC184-hhaSV. The point mutation *hha*D48N was introduced into pACYC184-hhaSV using the QuikChange site-directed mutagenesis kit (Stratagene) and the pair of oligonucleotides SalD48N_For/SalD48N_Rev, resulting in plasmid pACYC184-hhaD48NSV. The chromosomal *hha*D48N point mutant in strain SV5015 was constructed using the "stitch PCR" technique. This technique was used to "stitch" two DNA fragments together, the *hha*D48N and the kanamycin resistance genes. The *hha*D48N was PCR amplified from plasmid pACYC184-hhaD48NSV using primer pairs #1/#2. The kanamycin resistance cassette flanked by FRT sites was PCR amplified from plasmid pKD4 separately using oligonucleotides #3 and #4. The two PCR products were

TABLE 1 | Strains and plasmids used in this work.

Strains/Plasmids	Characteristics	Reference
Escherichia coli		
BL21 (DE3) Δhns	hsdS, gal, (λclts857, ind1, Sam7, nin5, lac-UV5-T7 gene1) Δhns::Km	Zhang et al. (1996)
DH5α	fhuA2 lac(del)U 169 phoA glnV44 Φ80' lacZ(del)M15 gyrA96 recA1 relA1 endA1 thi-1 hsdR17	Taylor et al. (1993)
Salmonella Typhimurium		
SV5015	SL1344 his+	Vivero et al. (2008)
SV5015H	SV5015 Δhha::Cm	Vivero et al. (2008)
SV5015S	SV5015 ΔstpA::Cm	Hüttener, M.
SV5015HY	SV5015 Δhha ΔydgT::Cm	This study
SV5015Y	SV5015 ΔydgT::Cm	This study
SV5015HYS	SV5015 Δhha ΔydgT ΔstpA::Cm	This study
SV5015D	SV5015 hhaD48N	This study
SV5015SY	SV5015 ΔstpA ΔydgT::Cm	This study
SV5015DY	SV5015 hhaD48N Δydgt::Cm	This study
SV5015DYS	SV5015 hhaD48N ΔydgT ΔstpA::Cm	This study
Plasmids		
pET15bHisHha	pET15b + 6xHis-hha; Apr	Cordeiro et al. (2008)
pET15bHisHhaD48N	pET15b + 6xHis-hhaD48N; Apr	de Alba et al. (2011)
pT7-stpA	pT7-5 + stpA; Apr	Zhang et al. (1995)
pLysS	T7 lysozyme; Cmr; ori p15A	Studier and Moffatt (1986)
pACYC184	Cloning vector Tcr, Cmr	Rose (1988)
pACYC184hhaSV	pACYC184 + hha; Cmr	This study
pACYC184hhaD48NSV	pACYC184 + hhaD48N; Cmr	This study
pIC-ssrA2	pIC552 ssrA::lacZ, Apr	Gaviria, T.
pGEM-T easy	Vector, Apr	Promega
pKD4	oriRγ Kmr Apr	Datsenko and Wanner (2000)
pKD3	oriRγ Cmr Apr	Datsenko and Wanner (2000)
pKD46	oriR 101, repA101 (ts), araBp-gam-bet-exo (Red helper plasmid,Ts; Apr)	Datsenko and Wanner (2000)
pCP20	λcI857 (ts), ts-rep (Recombinase FLP, Ts) Apr, Cmr	Cherepanov and Wackernagel (1995)

10 mM arabinose. Recombinant clones were selected at 37°C in LB medium containing kanamycin, tested for the absence of pKD46, and confirmed by PCR. The chromosomal deletion of *ydgT* gene was generated by the λ Red recombinant method as previously described (Datsenko and Wanner, 2000) resulting in strain SV5015Y. Oligonucleotides YdgT_P1 and YdgT_P2 were used to amplify chloramphenicol resistance gene from pKD3 plasmid. All the recombinants obtained were checked by PCR with YDGT-P1UP and YDGT-P2DOWN oligonucleotides.

When needed, the kanamycin and chloramphenicol resistance genes were removed using pCP20 plasmid (Cherepanov and Wackernagel, 1995). SV5015Y strain was used as a donor to transduce the *ydgT::Cm* mutation into strains SV5015H, SV5015D, and SV5015S using phage P22 HT (Sternberg and Maurer, 1991) generating strains SV5015HY, SV5015DY, and SV5015SY, respectively. Additionally, SV5015S strain was used to transfer the *stpA::Cm* mutation to the double mutant strains, obtaining strains SV5015HYS and SV5015DYS. The sequence of all oligonucleotides used in this work is indicated in **Table 2**.

Pull-Down Experiments

BL21 (DE3) Δ*hns* cells carrying plasmids pET15bHisHha, pET15bHisHhaD48N, or pLysS/pT7-stpA, were grown in 500 ml of LB at 37°C with shaking until OD$_{600nm}$ of 0.4. Following induction with 0.5 mM isopropyl-β-D-thiogalactopyranoside (IPTG), cells were grown for 2 h under the same conditions. Cells were then harvested by centrifugation at 11000 × *g* at 4°C for 30 min and resuspended in 20 ml of lysis buffer A (20 mM

TABLE 2 | Oligonucleotides used in this work.

Oligonucleotides	Sequence 5'–3'
HhaS_BamHI	GACGGATCCCAAAAATGGCGTAAATCGG
HhaS_HindIII	CGGAAGCTTGCCCGTTGTGTTATTAGCC
SalD48N_For	GTATTTTACTCAGCTGCGAATCACCGTCTTGCAGAATTG
SalD48N_Rev	CAATTCTGCAAGACGGTGATTCGCAGCTGAGTAAAATAC
#1	TTACAATCATAGGTAGAATTTATGTCTGATAAACCATTAACTAAAACTGATTATTTGATGC
#2	GAAGCAGCTCCAGCCTACACGAACGAGGAGGCAGATAACACCTGCGTGTTCTCTAAAAAG
#3	GTGTTATCTGCCTCCTCGTTCGTGTAGGCTGGAGCTGCTTCGAAGTTCCTATACTTTCTA
#4	CTATATCACTGTTCTATAATAGCCCGTTGTGTTATTAGCCACATATGAATATCCTCCTTAG
YdgT_P1	GTTTATTTTTTATCAGTGACTACTCCGTTGGCATTATATTTAATGTGTAGGCTGGAGCTGCTTC
YdgT_P2	GGGGCAAATATTATAAGGTTTTTGATGTTAAACGCTACTTTCTCATATGAATATCCTCCTTAGT
YdgT_P1UP	CCTGACTCTTTACCGGTAAG
YdgT_P2DOWN	GTAGTCATATCTTCTCCGGG
ssrA_qPCR_F	GCTCAATCTCAAGAATACGC
ssrA_qPCR_R	CTGCCGTTTCTGAACCATTG
sipB_qPCR_F	TTAGATAAGGCCACGGATGC
sipB_qPCR_R	CCTGGGAAACCTGATTCTGA
motB_qPCR_F	GATTTCCATCTCCAGCCCTA
motB_qPCR_R	GCTGTTGGGTGTAATCATCG

annealed at their overlapping regions and amplified as a single fragment using oligonucleotides pairs #1/#4 (for details see Supplementary Figure S1). The annealed product was purified, cloned into pGEM-T easy vector (Promega) after the addition of an A-tail and electroporated into *E. coli* DH5α. Sequencing analysis confirmed the correct amplification and the "stitch" fragment was amplified from the plasmid using oligonucleotides #1 and #4. Finally, the chromosomal *hhaD48N* point mutant (SV5015D) was obtained using the procedure described by (Datsenko and Wanner, 2000). Briefly, the PCR product was *Dpn*I-digested, purified and used to electroporate strain SV5015 carrying the plasmid pKD46 grown at 30°C in the presence of

HEPES pH 7.9, 100 mM KCl, 5 mM MgCl₂, 50 mM imidazole and 10 glycerol). Lysis was carried out by sonication. To obtain a clear lysate the extracts were centrifuged at 35000 × g at 4°C for 20 min. His-tagged Hha proteins were purified with 0.5 ml Ni²⁺-NTA beads (Qiagen) as previously described (Nieto et al., 2000).

Gel Electrophoresis and Western Blot

Protein samples were analyzed by SDS-PAGE and stained with Coomassie brilliant blue or immunoblotted by western blot upon transfer of proteins to PDVF membranes. Western blot analysis was performed with polyclonal antibodies raised against *E. coli* StpA protein [1:2000] (Johansson and Uhlin, 1999). Horseradish peroxidase-conjugated goat anti-rabbit IgG [1:100000] (Sigma) was used as secondary antibody. Immunodetection of the transferred proteins was performed by enhanced chemiluminescence using the software Quantity One (Bio-Rad).

Microarray Analysis

Total RNA was isolated from three independent cultures of the strains SV5015, SV5015H, SV5015D and SV5015S grown at 37°C in LB to an OD$_{600}$ of 0.6. The RNA was purified as previously described (Paytubi et al., 2014). Transcriptomic analyses were performed on a custom DNA microarray engineered by Nimblegen. The custom Nimblegen microarray contained 4941 probes (4519 SL1344, 103 SL1344_pSLT, 99 SL1344_pRSF, 14 SL1344_pCOL1B, 206 R27) from the genome sequence of *S. enterica* serovar Typhimurium SL1344 (Paytubi et al., 2014). Retrotranscription, labeling, hybridization, microarray scanning, and data analysis were performed as recommended by Nimblegen standard protocol. These transcriptomic experiments and the statistical analysis of the microarray data were carried out at Institute of Biomedical Research, Barcelona.

The complete data set has been deposited under accession number E-MTAB-3621 at http://www.ebi.ac.uk/arrayexpress.

qRT-PCR

qRT-PCR analysis was performed to corroborate the microarray data (Supplementary Table S3) using strains SV5015H, SV5015D and SV5015S, SV5015HY, and SV5015DY versus wild-type (WT) strain, SV5015. Real-time quantitative reverse transcription-PCR was used to confirm microarray results and analyze expression of the *ssrA*, *sipB*, and *motB*. Briefly, 1 μg of total RNA was reverse transcribed to generate cDNA using the "High-capacity cDNA Reverse Transcription kit" (Applied Biosystems) as recommended by the manufacturer. As a control, parallel samples were run in which reverse transcriptase was omitted from the reaction mixture. Real-time PCR using "Power SYBR Green PCR Master Mix kit" (Applied Biosystems) was carried out on the StepOne Real-Time PCR System Thermal Cycling Block (Applied Biosystems). Oligonucleotides complementary to the genes of interest were designed using Primer3 online tool provided by the Whitehead institute[1]. Expression levels of the

---

[1]http://bioinfo.ut.ee/primer3 (Rozen and Skaletsky, 2000).

tested genes were normalized to the reference strain (SV5015) as in Ali et al., 2013.

Siderophore Detection

A colorimetric assay, Siderotec Assay™ (Emergenbio), was used for the detection of the siderophores secreted by strain SV5015 and its derivatives. Cultures of each strain were grown in LB at 37°C until the beginning of stationary phase (OD$_{600nm}$ 2.0). One-milliliter of each culture was centrifuged and the supernatant was used for siderophores detection. The assay was performed as indicated by the supplier with the following modification. To enhance the detection, the standard solution was added to the recommended mix in the following proportion: 10 μl catalyst, 90 μl dye reagent, and 20 μl of the standard. Eighty-microliter of each strain supernatant was added to the mixture and the colorimetric changes were determined by measuring the OD$_{630nm}$ of the samples, as recommended by the supplier.

Swimming Motility

Swimming motility was performed on Tryptone broth (TB) plates (1% tryptone and 0.5% NaCl) containing 0.35% agar. Overnight bacterial cultures grown in LB at 37°C were spotted (5 μl) on the center of the plates. The colony diameter was measured after incubation for 8 h at 37°C.

Transmission Electron Microscopy

Bacterial strains used for flagella visualization were obtained from motility plates grown overnight at 37°C. A slice of the motility perimeter of the indicated strains was collected, resuspended in filtered Ringer ¼ solution and centrifuged at 1500 × g for 5 min. Cu-Carbon grid (CF200-Cu Carbon Film On 200 Mesh Copper Grids, Electron Microscopy Sciences) was soaked for 60 s on a 5 μl drop of each strain, washed three times with water for 20 s and stained for 60 s using a 2% (w/v) uranyl acetate solution (Polysciences). Once stained, the grids were dried for at least 24 h before visualization under a JEOL JEM1010 transmission electron microscope. Images were obtained using the software analysis (Soft Imaging System GmbH, Münster, Germany). Each sample was observed for at least 100 cells.

β-Galactosidase Activity

The reporter gene fusion *ssrA::lacZ* from plasmid pIC-ssrA2 was used to evaluate *ssrA* transcriptional expression. β-galactosidase activity assays were performed as previously described (Miller, 1993). Strains were grown at 37°C with shaking in LB until reaching an OD$_{600nm}$ of 0.6 (as in the transcriptomic experiments) or in LPM pH 5.8 culture medium, which induces SPI-2 gene expression as previously described (Coombes et al., 2004; Silphaduang et al., 2007).

Macrophage Survival Assay

RAW 264.7 murine macrophages (ATCCTIB-71), were grown in Dulbecco's modified Eagle medium (DMEM) containing 10% heat-inactivated fetal bovine serum (HI FBS), x1 MEM non-essential amino acids and 2 mM L-glutamine in a humidified atmosphere (37°C in 5% CO₂). S. Typhimurium strains were grown in LB to stationary phase and opsonized for 30 min in

DMEM containing 10% normal mouse serum (Charles River Laboratories; Balb/c mouse male). Bacteria were then centrifuged onto macrophages seeded in 6-well tissue culture plates at a multiplicity of infection (MOI) of 10:1 and incubated for 30 min. After infection, macrophages were washed twice with DPBS (Dulbecco's Phosphate-Buffered Saline) and incubated for 90 min more in medium containing 100 μg ml^{-1} gentamicin to kill the remaining extracellular bacteria (2 h post-infection, corresponding to time 0). For the remainder of the experiment, medium containing 10 μg ml^{-1} gentamicin was used to prevent extracellular bacterial replication. At 2 and 16 h post-infection time points, infected macrophages were washed twice with DPBS and lysed with 0.1% Triton X-100 in DPBS. Appropriate serial dilutions of the lysates were plated onto LB agar to enumerate colony-forming units. Intracellular replication ratio (16 h versus 2 h) was calculated.

Results

HhaD48N Interacts with StpA

In order to investigate the specific regulatory role of Hha protein independently of its interaction with H-NS, we decided to introduce the D48N mutation in the Hha protein of *S.* Typhimurium SV5015 strain, generating then a mutant derivative not capable to form Hha/H-NS complexes while maintaining the Hha structure (de Alba et al., 2011). Presumably, Hha residues involved in StpA interaction are the same as for the H-NS protein. However, to rule out that the effects of the *hha*D48N mutation might be caused by a possible interaction with StpA, pull-down experiments using His-tagged Hha and HhaD48N proteins were performed. To avoid the interaction of the WT Hha protein and H-NS, which could mask the interaction with StpA, an Δ*hns* genetic background was used. His-Hha, His-HhaD48N and StpA proteins were overexpressed using plasmids pET15bHisHha, pET15bHisHhaD48N, and pT7-stpA, respectively, in *E. coli* BL21 (DE3) Δ*hns* (His-Hha and His-HhaD48N) or *E. coli* BL21 (DE3) Δ*hns* pLysS (StpA). Ten-milliliter of clarified supernatants containing His-Hha or His-HhaD48N recombinant proteins were mixed with 10 ml of clarified supernatants containing overexpressed StpA and altogether was coated onto a Ni^{2+}-NTA matrix as described in the "Materials and Methods" section. His-tagged Hha variants were eluted with 200 mM imidazole and tested for the presence of the Hha variants and the StpA protein by SDS-PAGE followed by Coomassie staining or western blot, respectively (**Figure 1**). Unexpectedly, the results obtained showed that HhaD48N ("H-NS blind" Hha mutant) still supported the interaction with the StpA protein. The amount of immunodetected StpA protein that coeluted with HhaD48N was more than twofold higher than that detected for the WT Hha. Similar results were obtained when His-tagged proteins were purified directly from cellular extracts containing either overexpressed His-Hha, His-HhaD48N, or StpA. As shown in the StpA control lane, a low level of StpA was detected due to non-specific binding of the protein to the beads. However, when HhaD48N was overexpressed and purified, endogenous

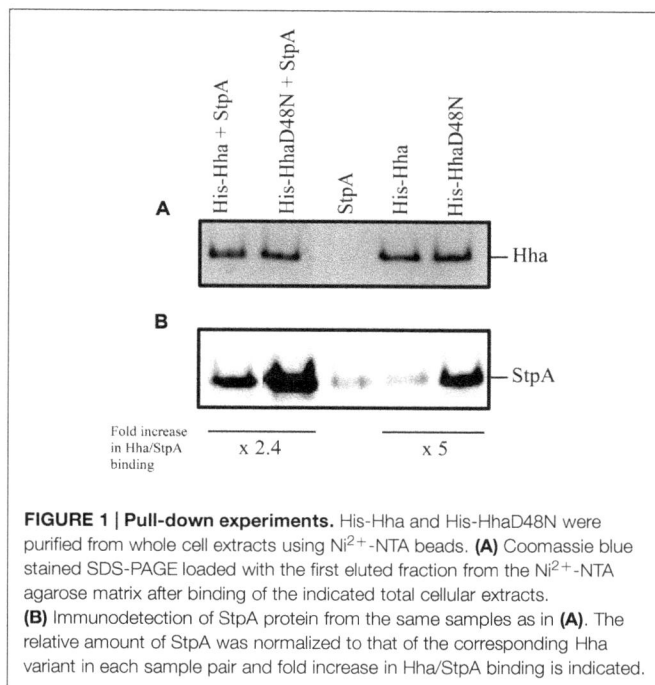

FIGURE 1 | Pull-down experiments. His-Hha and His-HhaD48N were purified from whole cell extracts using Ni^{2+}-NTA beads. **(A)** Coomassie blue stained SDS-PAGE loaded with the first eluted fraction from the Ni^{2+}-NTA agarose matrix after binding of the indicated total cellular extracts. **(B)** Immunodetection of StpA protein from the same samples as in **(A)**. The relative amount of StpA was normalized to that of the corresponding Hha variant in each sample pair and fold increase in Hha/StpA binding is indicated.

StpA clearly coeluted up to fivefold compared to the WT Hha.

Hha Protein Regulates a Set of Genes Independently of H-NS/StpA

The above reported data show that cells expressing the mutant HhaD48N protein can be used to assess whether Hha is able to modulate gene expression independently of its interaction with H-NS in *S.* Typhimurium. To investigate this, transcriptomic analyses of the WT strain *S.* Typhimurium SV5015, the Δ*hha* (SV5015H) and the point mutant *hha*D48N strain (SV5015D) were carried out. Having in mind that HhaD48N protein interacts with StpA, the Δ*stpA* strain (SV5015S) was included in the transcriptomic study to discard possible effects on gene regulation caused by this interaction. The transcriptome of each mutant was compared to that of the WT strain (**Table 3** and Supplementary Table S1). The *hha* mutation caused a more than twofold differential expression of 659 genes (406 up-regulated and 253 down-regulated). In the case of the *hha*D48N mutant, 499 genes were deregulated (310 up-regulated and 189 down-regulated). Finally, the largest effect was found in the Δ*stpA* mutant, which showed an altered expression of 783 genes (423 up-regulated and 360 down-regulated). For all mutant strains, more genes were up-regulated than down-regulated, indicating that these proteins play an important role in gene expression acting mainly as transcriptional repressors.

To further unravel the Hha regulon in *Salmonella*, we considered the following. Genes whose expression is altered in an Δ*hha* strain would correspond to both (i) genes whose expression is dependent of Hha through its interaction with H-NS, and (ii) the subset of genes regulated by Hha autonomously of this interaction. On the other hand, the genes showing an altered

TABLE 3 | Total and relative number of genes deregulated (more than twofold, *p*-value <0.05) in strains SV5015H, SV5015D and SV5015S, versus wild-type strain, SV5015.

	SV5015H vs. SV5015		SV5015D vs. SV5015		SV5015S vs. SV5015	
	Up-regulated	Down-regulated	Up-regulated	Down-regulated	Up-regulated	Down-regulated
Chromosome (4527)	386 (8.5%)	251 (5.5%)	293 (6.4%)	189 (4.1%)	409 (9%)	356 (7.8%)
pSLT (103)	11 (10.6%)	2 (1.9%)	9 (8.7%)	n.d	9 (8.7%)	n.d
pCollB (100)	9 (9%)	n.d	8 (8%)	n.d	5 (5%)	4 (4%)

n.d., not detected.

expression in an *hha*D48N strain are those regulated by the Hha/H-NS complex. Thus, the genes showing a deregulated expression in an *hha* mutant (Hha and Hha/H-NS dependent), but unaffected in an *hha*D48N mutant (Hha/H-NS dependent) or in an *stpA* mutant (Hha/StpA dependent), are good candidates to be regulated by Hha independently of its interaction with H-NS and StpA.

Consequently, when looking more deeply into the transcriptomic data, we were able to identify 120 genes as regulated by Hha independently of H-NS/StpA (Supplementary Table S2). Seventy-three of these genes were up-regulated and 47 down-regulated. In order to determine the global pattern of gene regulation held exclusively by Hha, the differentially expressed genes were grouped in their functional categories (J. Craig Venter Institute; **Figure 2**). The functional categories that showed the highest number of Hha-dependent genes correspond to genes

of unknown function, pathogenicity islands, cell envelope, transport, and binding of proteins and protein synthesis. It is noteworthy that the number of pathogenicity island genes affected represents more than 10% of the total number of genes that belong to this category, resulting in the category with the highest percentage of affected genes.

Regulation of Genes Associated to Iron Transport by the Hha Protein

fepG, *fepD*, *fepB*, and *entS* genes are presumably up-regulated, directly or indirectly, by Hha in an H-NS-independent fashion (**Table 4**). The products of all these genes are involved in iron transport. FepG and FepB form, together with FepC, an ABC transporter of siderophores (Zhu et al., 2005; Crouch et al., 2008). EntS is a transmembrane protein related to enterobactin secretion (Furrer et al., 2002; Methner et al., 2008).

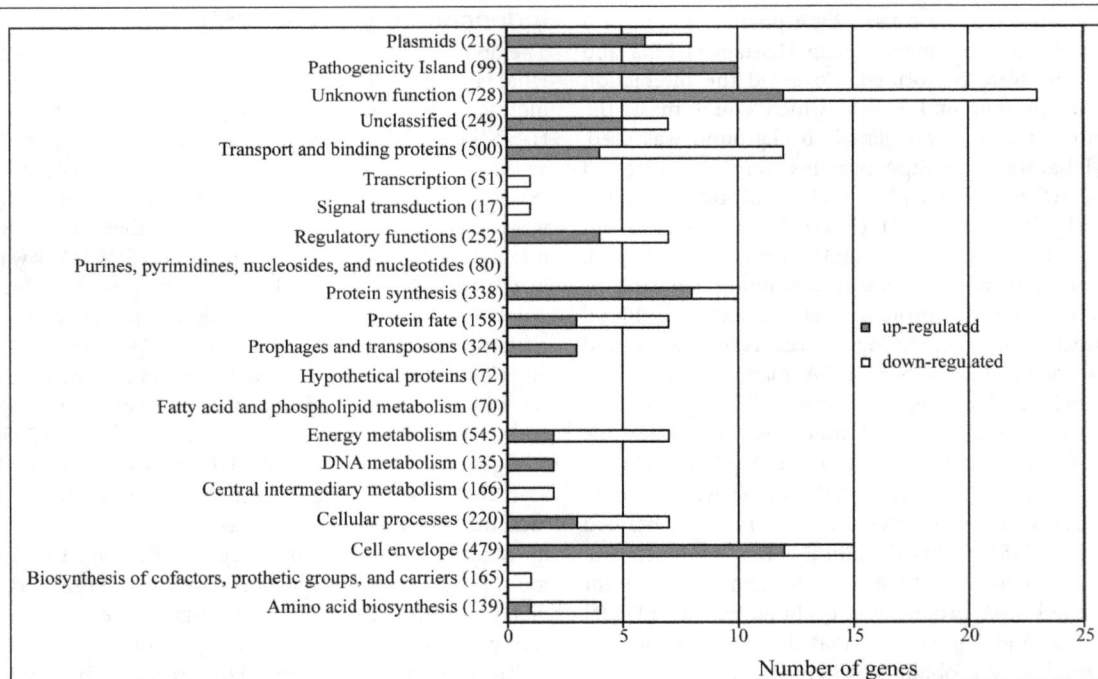

FIGURE 2 | The Hha regulon. The bars indicate the number of genes of each category that show an altered expression in an Δ*hha* mutant independently of its interaction with H-NS or StpA. The gray bars indicate the number of genes that are up-regulated (FC > 2, *p*-value <0.05) and the white bars indicate the down-regulated genes (FC < −2, *p*-value <0.05). The total number of genes belonging to each of the functional categories (JCV Institute) is specified in parenthesis.

TABLE 4 | Genes related to iron transport, motility, and chemotaxis and pathogenicity islands that are regulated by the Hha protein in an H-NS/StpA-independent manner.

Gene	ORF	Function	FC (SV5015H vs SV5015)
Iron transport			
fepG	SL0579	Ferric enterobactin transport system permease protein	−2.1
fepD	SL0580	Ferric enterobactin transport system permease protein	−2.6
fepB	SL0582	Ferrienterobactin-binding periplasmic protein	−2.3
entS	SL0581	Enterobactin exporter	−2.3
Motility and chemotaxis			
flgN	SL1108	Flagella synthesis protein	−2.3
motB	SL1857	Motility protein B	−2.1
tcp	SL3542	Methyl-accepting chemotaxis citrate transducer	−2.3
tsr	SL4464	Methyl-accepting chemotaxis protein I	−2.3
Pathogenicity islands			
pipA	SL1026	Pathogenicity island encoded protein (SPI-5)	2.8
sigE	SL1029	Pathogenicity island-encoded protein; cell invasion protein (SPI-5)	3.3
sigD	SL1030	Pathogenicity island-encoded protein; Type III secretion system effector protein (SPI-5)	3.2
ssrB	SL1325	Two-component response regulator (SPI-2)	2.7
ssrA	SL1326	Two-component sensor kinase (SPI-2)	2.7
sseG	SL1339	Type III secretion system effector protein-modulates the positioning of the SCV (SPI-2)	4.0
ssaQ	SL1352	Type III secretion system protein (SPI-2)	2.7
sipC	SL2863	Translocation machinery protein (SPI-1)	2.4
sipB	SL2864	Translocation machinery protein (SPI-1)	2.5
siiE	SL4197	Large repetitive protein (SPI-4)	3.5

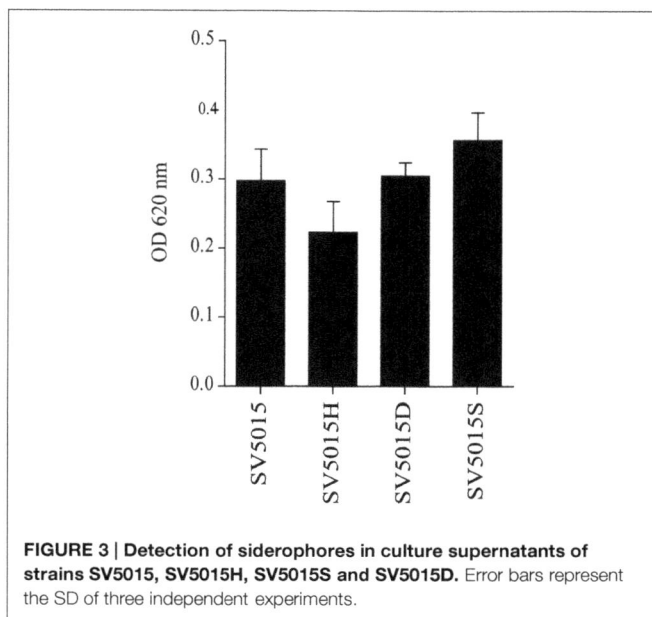

FIGURE 3 | Detection of siderophores in culture supernatants of strains SV5015, SV5015H, SV5015S and SV5015D. Error bars represent the SD of three independent experiments.

In light of the above, we tested the role of the Hha protein in the synthesis and transport of siderophores. To this end, the presence of siderophores in bacterial culture supernatants of strains SV5015, SV5015H, SV5015D, and SV5015S was determined (**Figure 3**). As expected, the level of siderophores detected in the supernatant of SV5015H strain was 25% lower compared to the WT strain. Despite these results are not statistically significant, the data are consistent with the changes observed in the transcriptomic data. In contrast, the level of siderophores detected in supernatants of strains SV5015D and SV5015S was similar or even higher than in the WT strain, indicating that the interaction of Hha with H-NS/StpA is not required to regulate the genes associated to iron transport.

Hha Controls Motility at the Motor Level

The transcriptomic data allowed us to identify four genes involved in motility and chemotaxis, *flgN*, *motB*, *tcp*, and *tsr*, that are down-regulated in an *hha* mutant (**Table 4**). FlgN is a chaperone related to the secretion of hook-associated proteins FlgK and FlgL. MotB forms, together with MotA, a transmembrane proton-channel that drives flagellar rotation (Morimoto and Minamino, 2014). Tcp and Tsr are chemoreceptors located in the cytoplasmic membrane

(Yamamoto and Imae, 1993; Okumura et al., 1998; Iwama et al., 2006).

The effect on genes contributing to bacterial chemotaxis and flagellar function was confirmed by a motility assay (**Figure 4**). YdgT, the paralog of Hha, appears to fulfill some of the functions of the Hha protein (Paytubi et al., 2004). Moreover, it has been previously described that the *hha ydgT* double mutant completely abolishes the motility of *S.* Typhimurium SV5015 (Vivero et al., 2008). Consequently, we decided to include *ydgT* mutant strains in the motility assays. By using strains deficient in YdgT, we intended to withdraw any possible effect of the YdgT-H-NS/StpA interaction that could potentially mask the effect of Hha on motility.

Cultures of SV5015 and its mutant derivatives were spotted on TB agar plates and the colony diameter was measured. The results obtained (**Figures 4A,B**) showed that the *hha* mutation does not cause a significant effect on motility, compared to the WT strain. In contrast, single SV5015D, SV5015Y, SV5015S, and double SV5015SY mutations cause a slight increase in motility compared to the SV5015 strain (1.2-, 1.7-, 1.6-, and 1.5-fold, respectively). As expected, the SV5015HY strain is totally impaired in motility as previously described (Vivero et al., 2008; Wallar et al., 2011), as well as the SV5015HYS triple mutant strain. Remarkably, the double mutant SV5015DY only shows a 1.7-fold decrease in motility compared to the WT strain. qRT-PCR analysis of strains SV5015H, SV5015D, SV5015S, SV5015HY, and SV5015DY versus WT strain SV5015 using primers against *motB* gene confirmed these results (Supplementary Table S3). Despite that this strain does not completely recover the motility showed by strain SV5015, the results suggest that Hha plays a role in motility in an H-NS-independent manner. The phenotype shown by this double mutant strain could still be due to the effect of the interaction between HhaD48N and StpA. To bypass this effect, a triple mutant *hha*D48N *ydgT stpA* (SV5015DYS) was

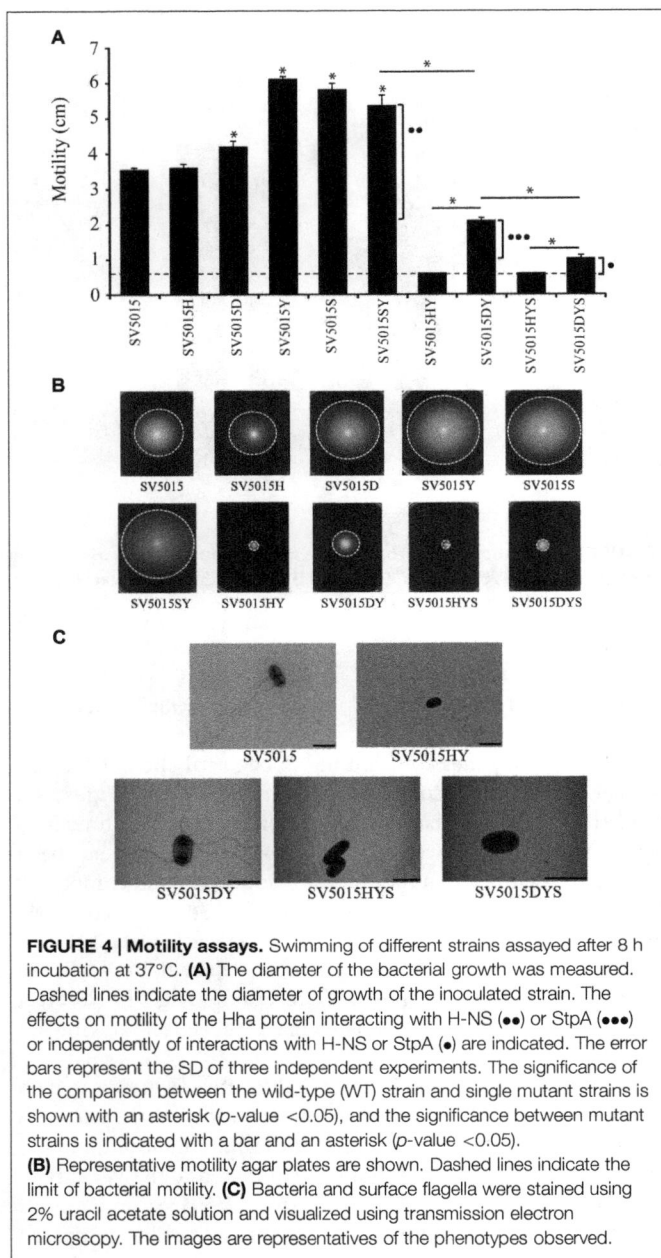

FIGURE 4 | Motility assays. Swimming of different strains assayed after 8 h incubation at 37°C. **(A)** The diameter of the bacterial growth was measured. Dashed lines indicate the diameter of growth of the inoculated strain. The effects on motility of the Hha protein interacting with H-NS (••) or StpA (•••) or independently of interactions with H-NS or StpA (•) are indicated. The error bars represent the SD of three independent experiments. The significance of the comparison between the wild-type (WT) strain and single mutant strains is shown with an asterisk (p-value <0.05), and the significance between mutant strains is indicated with a bar and an asterisk (p-value <0.05).
(B) Representative motility agar plates are shown. Dashed lines indicate the limit of bacterial motility. **(C)** Bacteria and surface flagella were stained using 2% uracil acetate solution and visualized using transmission electron microscopy. The images are representatives of the phenotypes observed.

Figure S2). The triple mutation SV5015HYS totally abrogates the motility of SV5015 strain which is completely recovered when complemented in *trans* with the WT *hha* gene cloned in plasmid pACYC184. Contrarily, when complementing the non-motile phenotype with the "H-NS blind" Hha mutant, a partial recovery of the motility is obtained, corresponding to the regulatory role in motility that Hha protein carries out independently of its interaction with H-NS or StpA.

To discern whether the absence of motility of strains SV5015HY and SV15015HYS was due to a defect on flagellar production or flagellar rotation, cells of the different strains grown on motility plates were observed using transmission electron microscopy. Strains SV5015HY and SV5015HYS (non-motile strains) presented flagella on their surface, although to a less extend than strains SV5015, SV5015DY, and SV5015DYS, which showed motile phenotype in different degrees (**Figure 4C**). Results suggest that the observed absence or decrease in motility might rely on genes involved in motility at the motor level, more than to a decrease of flagella.

Role of Hha Protein on the Regulation of Expression of SPI Genes

The transcriptomic data reveals that the Hha protein is engaged in the regulation of some genes related to pathogenicity islands independently of its interaction with H-NS or StpA (**Table 4**). Among these, *ssrA* and *ssrB* genes were found to be up-regulated. These genes encode the two-component regulatory system SsrA–SsrB responsible for the activation of expression of the SPI-2 genes (Tomljenovic-Berube et al., 2010; Xu and Hensel, 2010; Osborne and Coombes, 2011). It has been described that the expression of SPI-2 genes is essential for the intracellular survival and replication of *Salmonella* within macrophages (Schmidt and Hensel, 2004; Fàbrega and Vila, 2013). Moreover, Hha and YdgT proteins play a regulatory role in the expression of virulence factors encoded in SPI-2 (Coombes et al., 2004; Silphaduang et al., 2007; Vivero et al., 2008), although it is not clear whether this modulatory role depends upon their interaction with H-NS. The transcriptomic data revealed that the above mentioned role of Hha protein could be H-NS-independent, since *ssrA* and *ssrB* genes showed an altered expression in an *hha* mutant whereas the *hha*D48N mutation did not cause a significant deregulation of the gene expression when compared to the WT strain. A transcriptional fusion of the *ssrA* promoter with the *lacZ* reporter gene into the multicopy plasmid pIC552, pIC-*ssrA2*, was used to determine the transcriptional level of expression of the *ssrA* gene in different genetic backgrounds (**Figure 5A**). The strains were grown in LB medium at 37°C to an OD_{600} of 0.6, the same conditions as in the transcriptomic assay. In agreement with the transcriptomic data, the *hha* mutation causes an increase of the expression of the *lacZ* gene depending on the *ssrA* promoter (twofold). In contrast, expression from the *ssrA* promoter in strains SV5015D or SV5015Y was similar to the values obtained for the WT strain. Since the YdgT protein can compensate the effect of an *hha* mutation, we expected a more notorious regulatory role of Hha in a *ydgT* genetic background. Interestingly, the expression from the *ssrA* promoter in the double mutant strain

used. A twofold decrease in motility was observed, compared to the SV5015DY strain, although the motility is not completely abolished as it occurs in the *hha ydgT stpA* (SV5015HYS) strain. This result indicates that the interaction between Hha and H-NS/StpA is important for the regulation of expression of genes related to motility (as indicated in **Figure 4A**). Nevertheless, the fact that the presence of an "H-NS blind" Hha mutant causes an increased motility phenotype in the absence of YdgT (SV5015DYS compared to SV5015HYS), indicates that Hha also plays a role in motility in an H-NS/StpA-independent manner.

Trans-complementation assays were performed to confirm the role on motility of the Hha protein (Supplementary

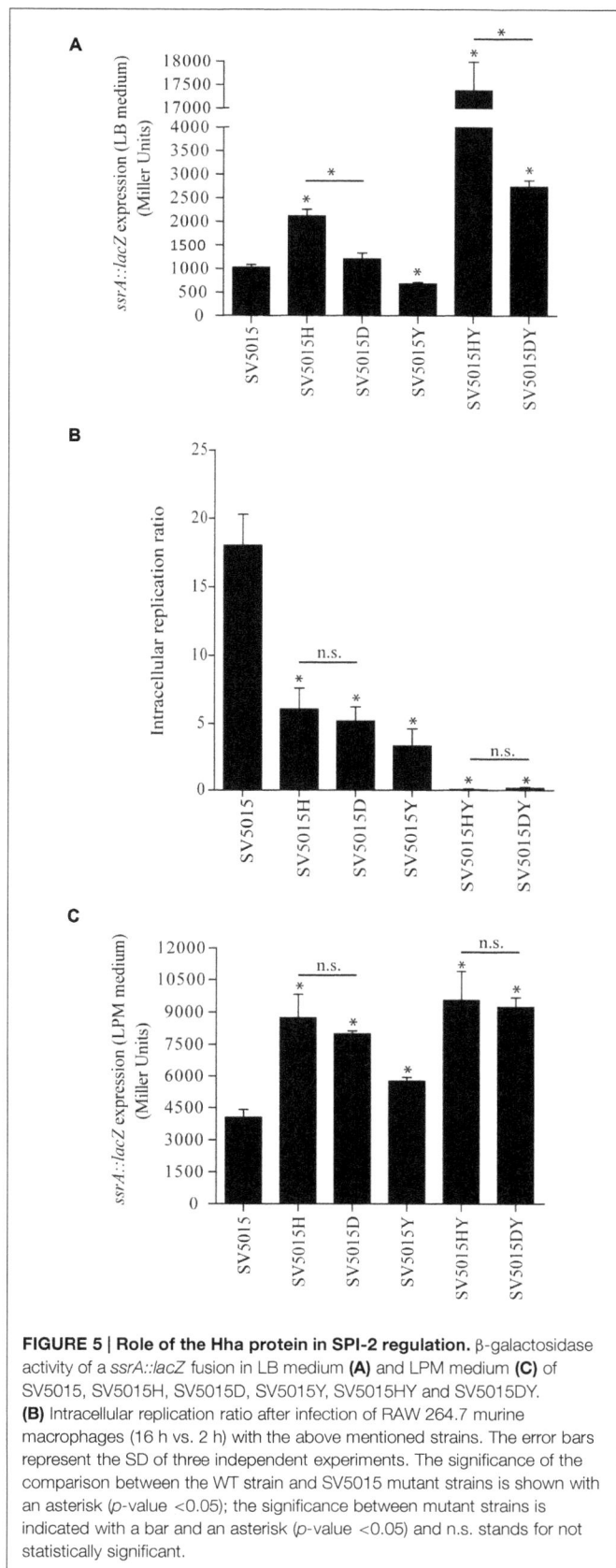

FIGURE 5 | Role of the Hha protein in SPI-2 regulation. β-galactosidase activity of a *ssrA::lacZ* fusion in LB medium **(A)** and LPM medium **(C)** of SV5015, SV5015H, SV5015D, SV5015Y, SV5015HY and SV5015DY. **(B)** Intracellular replication ratio after infection of RAW 264.7 murine macrophages (16 h vs. 2 h) with the above mentioned strains. The error bars represent the SD of three independent experiments. The significance of the comparison between the WT strain and SV5015 mutant strains is shown with an asterisk (*p*-value <0.05); the significance between mutant strains is indicated with a bar and an asterisk (*p*-value <0.05) and n.s. stands for not statistically significant.

SV5015HY, showed a drastic increase (17-fold) of the expression. In striking contrast, in the double mutant strain SV5015DY, the *lacZ* expression only showed a 2.6-fold increase compared to the WT strain. qRT-PCR analysis of strains SV5015H, SV5015D, SV5015S, SV5015HY, and SV5015DY compared to strain SV5015 to assess transcript levels of *ssrA* confirmed these results (Supplementary Table S3). Results showed that Hha protein might play a role by its own in the regulation of SsrA–SsrB two component system, and thus in the expression of SPI-2 genes.

The fine tuning of the SPI-2 gene expression contributes to the survival and replication of *Salmonella* within macrophages. Considering the results obtained, we would expect different behavior when testing replication of the different genetic background strains within macrophages. To verify the significance of the Hha protein on the regulation of the two component system SsrA–SsrB and thus on the pathogenicity of *Salmonella*, we assessed the ability of the same strains used in the β-galactosidase assay to replicate in macrophages (**Figure 5B**). All single mutant strains (SV5015H, SV5015Y and SV5015D) showed a reduced intracellular replication in the murine macrophage cell line RAW 264.7 when compared to the WT strain SV5015. This result is consistent with previous reported results concerning the role of Hha and YdgT in the SPI-2 regulation (Coombes et al., 2005; Silphaduang et al., 2007). However, these results are not consistent with the expression levels of *ssrA* observed both in the transcriptomic data and the β-galactosidase assay under the conditions used (LB medium). The data obtained in the macrophages replication assay indicates that the Hha protein does not play a role independent of H-NS in the replication of *Salmonella* within macrophages. It has been documented that the LPM medium (low phosphate and low magnesium, pH 5.8) mimics the physiological conditions that induce the expression of SPI-2 genes (Coombes et al., 2004; Silphaduang et al., 2007). To confirm whether the culture medium has an effect on the regulation of *ssrA*, held by Hha itself or with H-NS, we decided to analyze the transcriptional expression of *ssrA* in cells growing in SPI-2-inducing medium. **Figure 5C** shows the results obtained. When using LPM medium, the levels of β-galactosidase activity shown by strains *hha* and *hha*D48N are not significantly different, although in both cases are approximately twofold higher than the activity detected for the WT strain. In contrast to the data obtained when cells were grown in LB medium, the double mutant strains (SV5015HY and SV5015DY) do not show differences in the expression levels of *lacZ* compared to the respective single mutant strains (SV5015H and SV5015D). Strikingly, the β-galactosidase activity determined in the LPM medium was significantly similar in both double mutant strains. These results are consistent with the data obtained in the replication assay within macrophages.

In short, we suggest that Hha could play a "dual" regulatory role in the *ssrA* expression, based on its dependence of H-NS. The Hha-mediated repression of *ssrA* expression coulfd be H-NS-independent in extracellular conditions while H-NS-dependent during intercellular growth.

Discussion

Hha, a nucleoid- associated protein, was first identified as a modulator of the expression of hemolysin in *E. coli* (Nieto et al., 1991; Carmona et al., 1993; Fahlen et al., 2000; Sharma and Zuerner, 2004). Furthermore, Hha-like proteins have been related to the environmental regulation of virulence factors in several enterobacteria (Mouriño et al., 1996; Madrid et al., 2002a,b). It has been described that the Hha protein affects gene expression through its interaction with H-NS (Nieto et al., 2000; Forns et al., 2005; Ali et al., 2013; Aznar et al., 2013) and this interaction is common to other members of both families (Hha/StpA, YdgT/H-NS, YdgT/StpA, YmoA/H-NS; Nieto et al., 2002; Paytubi et al., 2004). As a result of the interaction with H-NS, Hha can alter the target specificity of H-NS playing a significant role in the recognition of HGT sequences (Baños et al., 2009; Ali et al., 2013; Aznar et al., 2013).

Several studies have suggested the modulatory role of Hha in enterobacteria through its interaction with DNA regulatory sequences (Fahlen et al., 2001; Sharma and Zuerner, 2004; Sharma et al., 2005; Sharma and Bearson, 2013; Sharma and Casey, 2014). It is worth to remark that only in one of these reports (Olekhnovich and Kadner, 2007) the Hha protein has been purified in an Δ*hns* genetic background. Bearing in mind that Hha interacts with H-NS, and that H-NS copurifies with Hha, it is likely that the DNA binding activity reported in most of these studies is due to Hha/H-NS complexes.

The H-NS protein has two different domains, the N-terminal domain involved in oligomerization, and the C-terminal domain responsible for its binding to the DNA. The Hha protein shows a high degree of homology with the N-terminal domain of H-NS and it has been proposed that Hha-like proteins might have evolved to mimic the N-terminal domain of H-NS (Madrid et al., 2007a). Despite the fact that the Hha protein does not have defined domains and most of the protein sequence contributes to its interactions with H-NS (Nieto et al., 2002), it shows a dipolar distribution. Positively and negatively charged residues cluster on opposing surfaces of the molecule showing an asymmetrical charge distribution that reveals specific functions (Paytubi et al., 2011; Ali et al., 2013). Conserved positively charged residues on the surface of Hha are positioned in the same orientation as predicted for the DNA binding domain of H-NS, whilst a combination of basic and acidic Hha residues interacts with H-NS. NMR analysis focused on the study of the function of the acidic residues of the Hha protein in the interaction with H-NS, allowed the identification of aspartic acid in position 48 as essential for this interaction (de Alba et al., 2011). Consistent with this finding, the structure resolved by Ali et al. (2013) confirmed that these residues are located within the Hha/H-NS (1–46) interaction interface.

StpA is the best characterized paralog of H-NS (Zhang and Belfort, 1992; Zhang et al., 1996; Johansson and Uhlin, 1999; Lucchini et al., 2009). Structurally, H-NS and StpA are similar and both have a tightly relationship with the Hha family of proteins. Although the specific residues of StpA involved in the interaction with Hha have not been identified, attending to its similarity with H-NS, it is reasonable to speculate that such

interactions might comprise the same residues described for the H-NS/Hha interaction (García et al., 2006; de Alba et al., 2011). Pull-down experiments of StpA with Hha or HhaD48N showed that the "H-NS-blind" Hha mutant binds with higher affinity to StpA than the WT Hha protein. This result suggests that the interaction of Hha with StpA involves different residues in Hha from the ones described for the H-NS/Hha complex. Although the high homology exhibited between H-NS and StpA, under some circumstances both proteins might have distinct biological functions (Deighan et al., 2000) and DNA-binding mechanisms (Lim et al., 2012). In this regard, it is tempting to speculate that several models of interaction of Hha with different members of the H-NS family might take place. Further structural studies will be required to resolve the Hha/StpA complex.

The use of the "H-NS-blind" Hha mutant permitted us to identify a group of genes regulated by Hha in an H-NS/StpA-independent manner. For this purpose, we compared the transcriptomic profiles of four different strains: WT, Δ*hha*, *hha*D48N, and Δ*stpA*. The results obtained drove us to the identification of a set of 120 genes regulated by Hha in an H-NS/StpA-independent manner. Among these 120 genes, 39 genes have been previously described to be regulated by H-NS and/or StpA (Supplementary Table S2; Baños et al., 2009; Lucchini et al., 2009). Nevertheless, this H-NS/StpA-dependent regulation does not necessarily imply protein–protein interactions with Hha. The 120 genes identified were grouped into the following functional categories: genes of unknown functions, pathogenicity islands, cell envelope, protein synthesis, and transport and binding of proteins. Several phenotypic studies were assayed to corroborate the transcriptomic data.

Since genes involved in iron uptake, *fepG*, *fepD*, *fepB*, and *entS*, were down-regulated in an *hha* mutant, we decided to check the siderophore content in the culture supernatant of different strains. The siderophore content was slightly lower in the Δ*hha* mutant compared to WT, *hha*D48N and Δ*stpA* strains, indicating that Hha could play a role in the expression of these genes. The down-regulation of *entS*, encoding the main enterobactin exporter protein in *Salmonella* and *E. coli* (Furrer et al., 2002; Methner et al., 2008) could be responsible for the faint difference of siderophores detected in the Δ*hha* mutant supernatant. Notably, secretion and transport of siderophores is a complex system, involving alternative transport systems. Although the detection of siderophores in the Δ*hha* mutant is lower than in any of the other strains under study, the mere presence of these iron-chelating compounds in Δ*hha* supernatants suggests the existence of another fully functional exporter system outside the Hha regulon. IroC, described as an enterobactin and salmochelin transporter (Hantke et al., 2003; Crouch et al., 2008), could replace EntS protein, therefore masking the decrease on the *entS* expression. Additionally, FepG, FepD and FepB proteins are associated with siderophore transport across the cytoplasmic membrane into the cytoplasm (Hantke et al., 2003; Zhu et al., 2005; Crouch et al., 2008). The down-regulated expression of these genes in an Δ*hha* mutant reveals that Hha, in addition to partially affect the siderophore exporter system, regulate the uptake of siderophores. However, siderophores that remain outside may be recognized by outer membrane receptors, such as IroN or FepA (Hantke et al., 2003) and stay in the periplasm waiting to be transported into the cytoplasm.

The modulatory role of Hha in the expression of the genes *flgN*, *motB*, *tcp*, and *tsr*, related to flagellar motility and chemotaxis, was assessed. The differences in motility exhibited by the strains used in the assay allowed us to discriminate between the regulation that could be held by Hha and different complexes (Hha/StpA and Hha/H-NS). The double mutant SV5015DY displayed a motile phenotype, which could be associated with the regulatory effects that can be played by Hha (in an H-NS independent-manner) and through its interaction with StpA. The existence of a modulatory function of Hha independently of H-NS and StpA in the expression of genes involved in motility was confirmed by using the SV5015DYS triple mutant. Complementation in *trans* of the *hha* mutation with the WT Hha or the "H-NS-blind" Hha mutant confirmed our hypothesis. Surprisingly, the *ydgT* mutation appears to cause an increase in motility compared to the WT strain. This result disagrees with previous studies that have reported a motile phenotype for the Δ*ydgT* mutant similar to the one exhibited by the WT strain (Wallar et al., 2011). However, differences in the composition of the motility plates could result in different swimming phenotypes. More studies will be required to understand the regulatory role of the YdgT protein in motility.

Transmission electron microscopy allowed us to determine that the non-motile SV5015HY and SV5015HYS mutant strains presented flagella on their surface, although the number of flagella was slightly lower than in strains SV5015DY and SV5015DYS. Among the Hha-dependent genes (H-NS/StpA-independent), *motB* was found to be down-regulated. This gene encodes a flagellar motor component that, with MotA, form the proton-channel complex of the proton-driven bacterial flagellar motor (Morimoto et al., 2010; Morimoto and Minamino, 2014). Previous studies determined that the non-motile phenotype observed in an Δ*hha* Δ*ydgT* mutant is due to a decrease in the expression of the genes encoding the transcriptional regulator $FlhD_4C_2$, mediated by PefI-SrgD. In other words, the expression of the PefI-SrgD repressor complex is under the negative control of Hha and YdgT. Wallar et al. (2011) showed that, in an Δ*hha* Δ*ydgT* double mutant, low levels of $FlhD_4C_2$ are translated in the down-regulation of flagellar promoters and this prompts the loss of surface flagella and motility. It is worth noting that in this work we have visualized cells grown on motility plates, whereas in the above mentioned work cells were grown in LB medium. Such differences in the methodology might explain the differences observed in the phenotype of the double mutant. Taking together, our data suggest that the loss of motility in SV5015HY and SV5015HYS strains could be explained by both: (i) the role accomplished by Hha and YdgT, through its interaction with H-NS, on the negative regulation of PefI-SrgD; (ii) the regulatory role that Hha, independently of H-NS, plays on flagella at the motor level.

The transcriptomic data obtained in this work suggest a role for the Hha protein in the regulation of expression of the *ssrA* and *ssrB* genes involved in SPI-2 regulation. Transcription assays in LB medium using the *ssrA::lacZ* transcriptional fusion allowed us to demonstrate that the *hha* mutant exhibits a higher expression of the *ssrA* gene compared to the wild type and the *hha*D48N mutant, suggesting a regulatory role of the Hha protein independently of H-NS. Moreover, this effect is exalted when the Δ*hha*Δ*ydgT* and *hha*D48N Δ*ydgT* genetic backgrounds are compared. In contrast, when the expression of *ssrA* was tested in cells grown under SPI-2-inducing conditions, LPM medium, no differences were observed between the Δ*hha* and *hha*D48N genetic backgrounds. This behavior is in good harmony with the survival and replication within macrophages assays. The results suggest that the regulation of SPI-2 genes in intracellular conditions is dependent on the complex Hha/H-NS, since the presence of a protein "H-NS blind" does not hold a modulatory effect, whereas a clear Hha repressor function of SPI-2 genes is observed in extracellular conditions. The role of Hha in SPI-2 genes expression has been previously reported although it is likely that H-NS is also involved in this regulation (Silphaduang et al., 2007). However, in this work we describe a role for Hha on SPI-2 regulation independently of its interaction with H-NS. Altogether these data highlight the role of the Hha protein as a repressor of SPI-2 genes when the cells are outside the intracellular environment. Furthermore, the results confirm that the Hha/H-NS regulatory complex is essential for the proper regulation of *ssrAB* during the intracellular phase of infection of S. Typhimurium (Coombes et al., 2005; Silphaduang et al., 2007).

In this work, we have described for the first time an Hha regulon independent of H-NS. It is tempting to speculate that, under specific environmental conditions, Hha, H-NS, and StpA could play different regulatory roles, acting as complexes or separately. By this means, regulation held by Hha, Hha/H-NS, and Hha/StpA guarantees the fine-tuning of the different regulons. Some of the genes identified in this work are good candidates for this dual regulation under different environmental conditions, as exemplified by the SPI-2 regulators SsrAB, as well as previously suggested for the regulation of *hlyABCD* or *hilA* gene expression (Nieto et al., 2000; Boddicker and Jones, 2004; Queiroz et al., 2011). In light of the above, the differential expression of *hns* and *hha* is likely to influence the abundance of the different complexes in the cell at any time. The expression of *hns* is relatively constant along the growth phase (Schröder and Wagner, 2002) and at different temperatures (Göransson et al., 1990), whereas *hha* expression is enhanced in LB medium at high temperature (37°C) and early stationary phase of growth (Paytubi et al., 2014). Therefore, Hha might become an important cellular protein under environmental conditions encountered by the bacterium during the infection process. Moreover, to accomplish the regulatory role described, we cannot rule out interactions of Hha with proteins other than H-NS or StpA.

The data obtained in this work indicates that 37.5% of the genes that belong to the Hha regulon correspond to HGT DNA (6.7% plasmid- and 30.8% chromosomally encoded genes). The percentage of HGT[2] DNA in *Salmonella* LT2 is approximately 10%. It is hence remarkable that the Hha regulon in *Salmonella* is enriched with HGT sequences. A recent study demonstrated that, in *E. coli* K-12, Hha/YdgT function in complexes with H-NS/StpA to regulate expression of HGT genes (Ueda et al., 2013).

---

[2] http://genomes.urv.es/HGT-DB (Garcia-Vallve et al., 2003).

Slight differences in the HGT content of both species, *E. coli* K-12 and *S.* Typhimurium, might account for the lack of identification of a specific Hha regulon in *E. coli*. Additionally, a significant proportion of the *Salmonella* "unique" genes are involved in virulence (Conner et al., 1998) where Hha plays a relevant role.

Future work will be aimed at precisely defining the molecular mechanism underlying the modulatory role of Hha protein.

Acknowledgments

We thank M. Hüttener and T. Gaviria for kindly providing SV5015S strain and pIC-ssrA2 plasmid. The authors are grateful to J. D. Hinton for his helpful comments. This work was supported by grants from the Spanish Ministry of Science and Innovation (CSD2008-00013, BFU2010-21836-C02-02), the Generalitat de Catalunya (2009SGR66) and the RecerCaixa program (2012/ACUP/00048). CS was recipient of a FPU grant from the Spanish Ministry.

References

Ali, S. S., Whitney, J. C., Stevenson, J., Robinson, H., Howell, P. L., and Navarre, W. W. (2013). Structural insights into the regulation of foreign genes in *Salmonella* by the Hha/H-NS complex. *J. Biol. Chem.* 288, 13356–13369. doi: 10.1074/jbc.M113.455378

Aznar, S., Paytubi, S., and Juárez, A. (2013). The Hha protein facilitates incorporation of horizontally acquired DNA in enteric bacteria. *Microbiology* 159, 545–554. doi: 10.1099/mic.0.062448-0

Badaut, C., Williams, R., Arluison, V., Bouffartigues, E., Robert, B., Buc, H., et al. (2002). The degree of oligomerization of the H-NS nucleoid structuring protein is related to specific binding to DNA. *J. Biol. Chem.* 277, 41657–41666. doi: 10.1074/jbc.M206037200

Baños, R. C., Vivero, A., Aznar, S., García, J., Pons, M., Madrid, C., et al. (2009). Differential regulation of horizontally acquired and core genome genes by the bacterial modulator H-NS. *PLoS Genet.* 5:e1000513. doi: 10.1371/journal.pgen.1000513

Boddicker, J. D., and Jones, B. D. (2004). Lon protease activity causes down-regulation of *Salmonella* pathogenicity island 1 invasion gene expression after infection of epithelial cells. *Infect. Immun.* 72, 2002–2013. doi: 10.1128/IAI.72.4.2002-2013.2004

Carmona, M., Balsalobre, C., Muñoa, F., Mouriño, M., Jubete, Y., Cruz, F., et al. (1993). *Escherichia coli* hha mutants, DNA supercoiling and expression of the haemolysin genes from the recombinant plasmid pANN202-312. *Mol. Microbiol.* 9, 1011–1018. doi: 10.1111/j.1365-2958.1993.tb01230.x

Cherepanov, P. P., and Wackernagel, W. (1995). Gene disruption in *Escherichia coli*: TcR and KmR cassettes with the option of Flp-catalyzed excision of the antibiotic-resistance determinant. *Gene* 158, 9–14. doi: 10.1016/0378-1119(95)00193-A

Conner, C. P., Heithoff, D. M., Julio, S. M., Sinsheimer, R. L., and Mahan, M. J. (1998). Differential patterns of acquired virulence genes distinguish *Salmonella* strains. *Proc. Natl. Acad. Sci. U.S.A.* 95, 4641–4645. doi: 10.1073/pnas.95.8.4641

Coombes, B. K., Brown, N. F., Valdez, Y., Brumell, J. H., and Finlay, B. B. (2004). Expression and secretion of *Salmonella* pathogenicity island-2 virulence genes in response to acidification exhibit differential requirements of a functional type III secretion apparatus and SsaL. *J. Biol. Chem.* 279, 49804–49815. doi: 10.1074/jbc.M404299200

Coombes, B. K., Wickham, M. E., Lowden, M. J., Brown, N. F., and Finlay, B. B. (2005). Negative regulation of *Salmonella* pathogenicity island 2 is required for contextual control of virulence during typhoid. *Proc. Natl. Acad. Sci. U.S.A.* 102, 17460–17465. doi: 10.1073/pnas.0505401102

Cordeiro, T. N., Garcia, J., Pons, J.-I., Aznar, S., Juárez, A., and Pons, M. (2008). A single residue mutation in Hha preserving structure and binding to H-NS results in loss of H-NS mediated gene repression properties. *FEBS Lett.* 582, 3139–3144. doi: 10.1016/j.febslet.2008.07.037

Crouch, M. L. V., Castor, M., Karlinsey, J. E., Kalhorn, T., and Fang, F. C. (2008). Biosynthesis and IroC-dependent export of the siderophore salmochelin are essential for virulence of *Salmonella enterica* serovar Typhimurium. *Mol. Microbiol.* 67, 971–983. doi: 10.1111/j.1365-2958.2007.06089.x

Cusick, M. E., and Belfort, M. (1998). Domain structure and RNA annealing activity of the *Escherichia coli* regulatory protein StpA. *Mol. Microbiol.* 28, 847–857. doi: 10.1046/j.1365-2958.1998.00848.x

Dame, R. T., Wyman, C., and Goosen, N. (2000). H-NS mediated compaction of DNA visualised by atomic force microscopy. *Nucleic Acids Res.* 28, 3504–3510. doi: 10.1093/nar/28.18.3504

Datsenko, K. A., and Wanner, B. L. (2000). One-step inactivation of chromosomal genes in *Escherichia coli* K-12 using PCR products. *Proc. Natl. Acad. Sci. U.S.A.* 97, 6640–6645. doi: 10.1073/pnas.120163297

de Alba, C. F., Solórzano, C., Paytubi, S., Madrid, C., Juarez, A., García, J., et al. (2011). Essential residues in the H-NS binding site of Hha, a co-regulator of horizontally acquired genes in Enterobacteria. *FEBS Lett.* 585, 1765–1770. doi: 10.1016/j.febslet.2011.05.024

Deighan, P., Free, A., and Dorman, C. J. (2000). A role for the *Escherichia coli* H-NS-like protein StpA in OmpF porin expression through modulation of micF RNA stability. *Mol. Microbiol.* 38, 126–139. doi: 10.1046/j.1365-2958.2000.02120.x

Dorman, C. J. (2007). H-NS, the genome sentinel. *Nat. Rev. Microbiol.* 5, 157–161. doi: 10.1038/nrmicro1598

Fàbrega, A., and Vila, J. (2013). *Salmonella enterica* serovar Typhimurium skills to succeed in the host: virulence and regulation. *Clin. Microbiol. Rev.* 26, 308–341. doi: 10.1128/CMR.00066-12

Fahlen, T. F., Mathur, N., and Jones, B. D. (2000). Identification and characterization of mutants with increased expression of hilA, the invasion gene transcriptional activator of *Salmonella typhimurium*. *FEMS Immunol. Med. Microbiol.* 28, 25–35. doi: 10.1111/j.1574-695X.2000.tb01453.x

Fahlen, T. F., Wilson, R. L., Boddicker, J. D., and Jones, B. D. (2001). Hha is a negative modulator of transcription of hilA, the *Salmonella enterica* serovar Typhimurium invasion gene transcriptional activator. *J. Bacteriol.* 183, 6620–6629. doi: 10.1128/JB.183.22.6620-6629.2001

Fang, F. C., and Rimsky, S. (2008). New insights into transcriptional regulation by H-NS. *Curr. Opin. Microbiol.* 11, 113–120. doi: 10.1016/j.mib.2008.02.011

Forns, I., Baños, R. C., Balsalobre, C., Juárez, A., and Madrid, C. (2005). Temperature-dependent conjugative transfer of R27: role of chromosome- and plasmid-encoded Hha and H-NS proteins. *J. Bacteriol.* 187, 3950–3959. doi: 10.1128/JB.187.12.3950-3959.2005

Free, A., Williams, R. M., and Dorman, C. J. (1998). The StpA protein functions as a molecular adapter to mediate repression of the bgl operon by truncated H-NS in *Escherichia coli*. *J. Bacteriol.* 180, 994–997.

Furrer, J. L., Sanders, D. N., Hook-Barnard, I. G., and McIntosh, M. A. (2002). Export of the siderophore enterobactin in *Escherichia coli*: involvement of a 43 kDa membrane exporter. *Mol. Microbiol.* 44, 1225–1234. doi: 10.1046/j.1365-2958.2002.02885.x

García, J., Cordeiro, T. N., Nieto, J. M., Pons, I., Juárez, A., and Pons, M. (2005). Interaction between the bacterial nucleoid associated proteins Hha and H-NS involves a conformational change of Hha. *Biochem. J.* 388, 755–762. doi: 10.1042/BJ20050002

García, J., Madrid, C., Juárez, A., and Pons, M. (2006). New roles for key residues in helices H1 and H2 of the *Escherichia coli* H-NS N-terminal domain:

H-NS dimer stabilization and Hha binding. *J. Mol. Biol.* 359, 679–689. doi: 10.1016/j.jmb.2006.03.059

Garcia-Vallve, S., Guzman, E., Montero, M. A., and Romeu, A. (2003). HGT-DB: a database of putative horizontally transferred genes in prokaryotic complete genomes. *Nucleic Acids Res.* 31, 187–189. doi: 10.1093/nar/gkg004

Göransson, M., Sondén, B., Nilsson, P., Dagberg, B., Forsman, K., Emanuelsson, K., et al. (1990). Transcriptional silencing and thermoregulation of gene expression in *Escherichia coli. Nature* 344, 682–685. doi: 10.1038/344682a0

Hantke, K., Nicholson, G., Rabsch, W., and Winkelmann, G. (2003). Salmochelins, siderophores of *Salmonella enterica* and uropathogenic *Escherichia coli* strains, are recognized by the outer membrane receptor IroN. *Proc. Natl. Acad. Sci. U.S.A.* 100, 3677–3682. doi: 10.1073/pnas.0737682100

Hommais, F., Krin, E., Laurent-Winter, C., Soutourina, O., Malpertuy, A., Le Caer, J. P., et al. (2001). Large-scale monitoring of pleiotropic regulation of gene expression by the prokaryotic nucleoid-associated protein, H-NS. *Mol. Microbiol.* 40, 20–36. doi: 10.1046/j.1365-2958.2001.02358.x

Iwama, T., Ito, Y., Aoki, H., Sakamoto, H., Yamagata, S., Kawai, K., et al. (2006). Differential recognition of citrate and a metal-citrate complex by the bacterial chemoreceptor Tcp. *J. Biol. Chem.* 281, 17727–17735. doi: 10.1074/jbc.M601038200

Johansson, J., and Uhlin, B. E. (1999). Differential protease-mediated turnover of H-NS and StpA revealed by a mutation altering protein stability and stationary-phase survival of *Escherichia coli. Proc. Natl. Acad. Sci. U.S.A.* 96, 10776–10781. doi: 10.1073/pnas.96.19.10776

Lim, C. J., Whang, Y. R., Kenney, L. J., and Yan, J. (2012). Gene silencing H-NS paralogue StpA forms a rigid protein filament along DNA that blocks DNA accessibility. *Nucleic Acids Res.* 40, 3316–3328. doi: 10.1093/nar/gkr1247

Lucchini, S., McDermott, P., Thompson, A., and Hinton, J. C. D. (2009). The H-NS-like protein StpA represses the RpoS (sigma 38) regulon during exponential growth of *Salmonella* Typhimurium. *Mol. Microbiol.* 74, 1169–1186. doi: 10.1111/j.1365-2958.2009.06929.x

Lucchini, S., Rowley, G., Goldberg, M. D., Hurd, D., Harrison, M., and Hinton, J. C. D. (2006). H-NS mediates the silencing of laterally acquired genes in bacteria. *PLoS Pathog.* 2:e81.

Madrid, C., Balsalobre, C., García, J., and Juárez, A. (2007a). The novel Hha/YmoA family of nucleoid-associated proteins: use of structural mimicry to modulate the activity of the H-NS family of proteins. *Mol. Microbiol.* 63, 7–14. doi: 10.1111/j.1365-2958.2006.05497.x

Madrid, C., García, J., Pons, M., and Juárez, A. (2007b). Molecular evolution of the H-NS protein: interaction with Hha-like proteins is restricted to enterobacteriaceae. *J. Bacteriol.* 189, 265–268. doi: 10.1128/JB.01124-06

Madrid, C., Nieto, J. M., and Juárez, A. (2002a). Role of the Hha/YmoA family of proteins in the thermoregulation of the expression of virulence factors. *Int. J. Med. Microbiol.* 291, 425–32. doi: 10.1078/1438-4221-00149

Madrid, C., Nieto, J. M., Paytubi, S., Falconi, M., Gualerzi, C. O., and Juárez, A. (2002b). Temperature- and H-NS-dependent regulation of a plasmid-encoded virulence operon expressing *Escherichia coli* hemolysin. *J. Bacteriol.* 184, 5058–5066. doi: 10.1128/JB.184.18.5058-5066.2002

Methner, U., Rabsch, W., Reissbrodt, R., and Williams, P. H. (2008). Effect of norepinephrine on colonisation and systemic spread of *Salmonella enterica* in infected animals: role of catecholate siderophore precursors and degradation products. *Int. J. Med. Microbiol.* 298, 429–39. doi: 10.1016/j.ijmm.2007.07.013

Miller, J. (1993). *A Short Course in Bacterial Genetics: A Laboratory Manual and Handbook for Escherichia coli and Related Bacteria.* New York: Cold Spring Harbor Laboratory Press.

Morimoto, Y. V., Che, Y.-S., Minamino, T., and Namba, K. (2010). Proton-conductivity assay of plugged and unplugged MotA/B proton channel by cytoplasmic pHluorin expressed in *Salmonella. FEBS Lett.* 584, 1268–1272. doi: 10.1016/j.febslet.2010.02.051

Morimoto, Y. V., and Minamino, T. (2014). Structure and function of the bi-directional bacterial flagellar motor. *Biomolecules* 4, 217–234. doi: 10.3390/biom4010217

Mouriño, M., Madrid, C., Balsalobre, C., Prenafeta, A., Muñoa, F., Blanco, J., et al. (1996). The Hha protein as a modulator of expression of virulence factors in *Escherichia coli. Infect. Immun.* 64, 2881–2884.

Navarre, W. W., Porwollik, S., Wang, Y., McClelland, M., Rosen, H., Libby, S. J., et al. (2006). Selective silencing of foreign DNA with low GC content by the H-NS protein in *Salmonella. Science* 313, 236–238. doi: 10.1126/science.1128794

Nieto, J. M., Carmona, M., Bolland, S., Jubete, Y., Cruz, F., and Juárez, A. (1991). The hha gene modulates haemolysin expression in *Escherichia coli. Mol. Microbiol.* 5, 1285–1293. doi: 10.1111/j.1365-2958.1991.tb01902.x

Nieto, J. M., Madrid, C., Miquelay, E., Parra, J. L., Rodríguez, S., and Juárez, A. (2002). Evidence for direct protein-protein interaction between members of the enterobacterial Hha/YmoA and H-NS families of proteins. *J. Bacteriol.* 184, 629–635. doi: 10.1128/JB.184.3.629-635.2002

Nieto, J. M., Madrid, C., Prenafeta, A., Miquelay, E., Balsalobre, C., Carrascal, M., et al. (2000). Expression of the hemolysin operon in *Escherichia coli* is modulated by a nucleoid-protein complex that includes the proteins Hha and H-NS. *Mol. Gen. Genet.* 263, 349–358. doi: 10.1007/s004380051178

Okumura, H., Nishiyama, S., Sasaki, A., Homma, M., and Kawagishi, I. (1998). Chemotactic adaptation is altered by changes in the carboxy-terminal sequence conserved among the major methyl-accepting chemoreceptors. *J. Bacteriol.* 180, 1862–1868.

Olekhnovich, I. N., and Kadner, R. J. (2007). Role of nucleoid-associated proteins Hha and H-NS in expression of *Salmonella enterica* activators HilD, HilC, and RtsA required for cell invasion. *J. Bacteriol.* 189, 6882–6890. doi: 10.1128/JB.00905-07

Ono, S., Goldberg, M. D., Olsson, T., Esposito, D., Hinton, J. C. D., and Ladbury, J. E. (2005). H-NS is a part of a thermally controlled mechanism for bacterial gene regulation. *Biochem. J.* 391, 203–213. doi: 10.1042/BJ20050453

Osborne, S. E., and Coombes, B. K. (2011). Transcriptional priming of *Salmonella* pathogenicity island-2 precedes cellular invasion. *PLoS ONE* 6:e21648. doi: 10.1371/journal.pone.0021648

Oshima, T., Ishikawa, S., Kurokawa, K., Aiba, H., and Ogasawara, N. (2006). *Escherichia coli* histone-like protein H-NS preferentially binds to horizontally acquired DNA in association with RNA polymerase. *DNA Res* 13, 141–153. doi: 10.1093/dnares/dsl009

Paytubi, S., Aznar, S., Madrid, C., Balsalobre, C., Dillon, S. C., Dorman, C. J., et al. (2014). A novel role for antibiotic resistance plasmids in facilitating *Salmonella* adaptation to non-host environments. *Environ. Microbiol.* 16, 950–962. doi: 10.1111/1462-2920.12244

Paytubi, S., García, J., and Juárez, A. (2011). Bacterial Hha-like proteins facilitate incorporation of horizontally transferred DNA. *Cent. Eur. J. Biol.* 6, 879–886. doi: 10.2478/s11535-011-0071-3

Paytubi, S., Madrid, C., Forns, N., Nieto, J. M., Balsalobre, C., Uhlin, B. E., et al. (2004). YdgT, the Hha paralogue in *Escherichia coli*, forms heteromeric complexes with H-NS and StpA. *Mol. Microbiol.* 54, 251–263. doi: 10.1111/j.1365-2958.2004.04268.x

Queiroz, M. H., Madrid, C., Paytubi, S., Balsalobre, C., and Juárez, A. (2011). Integration host factor alleviates H-NS silencing of the *Salmonella enterica* serovar Typhimurium master regulator of SPI1, hilA. *Microbiology* 157, 2504–2514. doi: 10.1099/mic.0.049197-0

Rose, R. E. (1988). The nucleotide sequence of pACYC184. *Nucleic Acids Res.* 16, 355. doi: 10.1093/nar/16.1.355

Rozen, S., and Skaletsky, H. (2000). Primer3 on the WWW for general users and for biologist programmers. *Methods Mol. Biol.* 132, 365–386.

Schmidt, H., and Hensel, M. (2004). Pathogenicity islands in bacterial pathogenesis. *Clin. Microbiol. Rev.* 17, 14–56. doi: 10.1128/CMR.17.1.14-56.2004

Schröder, O., and Wagner, R. (2002). The bacterial regulatory protein H-NS–a versatile modulator of nucleic acid structures. *Biol. Chem.* 383, 945–960. doi: 10.1515/BC.2002.101

Sharma, V. K., and Bearson, B. L. (2013). Hha controls *Escherichia coli* O157:H7 biofilm formation by differential regulation of global transcriptional regulators FlhDC and CsgD. *Appl. Environ. Microbiol.* 79, 2384–2396. doi: 10.1128/AEM.02998-12

Sharma, V. K., Carlson, S. A., and Casey, T. A. (2005). Hyperadherence of an hha mutant of *Escherichia coli* O157:H7 is correlated with enhanced expression of LEE-encoded adherence genes. *FEMS Microbiol. Lett.* 243, 189–196. doi: 10.1016/j.femsle.2004.12.003

Sharma, V. K., and Casey, T. A. (2014). Determining the relative contribution and hierarchy of hha and qseBC in the regulation of flagellar motility of *Escherichia coli* O157:H7. *PLoS ONE* 9:e85866. doi: 10.1371/journal.pone.0085866

Sharma, V. K., and Zuerner, R. L. (2004). Role of hha and ler in transcriptional regulation of the esp operon of enterohemorrhagic *Escherichia coli* O157:H7. *J. Bacteriol.* 186, 7290–7301. doi: 10.1128/JB.186.21.7290-7301.2004

Silphaduang, U., Mascarenhas, M., Karmali, M., and Coombes, B. K. (2007). Repression of intracellular virulence factors in *Salmonella* by the Hha and YdgT nucleoid-associated proteins. *J. Bacteriol.* 189, 3669–3673. doi: 10.1128/JB.00002-07

Spurio, R., Falconi, M., Brandi, A., Pon, C. L., and Gualerzi, C. O. (1997). The oligomeric structure of nucleoid protein H-NS is necessary for recognition of intrinsically curved DNA and for DNA bending. *EMBO J.* 16, 1795–1805. doi: 10.1093/emboj/16.7.1795

Stella, S., Spurio, R., Falconi, M., Pon, C. L., and Gualerzi, C. O. (2005). Nature and mechanism of the in vivo oligomerization of nucleoid protein H-NS. *EMBO J.* 24, 2896–2905. doi: 10.1038/sj.emboj.7600754

Sternberg, N. L., and Maurer, R. (1991). Bacteriophage-mediated generalized transduction in *Escherichia coli* and *Salmonella typhimurium*. *Methods Enzymol.* 204, 18–43. doi: 10.1016/0076-6879(91)04004-8

Studier, F. W., and Moffatt, B. A. (1986). Use of bacteriophage T7 RNA polymerase to direct selective high-level expression of cloned genes. *J. Mol. Biol.* 189, 113–130. doi: 10.1016/0022-2836(86)90385-2

Taylor, R. G., Walker, D. C., and McInnes, R. R. (1993). *E. coli* host strains significantly affect the quality of small scale plasmid DNA preparations used for sequencing. *Nucleic Acids Res.* 21, 1677–1678. doi: 10.1093/nar/21.7.1677

Tendeng, C., and Bertin, P. N. (2003). H-NS in Gram-negative bacteria: a family of multifaceted proteins. *Trends Microbiol.* 11, 511–518. doi: 10.1016/j.tim.2003.09.005

Tomljenovic-Berube, A. M., Mulder, D. T., Whiteside, M. D., Brinkman, F. S. L., and Coombes, B. K. (2010). Identification of the regulatory logic controlling *Salmonella* pathoadaptation by the SsrA-SsrB two-component system. *PLoS Genet.* 6:e1000875. doi: 10.1371/journal.pgen.1000875

Ueda, T. A., Takahashi, H. I., Uyar, E. B. R. U., Ishikawa, S. H. U., Ogasawara, N. A., and Oshima, T. A. K. U. (2013). Functions of the Hha and YdgT proteins in transcriptional silencing by the nucleoid proteins, H-NS and StpA, in *Escherichia coli*. *DNA Res* 20, 263–271. doi: 10.1093/dnares/dst008

Vivero, A., Baños, R. C., Mariscotti, J. F., Oliveros, J. C., García-del Portillo, F., Juárez, A., et al. (2008). Modulation of horizontally acquired genes by the Hha-YdgT proteins in *Salmonella enterica* serovar Typhimurium. *J. Bacteriol.* 190, 1152–1156. doi: 10.1128/JB.01206-07

Wallar, L. E., Bysice, A. M., and Coombes, B. K. (2011). The non-motile phenotype of *Salmonella* hha ydgT mutants is mediated through PefI-SrgD. *BMC Microbiol.* 11:141. doi: 10.1186/1471-2180-11-141

Williams, R. M., Rimsky, S., and Buc, H. (1996). Probing the structure, function, and interactions of the *Escherichia coli* H-NS and StpA proteins by using dominant negative derivatives. *J. Bacteriol.* 178, 4335–4343.

Xu, X., and Hensel, M. (2010). Systematic analysis of the SsrAB virulon of *Salmonella enterica*. *Infect. Immun.* 78, 49–58. doi: 10.1128/IAI.00931-09

Yamamoto, K., and Imae, Y. (1993). Cloning and characterization of the *Salmonella typhimurium*-specific chemoreceptor Tcp for taxis to citrate and from phenol. *Proc. Natl. Acad. Sci. U.S.A.* 90, 217–221. doi: 10.1073/pnas.90.1.217

Yee, A., Chang, X., Pineda-Lucena, A., Wu, B., Semesi, A., Le, B., et al. (2002). An NMR approach to structural proteomics. *Proc. Natl. Acad. Sci. U.S.A.* 99, 1825–1830. doi: 10.1073/pnas.042684599

Zhang, A., and Belfort, M. (1992). Nucleotide sequence of a newly-identified *Escherichia coli*. *Nucleic Acids Res.* 20, 6735. doi: 10.1093/nar/20.24.6735

Zhang, A., Derbyshire, V., Salvo, J. L., and Belfort, M. (1995). *Escherichia coli* protein StpA stimulates self-splicing by promoting RNA assembly in vitro. *RNA* 1, 783–793.

Zhang, A., Rimsky, S., Reaban, M. E., Buc, H., and Belfort, M. (1996). *Escherichia coli* protein analogs StpA and H-NS: regulatory loops, similar and disparate effects on nucleic acid dynamics. *EMBO J.* 15, 1340–1349.

Zhu, M., Valdebenito, M., Winkelmann, G., and Hantke, K. (2005). Functions of the siderophore esterases IroD and IroE in iron-salmochelin utilization. *Microbiology* 151, 2363–2372. doi: 10.1099/mic.0.27888-0

Conflict of Interest Statement: The authors declare that the research was conducted in the absence of any commercial or financial relationships that could be construed as a potential conflict of interest.

2

Impact of surface structure and feed gas composition on *Bacillus subtilis* endospore inactivation during direct plasma treatment

Christian Hertwig[1]*, Veronika Steins[2], Kai Reineke[1], Antje Rademacher[1], Michael Klocke[1], Cornelia Rauh[2] and Oliver Schlüter[1]*

[1] Leibniz Institute for Agricultural Engineering, Potsdam-Bornim, Germany, [2] Department of Food Biotechnology and Food Process Engineering, Berlin University of Technology, Berlin, Germany

Edited by:
Michael Gänzle,
University of Alberta, Canada

Reviewed by:
Eva-Guadalupe Lizárraga-Paulín,
Instituto Tecnológico y de Estudios
Superiores de Monterrey Campus
Estado de México, Mexico
Peter Setlow,
University of Connecticut Health
Center, USA

*Correspondence:
Oliver Schlüter and Christian Hertwig,
Leibniz Institute for Agricultural
Engineering, Max-Eyth-Allee 100,
14469 Potsdam-Bornim, Germany
oschlueter@atb-potsdam.de;
chertwig@atb-potsdam.de

This study investigated the inactivation efficiency of cold atmospheric pressure plasma treatment on *Bacillus subtilis* endospores dependent on the used feed gas composition and on the surface, the endospores were attached on. Glass petri-dishes, glass beads, and peppercorns were inoculated with the same endospore density and treated with a radio frequency plasma jet. Generated reactive species were detected using optical emission spectroscopy. A quantitative polymerase chain reaction (qPCR) based ratio detection system was established to monitor the DNA damage during the plasma treatment. Argon + 0.135% vol. oxygen + 0.2% vol. nitrogen as feed gas emitted the highest amounts of UV-C photons and considerable amount of reactive oxygen and nitrogen species. Plasma generated with argon + 0.135% vol. oxygen was characterized by the highest emission of reactive oxygen species (ROS), whereas the UV-C emission was negligible. The use of pure argon showed a negligible emission of UV photons and atomic oxygen, however, the emission of vacuum (V)UV photons was assumed. Similar maximum inactivation results were achieved for the three feed gas compositions. The surface structure had a significant impact on the inactivation efficiency of the plasma treatment. The maximum inactivation achieved was between 2.4 and 2.8 $\log_{10}$ on glass petri-dishes and 3.9 to 4.6 $\log_{10}$ on glass beads. The treatment of peppercorns resulted in an inactivation lower than 1.0 $\log_{10}$. qPCR results showed a significant DNA damage for all gas compositions. Pure argon showed the highest results for the DNA damage ratio values, followed by argon + 0.135% vol. oxygen + 0.2% vol. nitrogen. In case of argon + 0.135% vol. oxygen the inactivation seems to be dominated by the action of ROS. These findings indicate the significant role of VUV and UV photons in the inactivation process of *B. subtilis* endospores.

Keywords: cold plasma, spore inactivation, inactivation mechanism, DNA damage, qPCR

Introduction

In recent years, the application of cold atmospheric pressure plasma (CAPP) for the decontamination of food products, food packing material and/or food contact surfaces raised in attention (Pankaj et al., 2014; Schlüter and Fröhling, 2014). Plasma is in general an at least partially ionized gas, which contains charged particles such as ions and electrons as well as neutral species such as atoms, molecules, and radicals, furthermore also UV photons. Depending on their thermodynamic properties plasmas can be classified as thermal and non-thermal plasmas (Schlüter et al., 2013).

Thermal plasmas are characterized by a local thermodynamic equilibrium between the electrons, ions and neutral species, whereby the temperature of the plasma can reach several 1000 kelvins under atmospheric pressure (Moreau et al., 2008). In non-thermal plasma, there is a significant difference between the electron temperature and the temperature of the charged particles and bulk gas. The electron temperature can reach several 1000 kelvins, whereas the bulk gas temperature can be closed to ambient. These so called "cold" plasmas can be directly applied also to thermal sensitive surfaces (Ehlbeck et al., 2011).

Under atmospheric condition cold plasmas can be generated using different set-ups, such as dielectric barrier discharges or plasma jet systems (Surowsky et al., 2014). Various studies showed already the antimicrobial potential of different CAPP applications (Niemira, 2012), whereby the different reactive species inside the plasma, neutral and charges particles, UV photons and also irradiated heat, are responsible for the antimicrobial effect of the plasma application (Laroussi, 2002; Moisan et al., 2002). The obtained composition of the generated plasma depends on the plasma source, feed gas and also on operation conditions, e.g., energy input (Weltmann et al., 2008; Ehlbeck et al., 2011; Reineke et al., 2015).

The potential of several cold plasma applications to inactivate different endospores on various matrices was shown in other studies already (Lassen et al., 2005; Boudam et al., 2006; Deng et al., 2006; Brandenburg et al., 2009; Kim et al., 2014; Hertwig et al., 2015a,b; Reineke et al., 2015). Nevertheless, the mechanisms leading to the inactivation of endospores are not clear and are still under investigations. The inactivation behavior of endospores by plasma treatment is often described by biphasic inactivation kinetics (Moreau et al., 2000; Brandenburg et al., 2009; Reineke et al., 2015). These biphasic inactivation kinetics probably indicate the involvement of different inactivation effects of the cold plasma, like the inactivation by DNA damage due to emitted UV photons and the decomposition of microorganisms through photodesorption and etching. Whereas photodesorption is a UV-induced erosion of the cell, where UV photons break chemical bonds and lead to the formation of volatile compounds. Etching, however, is the adsorption of reactive species on microorganisms, leading to chemical reaction and the formation of volatile compounds (Moisan et al., 2002).

The main subject of the ongoing controversy regarding the CAPP based inactivation of microorganisms is the role of the generated UV and vacuum (V)UV photons. UV and VUV photons are known to induce strand breaks and other damages in DNA in the cell. Furthermore UV photons with wavelength below 275 nm can break C–C or C–H bonds (Brandenburg et al., 2009) and hence affect protein and other macromolecules structures and functioning. Most of the published studies claim that under atmospheric conditions UV photons play only a minor role in the inactivation process (Laroussi and Leipold, 2004; Perni

et al., 2007; Lu et al., 2008; Knorr et al., 2011), since major quantities of (V)UV photons are only emitted in low-pressure or vacuum plasma systems. Nevertheless, some groups showed that UV photons can dominate the inactivation process (Boudam et al., 2006; Reineke et al., 2015).

The structure of the contaminated surface has also a certain impact on the spore inactivation efficiency of the CAPP treatment, as rough surfaces with pits and cracks can hinder the inactivation of microorganisms (Surowsky et al., 2014). In most cases endospores were inoculated and plasma treated on smooth surfaces such as glass, polyethylene strips, polycarbonate membranes, or polymer foils (Heise et al., 2004; Deng et al., 2006; Brandenburg et al., 2009; Reineke et al., 2015). However, studies investigating the inactivation of endospores on different surfaces are scarce.

In this study the effect of different structured surfaces (glass petri-dishes, glass beads and peppercorns) concerning the inactivation of *Bacillus subtillis* endospores during CAPP treatment using a radio frequency plasma jet was investigated and to ensure comparable results all samples were inoculated with a similar endospore density. The selection of different surfaces, from a simple even glass surface via a spherical model (glass beads) to a real food matrix (peppercorns), in combination with an inoculation of comparable spore density enable a closer insight into the surface-related inactivation effect of *B. subtilis* endospores. The samples were treated using three different feed gas compositions, in order to vary the composition of the generated plasma and also the focus between the different involved mechanisms in endospore inactivation. Considering that the DNA play likely a considerable role during the inactivation process by CAPP, the quantification of DNA damage during the plasma exposure may help to closer understand the plasma based mechanisms responsible for the inactivation of endospores. Therefore a quantitative polymerase chain reaction (qPCR) based ratio detection system was established, which detected the degree of *B. subtilis* DNA damage during the CAPP treatment.

Materials and Methods

Bacillus subtilis Endospore Preparation

The endospore forming *B. subtilis* strain PS832 was used in this study. *B. subtilis* was sporulated using the method previously published by Nicholson and Setlow (1990). Sporulation was induced on 2x SG medium agar plates at 37°C without addition of antibiotics. After sporulation, the endospores were harvested with distilled water. The obtained suspension was washed and cleaned with cold distilled water by repeated centrifugation (threefold at 5000 g) and intermittently treated with ultrasonic (1 min). The cleaned endospore suspension contained $\geq$95% phase bright endospores and nearly no endospore agglomerates. The endospores were stored in the dark at 4°C, until needed.

Sample Preparation

In this study three different surfaces, such as glass petri-dishes (30 mm diameter), glass beads and whole black peppercorns

(*Piper nigrum*), were used. Peppercorns were purchased from JJ Albaracin (Murcia, Spain). To ensure comparable results between the different surfaces, all samples were inoculated with an endospore density of about $4*10^6$ endospores cm^{-2}, which is comparable to the native microbial load of black peppercorns (Hertwig et al., 2015a). Therefore, the surface of the peppercorns was measured using a particle analyzer PartAn 3001L (AnaTec GmbH, Duisburg, Germany). 1 g of peppercorns had a surface of 18.4 cm^2. The glass beads had a diameter of 5 mm. For the inoculation with *B. subtilis* endospores, 3.5 g of sterile peppercorns and 82 sterile glass beads, which had a similar surface of 64.4 cm^2, were placed into a sterile beaker and 175 µL stock endospore suspension was added. The beaker was placed on an automatic stirrer and shaken for 4 min at 400 rpm to obtain a homogenous coating of the microorganisms on the samples surface. The inoculated samples were placed under a clean bench for drying at room temperature for 30 min. Regarding the inoculation of the glass petri-dishes the stock endospore suspension was diluted 1:5 with ACES-buffer (pH 7). An aliquot of 300 µL diluted endospore suspension was mixed with 700 µL ethanol (96%). 35 µL of the ethanolic endospore suspension were spread on an area of 1.5 cm^2 of the glass petri-dishes.

Plasma Source and Plasma Treatment

The radio frequency (rf) plasma jet equipment used in this study was described elsewhere in detail (Brandenburg et al., 2007). The apparatus consists of a ceramic nozzle (nozzle tip diameter ~7 mm) with a needle electrode inside, a grounded ring electrode at the nozzle outlet, an rf-generator and a gas supply system. The rf-voltage is coupled with the needle electrode. The plasma is generated at the tip of this electrode and expands into the air outside the nozzle with a length of up to 15 mm. Prior to plasma treatment, the atmospheric pressure plasma jet was let run at experimental conditions for 15 min to allow preheating and passivation of the electrodes. Argon with additional mixing of 0.2% vol. nitrogen and/or 0.135% vol. oxygen was used as feed gas with a gas flow of 10 standard liter per minute and an operation power of 30 W. 1 g inoculated peppercorns and 23 inoculated glass beads were placed in individual sterile glass petri-dishes (30 mm diameter) and placed on an automatic stirrer below the collimated plasma beam with a distance of 12 mm to the nozzle outlet. The peppercorns were treated up to 15 min, glass beads up to 10 min, respectively. The inoculated glass petri-dishes were also plasma treated with a distance of 12 mm to the nozzle outlet, up to 5 min. A direct contact between the plasma and the surface of the glass petri-dishes was avoided, to prevent the endospores from abrasion by the plasma filaments. All treatments were done at least in quadruplicate.

Optical Emission Spectroscopy

A Black Comet UV-VIS Spectrometer (StellarNet, Inc., Tampa, FL, USA) equipped with a F400 UV-VIS-SR fiber optic and a quartz lens was used to measure the emission spectrum of the direct CAPP set-up. The spectrum was measured in the range from 190 to 850 nm. The distance from the middle of the nozzle outlet to the middle of the lens was 10 mm in vertical and 12 mm in the horizontal axis. The spectrum was measured 10 times with an integration time of 100 ms. The average spectrum was baseline corrected and normalized (between $\lambda = 450$–470 nm) using a self-written LabVIEW routine.

Viable Cell Counts

After the plasma treatment, the viable cell count was determined by standard cell culture methods. Therefore, the glass petri-dishes were filled with 1 ml ACES-buffer and four sterile glass beads were added. The *B. subtilis* endospores were resuspended by continuously shaking (250 rpm) for 30 min. The recovery of the endospores from the peppercorns and glass beads was carried out by shaking the samples in 4 ml ACES-buffer for 3 min at 400 rpm. The obtained suspensions were serially diluted in ACES-buffer and every dilution was plated on nutrient agar plates (Carl Roth GmbH, Karlsruhe, Germany) in duplicates. The plates were incubated at 37°C and the colony forming units (cfu) were counted after 24 and 48 h. The obtained inactivation kinetics were modeled with GInaFiT (Geeraerd and Van Impe Inactivation Model Fitting Tool), a freeware applet for Microsoft Excel, using a biphasic inactivation model (Cerf, 1977). In this model, the relation between the survival and exposure time is given by following equation:

$$\log_{10} S(t) = \varphi \cdot e^{k_1 \cdot t} + (1 - \varphi) \cdot e^{k_2 \cdot t}$$

where S(t) is N(t)/N0, with N(t) as the number of colony forming units at the time t and N0 as the initial number of colony forming units. φ is the fraction of the initial population in a major population and (1-φ) is the fraction of the initial population in a minor population; k_1 and k_2 are the specific inactivation rates of the two populations.

Infrared Temperature Imaging

During the plasma treatments, the surface temperature of the glass petri-dishes, glass beads and peppercorns was recorded by an infrared camera (ThermaCam 500, Flir, Frankfurt am Main, Germany) in triplicates. The emissivity of the glass petri-dishes and glass beads was set to 0.94 and 0.96 for peppercorns (BARTEC Messtechnik und Sensorik, 2001). The camera was installed from above at a distance of 1 m to the plasma treated sample; infrared images were taken at a frequency of 1 Hz. To exclude thermal inactivation effect, *B. subtilis* endospores inoculated on the three used material were thermal treated at the highest peak temperature measured during the plasma treatment, according to the corresponding plasma treatment time. The samples were placed in a heating and drying oven UT 20 (Heareus Instruments GmbH, Hanau, Germany) and the surface temperatures of the samples were measured using fiberglass-encased K-type thermocouples connected with a USB data acquisition system (Personal Daq/56, SynoTECH, Hückelhofen, Germany) and DASYLab 13.0 software.

Determination of Endospore DNA Damage

A qPCR based ratio detection system (Bauer et al., 2004; Roth et al., 2010) was used to determine the degree of DNA damage of the *B. subtilis* endospores after plasma treatment. The recovered endospore suspensions were pooled and collected by centrifugation (10 min; 10,000 *g*; 4°C). In case of the glass petri-dishes recovered endospore suspension of four replicates were pooled together. For chemical decoating, the pellet was suspended in 200 µL of 50 mmol L^{-1} Tris-HCl (pH 8.0), which contains 8 mol L^{-1} urea, 1% sodium dodecyl sulfate, 10 mmol L^{-1} EDTA and 50 mmol L^{-1} dithiothreitol. After 90 min incubation at 37°C, the decoated endospores were washed three times by repeated centrifugation (10 min; 10,000 *g*; 4°C) with cold, sterile water (Fairheadt et al., 1993). By suspending the endospores in 200 µL STE-buffer (150 mmol L^{-1} NaCl, 10 mmol L^{-1} Tris-HCl, pH 8.0; 10 mmol L^{-1} EDTA) containing 2 mg mL^{-1} lysozyme and incubating at 37°C for 60 min, the disruption of the endospores was accomplished. From the disrupted endospores, chromosomal DNA was purified using the High Pure PCR Template Preparation Kit (Roche, Penzberg, Germany). The concentration of the DNA was determined using a NanoDrop 3300 fluorospectrometer (Thermo Fisher Scientific, Inc., Waltham, MA, USA) applying the PicoGreen® dsDNA assay (Life Technologies GmbH, Darmstadt, Germany). The method used for the DNA damage determination was established by Roth et al. (2010) and optimized for the qPCR system used in this study. It can be assumed, that the plasma treatment causes randomly distributed defects along the DNA double strands, accordingly increases the probability of the detecting such defects with increasing length of the examined DNA fragment. Two PCR primer pairs Bs_dnaK855f (5′-CACAATGGGTCCTGTCCGTC-3′)/Bs_dnaK1254r (5′-AGACATTGGGCGCTCACCT-3′) and Bs_dnaK1154f (5′-ACACGACGATCCCAACAAGC-3′)/Bs_dnaK1254r were used to amplify a 400 bp reporter and an internal 101 bp fragment from the *dnaK* locus. The 101 bp fragment was used as an internal standard. Both fragments were amplified on a CFX96 Touch™ real-time PCR detection system (Bio-Rad Laboratories GmbH, München, Germany) in separate 20 µL volume reactions, each in triplicates. Per reaction 10 µL SYBR Green reagent (Quiagen, Hilden, Germany), 0.2 µmol L^{-1} of each primer and 0.10–0.15 ng templates DNA was used.

A 10-fold dilution series of a *B. subtilis* PS832 plasmid including the target fragment were used as an external standard and to determine the absolute copy number. The plasma treatment may cause cross-links between the DNA and other endospore components, thus influencing the quality and efficiency of DNA extraction and also unpredictably affecting the detection of the target fragments by PCR (Roth et al., 2010). The applied qPCR based ratio detection system (Roth et al., 2010) take this into account, whereby the target fragment copy numbers cn_{400} were normalized to those of the internal 101 bp fragment cn_{101}. Thus, the degree of DNA damage was expressed as a ratio (CN) between the detected 400 and 101 bp fragment copy numbers. For non-degraded DNA is the resulting ratio equal to 1 and decreases with increasing degrees of DNA damage. This ratio detection system can only be used for qualitative evaluation, because the correlation between actual degree of DNA damage and the ratio is unknown. This method allows the detection of various DNA damages like double or single strand breaks and thymidine dimers (Roth et al., 2010).

Results

Surface Temperature during CAPP Treatment

The average surface temperatures measured during the CAPP treatment are shown in **Table 1**. The CAPP treatment of glass beads leads to maximum local temperatures up to 90.1°C, the ones for the glass petri-dishes and peppercorns were slightly lower. The average surface temperature measured directly after the treatment was in the range of 56.9–75.2°C. Considering the measured surface temperatures, this CAPP application cannot be classified as a non-thermal treatment. To exclude thermal inactivation effects *B. subtilis* endospores inoculated on the three different sample types were thermal treated at 90°C in a heat and drying oven, according to the maximum CAPP treatment time. The thermal treatment resulted in no considerable inactivation; only for glass beads an inactivation of 0.2 log_{10} was obtained. However, the temperatures during the CAPP treatment may support the inactivation; nevertheless this effect should be comparable on the different treated surfaces due to the similar maximum local temperatures.

Characterization of Reactive Plasma Species

Three different feed gas compositions (1. pure argon, 2. argon + 0.135% vol. oxygen and 3. argon + 0.135% vol. oxygen + 0.2% vol. nitrogen) were used for detailed investigation of the involved inactivation mechanisms. Reineke et al. (2015) systematically investigated the emission intensity of argon plasma with the admixture of different oxygen and nitrogen concentration and showed that plasma running with argon + 0.135% vol. oxygen emitted a high amount of reactive oxygen species (ROS), whereas plasma running with argon + 0.135% vol. oxygen + 0.2% vol. nitrogen was characterized by the highest emission of UV-C photons. The emission spectra of the used plasmas, generated with the chosen gas compositions, are shown in **Figure 1**. The addition of oxygen and nitrogen causes significant changes in the emission spectra. In case of pure argon, molecular bands of oxygen, nitrogen, and other species were also detected due to interactions of the argon plasma with the surrounding air. The use of argon + 0.135% vol. oxygen + 0.2% vol. nitrogen resulted in considerable emission in the UV-C range (**Figure 1A**), whereas

TABLE 1 | Mean surface temperatures (±SD) before and after CAPP treatment.

	Starting temperature [°C]	Temperature after CAPP treatment [°C]	Maximum temperature during CAPP treatment [°C]
Glass petri-dishes	29.8 (±3.5)	56.9 (±1.0)	82.3 (±2.3)
Glass beads	27.7 (±0.4)	75.2 (±1.3)	90.1 (±0.2)
Peppercorns	28.3 (±1.7)	63.8 (±4.2)	88.3 (±0.6)

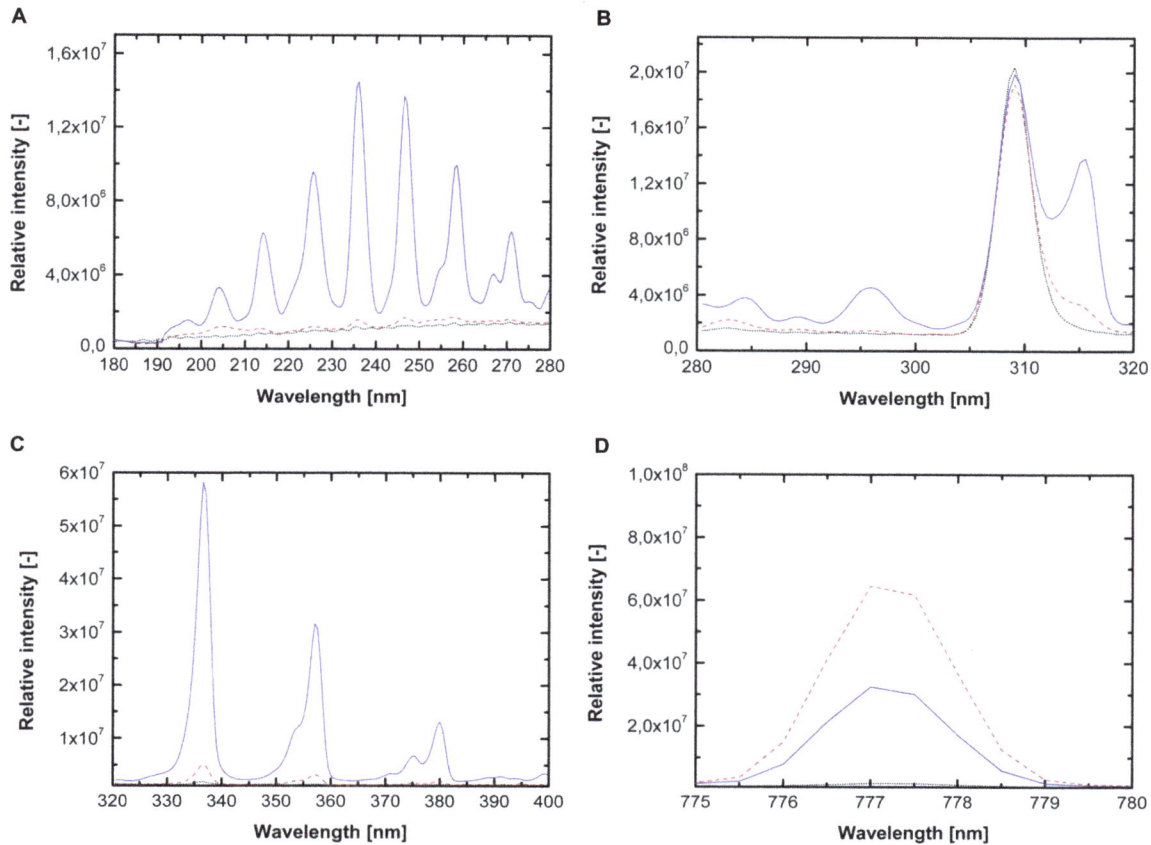

FIGURE 1 | Emission spectra for pure argon (dotted black line), argon + 0.135% vol. oxygen (dashed red line) and argon + 0.135% vol. oxygen + 0.2% vol. nitrogen (solid blue line) for wavelength from **(A)** 180–280 nm (UV-C light), **(B)** 280–320 nm (UV-B light), **(C)** 320–400 nm (UV-A light), and **(D)** 775–780 nm (atomic oxygen emission).

the emission of UV-C photons was negligible for the two other feed gas compositions. **Figure 1B** shows the emission intensity of the UV-B range, which was dominated by the signal of OH radicals with the maximum of 309 nm. The addition of oxygen and nitrogen resulted in no significant changes in the emission intensity of OH radicals. The emission spectrum from 320 to 400 nm (UV-A, **Figure 1C**) is dominated by molecular bands of the second positive system of N_2. The emissions for pure argon and argon + 0.135% vol. oxygen were negligible comparing to the use of argon + 0.135% vol. oxygen + 0.2% vol. nitrogen. **Figure 1D** shows the emission intensity from 775 to 780 nm, this wavelength range is characterized by the atomic oxygen band at 777 nm, and depicts considerable variations depending on the gas composition. The feed gas composition argon + 0.135% vol. oxygen emitted the highest photon intensity.

Effect of Feed Gas Composition and Surface Structure on the Inactivation of *B. subtilis* Endospores

To ensure comparable results between the three surfaces, all samples were inoculated with equal endospore densities of *B. subtilis* endospores. The obtained inactivation data were modeled using a biphasic inactivation model, which adequately described the inactivation behavior (**Table 2**). The resulting inactivation kinetics are shown in **Figure 2**. All inactivation kinetics showed an accelerated initial followed by a retarded inactivation for longer plasma exposure times, which can be also seen by inactivation rate constants (k). In general the k_1 (first inactivation phase) were higher than the k_2 (second inactivation phase) values. For each treated surface, the achieved maximum inactivation levels were relatively close together independent of the used feed gas composition. Nevertheless, the use of pure argon as feed gas for all three different surfaces resulted in the highest inactivation level, whereas the lowest inactivation was achieved by CAPP running with argon + 0.135% vol. oxygen in all cases.

In contrast to the feed gas composition, the treated surface had a tremendous impact on the inactivation efficiency of the CAPP plasma treatment. On the glass petri-dishes, the use of pure argon as feed gas for 5 min inactivated (**Figure 2A**) 2.7 $\log_{10}$ *B. subtilis* endospores, whereas the use of the second and third gas composition resulted in an inactivation of 2.2 and 2.5 $\log_{10}$, respectively. In contrast, the CAPP treatment of glass beads resulted in significantly higher maximum inactivation levels of

TABLE 2 | Statistical parameters and the corresponding standard error (in brackets) of the biphasic model obtained from GInaFit.

	Pure argon				Argon + 0.135% vol. oxygen				Argon + 0.135% vol. oxygen + 0.2% vol. nitrogen			
	Adj. R^2	φ [-]	k_1 [min^{-1}]	k_2 [min^{-1}]	Adj. R^2	φ [-]	k_1 [min^{-1}]	k_2 [min^{-1}]	Adj. R^2	φ [-]	k_1 [min^{-1}]	k_2 [min^{-1}]
Glass petri-dishes	1.00	0.93 (0.03)	2.54 (0.34)	0.71 (0.11)	0.98	0.86 (0.07)	3.28 (0.94)	0.62 (0.12)	0.99	0.77 (0.08)	3.85 (1.24)	0.87 (0.09)
Glass beads	0.99	1.00 (0.00)	4.15 (0.51)	0.42 (0.07)	0.99	1.00 (0.00)	4.04 (0.55)	0.32 (0.09)	1.00	1.00 (0.00)	4.48 (0.29)	0.19 (0.06)
Peppercorns	0.93	0.68 (0.12)	0.82 (0.43)	0.06 (0.03)	0.96	0.76 (0.27)	0.29 (0.16)	0.01 (0.07)	0.98	0.76 (0.04)	0.64 (0.12)	0.02 (0.02)

about 2 $\log_{10}$ (**Figure 2B**). CAPP generated by pure argon achieved an inactivation after 10 min treatment of 4.7 $\log_{10}$, followed by argon + 0.135% vol. oxygen + 0.2% vol. nitrogen with 4.5 $\log_{10}$ and argon + 0.135% vol. oxygen with 4.2 $\log_{10}$. The higher endospore inactivation after the CAPP treatment of glass beads cannot only be explained with the longer treatment time compared to the glass petri-dishes, because the inactivation obtained after a 4 min treatment were already higher than the maximum inactivation of the glass petri-dishes, i.e., 3.7, 3.5, and 4.0 $\log_{10}$ for the three feed gas compositions. The CAPP treatment of peppercorns resulted for all three used gas compositions in an inactivation less than 1.0 $\log_{10}$ after 15 min.

Endospore DNA Damage caused by Different Feed Gas Compositions

In case of CAPP treated peppercorns no assessment of DNA damage was conducted. Peppercorns are often highly spoiled with microorganisms. Even though they can be sterilized, the DNA material of the native microbial load is still present on the peppercorns surface and would falsify the results. For a better comparison between the inactivation and DNA damage ratio, only inactivation data of samples, which were also used for the analyzing of the DNA damage, were considered for the depiction of the inactivation behavior. Thus the inactivation kinetics shown in **Figures 3** and **4** may differ slightly from those shown in **Figure 2**. The DNA damage ratio values were also modeled using the biphasic equation (Cerf, 1977) to investigate if the damage of the *B. subtilis* DNA during the CAPP showed a similar course as the corresponding inactivation kinetics. Furthermore, the point of inflection (PI) of the biphasic kinetics (inactivation and DNA damage) was calculated, which describe the transition between the first and the second phase and can be calculated as the point of intersection between the two linear phases.

The results for the DNA damage of *B. subtilis* endospores on glass petri-dishes are depicted in **Table 3** and **Figure 3**. The maximum DNA damage ratio correlates with the corresponding maximum inactivation level. Comparing to the inactivation kinetics not all DNA damage kinetics could be adequately described using the biphasic model, with an accelerated initial DNA damage followed by a retarded damage of DNA. The use of pure argon resulted in the highest inactivation of 2.8 $\log_{10}$ and also in the highest DNA damage with a ratio of 0.51. The shift from the accelerated first inactivation phase to the second one was at 1.9 min, whereas the shift between these phases was already at 0.1 min for DNA damage kinetic. Plasma running with argon + 0.135% vol. oxygen + 0.2% vol. nitrogen achieved after 5 min treatment an inactivation and DNA damage ratio value of 2.5 $\log_{10}$ and 0.57, respectively. The DNA damage showed a linear behavior, because k_1 and k_2 had the same value. The shift to the second inactivation phase was after 0.4 min. Argon + 0.135% vol. oxygen as a feed gas composition inactivated up to 2.4 $\log_{10}$ endospores and the resulting DNA damage ratio was the highest with 0.63. For DNA damage kinetic no point of inflection could be calculated, since φ was smaller than 0.5. The shift between the two inactivation phases was at about 1.0 min.

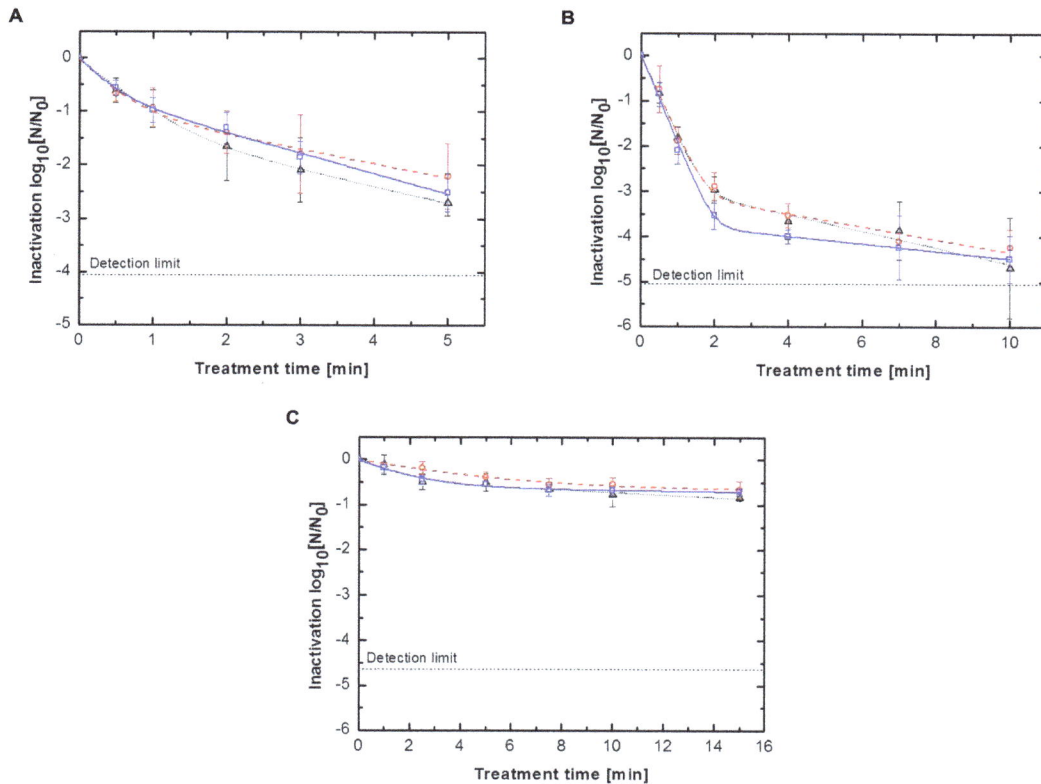

FIGURE 2 | Inactivation kinetics for *Bacillus subtilis* endospores for (△) pure argon, (○) argon + 0.135% vol. oxygen, (□) argon + 0.135% vol. oxygen + 0.2% vol. nitrogen inoculated on: (A) glass petri-dishes, (B) glass beads and (C) peppercorns, with biphasic fit.

As above mentioned, the inactivation efficiency of CAPP for the treatment of *B. subtilis* endospores on glass beads was significant higher than on the glass petri-dishes. The maximum DNA damage was also significant higher, indicated by ratio values between 0.46 and 0.25 (**Figure 4**). All DNA damage kinetics could be adequately described using the biphasic model (**Table 4**). CAPP generated with pure argon achieved again the highest inactivation level after 10 min with 4.6 $\log_{10}$, followed by the third and second gas composition with 4.0 and 3.9 $\log_{10}$ (**Figure 4**), respectively. The points of inflection for the biphasic inactivation kinetics of pure argon and argon + 0.135% vol. oxygen were relatively close together at 1.2 and 1.3 min. The accelerated first inactivation phase of argon + 0.135% vol. oxygen + 0.2% vol. nitrogen lasted until 1.9 min. The maximum values for the DNA damage showed the same trend as the maximum inactivation values. Pure argon had a ratio value of 0.25. The ones for the other two feed gas composition were considerably higher, 0.46 for argon + 0.135% vol. oxygen and 0.41 for argon + 0.135% vol. oxygen + 0.2% vol. nitrogen. No point of inflection could be calculated for the use of argon + 0.135% vol. oxygen, because φ was again smaller than 0.5. For pure argon, the shift of the two phases was already at 0.2 min, whereas the shift for the third gas composition was at 1.0 min.

Discussion

To study the inactivation mechanisms of CAPP on endospores, *B. subtilis* endospores inoculated on different surfaces were plasma treated using three different feed gas compositions. CAPP treatment is usually described as a non-thermal process, however, during the experiment in this study peak temperatures up to 90°C were reached (**Table 1**). However, a strong thermal inactivation effect on the *B. subtilis* endospores during the CAPP treatment could not be detected in this study. The three different used gas compositions were characterized by their different emission spectra. Plasma generated by pure argon showed less emission in the UV range and of atomic oxygen than the other used feed gas compositions. The use of argon + 0.135% vol. oxygen leads to a strong emission of atomic oxygen; however, the emission in the UV range was also negligible. Plasma running with argon + 0.135% vol. oxygen + 0.2% vol. nitrogen showed a significant emission in the UV-C range, furthermore it emitted also an significant amount of reactive oxygen and nitrogen species (RNS). The emission of OH radicals reached for all three gas compositions the same range. Reineke et al. (2015) showed similar results for the emission intensities of the three used gas composition. They reported a fourfold higher UV emission by plasma running with argon + 0.135% vol. oxygen + 0.2% vol. nitrogen compared

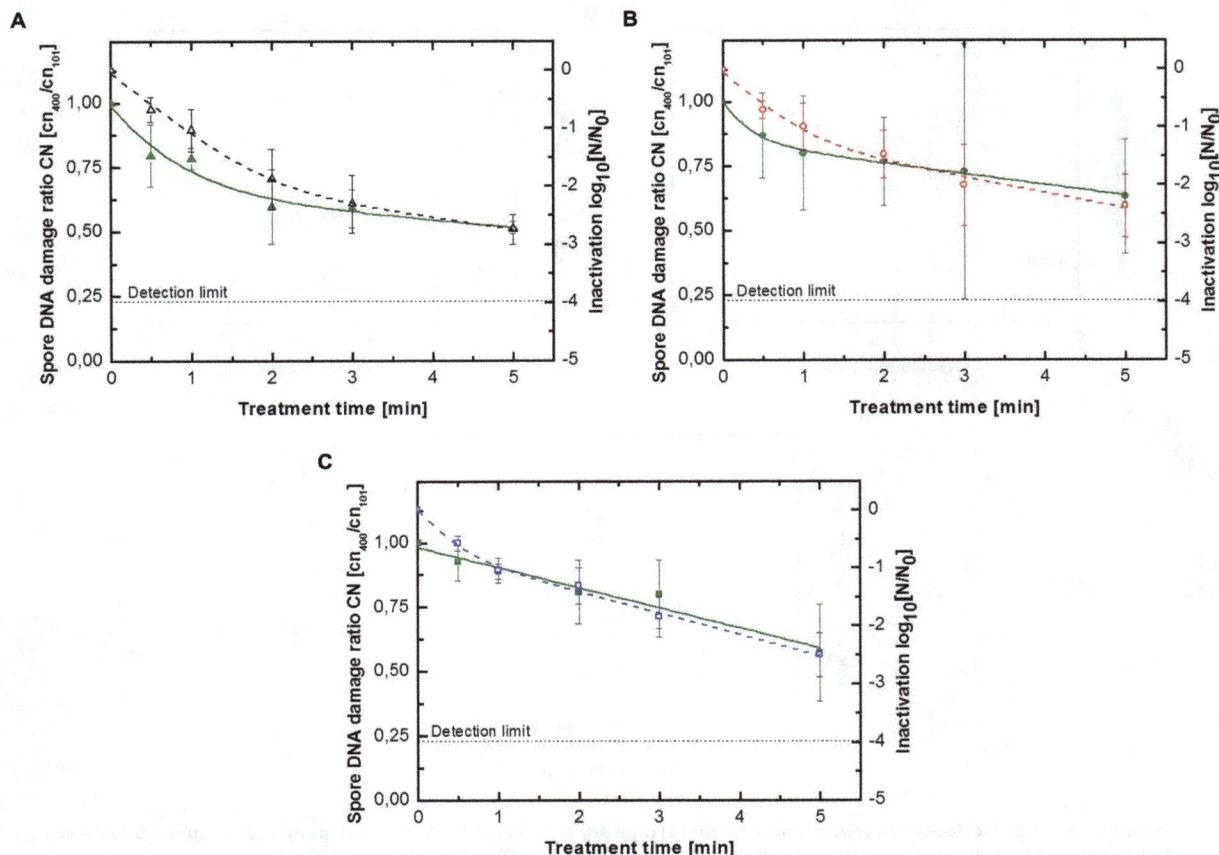

FIGURE 3 | Kinetics for *B. subtilis* endospores inoculated on glass petri-dishes for (A) pure argon with (△) inactivation and (▲)endospore DNA damage ratio, (B) argon + 0.135% vol. oxygen with (○) inactivation and (●)endospore DNA damage ratio, (C) argon + 0.135% vol. oxygen + 0.2% vol. nitrogen with (□) inactivation and (■) endospore DNA damage ratio. Solid lines represent the biphasic fit for the DNA damage and the dashed lines for endospore inactivation.

to the use of pure argon. Furthermore they also showed that the admixture of oxygen leads to an enhanced atomic oxygen emission. Regarding their different spectral intensities (**Figure 1**), different inactivation efficiencies could be expected and also different inactivation mechanisms.

In general the inactivation of endospores by CAPP treatment can be attributed to three different main mechanisms: (1) DNA damage due to UV photons, (2) intrinsic photodesorption, and (3) etching of organic molecules (Moisan et al., 2001). In case of plasma generated with argon + 0.135% vol. oxygen, which emitted the highest amount of ROS, the inactivation process should be dominated by the oxidation potential of the different ROS, namely OH radicals and atomic oxygen. In comparison, using plasma generated by argon + 0.135% vol. oxygen + 0.2% vol. nitrogen the inactivation process should be dominated by the damage of endospores DNA. A factor, which could contribute to the inactivation process are VUV photons, which can effectively inactivate *B. subtilis* endospores (Munakata et al., 1991). Argon driven plasma jets are well known to emit a certain amount of VUV light, dominated by argon excimer Ar^*_2 with an intensity maximum

at $\lambda = 126$ nm (Brandenburg et al., 2009; Ehlbeck et al., 2011). Brandenburg et al. (2009) characterized the VUV emission of an identically constructed argon driven plasma jet and showed that the absolute radiance in the VUV range (115–200 nm) did not change substantially up to a distance of 10 mm to the nozzle outlet and the irradiance at that distance can be estimated with about 2 Mw cm^{-2}. However, the emission of VUV photons could not be measured with the spectrometer used in this study.

The results depicted in **Figure 2** indicated that the inactivation efficiency is almost independent of the used feed gas composition. On all three CAPP treated surfaces the maximum achieved inactivation of the different gas compositions was relatively similar. In contrast, the treated surface had a significant impact on the inactivation efficiency of the CAPP treatment. All different surfaces were inoculated with a similar endospore density of about $4*10^6$ endospores cm^{-2} to ensure comparable results. Nevertheless, the distribution of the endospores on the surface can be different and can influence the inactivation efficiency of the plasma treatment. The significant lower inactivation of endospores on peppercorns can be explained by the structured and uneven surface. The surface of peppercorns is characterized

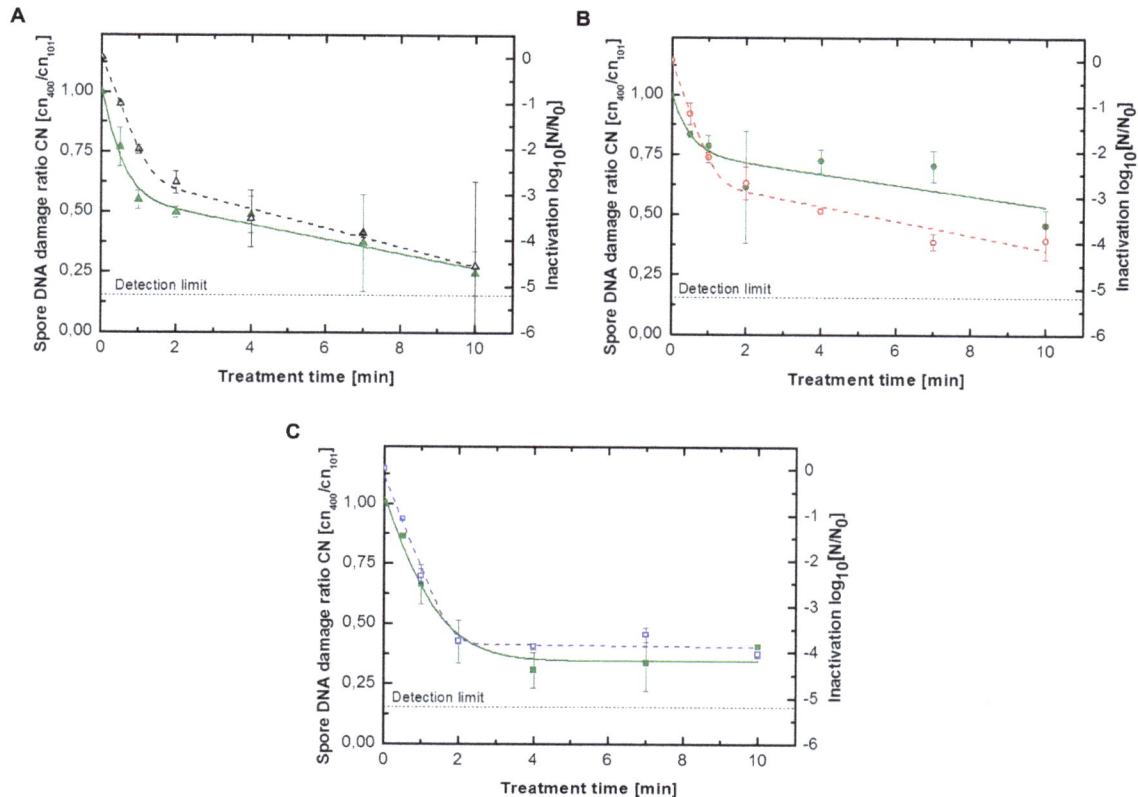

FIGURE 4 | Kinetics for *B. subtilis* endospores inoculated on glass beads for **(A)** pure argon with (Δ) inactivation and (▲)endospore DNA damage ratio, **(B)** argon + 0.135% vol. oxygen with (○) inactivation and (●)endospore DNA damage ratio, **(C)** argon + 0.135% vol. oxygen + 0.2% vol. nitrogen with (□) inactivation and (■) endospore DNA damage ratio. Solid lines represent the biphasic fit for the DNA damage and the dashed lines for endospore inactivation.

by cracks, grooves, and pits (Hertwig et al., 2015b) and cause probably shadow effects, for the emitted (V)UV photons, reactive species and charged particles, which reduce the efficiency of the CAPP treatment.

As described above, the higher endospore inactivation for the treatment of glass beads compared to the maximum achieved inactivation on glass petri-dishes cannot only be explained with the longer treatment time. The distribution of the inoculated *B. subtilis* endospores is presumably more homogenous on glass beads than on glass petri-dishes. The endospore inoculation on the petri-dishes causes probably the formation of agglomerates on the outer edge of the endospore suspension drop. Since the penetration depth of the (V)UV photons and reactive species is restricted to a few nm, the CAPP treatment would only effect the top layer of aggregated endospores (Hertwig et al., 2015b). Noticeable is that plasma generated by pure argon, which had the lowest emission of UV photons and other reactive species, achieved on all three treated surfaces the highest inactivation. The feed gas composition argon + 0.135% vol. oxygen + 0.2% vol. nitrogen, which emitted the highest amount of UV-C photons and considerable amounts of ROS and RNS, showed on all surfaces the second highest inactivation.

Reineke et al. (2015) reported similar results for the inactivation of *B. subtilis* endospores on glass petri-dishes and showed that the use of pure argon and argon + 0.135% vol. oxygen + 0.2% vol. nitrogen for the generation of plasma resulted in a similar inactivation efficiency after a treatment of 5 min. All inactivation kinetics showed a biphasic behavior, with an accelerated first inactivation phase and a slower second one, k_1 values were always higher than the k_2 values (**Table 2**). The biphasic inactivation curves pointed toward different inactivation mechanisms which may be involved during the treatment. Whereas the fast inactivation during the first phase can be attributed to the emitted (V)UV photons, which cause DNA damage, the slower inactivation during the second phase is presumably caused by a combination of DNA damage, photodesorption and etching (Knorr et al., 2011).

Reineke et al. (2015) treated UV-sensitive *B. subtilis* endospore mutant strains, FB122 and PS578, using plasmas with different UV-emission intensities and showed the significant impact of UV photons on the first inactivation phase. The strain FB122 is unable to synthesize DPA during the sporulation; whereas the strain PS578 lacks the genes encoding the spore's two major SASPs. Both DPA and SASPs contribute to the UV resistance

TABLE 3 | Statistical parameters and the corresponding standard error (in brackets) of biphasic model obtained regarding the inactivation and DNA damage of B. subtilis endospores inoculated on glass petri-dishes.

	Pure argon					Argon + 0.135% vol. oxygen					Argon + 0.135% vol. oxygen + 0.2% vol. nitrogen				
	Adj. R^2	φ [–]	k_1 [min^{-1}]	k_2 [min^{-1}]	PI [min]	Adj. R^2	φ [–]	k_1 [min^{-1}]	k_2 [min^{-1}]	PI [min]	Adj. R^2	φ [–]	k_1 [min^{-1}]	k_2 [min^{-1}]	PI [min]
Inactivation kinetic	0.99	0.98 (0.01)	2.52 (0.32)	0.48 (0.16)	1.9	0.97	0.90 (0.07)	2.92 (0.99)	0.62 (0.19)	1.0	0.99	0.79 (0.09)	4.02 (1.41)	0.85 (0.10)	0.4
DNA damage ratio kinetic	0.92	0.54 (0.16)	1.46 (0.92)	0.06 (0.09)	0.1	0.99	0.30 (0.04)	3.18 (1.05)	0.10 (0.01)	–	0.91	0.97 (8.0E+14)	0.18 (–)	0.18 (–)	–

Table 4 | Statistical parameters and the corresponding standard error in brackets of biphasic model obtained regarding the inactivation and DNA damage of B. subtilis endospores inoculated on glass beads.

	Pure argon					Argon + 0.135% vol. oxygen					Argon + 0.135% vol. oxygen + 0.2% vol. nitrogen				
	Adj. R^2	φ [–]	k_1 [min^{-1}]	k_2 min^{-1}]	PI [min]	Adj. R^2	φ [–]	k_1 [min^{-1}]	k_2 [min^{-1}]	PI [min]	Adj. R^2	φ [–]	k_1 [min^{-1}]	k_2 [min^{-1}]	PI [min]
Inactivation kinetic	0.99	1.00 (0.00)	4.81 (0.59)	0.48 (0.06)	1.3	0.98	1.00 (0.00)	5.20 (0.95)	0.36 (0.09)	1.2	0.98	1.00 (0.00)	4.59 (0.69)	0.02 (0.11)	1.9
DNA damage ratio kinetic	0.97	0.64 (0.06)	2.55 (0.73)	0.07 (0.02)	0.2	0.96	0.43 (0.19)	2.43 (3.08)	0.05 (0.04)	–	0.96	0.79 (0.05)	1.29 (0.32)	0.00 (0.03)	1.0

of *B. subtilis* endospores (Reineke et al., 2015). The different weighting between the involved inactivation mechanisms can be seen in the different shift between the two inactivation phases (**Tables 3 and 4**). To further investigate the inactivation mechanisms, a qPCR based ratio detection system (Bauer et al., 2004; Roth et al., 2010) was established to monitor the damage of endospore's DNA during the CAPP treatment. DNA is the primary target of emitted (V)UV photons during the CAPP treatment. ROS, namely atomic oxygen and OH radicals, are well known for their oxidation potential and can react with almost all cell components (Surowsky et al., 2014). Whereas OH radicals have the highest oxidation potential of all ROS, they are able to oxidize unsaturated fatty acids, proteins and to initiate DNA damage (Laroussi, 2002; Surowsky et al., 2014). However, the saturation of *B. subtilis* endospores DNA with α/β-type small acid soluble proteins (SASP) protects the DNA against damage caused by OH radicals (Setlow and Setlow, 1993). Hence, the measured DNA damage can be attributed to the emitted (V)UV photons. The DNA damage ratio values showed the same dependence on the used feed gas composition like the maximum achieved inactivation. The fact that plasma running with pure argon obtained on both surfaces the highest DNA damage showed that VUV photons can play a significant role in the inactivation of endospores at atmospheric pressure. All DNA damage kinetics, except the one of argon + 0.135% vol. oxygen + 0.2% vol. nitrogen on glass beads, showed a continuous increase of DNA damage during the entire inactivation process. The DNA damage kinetic for the treatment on glass beads using plasma running with argon + 0.135% vol. oxygen + 0.2% vol. nitrogen merge into a tailing after 3 min (**Figure 4C**). However, the corresponding inactivation curve showed the same tailing, this behavior support the finding, that the damage of DNA due to (V)UV photons is one of the main inactivation mechanisms.

The point of inflection between the two inactivation phases was achieved always significantly earlier for the DNA damage kinetics than for the corresponding inactivation kinetics (**Tables 3 and 4**). The accelerated first inactivation phase is often attributed to the damage of DNA by (V)UV photons (Moisan et al., 2002), however, the different shift between the two phases for the inactivation and DNA damage showed that the (V)UV based inactivation is presumably not the only mechanism contributing to the first fast inactivation phase. The linear behavior for the DNA damage kinetic of argon + 0.135% vol. oxygen + 0.2% vol. nitrogen (**Figure 3C**) can be explained by the high amount of emitted ROS and RNS. High amounts of reactive species are necessary for the decomposition of endospores due to etching, which lead to exposure of former covered endospores in, e.g., agglomerates and consequently to a further DNA damage. The early shift in the inactivation kinetic at 0.4 min may indicate the point, where the inactivation based on UV photons change to an inactivation process based on UV photons and supported by etching and photodesorption. For both DNA damage kinetics of argon + 0.135% vol. oxygen (glass beads and glass petri-dishes) the shift between the two phases could not be calculated, because both φ were under 0.5 and the treatment resulted in the highest DNA damage ratio value for both surfaces. Plasma generated with argon + 0.135% vol. oxygen emitted the highest amount of ROS, furthermore the admixture of air to argon decreases the VUV emission due to an increase of atomic oxygen emission (Brandenburg et al., 2009). Considering this, the inactivation process seems to be dominated by the action of ROS.

Conclusion

It can be stated that (V)UV photons emitted by CAPP play presumably a considerable role during the inactivation process of B. subtilis endospores, if the (V)UV intensity is high enough. Furthermore, the structure of the plasma treated surface as well as the distribution of the endospores on it affects the inactivation efficiency of CAPP treatment.

Acknowledgments

This study was funded by the research project "Plasma-based decontamination of dried plant related products for an enhancement of food safety (^{3}Plas)," which was financially supported by the German Federal Ministry of Food and Agriculture (2819102713).

References

BARTEC Messtechnik und Sensorik. (2001). *Emissionsfaktor-Tabelle*. Gotteszell: BARTEC Messtechnik und Sensorik. Available at: http://www.bartec.de/homepage/deu/downloads/produkte/19_temperatur/Ti_Tabelle_Emission_d.pdf (accessed September 14, 2014).

Bauer, T., Hammes, W. P., Haase, N. U., and Hertel, C. (2004). Effect of food components and processing parameters on DNA degradation in food. *Eur. Food Res. Technol.* 3, 215–223. doi: 10.1051/ebr:2005005

Boudam, M. K., Moisan, M., Saoudi, B., Popovici, C., Gherardi, N., and Massines, F. (2006). Bacterial spore inactivation by atmospheric-pressure plasmas in the presence or absence of UV photons as obtained with the same gas mixture. *J. Phys. D Appl. Phys.* 39, 3494–3507. doi: 10.1088/0022-3727/39/16/S07

Brandenburg, R., Ehlbeck, J., Stieber, M., Woedtke, T. V., Zeymer, J. S., and Schlüter, O. (2007). Antimicrobial treatment of heat sensitive materials by means of atmospheric pressure Rf-driven plasma jet. *Contribut. Plasma Phys.* 47, 72–79. doi: 10.1002/ctpp.200710011

Brandenburg, R., Lange, H., von Woedtke, T., Stieber, M., Kindel, E., and Ehlbeck, J. (2009). Antimicrobial Effects of UV and VUV Radiation of Nonthermal Plasma Jets. *IEEE Trans. Plasma Sci.* 37, 877–883. doi: 10.1109/TPS.2009.2019657

Cerf, O. (1977). Tailing of survival curves of bacterial spores. *J. Appl. Bacteriol.* 42, 1–19. doi: 10.1111/j.1365-2672.1977.tb00665.x

Deng, X., Shi, J., and Kong, M. (2006). Physical mechanisms of inactivation of *Bacillus subtilis* spores using cold atmospheric plasmas. *IEEE Trans. Plasma Sci.* 34, 1310–1316 doi: 10.1128/AEM.00583-12

Ehlbeck, J., Schnabel, U., Polak, M., Winter, J., Von Woedtke, T., Brandenburg, R., et al. (2011). Low temperature atmospheric pressure plasma sources for microbial decontamination. *J. Phys. D Appl. Phys.* 44, 1–18. doi: 10.1016/j.jhin.2012.02.012

Fairheadt, H., Setlow, B., and Setlow, P. (1993). Prevention of DNA damage in spores and in vitro by small, acid-soluble proteins from *Bacillus* species. *J. Bacteriol.* 175, 1367–1374.

Heise, M., Neff, W., Franken, O., Muranyi, P., and Wunderlich, J. (2004). Sterilization of polymer foils with dielectric barrier discharges at atmospheric pressure. *Plasmas Polym.* 9, 23–33. doi: 10.1023/B:PAPO.0000039814.70172.c0

Hertwig, C., Reineke, K., Ehlbeck, J., Knorr, D., and Schlüter, O. (2015a). Impact of remote plasma treatment on natural microbial load and quality parameters of selected herbs and spices. *J. Food Eng.* (in press). doi: 10.1016/j.jfoodeng.2014.12.017

Hertwig, C., Reineke, K., Ehlbeck, J., Knorr, D., and Schlüter, O. (2015b). Decontamination of whole black pepper using different cold atmospheric pressure plasma applications. *Food Control* 55, 221–229. doi: 10.1016/j.foodcont.2015.03.003

Kim, J. E., Lee, D.-U., and Min, S. C. (2014). Microbial decontamination of red pepper powder by cold plasma. *Food Microbiol.* 38, 128–136. doi: 10.1016/j.fm.2013.08.019

Knorr, D., Froehling, A., Jaeger, H., Reineke, K., and Schlueter, O. (2011). Emerging technologies in food processing. *Annu. Rev. Food Sci. Technol.* 2, 203–235. doi: 10.1146/annurev.food.102308.124129

Laroussi, M. (2002). Nonthermal decontamination of biological media by atmospheric-pressure plasmas: review, analysis, and prospects. *IEEE Trans. Plasma Sci.* 30, 1409–1415. doi: 10.1109/TPS.2002.804220

Laroussi, M., and Leipold, F. (2004). Evaluation of the roles of reactive species, heat, and UV radiation in the inactivation of bacterial cells by air plasmas at atmospheric pressure. *Int. J. Mass Spectrom.* 233, 81–86. doi: 10.1016/j.ijms.2003.11.016

Lassen, K. S., Nordby, B., and Grün, R. (2005). The dependence of the sporicidal effects on the power and pressure of RF-generated plasma processes. *J. Biomed. Mater. Res. Part B Appl. Biomater.* 74, 553–559. doi: 10.1002/jbm.b.30239

Lu, X., Ye, T., Cao, Y., Sun, Z., Xiong, Q., Tang, Z., et al. (2008). The roles of the various plasma agents in the inactivation of bacteria. *J. Appl. Phys.* 104, 1–5. doi: 10.1063/1.2977674

Moisan, M., Barbeau, J., Moreau, S., Pelletier, J., Tabrizian, M., and Yahia, L. H. (2001). Low-temperature sterilization using gas plasmas: a review of the experiments and an analysis of the inactivation mechanisms. *Int. J. Pharm.* 226, 1–21. doi: 10.1016/S0378-5173(01)00752-9

Moisan, M., Sakudo, A., Burke, P., and Mcdonnell, G. (2002). Plasma sterilization. Methods and mechanisms. *Pure Appl. Chem.* 74, 349–358. doi: 10.1016/S0378-5173(01)00752-9

Moreau, M., Orange, N., and Feuilloley, M. G. J. (2008). Non-thermal plasma technologies: new tools for bio-decontamination. *Biotechnol. Adv.* 26, 610–617. doi: 10.1016/j.biotechadv.2008.08.001

Moreau, S., Moisan, M., Tabrizian, M., Barbeau, J., Pelletier, J., Ricard, A., et al. (2000). Using the flowing afterglow of a plasma to inactivate *Bacillus subtilis* spores: influence of the operating conditions. *J. Appl. Phys.* 88, 1166–1174. doi: 10.1063/1.373792

Munakata, N., Saito, M., and Hieda, K. (1991). Inactivation action spectra of *Bacillus subtilis* spores in extended ultraviolet wavelengths(50–300nm) obtained with snchrotron radiation. *Photochem. Photobiol.* 54, 761–768. doi: 10.1111/j.1751-1097.1991.tb02087.x

Nicholson, W. L., and Setlow, P. (1990). "Sporulation, germination and outgrowth," in *Molecular Biological Methods for Bacillus*, eds C. R. Harwood, S. M. Cutting, and R. Chambert (Chichester: Wiley), 391–450. doi: 10.1111/j.1751-1097.1991.tb02087.x

Niemira, B. (2012). Cold plasma decontamination of foods. *Annu. Rev. Food Sci. Technol.* 3, 125–142. doi: 10.1146/annurev-food-022811-101132

Pankaj, S. K., Bueno-Ferrera, N. N., Misraa, V., Milosavljevića, B., O'Donnellb, C. P., Bourkea, P., et al. (2014). Applications of cold plasma technology in food packaging. *Trends. Food Sci. Technol.* 35, 5–17. doi: 10.1016/j.tifs.2013.10.009

Perni, S., Shama, G., Hobman, J. L., Lund, P. A., Kershaw, C. J., Hidalgo-Arroyo, G. A., et al. (2007). Probing bactericidal mechanisms induced by cold atmospheric plasmas with *Escherichia coli* mutants. *Appl. Phys. Lett.* 90:073902. doi: 10.1063/1.2458162

Reineke, K., Langera, K., Hertwig, C., Ehlbeck, J., and Schlüteret, O. (2015). The impact of different process gas compositions on the inactivation effect of an atmospheric pressure plasma jet on *Bacillus spores*. *Innov. Food Sci. Emerg. Technol.* 30, 112–118. doi: 10.1016/j.ifset.2015.03.019

Roth, S., Feichtinger, J., and Hertel, C. (2010). Characterization of *Bacillus subtilis* spore inactivation in low-pressure, low-temperature gas plasma sterilization processes. *J. Appl. Microbiol.* 108, 521–531. doi: 10.1111/j.1365-2672.2009.04453.x

Schlüter, O., Ehlbeck, J., Hertel, C., Habermeyer, M., Roth, A., Engel, K. H., et al. (2013). Opinion on the use of plasma processes for treatment of foods. *Mol. Nutr. Food Res.* 57, 920–927. doi: 10.1002/mnfr.201300039

Schlüter, O., and Fröhling, A. (2014). *Encyclopedia of Food Microbiology.* Oxford: Elsevier.

Setlow, B., and Setlow, P. (1993). Binding of small, acid-soluble spore proteins to DNA plays a significant role in the resistance of *Bacillus subtilis* spores to hydrogen peroxide. *Appl. Environ. Microbiol.* 59, 3418–3423.

Surowsky, B., Schlüter, O., and Knorr, D. (2014). Interactions of non-thermal atmospheric pressure plasma with solid and liquid food systems: a review. *Food Eng.* 7, 82–108. doi: 10.1007/s12393-014-9088-5

Weltmann, K.-D., Brandenburg, R., von Woedtke, T., Ehlbeck, J., Foest., R., Stieber, M., et al. (2008). Antimicrobial treatment of heat sensitive products by miniaturized atmospheric pressure plasma jets (APPJs). *J. Phys. D Appl. Phys.* 41, 1–6. doi: 10.1088/0022-3727/41/19/194008

Conflict of Interest Statement: The authors declare that the research was conducted in the absence of any commercial or financial relationships that could be construed as a potential conflict of interest.

3

Equol status and changes in fecal microbiota in menopausal women receiving long-term treatment for menopause symptoms with a soy-isoflavone concentrate

Lucía Guadamuro¹†, Susana Delgado¹†, Begoña Redruello², Ana B. Flórez¹, Adolfo Suárez³, Pablo Martínez-Camblor⁴,⁵ and Baltasar Mayo¹*

¹ Departamento de Microbiología y Bioquímica, Instituto de Productos Lácteos de Asturias – Consejo Superior de Investigaciones Científicas, Villaviciosa, Spain, ² Servicios Científico-Técnicos, Instituto de Productos Lácteos de Asturias – Consejo Superior de Investigaciones Científicas, Villaviciosa, Spain, ³ Servicio de Digestivo, Hospital de Cabueñes, Gijón, Spain, ⁴ Consorcio de Apoyo a la Investigación Biomédica en Red, Hospital Universitario Central de Asturias, Oviedo, Spain, ⁵ Facultad de Ciencias de la Educación, Universidad Autónoma de Chile, Santiago, Chile

Edited by:
Javier Carballo,
University of Vigo, Spain

Reviewed by:
Analia Graciela Abraham,
Centro de Investigacion y Desarrollo
en Criotecnologia de Alimentos,
Argentina
Amit Kumar Tyagi,
The University of Texas MD Anderson
Cancer Center, USA

*Correspondence:
Baltasar Mayo,
Departamento de Microbiologia y
Bioquimica, Instituto de Productos
Lácteos de Asturias – Consejo
Superior de Investigaciones
Cientificas, Paseo Río Linares s/n,
33300-Villaviciosa, Asturias, Spain
baltasar.mayo@ipla.csic.es

†These authors have contributed
equally to this work.

The knowledge regarding the intestinal microbial types involved in isoflavone bioavailability and metabolism is still limited. The present work reports the influence of a treatment with isoflavones for 6 months on the fecal bacterial communities of 16 menopausal women, as determined by culturing and culture-independent microbial techniques. Changes in fecal communities were analyzed with respect to the women's equol-producing phenotype. Compared to baseline, at 1 and 3 months the counts for all microbial populations in the feces of equol-producing women had increased strongly. In contrast, among the non-producers, the counts for all microbial populations at 1 month were similar to those at baseline, and decreased significantly by 3 and 6 months. Following isoflavone intake, major bands in the denaturing gradient gel electrophoresis (DGGE) profiles appeared and disappeared, suggesting important changes in majority populations. In some women, increases were seen in the intensity of specific DGGE bands corresponding to microorganisms known to be involved in the metabolism of dietary phytoestrogens (*Lactonifactor longoviformis*, *Faecalibacterium prausnitzii*, *Bifidobacterium* sp., *Ruminococcus* sp.). Real-Time quantitative PCR revealed that the *Clostridium leptum* and *C. coccoides* populations increased in equol producers, while those of bifidobacteria and enterobacteria decreased, and *vice versa* in the non-producers. Finally, the *Atopobium* population increased in both groups, but especially in the non-producers at three months. As the main findings of this study, (i) variations in the microbial communities over the 6-month period of isoflavone supplementation were large; (ii) no changes in the fecal microbial populations that were convincingly treatment-specific were seen; and (iii) the production of equol did not appear to be associated with the presence of, or increase in the population of, any of the majority bacterial types analyzed.

Keywords: soy isoflavone, equol, intestinal microbiology, fecal microbiota, menopause, probiotics

Introduction

Compared to Caucasian women, fewer Asian women suffer discomfort during menopause. They also have better intestinal health, and enjoy lower rates of cardiovascular disease (Bhupathy et al., 2010) and cancer (Virk-Baker et al., 2010). The better health of these women has been associated with a higher intake of soy foods (Messina, 2000). Soy contains many biologically active compounds (Kang et al., 2010), but its beneficial health effects have been attributed to its isoflavone content (Sánchez-Calvo et al., 2013). At the molecular level, the health benefits of dietary isoflavones appear to be mediated through their hormonal (Yuan et al., 2007), antioxidant (Wang et al., 2008), and enzyme-inhibitory (Crozier et al., 2009) activities.

In nature, isoflavones (daidzin, genistin, glycitin) mostly (>80%) appear conjugated with sugars as isoflavone-glycosides, the bioavailability and bioactivity of which are low (Crozier et al., 2009; de Cremoux et al., 2010). For their full activity to be realized, aglycones (daidzein, genistein, glycitein) need to be released from these isoflavone-glycosides and, occasionally, metabolized (Sánchez-Calvo et al., 2013). The transformations necessary are mostly performed by the enzymes of the gut microbiota. However, though improving, our knowledge of the gut microbes, their enzymes, and the pathways involved in the metabolism of isoflavones, is still limited (Atkinson et al., 2005; Yuan et al., 2007; Kemperman et al., 2010; Clavel and Mapesa, 2013). The metabolism of isoflavones is thought to occur in sequential steps involving several enzymes produced by a number of microbial types (Clavel and Mapesa, 2013; Sánchez-Calvo et al., 2013). The release of aglycones from conjugated isoflavones starts via the action of the widely distributed glycosyl hydrolases (members of the β-glucosidase family; Cantarel et al., 2009). The aglycones formed then undergo dehydroxylation, reduction, the breakage of the pyrone ring, and demethylation, etc., giving rise to compounds either of greater biological activity (such as equol and 5-hydroxy equol) or inactive molecules [such as o-demethylangolensin (O-DMA); Clavel and Mapesa, 2013]. Due to the inter-individual diversity in microbiota composition (Frankenfeld, 2011), only around 30–50% of Western women are capable of producing equol, while around 80–90% produce O-DMA. It may be that equol-producing women benefit more fully from the intake of isoflavones (Sánchez-Calvo et al., 2013).

Like other polyphenols, isoflavones have antimicrobial activity, which can modulate the diversity and composition of the gut microbiota after their consumption (Kemperman et al., 2010). The inhibition of pathogens or an increase in the size of beneficial populations might then contribute toward health benefits. However, studies on how isoflavones influence the composition and activity of the gut microbial community, and its effect on human health, are scarce (Clavel et al., 2005; Bolca et al., 2007; Nakatsu et al., 2014). Further, the results of the studies that are available are hard to compare, a consequence of differences in treatment regimen, target group, and the techniques employed. However, understanding how microorganisms and metabolites interact and elicit a physiological response (or lack thereof) is crucial if the results of observational and interventional studies

are to be properly interpreted. Investigations that assess in vivo the response of gut populations to isoflavone consumption are much needed.

The main aim of the present study was to determine the effect of long-term dietary supplementation with an isoflavone concentrate on the fecal microbial communities of menopausal women, via both culturing and culture-independent techniques. Women were entering a treatment of menopause symptoms with an isoflavone concentrate, which made unnecessary the design of an intervention study. The microbial results were correlated with equol production status in an attempt to identify those microbial populations and/or numbers linked to the production of this active, microbial-derived compound.

Materials and Methods

Human Volunteers and Urine and Stool Samples

This study was approved by the Research Ethics Committee of the Principado de Asturias, Spain. The selection of donors and later sampling was performed following standardized protocols recommended by the above committee. Sixteen menopausal women (age range 48–61; average 53.4) were recruited at the Obstetrics and Gynaecology Service of the Hospital de Cabueñes (Gijón, Spain). No participants suffered from any disease or intestinal disorder. Additionally, they had received no treatment with antibiotics or any other medication for at least 6 months prior to the start of the study. All participants consumed one tablet containing 80 mg of an isoflavone concentrate (Fisiogen; Zambon, Bresso, Italy) per day for 6 months. Urine and fecal samples were collected at four time points: before the start of the treatment ($t = 0$), and at one ($t = 1$), three ($t = 3$), and six ($t = 6$) months. Morning urine samples and freshly voided stools were collected by the volunteers themselves, the latter in sterile plastic containers, in which they were maintained under anaerobic conditions via the use of Anaerocult A (Merck, Darmstadt, Germany). All samples were a transported to the laboratory by courier. Stool samples were subjected to microbial analyses within 2 h of arrival; dilutions for culture-independent techniques were kept frozen at –80°C until use, as were urine samples for later equol and creatinine analysis.

Determination of Equol and Creatinine in Urine Samples

Three milliliters of thawed urine samples were diluted with 3 mL of 0.1 M acetate buffer (pH 4.5) and treated for 20 h at 37°C with 10 μL of extract type H-1 crude enzyme solution from Helix pomatia (Sigma–Aldrich, St. Louis, MO, USA). This has β-glucuronidase (100 units/μL) and sulphatase (7.5 units/μL) activities. Equol was extracted using Bond Elut-C18 200 mg solid-phase cartridges (Agilent Technologies, Santa Clara, CA, USA), pre-conditioned with 3 mL of methanol and 3 mL of acetate buffer 0.1 M pH 4.5. Treated urine was passed through the cartridges, which were then washed with 3 mL acetate buffer 0.1 M (pH 4.5). To remove any residual water, 200 μL of ethyl acetate were eluted through the cartridges and rejected; this roughly corresponded to the column dead volume. Equol was then eluted with 1 mL of ethyl acetate, filtered through a 0.2 μm PTFE membrane, and then evaporated to dryness under vacuum at room temperature. Prior to analysis, extracts were dissolved in 100 μL methanol [high-performance liquid chromatography (HPLC) grade] and kept at 4°C in opaque vials with screw caps. A 2 mM stock solution of equol (Sigma–Aldrich) in methanol was used to prepare a calibration curve covering six concentrations from 0.005 to 363.60 μM.

Equol in urine was determined using an H-Class Acquity UPLC™ ultra-high-performance liquid chromatography (UHPLC) system (Waters) equipped with a BEH reversed-phase C18 chromatographic column (1.7 μm, 2.1 mm × 100 mm; Waters, Palo Alto, CA, USA; Redruello et al., unpublished). The temperature of the column was set at 40°C. The mobile phase consisted of H_2O supplemented with 0.05% phosphoric acid (solvent A) and 100% methanol (solvent B). Samples were applied onto the column and eluted at a flow rate of 0.45 mL/min according to the following linear gradient: 0 min 90% A/10% B; 1 min 88% A/12% B; 3.5 min 75% A/25% B; 6.5 min 75% A/25% B; 10 min 50% A/50% B; 12 min 10% A/90% B; and 16 min 10% A/90% B. This was followed by washing with 10% A/90% B for 4 min and then 5 min with 90% A/10% B to re-equilibrate the column. Equol was measured using a fluorescence detector (excitation 280 nm, emission 310 nm).

As a single urine sample was analyzed, creatinine was determined to normalize equol excretion values. Equol was measured using a kinetic-photometric method based on the Jaffe reaction (Heinegård and Tiderström, 1973). For this, urine was treated with an alkaline picrate solution resulting in a bright orange–red complex. The formation rate of the complex over a prefixed interval of time (measured as an increase in absorbance) is proportional to the concentration of creatinine in the sample. Reactions were performed in 96-well microplates and measurements made using a Benchmark Plus microplate spectrophotometer (Bio-Rad, Richmond, CA, USA). Two hundred microliters of 10-fold diluted urine were added to a 2 mL solution of 25 mM picric acid prepared in 300 mM phosphate buffer (pH 12.1) containing 2 g/L SDS (to avoid protein precipitation). The reaction was allowed to proceed at 37°C and mixed every 5 s. The absorbance at 510 nm was measured over 6 min and the reaction rate determined as the tangent in the linear part of the kinetic curve between 0.08 and 5 min. Each sample was assayed in triplicate.

Fecal Microbiota Analyses
Microbial Counts
All fecal samples were processed in a Mac500 anaerobic chamber (Down Whitley Scientific, West Yorkshire, UK) containing a 10% H_2, 10% CO_2, 80% N_2 atmosphere. For microbial analyses, 1 g of feces was homogenized in 9 mL of sterile Maximum Recovery Diluent (Scharlab, Barcelona, Spain). The homogenates were then serially diluted and plated in duplicate onto general and selective agar media. Total and indicator bacterial populations were counted using the media as follows: for lactobacilli, de Man-Rogosa-Sharpe agar (Scharlab) supplemented with 0.25% cysteine (VWR International, Radnor, PA, USA; MRSC);

for clostridia, agarified Reinforced Clostridial Medium (RCM; Merck, Darmstadt, Germany); for *Bifidobacterium* sp., modified Columbia agar (BCCM™/LMG, Medium M144; BIF; Masco et al., 2003); for Enterobacteriaceae, Eosin Methylene Blue agar (EMB; Oxoid, Basingstoke Hampshire, UK); for *Veillonella* sp., *Veillonella* agar (VA; Merck); for *Bacteroides* and *Prevotella* (BP), specialized *Bacteroides* and *Prevotella* agar (Ly et al., 2007); and for total microorganisms, Medium for Colon Bacteria (MCB; Van der Meulen et al., 2006). Counting was performed after anaerobic incubation of the plates at 37°C for 72 h, with the exception of the EMB plates, for which enumeration analysis was performed after aerobic incubation for 24 h. Data were recorded as colony forming units (cfu)/g of feces, were transformed to logarithmic units before statistical analysis.

DNA Extraction from fecal Samples
Extraction of total bacterial DNA was based on the method of Zoetendal et al. (2006) using the QIAamp DNA Stool Minikit (Qiagen, Hilden, Germany) with some modifications of the manufacturer's protocol. Briefly, 0.2 g of thawed fecal samples were suspended in 1.8 mL of phosphate buffer saline solution (PBS; pH 7.4). The fecal suspension was homogenized by vortexing and centrifuged at 800 rpm at 4°C for 10 min to eliminate insoluble materials. Supernatants were transferred to new tubes and centrifuged at 14,000 rpm at 4°C for 5 min. Pellets were suspended in 200 μL of lysis solution (20 mM Tris-HCl pH 8.0, 2 mM EDTA, 1.20% Triton X-100, and 20 mg/mL lysozyme). Twenty units of mutanolysin (Merck, Darmstadt, Germany) were added to the mixture, which was then incubated at 37°C for 40 min. Immediately after, 1.2 mL of the lysis buffer from the DNA-isolation kit were added, the mixture placed in a screw-cap tube containing 0.3 and 0.1 g of 0.1 and 0.5 mm zirconia/silica beads, respectively, (BioSpec Products, Bartlesville, OK, USA) and subjected to mechanical breakage in a FastPrep FP120 Cell Disrupter (Qbiogene, Carlsbad, CA, USA; three cycles at 5.5 m s^{-1} for 30 s, cooling the samples on ice between cycles). Cell extracts were loaded onto the kit's column following the manufacturer's recommendations. Finally, the DNA was eluted with 150 μL sterile molecular biology grade water (Sigma–Aldrich) and stored at –20°C until required.

Denaturing Gradient Gel Electrophoresis (DGGE) Amplification
The variable V3 region of the 16S rRNA gene was amplified by PCR using the universal primers F357 (5′-TACGGGAGGCAGCAG-3′), to which a 39 bp GC sequence was linked to give rise to GC-F357 and R518 (5′-ATTACCGCGGCTGCTGG-3′), (Muyzer et al., 1993). The V2 and V4 regions were amplified with *Bifidobacterium*-specific primers F-Bif164 (5′-GGGTGGTAATGCCGGATG-3′) and R-Bif662 (5′-CCACCGTTACACCGGGAA-3′); a 40 bp GC sequence was linked to the latter to give rise to R-Bif662-GC (Satokari et al., 2001). Each reaction mixture consisted of 0.2 mM of each deoxynucleoside triphosphate (dNTPs), 0.24 μM of each forward and reverse primer, 2 U of 5 Prime Taq polymerase (VWR International), and between 100 and 150 ng of purified DNA. The PCR amplification programs were as follows: for

primers F357-GC and R518 – an initial denaturation step at 95°C for 5 min, followed by 35 cycles of 95°C for 30 s, 56°C for 30 s, and 72°C for 40 s, plus a final extension step at 72°C for 10 min; for primers F-Bif164 and R-Bif662-G – after an identical denaturation step, 35 cycles of 95°C for 30 s, 62°C for 40 s, and 72°C for 1 min, plus a final extension step at 72°C for 5 min.

DGGE Analysis

Denaturing gradient gel electrophoresis was undertaken in a DCode apparatus (Bio-Rad) at 60°C. Electrophoresis was performed at 200 V for 10 min in an 8% polyacrylamide stacking gel, followed by 75 V for 16 h in a denaturing gel in 0.5X Tris-acetate-EDTA (TAE) buffer. The urea-formamide denaturing ranges were 40–55% for amplicons obtained with primers F357-GC/R518 and 45–55% for amplicons obtained with primers F-Bif164/R-Bif662-GC. After staining in an ethidium bromide solution, the gels were visualized under UV light using a GBox system (Syngene, Cambridge, UK) equipped with GeneSys image acquisition software (Syngene). Selected bands were excised from the gels, suspended in sterile molecular grade water, and kept overnight at 4°C. Subsequently, the eluted DNA was used as a template in new amplification reactions involving the same primers without the GC-clamp. Finally, amplicons were purified using GenElute PCR Clean-Up columns (Sigma–Aldrich) and sequenced using an ABI Prism gene sequencer (Applied Biosystems, Foster City, CA, USA). Sequences were compared using the Blast program[1] and the Classifier tool provided by the Ribosomal Database Project[2]. Sequences with a percentage nucleotide match of 97% or higher to those in databases were assigned to the corresponding species.

The gels were digitalized and analyzed using Gene Tools v.4.03 software. For each DGGE profile, the Shannon–Weaver diversity index (H index) was estimated on the basis of the number

of bands (assuming them to be equivalent to the number of species) and their relative intensity. Additionally, the similarity between the DGGE profiles (presence or absence of bands and their intensities) was determined by calculating the Dice's coefficient. Clustering was performed using the unweighted pair group method with arithmetic averages (UPGMA), employing the DendroUPGMA computer program[3].

Real-Time quantitative PCR (qPCR)

Quantification of the different bacterial populations in feces was performed by qPCR using group-specific primers targeting the 16S rRNA gene (Table 1). Amplification reactions were performed in 96-well optical plates (Applied Biosystems) in a 7500 Fast Real-Time PCR System (Applied Biosystems). All amplifications were performed in triplicate in a final volume of 25 µL containing 2x SYBR Green PCR Master Mix (Applied Biosystems), 0.2 µM of each primer and 1 µL of template DNA (5–10 ng). The thermal cycling protocol followed consisted of an initial cycle at 95°C for 10 min, followed by 40 cycles at 95°C for 15 s, and 1 min at the appropriate primer-pair annealing temperature (Table 1). To check for specificity, melting curve analysis was performed, increasing the temperature from 60 to 95°C at a rate of 0.2°C per second with the continuous monitoring of fluorescence. Primer efficiency was calculated from the slope of the standard curve for each primer set ($E = 10^{-1/slope}$). The different bacterial groups were expressed as relative quantities (percentage of the total bacterial 16S rDNA in the sample) according to Vigsnæs et al. (2011).

Statistical Analysis

Statistical analysis of the culturing and qPCR data was performed using free R software[4]. The Shapiro–Wilk test was used to check for the normal distribution of the data. As several variables did

[1]http://www.ncbi.nlm.nih.gov/BLAST/
[2]http://rdp.cme.msu.edu/index.jsp
[3]http://genomes.urv.es/UPGMA/
[4]http://www.r-project.org

TABLE 1 | Bacterial target groups and characteristics of primers used for quantitative PCR (qPCR) in this study.

Microbial target	Primer	Sequence 5′–3′	Annealing (°C)	Efficiency[a]	Reference
Bifidobacterium sp.	F-bifido / R-bifido	CGCGTCYGGTGTGAAAG / CCCCACATCCAGCATCCA	60	1.90	Delroisse et al. (2008)
Lactobacillus sp.	Lacto-F / Lacto-R	AGCAGTAGGGAATCTTCCA / CACCGCTACACATGGAG	60	1.96	Heilig et al. (2002)
Clostridium coccoides group	g-Ccoc-F / g-Ccoc-R	AAATGACGGTACCTGACTAA / CTTTGAGTTTCATTCTTGCGAA	60	1.93	Matsuki et al. (2004)
Clostridium leptum group	g-Clept-F / g-Clept-R	GCACAAGCAGTGGAGT / CTTCCTCCGTTTTGTCAA	60	1.89	Matsuki et al. (2004)
Bacteroidetes phylum	Bact934F / Bact1060R	GGARCATGTGGTTTAATTCGATGAT / AGCTGACGACAACCATGCAG	60	1.92	Guo et al. (2008)
Atopobium cluster	c-Atopo-F / c-Atopo-R	GGGTTGAGAGACCGACC / CGGRGCTTCTTCTGCAGG	60	1.90	Matsuki et al. (2002)
Enterobacteriaceae	En-lsu3F / En-lsu3-R	TGCCGTAACTTCGGGAGAAGGCA / TCAAGGCTCAATGTTCAGTGTC	55	1.97	Matsuda et al. (2007)
Total bacteria[b]	TBA-F / TBA-R	CGGCAACGAGCGCAACCC / CCATTGTAGCACGTGTGTAGCC	60	1.93	Dennan and McSweeney (2006)

[a]Primer efficiency was calculated from the slope of the standard curve for each primer set as $E = 10^{-1/slope}$.
[b]TBA primers were used for quantification of total bacterial DNA. This value was used to normalize differences in DNA concentration between individual samples.

not follow a normal distribution, comparisons were performed by using non-parametric tests. The Mann–Whitney test for independent samples was performed to examine differences between equol producers and non-producers in terms of the microbial groups studied at every sampling point. The Wilcoxon signed-rank test for related samples was used to examine differences within microbial groups at all sampling points for both equol producers and non-producers. Two-tailed probability values of $P < 0.05$ were considered significant. Multivariate statistics was performed by principal coordinates analysis (PCoA) to search for associations between equol production status and the microbial groups determined by culturing and qPCR.

Results

Urine Equol after Isoflavone Supplementation

To test whether changes in the fecal microbial communities were associated with equol production phenotype, the equol status of the women was assessed by determining the concentration of this compound in their urine by UHPLC. Urine creatinine was also determined to normalize the equol values. Equol-producing women were defined as those responding to the soy challenge by showing an increase in urine equol to over 1000 nM (the cut-off defined by Rowland et al., 2000), and having an equol/creatinine ratio of >5.

Similar (and low) equol concentrations and equol/creatinine ratios were observed in the samples from all 16 women at $t = 0$ (**Table 2**). However, at $t = 1$, 100-fold increments in the ratios were obtained in the urine of four (25%) of the women (WA, WC, WG, and WP; Supplementary Table S1). In these subjects, the urinary excretion of equol at all post $t = 0$ sampling points (with the exception of $t = 3$ in woman WA) reached values above the stated cut-off. They were therefore considered equol producers. However, large variations in absolute equol concentrations were detected among their samples, as well as differences between samples for the same individual subject at different times (Supplementary Figure S1). In general, maximum equol production was observed at $t = 1$. After this point

production was either maintained (in WC and WP) or fell by $t = 3$ and $t = 6$ (in WA and WG). Low equol levels, sometimes close to the limit of detection and/or of the limit of quantification, were measured in most urine samples from the non-producer women (**Table 2**).

Microbial Counts

Wide inter-individual and inter-sample variations in counts for the different microbial populations were recorded over the supplementation period (**Table 3**). Although the response seemed erratic, some general trends were appreciated for equol producers and non-producers. At $t = 0$, counts for total and indicator microbial populations were generally slightly higher in the feces of the non-producers (**Table 3**). However, counts for all microbial populations increased strongly at $t = 1$ and $t = 3$ in the feces of the equol producers. In addition, in the equol producers, most cultivable populations decreased by $t = 6$, showing a trend toward the numbers recorded at $t = 0$. In contrast, in the non-producers, the counts for all microbial populations at $t = 1$ were similar to those at $t = 0$, but decreased significantly by $t = 3$ and $t = 6$ (**Table 3**). However, at the personal level, most microbial groups changed unpredictably in each of the women (Supplementary Table S2).

Community Profiling by DGGE

The community profiles showed 12–22 distinct DGGE bands of different intensity (**Figure 1**). Profiles contained between 3 (WP samples) through 8 (WF, WH samples) bands of a high intensity, being the rest of low or very low intensity. The appearance and disappearance of intense bands in the samples of individual subjects at consecutive sampling points (see **Figure 1**, WD-1, WF-3, and WH-1) indicates major changes in the majority bacterial populations. Since DGGE is a semi-quantitative technique, increases and reductions in the intensity of bands suggests corresponding changes in the fortune of the associated species. In general, isoflavone supplementation led to a reduction in the Shannon–Weaver diversity index (H index) compared to $t = 0$. The UPGMA comparison of Dice's coefficients showed a clear clustering of the profiles by subject (with only two exceptions, WC-0 and WD-1; **Figure 1**).

Eighty-two DGGE bands, the intensity of which varied over the supplementation period compared to $t = 0$, were assigned to bacterial types after DNA isolation, re-amplification, sequencing, and sequence comparison. The most common microbial types that responded to the presence of isoflavones were *Bifidobacterium adolescentis*, *Faecalibacterium prausnitzii*, *Lactonifactor longoviformis*, *Flavonifractor plautii*, *Coprococcus* sp., *Blautia* sp., *Oribacterium* sp., *Ruminococcus* sp., and members of the family Lachnospiraceae. However, the DGGE profiles from equol-producer women did not cluster together, and an apparent association between equol production and presence of specific bands was not observed.

Since the universal prokaryotic primers showed all bifidobacterial populations to increase over the supplementation period, DGGE analyses were performed using bifidobacteria-specific primers. Compared to the complex DGGE profiles for total bacteria, the bifidobacterial profiles were rather

---

TABLE 2 | Average equol production during isoflavone treatment in urine samples among equol producer and non-producer women of this study.

Equol status	Sample (month)	Parameter	
		Equol[a]	Equol/Creatinine
Producers ($n = 4$)	0	59.50	0.58
	1	18716.75	77.62
	3	8114.00	38.13
	6	5702.50	28.63
Non-producers ($n = 12$)	0	39.00	0.73
	1	19.42	0.12
	3	28.08	0.23
	6	15.58	0.10

[a]*Concentration of equol in nM.*

TABLE 3 | Viable counts of total and indicator fecal microbial populations in menopausal women treated with a soy isoflavone supplement over a 6-month period.

Equol	Month	Microbial counts[a]						
		MCB	MRSC	BIF	RCM	EMB	BP	VA
Producers (n = 4)	0	9.51 ± 1.12	9.29 ± 1.39	9.44 ± 1.11	9.21 ± 1.31	7.69 ± 1.19	9.87 ± 1.27	9.65 ± 1.35
	1	10.46 ± 0.68	10.55 ± 0.82	10.28 ± 0.75	9.79 ± 1.92	8.91 ± 1.53	11.16 ± 0.63	10.65 ± 0.29
	3	10.73 ± 0.75	10.37 ± 0.83	10.49 ± 0.95	10.27 ± 0.52	7.35 ± 1.01	10.87 ± 0.87	10.42 ± 0.60
	6	9.79 ± 0.25	8.96 ± 0.87	9.53 ± 0.19	9.32 ± 0.40	7.17 ± 0.57	10.08 ± 0.13	9.36 ± 0.22
Non-producers (n = 12)	0	10.32 ± 0.61	10.04 ± 0.81	10.21 ± 0.67	10.23 ± 0.70	8.01 ± 0.98	10.45 ± 0.60	10.16 ± 0.73
	1	10.42 ± 0.81	10.06 ± 1.00	10.30 ± 0.82	10.38 ± 0.87	7.82 ± 0.70	10.80 ± 0.71	10.64 ± 0.75
	3	9.75 ± 0.76	9.30 ± 1.15	9.65 ± 0.69	9.49 ± 0.92	7.32 ± 1.07*	10.13 ± 0.62	9.23 ± 0.85
	6	9.54 ± 0.54**	9.10 ± 0.72**	9.47 ± 0.49**	9.19 ± 0.68**	7.31 ± 1.03*	9.74 ± 0.57**	8.86 ± 0.75***

[a]Log_{10} cfu g^{-1} of feces (mean ± SD).
Key of media (target population): MCB, Medium for Colon Bacteria (total cultivable bacteria); MRSC; de Man, Rogosa, and Sharpe (lactobacilli); BIF, BCCMTM/LMG-Medium M144 (bifidobacteria); RCM, Reinforced Clostridium Medium (clostridia); EMB, Eosin Methylene Blue (enterobacterias); BP, Bacteroides and Prevotella medium (Bacteroides and Prevotella sp.); VA, Veillonella Agar (Veillonella sp.).
Key of statistical significance respect to basal time (t = 0): *$p < 0.05$; **$p < 0.01$; ***$p < 0.001$.

simple, involving just 2–5 bands per subject (partial results are provided in Supplementary Figure S2). In contrast to the general profiles, the bifidobacterial profiles for most of the women proved stable throughout the supplementation period. In total, 18 bands were isolated from the DGGE gels and identified as before. B. bifidum, B. longum, B. catenulatum/B. pseudocatenulatum, and B. adolescentis were the most common bifidobacterial species identified. Other species, such as B. saeculare and B. ruminantium, were occasionally identified. In two of the women (WB and WG), the bands corresponding to B. adolescentis increased markedly over the supplementation period.

qPCR Analysis

qPCR detected wide variations in bacterial population sizes among subjects, and between samples from the same subject at different sampling points (Supplementary Table S3). The majority populations were formed by Bacteroides and Clostridium. Bifidobacteria and lactobacilli made up <10 and <0.5%, respectively, of total bacterial numbers. Over the supplementation period, opposite trends were seen between the equol producers and non-producers in terms of the change in some other bacterial populations. For example, bifidobacterial populations decreased in size in the equol producers, but increased in the non-producers, while the population of Bacteroides increase in equol producers and decrease in the non-producers (**Table 4**). The population of enterobacterias was shown to be significantly lower in equol producers than in non-producers. In this last group of women, increased numbers of enterobacteria were seen at t = 3 and t = 6 (**Table 4**). The two Clostridium clusters (leptum and coccoides) targeted by specific primers were seen to increase in the equol producers, especially at t = 3 and t = 6, while their numbers remained similar in samples from the non-producers. Finally, the Atopobium population increased over the supplementation period in both groups of women, but more so in the non-producers (a significant increase was detected at t = 3).

Discussion

The metabolism of soy isoflavones, and therefore their bioavailability and activity in humans, strongly depends on the activity of the intestinal microbiota. The underlying interactions between microbial populations and isoflavone metabolites, however, remain poorly understood (Frankenfeld, 2011; Clavel and Mapesa, 2013).

Though several methods have been proposed to determine human equol production phenotype (Rowland et al., 2000; Setchell and Cole, 2006), the cut-off for the assignment of producer or non-producer status remains rather arbitrary. In the present work, producers and non-producers were identified based on a cut-off of 1000 nM equol in urine, as defined by Rowland et al. (2000). The same number of equol producers and non-producers were also obtained when considering an equol/creatinine ratio >5 as the cut-off (Rowland et al., 2000). The frequency of equol producers differ widely among human communities (Setchell et al., 2002). That reported in the present work agrees well with values reported by other authors for Western women (Atkinson et al., 2005; Setchell and Cole, 2006; Possemiers et al., 2007; Frankenfeld, 2011; Nakatsu et al., 2014). In contrast to that reported by Franke et al. (2012), equol production phenotype proved to be stable over the study period; no conversion from equol producer to non-producer or vice versa was observed.

The effects of isoflavones on the gut microbiota have been little examined (Clavel et al., 2005; Bolca et al., 2007; Nakatsu et al., 2014), and with the exception of the study by Clavel et al. (2005), which lasted two months, changes in bacterial populations have only ever been monitored over short periods. Further, apart from the work of Nakatsu et al. (2014), which involved phylogenetic/metagenomic analyses, the number of populations targeted has been very limited. In the present work, the effects of isoflavones on the fecal microbiota of healthy menopausal women were analyzed at 1, 3, and 6 months of supplementation, and the results compared to those at baseline. This design, however, may have overlooked significant changes

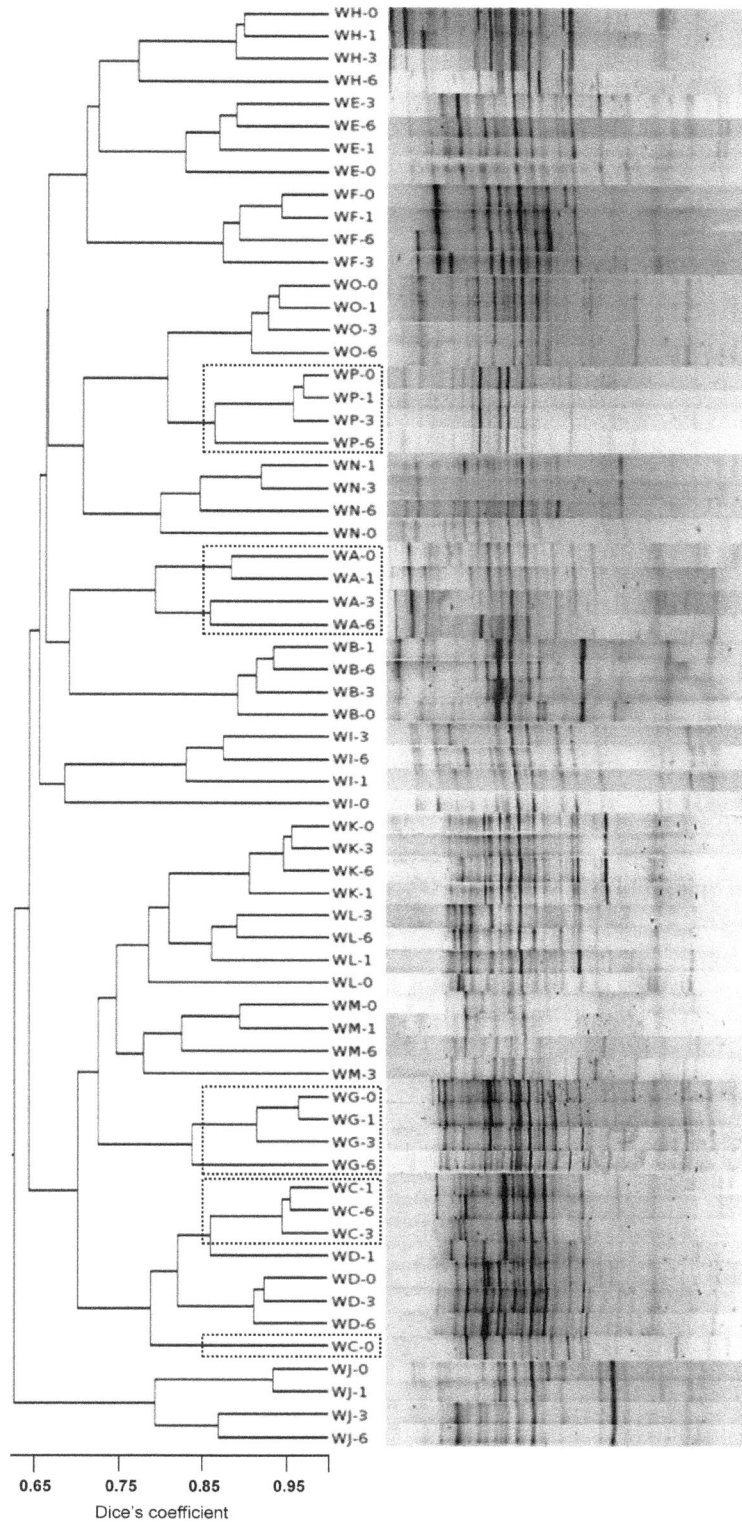

FIGURE 1 | Denaturing gradient gel electrophoresis (DGGE) profiles of the total microbial populations from fecal samples of menopausal women at different times during isoflavone treatment using universal prokaryotic primers amplifying the V2-V4 hypervariable regions of the 16S rRNA gene. Similarity of the profiles was calculated by the Dice's coefficient and clustered by the UPGMA method. Position of the profiles from equol producer women is indicated by a doted box.

TABLE 4 | Relative quantities of fecal microbial populations in the different equol status groups of the menopausal women treated with soy isoflavones of this study as determined by qPCR using universal and group-specific primers.

Equol status	Month	Microbial population[a]						
		Bifidobacteria	Lactobacilli	*Clostridium leptum*	*Clostridium coccoides*	*Bacteroides*	Enterobacteria	*Atopobium*
Producers (n = 4)	0	3.47 ± 5.74[a]	0.51 ± 0.91	28.70 ± 7.67	23.36 ± 14.31	73.6 ± 126.20	0.008 ± 0.001	4.87 ± 3.35
	1	1.00 ± 0.80	< 0.01 ± 0.01	19.64 ± 8.25	24.19 ± 18.58	104.9 ± 181.7	0.003 ± 0.004	5.64 ± 2.15
	3	2.05 ± 1.37	0.13 ± 0.20	35.69 ± 13.52	40.34 ± 24.18	45.31 ± 59.72	< 0.001 ± 0.001	5.96 ± 1.00
	6	1.43 ± 1.04	0.04 ± 0.03	33.14 ± 13.25	32.43 ± 24.08	113.3 ± 210.1	0.004 ± 0.006	4.70 ± 1.72
Non-producers (n = 12)	0	4.96 ± 6.19	0.31 ± 0.81	23.89 ± 8.75	25.33 ± 16.51	29.08 ± 63.03	0.20 ± 0.52	3.77 ± 1.97
	1	7.63 ± 7.87	0.31 ± 0.89	26.17 ± 17.11	26.97 ± 18.59	20.57 ± 31.07	0.14 ± 0.27	5.61 ± 4.25
	3	6.84 ± 9.17	0.05 ± 0.07	24.72 ± 10.89	25.15 ± 16.60	15.53 ± 14.42	0.83 ± 0.22	6.24 ± 4.90*
	6	9.27 ± 13.11	0.35 ± 0.68	19.98 ± 10.13	24.26 ± 16.80	12.55 ± 9.46	0.49 ± 0.13	4.85 ± 2.88

[a]*Percentage of the total bacterial 16S rDNA (mean ± SD), as determined using the universal prokaryotic primers TBA-F and TBA-R* (**Table 1**).
*Key of statistical significance: *p < 0.05 versus basal sample (t = 0).*

occurring soon after the start of isoflavone supplementation (Bolca et al., 2007; Nakatsu et al., 2014).

In general, the community structure and composition of the fecal bacterial populations changed significantly with the isoflavone supplementation. However, wide inter- and intra-individual (at different times) variations were detected by the different techniques employed, making it very difficult to correlate isoflavone intake with any changes in the bacterial community structure/population size, or changes in the latter with equol production. Thus, not surprisingly, PCoA of culturing and qPCR results showed no association between microbial communities and equol production status (Supplementary Figure S3). It is conceivable that isoflavones directly or indirectly affect members of the dominant microbiota (Clavel and Mapesa, 2013). In addition to the selective pressure they may exert on isoflavone-utilizing microorganisms, isoflavones or their metabolites (aglycones, equol, O-DMA, etc.) might modify conditions in the intestinal tract with an ensuing effect on susceptible bacterial communities. A bifidogenic effect of isoflavones has been reported in some studies (Clavel et al., 2005; Nakatsu et al., 2014), and a modest effect of this kind for most of the samples at $t = 1$ was confirmed by culturing. However, it did not persist through to $t = 3$ and $t = 6$. In equol producers, increases in population size within *Clostridium* clusters have been observed before (Clavel et al., 2005; Possemiers et al., 2007). However, such stimulatory effects may depend on the baseline sizes of these populations, which can vary widely between subjects. In contrast, a reduction was seen in the number of enterobacterias in the samples from most women (Supplementary Table S2). Since several members of the Enterobacteriaceae, including *Escherichia coli* phylotypes, harbor a ~54 kb polyketide synthase (*pks*) pathogenicity island that encodes multi-enzymatic machinery for synthesizing a genotoxin that promotes tumorigenesis (Arthur et al., 2012), a fall in their numbers might be beneficial. However, no changes in the fecal microbial populations that were convincingly supplementation-specific were seen, suggesting that (at least some) changes might

be due to normal variations caused by diet or other uncontrolled environmental factors. Together, the present enumeration results suggest that the consumption of isoflavones induces changes in the majority of microbial populations in both equol producers and non-producers, although these might be opposite.

The inadequacy of conventional culturing techniques for reflecting the microbial diversity of the intestinal ecosystem (Leser et al., 2002; Kemperman et al., 2010; Li et al., 2014) prompted the use of two culture-independent methods: DGGE for profiling the majority microbial species, and qPCR for targeting specific microbial populations. In some woman, specific bands corresponding to microorganisms previously associated with the metabolism of isoflavones and other dietary phytoestrogens (*L. longoviformis, F. prausnitzii, Bifidobacterium* sp., *Ruminococcus* sp.; Bolca et al., 2007; Clavel et al., 2007; Nakatsu et al., 2014) were enhanced in intensity. As suggested by Nakatsu et al. (2014), such increases may argue for isoflavones providing a chemical environment that selects for a subset of the initial bacterial community (Clavel and Mapesa, 2013). Alternatively, isoflavones might have antimicrobial effects on certain bacterial populations, as reported for flavonoid compounds (Orhan et al., 2010). In either case, the UPGMA analysis grouped the DGGE profiles by individual rather than by treatment (isoflavones, non-isoflavones), suggesting the changes detected by this technique depend on the initial microbial profile (personal microbiota). Furthermore, DGGE only detects changes occurring within the dominant bacterial populations (Muyzer et al., 1993); subtle changes in subdominant or minority species also occur but go unseen, and equol production might result from interactions between dominant and subdominant (or even minority) intestinal populations. In contrast, qPCR involving universal and group-specific primers produces a general microbial picture similar to that produced by state-of-the-art metagenomic techniques (Qin et al., 2010; Li et al., 2014).

Discrepancies between culturing and culture-independent methods may result from differences in lysis efficiency during DNA isolation, preferential PCR amplification (which may be different for each primer pair), or interspecies differences in 16S rRNA operon copy number (Kemperman et al., 2010). The detection of non-cultivatable and dead cells by DNA-based techniques may account for further differences. In this sense, low recoveries of Bacteroidetes from frozen fecal samples have been reported (Bahl et al., 2012). However, in the present work *Bacteroides* sp. proved to be majority population in both equol producers and non-producers.

Conclusion

Both the culturing and the culture-independent methods used in this work detected wide microbial diversity in the studied menopausal women at baseline. Variations in the microbial communities over the six-month period of isoflavone supplementation were also large. No general patters of change due to isoflavone ingestion, or associated with equol production, were observed. The production of equol could not be correlated to the presence of, or increase in, any of the bacterial populations analyzed. This suggests isoflavone metabolism may differ in different people depending on their personal gut microbiota. Metagenomics, metabolomics, and metatranscriptomics analyses will be required to uncover

the relationships between the structure and composition of the intestinal microbial communities in response to soy isoflavones and their metabolic compounds. Further research in this area could ultimately lead to the modulation of numbers of equol-producing bacteria, and thus equol producer-status, through the use of specific prebiotics and/or probiotics.

Acknowledgments

This research was funded by projects from the Spanish Ministry of Economy and Competitiveness (MINECO; Ref. AGL2011-24300 and AGL-2014-57820-R). LG was supported by a research fellowship of the FPI Program from MINECO (Ref. BES-2012-062502). AF was supported by a research contract under CSIC JAE-Doc Program.

References

Arthur, J. C., Perez-Chanona, E., Mühlbauer, M., Tomkovich, S., Uronis, J. M., Fan, T. J., et al. (2012). Intestinal inflammation targets cancer-inducing activity of the microbiota. *Science* 338, 120–123. doi: 10.1126/science.1224820

Atkinson, C., Frankenfeld, C. L., and Lampe, J. W. (2005). Gut bacterial metabolism of the soy isoflavone daidzein: exploring the relevance to human nutrition. *Exp. Biol. Med.* 230, 155–170.

Bahl, M. I., Bergström, A., and Licht, R. (2012). Freezing fecal samples prior to DNA extraction affects the *Firmicutes* to *Bacteroidetes* ratio determined by downstream quantitative PCR analysis. *FEMS Microbiol. Lett.* 329, 193–197. doi: 10.1111/j.1574-6968.2012.02523.x

Bhupathy, P., Haines, C. D., and Leinwand, L. A. (2010). Influence of sex hormones and phytoestrogens on heart disease in men and women. *Womens Health* 6, 77–95. doi: 10.2217/whe.09.80

Bolca, S., Possemiers, S., Herregat, A., Huybrechts, I., Heyerick, A., De Vriese, S., et al. (2007). Microbial and dietary factors are associated with the equol producer phenotype in healthy postmenopausal women. *J. Nutr.* 137, 2242–2246.

Cantarel, B. L., Coutinho, P. M., Rancurel, C., Bernard, T., Lombard, V., and Henrissat, B. (2009). The Carbohydrate-Active Enzymes database (CAZy): an expert resource for Glycogenomics. *Nucleic Acids Res.* 37, D233–D238. doi: 10.1093/nar/gkn663

Clavel, T., Fallani, M., Lepage, P., Levenez, F., Mathey, J., Rochet, V., et al. (2005). Isoflavones and functional foods alter the dominant intestinal microbiota in postmenopausal women. *J. Nutr.* 135, 2786–2792.

Clavel, T., Lippman, R., Gavini, F., Doré, J., and Blaut, M. (2007). *Clostridium saccharogumia* sp. nov. and *Lactonifactor longoviformis* gen. nov., two novel human faecal bacteria involved in the conversion of the dietary phytoestrogen secoisolariciresinol diglucoside. *Syst. Appl. Microbiol.* 30, 16–26. doi: 10.1016/j.syapm.2006.02.003

Clavel, T., and Mapesa, J. O. (2013). "Phenolics in human nutrition: importance of the intestinal microbiome for isoflavone and lignan bioavailability," in *Handbook of Natural Products*, eds K. G. Ramawat and J. M. Merillon (Amsterdam: Elsevier).

Crozier, A., Jaganath, I. B., and Clifford, M. N. (2009). Dietary phenolics: chemistry, bioavailability and effects on health. *Nat. Prod. Rep.* 26, 1001–1043. doi: 10.1039/b802662a

de Cremoux, P., This, P., Leclercq, G., and Jacquot, Y. (2010). Controversies concerning the use of phytoestrogens in menopause management: bioavailability and metabolism. *Maturitas* 65, 334–339. doi: 10.1016/j.maturitas.2009.12.019

Delroisse, J. M., Boulvin, A. L., Parmentier, I., Dauphin, R. D., Vandenbol, M., and Portetelle, D. (2008). Quantification of *Bifidobacterium* spp. and *Lactobacillus* spp. in rat fecal samples by real-time PCR. *Microbiol. Res.* 163, 663–670. doi: 10.1016/j.micres.2006.09.004

Dennan, S. E., and McSweeney, C. S. (2006). Development of a real-time PCR assay for monitoring anaerobic fungal and cellulolytic bacterial populations within the rumen. *FEMS Microbiol. Ecol.* 58, 572–582. doi: 10.1111/j.1574-6941.2006.00190.x

Franke, A. A., Lai, J. F., Halm, B. M., Pagano, I., Kono, N., Mack, W. J., et al. (2012). Equol production changes over time in postmenopausal women. *J. Nutr. Biochem.* 23, 573–579. doi: 10.1016/j.jnutbio.2011.03.002

Frankenfeld, C. L. (2011). O-desmethylangolensin: the importance of equol's lesser known cousin to human health. *Adv. Nutr. Inter. Rev. J.* 2, 317–324. doi: 10.3945/an.111.000539

Guo, X., Xia, X., Tang, R., Zhou, J., Zhao, H., and Wang, K. (2008). Development of a real-time PCR method for *Firmicutes* and *Bacteroidetes* in faeces and its application to quantify intestinal population of obese and lean pigs. *Lett. Appl. Microbiol.* 47, 367–373. doi: 10.1111/j.1472-765X.2008.02408.x

Heilig, H. G. H. J., Zoetendal, E. G., Vaughan, E. E., Marteau, P., Akkermans, A. D., and de Vos, W. M. (2002). Molecular diversity of *Lactobacillus* spp. and other lactic acid bacteria in the human intestine as determined by specific amplification of 16S ribosomal DNA. *Appl. Environ. Microbiol.* 68, 114–123. doi: 10.1128/AEM.68.1.114-123.2002

Heinegård, D., and Tiderström, G. (1973). Determination of serum creatinine by a direct colorimetric method. *Clin. Chem. Acta* 43, 305–310. doi: 10.1016/0009-8981(73)90466-X

Kang, J., Badger, T. M., Ronis, M. J., and Wu, X. (2010). Non-isoflavone phytochemicals in soy and their health effects. *J. Agric. Food Chem.* 58, 8119–8133. doi: 10.1021/jf100901b

Kemperman, R. A., Bolca, S., Loger, L. C., and Vaughan, E. E. (2010). Novel approaches for analyzing gut microbes and dietary polyphenols: challenges and opportunities. *Microbiology* 156, 3224–3231. doi: 10.1099/mic.0.042127-0

Leser, T. D., Amenuvor, J. Z., Jensen, T. K., Lindecrona, R. H., Boye, M., and Møller, K. (2002). Culture-independent analysis of gut bacteria: the pig gastrointestinal tract microbiota revisited. *Appl. Environ. Microbiol.* 68, 673–690. doi: 10.1128/AEM.68.2.673-690.2002

Li, J., Jia, H., Cai, X., Zhong, H., Feng, Q., Sunagawa, S., et al. (2014). An integrated catalog of reference genes in the human gut microbiome. *Nat. Biotechnol.* 32, 834–841. doi: 10.1038/nbt.2942

Ly, N. P., Rifas-Shiman, S. L., Litonjua, A. A., Tzianabos, A. O., Schaub, B., Ruiz-Pérez, B., et al. (2007). Cord blood cytokines and acute lower respiratory illnesses in the first year of life. *Pediatrics* 119, e171–e178. doi: 10.1542/peds.2006-0524

Masco, L., Huys, G., Gevers, D., Verbrugghen, L., and Swings, J. (2003). Identification of Bifidobacterium species using rep-PCR fingerprinting. *Syst. Appl. Microbiol.* 26, 557–563. doi: 10.1078/072320203770865864

Matsuda, K., Tsuji, H., Asahara, T., Kado, Y., and Nomoto, K. (2007). Sensitive quantitative detection of commensal bacteria by rRNA-targeted reverse transcription-PCR. *Appl. Environ. Microbiol.* 73, 32–39. doi: 10.1128/AEM.01224-06

Matsuki, T., Watanabe, K., Fujimoto, J., Miyamoto, Y., Takada, T., Matsumoto, K., et al. (2002). Development of 16S rRNA-gene-targeted group-specific primers for the detection and identification of predominant bacteria in human feces. *Appl. Environ. Microbiol.* 68, 5445–5451. doi: 10.1128/AEM.68.11.5445-5451.2002

Matsuki, T., Watanabe, K., Fujimoto, J., Takada, T., and Tanaka, R. (2004). Use of 16S rRNA gene-targeted group-specific primers for real-time PCR analysis of predominant bacteria in human feces. *Appl. Environ. Microbiol.* 70, 7220–7228. doi: 10.1128/AEM.70.12.7220-7228.2004

Messina, M. (2000). Soyfoods and shoybean phyto-oestrogens (isoflavones) as possible alternatives to hormone replacement therapy. *Eur. J. Cancer* 36, S71–S77. doi: 10.1002/1521-3765(20001002)6:19<3495::AID-CHEM3495>3.0.CO;2-1

Muyzer, G., de Waal, E. C., and Uitterlinden, A. G. (1993). Profiling of complex microbial populations by denaturing gradient gel electrophoresis analysis of polymerase chain reaction-amplified genes encoding for 16S rRNA. *Appl. Environ. Microbiol.* 59, 695–700.

Nakatsu, C. H., Arsmstrong, A., Cavijo, A. P., Martin, B. R., Barnes, S., and Weaver, C. M. (2014). Fecal bacterial community changes associated with isoflavone metabolites in postmenopausal women after soy bar consumption. *PLoS ONE* 9:e108924. doi: 10.1371/journal.pone.0108924

Orhan, D. D., Özçelik, B., Ozgen, S., and Ergun, F. (2010). Antibacterial, antifungal, and antiviral activities of some flavonoids. *Microbiol. Res.* 165, 496–504. doi: 10.1016/j.micres.2009.09.002

Possemiers, S., Bolca, S., Eeckhaut, E., Depypere, H., and Verstraete, W. (2007). Metabolism of isoflavones, lignans and prenylflavonoids by intestinal bacteria: producer phenotyping and relation with intestinal community. *FEMS Microbiol. Ecol.* 61, 372–383. doi: 10.1111/j.1574-6941.2007.00330.x

Qin, J., Li, R., Raes, J., Arumugam, M., Burgdorf, K. S., Manichanh, C., et al. (2010). A human gut microbial gene catalogue established by metagenomic sequencing. *Nature* 464, 59–65. doi: 10.1038/nature08821

Rowland, I. R., Wiseman, H., Sanders, T. A., Adlercreutz, H., and Bowey, E. A. (2000). Interindividual variation in metabolism of soy isoflavones and lignans: influence of habitual diet on equol production by the gut microflora. *Nutr. Cancer* 36, 27–32. doi: 10.1207/S15327914NC3601_5

Sánchez-Calvo, J. M., Rodríguez-Iglesias, M. A., Molinillo, J. M. G., and Macías, F. A. (2013). Soy isoflavones and their relationship with microflora: beneficial effects on human health in equol producers. *Phytochem. Rev.* 12, 979–1000. doi: 10.1007/s11101-013-9329-x

Satokari, R. M., Vaughan, E. E., Akkermans, A. D., Saarela, M., and de Vos, W. M. (2001). Bifidobacterial diversity in human feces detected by genus-specific PCR and denaturing gradient gel electrophoresis. *Appl. Environ. Microbiol.* 67, 504–513. doi: 10.1128/AEM.67.2.504-513.2001

Setchell, K. D. R., Brown, N. M., and Lydeking-Olsen, E. (2002). Clinical importance of the metabolite equol. A clue to the effectiviness of soy and its isoflavones. *J. Nutr.* 132, 3577–3584.

Setchell, K. D. R., and Cole, S. J. (2006). Method of defining equol producer status and its frequency among vegetarians. *J. Nutr.* 136, 2188–2193.

Van der Meulen, R., Makras, L., Verbrugghe, K., Adriany, T., and De Vuyst, L. (2006). In vitro kinetic analysis of oligofructose consumption by *Bacteroides* and *Bifidobacterium* spp. indicates different degradation mechanisms. *Appl. Environ. Microbiol.* 72, 1006–1012. doi: 10.1128/AEM.72.2.1006-1012.2006

Vigsnæs, L. K., Holck, J., Meyer, A. S., and Licht, T. R. (2011). In vitro fermentation of sugar beet arabino-oligosaccharides by fecal microbiota obtained from patients with ulcerative colitis to selectively stimulate the growth of *Bifidobacterium* spp. and *Lactobacillus* spp. *Appl. Environ. Microbiol.* 77, 8336–8344. doi: 10.1128/AEM.05895-11

Virk-Baker, M. K., Nagy, T. R., and Barnes, S. (2010). Role of phytoestrogens in cancer therapy. *Planta Med.* 76, 1132–1142. doi: 10.1055/s-0030-1250074

Wang, D., Wang, L., Zhu, F., Zhu, J., Chen, X. D., Zou, L., et al. (2008). In vitro and in vivo studies on the antioxidant activities of the aqueous extracts of Douchi (a traditional Chinese slat fermented soybean food). *Food Chem.* 17, 1421–1428. doi: 10.1016/j.foodchem.2007.09.072

Yuan, J.-P., Wang, J.-H., and Liu, X. (2007). Metabolism of dietary soy isoflavones to equol by human intestinal microbiota. Implications for health. *Mol. Nutr. Food Res.* 51, 765–781. doi: 10.1002/mnfr.200600262

Zoetendal, E. G., Heilig, H. G., Klaassens, E. S., Booijink, C. C., Kleerebezem, M., Smidt, H., et al. (2006). Isolation of DNA from bacterial samples of the human gastrointestinal tract. *Nat. Protoc.* 1, 870–873. doi: 10.1038/nprot.2006.142

Conflict of Interest Statement: The authors declare that the research was conducted in the absence of any commercial or financial relationships that could be construed as a potential conflict of interest.

4

Defects in polynucleotide phosphorylase impairs virulence in *Escherichia coli* O157:H7

Jia Hu[1,2] and Mei-Jun Zhu[1]*

[1] School of Food Science, Washington State University, Pullman, WA, USA, [2] Department of Animal Science, University of Wyoming, Laramie, WY, USA

Edited by:
Dongsheng Zhou,
Beijing Institute of Microbiology
and Epidemiology, China

Reviewed by:
Yufeng Yao,
Shanghai Jiao Tong University School
of Medicine, China
Grzegorz Wegrzyn,
University of Gdansk, Poland

*Correspondence:
Mei-Jun Zhu,
School of Food Science, Washington
State University, Pullman, WA 99164,
USA
meijun.zhu@wsu.edu

Polynucleotide phosphorylase (PNPase) is reported to regulate virulence in *Salmonella, Yersinia* sp. and *Campylobacter jejuni,* yet its role in *Escherichia coli* O157:H7 has not been investigated. To gain insights into its roles in *E. coli* O157:H7 virulence, *pnp* deletion mutants were generated and the major virulence factors were compared to their parental wild type strains. Deletion of *pnp* in *E. coli* O157:H7 dramatically decreased *stx2* mRNA expression and Stx2 protein production, and impaired lambdoid prophage activation in *E. coli* O157:H7. Quantitative PCR further confirmed that the Stx2 phage lytic growth was repressed by *pnp* deletion. Consistent with reduced Stx2 production and Stx2 phage activation, the transcriptional levels of genes involved in phage lysis and replication were down-regulated. In addition, disruption of *pnp* in *E. coli* O157:H7 decreased its adhesion to intestinal epithelial cells as well as cattle colonic explant tissues. On the other hand, PNPase inactivation in *E. coli* O157:H7 enhanced Tir protein content and the transcription of type three secretion system components, including genes encoding intimin, Tir, and EspB as well as locus of enterocyte and effacement positive regulator, Ler. Collectively, data indicate that PNPase has pleiotropic effects on the virulence of *E. coli* O157:H7.

Keywords: *E. coli* O157:H7, PNPase, Shiga toxin 2, prophage, Type three secretion system, intestine, epithelium, adhesion

Introduction

Shiga toxin (Stx) producing *Escherichia coli* O157:H7 is a major food safety threat that results in significant economic losses, especially in the beef industry. Stx is the major virulence factor in *E. coli* O157:H7, which causes bloody diarrhea and life threatening hemolytic-uremic syndrome (Mainil and Daube, 2005). The mortality associated with *E. coli* O157:H7 infection is due to the production and release of Stx, which is composed of a single 32-kDa A subunit and five 7.7-kDa B subunits (Paton and Paton, 1998). Stx binds to receptors on cell surface and is internalized through endocytosis, which inhibits protein synthesis and causes complications associated with *E. coli* O157:H7 infection (Johannes and Romer, 2010). In addition, *E. coli* O157:H7 has a chromosomal pathogenicity island, i.e., locus of enterocyte and effacement (LEE; McDaniel et al., 1995), which mediates intimate contact to epithelial cells, further leading to attaching and effacing (AE) lesions (McDaniel et al., 1995). The LEE is composed of five major operons (LEE1-5) that encode type three secretion system (T3SS) apparatus and effector proteins (Deng et al., 2004).

There is a plethora of studies on the regulation of LEE in *E. coli* O157:H7, which is regulated by multiple proteins such as Ler, H-NS, GrlA, CsrA through direct or indirect interaction (Bhatt et al., 2011). Compared with LEE, the regulation of Stx 2 production in *E. coli* O157:H7 has only been sparsely studied. Gene *stx2* is localized on Stx2 prophage, a lambdoid like bacteriophage, DNA that inserted to *E. coli* O157:H7 genome (Herold et al., 2004; Allison, 2007; Los et al., 2011). Stx production is generally linked to the expression of induced Stx-phage's late genes, which results in phage lytic cycle induction and release of Stx (Wagner et al., 2001). The life cycle of prophages is controlled by phage repressor CI proteins in lambdoid phage (Ptashne and Hopkins, 1968). Mutations in the *cI* gene make a non-cleavable repressor and an uninducible 933W Stx2 prophage (Tyler et al., 2013). RecA can stimulate the self-cleavage of CI further allowing initiation of transcription from the early P_L and P_R promoters (Little, 2005). The transcription from P_L results in expression of *N* protein, which modified RNA polymerase initiating at P_L and P_R to form resistant to downstream terminators (Friedman Di, 2006). The transcription from P_R results in Q expression, which initiated the transcription from P_R' and the downstream genes expression including *stx* A and B (Karch et al., 1999). SOS response due to DNA damage and replication arrest can enhance *recA* expression and further stimulate Stx production (Kimmitt et al., 2000). There is also a RecA independent lambdoid prophage activation pathway for Stx production in *E. coli* O157:H7 (Imamovic and Muniesa, 2012), which is through regulation of RcsA (Rozanov et al., 1998). However, the mechanism involved in RecA-independent Stx2 933W prophage induction remains unknown (Imamovic and Muniesa, 2012). Stx prophage is also regulated by poly (A) polymerase I (PAP I), and PAP I-deficient cells showed a significant impairment of lysogenization and lytic development by Stx phages (Nowicki et al., 2015).

Polynucleotide phosphorylase (PNPase) has both 3'-5'exoribonuclease activity and 3' terminal oligonucleotide polymerase activity, which is involved in mRNA degradation and small RNA turnover (Carpousis et al., 1999; Arraiano et al., 2010) and also serves as PAP II (Kushner, 2015). PNPase regulates T3SS expression in *Yersinia* sp. and *Salmonella* Typhimurium (Ygberg et al., 2006; Rosenzweig et al., 2007); acts as a global regulator of virulence genes in *Salmonella*, which repressed invasion and intracellular replication of *Salmonella* (Clements et al., 2002), but was required for *S.* Typhimurium gut colonization in swine (Bearson et al., 2013). It enhanced *Campylobacter jejuni* motility and its chicken gut colonization (Haddad et al., 2012). However, roles of PNPase in the colonization and virulence of *E. coli* O157:H7 had not been explored, which were examined in the current study.

Materials and Methods

Cell Line, Media, Bacterial Strains, and Plasmids

The human colonic epithelial cell line HT-29 was obtained from the American Type Culture Collection (ATCC® HTB-38™,

Manassas, VA, USA). HT-29 cells were routinely cultured in Dulbecco's Modified Eagle's medium (DMEM; Sigma, St. Louis, MO, USA) supplemented with 10% fetal bovine serum (Sigma), 100 units/ml penicillin G, and 100 µg/ml of streptomycin (Sigma).

The *E. coli* O157:H7 EDL933, Sakai, and 86-24 strains were obtained from the STEC center at Michigan State University. *E. coli* O157:H7 was routinely grown in LB broth at 37°C with aeration. *E. coli* O157:H7 *pnp* deletion mutant was generated per the published method (Datsenko and Wanner, 2000). Briefly, using the forward primer GGCTTTACCCACATAGAGCTGGGTT AGGGTTGTCATTAGTCGCGAGGATGattccggggatccgtcgacc, and the reverse primer: CCGCCGCAGCGGAYGGCAAATGGC AACCTTACTCGCCCTGTTCAGCAGC tgtaggctggagctgcttcg, Kanamycin resistance cassette (lowercase letter) with *pnp* homology flanking sequence (uppercase letters) was PCR-amplified from pKD13 and electroporated into *E. coli* O157:H7 containing the λ-Red recombinase plasmid. The deletion of *pnp* was confirmed by PCR amplification using primers external to the disrupted gene (forward primer: TGTCA TTAGTCGCGAGGATG; and reverse primer: GCGGAYGGCA AATGGCAACC). To construct the *pnp* complementary plasmid, the *pnp* gene was PCR-amplified from *E. coli* O157:H7 genomic DNA using primers flanking with *Kpn*I and *Xba*I restriction enzyme sites, respectively (underlined letters): ATGGTACCACGCAAACGACACCGGTTCT and AG TCTAGATTACTCGCCCTGTTCAGCAG, and cloned into pBBR1-MCS3 (Kovach et al., 1995), which was kindly provided by Dr. Mark Gomelsky. The resulting plasmid pBBR1MCS3::*pnp* and its control vector were electroporated into *E. coli* O157:H7 EDL933 *pnp* mutant strains, which resulted in six strains: (1) EDL933; (2) EDL933 Δ*pnp*; (3) EDL933 Δ*pnp* pBBR1; (4) EDL933Δ*pnp* pBBR1::*pnp*; (5) 86-24 Δ*pnp*; (6) Sakai Δ*pnp*. Given that *pnp* and its downstream genes, *yhbM* or *nlpI* (Lipoprotein NlpI), *deaD* (ATP-dependent RNA helicase) and *mtr* (tryptophan permease) are located in one operon, to confirm that *pnp* deletion did not impact transcription of its downstream genes, we further compared the transcription of *yhbM*, the nearest downstream gene of *pnp*, between *pnp* deletion mutant and wild-type strain using forward primer: CAGTTTGAACAGTGCCGTGG and reverse primer: GATAACACCTCGCTCGCTGA.

Immunoblotting

Supernatant components: overnight *E. coli* O157:H7 cultures were subcultured at 1:1,000 in LB broth for 14 h at 37°C under shaking at 200 rpm, when bacteria were in stationary phase. Bacterial cultures were centrifuged at 5,000 × *g* at room temperature for 5 min, and the supernatant was then filtered through 0.22 µm filter (Millipore, Bedford, MA, USA). Trichloroacetic acid was added to a final concentration of 10% to the supernatant and incubated overnight at 4°C to precipitate proteins.

Whole bacteria: overnight *E. coli* O157:H7 cultures were subcultured at 1:1,000 in LB broth for 14 h at 37°C under shaking at 200 rpm, when bacterial culture were used for protein extraction.

Protein extracts were separated by 10% SDS-PAGE, transferred to nitrocellulose membranes and assayed with antibodies specific to Stx2A (Toxin Technology Inc., Sarasota, FL, USA) and Tir (a generous gift from Dr. John Leong, Tufts University). Blotted membranes were visualized using ECL^{TM} Western blotting detection reagents (Amersham Bioscience, Piscataway, NJ, USA).

Phage Spontaneous Induction and Enumeration

Phage enumeration assay was conducted according to our published method (Harris et al., 2012; Yue et al., 2012). Briefly, E. coli O157 cells were activated from frozen glycerol stock and grown in LB broth at 37°C with aeration for 8 h, which were sub-cultured at 1:1,000 in LB broth for 14 h at 37°C under shaking at 200 rpm when cultures were used to assess phage spontaneous induction. Activated bacterial culture was centrifuged at 5,000 × g at 4°C for 5 min. The resulting supernatant was serially diluted in a phage solution containing 10 mM CaCl$_2$ (Sigma) and 5 mM MgSO$_4$ (Sigma). Proper phage titration (200 μl) was mixed with 0.9 ml of the sensitive E. coli strain MG1655 culture and 5 ml of tempered top LB agar (0.7%, W/V) supplemented with 10 mM CaCl$_2$ and 10 mM MgSO$_4$. They were then mixed and poured immediately on top of the bottom LB agar plates (1.5%, W/V) containing 3 μg/ml chloramphenicol (Sigma). Plates were incubated at 37°C and plaque forming units (Pfu) were counted after 36 h incubation. Measurements for each strain had four independent replicates.

Quantitative PCR (qPCR) Analysis of Stx2 Prophage

Shiga toxin2 prophage qPCR was conducted per published methods (Harris et al., 2012; Yue et al., 2012). Briefly, overnight bacterial cultures were centrifuged at 5,000 × g at 4°C for 5 min. The supernatant was then filtered through 0.22 μm filter (Millipore, Bedford, MA, USA) to ensure complete removal of bacteria. The filtrate was centrifuged at 35,000 × g at 4°C for 2 h. The resulting phage pellet was dissolved in sterile ddH$_2$O and treated with DNase I (2 units/μl) at 37°C for 2 h to hydrolyze any remaining contaminated E. coli O157:H7 genomic DNA. The above phage preparation was boiled at 100°C for 10 min to denature DNase I and release phage DNA, and then used as a template for qPCR. PCR analysis of stx2 gene was conducted using stx2A primers and SYBR Green master mix (Bio-Rad, Hercules, CA, USA). Absence of bacterial DNA in phage lysates was confirmed by the absence of tufA DNA as assayed by PCR. Stx2 phages are known to carry only one copy of the stx2 gene, therefore the stx2 gene copy number can be extrapolated to quantify Stx2 prophages.

Adhesion of E. coli O157:H7 to Colonic Epithelial Cells

HT-29 epithelial cells were seeded in 24 well plates and cultured in DMEM media (Sigma) with 10 % FBS (Sigma) until ~90 % confluence. The medium was removed and each well was washed three times with PBS (pH 7.4). The cells in each well were challenged with 10^7 CFU/well E. coli O157:H7. The HT-29 cells and E. coli O157:H7 were co-cultured at 37°C with 5% CO$_2$ for 2 h. Then the HT-29 cell monolayers were washed three times with PBS, and lysed with 0.2% Triton X-100. Lysates were serially diluted, plated on LB, and bacterial colonies were counted after 18 h incubation (Xicohtencatl-Cortes et al., 2007; Torres et al., 2008).

E. coli O157:H7 Adhesion to Cattle Colonic Explants

Terminal colons (5–10 cm from anus) were aseptically collected from E. coli O157:H7 free beef cattle slaughtered in the University of Wyoming Meat Laboratory and transferred to the Microbiology Lab within 10 min. Specimens were washed five times with 0.9% (W/V) NaCl and the mucosa was dissected from the underlying tissues using sterile dissecting scissors. The trimmed specimens were reduced to 8 mm round pieces using Miltex disposable biopsy punches (Cardinal Health, Cleveland, OH, USA), and transferred to the individual well of 24 well plates and cultured in DMEM medium with 5% FBS for 1 h. E. coli O157:H7 was added to these wells at 10^7 CFU/well, and cultured at 37°C for 2 h (Xicohtencatl-Cortes et al., 2007). The attached E. coli O157:H7 cells were serially diluted, plated and enumerated.

Quantitative Reverse Transcription PCR (qRT-PCR) Analyses

Total RNA was extracted from E. coli O157:H7 grown in LB broth using RNeasy Protect Bacteria Mini Kit (Qiagen, Valencia, CA, USA) and reverse transcribed using a QuantiTect Reverse Transcription Kit (Qiagen). cDNAs were used as a template for qRT-PCR analysis of selected genes using a CFX96^{TM} Real-Time PCR Detection System (Bio-Rad, Hercules, CA, USA). SYBR Green Master Mix (Bio-Rad, Hercules, CA, USA) was used for all qRT-PCR reactions. Primers for qRT-PCR are listed in **Tables 1** and **2**. Gene gapA was used as the housekeeping gene. Amplification efficiency was 0.90–0.99.

Statistical Analysis

Data were analyzed as a randomized design using GLM (General Linear Model of Statistical Analysis System, SAS, 2000). Differences between means were determined using Student's t-test followed by Duncan's multiple test when appropriate. P < 0.05 was considered to be statistically significant.

Results

PNPase is Essential for Stx2 Production in E. coli O157:H7 EDL933 Strain

In transcriptional level, deletion of pnp completely eliminated stx2 mRNA expression in E. coli O157:H7 EDL933 strain without affecting stx1 expression (**Figure 1A**). We further compared the Stx2 production. In wild-type strain, a 32 kDa band Stx2 A subunit was detected in both supernatant fraction and

TABLE 1 | Primer sets used for quantitative Reverse Transcription PCR (qRT-PCR) for Shiga toxin (Stx) and phage related genes.

Gene name	Product size	Direction	Sequence
cI	80 bp	Forward	TCACACGAAGACCAAAGGCA
		Reverse	TGCCATCGAGATGACCGAAG
cII	118 bp	Forward	ACCTGTCAACGCTTACCCAG
		Reverse	GATGCCATGCCAAAAGCACA
cIII	101 bp	Forward	TTCTTTGGGACTCCTGGCTG
		Reverse	GGCTGCCCTTCCGAATCTTT
cro	75 bp	Forward	GAAGTTGCGAAGGCTTGTGG
		Reverse	AGTCTTAGGGAGGAAGCCGT
gapA	103 bp	Forward	CCAGGACATCGTTTCCAAC
		Reverse	GGTGGTCATCAGACCTTCG
O	104 bp	Forward	GATGCTGCAATTCAGAGCGG
		Reverse	TTTCTGGCTGATGGTGCGAT
rcsA	124 bp	Forward	CTCGACGATATCCTTGGCGA
		Reverse	CCTGACCTGCCATCCACATT
recA	70 bp	Forward	CTGGGAGCACAGCAGGTTGTCG
		Reverse	TGAACACGCGCTGGACCCAATC
S	124 bp	Forward	AGTTGCTGGACAGGGTTTCC
		Reverse	GCCTTACGCCGGTCTTCTTT
stx1	180 bp	Forward	ATAAATCGCCATTCGTTGACTAC
		Reverse	AGAACGCCCACTGAGATCATC
stx2A	133 bp	Forward	CGTCACTCACTGGTTTCATCAT
		Reverse	TCTGTATCTGCCTGAAGCGTAA

TABLE 2 | Primer sets used for qRT-PCR for T3SS related genes.

Gene name	Product size	Direction	Sequence
csrA	102 bp	Forward	TCCTTCGGGGCATTTACGCCA
		Reverse	TCGTCGAGTTGGTGAGACCCTC
eae	139 bp	Forward	TCACTTTGAATGGTAAAGGCAGT
		Reverse	CAAATGGACATAGCATCAGCATA
espB	122 bp	Forward	TAAAGGGGCTGGTGAGATTG
		Reverse	CTGCCGACATCAGCAACACTT
gapA	103 bp	Forward	CCAGGACATCGTTTCCAAC
		Reverse	GGTGGTCATCAGACCTTCG
grlA	71 bp	Forward	TAGAAAGTCCTGGAACAAC
		Reverse	AGACTGTCCCACAATACC
hns	150 bp	Forward	GCCGGACGCTGAGCACGTTT
		Reverse	CGCGGCTGCTGCTGAAGTTG
ler	71 bp	Forward	CCCGACCAGGTCTGCCCTTCT
		Reverse	GATGGACTCGCTCGCCGGAAC
tir	128 bp	Forward	AACGAAAGAAGCGTTCCAGA
		Reverse	CTGCTGCTTTAGCCTGCTCT

whole cell lysis (**Figure 1B**). Deletion of *pnp* resulted in no detection of Stx2A (**Figure 1B**). This defective in Stx2 protein production was not restored by complementation of pBBR1::*pnp* (**Figure 1B**).

PNPase is Indispensable for Stx2 Phage Lytic Growth in *E. coli* O157:H7

Shiga toxin production is mediated by Stx2 phage activation, thus we further compared phage activation in *E. coli* O157:H7 EDL933 and its *pnp* mutant strains. Deletion of *pnp* completely abolished lambdoid phage progeny production in *E. coli* O157:H7 (**Figure 2A**). Quantitative PCR further confirmed that the *stx2* phage DNA accumulation was repressed in PNPase defective *E. coli* O157:H7 strain (**Figure 2B**). To gain insights into molecular mechanisms of PNPase mediated impairment in Stx phage activation, we further analyzed expression of phage genes regulating prophage activation. Deletion of *pnp* did not affect *recA* and *rcsA* expressions (**Figure 2C**). We could not detect *cro* mRNA expression and only detected a very low level of *cI* expression in Δ*pnp* strain (**Figure 2C**). In addition, the transcriptional levels of *cII* and *cIII*, phage lysis *S* gene and phage replication gene *O* were all diminished in Δ*pnp* strain (**Figure 2C**). Altogether, these data indicated that PNPase in *E. coli* O157:H7 plays an indispensable role in Stx2 prophage lysis.

Deletion of PNPase in *E. coli* O157:H7 Impairs Epithelial Adhesion but Enhances T3SS Encoding Transcripts and Proteins

In addition to phage encoding Stx2 production, *E. coli* O157:H7 has a LEE chromosomal pathogenicity island that encoding T3SS apparatus and effectors, mediating intimate contact to epithelial cells (McDaniel et al., 1995). We further evaluated the significance of *pnp* deletion in *E. coli* O157: H7 epithelial adhesion. Deletion of *pnp* in *E. coli* O157: H7 EDL933 strain decreased its adhesion to colonic epithelial HT-29 cells by approximately twofold (**Figure 3A**). Consistently, adhesion to cattle colonic gut explant was reduced about fourfold in EDL933Δ*pnp* (**Figure 3B**).

To understand possible factors limiting *E. coli* O157:H7 epithelial colonization in the absence of PNPase, we further analyzed T3SS apparatus and effectors transcripts and proteins in EDL933Δ*pnp* and its wild type strains. Surprisingly, the transcriptional levels of genes located on LEE region including *tir*, *eae*, and *espB* were enhanced in EDL933Δ*pnp* strain (**Figure 4A**). Tir protein production was also increased in both supernatant fraction and whole cell lysis of EDL933Δ*pnp* strain compared to the wild-type EDL933 strain (**Figure 4B**). In line with, the transcriptional level of LEE positive regulators, the *ler* transcript level, was significantly enhanced in EDL933Δ*pnp* strain (**Figure 4C**), though other regulators of LEE (*hns*, *csrA*, and *grlA*) did not differ between EDL933Δ*pnp* and its isogenic wild type strain.

Impact of PNPase in Other *E. coli* O157:H7 Strains

To confirm the results from *E. coli* O157:H7 strain EDL933, we further examined the Stx2 production and Stx2 phage lytic activation in *E. coli* O157:H7 strain Sakai and 86-24. Consistent to the finding in EDL933 strain, the *pnp* deletion decreased Stx2 production (**Figure 5A**) and impaired Stx2 phage lytic growth (**Figure 5B**) in Sakai and 86-24. As in strain EDL933, Tir protein content was enhanced in both *E. coli* O157:H7 strain Sakai and 86-24 (**Figure 5A**).

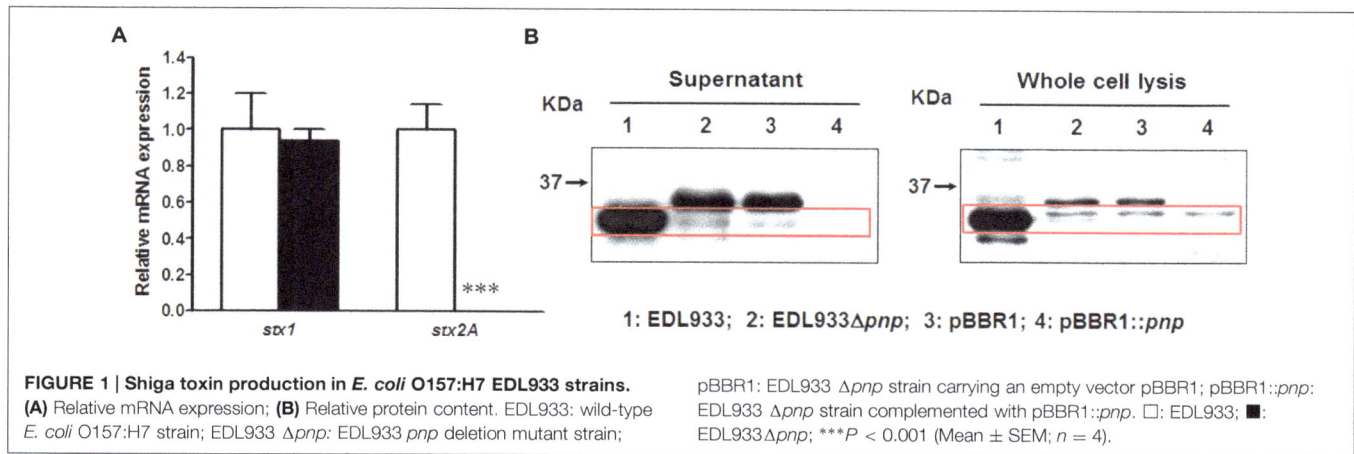

FIGURE 1 | Shiga toxin production in *E. coli* O157:H7 EDL933 strains.
(A) Relative mRNA expression; **(B)** Relative protein content. EDL933: wild-type *E. coli* O157:H7 strain; EDL933 Δ*pnp*: EDL933 *pnp* deletion mutant strain; pBBR1: EDL933 Δ*pnp* strain carrying an empty vector pBBR1; pBBR1::*pnp*: EDL933 Δ*pnp* strain complemented with pBBR1::*pnp*. □: EDL933; ■: EDL933Δ*pnp*; ***$P < 0.001$ (Mean ± SEM; $n = 4$).

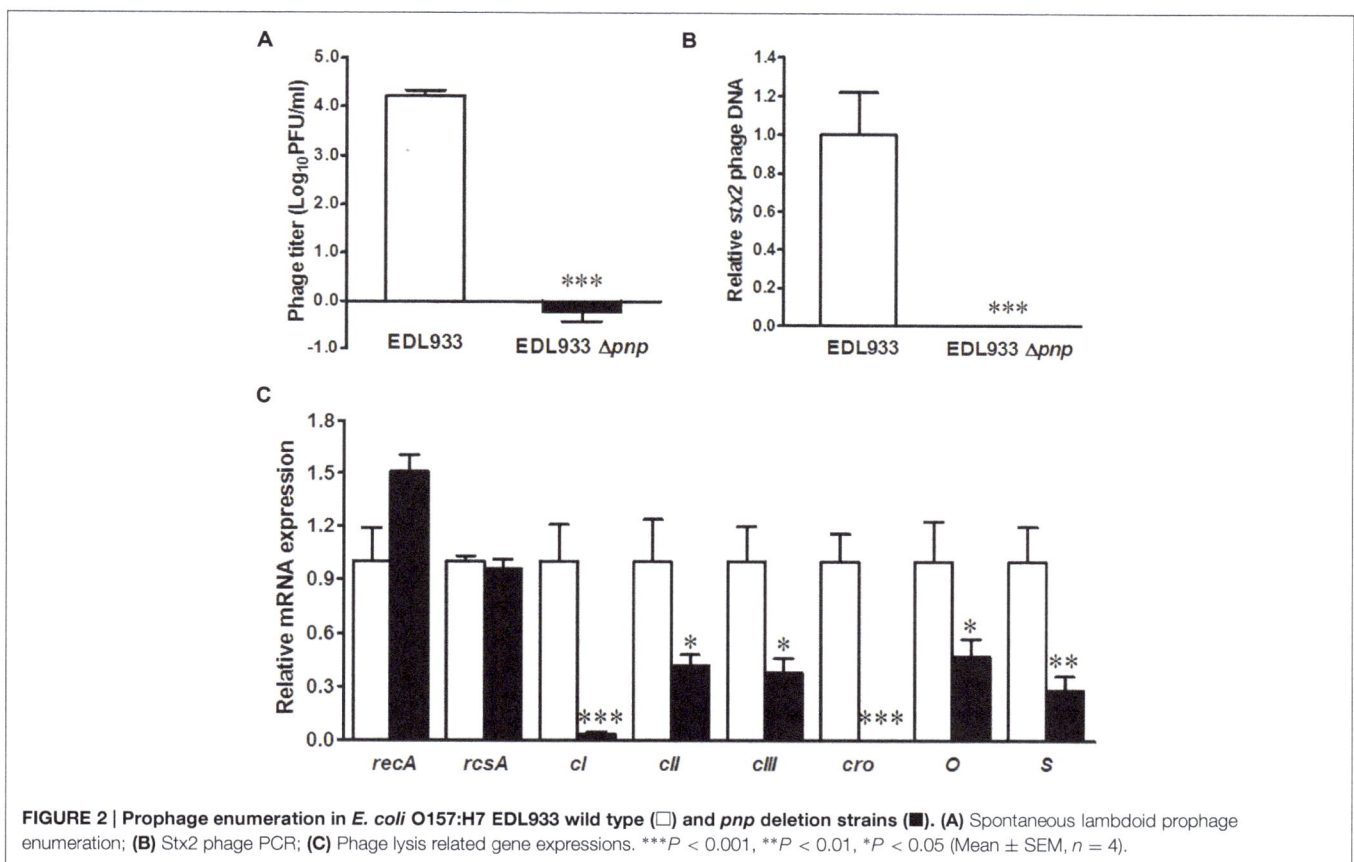

FIGURE 2 | Prophage enumeration in *E. coli* O157:H7 EDL933 wild type (□) and *pnp* deletion strains (■). (A) Spontaneous lambdoid prophage enumeration; **(B)** Stx2 phage PCR; **(C)** Phage lysis related gene expressions. ***$P < 0.001$, **$P < 0.01$, *$P < 0.05$ (Mean ± SEM, $n = 4$).

Discussion

PNPase Regulates Stx2 Production Through Repressing Stx2 Phage Lytic Cycle

Shiga toxin is mainly responsible for the mortality associated with *E. coli* O157:H7 infection. PNPase is a bifunctional enzyme with an exoribonuclease activity and a poly (A) polymerase activity. Study herein demonstrated that PNPase plays a vital role in Stx production. The deletion of *pnp* in *E. coli* O157:H7 completely eliminated *stx2* mRNA expression and Stx2 production, which could not be recovered by complementary *pnp* overexpression. Thus we speculated that the observed effect of PNPase on Stx2 production in EDL933 strain might be due to a secondary mutation in *stx2* gene during generation of *pnp* deletion mutant. Therefore, we did three independent *pnp* deletion experiments. Different colony from three independent knock out experiments had the same Stx2 profile (data not shown). We further sequenced the promoter region of *stx2*, which revealed no alteration in the sequence of *stx2* promoter region (data not shown). Given

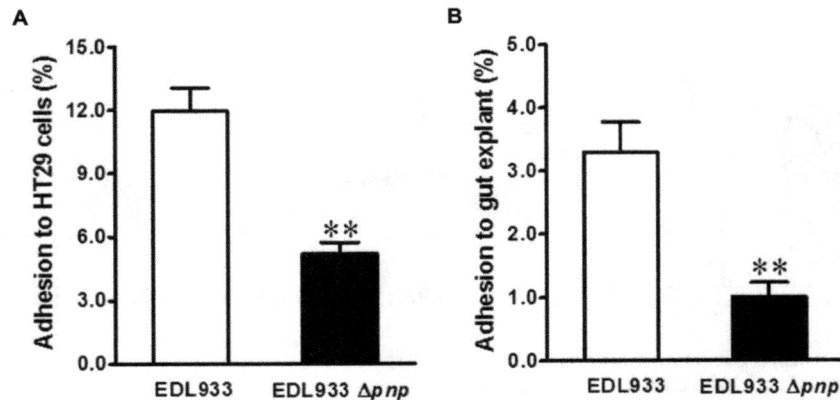

FIGURE 3 | Adhesion of _E. coli_ O157:H7 to intestinal epithelium. (A) HT-29 colonic epithelial cell line; **(B)** Cattle colonic explant tissues. EDL933: wild-type _E. coli_ O157:H7 strain; EDL933 Δ_pnp_: EDL933 _pnp_ deletion mutant strain, ****P** < 0.01 (Mean ± SEM; _n_ = 8).

that _pnp_ and its downstream genes, _yhbM, deaD,_ and _mtr_ are in one operon, we questioned whether disruption of _pnp_ might change the mRNA levels of _pnp_ downstream genes. Thus, we further analyzed the transcription of the _pnp_ nearest downstream gene, _yhbM,_ and qRT-PCR analysis indicated that it was not affected by _pnp_ deletion (data not shown). These testimonies showed that the decreased Stx2 production was due to _pnp_ deletion. Although the polar effect of _pnp_ deletion would be low in our experiment (Our experiment is in frame deletion), we can't rule out the possibility that the polar effect of the _pnp_ gene deletion cause the deficiency of Stx production.

Effective production of Stx requires phage lytic status development. Consistent with decreased Stx production, _pnp_ deletion in _E. coli_ O157:H7 repressed Stx2 phage lytic status, indicating that the lytic growth of 933w phage in _E. coli_ O157:H7 was regulated by PNPase. CI protein is required for lysogenic establishment and has self-feedback inhibition activity, which represses prophage replication and lysis (Echols and Green, 1971). _cro_ is the early lytic gene antagonizes CI and regulates phage switch from lysogenic to lytic growth during prophage induction (Svenningsen et al., 2005). Consistent with impaired Stx2 phage lytic growth, transcriptional level of _cro_ were decreased in the EDL933Δ_pnp_ strain. Additionally, transcriptional levels of _cII_ and _cIII_, phage lysis gene, _S_, and replication genes, _O_, were also repressed in Δ_pnp_ compared to the wild type host. These data demonstrated that the lytic cycle of the Stx2 phage in EDL933Δ_pnp_ strain was repressed, which at least partially through influencing transcripts of gene encoding proteins vital for lytic phage growth. In line with our findings, dysfunction of PAP I in _E. coli_ resulted in impaired Stx phage lytic development associated with reduced transcripts of _cII, cIII, cro, and O_ at later time points of induction (Nowicki et al., 2015). Of note, the transcript level of _O_ and _cro_ are altered differently in Δ_pnp_ relative to wild-type strain, which might be due to the different post-transcriptional regulation (e.g., different sRNA regulation) and warrants further research.

Similar to what observed in strain EDL933, deletion of _pnp_ significantly decreased Stx2 production in Sakai and 86-24 strains. EDL933 and Sakai strains have both _stx1_ and _stx2_ genes, while 86-24 only has _stx2_ gene. 933W prophage from EDL933 and Sakai Stx2 prophage have different sequence in early gene regulators and replication proteins (Makino et al., 1999), but almost identical sequence for structural genes (Makino et al., 1999). Compared to the genome of 86-24 Stx2 producing phage, there is a 1439 bp insertion replaced a 2091 bp sequence in 933W phage (Kudva et al., 2002). In agreement with diminished Stx2 production, Stx2 phage spontaneous lytic growth was repressed in all tested PNPase deficient strains, indicating that PNPase is important in lytic cycle development of Stx2 phage in _E. coli_ O157:H7 strains.

PNPase Reduced _E. coli_ O157:H7 Intestinal Epithelial Colonization Independent of Expression Levels of T3SS

Deletion of _pnp_ in _E. coli_ O157:H7 decreased its adhesion to epithelial cells and colonic epithelial tissues. In agreement, _C. jejun_ deficient in PNPase has reduced ability to adhere and invade to HT-29 cells, and colonize chick gut (Haddad et al., 2012). PNPase was required for _S._ Typhimurium gut colonization in swine, PNPase deficiency was associated with reduced fecal _S._ Typhimurium shedding and intestinal colonization (Bearson et al., 2013). These data collectively indicated PNPase plays a critical role in gut colonization.

Escherichia coli O157:H7 possess LEE pathogenicity island that encode T3SS apparatus and effector proteins, enabling its intimate contact to epithelial cells and further forming AE lesions. In contrast to decreased epithelial adhesion, we revealed that _pnp_ deletion elevated the transcription level of genes located in the LEE region as well as Tir protein content, which is possibly through enhancing _ler_ transcription. These data indicated that PNPase in _E. coli_ O157:H7 might regulate expression and functional activity

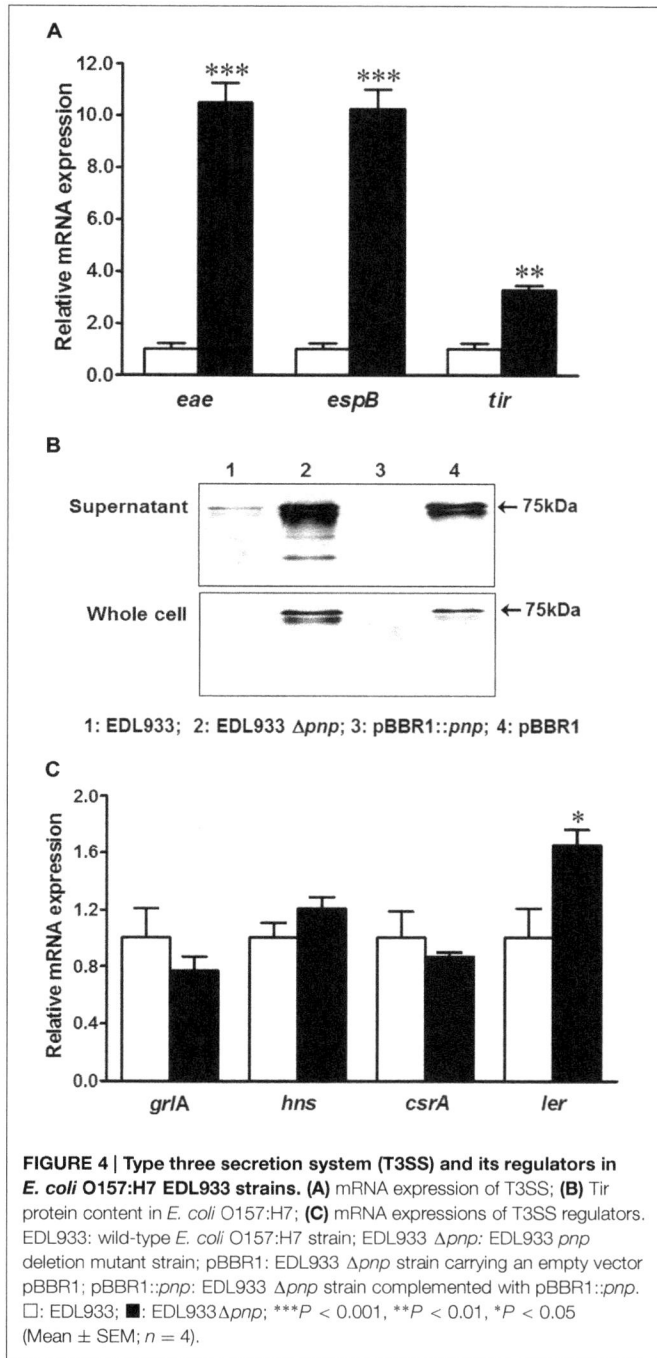

FIGURE 4 | Type three secretion system (T3SS) and its regulators in
E. coli O157:H7 EDL933 strains. (A) mRNA expression of T3SS; (B) Tir
protein content in *E. coli* O157:H7; (C) mRNA expressions of T3SS regulators.
EDL933: wild-type *E. coli* O157:H7 strain; EDL933 Δ*pnp*: EDL933 *pnp*
deletion mutant strain; pBBR1: EDL933 Δ*pnp* strain carrying an empty vector
pBBR1; pBBR1::*pnp*: EDL933 Δ*pnp* strain complemented with pBBR1::*pnp*.
□: EDL933; ■: EDL933Δ*pnp*; ***$P < 0.001$, **$P < 0.01$, *$P < 0.05$
(Mean ± SEM; $n = 4$).

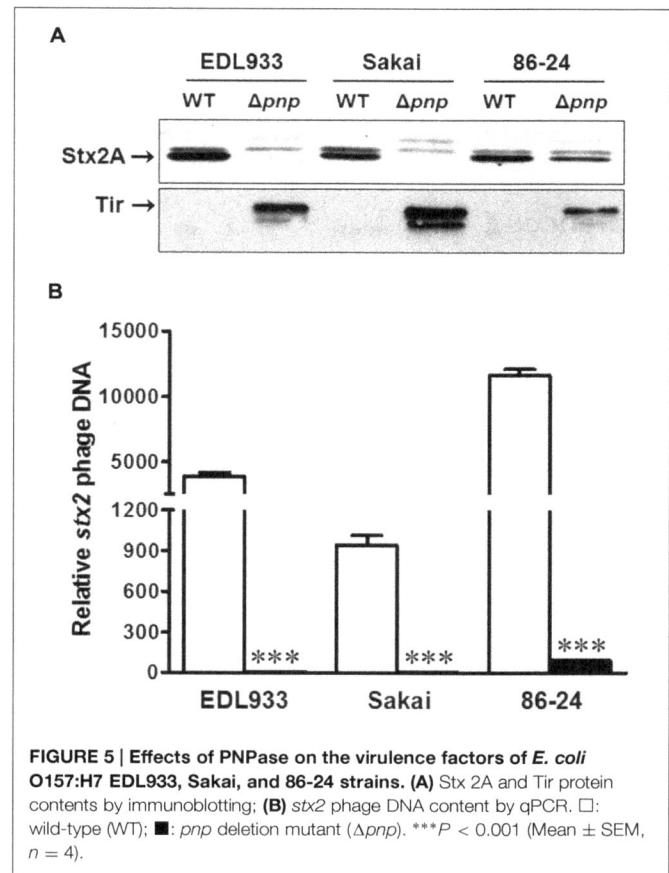

FIGURE 5 | Effects of PNPase on the virulence factors of *E. coli*
O157:H7 EDL933, Sakai, and 86-24 strains. (A) Stx 2A and Tir protein
contents by immunoblotting; (B) *stx2* phage DNA content by qPCR. □:
wild-type (WT); ■: *pnp* deletion mutant (Δ*pnp*). ***$P < 0.001$ (Mean ± SEM,
$n = 4$).

of T3SS independently, which warrants future research. The
reduced T3SS functional activity as indicated by reduced
epithelial adhesion might be due to the defectiveness in T3SS
apparatus organization or effector secretion. Meanwhile, we
speculated that the decreased intestinal adhesion might be
partially due to the reduced Stx2 production. Stx regulates
the distribution of nucleolin, a host cell protein that is a
receptor for intimin in *E. coli* O157:H7 (Robinson et al., 2006).
PNPase deletion decreased Stx2, which might disturb nucleolin

distribution and further impair the adherence capacity of *E. coli*
O157:H7.

Our observation was supported by previous studies in
Yersinia, where PNPase is required for the proper T3SS
function. The *pnp* mutant strain was defective in rapidly
exporting T3SS effector proteins (Rosenzweig et al., 2005), less
virulent in mouse, but enhanced T3SS encoding transcripts
and proteins compared with its isogenic wild-type *Yersinia*
strain (Rosenzweig et al., 2007). Similarly, in *S.* Typhimurium,
despite of decreased swine intestinal colonization in response
to PNPase deficiency (Bearson et al., 2013), T3SS-encoding
transcripts and proteins were expressed at higher levels in
PNPase inactivated strain (Ygberg et al., 2006). On the other
hand, inactivation of PNPase in *Salmonella* increased invasion
and intracellular replication as well as the establishment of
persistent infection in mice (Clements et al., 2002). These
implicated a complex nature of PNPases in regulating bacterial
virulence.

In summary, PNPase has pleiotropic effects on the virulence of
E. coli O157:H7. It is not only indispensable for Stx2 phage lytic
cycle growth and associated Stx2 production in *E. coli* O157:H7,
but also required for *E. coli* O157:H7 intestinal epithelial adhesion
and regulation of type III secretion system. Alteration in PNPase
activity might provide a potential strategy to reduce the virulence
of *E. coli* O157:H7.

Acknowledgments

We thank Dr. Xin Fang and Dr. Mark Gomelsky at Department of Molecular Biology, University of Wyoming for their helpful suggestions. We acknowledge Dr. John Leong at Tufts University and Dr. Mark Gomelsky at University of Wyoming for their generous gifts of Tir antibodies and pBBR1-MCS3 plasmid, respectively. This work was supported by USDA-AFRI 2010-65201-20599 and Washington State University Emerging Research Issues Competitive Grants.

References

Allison, H. E. (2007). Stx-phages: drivers and mediators of the evolution of STEC and STEC-like pathogens. *Future Microbiol.* 2, 165–174. doi: 10.2217/17460913.2.2.165

Arraiano, C. M., Andrade, J. M., Domingues, S., Guinote, I. B., Malecki, M., Matos, R. G., et al. (2010). The critical role of RNA processing and degradation in the control of gene expression. *FEMS Microbiol. Rev.* 34, 883–923. doi: 10.1111/j.1574-6976.2010.00242.x

Bearson, S. M., Bearson, B. L., Lee, I. S., and Kich, J. D. (2013). Polynucleotide phosphorylase (PNPase) is required for *Salmonella enterica* serovar Typhimurium colonization in swine. *Microb. Pathog.* 65, 63–66. doi: 10.1016/j.micpath.2013.10.001

Bhatt, S., Romeo, T., and Kalman, D. (2011). Honing the message: post-transcriptional and post-translational control in attaching and effacing pathogens. *Trends Microbiol.* 19, 217–224. doi: 10.1016/j.tim.2011.01.004

Carpousis, A. J., Vanzo, N. F., and Raynal, L. C. (1999). mRNA degradation. A tale of poly(A) and multiprotein machines. *Trends Genet.* 15, 24–28. doi: 10.1016/S0168-9525(98)01627-8

Clements, M. O., Eriksson, S., Thompson, A., Lucchini, S., Hinton, J. C., Normark, S., et al. (2002). Polynucleotide phosphorylase is a global regulator of virulence and persistency in *Salmonella enterica. Proc. Natl. Acad. Sci. U.S.A.* 99, 8784–8789. doi: 10.1073/pnas.132047099

Datsenko, K. A., and Wanner, B. L. (2000). One-step inactivation of chromosomal genes in *Escherichia coli* K-12 using PCR products. *Proc. Natl. Acad. Sci. U.S.A.* 97, 6640–6645. doi: 10.1073/pnas.120163297

Deng, W., Puente, J. L., Gruenheid, S., Li, Y., Vallance, B. A., Vazquez, A., et al. (2004). Dissecting virulence: systematic and functional analyses of a pathogenicity island. *Proc. Natl. Acad. Sci. U.S.A.* 101, 3597–3602. doi: 10.1073/pnas.0400326101

Echols, H., and Green, L. (1971). Establishment and maintenance of repression by bacteriophage lambda: the role of the cI, cII, and c3 proteins. *Proc. Natl. Acad. Sci. U.S.A.* 68, 2190–2194. doi: 10.1073/pnas.68.9.2190

Friedman Di, C. D. (2006). "Regulation of lambda gene experssion by transcription termination and antitermination," in *The Bacteriophages*, ed. R. Calendar (Oxford: Oxford Press.), 83–103.

Haddad, N., Tresse, O., Rivoal, K., Chevret, D., Nonglaton, Q., Burns, C. M., et al. (2012). Polynucleotide phosphorylase has an impact on cell biology of *Campylobacter jejuni. Front. Cell. Infect. Microbiol.* 2:30. doi: 10.3389/fcimb.2012.00030

Harris, S. M., Yue, W. F., Olsen, S. A., Hu, J., Means, W. J., Mccormick, R. J., et al. (2012). Salt at concentrations relevant to meat processing enhances Shiga toxin 2 production in *Escherichia coli* O157:H7. *Int. J. Food Microbiol.* 159, 186–192. doi: 10.1016/j.ijfoodmicro.2012.09.007

Herold, S., Karch, H., and Schmidt, H. (2004). Shiga toxin-encoding bacteriophages–genomes in motion. *Int. J. Med. Microbiol.* 294, 115–121. doi: 10.1016/j.ijmm.2004.06.023

Imamovic, L., and Muniesa, M. (2012). Characterizing RecA-independent induction of Shiga toxin2-encoding phages by EDTA treatment. *PLoS ONE* 7:e32393. doi: 10.1371/journal.pone.0032393

Johannes, L., and Romer, W. (2010). Shiga toxins–from cell biology to biomedical applications. *Nat. Rev. Microbiol.* 8, 105–116. doi: 10.1038/nrmicro2279

Karch, H., Schmidt, H., Janetzki-Mittmann, C., Scheef, J., and Kroger, M. (1999). Shiga toxins even when different are encoded at identical positions in the genomes of related temperate bacteriophages. *Mol. Gen. Genet.* 262, 600–607. doi: 10.1007/s004380051122

Kimmitt, P. T., Harwood, C. R., and Barer, M. R. (2000). Toxin gene expression by shiga toxin-producing *Escherichia coli*: the role of antibiotics and the bacterial SOS response. *Emerg. Infect. Dis.* 6, 458–465. doi: 10.3201/eid0605.000503

Kovach, M. E., Elzer, P. H., Hill, D. S., Robertson, G. T., Farris, M. A., Roop, R. M. II, et al. (1995). Four new derivatives of the broad-host-range cloning vector pBBR1MCS, carrying different antibiotic-resistance cassettes. *Gene* 166, 175–176. doi: 10.1016/0378-1119(95)00584-1

Kudva, I. T., Evans, P. S., Perna, N. T., Barrett, T. J., Ausubel, F. M., Blattner, F. R., et al. (2002). Strains of *Escherichia coli* O157:H7 differ primarily by insertions or deletions, not single-nucleotide polymorphisms. *J. Bacteriol.* 184, 1873–1879. doi: 10.1128/JB.184.7.1873-1879.2002

Kushner, S. R. (2015). Polyadenylation in *E. coli*: a 20 year odyssey. *RNA* 21, 673–674. doi: 10.1261/rna.049700.115

Little, J. (2005). "Lysogeny, prophage induction, and lysogenic conversion," in *Phages: Their Role in Bacterial Pathogenesis and Biotechnology*, eds D. I. Friedman, M. K. Waldor, and S. L. Adhya (Washington, DC: ASM Press), 37–54.

Los, J. M., Los, M., and Wegrzyn, G. (2011). Bacteriophages carrying Shiga toxin genes: genomic variations, detection and potential treatment of pathogenic bacteria. *Future Microbiol.* 6, 909–924. doi: 10.2217/fmb.11.70

Mainil, J. G., and Daube, G. (2005). Verotoxigenic *Escherichia coli* from animals, humans and foods: who's who? *J. Appl. Microbiol.* 98, 1332–1344. doi: 10.1111/j.1365-2672.2005.02653.x

Makino, K., Yokoyama, K., Kubota, Y., Yutsudo, C. H., Kimura, S., Kurokawa, K., et al. (1999). Complete nucleotide sequence of the prophage VT2-Sakai carrying the verotoxin 2 genes of the enterohemorrhagic *Escherichia coli* O157:H7 derived from the Sakai outbreak. *Genes Genet. Syst.* 74, 227–239. doi: 10.1266/ggs.74.227

McDaniel, T. K., Jarvis, K. G., Donnenberg, M. S., and Kaper, J. B. (1995). A genetic locus of enterocyte effacement conserved among diverse enterobacterial pathogens. *Proc. Natl. Acad. Sci. U.S.A.* 92, 1664–1668. doi: 10.1073/pnas.92.5.1664

Nowicki, D., Bloch, S., Nejman-Falenczyk, B., Szalewska-Palasz, A., Wegrzyn, A., and Wegrzyn, G. (2015). Defects in RNA polyadenylation impair both lysogenization by and lytic development of Shiga toxin-converting bacteriophages. *J. Gen. Virol.* doi: 10.1099/vir.0.000102 [Epub ahead of print].

Paton, J. C., and Paton, A. W. (1998). Pathogenesis and diagnosis of Shiga toxin-producing *Escherichia coli* infections. *Clin. Microbiol. Rev.* 11, 450–479.

Ptashne, M., and Hopkins, N. (1968). The operators controlled by the lambda phage repressor. *Proc. Natl. Acad. Sci. U.S.A.* 60, 1282–1287. doi: 10.1073/pnas.60.4.1282

Robinson, C. M., Sinclair, J. F., Smith, M. J., and O'Brien, A. D. (2006). Shiga toxin of enterohemorrhagic *Escherichia coli* type O157 : H7 promotes intestinal colonization. *Proc. Natl. Acad. Sci. U.S.A.* 103, 9667–9672. doi: 10.1073/pnas.0602359103

Rosenzweig, J. A., Chromy, B., Echeverry, A., Yang, J., Adkins, B., Plano, G. V., et al. (2007). Polynucleotide phosphorylase independently controls virulence factor expression levels and export in *Yersinia* spp. *FEMS Microbiol. Lett.* 270, 255–264. doi: 10.1111/j.1574-6968.2007.00689.x

Rosenzweig, J. A., Weltman, G., Plano, G. V., and Schesser, K. (2005). Modulation of *Yersinia* type three secretion system by the S1 domain of polynucleotide phosphorylase. *J. Biol. Chem.* 280, 156–163. doi: 10.1074/jbc.M405662200

Rozanov, D. V., D'Ari, R., and Sineoky, S. P. (1998). RecA-independent pathways of lambdoid prophage induction in *Escherichia coli. J. Bacteriol.* 180, 6306–6315.

Svenningsen, S. L., Costantino, N., Court, D. L., and Adhya, S. (2005). On the role of Cro in lambda prophage induction. *Proc. Natl. Acad. Sci. U.S.A.* 102, 4465–4469. doi: 10.1073/pnas.0409839102

Torres, A. G., Slater, T. M., Patel, S. D., Popov, V. L., and Arenas-Hernandez, M. M. (2008). Contribution of the Ler- and H-NS-regulated long polar fimbriae of

Escherichia coli O157:H7 during binding to tissue-cultured cells. *Infect. Immun.* 76, 5062–5071. doi: 10.1128/IAI.00654-08

Tyler, J. S., Beeri, K., Reynolds, J. L., Alteri, C. J., Skinner, K. G., Friedman, J. H., et al. (2013). Prophage induction is enhanced and required for renal disease and lethality in an EHEC mouse model. *PLoS Pathog.* 9:e1003236. doi: 10.1371/journal.ppat.1003236

Wagner, P. L., Neely, M. N., Zhang, X., Acheson, D. W., Waldor, M. K., and Friedman, D. I. (2001). Role for a phage promoter in Shiga toxin 2 expression from a pathogenic *Escherichia coli* strain. *J. Bacteriol.* 183, 2081–2085. doi: 10.1128/JB.183.6.2081-2085.2001

Xicohtencatl-Cortes, J., Monteiro-Neto, V., Ledesma, M. A., Jordan, D. M., Francetic, O., Kaper, J. B., et al. (2007). Intestinal adherence associated with type IV pili of enterohemorrhagic *Escherichia coli* O157:H7. *J. Clin. Invest.* 117, 3519–3529. doi: 10.1172/JCI 30727

Ygberg, S. E., Clements, M. O., Rytkonen, A., Thompson, A., Holden, D. W., Hinton, J. C., et al. (2006). Polynucleotide phosphorylase negatively controls spv virulence gene expression in *Salmonella enterica*. *Infect. Immun.* 74, 1243–1254. doi: 10.1128/IAI.74.2.1243-1254.2006

Yue, W. F., Du, M., and Zhu, M. J. (2012). High temperature in combination with UV irradiation enhances horizontal transfer of stx2 gene from *E. coli* O157:H7 to non-pathogenic *E. coli*. *PLoS ONE* 7:e31308. doi: 10.1371/journal.pone.0031308

Conflict of Interest Statement: The authors declare that the research was conducted in the absence of any commercial or financial relationships that could be construed as a potential conflict of interest.

5

The two-component system CpxR/A represses the expression of *Salmonella* virulence genes by affecting the stability of the transcriptional regulator HilD

Miguel A. De la Cruz[1†], Deyanira Pérez-Morales[2†], Irene J. Palacios[2], Marcos Fernández-Mora[2], Edmundo Calva[2] and Víctor H. Bustamante[2*]

[1] Unidad de Investigación Médica en Enfermedades Infecciosas y Parasitarias, Centro Médico Nacional Siglo XX1-IMSS, México DF, Mexico, [2] Departamento de Microbiología Molecular, Instituto de Biotecnología, Universidad Nacional Autónoma de México, Cuernavaca, Morelos, Mexico

Edited by:
Dongsheng Zhou,
Beijing Institute of Microbiology and
Epidemiology, China

Reviewed by:
Weili Liang,
National Institute for Communicable
Disease Control and Prevention, China
Tracy Raivio,
University of Alberta, Canada

*Correspondence:
Víctor H. Bustamante,
Departamento de Microbiología
Molecular, Instituto de Biotecnología,
Universidad Nacional Autónoma de
México, Avenida Universidad 2001,
Cuernavaca, Mor 62210, Mexico
victor@ibt.unam.mx

[†] These authors have contributed
equally to this work.

Salmonella enterica can cause intestinal or systemic infections in humans and animals mainly by the presence of pathogenicity islands SPI-1 and SPI-2, containing 39 and 44 genes, respectively. The AraC-like regulator HilD positively controls the expression of the SPI-1 genes, as well as many other *Salmonella* virulence genes including those located in SPI-2. A previous report indicates that the two-component system CpxR/A regulates the SPI-1 genes: the absence of the sensor kinase CpxA, but not the absence of its cognate response regulator CpxR, reduces their expression. The presence and absence of cell envelope stress activates kinase and phosphatase activities of CpxA, respectively, which in turn controls the level of phosphorylated CpxR (CpxR-P). In this work, we further define the mechanism for the CpxR/A-mediated regulation of SPI-1 genes. The negative effect exerted by the absence of CpxA on the expression of SPI-1 genes was counteracted by the absence of CpxR or by the absence of the two enzymes, AckA and Pta, which render acetyl-phosphate that phosphorylates CpxR. Furthermore, overexpression of the lipoprotein NlpE, which activates CpxA kinase activity on CpxR, or overexpression of CpxR, repressed the expression of SPI-1 genes. Thus, our results provide several lines of evidence strongly supporting that the absence of CpxA leads to the phosphorylation of CpxR via the AckA/Pta enzymes, which represses both the SPI-1 and SPI-2 genes. Additionally, we show that in the absence of the Lon protease, which degrades HilD, the CpxR-P-mediated repression of the SPI-1 genes is mostly lost; moreover, we demonstrate that CpxR-P negatively affects the stability of HilD and thus decreases the expression of HilD-target genes, such as *hilD* itself and *hilA*, located in SPI-1. Our data further expand the insight on the different regulatory pathways for gene expression involving CpxR/A and on the complex regulatory network governing virulence in *Salmonella*.

Keywords: *Salmonella*, SPI, CpxR/A, HilD, Lon, regulation, RpoH, virulence

Introduction

Salmonella enterica groups Gram-negative bacteria comprising around 2500 serotypes, which can infect a wide variety of hosts ranging from humans to birds (Haraga et al., 2008; Sánchez-Vargas et al., 2011). Acquisition of DNA fragments by horizontal transfer and the ensuing adaptation of regulatory mechanisms to control the expression of the newly acquired genes have been pivotal events in the *Salmonella* pathogenicity evolution (Schmidt and Hensel, 2004; Fàbrega and Vila, 2013). Around 30% of the *S. enterica* genome has been shaped by horizontal transfer events; most of the acquired genes are clustered in regions denominated islands (Mcclelland et al., 2001; Porwollik and Mcclelland, 2003). *Salmonella* pathogenicity islands 1 and 2 (SPI-1 and SPI-2), which are chromosomal regions composed of 39 and 44 genes, respectively, have crucial roles in the pathogenesis of *Salmonella* (Hansen-Wester and Hensel, 2001; Haraga et al., 2008; Fàbrega and Vila, 2013). SPI-1 is conserved in the two *Salmonella* species, *enterica* and *bongori*, whereas SPI-2 is only present in *S. enterica*, suggesting that SPI-1 was acquired earlier than SPI-2 during the evolution of *Salmonella* pathogenicity (Groisman and Ochman, 1997; Porwollik and Mcclelland, 2003). Both SPI-1 and SPI-2 encode a type 3 secretion system (T3SS), different effector proteins, chaperones, and transcriptional regulators that control the expression of the genes within each island (Hansen-Wester and Hensel, 2001; Haraga et al., 2008; Fàbrega and Vila, 2013). The T3SSs are highly complex needle-like nanomachines formed by more than 20 proteins, which span the inner and outer membrane of the bacteria and thus are able to inject effector proteins from the bacterial cytoplasm into the eukaryotic cytosol; once inside the host cell, effector proteins translocated by their cognate T3SS manipulate different signal transduction pathways and induce rearrangement of the host cell cytoskeleton (Moest and Méresse, 2013; Abrusci et al., 2014; Diepold and Wagner, 2014). The T3SS-1 and effector proteins encoded in SPI-1 are necessary for *Salmonella* invasion of intestinal epithelial cells, and thus for the intestinal colonization leading to enteritis; whereas the T3SS-2 and effector proteins encoded in SPI-2 are mainly required for *Salmonella* survival and replication inside macrophages, and hence for the systemic disease (Hansen-Wester and Hensel, 2001; Haraga et al., 2008; Fàbrega and Vila, 2013; Moest and Méresse, 2013). Different studies have shown that the SPI-2 genes also induce a *Salmonella* non-proliferative life style inside phagocytes and non-phagocytic cells (Grant et al., 2012; Núñez-Hernández et al., 2014) and that they contribute to the development of the intestinal inflammatory disease (Bispham et al., 2001; Coburn et al., 2005; Coombes et al., 2005). *S. enterica* serovar Typhimurium (*S.* Typhimurium) can cause self-limiting enteritis in humans, chickens and calves, while in mice it produces a systemic infection similar to the typhoid fever caused by *S.* Typhi in humans (Haraga et al., 2008; Sánchez-Vargas et al., 2011). Since *S.* Typhimurium can cause both intestinal and systemic infections in different hosts, it is widely used as a model to study the molecular virulence mechanisms of *Salmonella*.

The expression of the SPI-1 and SPI-2 genes is induced in different *in vivo* and *in vitro* growth conditions. *In vivo*, the SPI-1 genes are mainly expressed when *Salmonella* is in the intestinal lumen, associated with the epithelium or with extruding enterocytes (Laughlin et al., 2014), and also in a *Salmonella* subpopulation that replicates in the cytosol of epithelial cells (Knodler et al., 2010). In contrast, the SPI-2 genes are mainly expressed when *Salmonella* is inside epithelial cells or macrophages, within vacuoles (Cirillo et al., 1998; Deiwick et al., 1999; Eriksson et al., 2003; Knodler et al., 2010), and also when *Salmonella* is in the intestinal lumen (Brown et al., 2005), in the lamina propria or in the underlying mucosa (Laughlin et al., 2014). *In vitro*, the SPI-1 and SPI-2 genes are both expressed when *Salmonella* is grown in nutrient-rich media, such as the Luria-Bertani (LB) medium, albeit they are differentially regulated by growth phase (Lundberg et al., 1999; Miao and Miller, 2000; Bustamante et al., 2008; Kröger et al., 2013). Moreover, the expression of the SPI-2 genes is also induced when *Salmonella* is grown in acidic minimal media containing low concentrations of phosphate, calcium, and magnesium (Deiwick et al., 1999; Miao and Miller, 2000; Kröger et al., 2013).

SPI-1 encodes the transcriptional regulators HilD, HilA and InvF, which induce the expression of the genes within this island in a cascade fashion (Golubeva et al., 2012; Fàbrega and Vila, 2013). HilD, a member of the AraC family of transcriptional regulators, induces the expression of HilA (Schechter et al., 1999; Schechter and Lee, 2001; Ellermeier et al., 2005), an OmpR/ToxR-like transcriptional regulator, which in turn, activates the expression of InvF (Lostroh et al., 2000; Lostroh and Lee, 2001), another AraC-like regulator. HilA directly activates the expression of genes encoding T3SS-1 components, whereas InvF induces the expression of SPI-1 genes encoding effector proteins (Golubeva et al., 2012; Fàbrega and Vila, 2013). Furthermore, HilD regulates directly, or indirectly, through HilA and InvF, the expression of several other genes located outside SPI-1, including acquired and ancestral genes (Bustamante et al., 2008; Golubeva et al., 2012; Fàbrega and Vila, 2013; Petrone et al., 2014; Singer et al., 2014). Interestingly, when *S.* Typhimurium is grown to late stationary phase in LB medium, HilD directly induces the expression of the *ssrAB* operon that is located in SPI-2 and codes for the SsrA/B two-component system, the central positive regulator of the SPI-2 genes, thus establishing a transcriptional cross talk between SPI-1 and SPI-2 (Bustamante et al., 2008; Martínez et al., 2014).

Many *Salmonella*-specific and global regulators have been involved in the expression of the SPI-1 and SPI-2 genes, which mainly act on the expression of *hilD*, *hilA*, or *ssrAB* (Fass and Groisman, 2009; Martínez et al., 2011; Golubeva et al., 2012; Fàbrega and Vila, 2013). Notably, according to its role as a central regulator of the SPI-1 and several other virulence genes, the expression, concentration and activity of HilD is highly controlled. At the transcriptional level, the expression of *hilD* is positively autoregulated and modulated by a feed-forward regulatory loop involving HilD itself and the AraC-like regulators HilC and RtsA (Olekhnovich and Kadner, 2002; Ellermeier et al., 2005; Golubeva et al., 2012), while post-transcriptionally it is positively controlled by a regulatory cascade integrated by the SirA/BarA and Csr global regulatory systems (Martínez et al., 2011). On the other hand, HilD activity is positively regulated by FliZ and Fur, through still unknown mechanisms (Ellermeier and Slauch, 2008; Chubiz et al., 2010), as well as negatively regulated by HilE, through protein-protein interactions (Baxter et al., 2003). Moreover, the cellular concentration of HilD is controlled by the Lon protease (Takaya et al., 2005).

One of the regulators that have been involved in the expression of the SPI-1 genes is CpxA, the sensor histidine kinase of the Cpx-envelope stress two-component system (Nakayama et al., 2003). CpxA phosphorylates its cognate response regulator CpxR in response to a broad range of stimuli that cause perturbations in the cell envelope, such as pH, salt, metals, lipids and misfolded proteins; whereas in the absence of these activating signals CpxA has phosphatase activity on CpxR (Hunke et al., 2012; Vogt and Raivio, 2012; Raivio, 2014). CpxR can also be phosphorylated independently of CpxA by acetyl phosphate, which is generated *in vivo* from acetyl-CoA by the phosphotransacetylase (Pta) and acetate kinase (AckA) enzymes (Raivio and Silhavy, 1997; Wolfe et al., 2008). Phosphorylated CpxR (CpxR-P) positively or negatively regulates many genes encoding protein folding and degrading factors, peptidoglycan metabolic enzymes, inner membrane proteins, envelope-localized protein complexes, and other cellular regulators (Hunke et al., 2012; Vogt and Raivio, 2012; Raivio, 2014). Additionally, the Cpx system has been involved in the expression of virulence genes in different pathogenic bacteria, such as enteropathogenic and uropathogenic *Escherichia coli*, *Yersinia*, *Shigella*, *Legionella*, *Haemophilus* and *Salmonella* (Hunke et al., 2012; Vogt and Raivio, 2012; Raivio, 2014). In *S.* Typhimurium, deletion of *cpxA*, but not *cpxR*, decreases the expression of the SPI-1 genes and, as a consequence, reduces *Salmonella* invasion into host cells (Nakayama et al., 2003). Therefore, on the basis of these results, it was suggested that CpxA positively regulates the SPI-1 genes through regulator(s) other than CpxR (Nakayama et al., 2003).

In this work, we determined that the absence of CpxA renders activation of CpxR via the AckA-Pta pathway, which represses the expression of the SPI-1 genes. Consistently, it was found that CpxR-P generated by the activation of CpxA, or the overexpression of CpxR, also represses the expression of these genes. Our results indicate that CpxR negatively controls the expression of the SPI-1 genes, as well as genes located in SPI-2, by repressing the autoregulation of HilD, a central positive regulator for the expression of the genes within SPI-1 and SPI-2 and other virulence genes. Furthermore, we found that activation of CpxR decreases the stability of HilD and that, in the absence of the Lon protease, which degrades HilD, the CpxR-mediated repression of the SPI-1 genes is mostly lost. Thus, our data clarify and expand the regulatory role of the two-component system CpxR/A for the expression of *S.* Typhimurium virulence genes.

Materials and Methods

Bacterial Strains, Media, and Culture Conditions

Bacterial strains used in this study are listed in **Table 1**. Bacterial cultures were grown at 37°C in LB medium containing 1% tryptone, 0.5% yeast agar and 1% NaCl, pH 7.5. When necessary, media were supplemented with ampicillin (200 μg ml^{-1}), kanamycin (20 μg ml^{-1}) or chloramphenicol (30 μg ml^{-1}). Cultures for chloramphenicol acetyltransferase (CAT), Western blot and protein secretion assays were performed as we described previously (Bustamante et al., 2008; Martínez et al., 2011).

Construction of Plasmids

Plasmids and primers used in this study are listed in **Tables 1, 2**, respectively. To construct the plasmids containing the transcriptional fusions *lon-cat*, *clpX-cat*, *clpP-cat* and *cpxRA-cat*, the regulatory regions of *lon*, *clpX*, *clpP*, and *cpxRA* were amplified by PCR with the primer pairs promlon-Fw1/promlon-Rv1, pClpX-Bam/pClpX-Hind, pClpXP-Bam/pClpXP-Hind and CpxR-Bam-5′/CpxR-Hind-3′, respectively. The PCR products were digested with BamHI and HindIII restriction enzymes and then cloned into the BamHI and HindIII sites of the vector pKK232-8, which carries a promoterless *cat* gene (Amersham Pharmacia LKB Biotechnology), generating plasmids plon-cat, pclpX-cat, pclpP-cat, and pcpxRA-cat. To construct the plasmids pK3-CpxR and pK3-RpoH, the *cpxR* and *rpoH* genes were amplified by PCR using primer pairs CpxR-Fw1/CpxR-Rv1 and RpoH-FwKpn/RpoH-RvBam, respectively. The PCR products were digested with HindIII and BamHI (*cpxR* gene) or KpnI and BamHI (*rpoH* gene) restriction enzymes and then cloned into the vector pMPM-K3 (Mayer, 1995) digested with the respective restriction enzymes. pK3-CpxR and pK3-RpoH constitutively express CpxR and RpoH, respectively, under a *lac* promoter, since *Salmonella* and the vector pMPM-K3 lack the gene encoding LacI, the repressor of *lac*.

Construction of Deletion Mutants and Strains Expressing FLAG-tagged Proteins

Non-polar gene-deletion mutant strains were generated by the λRed recombinase system, as reported previously (Datsenko and Wanner, 2000), using the respective primers described in **Table 2**. The genes *cpxR*, *cpxA*, *cpxRA*, *ackA-pta*, *hilE*, or *lon* were replaced with a selectable kanamycin resistance cassette in the *S.* Typhimurium strain 14028s, generating the Δ*cpxR::kan* (DTM48), Δ*cpxA::kan* (DTM50), Δ*cpxRA::kan* (DTM52), Δ*ackA-pta::kan* (DTM54), Δ*hilE::kan* (DTM56) and Δ*lon::kan* (DTM60) mutants, respectively. The kanamycin resistance cassette was excised from the Δ*cpxR::kan* (DTM48), Δ*cpxA::kan* (DTM50), Δ*cpxRA::kan* (DTM52), Δ*hilE::kan* (DTM56), Δ*cpxA* Δ*hilE::kan* (DTM58), Δ*lon::kan* (DTM60), Δ*cpxA* Δ*lon::kan* (DTM62) and Δ*hilD::kan* (DTM64) mutants, by using helper plasmid pCP20, expressing the FLP recombinase, as described previously (Datsenko and Wanner, 2000), generating the Δ*cpxR* (DTM49), Δ*cpxA* (DTM51), Δ*cpxRA* (DTM53), Δ*hilE* (DTM57), Δ*cpxA* Δ*hilE* (DTM59), Δ*lon* (DTM61), Δ*cpxA* Δ*lon* (DTM63) and Δ*hilD* (DTM65) mutants, respectively. P22 transduction was used to transfer the Δ*hilD::kan* allele from strain JPTM5 into strain 14028s, generating strain DTM64, to transfer the Δ*ackA-pta::kan*, Δ*hilE::kan* or Δ*lon::kan* alleles from strains DTM54, DTM56 and DTM60 into strain DTM51, generating the Δ*cpxA* Δ*ackA-pta::kan* (DTM55), Δ*cpxA* Δ*hilE::kan* (DTM58) and Δ*cpxA* Δ*lon::kan* (DTM62) mutants, respectively, to transfer the Δ*cpxA::kan* or Δ*lon::kan* alleles from

TABLE 1 | Bacterial strains and plasmids used in this study.

Strain or plasmid	Genotype or description	References or sources
S. TYPHIMURIUM STRAINS		
14028s	Wild-type	ATCC
DTM48	ΔcpxR::kan	This study
DTM49	ΔcpxR	This study
DTM50	ΔcpxA::kan	This study
DTM51	ΔcpxA	This study
DTM52	ΔcpxRA::kan	This study
DTM53	ΔcpxRA	This study
DTM54	ΔackA-pta::kan	This study
DTM55	ΔcpxA ΔackA-pta::kan	This study
DTM56	ΔhilE::kan	This study
DTM57	ΔhilE	This study
DTM58	ΔcpxA ΔhilE::kan	This study
DTM59	ΔcpxA ΔhilE	This study
DTM60	Δlon::kan	This study
DTM61	Δlon	This study
DTM62	ΔcpxA Δlon::kan	This study
DTM63	ΔcpxA Δlon	This study
JPTM5	SL1344 ΔhilD::kan	Bustamante et al., 2008
DTM64	14028s ΔhilD::kan	This study
DTM65	ΔhilD	This study
DTM66	ΔhilD ΔcpxA::kan	This study
DTM67	ΔhilD Δlon::kan	This study
JPTM7	SL344 hilA::3XFLAG-kan	Bustamante et al., 2008
DTM68	14028s hilA::3XFLAG-kan	This study
DTM69	ΔcpxR hilA::3XFLAG-kan	This study
DTM70	ΔcpxA hilA::3XFLAG-kan	This study
DTM71	ΔcpxRA hilA::3XFLAG-kan	This study
JPTM30	SL1344 ssrB::3XFLAG-kan	Martínez et al., 2011
DTM72	14028s ssrB::3XFLAG-kan	This study
DTM73	ΔcpxR ssrB::3XFLAG-kan	This study
DTM74	ΔcpxA ssrB::3XFLAG-kan	This study
DTM75	ΔcpxRA ssrB::3XFLAG-kan	This study
DTM76	14028s invF::3XFLAG-kan	This study
DTM77	ΔcpxR invF::3XFLAG-kan	This study
DTM78	ΔcpxA invF::3XFLAG-kan	This study
DTM79	ΔcpxRA invF::3XFLAG-kan	This study
DTM80	ΔhilE invF::3XFLAG-kan	This study
DTM81	ΔcpxA ΔhilE invF::3XFLAG-kan	This study
DTM82	Δlon invF::3XFLAG-kan	This study
DTM83	ΔcpxA Δlon invF::3XFLAG-kan	This study
MF100	14028s ΔCthns::kan (lacking codons 99–136 of hns)	Fernández-Mora, personal communication
DTM84	ΔhilD ΔCthns::kan	This study
E.COLI K12 STRAIN		
DH10β	Laboratory strain	Invitrogen
PLASMIDS		
pKK232-8	pBR322 derivative containing a promoterless chloramphenicol acetyltransferase (cat) gene, ApR	Brosius, 1984

(Continued)

TABLE 1 | Continued

Strain or plasmid	Genotype or description	References or sources
philD-cat	pKK232-8 derivative containing a hilD-cat transcriptional fusion from nucleotides −364 to +88	Bustamante et al., 2008
philA-cat	pKK232-8 derivative containing a hilA-cat transcriptional fusion from nucleotides −410 to +446	Bustamante et al., 2008
pinvF-cat	pKK232-8 derivative containing a invF-cat transcriptional fusion from nucleotides −306 to +213	Bustamante et al., 2008
psirA-cat	pKK232-8 derivative containing a sirA-cat transcriptional fusion from nucleotides −563 to +98	Martínez et al., 2011
plon-cat	pKK232-8 derivative containing a lon-cat transcriptional fusion from nucleotides −296 to +61	This study
pclpX-cat	pKK232-8 derivative containing a clpX-cat transcriptional fusion from nucleotides −330 to +76	This study
pclpP-cat	pKK232-8 derivative containing a clpP-cat transcriptional fusion from nucleotides −335 to +57	This study
pcpxRA-cat	pKK232-8 derivative containing a cpxRA-cat transcriptional fusion from nucleotides −544 to +57	This study
pCA24N	High-copy-number cloning vector, lac promoter, lacIq, CmR	Kitagawa et al., 2005
pCA-NlpE	pCA24N derivative expressing E. coli K12 NlpE from the lac promoter	Kitagawa et al., 2005
pCA-CpxR	pCA24N derivative expressing E. coli K12 CpxR from the lac promoter	Kitagawa et al., 2005
pMPM-K3	p15A derivative low-copy-number cloning vector, lac promoter, KanR	Mayer, 1995
pK3-CpxR	pMPM-K3 derivative expressing S. Typhimurium 14028s CpxR from the lac promoter	This study
pK3-RpoH	pMPM-K3 derivative expressing S. Typhimurium 14028s RpoH from the lac promoter	This study
pBAD-HilD	pBADMycHis derivative expressing HilD-MycHis from the ara promoter	Martínez et al., 2011
pKD46	pINT-ts derivative containing red recombinase system under an arabinose-inducible promoter, ApR	Datsenko and Wanner, 2000
pKD4	pANTsγ derivative template plasmid containing the kanamycin cassette for λ Red recombination, ApR	Datsenko and Wanner, 2000
pCP20	Plasmid expressing FLP recombinase from a temperature-inducible promoter, ApR	Datsenko and Wanner, 2000
pSUB11	pGP704 derivative template plasmid for FLAG epitope tagging	Uzzau et al., 2001

The coordinates for the cat fusions are indicated with respect to the transcriptional start site for each gene. ApR, ampicillin resistance; CmR, chloramphenicol resistance; KanR, kanamycin resistance.

strains DTM50 or DTM60 into strain DTM65, generating the ΔhilD ΔcpxA::kan (DTM66) and ΔhilD Δlon::kan (DTM67) mutants, respectively, and to transfer the ΔCthns::kan allele

TABLE 2 | Primers used in this study.

Primer	Sequence (5'-3')	Target gene	RE
FOR *CAT* TRANSCRIPTIONAL FUSIONS			
promIon-Fw1	GTC*GGATCC*TGCCGGTCAGAGTAAGCCG	*lon*	BamHI
promIon-Rv1	TAC*AAGCTT*GGGTATGACCATGTGCGG		HindIII
pClpX-Bam	TCA*GGATCC*TGAGCAGATTGAACGTGATAC	*clpX*	BamHI
pClpX-Hind	GA*AAGCTT*CGGATGGACCGGCAATCAG		HindIII
pClpXP-Bam	GAT*GGATCC*TATGCGTAACGTCGCTCTGG	*clpP*	BamHI
pClpXP-Hind	GAG*AAGCTT*TATCAAAAGAGCGCTCACCG		HindIII
CpxR-Bam-5'	TTG*GGATCCC*GCCACGTCGCGC	*cpxRA*	BamHI
CpxR-Hind-3'	ATC*AAGCTT*CTTTAACAGGATTTTATTC		HindIII
FOR GENE CLONING			
CpxR-Fw1	TGA*AAGCTT*TTTCTGCCTCGGAGGTACG	*cpxR*	HindIII
CpxR-Rv1	CAG*GGATCCC*GTCAACCAGAAGATGGCG		BamHI
RpoH-FwKpn	ACG*GTACC*AGGCAATACTGATTGA	*rpoH*	KpnI
RpoH-RvBam	CAT*GGATCC*AACAGATTTGTGTCGGTGGG		BamHI
FOR GENE DELETIONS			
ScpxR-H1PI	GATGACCGAGAGCTGACTTCCCTGTT AAAAGAGCTCCTCGAA<u>TGTAGGCTGG AGCTGCTTCG</u>	*cpxR*	
ScpxR-H2P2	TGTTTTAAACCACGGGTGACCGTCTTT GCGTTCCGGCAGTTT<u>CATATGAATATC CTCCTTAG</u>		
ScpxA-H1P1	CTATCTGATGGTTTCCGCTTCATGATAG GAAGTTTAACCGCG<u>TGTAGGCTGGAGC TGCTTCG</u>	*cpxA*	
ScpxA-H2P2	GCATTCGCAGGCCGATGGTTTTTAGGTT CGCTTGTACAGCGG<u>CATATGAATATCCT CCTTAG</u>		
SackA-pta-H1P1	GTATCATAAATAGGTACTTCCATGTCGA GTAAGTTAGTACTGT<u>GTAGGCTGGAGCT GCTTCG</u>	*ackA-pta*	
SackA-pta-H2P2	ATCCGGCATTAGCTTTTACTGTTACTGC TGCTGCTGAGAAGC<u>CATATGAATATCCT CCTTAG</u>		
ShilE-H1P1	TACAGAGACACCAACGAAATGGCTGG AAAATGGAACGTTCTT<u>TGTAGGCTGGA GCTGCTTCG</u>	*hilE*	
ShilE-H2P2	CGCAAGCTTGTTTTGTCCTCATCGCCA CAGCGCCTGTCGGTG<u>CATATGAATATC CTCCTTAG</u>		
SIonH1P1	AAACTAAGAGAGAGCTCTATGAATCCT GAGCGTTCTGAACGC<u>TGTAGGCTGGA GCTGCTTCG</u>	*lon*	
SIonH2P2	GTCATTTGCGCGAGGTCACTATTTTGC GGTTACAACCTGCAT<u>CATATGAATATC CTCCTTAG</u>		

(Continued)

TABLE 2 | Continued

Primer	Sequence (5'-3')	Target gene	RE
FOR GENE FLAG TAGGING			
invFflag-F	CCGCGGAAATTATCAAATATTATTCAAT TGGCAGACAAA<u>GACTACAAAGACCATG ACGGT</u>	*invF*	
invFflag-R	CGGCACATGCCAGCACTCTGGCCAAAA GAATATGTGTCT<u>CATATGAATATCCTCCT TAGTTC</u>		

Italic letters indicate the respective restriction enzyme site in the primer. The sequence corresponding to the template plasmids pKD4 or pSUB11 is underlined. RE, restriction enzyme for which a site was generated in the primer.

from strain MF100 into strain DTM65, generating the Δ*hilD* Δ*Cthns::kan* (DTM84) mutant.

The chromosomal *invF* gene was FLAG-tagged in *S.* Typhimurium strain 14028s, using a modification of the λ Red recombinase system for gene replacement, as described previously (Uzzau et al., 2001), and the respective primers described in **Table 2**, generating strain DTM76. P22 transduction was used to transfer the *invF::3XFLAG-kan* allele from strain DTM76 into strains DTM49, DTM51, DTM53, DTM57, DTM59, DTM61 and DTM63, generating the Δ*cpxR invF::3XFLAG-kan* (DTM77), Δ*cpxA invF::3XFLAG-kan* (DTM78), Δ*cpxRA invF::3XFLAG-kan* (DTM79), Δ*hilE invF::3XFLAG-kan* (DTM80), Δ*cpxA* Δ*hilE invF::3XFLAG-kan* (DTM81), Δ*lon invF::3XFLAG-kan* (DTM82) and Δ*cpxA* Δ*lon invF::3XFLAG-kan* (DTM83) mutants, respectively, to transfer the *hilA::3XFLAG-kan* allele from strain JPTM7 into strains 14028s, DTM49, DTM51 and DTM53, generating the *hilA::3XFLAG-kan* (DTM68), Δ*cpxR hilA::3XFLAG-kan* (DTM69), Δ*cpxA hilA::3XFLAG-kan* (DTM70) and Δ*cpxRA hilA::3XFLAG-kan* (DTM71) mutants, respectively, and to transfer the *ssrB::3XFLAG-kan* allele from strain JPTM30 into strains 14028s, DTM49, DTM51 and DTM53, generating the *ssrB::3XFLAG-kan* (DTM72), Δ*cpxR ssrB::3XFLAG-kan* (DTM73), Δ*cpxA ssrB::3XFLAG-kan* (DTM74) and Δ*cpxRA ssrB::3XFLAG-kan* (DTM75) mutants, respectively. All mutant strains were verified by PCR amplification and sequencing.

CAT Assays

The CAT assays and protein quantification to calculate CAT specific activities were performed as described previously (Puente et al., 1996).

Statistical Analysis

Results from chloramphenicol acetyltransferase (CAT) assays were analyzed using One-Way analysis of variance (ANOVA) with the Dunnett multiple comparison test for **Figures 1A,D**, or *t*-Test with the Mann–Whitney test for **Figures 4A–D, 7B,C**. A *P*-value of <0.05 was considered significant. This statistical analysis was performed using Prism 5 program version 5.04 (GraphPad Software, San Diego, CA).

Protein Secretion Analysis

Protein secretion assays were performed as we described previously (Martínez et al., 2011). Samples were subjected to SDS-PAGE analysis using 12% polyacrylamide gels and stained with Coomassie Brilliant Blue R-250.

Western Blotting

Whole-cell extracts were prepared from samples collected at the indicated time points of bacterial cultures. Ten micrograms of each extract were subjected to electrophoresis in SDS-12% polyacrylamide gels, and then transferred to 0.45 μm pore size nitrocellulose membranes (Bio-Rad), using a semidry transfer apparatus (Bio-Rad). Membranes were blocked with 5% nonfat milk and then incubated with anti-c-Myc (Sigma), anti-FLAG M2 (Sigma) or anti-DnaK (StressGen) monoclonal antibodies, or anti-SseB polyclonal antibody (Coombes et al., 2004), at 1:3000, 1:4000, 1:20,000 and 1:2000 dilutions, respectively. Horseradish peroxidase-conjugated anti-mouse or anti-rabbit (Pierce), at a dilution of 1:10,000, were used as the secondary antibodies. Bands on the blotted membranes were developed by incubation with the Western Lightning Chemiluminescence Reagent Plus (Perkin-Elmer) and exposed to Kodak X-Omat films.

HilD Protein Stability Assays

Bacterial strains were grown in LB medium at $37^{\circ}C$ to an OD_{600} equal to 0.8. Then, the expression of HilD-Myc from plasmid pBAD-HilD was induced by adding 0.05% L-arabinose for 45 min. After this time, antibiotics streptomycin, rifampicin and chloramphenicol, at final concentrations of 200, 100 and 200 μg ml^{-1}, respectively, were added to the cultures to prevent transcription and translation. To ensure repression of the ara promoter expressing HilD-Myc, 2% glucose was also added. The bacterial cultures were further incubated at $37^{\circ}C$ and samples were taken at 0, 15, 30, 60, 90 and 120 min, and analyzed by Western blotting as described above. Intensity of protein bands from the blots was quantified by using ImageJ software (Image Processing and Analysis in Java), version 1.48 (National Institutes of Health, USA). Values for HilD-Myc bands were normalized with those respective of DnaK bands and then the relative percentage of HilD-Myc at each time with respect to time 0 was calculated. The HilD half-life ($t_{1/2}$) was calculated by linear regression.

Results

AckA-Pta-dependent Activation of CpxR Represses the SPI-1 Genes

Intriguingly, the absence of the histidine kinase CpxA, but not its cognate response regulator CpxR, negatively affects the expression of SPI-1 genes, which could suggest that CpxA positively regulates the SPI-1 genes by interacting with regulator(s) other than CpxR (Nakayama et al., 2003). However, there was the possibility that the absence of CpxA, and thus of its phosphatase activity, could lead to the phosphorylation of CpxR, mainly by acetyl phosphate produced by the AckA and Pta enzymes, as described previously (Batchelor et al., 2005; Spinola et al., 2010; Liu et al., 2012). Therefore, an alternative to the

proposed positive regulatory role of CpxA on SPI-1 genes, was that CpxR-P generated in the absence of CpxA could actually repress these genes. To investigate this possibility, we tested the expression of a transcriptional fusion of the SPI-1 gene $invF$ with the cat (chloramphenicol acetyl transferase) reporter gene, in wild-type (WT) $S.$ Typhimurium strain 14028s, as well as in different derivative mutants containing single, double or triple deletions of $cpxA$, $cpxR$, $ackA$, or pta genes. As a control, the expression of a cat transcriptional fusion of $sirA$, which encodes a positive regulator of the SPI-1 genes that is located outside SPI-1, was also assessed. In agreement with the results reported previously (Nakayama et al., 2003), the expression of the $invF$-cat fusion was reduced in the $\Delta cpxA$ mutant, but was not affected in the $\Delta cpxR$ mutant (**Figure 1A**). Additionally, the expression of this fusion was not affected in the $\Delta cpxRA$ or $\Delta ackA$-pta double mutants neither in the $\Delta cpxA$ $\Delta ackA$-pta triple mutant, whereas the $sirA$-cat fusion was expressed at a similar level in all strains tested (**Figure 1A**). Consistently, protein secretion analyses showed that the secretion/expression of the SPI-1-encoded proteins SipA, SipB, SipC, and SipD was drastically diminished in the $\Delta cpxA$ mutant, but not in the $\Delta cpxR$, $\Delta cpxRA$, $\Delta ackA$-pta or $\Delta cpxA$ $\Delta ackA$-pta mutants (**Figure 1B**). Furthermore, the expression of the 3xFLAG-tagged regulators HilA (HilA-FLAG) and InvF (InvF-FLAG), which are encoded in SPI-1, was reduced in the $\Delta cpxA$ mutant, but not in the $\Delta cpxR$ or $\Delta cpxRA$ mutants (**Figure 1C**). These results show that deletion of $cpxR$ or the $ackA$-pta genes restores the expression of the SPI-1 genes in the $\Delta cpxA$ mutant, indicating that the absence of CpxA actually represses these genes through CpxR and the AckA/Pta enzymes.

Previous studies have shown that CpxR-P activates the expression of the $cpxRA$ operon (De Wulf et al., 1999; Raivio et al., 1999, 2013; Price and Raivio, 2009). Therefore, to further investigate whether the absence of CpxA turns on CpxR-mediated gene regulation in $S.$ Typhimurium, in the growth conditions tested, we determined the expression of a $cpxRA$-cat transcriptional fusion in the WT $S.$ Typhimurium strain and its derivative $\Delta cpxR$, $\Delta cpxA$, and $\Delta cpxRA$ mutants. As shown in **Figure 1D**, the expression of the $cpxRA$-cat fusion was increased in the $\Delta cpxA$ mutant, but not in the $\Delta cpxR$ and $\Delta cpxRA$ mutants, indicating that the absence of CpxA induces the expression of the $cpxRA$ operon through CpxR.

Together, these results strongly support that the absence of CpxA leads to the phoshorylation of CpxR via the AckA-Pta pathway, which in turn represses the expression of the SPI-1 genes and probably induces the positive or negative regulation of the whole CpxR regulon.

CpxA-dependent Activation or Overexpression of CpxR Represses the SPI-1 Genes

Overproduction of the lipoprotein NlpE activates the kinase activity of CpxA and thus the CpxA-dependent phosphorylation of CpxR (Snyder et al., 1995; Hunke et al., 2012; Vogt and Raivio, 2012). Hence, to determine whether the CpxA-mediated activation of CpxR also represses the expression of the SPI-1 genes, we examined the effect of the overexpression of NlpE, from an IPTG-inducible promoter, on the protein secretion profiles

FIGURE 1 | The absence of CpxA represses SPI-1 genes by the activation of CpxR via the AckA and Pta enzymes. (A) Expression of the *invF-cat* and *sirA-cat* transcriptional fusions, carried by plasmids pinvF-cat and psirA-cat, respectively, was tested in the WT *S.* Typhimurium 14028s strain and its isogenic ΔcpxA, ΔcpxR, ΔcpxRA, ΔackA-pta, and ΔcpxA ΔackA-pta mutants. **(B)** Secretion analysis of the SPI-1-encoded proteins SipA, SipB, SipC, and SipD was tested in the WT *S.* Typhimurium strain 14028s and its isogenic ΔcpxA, ΔcpxR, ΔcpxRA, ΔackA-pta, and ΔcpxA ΔackA-pta mutants grown for 9 h in LB medium at 37°C. FliC is a flagellar protein whose secretion is SPI-1-independent. **(C)** Expression of the SPI-1-encoded HilA-FLAG and InvF-FLAG and the SPI-2-encoded SsrB-FLAG and SseB, in the WT *S.* Typhimurium strain and its isogenic ΔcpxR, ΔcpxA and ΔcpxRA mutants, carrying the respective chromosomal

FLAG-tagged gene, was analyzed by Western blotting using monoclonal anti-FLAG or polyclonal anti-SseB antibodies. Whole cell lysates were prepared from samples of bacterial cultures grown in LB medium at 37°C for 5 or 9 h, for detection of SPI-1- and SPI-2-encoded proteins, respectively. As a loading control, the expression of DnaK was also determined using monoclonal anti-DnaK antibodies. **(D)** Expression of the *cpxRA-cat* transcriptional fusion, carried by plasmid pcpxRA-cat, was tested in the WT *S.* Typhimurium 14028s strain and its isogenic ΔcpxR, ΔcpxA, ΔcpxRA, and ΔhilD ΔcpxA mutants. CAT-specific activity of *cat* fusions was determined from samples collected of bacterial cultures grown for 5 h in LB medium at 37°C. The data are the averages of three different experiments performed in duplicate. Bars represent the standard deviations. *Expression statistically different with respect to that shown by the same fusion in the WT strain.

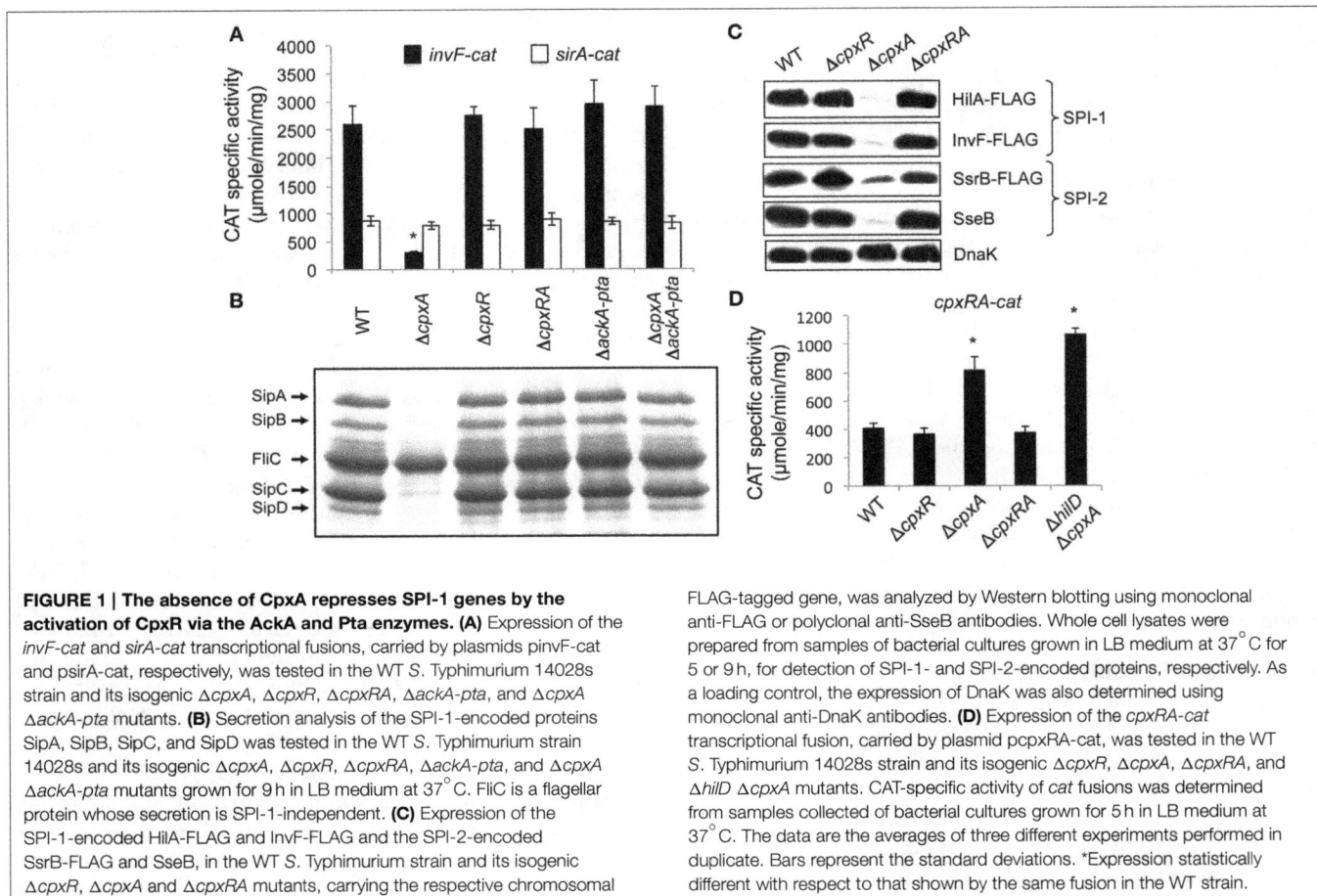

of the WT *S.* Typhimurium strain and its derivative ΔcpxR mutant. Since *Salmonella* lacks NlpE, the *E. coli* K12 NlpE was used in these assays. As shown in **Figure 2**, the induction of the NlpE expression by the presence of IPTG decreased the secretion/expression of the SipA-D proteins in the WT strain but not in its derivative ΔcpxR mutant, indicating that the activation of CpxA represses the secretion/expression of SPI-1-encoded proteins through CpxR. To further confirm the regulatory role of CpxR on the SPI-1 genes, we determined the effect of its overexpression on the protein secretion profile of the WT *S.* Typhimurium strain, since the overexpression can bypass the need for phosphorylation of CpxR to regulate target genes (Macritchie et al., 2008; Acosta et al., 2015; Yun et al., 2015). The *E. coli* K12 CpxR is 97% identical to that of *S.* Typhimurium 14028s; thus, the plasmid pCA-CpxR from the ASKA library (Kitagawa et al., 2005), which expresses the *E. coli* K12 CpxR from an IPTG-inducible promoter, was used in these assays. As shown in **Figure 3A**, the overexpression of CpxR reduced the secretion/expression of the SipA-D proteins. Furthermore, the overexpression of CpxR repressed the expression of HilA-FLAG and InvF-FLAG in the WT *S.* Typhimurium strain (**Figure 3B**). In all, these results indicate that CpxA-mediated phosphorylation of CpxR or the overexpression of CpxR represses the expression of the SPI-1 genes.

FIGURE 2 | NlpE-mediated activation of CpxA represses SPI-1 through CpxR. Secretion analysis of the SPI-1-encoded proteins SipA, SipB, SipC, and SipD was tested in the WT *S.* Typhimurium strain and its isogenic ΔcpxR mutant carrying plasmid pCA-NlpE, grown for 9 h in LB medium at 37°C. FliC is a flagellar protein whose secretion is SPI-1-independent. Expression (+) of NlpE from the *T5-lac* promoter of plasmid pCA-NlpE was induced by adding 50 μM IPTG at the beginning of the bacterial cultures.

CpxR Represses *hilD* and thus Indirectly Affects HilD-regulated Genes

Several global regulators control the expression of the SPI-1 genes by directly affecting the expression, activity or concentration of

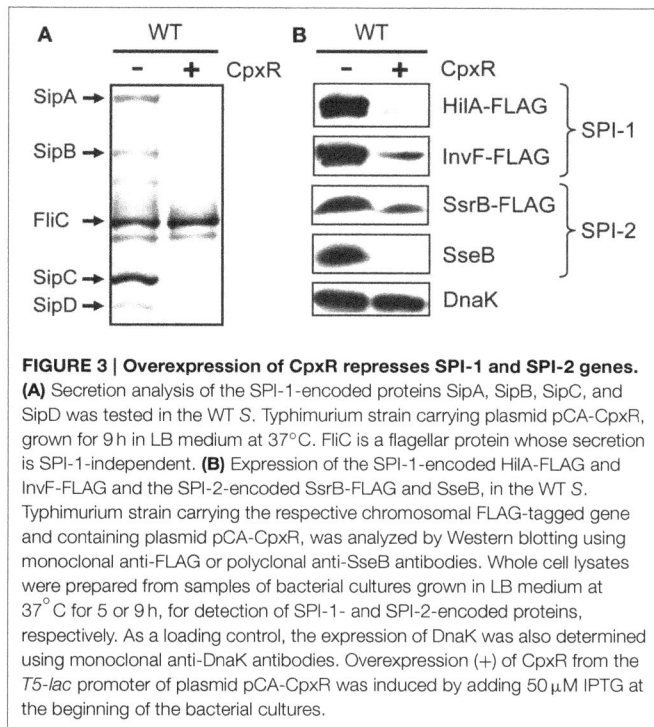

FIGURE 3 | Overexpression of CpxR represses SPI-1 and SPI-2 genes.
(A) Secretion analysis of the SPI-1-encoded proteins SipA, SipB, SipC, and SipD was tested in the WT *S.* Typhimurium strain carrying plasmid pCA-CpxR, grown for 9 h in LB medium at 37°C. FliC is a flagellar protein whose secretion is SPI-1-independent. **(B)** Expression of the SPI-1-encoded HilA-FLAG and InvF-FLAG and the SPI-2-encoded SsrB-FLAG and SseB, in the WT *S.* Typhimurium strain carrying the respective chromosomal FLAG-tagged gene and containing plasmid pCA-CpxR, was analyzed by Western blotting using monoclonal anti-FLAG or polyclonal anti-SseB antibodies. Whole cell lysates were prepared from samples of bacterial cultures grown in LB medium at 37°C for 5 or 9 h, for detection of SPI-1- and SPI-2-encoded proteins, respectively. As a loading control, the expression of DnaK was also determined using monoclonal anti-DnaK antibodies. Overexpression (+) of CpxR from the *T5-lac* promoter of plasmid pCA-CpxR was induced by adding 50 μM IPTG at the beginning of the bacterial cultures.

HilD or HilA, the central regulators of these genes (Golubeva et al., 2012; Fàbrega and Vila, 2013). Our results indicate that CpxR represses the expression of HilA (**Figure 1C**). To start to define whether CpxR affects *hilA* directly or through HilD, which positively regulates *hilA*, we determined the effect of the overexpression of CpxR on the activity of a *hilD-cat* transcriptional fusion in the WT *S.* Typhimurium strain. Since plasmid pCA-CpxR, expressing the *E. coli* K12 CpxR, is incompatible with the vector carrying the *cat* fusions tested, for the next assays we constructed and used the plasmid pK3-CpxR, which constitutively expresses CpxR of *S.* Typhimurium 14028s. The overexpression of CpxR reduced 50% the expression of the *hilD-cat* fusion (**Figure 4A**), revealing that CpxR represses *hilD*. CpxR could directly repress the transcription of *hilD* or reduce post-transcriptionally the concentration of HilD and thus affect its positive autoregulation. To determine if CpxR affects the autoregulation of *hilD*, the expression of the *hilD-cat* fusion was determined in the WT *S.* Typhimurium strain and its derivatives Δ*cpxA*, Δ*hilD*, and Δ*hilD* Δ*cpxA* mutants. As shown in **Figure 4B**, the expression of the *hilD-cat* fusion was similarly reduced in these three mutants, indicating that the absence of CpxA or HilD has the same effect on the expression of *hilD*, and that, when HilD is not present, the absence of CpxA does not longer repress *hilD*. In contrast, the expression of the *cpxRA-cat* fusion was similarly increased in the Δ*cpxA* and Δ*hilD* Δ*cpxA* mutants (**Figure 1D**), indicating that the absence of CpxA activates the expression of *cpxRA* independently of HilD. These results suggest that CpxR represses *hilA* and thus the other SPI-1 genes by affecting the autoregulation of HilD. To confirm that CpxR regulates *hilA* through HilD and not directly, we analyzed

the effect of CpxR on the expression of *hilA* in the presence or not of HilD. Previous studies indicate that HilD induces the expression of *hilA* by counteracting the repression exerted by the nucleoid protein H-NS on the promoter of this gene (Schechter et al., 1999; Schechter and Lee, 2001; Olekhnovich and Kadner, 2006); thus, in the absence of H-NS activity *hilA* can be expressed independently of HilD. Full-length deletion of *hns* produces severe growth defects in *S.* Typhimurium (Lucchini et al., 2006; Navarre et al., 2006). However, deletion of the sequence encoding the C-terminal region of H-NS (Δ*Cthns*), which contains its DNA-binding domain, has only a minor effect on *S.* Typhimurium fitness (Fernández-Mora, personal communication), probably because the N-terminal of H-NS can still repress some of its target genes by interacting with StpA, another nucleoid protein (Free et al., 2001). Therefore, we constructed and tested a *S.* Typhimurium 14028s Δ*hilD* Δ*Cthns* mutant. The overexpression of CpxR, from plasmid pK3-CpxR, reduced five-fold the expression of a *hilA-cat* transcriptional fusion in the WT strain (**Figure 4C**), but did not affect the high levels of expression showed by this fusion in the Δ*hilD* Δ*Cthns* mutant (**Figure 4D**), indicating that CpxR regulates *hilA* through HilD and not directly.

In agreement with our results indicating that CpxR represses the HilD-dependent expression of *hilD* and *hilA*, both the overexpression of CpxR and the absence of CpxA drastically reduced the production of SsrB-FLAG and SseB proteins (**Figures 1C**, **3B**), which are encoded in SPI-2 and whose expression is also dependent of HilD in the condition tested.

Taken together, these results show that CpxR represses the autoregulation of HilD, which in turn affects the expression of *hilA* and thus the SPI-1 genes, as well as of other virulence genes regulated by HilD, such as *ssrB* and *sseB* located in SPI-2.

CpxR-mediated Repression of the SPI-1 Genes Is lost in the Absence of the Lon Protease

A previous study shown that the overexpression of the sigma factor RpoH represses the SPI-1 genes through the Lon protease that degrades HilD (Matsui et al., 2008). On the other hand, it was reported that CpxR positively regulates *rpoH* in *E. coli* (Zahrl et al., 2006). Interestingly, we observed a very similar effect with the overexpression of CpxR or RpoH on the activity of *hilD-cat* and *hilA-cat* fusions (**Figures 4A,C,D**), which was initially tested as an expression control of the promoter expressing CpxR from plasmid pMPM-K3. Therefore, we thought that CpxR could repress the SPI-1 genes through RpoH and Lon. To investigate this, we sought to determine the effect of CpxR on the SPI-1 genes in the absence of RpoH or Lon. After several attempts, we were unable to delete *rpoH* in the *S.* Typhimurium 14028s strain by the λRed recombination method (Datsenko and Wanner, 2000), which could suggest that the absence of RpoH affects *Salmonella* fitness; although, a *S.* Typhimurium 14028s Δ*rpoH* mutant was reported previously (Bang et al., 2005). In contrast, a *lon* deletion strain was successful; thus, we constructed and analyzed Δ*lon* and Δ*cpxA* Δ*lon* mutants. Furthermore, since HilE regulates the activity of HilD by protein-protein interaction (Baxter et al., 2003), Δ*hilE* and Δ*cpxA* Δ*hilE* mutants were also constructed and used as controls. Interestingly, the expression

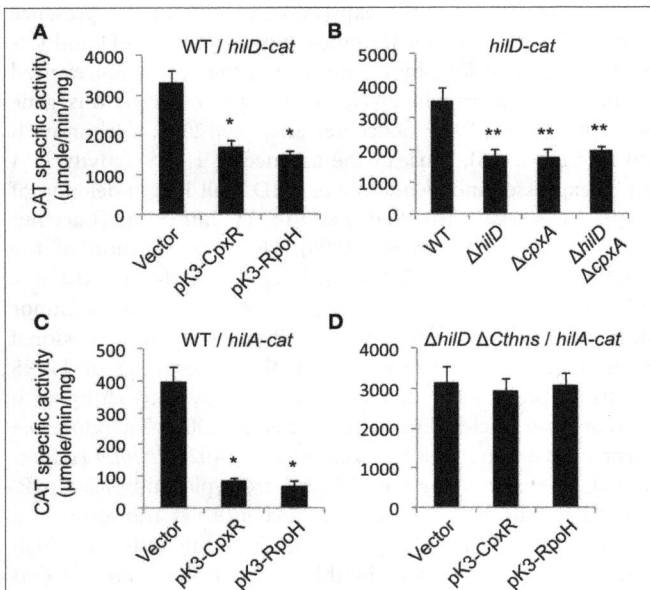

FIGURE 4 | CpxR represses the autoregulation of *hilD* and thus negatively affects the expression of *hilA*. Expression of the *hilD-cat* transcriptional fusion carried by plasmid philD-cat was tested in the WT *S.* Typhimurium strain carrying plasmid pK3-CpxR or pK3-RpoH, or the vector pMPM-K3 **(A)**, as well as in the WT *S.* Typhimurium strain and its isogenic Δ*hilD*, Δ*cpxA*, and Δ*hilD* Δ*cpxA* mutants **(B)**. Expression of the *hilA-cat* transcriptional fusion carried by plasmid philA-cat was tested in the WT *S.* Typhimurium strain **(C)**, or in its isogenic Δ*hilD* Δ*Cthns* mutant **(D)**, containing plasmid pK3-CpxR or pK3-RpoH, or the vector pMPM-K3. Plasmids pK3-CpxR and pK3-RpoH, as well as *Salmonella*, lack the gene encoding the repressor LacI and thus they constitutively express CpxR and RpoH, respectively, from a *lac* promoter. CAT-specific activity was determined from samples collected of bacterial cultures grown for 5 h in LB medium at 37°C. The data are the averages of three different experiments performed in duplicate. Bars represent the standard deviations. *Expression statistically different with respect to that shown by the same fusion in the WT strain containing the vector. **Expression statistically different with respect to that shown by the same fusion in the WT strain.

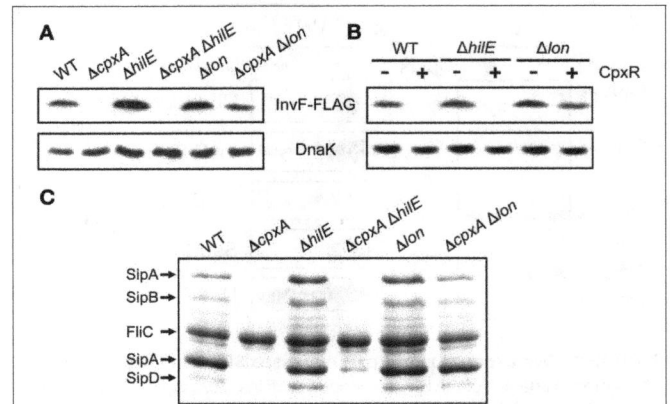

FIGURE 5 | Repression of the SPI-1 genes by CpxR is lost in the absence of the Lon protease. Expression of the SPI-1-encoded InvF-FLAG in the WT *S.* Typhimurium strain and its isogenic Δ*cpxA*, Δ*hilE*, Δ*cpxA* Δ*hilE*, Δ*lon* and Δ*cpxA* Δ*lon* mutants **(A)**, or in the WT *S.* Typhimurium strain and its isogenic Δ*hilE* and Δ*lon* mutants carrying plasmid pCA-CpxR **(C)**, was analyzed by Western blotting using monoclonal anti-FLAG antibodies. Whole cell lysates were prepared from samples of bacterial cultures grown for 5 h in LB medium at 37°C. As a loading control, the expression of DnaK was also determined using monoclonal anti-DnaK antibodies. Overexpression (+) of CpxR from the *T5-lac* promoter of plasmid pCA-CpxR was induced by adding 50 μM IPTG at the beginning of the bacterial cultures. **(B)** Secretion analysis of the SPI-1-encoded proteins SipA, SipB, SipC, and SipD was tested in the WT *S.* Typhimurium strain and its isogenic Δ*cpxA*, Δ*hilE*, Δ*cpxA* Δ*hilE*, Δ*lon*, and Δ*cpxA* Δ*lon* mutants grown for 9 h in LB medium at 37°C. FliC is a flagellar protein whose secretion is SPI-1-independent.

of InvF-FLAG, as well as the secretion/expression of the SipA-D proteins, was drastically reduced in the Δ*cpxA* and Δ*cpxA* Δ*hilE* mutants, but not in the Δ*hilE*, Δ*lon*, and Δ*cpxA* Δ*lon* mutants (**Figures 5A,B**). Consistently, the overexpression of CpxR clearly repressed the expression of InvF-FLAG in the WT strain and the Δ*hilE* mutant, but only slightly in the Δ*lon* mutant (**Figure 5C**). Therefore, these results show that deletion of *lon* counteracts repression exerted by CpxR on the SPI-1 genes, which supports that CpxR acts through the Lon protease to repress these genes. Nevertheless, even in the absence of Lon, either the absence of CpxA or the overexpression of CpxR slightly repressed the SPI-1 genes (**Figure 5**), revealing an additional Lon-independent mechanism for the repression of these genes by CpxR.

CpxR Affects Stability of HilD

On the basis of our results indicating that repression of the SPI-1 genes by CpxR is mostly lost in the absence of the Lon protease, which degrades HilD, we hypothesized that CpxR should reduce the stability of HilD. To investigate this, we determined the *in vivo* half-life of HilD in the presence or absence of CpxA or Lon. The cellular levels of Myc-tagged HilD (HilD-Myc) expressed from plasmid pBAD-HilD, under an arabinose-inducible promoter, were monitored in the Δ*hilD*, Δ*hilD* Δ*cpxA* and Δ*hilD* Δ*lon* mutants, at indicated times after adding a cocktail of transcription and translation inhibitors. As shown in **Figure 6A**, the levels of HilD-Myc were reduced faster in the Δ*hilD* Δ*cpxA* mutant than in the Δ*hilD* mutant, whereas, as expected, the stability of HilD-Myc was drastically increased in the Δ*hilD* Δ*lon* mutant. In these assays, the half-life of HilD-Myc in the presence and absence of CpxA was 38 and 20 min, respectively (**Figure 6B**), supporting the notion that the activation of CpxR by the absence of CpxA decreases the stability of HilD.

CpxR does not Affect the Transcription of *lon*

As most response regulators, CpxR directly controls gene expression at transcriptional level (Hunke et al., 2012; Vogt and Raivio, 2012; Raivio, 2014). Thus, we tested if CpxR affects the transcription of *lon*. In *E. coli*, *lon* seems to be transcribed from promoters located upstream of *lon*, or from those of neighboring genes *clpX* and *clpP* (RegulonDB database, www.regulondb.ccg.unam.mx). Therefore, to monitor the promoters expressing *lon*, we constructed *lon-cat*, *clpX-cat* and *clpP-cat* transcriptional fusions, which contain the full intergenic region upstream of the respective gene (**Figure 7A**). Each of these fusions showed similar levels of expression in the WT strain and its derivative Δ*cpxA* mutant (**Figure 7B**).

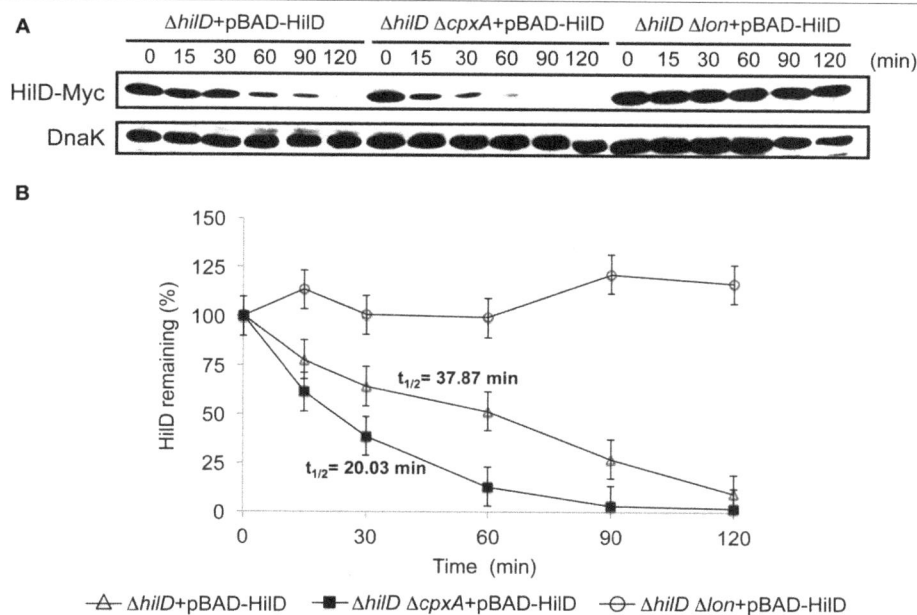

FIGURE 6 | CpxR reduces the stability of HilD. (A) Stability of HilD-Myc was determined in the ΔhilD, ΔhilD ΔcpxA, and ΔhilD Δlon mutants carrying plasmid pBAD-HilD, which were grown in LB medium at 37°C. Expression of HilD-Myc, from the arabinose-inducible promoter of plasmid pBAD-HilD, was induced with 0.05% L-arabinose for 45 min; then, transcription and translation were halted by the addition of a cocktail of antibiotics and glucose, and samples of bacterial cultures were taken at indicated times. HilD-Myc was detected from whole cell lysates of the samples by Western blotting using monoclonal anti-Myc antibodies. As a loading control, the expression of DnaK was also determined using monoclonal anti-DnaK antibodies. A representative Western blot of three independent experiments is shown. **(B)** Densitometric analysis of the HilD-Myc bands from the Western blots is indicated as the relative percentage of HilD-Myc at each time with respect to time 0. Intensity values of HilD-Myc bands were normalized with those respective of DnaK bands. The data are the averages of three independent experiments. Bars represent the standard deviations and $t_{1/2}$ indicates the half-life of HilD.

Furthermore, the expression of the *lon-cat*, *clpX-cat*, and *clpP-cat* fusions was not affected in the WT strain by the overexpression of CpxR; in contrast, their expression was increased by the overproduction of RpoH (**Figure 7C**). These results indicate that CpxR does not affect the transcription of *lon* and demonstrate that RpoH positively regulates *lon*, *clpX*, and *clpP* in *S.* Typhimurium.

Discussion

Previous studies have shown that deletion of *cpxA* or mutations in *cpxA* that activate the Cpx system reduce *Salmonella* adherence and invasion to eukaryotic cells, as well as the ability of *Salmonella* to infect mice (Leclerc et al., 1998; Nakayama et al., 2003; Humphreys et al., 2004). Furthermore, it was shown that deletion of *cpxA* decreases the expression of *hilA* and thus the SPI-1 genes (Nakayama et al., 2003), which code for the T3SS-1 and their cognate effector proteins that are required for *Salmonella* invasion into the intestinal epithelium of hosts (Haraga et al., 2008; Fàbrega and Vila, 2013). In this study, we show that deletion of *cpxA* negatively affects the expression of the SPI-1 genes when *S.* Typhimurium is grown in LB medium, but only in the presence of *cpxR*, indicating that the absence of CpxA leads to the repression of the SPI-1 genes through CpxR. Furthermore, we show that deletion of

cpxA increases the expression of the *cpxRA* operon through CpxR, which is in agreement with previous studies indicating that CpxR-P induces the expression of the *cpxRA* operon (De Wulf et al., 1999; Raivio et al., 1999, 2013; Price and Raivio, 2009). Thus, our results could suggest that the absence of CpxA turns on the regulation of the whole CpxR regulon in *S.* Typhimurium.

Our data provide several lines of evidence strongly supporting that CpxR-P represses the SPI-1 genes. First, the AckA and Pta enzymes, which generate acetyl phosphate that phosphorylates CpxR (Raivio and Silhavy, 1997; Wolfe et al., 2008), are also required for the repression of the SPI-1 genes mediated by deletion of *cpxA*. Second, overexpression of the lipoprotein NlpE, which activates the kinase activity of CpxA on CpxR (Snyder et al., 1995; Hunke et al., 2012; Vogt and Raivio, 2012), represses the SPI-1 genes via CpxR. Third, the overexpression of CpxR, which can bypass the need for phosphorylation of CpxR to regulate target genes (Macritchie et al., 2008; Acosta et al., 2015; Yun et al., 2015), has the same effect on the expression of the SPI-1 genes than the absence of CpxA or the overexpression of NlpE. Furthermore, a previous study showed accumulation of CpxR-P in a *cpxA* deletion mutant of *Yersinia pseudotuberculosis* grown in LB medium, which was generated through the AckA-Pta pathway (Liu et al., 2012). However, since CpxR-P induces its own expression, as mentioned above, both phosphorylation

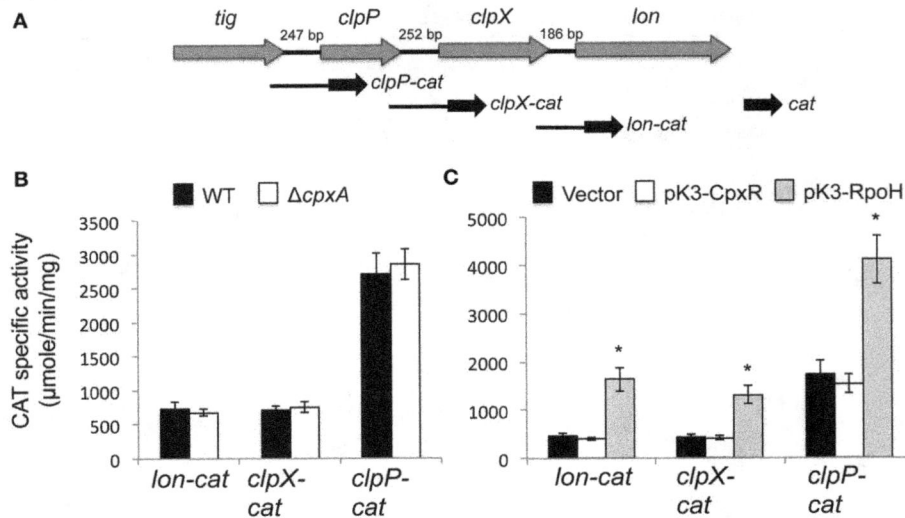

FIGURE 7 | RpoH, but not CpxR, induces the transcription of *lon*, *clpX*, and *clpP*. (A) Genetic organization of the *lon*, *clpX*, and *clpP* genes and schematic representation of the *cat* transcriptional fusions of these genes. Expression of the *lon-cat*, *clpX-cat* and *clpP-cat* transcriptional fusions, carried by plasmids plon-cat, pclpX-cat and pclpP-cat, respectively, was tested in the WT *S.* Typhimurium strain and its isogenic Δ*cpxA* mutant **(B)**, as well as in the WT *S.* Typhimurium strain carrying plasmid pK3-CpxR or pK3-RpoH, which constitutively express CpxR and RpoH, respectively, or carrying the vector pMPM-K3 **(C)**. CAT-specific activity was determined from samples collected of bacterial cultures grown for 5 h in LB medium at 37°C. The data are the averages of three different experiments performed in duplicate. Bars represent the standard deviations. *Expression statistically different with respect to that shown by the same fusion in the WT strain containing the vector.

and a higher concentration of CpxR would be involved in the repression of the SPI-1 genes.

Our data indicate that CpxR-P decreases the stability of HilD, the regulator that is at the apex of a regulatory cascade controlling the expression of the SPI-1 genes, as well as other virulence genes, such as those located in SPI-2 (Bustamante et al., 2008; Golubeva et al., 2012; Fàbrega and Vila, 2013; Martínez et al., 2014). Consistently, we show that CpxR represses the expression of both SPI-1 and SPI-2 (*ssrAB* and *sseB*) virulence genes when *S.* Typhimurium is grown in LB medium. Furthermore, we demonstrate that CpxR-P negatively affects the transcription of the SPI-1 genes *hilD* and *hilA*, but only in the presence of HilD, which would be expected, since the expression of HilD is autoregulated and HilD directly regulates *hilA* (Golubeva et al., 2012; Fàbrega and Vila, 2013). Therefore, the effect of CpxR-P on the expression of the SPI-1 and SPI-2 genes could be the result of its negative control on the stability of HilD and, as a consequence, on the transcription of *hilD*, which, in an additive manner, would decrease the concentration of HilD. In agreement with this conclusion, we did not find any putative CpxR binding-site in the regulatory regions of *hilD*, *hilA*, *ssrAB* and *sseB*, using the Virtual Footprint tool (Munch et al., 2005) (http://prodoric.tu-bs.de/vfp/) with the Position Weight Matrix for the binding-consensus sequence of *E. coli* K12 CpxR, 5′-GTAAA(N)5GTAA(A/G)-3′ (De Wulf et al., 2002), supporting that these genes are not directly regulated by CpxR. In contrast, these analyses revealed CpxR binding-sites in the regulatory regions of the *S.* Typhimurium *cpxR* and *cpxP* genes (data not shown). In *E. coli*, *cpxR* and *cpxP* belong to the CpxR regulon (Raivio et al., 1999, 2013; Price and Raivio, 2009).

CpxR has been shown to directly act as a transcriptional regulator (Hunke et al., 2012; Vogt and Raivio, 2012; Raivio, 2014). However, deletion of *cpxA* represses T3SS genes in *Shigella sonnei* through posttranscriptional processing of the regulator InvE (Mitobe et al., 2005). Furthermore, activation of the CpxR/A system reduces the stability of the *E. coli* F plasmid regulator TraJ via the HsIVU protease-chaperone pair (Gubbins et al., 2002; Lau-Wong et al., 2008). These latter studies indicate that CpxR can indirectly control protein stability by activating proteases. Interestingly, we found that the absence of the Lon protease, which has been shown to degrade HilD (Takaya et al., 2005), severely affects the repression of the SPI-1 genes mediated by CpxR. In contrast, the absence of HilE, a regulator which negatively controls HilD activity by protein-protein interactions (Baxter et al., 2003), does not affect this repression by CpxR. Taken together, our results support that CpxR-P represses the expression of the SPI-1 and SPI-2 genes mainly by reducing the stability of HilD through the Lon protease. However, our data also show that both deletion of *cpxA* and overexpression of CpxR slightly repress the expression of the SPI-1 genes in the absence of the Lon protease, suggesting an additional minor Lon-independent mechanism for the repression of the SPI-1 genes by CpxR-P. Alternatively, this could suggest that CpxR-P actually controls the stability of HilD through another protease, not involving Lon, which could be obfuscated by the extremely high stability of HilD in the absence of Lon. How CpxR-P reduces the stability of HilD or whether there is another mechanism by which CpxR-P represses the SPI-1 and SPI-2 genes is a matter of our current and future studies.

Overexpression of the heat shock sigma factor RpoH represses the SPI-1 genes only in the presence of the Lon protease (Matsui et al., 2008). Our results show that RpoH, but not CpxR, induces the transcription of *lon*, as well as of the *clpX* and *clpP* neighbor genes encoding the ClpXP protease. Accordingly, previous studies indicate that *lon*, *clpX* and *clpP* belong to the RpoH regulon (Nonaka et al., 2006; Wade et al., 2006), but not to the CpxR regulon of *E. coli* (Bury-Mone et al., 2009; Price and Raivio, 2009; Raivio et al., 2013). Therefore, CpxR and RpoH seem to affect HilD concentration and thus repress the SPI-1 genes differentially; RpoH by inducing transcription of *lon* and CpxR by probably affecting the posttranscriptional expression or activity of Lon, or through another protease. Anyway, HilD would integrate the regulation of *Salmonella* virulence genes to the stresses sensed by CpxR/A and RpoH.

The two-component system CpxR/A regulates virulence in many bacteria, mostly by inhibiting the production of secretion systems, pili, flagella, fimbriae and curli, which are required for bacteria interaction with host cells; furthermore, several studies have shown that biogenesis of these envelope-localized multiprotein complexes activates the Cpx response (Hunke et al., 2012; Vogt and Raivio, 2012; Raivio, 2014). In this study, we demonstrate that activation of the CpxR/A system represses the expression of the genes encoding the T3SS-1 and T3SS-2,

and their respective effector proteins, in *S.* Typhimurium. The activation of the CpxR/A system also represses the expression of T3SS genes in enteropathogenic *Escherichia coli* (Macritchie et al., 2008), *Yersinia pseudotuberculosis* (Carlsson et al., 2007; Liu et al., 2012) and *Shigella sonnei* (Mitobe et al., 2005, 2011). Therefore, it is tempting to speculate that the CpxR/A system controls biogenesis of T3SSs by sensing misfolded proteins generated during their production.

The insight from this study better explains the mechanism by which the CpxR/A system regulates the expression of the SPI-1 genes and further increases the current knowledge about the complex regulatory network governing virulence in *Salmonella*. Additionally, it reveals that deletion of *cpxA* activates CpxR-mediated gene regulation in *S.* Typhimurium.

Acknowledgments

We thank F.J. Santana for technical assistance, L.C. Martínez and A. Vázquez for constructing strains DTM76 and DTM60, respectively, B.B. Finlay and J.L. Puente for providing the anti-SseB polyclonal antibody and I. Martínez-Flores for critical reading of the manuscript. This work was supported by grants from the Dirección General de Asuntos del Personal Académico de la UNAM (IN205512 and IN203415 to VB and IN201513 to EC) and from the Consejo Nacional de Ciencia y Tecnología (179071 to VB and 179946 to EC).

References

Abrusci, P., Mcdowell, M. A., Lea, S. M., and Johnson, S. (2014). Building a secreting nanomachine: a structural overview of the T3SS. *Curr. Opin. Struct. Biol.* 25, 111–117. doi: 10.1016/j.sbi.2013.11.001

Acosta, N., Pukatzki, S., and Raivio, T. L. (2015). The *Vibrio cholerae* Cpx envelope stress response senses and mediates adaptation to low iron. *J. Bacteriol.* 197, 262–276. doi: 10.1128/JB.01957-14

Bang, I. S., Frye, J. G., Mcclelland, M., Velayudhan, J., and Fang, F. C. (2005). Alternative sigma factor interactions in *Salmonella*: σ^E and σ^H promote antioxidant defences by enhancing σ^S levels. *Mol. Microbiol.* 56, 811–823. doi: 10.1111/j.1365-2958.2005.04580.x

Batchelor, E., Walthers, D., Kenney, L. J., and Goulian, M. (2005). The *Escherichia coli* CpxA-CpxR envelope stress response system regulates expression of the porins OmpF and OmpC. *J. Bacteriol.* 187, 5723–5731. doi: 10.1128/JB.187.16.5723-5731.2005

Baxter, M. A., Fahlen, T. F., Wilson, R. L., and Jones, B. D. (2003). HilE interacts with HilD and negatively regulates *hilA* transcription and expression of the *Salmonella enterica* serovar Typhimurium invasive phenotype. *Infect. Immun.* 71, 1295–1305. doi: 10.1128/IAI.71.3.1295-1305.2003

Bispham, J., Tripathi, B. N., Watson, P. R., and Wallis, T. S. (2001). *Salmonella* pathogenicity island 2 influences both systemic salmonellosis and *Salmonella*-induced enteritis in calves. *Infect. Immun.* 69, 367–377. doi: 10.1128/IAI.69.1.367-377.2001

Brosius, J. (1984). Plasmid vectors for the selection of promoters. *Gene* 27, 151–160. doi: 10.1016/0378-1119(84)90136-7

Brown, N. F., Vallance, B. A., Coombes, B. K., Valdez, Y., Coburn, B. A., and Finlay, B. B. (2005). *Salmonella* pathogenicity island 2 is expressed prior to penetrating the intestine. *PLoS Pathog.* 1:e32. doi: 10.1371/journal.ppat.0010032

Bury-Mone, S., Nomane, Y., Reymond, N., Barbet, R., Jacquet, E., Imbeaud, S., et al. (2009). Global analysis of extracytoplasmic stress signaling in *Escherichia coli*. *PLoS Genet.* 5:e1000651. doi: 10.1371/journal.pgen.1000651

Bustamante, V. H., Martínez, L. C., Santana, F. J., Knodler, L. A., Steele-Mortimer, O., and Puente, J. L. (2008). HilD-mediated transcriptional cross-talk between SPI-1 and SPI-2. *Proc. Natl. Acad. Sci. U.S.A.* 105, 14591–14596. doi: 10.1073/pnas.0801205105

Carlsson, K. E., Liu, J., Edqvist, P. J., and Francis, M. S. (2007). Extracytoplasmic-stress-responsive pathways modulate type III secretion in *Yersinia pseudotuberculosis*. *Infect. Immun.* 75, 3913–3924. doi: 10.1128/IAI.01346-06

Chubiz, J. E., Golubeva, Y. A., Lin, D., Miller, L. D., and Slauch, J. M. (2010). FliZ regulates expression of the *Salmonella* pathogenicity island 1 invasion locus by controlling HilD protein activity in *Salmonella enterica* serovar typhimurium. *J. Bacteriol.* 192, 6261–6270. doi: 10.1128/JB.00635-10

Cirillo, D. M., Valdivia, R. H., Monack, D. M., and Falkow, S. (1998). Macrophage-dependent induction of the *Salmonella* pathogenicity island 2 type III secretion system and its role in intracellular survival. *Mol. Microbiol.* 30, 175–188. doi: 10.1046/j.1365-2958.1998.01048.x

Coburn, B., Li, Y., Owen, D., Vallance, B. A., and Finlay, B. B. (2005). *Salmonella enterica* serovar Typhimurium pathogenicity island 2 is necessary for complete virulence in a mouse model of infectious enterocolitis. *Infect. Immun.* 73, 3219–3227. doi: 10.1128/IAI.73.6.3219-3227.2005

Coombes, B. K., Brown, N. F., Valdez, Y., Brumell, J. H., and Finlay, B. B. (2004). Expression and secretion of *Salmonella* pathogenicity island-2 virulence genes in response to acidification exhibit differential requirements of a functional type III secretion apparatus and SsaL. *J. Biol. Chem.* 279, 49804–49815. doi: 10.1074/jbc.M404299200

Coombes, B. K., Coburn, B. A., Potter, A. A., Gomis, S., Mirakhur, K., Li, Y., et al. (2005). Analysis of the contribution of *Salmonella* pathogenicity islands 1 and 2 to enteric disease progression using a novel bovine ileal loop model and a murine model of infectious enterocolitis. *Infect. Immun.* 73, 7161–7169. doi: 10.1128/IAI.73.11.7161-7169.2005

Datsenko, K. A., and Wanner, B. L. (2000). One-step inactivation of chromosomal genes in *Escherichia coli* K-12 using PCR products. *Proc. Natl. Acad. Sci. U.S.A.* 97, 6640–6645. doi: 10.1073/pnas.120163297

Deiwick, J., Nikolaus, T., Erdogan, S., and Hensel, M. (1999). Environmental regulation of *Salmonella* pathogenicity island 2 gene expression. *Mol. Microbiol.* 31, 1759–1773. doi: 10.1046/j.1365-2958.1999.01312.x

De Wulf, P., Kwon, O., and Lin, E. C. (1999). The CpxR/A signal transduction system of *Escherichia coli*: growth-related autoactivation and control of unanticipated target operons. *J. Bacteriol.* 181, 6772–6778.

De Wulf, P., Mcguire, A. M., Liu, X., and Lin, E. C. (2002). Genome-wide profiling of promoter recognition by the two-component response regulator CpxR-P in *Escherichia coli*. *J. Biol. Chem.* 277, 26652–26661. doi: 10.1074/jbc.M203487200

Diepold, A., and Wagner, S. (2014). Assembly of the bacterial type III secretion machinery. *FEMS Microbiol. Rev.* 38, 802–822. doi: 10.1111/1574-6976.12061

Ellermeier, C. D., Ellermeier, J. R., and Slauch, J. M. (2005). HilD, HilC and RtsA constitute a feed forward loop that controls expression of the SPI1 type three secretion system regulator *hilA* in *Salmonella enterica* serovar Typhimurium. *Mol. Microbiol.* 57, 691–705. doi: 10.1111/j.1365-2958.2005.04737.x

Ellermeier, J. R., and Slauch, J. M. (2008). Fur regulates expression of the *Salmonella* pathogenicity island 1 type III secretion system through HilD. *J. Bacteriol.* 190, 476–486. doi: 10.1128/JB.00926-07

Eriksson, S., Lucchini, S., Thompson, A., Rhen, M., and Hinton, J. C. (2003). Unravelling the biology of macrophage infection by gene expression profiling of intracellular *Salmonella enterica*. *Mol. Microbiol.* 47, 103–118. doi: 10.1128/JB.2003.03313.x

Fàbrega, A., and Vila, J. (2013). *Salmonella enterica* serovar Typhimurium skills to succeed in the host: virulence and regulation. *Clin. Microbiol. Rev.* 26, 308–341. doi: 10.1128/CMR.00066-12

Fass, E., and Groisman, E. A. (2009). Control of *Salmonella* pathogenicity island-2 gene expression. *Curr. Opin. Microbiol.* 12, 199–204. doi: 10.1016/j.mib.2009.01.004

Free, A., Porter, M. E., Deighan, P., and Dorman, C. J. (2001). Requirement for the molecular adapter function of StpA at the *Escherichia coli bgl* promoter depends upon the level of truncated H-NS protein. *Mol. Microbiol.* 42, 903–917. doi: 10.1046/j.1365-2958.2001.02678.x

Golubeva, Y. A., Sadik, A. Y., Ellermeier, J. R., and Slauch, J. M. (2012). Integrating global regulatory input into the *Salmonella* pathogenicity island 1 type III secretion system. *Genetics* 190, 79–90. doi: 10.1534/genetics.111.132779

Grant, A. J., Morgan, F. J., Mckinley, T. J., Foster, G. L., Maskell, D. J., and Mastroeni, P. (2012). Attenuated *Salmonella* Typhimurium lacking the pathogenicity island-2 type 3 secretion system grow to high bacterial numbers inside phagocytes in mice. *PLoS Pathog.* 8:e1003070. doi: 10.1371/journal.ppat.1003070

Groisman, E. A., and Ochman, H. (1997). How *Salmonella* became a pathogen. *Trends Microbiol.* 5, 343–349. doi: 10.1016/S0966-842X(97)01099-8

Gubbins, M. J., Lau, I., Will, W. R., Manchak, J. M., Raivio, T. L., and Frost, L. S. (2002). The positive regulator, TraJ, of the *Escherichia coli* F plasmid is unstable in a *cpxA** background. *J. Bacteriol.* 184, 5781–5788. doi: 10.1128/JB.184.20.5781-5788.2002

Hansen-Wester, I., and Hensel, M. (2001). *Salmonella* pathogenicity islands encoding type III secretion systems. *Microbes Infect.* 3, 549–559. doi: 10.1016/S1286-4579(01)01411-3

Haraga, A., Ohlson, M. B., and Miller, S. I. (2008). *Salmonellae* interplay with host cells. *Nat. Rev. Microbiol.* 6, 53–66. doi: 10.1038/nrmicro1788

Humphreys, S., Rowley, G., Stevenson, A., Anjum, M. F., Woodward, M. J., Gilbert, S., et al. (2004). Role of the two-component regulator CpxAR in the virulence of *Salmonella enterica* serotype Typhimurium. *Infect. Immun.* 72, 4654–4661. doi: 10.1128/IAI.72.8.4654-4661.2004

Hunke, S., Keller, R., and M uller, V. S. (2012). Signal integration by the Cpx-envelope stress system. *FEMS Microbiol. Lett.* 326, 12–22. doi: 10.1111/j.1574-6968.2011.02436.x

Kitagawa, M., Ara, T., Arifuzzaman, M., Ioka-Nakamichi, T., Inamoto, E., Toyonaga, H., et al. (2005). Complete set of ORF clones of *Escherichia coli* ASKA library (a complete set of *E. coli* K-12 ORF archive): unique resources for biological research. *DNA Res.* 12, 291–299. doi: 10.1093/dnares/dsi012

Knodler, L. A., Vallance, B. A., Celli, J., Winfree, S., Hansen, B., Montero, M., et al. (2010). Dissemination of invasive *Salmonella* via bacterial-induced extrusion of mucosal epithelia. *Proc. Natl. Acad. Sci. U.S.A.* 107, 17733–17738. doi: 10.1073/pnas.1006098107

Kröger, C., Colgan, A., Srikumar, S., Händler, K., Sivasankaran, S. K., Hammarlöf, D. L., et al. (2013). An infection-relevant transcriptomic compendium for

Salmonella enterica Serovar Typhimurium. *Cell Host Microbe* 14, 683–695. doi: 10.1016/j.chom.2013.11.010

Laughlin, R. C., Knodler, L. A., Barhoumi, R., Payne, H. R., Wu, J., Gomez, G., et al. (2014). Spatial segregation of virulence gene expression during acute enteric infection with *Salmonella enterica* serovar Typhimurium. *MBio* 5, e00946–e00913. doi: 10.1128/mBio.00946-13

Lau-Wong, I. C., Locke, T., Ellison, M. J., Raivio, T. L., and Frost, L. S. (2008). Activation of the Cpx regulon destabilizes the F plasmid transfer activator, TraJ, via the HslVU protease in *Escherichia coli*. *Mol. Microbiol.* 67, 516–527. doi: 10.1111/j.1365-2958.2007.06055.x

Leclerc, G. J., Tartera, C., and Metcalf, E. S. (1998). Environmental regulation of *Salmonella typhi* invasion-defective mutants. *Infect. Immun.* 66, 682–691.

Liu, J., Thanikkal, E. J., Obi, I. R., and Francis, M. S. (2012). Elevated CpxR~P levels repress the Ysc-Yop type III secretion system of *Yersinia pseudotuberculosis*. *Res. Microbiol.* 163, 518–530. doi: 10.1016/j.resmic.2012.07.010

Lostroh, C. P., Bajaj, V., and Lee, C. A. (2000). The *cis* requirements for transcriptional activation by HilA, a virulence determinant encoded on SPI-1. *Mol. Microbiol.* 37, 300–315. doi: 10.1046/j.1365-2958.2000.01991.x

Lostroh, C. P., and Lee, C. A. (2001). The HilA box and sequences outside it determine the magnitude of HilA-dependent activation of P(*prgH*) from *Salmonella* pathogenicity island 1. *J. Bacteriol.* 183, 4876–4885. doi: 10.1128/JB.183.16.4876-4885.2001

Lucchini, S., Rowley, G., Goldberg, M. D., Hurd, D., Harrison, M., and Hinton, J. C. (2006). H-NS mediates the silencing of laterally acquired genes in bacteria. *PLoS Pathog.* 2:e81. doi: 10.1371/journal.ppat.0020081

Lundberg, U., Vinatzer, U., Berdnik, D., Von Gabain, A., and Baccarini, M. (1999). Growth phase-regulated induction of *Salmonella*-induced macrophage apoptosis correlates with transient expression of SPI-1 genes. *J. Bacteriol.* 181, 3433–3437.

Macritchie, D. M., Ward, J. D., Nevesinjac, A. Z., and Raivio, T. L. (2008). Activation of the Cpx envelope stress response down-regulates expression of several locus of enterocyte effacement-encoded genes in enteropathogenic *Escherichia coli*. *Infect. Immun.* 76, 1465–1475. doi: 10.1128/IAI.01265-07

Martínez, L. C., Banda, M. M., Fernández-Mora, M., Santana, F. J., and Bustamante, V. H. (2014). HilD induces expression of *Salmonella* pathogenicity island 2 genes by displacing the global negative regulator H-NS from *ssrAB*. *J. Bacteriol.* 196, 3746–3755. doi: 10.1128/JB.01799-14

Martínez, L. C., Yakhnin, H., Camacho, M. I., Georgellis, D., Babitzke, P., Puente, J. L., et al. (2011). Integration of a complex regulatory cascade involving the SirA/BarA and Csr global regulatory systems that controls expression of the *Salmonella* SPI-1 and SPI-2 virulence regulons through HilD. *Mol. Microbiol.* 80, 1637–1656. doi: 10.1111/j.1365-2958.2011.07674.x

Matsui, M., Takaya, A., and Yamamoto, T. (2008). σ^{32}-mediated negative regulation of *Salmonella* pathogenicity island 1 expression. *J. Bacteriol.* 190, 6636–6645. doi: 10.1128/JB.00744-08

Mayer, M. P. (1995). A new set of useful cloning and expression vectors derived from pBlueScript. *Gene* 163, 41–46. doi: 10.1016/0378-1119(95)00389-N

Mcclelland, M., Sanderson, K. E., Spieth, J., Clifton, S. W., Latreille, P., Courtney, L., et al. (2001). Complete genome sequence of *Salmonella enterica* serovar Typhimurium LT2. *Nature* 413, 852–856. doi: 10.1038/35101614

Miao, E. A., and Miller, S. I. (2000). A conserved amino acid sequence directing intracellular type III secretion by *Salmonella typhimurium*. *Proc. Natl. Acad. Sci. U.S.A.* 97, 7539–7544. doi: 10.1073/pnas.97.13.7539

Mitobe, J., Arakawa, E., and Watanabe, H. (2005). A sensor of the two-component system CpxA affects expression of the type III secretion system through posttranscriptional processing of InvE. *J. Bacteriol.* 187, 107–113. doi: 10.1128/JB.187.1.107-113.2005

Mitobe, J., Yanagihara, I., Ohnishi, K., Yamamoto, S., Ohnishi, M., Ishihama, A., et al. (2011). RodZ regulates the post-transcriptional processing of the *Shigella sonnei* type III secretion system. *EMBO Rep.* 12, 911–916. doi: 10.1038/embor.2011.132

Moest, T. P., and Méresse, S. (2013). *Salmonella* T3SSs: successful mission of the secret(ion) agents. *Curr. Opin. Microbiol.* 16, 38–44. doi: 10.1016/j.mib.2012.11.006

Munch, R., Hiller, K., Grote, A., Scheer, M., Klein, J., Schobert, M., et al. (2005). Virtual Footprint and PRODORIC: an integrative framework for regulon prediction in prokaryotes. *Bioinformatics* 21, 4187–4189. doi: 10.1093/bioinformatics/bti635

Nakayama, S., Kushiro, A., Asahara, T., Tanaka, R., Hu, L., Kopecko, D. J., et al. (2003). Activation of *hilA* expression at low pH requires the signal sensor CpxA, but not the cognate response regulator CpxR, in *Salmonella enterica* serovar Typhimurium. *Microbiology* 149, 2809–2817. doi: 10.1099/mic.0.26229-0

Navarre, W. W., Porwollik, S., Wang, Y., Mcclelland, M., Rosen, H., Libby, S. J., et al. (2006). Selective silencing of foreign DNA with low GC content by the H-NS protein in *Salmonella*. *Science* 313, 236–238. doi: 10.1126/science.1128794

Nonaka, G., Blankschien, M., Herman, C., Gross, C. A., and Rhodius, V. A. (2006). Regulon and promoter analysis of the *E. coli* heat-shock factor, σ^{32}, reveals a multifaceted cellular response to heat stress. *Genes Dev.* 20, 1776–1789. doi: 10.1101/gad.1428206

Núñez-Hernández, C., Alonso, A., Pucciarelli, M. G., Casadesús, J., and García-Del Portillo, F. (2014). Dormant intracellular *Salmonella enterica* serovar Typhimurium discriminates among Salmonella pathogenicity island 2 effectors to persist inside fibroblasts. *Infect. Immun.* 82, 221–232. doi: 10.1128/IAI.01304-13

Olekhnovich, I. N., and Kadner, R. J. (2002). DNA-binding activities of the HilC and HilD virulence regulatory proteins of *Salmonella enterica* serovar Typhimurium. *J. Bacteriol.* 184, 4148–4160. doi: 10.1128/JB.184.15.4148-4160.2002

Olekhnovich, I. N., and Kadner, R. J. (2006). Crucial roles of both flanking sequences in silencing of the *hilA* promoter in *Salmonella enterica*. *J. Mol. Biol.* 357, 373–386. doi: 10.1016/j.jmb.2006.01.007

Petrone, B. L., Stringer, A. M., and Wade, J. T. (2014). Identification of HilD-regulated genes in *Salmonella enterica* serovar Typhimurium. *J. Bacteriol.* 196, 1094–1101. doi: 10.1128/JB.01449-13

Porwollik, S., and Mcclelland, M. (2003). Lateral gene transfer in *Salmonella*. *Microbes Infect.* 5, 977–989. doi: 10.1016/S1286-4579(03)00186-2

Price, N. L., and Raivio, T. L. (2009). Characterization of the Cpx regulon in *Escherichia coli* strain MC4100. *J. Bacteriol.* 191, 1798–1815. doi: 10.1128/JB.00798-08

Puente, J. L., Bieber, D., Ramer, S. W., Murray, W., and Schoolnik, G. K. (1996). The bundle-forming pili of enteropathogenic *Escherichia coli*: transcriptional regulation by environmental signals. *Mol. Microbiol.* 20, 87–100. doi: 10.1111/j.1365-2958.1996.tb02491.x

Raivio, T. L. (2014). Everything old is new again: an update on current research on the Cpx envelope stress response. *Biochim. Biophys. Acta* 1843, 1529–1541. doi: 10.1016/j.bbamcr.2013.10.018

Raivio, T. L., Leblanc, S. K., and Price, N. L. (2013). The *Escherichia coli* Cpx envelope stress response regulates genes of diverse function that impact antibiotic resistance and membrane integrity. *J. Bacteriol.* 195, 2755–2767. doi: 10.1128/JB.00105-13

Raivio, T. L., Popkin, D. L., and Silhavy, T. J. (1999). The Cpx envelope stress response is controlled by amplification and feedback inhibition. *J. Bacteriol.* 181, 5263–5272.

Raivio, T. L., and Silhavy, T. J. (1997). Transduction of envelope stress in *Escherichia coli* by the Cpx two-component system. *J. Bacteriol.* 179, 7724–7733.

Sánchez-Vargas, F. M., Abu-El-Haija, M. A., and Gómez-Duarte, O. G. (2011). *Salmonella* infections: an update on epidemiology, management, and prevention. *Travel Med. Infect. Dis.* 9, 263–277. doi: 10.1016/j.tmaid.2011.11.001

Schechter, L. M., Damrauer, S. M., and Lee, C. A. (1999). Two AraC/XylS family members can independently counteract the effect of repressing sequences upstream of the *hilA* promoter. *Mol. Microbiol.* 32, 629–642. doi: 10.1046/j.1365-2958.1999.01381.x

Schechter, L. M., and Lee, C. A. (2001). AraC/XylS family members, HilC and HilD, directly bind and derepress the *Salmonella typhimurium hilA* promoter. *Mol. Microbiol.* 40, 1289–1299. doi: 10.1046/j.1365-2958.2001.02462.x

Schmidt, H., and Hensel, M. (2004). Pathogenicity islands in bacterial pathogenesis. *Clin Microbiol. Rev.* 17, 14–56. doi: 10.1128/CMR.17.1.14-56.2004

Singer, H. M., Kühne, C., Deditius, J. A., Hughes, K. T., and Erhardt, M. (2014). The *Salmonella* Spi1 virulence regulatory protein HilD directly activates transcription of the flagellar master operon *flhDC*. *J. Bacteriol.* 196, 1448–1457. doi: 10.1128/JB.01438-13

Snyder, W. B., Davis, L. J., Danese, P. N., Cosma, C. L., and Silhavy, T. J. (1995). Overproduction of NlpE, a new outer membrane lipoprotein, suppresses the toxicity of periplasmic LacZ by activation of the Cpx signal transduction pathway. *J. Bacteriol.* 177, 4216–4223.

Spinola, S. M., Fortney, K. R., Baker, B., Janowicz, D. M., Zwickl, B., Katz, B. P., et al. (2010). Activation of the CpxR/A system by deletion of *cpxA* impairs the ability of *Haemophilus ducreyi* to infect humans. *Infect. Immun.* 78, 3898–3904. doi: 10.1128/IAI.00432-10

Takaya, A., Kubota, Y., Isogai, E., and Yamamoto, T. (2005). Degradation of the HilC and HilD regulator proteins by ATP-dependent Lon protease leads to downregulation of *Salmonella* pathogenicity island 1 gene expression. *Mol. Microbiol.* 55, 839–852. doi: 10.1111/j.1365-2958.2004.04425.x

Uzzau, S., Figueroa-Bossi, N., Rubino, S., and Bossi, L. (2001). Epitope tagging of chromosomal genes in *Salmonella*. *Proc. Natl. Acad. Sci. U.S.A.* 98, 15264–15269. doi: 10.1073/pnas.261348198

Vogt, S. L., and Raivio, T. L. (2012). Just scratching the surface: an expanding view of the Cpx envelope stress response. *FEMS Microbiol. Lett.* 326, 2–11. doi: 10.1111/j.1574-6968.2011.02406.x

Wade, J. T., Castro Roa, D., Grainger, D. C., Hurd, D., Busby, S. J., Struhl, K., et al. (2006). Extensive functional overlap between sigma factors in *Escherichia coli*. *Nat. Struct. Mol. Biol.* 13, 806–814. doi: 10.1038/nsmb1130

Wolfe, A. J., Parikh, N., Lima, B. P., and Zemaitaitis, B. (2008). Signal integration by the two-component signal transduction response regulator CpxR. *J. Bacteriol.* 190, 2314–2322. doi: 10.1128/JB.01906-07

Yun, S., Lee, E. G., Kim, S. Y., Shin, J. M., Jung, W. S., Oh, D. B., et al. (2015). The CpxR/A two-component system is involved in the maintenance of the integrity of the cell envelope in the rumen bacterium *Mannheimia succiniciproducens*. *Curr. Microbiol.* 70, 103–109. doi: 10.1007/s00284-014-0686-5

Zahrl, D., Wagner, M., Bischof, K., and Koraimann, G. (2006). Expression and assembly of a functional type IV secretion system elicit extracytoplasmic and cytoplasmic stress responses in *Escherichia coli*. *J. Bacteriol.* 188, 6611–6621. doi: 10.1128/JB.00632-06

Conflict of Interest Statement: The authors declare that the research was conducted in the absence of any commercial or financial relationships that could be construed as a potential conflict of interest.

6

Wine fermentation microbiome: a landscape from different Portuguese wine appellations

Cátia Pinto[1†], Diogo Pinho[1†], Remy Cardoso[1], Valéria Custódio[1], Joana Fernandes[1], Susana Sousa[1], Miguel Pinheiro[2], Conceição Egas[2] and Ana C. Gomes[1]*

[1] Genomics Unit, Biocant – Biotechnology Innovation Center, Cantanhede, Portugal, [2] GenoInSeq Unit, Biocant – Biotechnology Innovation Center, Cantanhede, Portugal

Edited by:
Sandra Torriani,
Università degli Studi di Verona, Italy

Reviewed by:
Giuseppe Spano,
University of Foggia, Italy
Maret Du Toit,
Stellenbosch University, South Africa
Sandra Torriani,
Università degli Studi di Verona, Italy
Gianluca Bleve,
Istituto di Scienze delle Produzioni
Alimentari – CNR, Italy

*Correspondence:
Ana C. Gomes,
Genomics Unit, Biocant –
Biotechnology Innovation Center,
Biocant -Parque Tecnológico
de Cantanhede, Núcleo 04, Lote 8,
3060-197 Cantanhede, Portugal
acgomes@biocant.pt

[†] These authors have contributed
equally to this work.

Grapes and wine musts harbor a complex microbiome, which plays a crucial role in wine fermentation as it impacts on wine flavour and, consequently, on its final quality and value. Unveiling the microbiome and its dynamics, and understanding the ecological factors that explain such biodiversity, has been a challenge to oenology. In this work, we tackle this using a metagenomics approach to describe the natural microbial communities, both fungal and bacterial microorganisms, associated with spontaneous wine fermentations. For this, the wine microbiome, from six Portuguese wine appellations, was fully characterized as regards to three stages of fermentation – Initial Musts (IM), and Start and End of alcoholic fermentations (SF and EF, respectively). The wine fermentation process revealed a higher impact on fungal populations when compared with bacterial communities, and the fermentation evolution clearly caused a loss of the environmental microorganisms. Furthermore, significant differences ($p < 0.05$) were found in the fungal populations between IM, SF, and EF, and in the bacterial population between IM and SF. Fungal communities were characterized by either the presence of environmental microorganisms and phytopathogens in the IM, or yeasts associated with alcoholic fermentations in wine must samples as *Saccharomyces* and non-*Saccharomyces* yeasts (as *Lachancea, Metschnikowia, Hanseniaspora, Hyphopichia, Sporothrix, Candida,* and *Schizosaccharomyces*). Among bacterial communities, the most abundant family was Enterobacteriaceae; though families of species associated with the production of lactic acid (Lactobacillaceae, Leuconostocaceae) and acetic acid (Acetobacteriaceae) were also detected. Interestingly, a biogeographical correlation for both fungal and bacterial communities was identified between wine appellations at IM suggesting that each wine region contains specific and embedded microbial communities which may contribute to the uniqueness of regional wines.

Keywords: grape microbiology, wine spontaneous fermentation microbiome, industrial metagenomics

Introduction

The knowledge and the understanding of the microbial *terroir* – how the microbiome contributes to the natural environment of grapes and to the identity of wine, is a process that starts at the vineyards, at the harvest of grapes, and then evolves along the different stages of fermentation (Van Leeuwen and Seguin, 2006; Bokulich et al., 2013). Indeed, it is known that grapes harbor a complex microbiome, including a high range of filamentous fungi, yeasts and bacteria with different physiological and metabolic characteristics (Pretorius, 2000; Fleet, 2003; Barata et al., 2012). The microflora of the grapes is highly variable, mostly due to the influence of external factors as environmental parameters, geographical location, grape cultivars and application of phytochemicals on the vineyards (Pretorius, 2000; Cadez et al., 2010; Pinto et al., 2014). These microbial communities play an important role during the winemaking process, as they metabolize the sugars from the grapes and produce a whole set of secondary metabolites that influence the wine aromatic quality (Fleet, 2003). In fact, the natural diversity of those metabolic pathways, and the contribution of the different microorganisms involved on the fermentation process, is well documented (Setati et al., 2012). Therefore, unveiling the microbial biodiversity of grapes and during their fermentation will expand our understanding on fermentation dynamics, on its control (Bisson, 1999; Bisson and Butzke, 2000) and may also contribute to the identification of novel starter cultures (Fleet, 2008; Ciani et al., 2010).

The spontaneous wine fermentation is carried out by indigenous microbiota (Heard, 1999; Pretorius, 2000; Ciani et al., 2006; Renouf et al., 2007). Species of *Metschnikowia, Candida, Hanseniaspora, Pichia, Lachancea* (*Kluyveromyces*), and *Saccharomyces* are often present at the initial stages of wine fermentations and form the dominant consortium (Cocolin et al., 2000; Mills et al., 2002; Fleet, 2008). However, during the wine fermentation, the ethanol content increases and *Saccharomyces cerevisiae* strains dominate the alcoholic fermentation (AF; Fleet, 2008). Additionally, a deacidification may occur, by conversion of malic acid into lactic acid. This process is known as malolactic fermentation (MLF) and is due to the activity of lactic acid bacteria (LAB; Lonvaud-Funel, 1999; Lerm et al., 2011). The LAB species associated with MLF generally belong to the *Oenococcus, Pediococcus, Lactobacillus,* and *Leuconostoc* genera (Lonvaud-Funel, 1999). Indeed, MLF mainly influences the organoleptic characteristics and the aging of wines (Lonvaud-Funel, 1999). On the other hand, acetic acid bacteria (AAB) may cause a negative impact on the winemaking process, due to the production of undesirable metabolites, as acetic acid, thus affect negatively the quality of wine and so are considered spoilage microorganisms (Zoecklein et al., 2000).

The majority of the wine microbiology studies focus on the characterization of *S. cerevisiae* strains (Pretorius, 2000; Fleet, 2008; Nisiotou et al., 2011). Nevertheless, recent studies based on culture-independent methods, started to explore the microbial communities associated with wine grapes (Bokulich et al., 2013; Taylor et al., 2014). It is widely accepted that unveiling the indigenous microbial community associated with particular grape varieties, from specific locations, could represent an important source of distinctive metabolites and introduce an authenticity *terroir* to the region (Heard, 1999; Jolly et al., 2006; Fleet, 2008). The biogeographical distribution of the wine associated microorganisms has been recently investigated in vineyards from different regions of California (Bokulich et al., 2013), New Zealand (Taylor et al., 2014), and in conventional, biodynamic, and integrated vineyards of South Africa (Setati

et al., 2012). These studies allowed for a better spatial and temporal characterization of the wine grapes microbiome and brought new insights of its dynamics and biodiversity. Also, other biogeography wine studies have been previously published focusing on *S. cerevisiae* (Schuller et al., 2012). Nevertheless, there is still a lack of knowledge on the diversity and the dynamics of microbial communities as a whole– from the wine grapes until the wine fermentation, which can now be obtained using high-throughput sequencing technologies and metagenomics approaches that allow for the identification of both non-cultivable microorganisms, and of less represented species.

In this work, a total of six different Portuguese wine appellations were considered to analysis and high-throughput sequencing was used to unveil the wine microbiome present at initial musts (IM), and start and end of alcoholic fermentations (SF and EF, respectively). This work aims to understand the dynamics of microbial communities across spontaneous wine fermentations and also to reveal the biogeographic distribution of grape and wine microbiomes of Portuguese wine appellations.

Materials and Methods

Grape Sampling, Laboratory-Scale Fermentation, and DNA Extraction

The grape samples were collected during the 2010 vintage, from six different Portuguese appellations, namely, Minho (Mi), Douro (Dr), Dão (D), Bairrada (B), Estremadura (E), and Alentejo (Al). For each appellation, the three most representative grape varieties were considered for sampling, with exception of Minho where only two grape varieties were considered (**Supplementary Figure S1**). For all regions, the sampling was carried out 1 day prior the harvest. The sampling was authorized by private wine producers, who are fully acknowledged in this paper, and no specific permissions were required for this activity. Also, the field study did not involve any endangered or protected species.

For each appellation, one vineyard (farm) with different grape varieties was selected, and for each grape variety, 2 kg of healthy and undamaged grapes were collected. Grapes were collected from multiple bunches of different grapevines, randomly distributed across the vineyard in order to assure the representativeness of the sampling. These samples were collected into sterile plastic bags and transported to the laboratory chilled on ice. In total, 17 grape samples were collected, crushed and allowed for laboratory-scale fermentation (spontaneous AF) under aseptic conditions and acclimatised at 21°C, at the Genomics Unit from Biocant. For each sample, the microbial diversity was analyzed at three stages: IM, corresponding to the juice of crushed grapes; start of alcoholic fermentation (SF) and end of alcoholic fermentation (EF), which corresponded to the weight loss of 5 and 70 g/L of sugar, respectively. The SF and EF where daily monitored through weighting. At each stage, 50 mL of wine must were collected and centrifuged at 4000 rpm for 10 min. The respective microbial pellets were collected, washed twice with 0.9% NaCl and re-suspended with glycerol. A total of 51 samples ($n = 17 \times 3$ fermentation stages) were stored at $-80°C$ for DNA extraction. The DNA from each individual sample was extracted using the DNeasy Plant mini kit (QIAGEN, USA), according to the manufacturer's instructions, with a prior cell rupture using glass beads in Tissue Lyser (Qiagen, USA), to assure full disruption of microbial cells.

rDNA Library Construction and Pyrosequencing

A PCR amplicon library was built for each individual sample. For a better discrimination of the entire microbial community present during the fermentation process, rDNA sequences from both prokaryotic and eukaryotic microorganisms were amplified, using PCR primers that were designed to target three distinct regions. The V6 hypervariable region of the 16S rRNA was used for the identification of prokaryotic microorganisms (Sogin et al., 2006) and the D2, from the 26S rRNA, and ITS2 regions (White et al., 1990) for eukaryotic identification. The sequence-specific portions of the used primers were: V6_F 5′-ATGCAACGCGAAGAACCT-3′ and V6_R 5′-TAGCGATTCCGACTTCA-3′ of V6 region; D2_F 5′-AAGMACTTTGRAAAGAGAG-3′ and D2_R 5′-GGTCCGTGT TTCAAGACG-3′ of D2 region; and ITS2_F 5′-GCATCGATG AAGAACGC-3′ and ITS2_R 5′-CCTCCGCTTATTGATAT GC-3′ of ITS2 region. Additionally, the fusion primers also contained a specific Roche 454 adaptor sequence and a multiplex identifier sequence with eight nucleotides, which allows the pooling of amplicons.

All PCR reactions were carried out in 30 μL reaction mix containing 2 μL of DNA template, 1.5 units of FastStart High Fidelity Taq DNA polymerase (Roche, USA), 1x reaction buffer with $MgCl_2$ (1.8 mM) incorporate (Roche, USA), 0.2 mM dNTPs (Bioron, Germany) and 0.8 μM of the forward and reverse primers for V6 region or 0.4 μM of forward and reverse primers for D2 and ITS2 regions. For prokaryotes amplification, cycling conditions consisted in a first denaturation step at 94°C for 5 min followed by 20 cycles with a denaturation step at 94°C for 35 s, annealing at 50°C for 35 s and an extension at 72°C for 40 s. A final extension cycle at 72°C for 5 min was applied. The cycling conditions applied for eukaryotic microorganisms were the same, but the PCR consisted in 25 cycles. The amplification success was assessed by electrophoresis using the HT DNA 5K/RNA LabChip for the LabChip 90 (Caliper Life Sciences, USA). The PCR reaction products were then purified with the High Pure 96 UF Cleanup Plates (Roche, USA) and quantified using the PicoGreen® dsDNA quantitation kit (Invitrogen, USA). Samples were pooled together according to the number of DNA molecules, in equimolar concentrations and submitted for pyrosequencing using the GS FLX Titanium platform (454 Life Sciences, Roche) at Biocant, Portugal. The raw data obtained was deposited in NCBI platform with the accession number SRA097159.

Bioinformatic Data Analysis

Raw sequence reads were processed with MetaBiodiverse, an automatic annotation pipeline fully implemented at Genoinseq of Biocant (Vaz-Moreira et al., 2011; Egas et al., 2012; Pinto et al., 2014). Briefly, the raw data obtained was split through the identification of barcode sequences and quality filters were applied to remove low quality reads. Thus, (i) short sequences (<120 bp), (ii) sequences containing more than two undetermined nucleotides (N), (iii) masked sequences with more 50% of low complexity areas (Sogin et al., 2006) and (iv) chimera sequences, detected using UChime were removed (Edgar et al., 2011). All sequences with a distance value below 0.03, which corresponds to the species-level threshold (Sharpton et al., 2011), were grouped in operational taxonomic units (OTUs) through USearch, version 6.0.307 (Edgar, 2010). The Mothur package (Schloss et al., 2009) was used to generate rarefaction curves (richness of population analysis) and to calculate the population diversity analysis estimator Chao1 (α diversity). For the taxonomic annotation, each generated consensus sequences were queried by BLAST on curated databases. The Ribosomal Database Project II (RDP; Cole et al., 2009) was used for prokaryotic microorganisms assignment and the nt@ncbi/SILVA database for eukaryotic classification. After BLAST, the best hits were selected and subjected to another quality control. All sequences with an alignment of less than 40% or with an E-value greater than $1e^{-50}$ were rejected. Sequences that passed the quality check were subjected to a bootstrap test with 100 replicates, using the seqBoot application from the Phylip package (Felsenstein, 1989). The OTU identification process implemented provided a high level of confidence in taxon assignment of each sequence. The process assessed the correct E-values scores, went through the taxonomy path and identified the lowest common taxonomy level in the bootstrap process. Only those sequences with an identity greater than 70% were reported, while all the others went up the taxonomy levels until reached 70%.

Statistical Analyses

To determine the minimum significant difference ($p < 0.05$) in the biodiversity (Chao1) of IM, SF and EF samples, one-way analysis of variance (ANOVA) was performed using SPSS 20.0 (IBM, US). Shapiro-Wilk normality tests were carried out for each eukaryotic and prokaryotic phylogenetic group. As most groups did not follow the normal distribution, Friedman and Sign tests (pairwise comparisons) were used. The microbial communities were compared at family level for prokaryotic population and at genus level for eukaryotic population through the sequence reads analysis. Thus, microbial population comparisons were carried out using these taxa.

Sequence reads data matrixes of the 97% similarity grouped bacterial and fungal OTUs, produced by Metabiodiverse, were normalized by the total reads obtained for each analyzed sample, and then log(X+1) transformed and used to calculate a Bray–Curtis resemblance matrixes. The data obtained for the three fermentation stages were (i) explored by principal coordinate analysis (PCO), (ii) tested by Analysis of Similarities (ANOSIM) for significant differences and (iii) analyzed by SIMPER to identify the taxa responsible for similarity between samples within each group and dissimilarities between groups, using Primer E software version 6 (Clarke and Gorley, 2006). The same analyses were performed to explore and test the influence of wine appellations on microbiome although, for each fermentation stage, individual matrixes were created in order to remove the "fermentation stage" variable.

Results

Diversity and Richness of Microbial Communities

In this study, we assessed and compared the microbial community of IM, and the Start and End of wine alcoholic fermentations (SF and EF, respectively), from six Portuguese appellations by DNA massive parallel sequencing of 16S rDNA for bacteria, and both, ITS2 and D2 for fungal analysis. Two target regions were used for the fungal population identification as previous experiments demonstrated that these combination would allow for the highest coverage of eukaryotic organisms (Pinto et al., 2014).

The deep sequencing of microbial communities generated a total of 1,180,106 sequences of ITS2, D2, and V6 regions from IM, SF, and EF (**Table 1** and **Supplementary Table S1**). A total of 1,160,482 sequences passed the quality control parameters, representing an acceptance of 98.3% of high quality sequences (723,474 eukaryotic sequences: 313,919 reads for ITS2 region and 409,555 for D2 region; and 437,008 prokaryotic sequences). The clustering of the sequences at a phylogenetic distance of 3% generated a total of 1,034 OTUs for ITS2, 1,099 for D2, and 1,461 for V6. The number of OTUs from both eukaryotic and prokaryotic communities decreased along the fermentation.

The diversity of microbial community was compared by rarefaction curve analysis (**Supplementary Figure S2**) and the ratio between the number of the obtained and the expected OTUs (predicted by Chao1) was used to determine the coverage for the microbial communities: it was of 73.7 ± 2.0% for ITS2 region,

71.7 ± 1.9% for D2 region and 65.1 ± 1.9% for V6 region (**Supplementary Table S1**).

In order to assess the variations of microbial biodiversity, the Chao1 richness estimator was used to compare the three fermentation stages at both domain and phylum levels. In general, and as expected, a decrease of richness was observed over the spontaneous wine fermentation for both fungi and bacteria, at the analyzed taxonomical levels (domain and phylum; **Figure 1**). Considering the domain (**Figure 1A**), no significant differences were found for the three rDNA regions. At the phylum level, significant differences ($p < 0.05$) in the Basidiomycota between all stages of fermentation were observed (both for ITS2 and D2 regions), and in the Ascomycota population differences were between SF and EF, but not between IM and SF (**Figure 1B**). For the bacterial population, a decrease in biodiversity was observed but no significant differences were detected (V6 rDNA region). A clear relationship was observed between the microbial community biodiversity and the stage of fermentation. Interestingly, the variations of biodiversity, which were observed along the fermentation stages, revealed a higher impact on the structure of the eukaryotic population, when compared with the prokaryotic communities. Moreover, regarding the microbial biodiversity, the prokaryotic population was richer than the eukaryotic population.

General Characterization of Microbial Communities

The dominant phylum across the entire eukaryotic population was Ascomycota (42.4%), though it also contained Basidiomycota (17.7%), and other fungi, as Chytridiomycota phylum (0.2%) and *basal fungal lineages* (5.6%). Also, a considerable number of unidentified microorganisms (34.1%) were mostly present at IM (**Figure 2A**).

TABLE 1 | Total sequences obtained for eukaryotic (ITS2 and D2) and prokaryotic (V6) microbial community for IM, SF, and EF samples.

Sampling point	Target region	No. Reads		0.03 distance		
		Total	High quality	OTU obtained (mean ± SEM)	Estimated species (mean ± SEM)	Coverage (mean ± SEM)
IM	ITS2	119876	116064	68 ± 6	100 ± 9	68.83 ± 2.26%
	D2	131837	129652	71 ± 6	110 ± 10	66.54 ± 2.52%
	V6	145796	145051	78 ± 12	134 ± 21	60.30 ± 3.19%
SF	ITS2	114993	111075	33 ± 3	47 ± 5	74.44 ± 3.62%
	D2	145559	143100	36 ± 3	56 ± 7	68.63 ± 3.29%
	V6	159940	159054	56 ± 9	83 ± 13	66.92 ± 3.28%
EF	ITS2	90207	86780	20 ± 1	29 ± 4	77.74 ± 4.10%
	D2	138156	136803	19 ± 2	25 ± 2	79.82 ± 3.23%
	V6	133742	132903	54 ± 9	81 ± 12	68.15 ± 3.48%
	Eukaryotic	740628	723474			
	Prokaryotic	439478	437008			
	Total	**1180106**	**1160482**			

*Operational taxonomic units (OTUs) and estimated species (chao1) were determined at a genetic distance of 3% using Mothur. The coverage obtained was also determined as being the ratio between the observed OTUs and estimated Chao1 (OTUs/Chao1). A detailed table with indication of the samples origin is provided as **Supplementary Table S1**.*

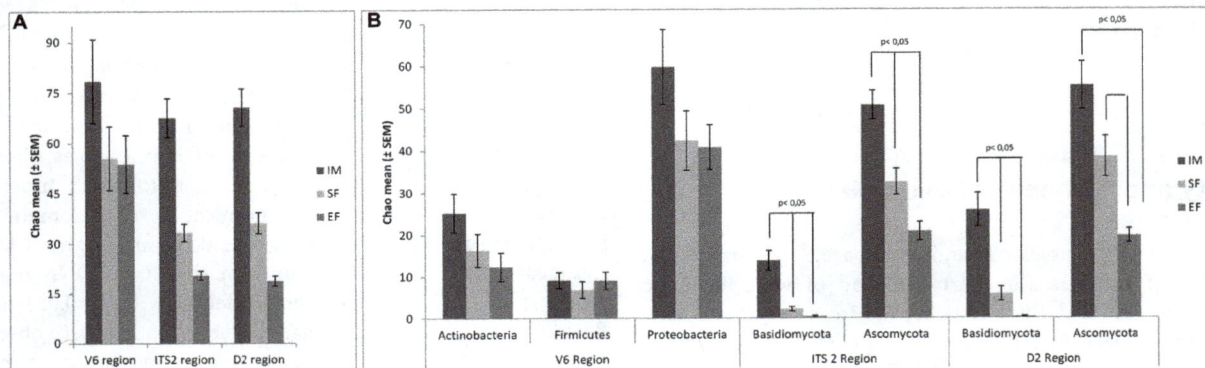

FIGURE 1 | Biodiversity dynamics associated with V6, ITS2, and D2 region, at domain (A) and phylum level (B). The mean of Chao1 index ± SEM are represented in the graph. Significance was assessed with Friedman test and signal test. $p < 0.05$ was set as statistic significant level.

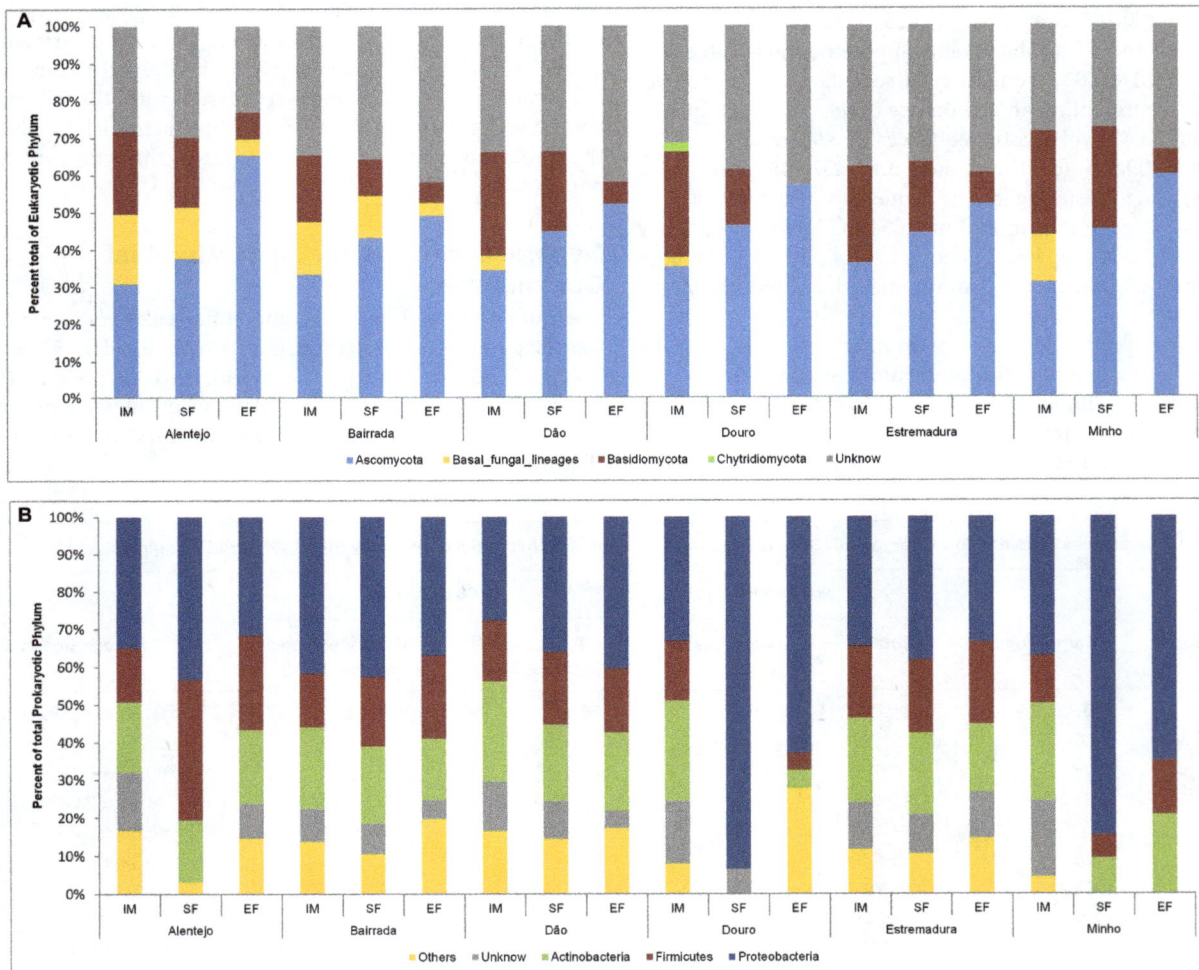

FIGURE 2 | Eukaryotic (A) and prokaryotic (B) community distribution over IM, SF, and EF from Portuguese appellations at the phylum level. Relative abundance of the eukaryotic **(A)** and prokaryotic **(B)** community through phylum analysis. For the whole figure, "Unknown" represents unclassified sequences. The prokaryotic members of rare population phyla were placed in an artificial group designed as "Others" and included Acidobacteria, Bacteroidetes, Chloroflexi, Cyanobacteria, Deinococcus-Thermus, Gemmatimonadetes, Nitrospirae, Planctomycetes, Tenericutes, and Verrumicrobia.

In all samples, the dynamics of microbial populations at phylum level were very similar. Nevertheless, the relative abundances varied along the fermentation and across Portuguese appellations (**Figure 2A**). Microorganisms belonging to Basidiomycota phylum decreased during the fermentation process. To better understand such population dynamics, the relative abundance at class level was analyzed. The entire microbial community was mostly characterized by Saccharomycetes (22.9%), Dothideomycetes (16.2%), Leotiomycetes (12.9%), Microbotryomycetes (9.6%), and Schizosaccharomycetes (7.7%; **Figure 3A**).

Concerning the prokaryotic communities, the dominant phyla were Proteobacteria (41.6%), Actinobacteria (19.2%), and Firmicutes (17.9%; **Figure 2B**). The members of under-represented phyla were grouped together in the artificial group "Other" (12.4%) and included Acidobacteria, Bacteroidetes, Chloroflexi, Cyanobacteria, Deinococcus-Thermus, Gemmatimonadetes, Nitrospirae, Planctomycetes, Tenericutes, and Verrumicrobia. As a reflection of the microbial community dynamics, and as seen in eukaryotic microorganisms, the relative abundances of all prokaryotic communities varied in both time and space. Along the spontaneous wine fermentations, it was possible to observe an increase of microorganisms belonging to the Proteobacteria phylum (**Figure 2B**), thus indicating that samples are losing their environmental characteristics. Regarding the prokaryotic classes, microorganisms from Gammaproteobacteria (27.9%), Betaproteobacteria (15.9%), Alphaproteobacteria (14.8%), Actinobacteria (13.2%), and Bacilli (11.5%) were identified (**Figure 3B**).

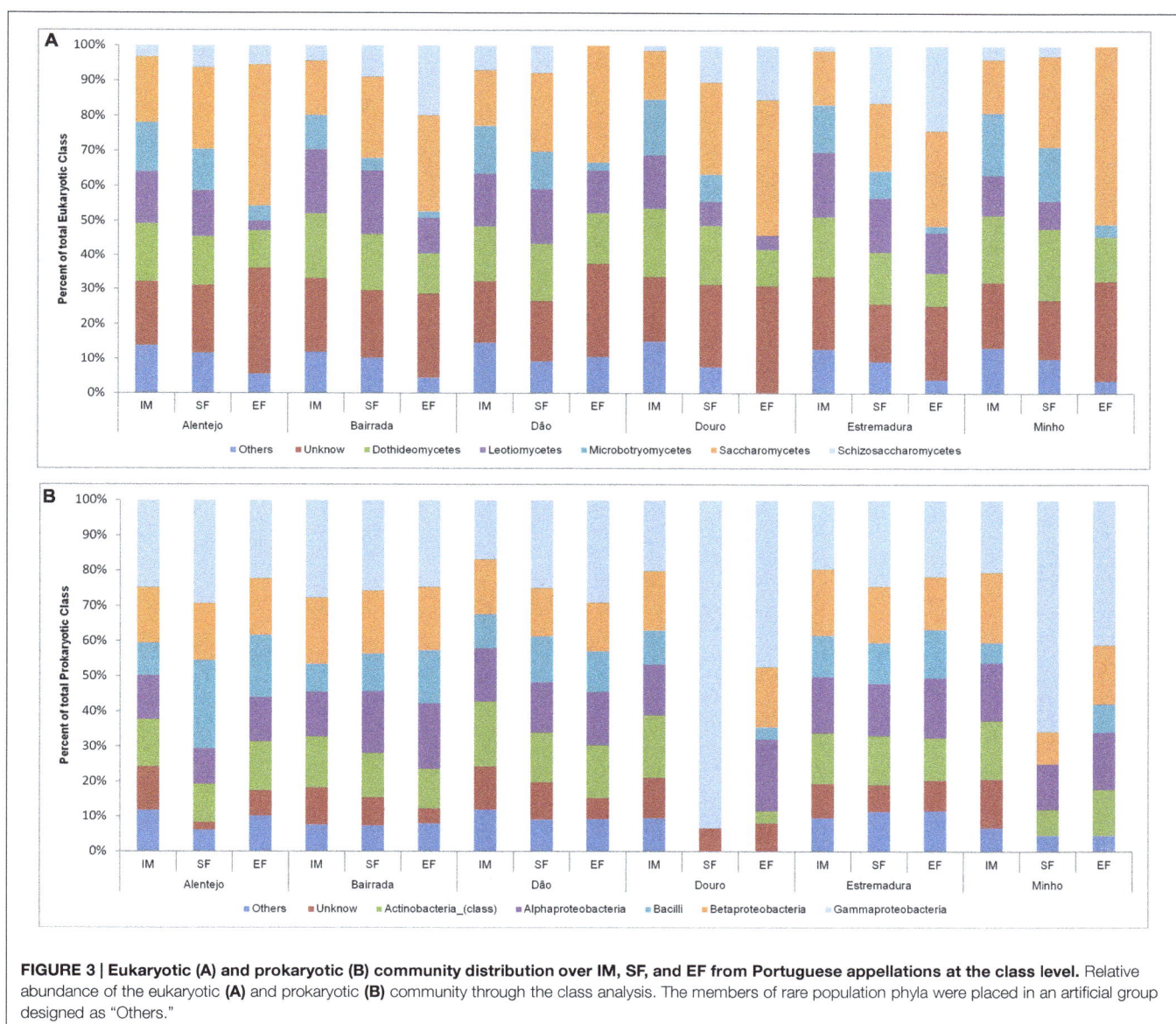

FIGURE 3 | Eukaryotic (A) and prokaryotic (B) community distribution over IM, SF, and EF from Portuguese appellations at the class level. Relative abundance of the eukaryotic **(A)** and prokaryotic **(B)** community through the class analysis. The members of rare population phyla were placed in an artificial group designed as "Others."

The Landscape of Microbial Communities Throughout Wine Fermentation

The dynamics of microbial communities present at IM, SF, and EF of samples from different Portuguese wine appellations were explored by principal coordinates analysis (PCO; **Figure 4**). For both fungal (**Figure 4A**) and bacterial communities (**Figure 4B**), samples were grouped according to their fermentative stage, where the first axis explains 48 and 52.3% of the total variation, respectively. Interestingly, SF samples were mixed with both IM and EF and, indeed this stage is a transition between IM and EF. As expected, the distribution of the microbial community composition is affected by fermentation. Significant differences (Fungi: $R_{ANOSIM} = 0.512$, $p = 0.001$; Bacteria: $R_{ANOSIM} = 0.170$, $p = 0.002$) between IM, SF, and EF samples were observed for a global test. Conversely, no significant differences were observed between SF and EF samples of the bacterial communities ($R_{ANOSIM} = 0.155$, $p = 0.954$) when analyzed by pairwise tests.

The fungal and bacterial microorganisms responsible for the similarities within each group, and the dissimilarity between different stages of fermentation, were analyzed using SIMPER analysis (**Supplementary Table S2**). The average of similarity within each group increased over the fermentation process for both fungal (IM: 39.84%; SF: 42.27%; EF: 64.19%) and bacterial community (IM: 42.64%; SF: 48.36%; EF: 46.96%). Further, the fungal communities of IM samples were mainly characterized by the environmental yeasts *Aureobasidium* and *Rhodotorula*, which contributed with 64.55% for the group similarity. Other microorganisms, such as *Hanseniaspora*,

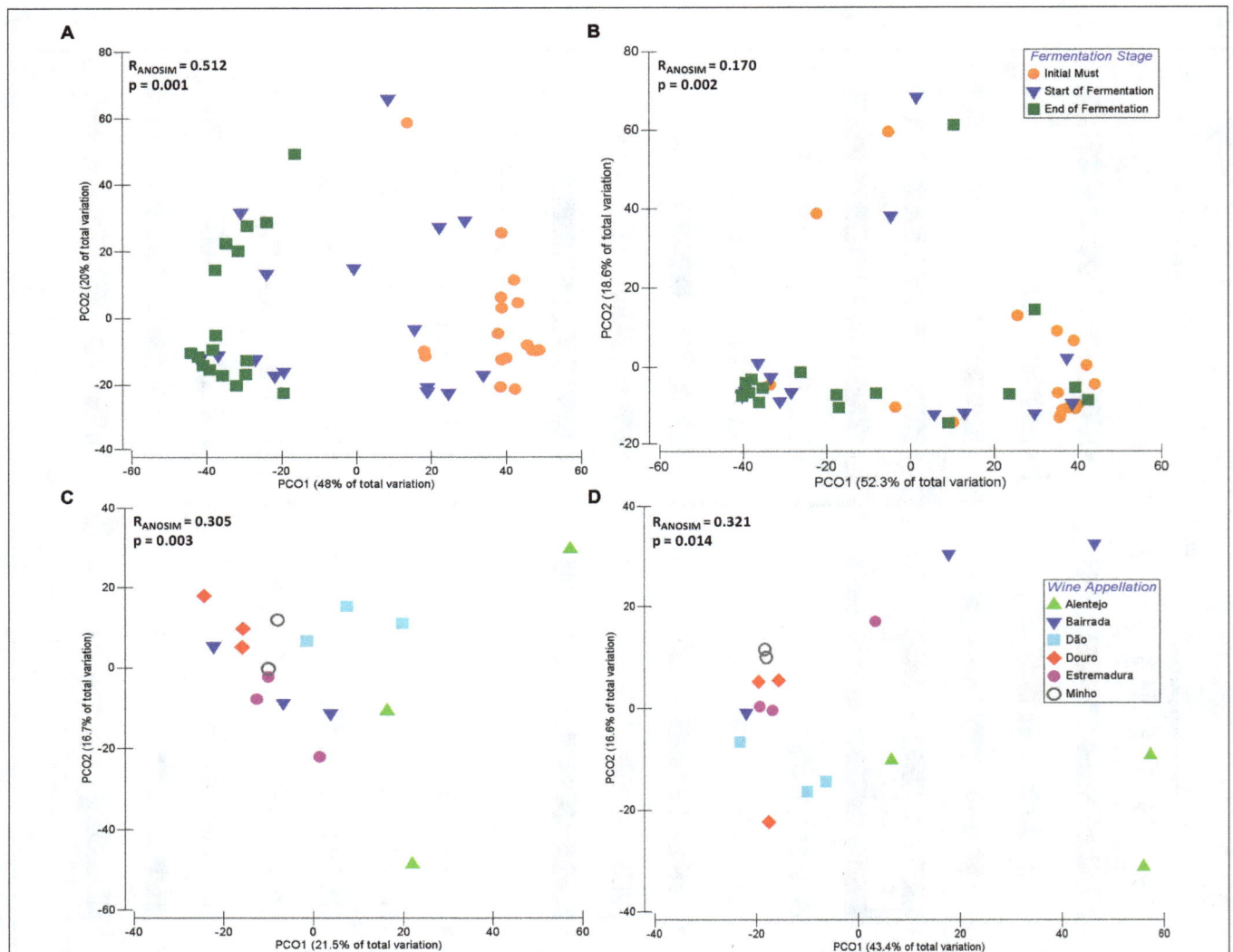

FIGURE 4 | Principal coordinate analysis (PCO) biplot diagram of microbial community during fermentation process, based on sequence abundance of eukaryotic genus and bacterial family. Principal coordinates analysis (showing the first and second components) of fungal (A) and bacterial (B) communities across the fermentation stage namely, initial musts (IM), start of fermentation (SF) and end of fermentation (EF) for Portuguese appellations. Biogeographical distribution of fungal (C) and bacterial (D) microorganisms at IM across the six Portuguese wine appellations namely, Alentejo, Bairrada, Dão, Douro, Estremadura, and Minho.

Saccharomyces, Lachancea, Botryotinia, Alternaria, Aspergillus, Metschnikowia, Filobasidiella, and *Candida* contributed with 25.80% for the group similarity. Regarding the bacterial community at IM, Enterobacteriaceae, Pseudomonadaceae, Microbacteriaceae, Comamonadaceae families contribuited with 52.68% for group similarity, followed by Oxalobacteraceae, Sphingomonadaceae, Xanthomonadaceae, Nocardioidaceae, Methylobacteriaceae, Halomonadaceae, Propionibacteriaceae, Rhodobacteraceae, Micrococcaceae, Acetobacteraceae, which all together contributed with 38.25%.

The analysis of similarity of the fungal community at SF and EF revealed that fewer microorganisms contributed to the similarity of groups when compared with IM, which is explained by the evolution of the fermentative process. In fact, the microbial community tended to be more similar and less diverse at EF. At SF, the microorganisms *Saccharomyces, Hanseniaspora, Aureobasidium,* and *Lachancea* contributed with 91.91% for group similarity, and at EF the *Saccharomyces* and *Hanseniaspora* microorganisms contributed with 91.19%. The same behavior was observed for bacterial communities where Enterobacteriaceae, Halomonadaceae, Comamonadaceae, Pseudomonadaceae, and Xanthomonadaceae families contributed with 91.44% of similarity for SF group, whereas Enterobacteriaceae, Comamonadaceae, Acetobacteraceae, Xanthomonadaceae, Pseudomonadaceae, and Oxalobacteraceae families contributed with 91.44% for EF group similarity.

Regarding the comparison between IM, SF, and EF groups of fungal communities, a higher dissimilarity value was obtained for IM vs. EF (86.53%) followed by IM vs. SF (73.84%) and SF vs. EF (53.44%), where microorganisms belonging to the *Lachancea, Saccharomyces, Hanseniaspora, Aureobasidium, Schizosaccharomyces, Candida, Metschnikowia, Torulaspora, Rhodotorula,* and *Alternaria* genera contributed for the dissimilarity of the groups. Furthermore, the diferences of the dissimilary were less pronounced for the bacterial community when compared with fungal population: IM vs EF (66.09%), IM vs SF (66.05%), and SF vs EF (50.51%). Micoorganisms belonging to the Halomonadaceae, Enterobacteriaceae, Pseudomonadaceae, Comamonadaceae, Oxalobacteraceae, Microbacteriaceae, Sphingomonadaceae, Acetobacteraceae, and Xanthomonadaceae familes were those that mostly contributed for the dissimilarity of groups (**Supplementary Table S2**).

Microbiome of Wine Appellations

In order to understand the biogeographical distribution of microbial populations, the microbiome associated with the six Portuguese appellations was individually compared for IM, SF, and EF, for both bacterial and fungal communities (**Figures 4C,D**). Significant differences were observed across wine appellations for IM samples (Fungi: $R_{ANOSIM} = 0.305$, $p = 0.003$; Bacteria: $R_{ANOSIM} = 0.321, p = 0.014$). For both fungal (**Figure 4C**) and bacterial communities (**Figure 4D**), samples were grouped according to their similarity, where the first axis explain 21.5 and 43.4% of the total variation, respectively. The SIMPER analysis (**Supplementary Table S2**) revealed that the average of similarity within each wine appellation was higher at Minho for both bacterial (76.20%) and fungal (63.21%)

communities, followed by Estremadura (50.49 and 51.99% for bacterial and fungal populations, respectively), Bairrada (40.81 and 51.77%), Douro (49.68 and 50.68%), Dão (59.74 and 45.29%), and Alentejo (51.54 and 23.98%). The SF samples (fungi: $R_{ANOSIM} = 0.060$, $p = 0.320$; bacteria: $R_{ANOSIM} = 0.073$, $p = 0.271$) and EF samples (fungi: $R_{ANOSIM} = -0.039, p = 0.596$; bacteria: $R_{ANOSIM} = 0.093$, $p = 0.199$) did not show any significant differences.

Regarding the fungal microorganisms that contributed for each wine appellation, the genus *Aureobasidium* dominated and contributed for an average of 44.39% appellations similarity (**Supplementary Table S2**). Interestingly, it was observed a regional effect on the contribution of other microorganisms: at Alentejo appellation *Lachancea* prevailed, contributing for 21.44% of region's similarity; in the Estremadura appellation *Rhodotorula* and *Botryotinia* contributed for 37.96% of the similarity; the Bairrada appellation was characterized by the presence of *Hanseniaspora* and *Ramularia,* who contributed for 18.86% of the regional similarity; the Dão appellation was characterized by the presence of microorganisms from the *Lachancea* and *Rhodotorula* genera (29.07% of similarity); within Douro appellation, *Rhodotorula* and *Erysiphe* contributed with 21.29% for the similarity; and finally, the Minho appellation was characterized by *Rhodotorula* and *Alternaria* (40% of similarity; **Supplementary Table S2**). In general, the fungal populations of IM were characterized by ubiquitous genera as *Aureobasidium, Rhodotorula, Hanseniaspora, Alternaria, Metschnikowia, Saccharomyces, Candida, Ramularia, Penicillium, Lewia, Filobasidiella, Leptosphaerulina,* and *Schizosaccharomyces,* forming the principal structure of the microbial populations (**Figure 5A**).

In SF samples, an increase of *Saccharomyces* population was observed in all regions. Nevertheless, Alentejo had the highest abundance of *Lachancea* and Minho was characterized by having the richest biodiversity, which included *Hanseniaspora, Lachancea, Metschnikowia,* and *Aureobasidium.* Expectedly at EF the dominant genus was *Saccharomyces,* but still some regional differences were observed: samples from Alentejo, Douro, and Minho presented a similar composition (*Saccharomyces* and *Lachancea*), while Bairrada and Dão were mostly composed by *Saccharomyces.* Samples from Estremadura region contained high amounts of both *Saccharomyces* and *Schizosaccharomyces.*

Regarding the bacterial community, the families of Halomonadaceae and Enterobacteriaceae contributed with 91.93% for the Alentejo appellation similarity whereas at Bairrada region, Enterobacteriaceae and Pseudomonadaceae contributed with 75.78%. At Dão appellation, Microbacteriaceae, Oxalobacteraceae, and Enterobacteriaceae contributed with 36.83% and Comamonadaceae, Enterobacteriaceae, Oxalobacteraceae, and Microbacteriaceae families with 52.35% for Douro region similarity. Finally, at Estremadura, Enterobacteriaceae, contributed with 22.47% and at Minho appellation, Oxalobacteraceae, Pseudomonadaceae, and or Enterobacteriaceae with 45.39% for the similarity. It is interesting to notice that the bacterial families responsible for the regional similarities were mostly environmental, and are not related with the oenological process.

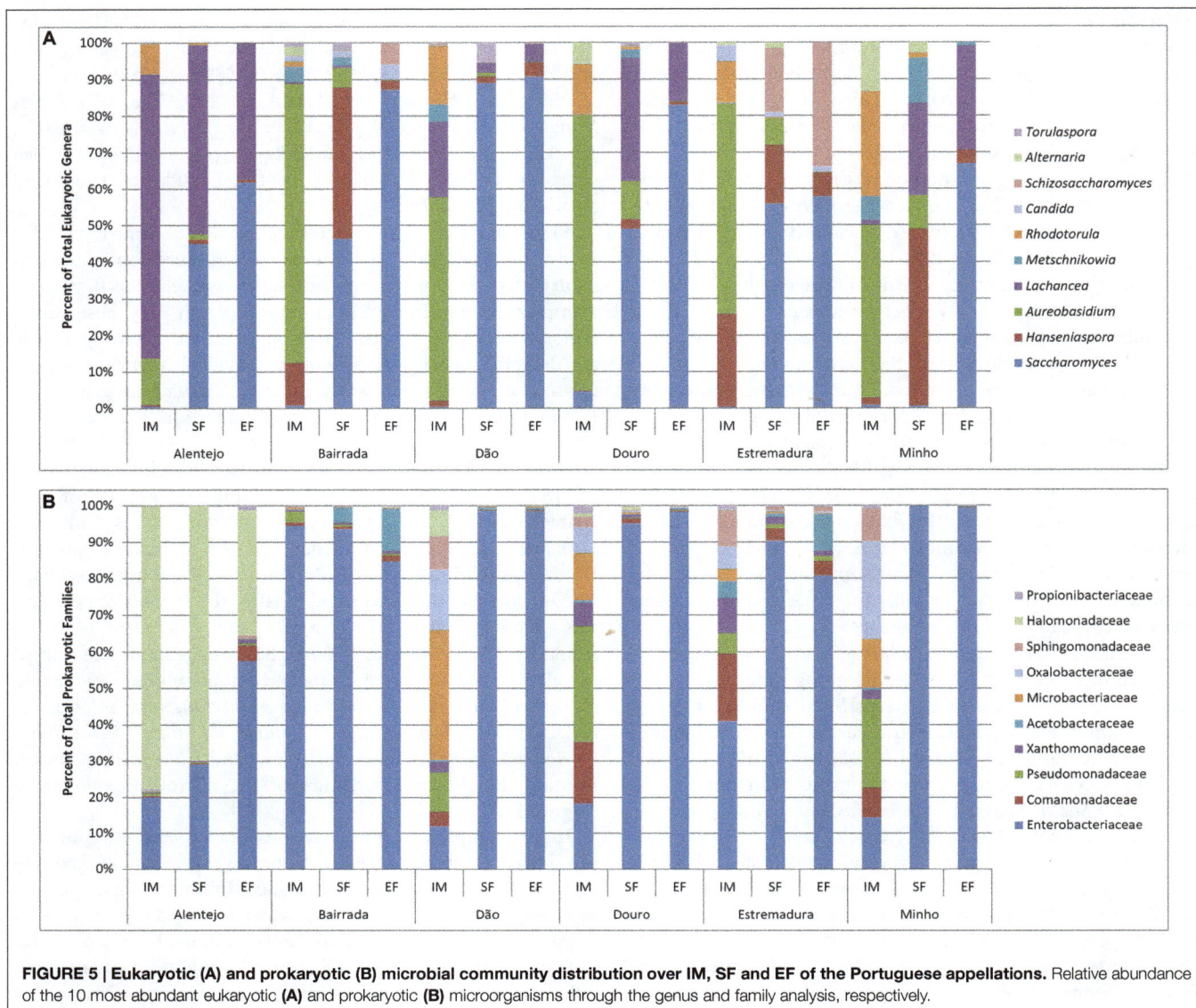

FIGURE 5 | Eukaryotic (A) and prokaryotic (B) microbial community distribution over IM, SF and EF of the Portuguese appellations. Relative abundance of the 10 most abundant eukaryotic **(A)** and prokaryotic **(B)** microorganisms through the genus and family analysis, respectively.

In general, the bacterial community was observed to differ across the appellations at IM samples. Additionally, grapes from Alentejo and Bairrada appellations presented the most distinct bacterial profiles (**Figure 5B**). Regarding SF and EF samples, Enterobacteriaceae was ubiquitous to all appellations. Bairrada and Estremadura were also characterized by high amounts of Acetobacteriaceae, while samples from Alentejo presented a unique microbiome characterized by the Halomonadaceae family (**Figure 5B**).

Regarding the most abundant bacterial family, Enterobacteriaceae, microorganisms from the genus *Pantoea* were found in all samples, whereas *Klebsiella* was only detected at IM and SF, and *Tatumella* was only identified at SF and EF samples. Also, bacteria belonging to the Microbacteriaceae family as *Curtobacterium* and *Frigobacterium* were detected in all samples and *Leifsonia* only at IM samples. Concerning all samples, the bacterial genera *Gluconobacter* (Acetobacteraceae)

and *Leuconostoc* (Leuconostocaceae) were also abundant, which was expected as they have been long related with wine fermentations. *Variovorax* (Comamonadaceae); *Carnimonas, Halotalea,* and *Zymobacter* (Halomonadaceae); *Massilia* (Oxalobacteraceae); *Pseudomonas* (Pseudomonadaceae); and *Sphingomonas* (Sphingomonadaceae) were also extensively detected in all samples.

Discussion

The aims of this work were to characterize and to compare the diversity of the microbial communities during spontaneous wine fermentations and across different wine Portuguese appellations. To achieve this, high-throughput sequencing was used to fully characterize both eukaryotic and prokaryotic communities from samples collected from six Portuguese wine regions.

Wine fermentations are known to harbor a heterogeneous population of microorganisms. In this work, a diverse set of microbial communities was identified, where the most abundant phyla were Proteobacteria and Ascomycota from prokaryotic and eukaryotic populations, respectively. As expected, a clear relationship was observed between the microbial community and fermentation stage. The biodiversity across the fermentation process decreased for both prokaryotic and eukaryotic communities as a result of the selective environment created over the spontaneous wine fermentation. Interestingly, the variations of biodiversity along this process revealed a higher impact on the fungal community structure, when compared with the bacterial populations. Furthermore, the prokaryotic populations were more diverse than the eukaryotic populations.

In this study, the most abundant eukaryotic microorganisms at IMs were *Aureobasidium* (*A. pullulans*), *Rhodothorula* (*R. nothofagi*), *Hanseniaspora* (*H.uvarum*), and *Lachancea* (*L. thermotolerans*). A diverse set of bacterial population was also uncovered, where Enterobacteriaceae (namely, *Pantoea*, and *Klebsiella*) and Pseudomonadaceae (namely, *Cellvibrio*, and *Pseudomonas*) were the most abundant families. This is in line with the previous reported by Bokulich et al. (2013), where microorganisms as *Cladosporium* spp., *A. pullulans*, *H. uvarum* were detected as the major eukaryotic population in the IMs, and as regards to prokaryotic population, Lactobacillales, Pseudomonadales, or Enterobacteriales were also identified.

The high microbial biodiversity within IM samples was mostly due to environmental microorganisms derived from vineyard. Indeed, several detected microorganisms, namely, *Botryotinia*, *Phomopsis*, *Aspergillus*, *Penicillium*, *Aureobasidium*, *Rhodotorula*, Enterobacteriaceae, or *Sphingomonas*, were previously described on grapevine leafs and grape surfaces and some of them are even refereed as inhabitant of grapes (Mills et al., 2008; Martins, 2012; Bokulich et al., 2013; Pinto et al., 2014). Also, *Saccharomyces* was detected at IMs, which suggests that this community comes from grapes, reinforcing findings from Bokulich et al. (2013), Pinto et al. (2014), and Taylor et al., 2014.

Regarding the origin of spoilage microorganisms, there has been a vivid discussion on whether or not these are present at the vineyards, where grapes are the principal source for wine contamination and deterioration (Renouf et al., 2005), or otherwise, winemaking equipment is the source of spoilage microorganisms (Couto et al., 2005). For instance, it is considered that *Dekkera/Brettanomyces*, the lactic and AAB are the most important wine spoilage microorganisms (Bartowsky et al., 2003; Beneduce et al., 2004; Cocolin et al., 2004). In this study, *Dekkera/Brettanomyces bruxellensis* was not detected, which is in line with the study of Suárez et al. (2007), who reported that this spoilage yeast is mainly present in winemaking equipment with deficient cleaning; and is opposed to the findings reported by Renouf and Lonvaud-Funel (2007). Still, these results *per se* do not yet allow for a clear conclusion on their origin. In the other hand, LAB and AAB were detected at low abundances, but *Oenococcus oeni*, a LAB extensively used to carry out the MLF, was not detected. Additionally, filamentous fungi (molds) were identified on IMs: *Alternaria*, *Aspergillus*, *Botrytis*, *Cladosporium*, *Penicillium*, or *Rhizopus*, which are undesirable

for wine quality (Toit and Pretorius, 2000). *Aspergillus* (*A. niger*) and *Penicillium* (*P. glabrum* and *P. brevicompactum*) were found in all the appellations considered in this work. However, and along fermentations, these molds disappeared, which supports the observations that they are sensitive to the wine fermentation conditions (Blesa et al., 2006).

From the IM to the wine, sequential stages of microbial development were observed, as result of fermentation activities (Fleet et al., 1984; Jolly et al., 2003). An initial growth of non-*Saccharomyces*, such as *Hanseniaspora*, *Torulaspora*, *Metschnikowia*, and *Pichia* at SF was followed by a decrease or even a disappearance of these yeasts at the EF and, conversely, the increase of *S. cerevisiae* was evidenced. A similar kinetic pattern was also observed on prokaryotic community, where in transition from IM to SF, Enterobacteriaceae family increased, and then decreased from SF to EF, specifically in Bairrada, Dão, and Estremadura appellations.

In spontaneous wine fermentations, *S. cerevisiae* was dominant despite the high abundance of *Hanseniaspora* and *Lachancea*. Yeasts associated with wine fermentation such as *Metschnikowia* (*M. pulcherrima and M. viticola*), *Torulaspora* (*T. delbrueckii*), *Schizosaccharomyces* (*S. japonicus*), *Candida* (*C. zemplinina*), *Issatchenkia* (*I. terricola*), and, less frequently, *Pichia* (*P. kluyveri* and *P. kudriavzevii*) were also detected. However, their relative abundances varied according to their appellation of origin. Indeed, each appellation presented characteristic microbial communities, with different abundances of non-*Saccharomyces* and specific patterns of microbial communities. Interestingly, *Schizosaccharomyces* (*S. japonicus*) was also detected, even at later stages, and was present at higher abundances in the Estremadura region. This yeast is characterized by having a high fermentative capacity at high temperatures (optimal growth around 30°C), and by being resistant to SO_2 and to the stringent conditions of fermentation (Torija et al., 2001). Regarding *Torulaspora delbrueckii*, it was found until EF, and it has been previously reported to survive until later stages of fermentation and to produce lower levels of acetic acid (Ciani et al., 2006). Interestingly, samples which presented higher abundance of this microorganism also generally had higher abundance of AAB namely, *Gluconobacter* (*G. oxydans*).

Among bacterial communities, during the fermentation, Enterobacteriaceae was the most abundant family (namely, *Tatumella* sp.). Nisiotou et al. (2011) also showed that Enterobacteriaceae persists in fermentation, and Ruiz et al. (2010) also confirmed its prevalence at beginning, mid and final stages of MLFs in different Spanish wineries. This raises the question if these bacteria interact with fermenting yeasts and, if so, in what degree can this microbial population influence (negatively or positively) the organoleptic proprieties of wine. The bacterial populations were found to be less dynamic than the eukaryotic populations in the later stages of fermentation process, and their geographic profiles were more similar: it was observed a clear dominance of Enterobactereaceae family at all appellations but Alentejo, where microorganisms from Halomonadaceae family were also presented with high abundance. The Bairrada and Estremadura appellations were also characterized by the presence of microorganisms from the Acetobacteraceae family. Among the LAB, high amounts of *Lactobacillus* (Lactobacillaceae), *Leuconostoc* (Leuconostocaceae), *Lactococcus*, and *Streptococcus* (Streptococcaceae) were detected. Additionally, *Facklamia* (Aerococcaceae), *Carnobacterium, Dolosigranulum, Granulicatella,* and *Trichococcus* from Carnobacteriaceae family, *Enterococcus* (Enterococcaceae) and *Weisella* as *W. cibaria* (Leuconostocaceae) were also detected, but at lower abundances. Interestingly, and with exception of *Weisella*, those specific microorganisms had not been previously isolated from musts and wines (König and Fröhlich, 2009).

To investigate whether or not there is a geographic imprint on the wine fermentation microbiome, a PCO was performed for each fermentation stage in order to evaluate differences according to wine appellation. Interestingly, significant differences ($p < 0.05$) were observed for both fungal and bacterial microbial communities at IM between wine appellations. These results are consistent with those reported by Bokulich et al. (2013), who observed differences in the microbial community structure across wine appellations from California. Over the fermentation process, the initial microbiome associated with each wine appellation disappears and, as a consequence, the biogeographic profile was lost (no significant differences were observed for SF and EF). As observed, this microbiome is characterized by the presence of environmental microorganisms, which constituted a signature of each Portuguese wine regions. Moreover, these results also suggested that the initial microbial community could strongly contribute to the uniqueness of the wines derived from each specific wine appellation. Furthermore, each wine appellation presented its own pattern of biodiversity that varied in terms of the microbial abundance. This finding is of special interest when considering the non-saccharomyces population at the SF, whom have been acknowledged for their metabolic contribution to the final wine sensorial properties (Romano et al., 2003; Jolly et al., 2014), which reinforces their role on the regional attributes of wines. These findings open new horizons to dissect how microbiomes affect wine properties and support the need to unveil the endogenous microflora of such regions and explore its natural microbial populations in order to produce valuable wines styles.

Acknowledgments

This work was carried out within the "O Microbioma índigeno das fermentações vínicas Portuguesas" and "Holiwine" projects, which were financed with funds from FEDER through "Programa Operacional Factores de Competitividade" – COMPETE and FCT – "Fundação para a Ciência e Tecnologia," with the references FCOMP-01-0124-FEDER-008749 and FCOMP-01-0124-FEDER-027411. CP is supported by a Ph.D grant from FCT, with the reference SFRH/BD/84197/2012. We are also grateful to all wine producers that collaborate with this study.

References

Barata, A., Malfeito-Ferreira, M., and Loureiro, V. (2012). The microbial ecology of wine grape berries. *Int. J. Food Microbiol.* 153, 243–259. doi: 10.1016/j.ijfoodmicro.2011.11.025

Bartowsky, E., Xia, D., Gibson, R., Fleet, G., and Henschke, P. (2003). Spoilage of bottled red wine by acetic acid bacteria. *Lett. Appl. Microbiol.* 36, 307–314. doi: 10.1046/j.1472-765X.2003.01314.x

Beneduce, L., Spano, G., Vernile, A., Tarantino, D., and Massa, S. (2004). Molecular characterization of lactic acid populations associated with wine spoilage. *J. Basic Microbiol.* 44, 10–16. doi: 10.1002/jobm.200310281

Bisson, L. (1999). Stuck and sluggish fermentations. *Am. J. Enol. Viticulture* 50, 107–119.

Bisson, L., and Butzke, C. (2000). Diagnosis and rectification of stuck and sluggish fermentations. *Am. J. Enol. Viticulture* 51, 168–177.

Blesa, J., Soriano, J., Moltó, J., and Mañes, J. (2006). Factors affecting the presence of ochratoxin a in wines. *Crit. Rev. Food Sci. Nutr.* 46, 473–478. doi: 10.1080/10408390500215803

Bokulich, N. A., Thorngate, J. H., Richardson, P. M., and Mills, D. A. (2013). Microbial biogeography of wine grapes is conditioned by cultivar, vintage, and climate. *Proc. Natl. Acad. Sci. U.S.A.* 11, E139–E148. doi: 10.1073/pnas.1317377110

Cadez, N., Zupan, J., and Raspor, P. (2010). The effect of fungicides on yeast communities associated with grape berries. *FEMS Yeast Res.* 10, 619–630. doi: 10.1111/j.1567-1364.2010.00635.x

Ciani, M., Beco, L., and Comitini, F. (2006). Fermentation behaviour and metabolic interactions of multistarter wine yeast fermentation. *Int. J. Food Microbiol.* 108, 239–245. doi: 10.1016/j.ijfoodmicro.2005.11.012

Ciani, M., Comitini, F., Mannazzu, I., and Domizio, P. (2010). Controlled mixed culture fermentation: a new perspective on the use of non-*Saccharomyces* yeasts in winemaking. *FEMS Yeast Res.* 10, 123–133. doi: 10.1111/j.1567-1364.2009.00579.x

Clarke, K. R., and Gorley, R. N. (2006). *PRIMER v6: User Manual/Tutorial.* Plymouth: PRIMER-E, 192.

Cocolin, L., Bisson, L., and Mills, D. (2000). Direct profiling of the yeast dynamics in wine fermentations. *FEMS Microbiol. Lett.* 189, 81–87. doi: 10.1016/S0378-1097(00)00257-3

Cocolin, L., Rantsiou, K., Iacumin, L., Zironi, R., and Comi, G. (2004). Molecular detection and identification of *Brettanomyces/Dekkera bruxellensis* and *Brettanomyces/Dekkera anomalus* in spoiled wines. *Appl. Environ. Microbiol.* 70, 1347–1355. doi: 10.1128/AEM.70.3.1347-1355.2004

Cole, J., Wang, Q., Cardenas, E., Fish, J., Chai, B., Farris, R. J., et al. (2009). The ribosomal database project: improved alignments and new tools for rrna analysis. *Nucleic Acids Res.* 37, D141–D145. doi: 10.1093/nar/gkn879

Couto, J., Neves, F., Campos, F., and Hogg, T. (2005). Thermal inactivation of the wine spoilage yeasts *Dekkera/Brettanomyces. Int. J. Food Microbiol.* 104, 337–344. doi: 10.1016/j.ijfoodmicro.2005.03.014

Edgar, R. (2010). Search and clustering orders of magnitude faster than BLAST. *Bioinformatics* 26, 2460–2461. doi: 10.1093/bioinformatics/btq461

Edgar, R., Haas, B., Clemente, J., Quince, C., and Knight, R. (2011). UCHIME improves sensitivity and speed of chimera detection. *Bioinformatics* 27, 2194–2200. doi: 10.1093/bioinformatics/btr381

Egas, C., Pinheiro, M., Gomes, P., Barroso, C., and Bettencourt, R. (2012). The transcriptome of bathymodiolus azoricus gill reveals expression of genes from endosymbionts and free-living deep-sea bacteria. *Mar. Drugs* 10, 1765–1783. doi: 10.3390/md10081765

Felsenstein, J. (1989). PHYLIP-Phylogeny inference package (version 3.2). *Cladistics* 5, 164–166.

Fleet, G. (2003). Yeast interactions and wine flavour. *Int. J. Food Microbiol.* 86, 11–22. doi: 10.1016/S0168-1605(03)00245-9

Fleet, G. (2008). Wine yeasts for the future. *FEMS Yeast Res.* 8, 979–995. doi: 10.1111/j.1567-1364.2008.00427.x

Fleet, G., Lafon-Lafourcade, S., and Ribéreau-Gayon, P. (1984). Evolution of yeasts and lactic acid bacteria during fermentation and storage of *Bordeaux* wines. *Appl. Environ. Microbiol.* 48, 1034–1038.

Heard, G. (1999). Novel yeasts in winemaking - looking to the future. *Food Aust.* 51, 347–352.

Jolly, N., Augustyns, O., and Pretorius, I. (2003). The effect of non-Saccharomyces on fermentation and wine quality. *S. Afr. J. Enol. Vitic.* 24, 55–62.

Jolly, N., Augustyn, O., and Pretorius, I. (2006). The role and use of non-*Saccharomyces* yeasts in wine production. *S. Afr. J. Enol. Vitic.* 27, 15–39.

Jolly, N. P., Varela, C., and Pretorius, I. S. (2014). Not your ordinary yeast: non-*Saccharomyces* yeasts in wine production uncovered. *FEMS Yeast Res.* 14, 215–237. doi: 10.1111/1567-1364.12111

König, H., and Fröhlich, J. (2009). "Lactic acid bacteria," in *Biology of Microorganisms on Grapes, in Must and in Wine*, eds H. König, G. Unden, and J. Fröhlich (Verlag Berlin Heidelberg: Springer), 3–30.

Lerm, E., Engelbrecht, L., and Du Toit, M. (2011). Selection and characterisation of *Oenococcus oeni* and *Lactobacillus plantarum* south african wine isolates for use as malolactic fermentation starter cultures. *S. Afr. J. Enol. Viticulture* 32, 280–295.

Lonvaud-Funel, A. (1999). Lactic acid bacteria in the quality improvement and depreciation of wine. *Antonie Van Leeuwenhoek* 76, 317–331. doi: 10.1023/A:1002088931106

Martins, G. (2012). *Communautés Microbiennes de la Baie de Raisin: Incidence des Facteurs Biotiques et Abiotiques.* Ph.D. theis, Université de Bourdeaux, Bourdeaux.

Mills, D., Johannsen, E., and Cocolin, L. (2002). Yeast diversity and persistence in botrytis-affected wine fermentations. *Appl. Environ. Microbiol.* 68, 4884–4893. doi: 10.1128/AEM.68.10.4884-4893.2002

Mills, D., Phister, T., Neeley, E., and Johannsen, E. (2008). "Wine fermentation," in *Molecular Techniques in the Microbial Ecology of Fermented Foods*, eds L. Cocolin and D. Ercolini (New York, NY: Springer), 162–192. doi: 10.1007/978-0-387-74520-6_6

Nisiotou, A., Rantsiou, K., Iliopoulos, V., Cocolin, L., and Nychas, G.-J. (2011). Bacterial species associated with sound and *Botrytis*-infected grapes from a Greek vineyard. *Int. J. Food Microbiol.* 145, 432–436. doi: 10.1016/j.ijfoodmicro.2011.01.017

Pinto, C., Pinho, D., Sousa, S., Pinheiro, M., Egas, C., and Gomes, A. (2014). Unravelling the diversity of grapevine microbiome. *PLoS ONE* 9:e85622. doi: 10.1371/journal.pone.0085622

Pretorius, I. (2000). Tailoring wine yeast for the new millennium: novel approaches to the ancient art of winemaking. *Yeast* 16, 675–729. doi: 10.1002/1097-0061(20000615)16:8<675::AID-YEA585>3.0.CO;2-B

Renouf, V., Claisse, O., and Lonvaud-Funel, A. (2005). Understanding the microbial ecosystem on the grape berry surface through numeration and identification of yeast and bacteria. *Aust. J. Grape Wine Res.* 11, 316–327. doi: 10.1111/j.1755-0238.2005.tb00031.x

Renouf, V., Claisse, O., and Lonvaud-Funel, A. (2007). Inventory and monitoring of wine microbial consortia. *Appl. Microbiol. Cell Physiol.* 75, 149–164. doi: 10.1007/s00253-006-0798-3

Renouf, V., and Lonvaud-Funel, A. (2007). Development of an enrichment medium to detect *Dekkera/Brettanomyces bruxellensis*, a spoilage wine yeast, on the surface of grape berries. *Microbiol. Res.* 162, 154–167. doi: 10.1016/j.micres.2006.02.006

Romano, P., Fiore, C., Paraggio, M., Caruso, M., and Capece, A. (2003). Function of yeast species and strains in wine flavour. *Int. J. Food Microbiol.* 86, 169–180. doi: 10.1016/S0168-1605(03)00290-3

Ruiz, P., Seseña, S., Izquierdo, P. M., and Palop, M. L. (2010). Bacterial biodiversity and dynamics during malolactic fermentation of Tempranillo wines as determined by a culture-independent method (PCR-DGGE). *Appl. Microbiol. Biotechnol.* 86, 1555–1562. doi: 10.1007/s00253-010-2492-8

Schloss, P., Westcott, S., Ryabin, T., Hall, J. R., Hartmann, M., Hollister, E. B., et al. (2009). Introducing mothur: open-Source, platform-independent, community-supported software for describing and comparing microbial communities. *Appl. Environ. Microbiol.* 75, 7537–7541. doi: 10.1128/AEM.01541-09

Schuller, D., Cardoso, F., Sousa, S., Gomes, P., Gomes, A. C., Santos, M. A. S., et al. (2012). Genetic diversity and population structure of *Saccharomyces cerevisiae* strains isolated from different grape varieties and winemaking regions. *PLoS ONE* 7:e32507. doi: 10.1371/journal.pone.0032507

Setati, M., Jacobson, D., Andong, U., and Bauer, F. (2012). The vineyard yeast microbiome, a mixed model microbial map. *PLoS ONE* 7:e52609. doi: 10.1371/journal.pone.0052609

Sharpton, T., Riesenfeld, S., Kembel, S., Ladau, J., O'Dwyer, J. P., Green, J. L., et al. (2011). PhylOTU: a high-throughput procedure quantifies microbial community diversity and resolves novel taxa from metagenomic data. *PLoS Comput. Biol.* 7:e1001061. doi: 10.1371/journal.pcbi.1001061

Sogin, M., Morrison, H., Huber, J., Mark Welch, D., Huse, S. M., Neal, P. R., et al. (2006). Microbial diversity in the deep sea and the underexplored "rare biosphere." *Proc. Natl. Acad. Sci. U.S.A.* 103, 12115–12120. doi: 10.1073/pnas.0605127103

Suárez, R., Suárez-Lepe, J. A., Morata, A., and Calderón, F. (2007). The production of ethylphenols in wine by yeasts of the genera *Brettanomyces and Dekkera*: a review. *Food Chemis.* 102, 10–21. doi:10.1016/j.foodchem.2006.03.030

Taylor, M. W., Tsai, P., Anfang, N., Ross, H. A., and Goddard, M. R. (2014). Pyrosequencing reveals regional differences in fruit-associated fungal communities. *Environ. Microbiol.* 16, 2848–2858. doi: 10.1111/1462-2920.12456.

Toit, M., and Pretorius, I. (2000). Microbial spoilage and preservation of wine: using weapons from nature's own arsenal - a review. *S. Afr. J. Enol. Viticulture* 21, 74–96.

Torija, M., Rozès, N., Poblet, M., Guillamón, J., and Mas, A. (2001). Yeast population dynamics in spontaneous fermentations: comparison between two different wine-producing areas over a period of three years. *Antonie Van Leeuwenhoek* 79, 345–352. doi: 10.1023/A:1012027718701

Van Leeuwen, C., and Seguin, G. (2006). The concept of terroir in viticulture. *J. Wine Res.* 17, 1–10. doi: 10.1080/095712 6060063

Vaz-Moreira, I., Egas, C., Nunes, O., and Manaia, C. (2011). Culture-dependent and culture-independent diversity surveys target different bacteria: a case study in a freshwater sample. *Antonie Van Leeuwenhoek* 100, 245–257. doi: 10.1007/s10482-011-9583-0

White, T., Bruns, T., Lee, S., and Taylor, J. (1990). "Amplification and direct sequencing of fungal ribosomal RNA genes for phylogenetics," in *Pcr Protocols: A Guide to Methods and Applications*, eds M. Innis, D. Gelfand, J. Sninsky, and T. White (New York, NY: Academic Press), 315–322.

Zoecklein, B., Williams, J., and Duncan, S. (2000). Effect of sour rot on the composition of white riesling (*Vitis vinifera* l.) grapes. *Small Fruits Rev.* 1, 63–77. doi: 10.1300/J301v01n01_08

Conflict of Interest Statement: The authors declare that the research was conducted in the absence of any commercial or financial relationships that could be construed as a potential conflict of interest.

7

Staphylococcus aureus strains associated with food poisoning outbreaks in France: comparison of different molecular typing methods, including MLVA

Sophie Roussel*, Benjamin Felix, Noémie Vingadassalon, Joël Grout, Jacques-Antoine Hennekinne, Laurent Guillier, Anne Brisabois and Fréderic Auvray

Université Paris-Est, ANSES, Food Safety Laboratory, European Union Reference Laboratory for Coagulase Positive Staphylococci, Maisons-Alfort, France

Edited by:
Javier Carballo,
University of Vigo, Spain

Reviewed by:
Kiiyukia Matthews Ciira,
Jomo Kenyatta University
of Agriculture and Technology, Kenya
Roberto Mauricio Vidal,
Universidad de Chile, Chile

*Correspondence:
Sophie Roussel,
ANSES, Laboratoire de sécurité des
aliments, 14 Rue Pierre et Marie
Curie, 94701 Maisons-Alfort-Cedex,
France
sophie.roussel@anses.fr

Staphylococcal food poisoning outbreaks (SFPOs) are frequently reported in France. However, most of them remain unconfirmed, highlighting a need for a better characterization of isolated strains. Here we analyzed the genetic diversity of 112 *Staphylococcus aureus* strains isolated from 76 distinct SFPOs that occurred in France over the last 30 years. We used a recently developed multiple-locus variable-number tandem-repeat analysis (MLVA) protocol and compared this method with pulsed field gel electrophoresis (PFGE), *spa*-typing and carriage of genes (*se* genes) coding for 11 staphylococcal enterotoxins (i.e., SEA, SEB, SEC, SED, SEE, SEG, SEH, SEI, SEJ, SEP, SER). The strains known to have an epidemiological association with one another had identical MLVA types, PFGE profiles, spa-types or *se* gene carriage. MLVA, PFGE and *spa*-typing divided 103 epidemiologically unrelated strains into 84, 80, and 50 types respectively demonstrating the high genetic diversity of *S. aureus* strains involved in SFPOs. Each MLVA type shared by more than one strain corresponded to a single spa-type except for one MLVA type represented by four strains that showed two different-but closely related-spa-types. The 87 enterotoxigenic strains were distributed across 68 distinct MLVA types that correlated all with *se* gene carriage except for four MLVA types. The most frequent *se* gene detected was *sea*, followed by *seg* and *sei* and the most frequently associated *se* genes were *sea-seh* and *sea-sed-sej-ser*. The discriminatory ability of MLVA was similar to that of PFGE and higher than that of spa-typing. This MLVA protocol was found to be compatible with high throughput analysis, and was also faster and less labor-intensive than PFGE. MLVA holds promise as a suitable method for investigating SFPOs and tracking the source of contamination in food processing facilities in real time.

Keywords: food poisoning outbreaks, *Staphylococcus aureus*, MLVA, PFGE, spa-typing, enterotoxin genes, genetic diversity

Abbreviations: CPS, coagulase-positive staphylococci; EURL, European Union Reference Laboratory; NRL, National Reference Laboratory; SEs, staphylococcal enterotoxins; SFPO, staphylococcal food poisoning outbreaks.

Introduction

Staphylococcal food poisoning is one of the most common food-borne diseases worldwide (Yu et al., 2007; Kadariya et al., 2014). It results from the ingestion of SEs preformed in food and produced by enterotoxigenic strains of CPS (Argudin et al., 2010). SEs are frequent causes of food-borne outbreaks in Europe (EFSA-ECDC, 2013). Two types of SFPOs can be differentiated. Outbreaks for which the evidence implicating a particular food vehicle is strong, based on the assessment of all available data, are referred to as "strong-evidence SFPO," whereas outbreaks for which no particular food vehicle is suspected or where the evidence implicating a particular food vehicle is weak are referred to as "weak-evidence SFPO" (EFSA, 2011).

Among the seven described species belonging to the CPS group, *Staphylococcus aureus* ssp. *aureus* is the main causative agent of SFPOs. To date, 21 SEs have been described: SEA to SElV all possess superantigenic activity whereas only a subset of SEs (i.e., SEA to SEI, SER, SES, and SET) are emetic (Ono et al., 2008). Out of the 21 SEs, 11 (i.e., SEA, SEB, SEC, SED, SEE, SEG, SEH, SEI, SEJ, SEP, SER) are suspected to cause SFPOs (Hennekinne et al., 2011).

Few data is available on the genetic diversity of the strains isolated from SFPOs. Among the molecular methods available, pulsed field gel electrophoresis (PFGE) and *Staphylococcus* protein A gene (*spa*) typing have been extremely helpful in short-term investigations and identification of SFPOs (Chiou et al., 2000; Shimizu et al., 2000; Wei and Chiou, 2002; Strommenger et al., 2006; Dyer et al., 2007; Hallin et al., 2007; Kerouanton et al., 2007; Kellermann et al., 2008; Ostyn et al., 2010; Wattinger et al., 2012; Chiang et al., 2014). Although PFGE is highly discriminatory, it remains a time-consuming and labor intensive method. It also requires highly skilled operators and there are no standardized reagents. Moreover, profile interpretation requires several subjective decisions, increasing the variability of the profiles and leading to possible uncertainty about profiles relatedness (Cookson et al., 1996; Murchan et al., 2003).The advantages of *spa*-typing are its excellent inter-laboratory reproducibility, the portability of the data and its flexible analysis throughput (Koreen et al., 2004; Aires-de-Sousa et al., 2006; Cookson et al., 2007). However, this method is less discriminatory than PFGE for the characterization of food isolates (Babouee et al., 2011).

Identification of SE-encoding genes (*se* genes) in isolated strains represents a complementary approach for investigating SFPOs. All known *se* genes are located on mobile genetic elements, including the νSaβ genomic island which contains the enterotoxin gene cluster known as *egc* (carrying *seg* and *sei*), *S. aureus* pathogenicity islands (SaPIs; carrying *seb* and *sec*), prophages (carrying *sea*, *see,* and *sep*), and plasmids (carrying *sed*, *sej*, and *ser*; Argudin et al., 2012). Many PCR assays have been developed to detect *se* genes in *S. aureus* strains isolated from contaminated foods (Martin et al., 2004; Morandi et al., 2007; Kadariya et al., 2014). Screening for *se* genes in the strains involved in SFPOs is useful in two ways. First, the identified *se* gene may correspond to the type of SE detected in food, thus confirming the result obtained by an immuno-enzymatic method (Ostyn et al., 2010). Second, the *se* gene identified may correspond to a type of SE known to be emetic, but for which no detection method is available, suggesting the involvement of the corresponding toxin in the outbreak (Kerouanton et al., 2007).

The ANSES Laboratory for Food Safety is the French NRL and the EURL for CPS, including *S. aureus* and their toxins. One of the EURL activities is to develop and evaluate new molecular methods for bacterial typing and to transfer them to the European NRL network. Simultaneously to the screening for enterotoxins in suspected food, staphylococcal isolates are characterized using (i) spa-typing (ii) PFGE and (iii) a multiplex PCR assay for the detection of *se* genes coding for 11 SEs. Given the limitations described above, there is still a need for an alternative typing method that would be as discriminatory as PFGE and as portable as *spa*-typing, at a low cost.

Multiple-locus variable-number tandem-repeat analysis (MLVA) is based on PCR amplification and size analysis of DNA regions containing variable numbers of tandem repeats (VNTRs). MLVA assays offer fast typing of various bacteria, with high resolution (Lindstedt et al., 2013). An assay based on eight VNTR loci was applied on a panel of (i) 1781 *S. aureus* strains isolated from animal and patients (Schouls et al., 2009) and (ii) 78 strains related to SFPOs, in China, between 2010 and 2012 (Lv et al., 2014). Another MLVA assay targeting 14 loci was used in a survey of 309 strains including clinical methicillin-resistant *S. aureus* (MRSA) isolates and nasal carriage staphylococcal isolates (Pourcel et al., 2009). Finally, Sobral et al. (2012) proposed a third MLVA protocol based on the detection of 16 VNTR loci, including eight from Schouls et al. (2009) and eight from Pourcel et al. (2009). This protocol was implemented for the characterization of a panel (i) of 251 strains isolated primarily from humans and animals and also, to a lesser extent, from food and food poisoning samples (Sobral et al., 2012) and (ii) of 152 strains isolated from cases of bovine, ovine and caprine mastitits in France (Bergonier et al., 2014).

The aim of this study was to analyze the genetic diversity of a panel of *S. aureus* strains associated with SFPOs that occurred in France over the past 30 years. More specifically, we assessed the diversity of strains implicated in each outbreak and compared strains obtained from distinct outbreaks. MLVA data generated using the recent protocol of Sobral et al. (2012) were compared with those obtained by PFGE, spa-typing, and *se* gene detection. In light of our results, we discuss the usefulness of MLVA for routine typing of *S. aureus*, in terms of discriminatory power, and for investigating SFPOs.

Materials and Methods

Description of the Molecular Database

The French NRL has established a large collection of strains isolated from the main food production sectors throughout various French regions, over the past 30 years. This collection also includes clinical strains mainly obtained during collaborative research projects. All the strains have been typed by spa-typing and PFGE and characterized with regard to their *se* genes. The NRL molecular typing database (BioNumerics software, V 7.1,

Applied Maths, Sint-Martens-Latem, Belgium) centralizes the epidemiological information, genotype and phenotype data for all the strains.

Strain Panel

A panel of 112 strains isolated from 76 distinct SFPOs that occurred in France from 1981 to 2009 was selected for this study (**Table 1**). Out of these 112 strains, 13 strains were considered as epidemiologically related because they originated from four distinct "strong evidence" SFPOs (no "3," "8," "20," "102"; **Table 2**). The epidemiological data regarding these four SFPOs were collected by the local health authorities using interviews or questionnaires. At the same time, tracing back of incriminated food was performed by the local services of the French Ministry in charge of agriculture and food. Three SFPOs ("3," "8," "20") included food and clinical strains isolated within one region. The fourth SFPO ("102") included only food strains isolated from three different regions in France. For this latter SFPO, a soft cheese made from unpasteurised cow's milk was identified as the common and single source (Ostyn et al., 2010).

The 99 remaining strains were isolated from 72 different "weak evidence" SFPOs that occurred in different locations in France between 1981 and 2009 (**Table 1**). Out of these 72 "weak evidence" SFPOs, 53 involved only one strain, and the remaining 19 involved between two and four strains isolated from different food samples (**Table 3**).

To compare the discriminatory power of spa-typing, MLVA and PFGE, 43 additional strains were included in the panel. These 43 strains were selected according to their PFGE profiles to represent the genetic diversity observed in the molecular database. Then, we selected a total panel of 146 epidemiogically unrelated isolates comprising one strain of each of the four "strong evidence" SFPO (i.e., four isolates), 99 strains related to each of the "weak evidence" SFPOs (99 isolates), four strains related to four outbreaks that occurred in Belgium, one strain related to one outbreak in Japan (Omoe et al., 2003; Ono et al., 2008), 31 strains isolated from research projects and seven strains isolated from monitoring programs. This panel included food strains ($n = 122$ with 112 isolated from SFPOs) and strains isolated from either human cases ($n = 19$) or animals ($n = 5$).

DNA Extraction

All strains were stored in cryobeads at $-80°C$. They were cultured overnight at $37°C$ in brain heart infusion (BHI), isolated on a non-selective medium (Milk Plate Count Agar) and incubated at $37°C$ for 24 h, prior to extraction of total DNA. DNA extraction was performed using the InstaGene kit (Bio-Rad, Marnes-la-Coquette, France) according to the manufacturer's recommendations. DNA concentrations were adjusted approximately to 100 ng/μl using a Nanodrop1000 spectrophotometer (spectrophotometer, Thermo scientific, Wilmington, DE, USA).

Detection of *se* Genes Using the EURL Multiplex PCR Assay

Two multiplex PCR systems were used to detect the genes of 11 types of enterotoxins, i.e., *sea, seb, sec, sed, see, seg, seh,*

sei, sej, sep, and *ser*. The primers used to amplify the *sea, seb, sec,* and *sed* genes were designed by Mehrotra et al. (2000), those for *seg, seh, sei, sej,* and *sep* by Bania et al. (2006), those for *see* by Sharma et al. (2000), and those for *ser* by Chiang et al. (2008). They were synthesized by Eurofins (MWG Operon, France).

For the multiplex reaction targeting *sea, seb, sec, sed, see,* and *ser* genes, the 25 μl reaction mixture contained 1 U Fast Start Taq DNA polymerase (Roche, Diagnostics, Meylan, France), 2.5 mM MgCl$_2$, 0.2 mM dNTPs, 1X PCR buffer, 0.2 μM of each primer for *sea, seb, sec, ser*, 0.8 μM of each primer for *sed*, 0.6 μM of each primer for *see* and 2 μl of DNA. PCR was performed on a Veriti® PCR cycler (Applied Biosystems, Courtaboeuf, France). The thermal cycle included an initial denaturation at $94°C$ for 3 min, followed by 35 cycles of denaturation at $94°C$ for 30 s, annealing at $56°C$ for 40 s, extension at $72°C$ for 1 min 30, and a final extension at $72°C$ for 7 min. The conditions for the multiplex targeting *seg, seh, sei, sej,* and *sep* genes were as described above, except that the annealing step was performed at $53°C$ for 40 s and the reaction mixture included 0.8 μM of each primer for *seg, sei, sej, sep* and 0.4 μM of each primer for *seh*. DNA of each isolate was tested by polymerase chain reaction (PCR) targeting the ribosomal RNA 23S gene region specific for *S. aureus* (Kerouanton et al., 2007).

Five reference *S. aureus* strains (i.e., FRIS6, 374F, FRI137, HMPL280, FRI326) were used as positive controls. The PCR products were separated by electrophoresis in a 2% agarose gel and visualized using the Gel Doc EQ apparatus (Bio-Rad).

MLVA Typing

DNA samples were diluted in molecular grade water to obtain solutions at 10 ng/μl. They were used as DNA templates for PCR amplification according to the protocol described by Sobral et al. (2012) with minor modifications to the amplification program: DNA template concentration was set at 10 ng/μl (instead of 5 ng/μl in Sobral's protocole), touchdown PCR and long range PCR were both set at 18 thermal cycles (instead of 15). Samples were loaded onto an ABI3500® capillary sequencer using a 50 cm capillary filled with performance-optimized polymer 7 (Applied Biosystems) at $60°C$ for 6200 s with a running voltage of 12 kV, and an injection time and voltage of 10 s and 1.6 kV, respectively.

For each multiplex reaction, 2 μl of purified PCR product was combined with 7.75 μl HiDi formamide and 0.25 μl GS1200LIZ (Applied Biosystems). Samples were loaded onto an ABI3500® capillary sequencer using a 50 cm capillary filled with performance-optimized polymer 7 (Applied Biosystems) at $60°C$ for 6200 s with a running voltage of 12 kV, and an injection time and voltage of 10 s and 1.6 kV, respectively.

Each run included a negative (water) control to ensure the absence of contamination and a positive control to verify the PCR reaction.

From the panel of 112 strains, 12 strains related to "weak evidence" SFPOs (no 431G, 360F, 338E, 419G, 372F, 402F, 353E, 301E, 384F, 339E, 363F, 399F) that had previously been tested by MLVA by Sobral et al. (2012) were used here as positive controls.

TABLE 1 | Description of the 76 SFPOs that occurred in France from 1981 to 2009.

SFPO no.	Year	Incriminated food	Number of cases	SE detected in food	Number of strains analyzed per SFPO	SFPO assessment	SFPO reference in Kerouanton et al. (2007)
3	1999	Chocolate milk	NA[a]	SEA	3	Strong evidence	18
4	2000	Raw sheeps' milk cheese	NA	SEA	1	Weak evidence	20
5	1985	Soft cheese	NA	SEA, SEB, SED	1	Weak evidence	7
6	1985	Soft cheese	NA	SEB	1	Weak evidence	8
7	1989	Chicken	NA	SEC	1	Weak evidence	12
8	2000	Sliced pork	NA	SEA, SED	3	Strong evidence	21
9	1981	Fresh cheese	NA	NT	2	Weak evidence	
10	1981	Raw milk semi-soft cheese	NA	SEA, SED	1	Weak evidence	1
11	2000	Soft cheese	NA	NT[b]	1	Weak evidence	
12	1986	Sheeps' cheese	NA	NT	1	Weak evidence	
15	1994	Lasagna	NA	NT	1	Weak evidence	
16	1987	Unknown	NA	NT	1	Weak evidence	
17	1987	Milk	NA	NT	1	Weak evidence	
18	1992	Potato and rice salad	NA	SEA	1	Weak evidence	13
19	2001	Minced lamb meat	NA	NT	1	Weak evidence	
20	2000	Mixed salad	NA	SEC	3	Strong evidence	19
21	2001	Sliced soft cheese	NA	NT	1	Weak evidence	27
23	2001	Roasted pork	NA	SED	3	Weak evidence	28
24	2001	Leg of lamb	NA	SEA	1	Weak evidence	26
25	1983	Raw milk soft cheese	NA	ND[c]	1	Weak evidence	5
26	1983	Meat	NA	SEA, SED	1	Weak evidence	6
28	1987	Cake	NA	SEA	1	Weak evidence	10
29	1987	Strawberry tart	NA	NT	1	Weak evidence	
32	1983	Cooked beef	NA	SEA	1	Weak evidence	2
33	1983	Raw milk semi-soft cheese	NA	SEA, SED	1	Weak evidence	3
35	2001	Cream	NA	SEA	1	Weak evidence	25
36	1997	Nougatine	NA	SEA	1	Weak evidence	14
37	1988	Spaghetti	NA	SEA	1	Weak evidence	11
39	2001	Pancakes	NA	SEA, SED	2	Weak evidence	23
40	2001	Chocolate cake	NA	SEA, SED	1	Weak evidence	24
41	2001	Cooked rice	NA	NT	1	Weak evidence	
42	2001	Raw milk semi-soft cheese	NA	ND	4	Weak evidence	29
43	1997	Raw milk cheese	NA	SEA, SED	2	Weak evidence	15
45	1998	Raw milk semi-soft cheese	NA	ND	1	Weak evidence	17
46	2002	Raw sheep milk cheese	NA	SEA	1	Weak evidence	30
47	2002	Potted meat	NA	SEA	2	Weak evidence	31
49	2002	Cheese	NA	SEA	2	Weak evidence	
50	2002	Raw ham smoked	4	ND	1	Weak evidence	
52	2003	Custard topped with caramelized sugar	20	SEA, SED	1	Weak evidence	
53	2002	Custard topped with caramelized sugar	NA	NT	1	Weak evidence	
55	2003	Hard cheese made from raw milk	NA	NT	3	Weak evidence	
56	2004	Semi-solft cheese made from raw milk	5	ND	1	Weak evidence	
57	2004	Cheese	NA	SEA[d]	1	Weak evidence	
58	2004	Hard cheese made with raw milk	3	ND	1	Weak evidence	
59	2005	Strawberry shortcake	NA	SEA	1	Weak evidence	
60	2005	Potato salad	NA	SEA	2	Weak evidence	
61	2005	Chicken drumstick; Chicken drumstick Indian style	> 2	SEA	2	Weak evidence	

TABLE 1 | Continued

SFPO no.	Year	Incriminated food	Number of cases	SE detected in food	Number of strains analyzed per SFPO	SFPO assessment	SFPO reference in Kerouanton et al. (2007)
62	2005	Cake	NA	SEA	2	Weak evidence	
63	2005	Mussels and shrimp	NA	SEA	1	Weak evidence	
64	2006	Cheese	4	ND	3	Weak evidence	
65	2006	Coconut pastry	11	SEA, SED	2	Weak evidence	
66	2006	Smoked dry sausage	NA	ND	1	Weak evidence	
69	2007	Corned beef hash, cottage pie	3	ND	1	Weak evidence	
70	2007	Sauerkraut	3	NT	1	Weak evidence	
71	2007	Meat raviolis	NA	NT	1	Weak evidence	
72	2007	Minced chicken vetetable mix for sandwich	10	SEA, SEC	1	Weak evidence	
73	2007	Semolina, vegetables and sweet red peper; pepper and vegetables	22	SEA	2	Weak evidence	
80	2007	Pasta	NA	NT	1	Weak evidence	
81	2007	Semi-solft cheese	NA	NT	2	Weak evidence	
82	2007	Cheese made from raw milk	NA	ND	1	Weak evidence	
83	2007	Coconut pastry frozen	3	SEA	3	Weak evidence	
84	2007	Goat milk	NA	NT	2	Weak evidence	
85	2007	Mixed salad	20	SEA, SEC	1	Weak evidence	
86	2008	Duck liver	NA	NT	1	Weak evidence	
87	2008	Soft cheese	5	SED	4	Weak evidence	
88	2008	Ham hock leftovers	2	ND	1	Weak evidence	
91	2008	Pasta salad	100	ND	1	Weak evidence	
92	2008	Baked salmon in foil	47	SEA	2	Weak evidence	
94	2008	Tuna, mayonnaise, salad	15	SEA	1	Weak evidence	
96	2009	Ripened sheeps' cheese made with raw milk	27	ND	1	Weak evidence	
97	2009	Chicken filet	10	SEA	1	Weak evidence	
98	2009	Fruit pie with cream	4	SEA	1	Weak evidence	
99	2009	Mackerel mustard sauce	21	SEA, SEC	1	Weak evidence	
100	2009	Boned ham	3	SEA	1	Weak evidence	
101	2009	Spare rib smoked	NA	NT	1	Weak evidence	
102[e]	2009	Soft cheese	23	SEE	4	Strong evidence	

[a]NA, not available; [b]NT, not tested; ND, not detected (Se detected at a concentration inferior to 20 ng/ml); [c]ND, not detected; [d]SEs were detected at a concentration of between 20 and 100 ng/ml; [e](Ostyn et al., 2010 #299).

Amplification products were electrophoresed twice in independent runs. At least two independent PCRs were performed from a given DNA extract of the reference strains.

Data Analysis

The products of both multiplex PCR amplifications were resolved by capillary electrophoresis, and the alleles from each of the 16 targeted loci were automatically identified. Each VNTR locus was identified according to specific fluorescent dyes and automatically assigned to a DNA fragment size by the GeneMapper software (Applied Biosystems). This size was then converted into an allele designation according to the number of repeats found on the fragment, in associated with a quality index. The typing data file was imported into the NRL molecular database.

Minimum spanning trees were constructed using a categorical coefficient and unweighted pair group method with arithmetic mean (UPGMA) clustering. Allele designations and nomenclature were used according to Sobral et al. (2012). Partial repeats were rounded down to the closest half decimal (e.g., 1.0, 1.5, 2.0).

Sequence Verification

Any new alleles of unexpected size were sequenced. The loci and flanking regions were amplified in both directions with high-fidelity HotStart Taq Polymerase (Roche Diagnostics). Amplification products were sequenced by Eurofins (MWG Operon, France). The sequence analysis was performed into the NRL molecular typing database.

PFGE Typing

The initial step of PFGE as defined by the EURL for CPS protocol was performed according to the protocol described by Chung et al. (2000) with minor modifications. Briefly, strains were cultured in liquid BHI (instead of liquid TSB medium). Cell density was determined from a 2 ml suspension (instead

TABLE 2 | Typing data on the 13 strains related to four distinct "strong evidence" SFPOs.

SFPO no.	Strain no.	Strain origin	Food product and batch number	*Spa* type	Smal-pulsotype	PCR *se* genes	MLVA profile
3	255D	Clinical	Chocolate milk	t127	21	*sea, seh*	11
	256D	Clinical		t127	21	*sea, seh*	11
	257D	Food		t127	21	*sea, seh*	11
8	355E	Clinical	Pork ribs	t008	91	*sea, sed, sej*	43
	354E	Clinical		t008	91	*sea, sed, sej*	43
	349E	Food		t008	91	*sea, sed, sej*	43
20	377F	Clinical	Mixed salad	t156	44	*sec*	79
	378F	Clinical		t156	44	*sec*	79
	293E	Food		t156	44	*sec*	79
102	09CEB319STA	Food	Soft cheese 3412	t4461	155	*se*	55
	09CEB319STA	Food	Soft cheese 3402	t4461	155	*se*	55
	09CEB314STA	Food	Soft cheese 3403	t4461	155	*se*	55
	09CEB329STA	Food	Soft cheese 3405	t4461	155	*se*	55

TABLE 3 | Multiple-locus variable-number tandem-repeat analysis (MLVA) and pulsed field gel electrophoresis (PFGE) typing data for the 46 strains originating from the 19 "weak evidence" SFPOs including several strains.

SFPO no.	Number of strains	Total number of distinct MLVA types	Total number of distinct PFGE types
42	4	4	2
87	4	4	4
23	3	1	1
55	3	2	2
83	3	3	3
64	3	2	1
81	2	2	2
65	2	1	1
49	2	2	1
62	2	2	1
47	2	1	1
73	2	1	1
61	2	1	1
84	2	2	2
60	2	2	2
43	2	1	1
9	2	2	2
92	2	2	2
39	2	2	2
Total	**46**	**37**	**32**

of 0.5 ml) at 600 nm (instead of 620 nm). The agarose plugs were prepared in TE buffer (instead of PIV) and the PIV buffer contained 2 mM Tris-HCL and 1 M NaCl (instead of 10 mM Tris, 1 M NaCl). Plug shape was cubic (instead of circular) and EC buffer was incubated for 3 h (instead of 5 h).

The protocol for the subsequent steps of PFGE was based on the recommendations of the Harmony typing group (Murchan et al., 2003). The reference standard, *S. aureus* NCTC 8325 *Sma*I profile, was loaded in every fifth or sixth lane. The total running time was 20 h, the first-block switch time was 5–15 s for 8.5 h, and the second-block switch time was 15–60 s for 11.5 h. The voltage applied for the run was 6 V/cm. The CHEF DRIII system (Bio-Rad) was used, with an included angle of 120° and a linear ramp factor.

Gels were stained for 30 min in a 400 ml ultra- pure sterile water solution containing ethidium bromide at 10 mg/ml and banding profiles were visualized under UV light, using the Gel Doc Eq system and Quantity One software (Bio-Rad). DNA profiles were analyzed in the NRL molecular typing database. PFGE pulsotypes were considered as different if there was at least one band different between them (Barrett et al., 2006). Each PFGE profile was arbitrarily assigned to a pulsotype number.

Spa-Typing

Spa-typing was performed as previously described by Shopsin et al. (1999) and Aires-de-Sousa et al. (2006).

A strain already known for its *spa*-type (*S. aureus* Mu50, spa-type t002, Kuroda et al., 2001) was used as a positive control and a reaction without DNA was included within each run as a negative control. The PCR products were migrated on a 2% agarose gel and visualized using the Gel Doc EQ apparatus (Bio-Rad). They were sequenced by Eurofins (Esberg, Germany), on both DNA strands. The sequences were analyzed using BioNumerics software which provides a fully automated workflow, from import of raw sequencer trace files to assignment of repeat codes and *spa* types using the plug-in *spa*-typing which connects to SeqNet/Ridom Spa Server[1]. Each new base composition of the polymorphic repeat found in a strain was assigned a unique repeat code. The succession of repeats in a given strain determines the strain's *spa* type. New *spa*-types were submitted to the SeqNet server[2].

Epidemiological Concordance, Discriminatory Power, and Congruence of the Typing Methods

The epidemiological concordance of PFGE, *spa*-typing and MLVA was assessed by testing their capacity to recognize the homogeneity of 13 strains related to four distinct "strong evidence" SFPOs (**Table 2**) in the same epidemiological groups.

---

[1]http://spaserver.ridom.de/

[2]http://www.seqnet.org/spaserver.shtml

The ability of the methods to discriminate *S. aureus* strains (i.e., unrelated strains) was assessed by calculating Simpson's index of diversity (ID; Hunter and Gaston, 1988) with confidence interval calculated according to (Carriço et al., 2006). The ID was calculated from PFGE, spa-typing and MLVA results obtained from the panel of 146 epidemiologically unrelated isolates.

This strain panel was reduced of four strains not typable by spa-typing, and used for congruence test between spa-typing, MLVA and PFGE. The congruence assessments were performed using the adjusted Rand's coefficient (Carriço et al., 2006, #1093). Adjusted rand coefficient consider (i) the probability that a pair of isolates which is assigned to the same type by one typing method is also typed as identical by the other method, (ii) the probability that a pair of isolates which is assigned to two types by one typing method is also typed as different by the other method and corrects the typing concordance for chance agreement, avoiding the overestimation of congruence between typing methods.

For a finer comparison the adjusted Wallace (AW) coefficients (Severiano et al., 2011, #1095) were also performed. Statistics analysis were performed in BioNumerics software, V 7.1, (Applied Maths) using a script developed by Ana Severiano and João André Carriço available online at http://darwin.phyloviz.net/ComparingPartitions/. The AW coefficient indicates the probability that pairs of isolates which are assigned to the same type by one typing method are also typed as identical by the other and corrects the typing concordance for chance agreement. The AW coefficient is directional, i.e., given a standard method. It considers the probability of two strains having the same type of standard method also sharing the same type of the compared method.

Results

Discriminatory Power and Concordance of MLVA, PFGE, and Spa-Typing

All the strains were identified as *S. aureus* by the 23S rDNA PCR assay specific for this bacterial species. The ID diversity index was assessed on a panel of 146 epidemiologically unrelated strains, which included 103 strains associated with various French SFPOs. MLVA and PFGE separated the 146 strains into 125 and 118 distinct groups, respectively. *Spa*-typing separated 142 out of the 146 strains into 71 groups only (**Table 4**). The four remaining strains could not be typed using this technique. High ID values were observed for MLVA (0.997) and PFGE (0.995) indicating that almost each strain can be distinguished from all other members of the strain panel by these two typing methods. PFGE and MLVA were found to be more discriminatory than *spa*-typing whose ID value was 0.970 (**Table 4**).

Four PFGE groups and five MLVA groups were shared mainly into two different spa-types. Eight MLVA groups were shared mainly into two PFGE types. Eleven PFGE groups were shared mainly into two MLVA types (data not shown).

Quantitative determination of concordance of the three methods was calculated (**Table 5**). Congruence between the three methods was found to be low with adjusted Rand coefficient values ranging from 0.131 to 0.269 (**Table 5**). The AWMLVA→spa-typing and AWPFGE→spa-typing were respectively higher than AWspa-typing→MLVA and AWspa-typing→PFGE (**Table 5**) indicating that partitions defined by spa-typing could have been predicted from the results of MLVA or PFGE.

TABLE 4 | Multiple-locus variable-number tandem-repeat analysis, PFGE and spa-typing results from a panel of 146 not epidemiologically related strains.

	Number of analyzed strains	Number of successfully typed strains	ID and CI 95%	Total number of distinct types	Total number of unique types
MLVA	146	146	0.997 (0.995–0.999)	125	109
PFGE	146	146	0.995 (0.992–0.998)	118	100
spa-typing	146	142	0.970 (0.959–0.982)	71	50

ID, Simpson's index of diversity (Hunter and Gaston, 1988, #1092).
CI, confidence interval.

TABLE 5 | Congruence between the typing methods using adjusted Rand and adjusted Wallace coefficients.

Method	Adjusted Rand coefficient			Adjusted Wallace coefficient 95% CI		
	Spa-typing	MLVA	PFGE	Spa-typing	MLVA	PFGE
spa-typing					0.072 (0.021–0.123)	0.104 (0.043–0.166)
MLVA	0.131			0.725 (0.621–0.830)		0.363 (0.212–0.515)
PFGE	0.178	0.269		0.616 (0.429–0.804)	0.213 (0.085–0.342)	

CI, confidence interval.

Typing of 112 French SFPO Strains

Typing Data from Four "Strong Evidence" SFPOs (13 Strains)

Typing of 13 *S. aureus* strains from four "strong evidence" SFPOs showed that the *se* gene profiles and the MLVA-, PFGE-, and spa- types were indistinguishable within each set of epidemiologically related strains (**Table 2**). This demonstrated the good epidemiological concordance of the four methods.

Typing Data from 76 "Weak Evidence" SFPOs (103 Strains)

MLVA

A panel of 103 epidemiologically unrelated strains from 76 French SFPOs was selected, containing one strain of each of the four "strong evidence" SFPOs described above (**Table 2**) and 99 strains from 72 "weak evidence" SFPOs (**Table 1**). For 12 strains previously tested elsewhere (i.e., 431G, 360F, 338E, 419G, 372F, 402F, 353E, 301E, 384F, 339E, 363F, 399F), the MLVA profiles obtained here were similar to those obtained by Sobral et al. (2012).

Multiple-locus variable-number tandem-repeat analysis separated the 103 strains into 84 different types. The most prevalent MLVA types were "20" and "1," with each of these two types containing four strains isolated from four different SFPOs that occurred over the periods 1983–2007 and 2000–2008, respectively. The incriminated food was different for each of these SFPOs.

PFGE compared with MLVA

Pulsed field gel electrophoresis separated the 103 strains into 80 pulsotypes. The most prevalent PFGE type was "18" and contained five strains isolated from four distinct SFPOs that occurred between 1999 and 2008.

Out of the 76 different "weak evidence" SFPOs, 19 included several strains (i.e., between two and four strains; **Table 3**). For 15 of these SFPOs, strains that displayed distinct MLVA profiles also displayed distinct PFGE pulsotypes, and strains with similar MLVA profiles also showed similar PFGE profiles. MLVA data were therefore in agreement with those of PFGE. For three of the 19 SFPOs (i.e., "62," "49," "64"), the strains had different MLVA profiles but showed indistinguishable PFGE pulsotypes (**Table 3**). However, for the SFPOs "49" and "64," the MLVA profiles obtained were very similar, differing by only 1.5 repeat. For the remaining SFPO (i.e., "42"), the four strains displayed four different MLVA profiles but showed only two distinct PFGE pulsotypes (**Table 3**).

Spa-typing compared with MLVA

Spa-typing separated 102 out of the 103 strains into 50 different types; one strain could not be typed. The most frequent spa-types observed were t127 ($n = 14$) and t008 ($n = 10$; **Figure 1**). Five new spa-types were observed for the first time here (i.e., t7667, t7668, t7669, t7671, and t7674). Except for one MLVA type ("20"), all the MLVA types that included several strains corresponded to the same spa-type. The type "20" included one and three strains with two different spa-types, t656 and t008, respectively (**Figure 1**). These two spa-types were close: they had the same repeat number

and only differed at two single nucleotide positions in the second *spa* repeat.

PCR se genes compared with MLVA

Out of the 103 strains tested, 16 strains did not carry any *se* gene and 87 strains carried at least one *se* gene. These 87 strains were divided into 20 distinct *se* gene profiles. The most frequently occurring gene was *sea* ($n = 58$), followed by *seg* ($n = 23$) and *sei* ($n = 23$), by *seh* ($n = 19$) and then by *sed, sej, ser, sec, sep, seb,* and *see*. Several *se* genes could be present, and the most frequent associations of *se* genes detected were *sea-seh* ($n = 15$), *sea-sed-sej-ser* ($n = 13$) and *seg-sei* ($n = 11$). Moreover, 65 strains carried genes corresponding to 'new' enterotoxins, i.e., *seg, seh, sei, sej, sep,* and *ser*. Eleven strains carried the *seg* and *sei* genes and one strain the *seg, sei, sej, sep,* and *ser* genes.

The 16 strains for which no *se* gene were detected were divided into 16 unique MLVA types. The 87 *se* gene-carrying strains were separated into 68 distinct MLVA-types: 64 MLVA-types correlated with *se* gene carriage (**Figure 2**). The four remaining MLVA-types included strains with different *se* genes as indicated by circles including various colors in the **Figure 2**. The 21 strains that carried *seh* either alone or in combination with *sea* were divided into 19 very close MLVA types that clustered within the same MLVA subgroup (**Figure 2**). 14 out of the 21 *seh*-carrying strains that clustered within the same MLVA sub-group possessed the spa type t127 (**Figures 1** and **2**).

Discussion

Comparison MLVA with PFGE

The aim of this study was to investigate the genetic diversity of 112 *S. aureus* strains involved in SFPOs in France between 1981 and 2009. The MLVA protocol developed by Sobral et al. (2012) was compared here for the first time with the EURL PFGE protocol. The analysis of a panel of 146 *S. aureus* strains isolated from different sources (clinical cases, food, animal cases, or SFPOs) showed that MLVA and PFGE have equal discriminatory power.

For the strains related to the strong evidence SFPOs, MLVA, and PFGE typing data demonstrated the good epidemiological concordance of each of the two methods. Moreover, for the strains which were involved in the same "weak-evidence" SFPO MLVA could discriminate strains which had similar PFGE types.

With the MLVA protocol used here, results could be obtained within 48 h, i.e., faster than PFGE analysis which is usually completed in 3 days from receipt of pure culture. Moreover, the method is easy to perform, is readily automatable, and allows high sample throughput. The use of commercially available reagents can help foster standardization. MLVA data are also suitable for electronic transmission between laboratories and are not prone to subjective interpretation. Moreover, this MLVA method benefits from the higher resolution of using a capillary electrophoresis methodology over standard gel-based techniques. The method can therefore be easily implemented by NRLs equipped with a capillary electrophoresis system. One inconvenience could

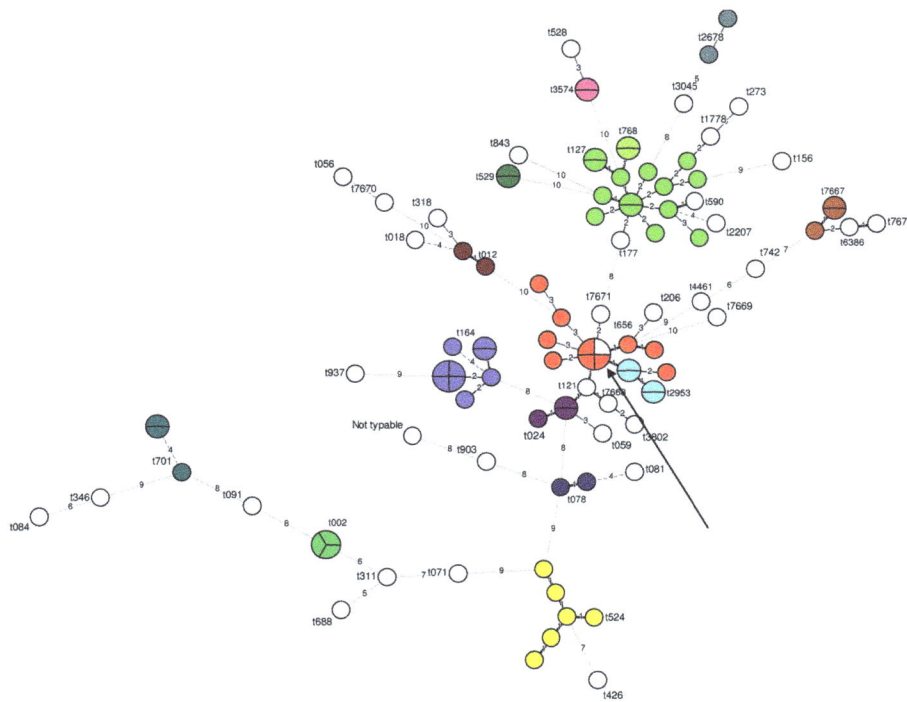

FIGURE 1 | Minimum spanning tree of spa-types according to the multiple-locus variable-number tandem-repeat analysis (MLVA) types determined on the 103 strains tested. 102 strains were successfully typed using spa-typing. Each circle represents a particular MLVA type. The size of each circle is proportional to the number of isolates within the MLVA type. The distance between the circles represents the genetic divergence. The divergence is given in number of mutations and is indicated on the branch. Each color represents a different spa-type (n > 1 isolate). All unique (n = 1 isolate) spa-type are shown in white. The arrow shows the single MLVA type including two different spa-types.

be the use of an expensive system. Nevertheless, this MLVA protocol is also suitable for agarose-gel electrophoresis. Indeed, for each of the 16 VNTRs, repeat size was sufficiently high for accurate band sizing on agarose gels. High-throughput typing on agarose gels can also be facilitated by the use of an automated flow-through gel electrophoresis system (Reskova et al., 2014).

The major drawback of this MLVA assay is the lack of free internet-accessible databases for comparative purposes. To date, the databases used worldwide[3] (MLVA.net) centralize profiles obtained with the protocols of Pourcel et al. (2009) and Schouls et al. (2009) using subset of loci. However, the development of an easily accessible database centralizing profiles obtained from the 16 VNTRs with this protocol is in progress and should be available in the coming years.

Prevalence of se Genes among the SFPO Strains

In this study, 65 out of 87 enterotoxigenic strains carried more than one gene coding for enterotoxins, illustrating the importance of searching for se genes in the strains involved in SFPOs. Our results confirmed the clear predominance of the sea gene (67%) among the SFPO strains and its frequent association

with other genes such as seh or sed-sej-ser. The prevalence of the sea gene in the strains linked to SPFOs has been already observed in Spain (Dyer et al., 2007; Sobral et al., 2007; Kellermann et al., 2008), Italy (Morandi et al., 2007), and in Japan (Sato'o et al., 2014). The association of sea with seh has also been observed for strains linked to SFPOs in France (Kerouanton et al., 2007) and recently in Japan (Sato'o et al., 2014).

Genetic Diversity and Origin of Strains Involved in the French SFPOs

The three typing method, MLVA, PFGE and spa-typing, show ID greater than 0.90. This value is considered as a cut-off in order to interpret the typing results with confidence (Hunter and Gaston, 1988, #1092; Maâtallah et al., 2013, #1096). PFGE and MLVA bring more information than spa-typing.

Whatever the typing method used here for the analysis of 112 French SFPOs-isolates, i.e., MLVA, PFGE, spa-typing or se genes detection, a large number of different molecular types was found, highlighting the high genetic diversity of the tested isolates. In this study, associations between MLVA types, se gene combinations and spa-types were identified. All the 21 seh-carrying strains clustered within the same MLVA sub-group, and 14 of these showed spa type t127. These results demonstrated that the strains carrying the seh gene have a genetically related background, as previously suggested by Ruzickova et al. (2008)

---

[3]http://minisatellites.u-psud.fr

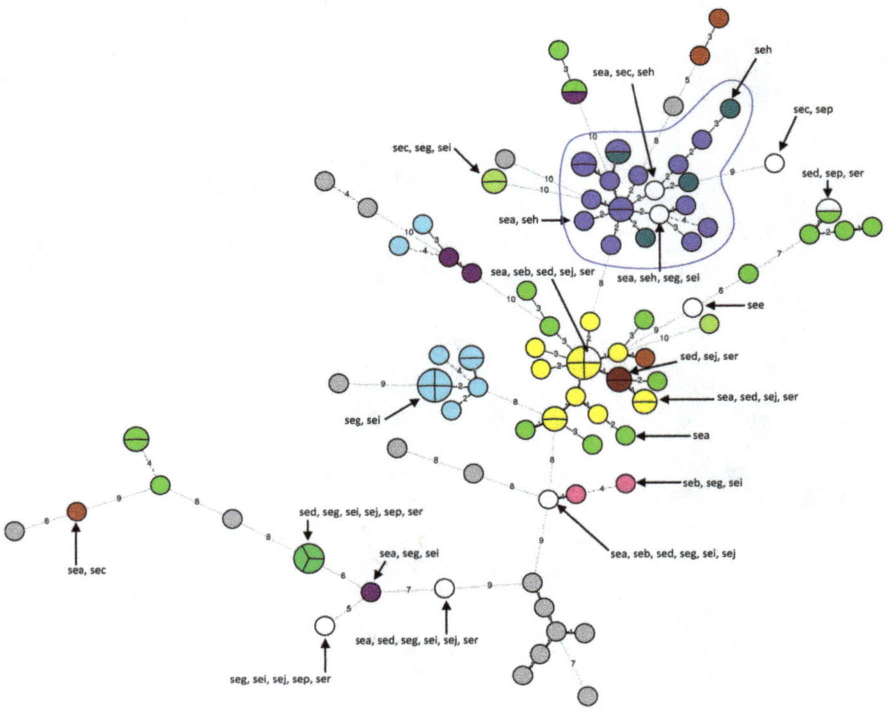

FIGURE 2 | Minimum spanning tree of 103 *Staphylococcus aureus se* gene-carrying strains according to their MLVA types. Each circle represents a particular MLVA type. The size of each circle is proportional to the number of isolates within this MLVA type. The distance between the circles represents the genetic divergence. The divergence is given in number of mutations and is indicated on the branch. Each color represents a particular *se* gene combination (*n* > 1 isolate), with unique *se* combination (*n* = 1 isolate) shown in white and no *se* gene shown in gray. Four MLVA-types had strains with different *se* genes combinations (circles with several colors). The MLVA subgroup that includes seh-carrying strains is highlighted in blue and circled. A subgroup groups strains with MLVA types that differ by a maximum of five mutations.

for 28 enterotoxin H-positive *S. aureus* strains isolated from food samples in the Czech Republic.

Previously, a high concordance between the MLVA protocol and the MLST method was demonstrated and a distribution signature of clonal complexes typical of human isolates was found for 13 SFPO strains (Sobral et al., 2012). Further comparison of the MLVA data obtained here from 112 SFPO strains with those from a representative panel of human strains could be useful to confirm the hypothesis of a human origin for SFPO strains.

Another investigation could be useful to explore the genetic structure of all the populations of food *S. aureus* strains associated to SFPO isolated in France over the past 30° years.

Conclusion

By combining various molecular typing methods, we highlighted the high genetic diversity of *S. aureus* strains involved in SFPO in France over the past 30 years. In addition, the MLVA protocol developed by Sobral et al. (2012) was found highly discriminatory, therefore representing a very interesting alternative to PFGE for establishing epidemiological links during SFPOs investigations. This MLVA method could be used also to better understand the population biology of *S. aureus*. Before transferring this MLVA protocol through the European NRL network, its reproducibility will need to be assessed in proficiency testing trials, in order to compare and interpret MLVA data and harmonize assignment of MLVA types.

Author Contributions

SR participated in the design and coordination of the study, the data interpretation and the draft of all the manuscript. BF participated to the design of the study, the data interpretation under BioNumerics software and the strain typing by PFGE, spa-typing and *se* genes. NV was in charge of the strain typing by MLVA and MLVA data analysis. JG carried out all the PCR tests at EURL. J-AH took part to the draft of the manuscript. LG carried all the congruence analysis and took part to the draft of the manuscript. AB and FA participated in the design and coordination of the study, and the draft of all the manuscript. All authors read and approved the final manuscript.

Acknowledgments

This work was conducted as part of the activities of the EURL and the French NRL for CPS. It was supported by a grant from the Directorate-General for Heath and Consumers (DG SANCO) of the European Commission and from ANSES. We thank Dr. Bertrand Lombard (ANSES) for his helpful advice on writing this manuscript.

References

Aires-de-Sousa, M., Boye, K., De Lencastre, H., Deplano, A., Enright, M. C., Etienne, J., et al. (2006). High interlaboratory reproducibility of DNA sequence-based typing of bacteria in a multicenter study. *J. Clin. Microbiol.* 44, 619–621. doi: 10.1128/JCM.44.2.619-621.2006

Argudin, M. A., Mendoza, M. C., Gonzalez-Hevia, M. A., Bances, M., Guerra, B., and Rodicio, M. R. (2012). Genotypes, exotoxin gene content, and antimicrobial resistance of *Staphylococcus aureus* strains recovered from foods and food handlers. *Appl. Environ. Microbiol.* 78, 2930–2935. doi: 10.1128/AEM.07 487-11

Argudin, M. A., Mendoza, M. C., and Rodicio, M. R. (2010). Food poisoning and *Staphylococcus aureus* enterotoxins. *Toxins (Basel)* 2, 1751–1773. doi: 10.3390/toxins2071751

Babouee, B., Frei, R., Schultheiss, E., Widmer, A. F., and Goldenberger, D. (2011). Comparison of the DiversiLab repetitive element PCR system with spa typing and pulsed-field gel electrophoresis for clonal characterization of methicillin-resistant *Staphylococcus aureus*. *J. Clin. Microbiol.* 49, 1549–1555. doi: 10.1128/JCM.02254-10

Barrett, T. J., Gerner-Smidt, P., and Swaminathan, B. (2006). Interpretation of pulsed-field gel electrophoresis patterns in foodborne disease investigations and surveillance. *Foodborne Pathog. Dis.* 3, 20–31. doi: 10.1089/fpd.2006.3.20

Bania, J., Dabrowska, A., Korzekwa, K., Zarczynska, A., Bystron, J., Chrzanowska, J., et al. (2006). The profiles of enterotoxin genes in *Staphylococcus aureus* from nasal carriers. *Lett. Appl. Microbiol.* 42, 315–320. doi: 10.1111/j.1472-765X.2006.01862.x

Bergonier, D., Sobral, D., Fessler, A. T., Jacquet, E., Gilbert, F. B., Schwarz, S., et al. (2014). *Staphylococcus aureus* from 152 cases of bovine, ovine and caprine mastitis investigated by Multiple-locus variable number of tandem repeat analysis (MLVA). *Vet. Res.* 45, 97. doi: 10.1186/s13567-014-0097-4

Carriço, J. A., Silva-Costa, C., Melo-Cristino, J., Pinto, F. R., De Lencastre, H., Almeida, J. S., et al. (2006). Illustration of a common framework for relating multiple typing methods by application to macrolide-resistant streptococcus pyogenes. *J. Clin. Microbiol.* 44, 2524–2532. doi: 10.1128/JCM.02536-05

Chiang, Y. C., Lai, C. H., Lin, C. W., Chang, C. Y., and Tsen, H. Y. (2014). Improvement of strain discrimination by combination of superantigen profiles, PFGE, and RAPD for *Staphylococcus aureus* isolates from clinical samples and food-poisoning cases. *Foodborne Pathog. Dis.* 11, 468–477. doi: 10.1089/fpd.2013.1708

Chiang, Y. C., Liao, W. W., Fan, C. M., Pai, W. Y., Chiou, C. S., and Tsen, H. Y. (2008). PCR detection of Staphylococcal enterotoxins (SEs) N, O, P, Q, R, U, and survey of SE types in *Staphylococcus aureus* isolates from food-poisoning cases in Taiwan. *Int. J. Food Microbiol.* 121, 66–73. doi: 10.1016/j.ijfoodmicro.2007.10.005

Chiou, C. S., Wei, H. L., and Yang, L. C. (2000). Comparison of pulsed-field gel electrophoresis and coagulase gene restriction profile analysis techniques in the molecular typing of *Staphylococcus aureus*. *J. Clin. Microbiol.* 38, 2186–2190.

Chung, M., De Lencastre, H., Matthews, P., Tomasz, A., Adamsson, I., Aires De Sousa, M., et al. (2000). Molecular typing of methicillin-resistant *Staphylococcus aureus* by pulsed-field gel electrophoresis: comparison of results obtained in a multilaboratory effort using identical protocols and MRSA strains. *Microb. Drug Resist.* 6, 189–198. doi: 10.1089/mdr.2000.6.189

Cookson, B. D., Aparicio, P., Deplano, A., Struelens, M., Goering, R., and Marples, R. (1996). Inter-centre comparison of pulsed-field gel electrophoresis for the typing of methicillin-resistant *Staphylococcus aureus*. *J. Med. Microbiol.* 44, 179–184. doi: 10.1099/00222615-44-3-179

Cookson, B. D., Robinson, D. A., Monk, A. B., Murchan, S., Deplano, A., De Ryck, R., et al. (2007). Evaluation of molecular typing methods in characterizing a European collection of epidemic methicillin-resistant *Staphylococcus aureus* strains: the HARMONY collection. *J. Clin. Microbiol.* 45, 1830–1837. doi: 10.1128/JCM.02402-06

Dyer, M. D., Murali, T. M., and Sobral, B. W. (2007). Computational prediction of host-pathogen protein-protein interactions. *Bioinformatics* 23, i159–i166. doi: 10.1093/bioinformatics/btm208

EFSA. (2011). Updated technical specifications for harmonised reporting of foodborneoutbreaks through the European Union reporting system in accordance with Directive 2003/99/EC1. *EFSA J.* 9, 2101.

EFSA-ECDC. (2013). The European Union summary report on trends and sources of zoonoses, zoonotic agents and food-borne outbreaks in 2011. *EFSA J.* 11, 250. doi: 10.2903/j.efsa.2013.3129

Hallin, M., Deplano, A., Denis, O., De Mendonca, R., De Ryck, R., and Struelens, M. J. (2007). Validation of pulsed-field gel electrophoresis and spa typing for long-term, nationwide epidemiological surveillance studies of *Staphylococcus aureus* infections. *J. Clin. Microbiol.* 45, 127–133. doi: 10.1128/JCM. 01866-06

Hennekinne, J. A., De Buyser, M. L., and Dragacci, S. (2011). *Staphylococcus aureus* and its food poisoning toxins: characterization and outbreak investigation. *FEMS Microbiol Rev.* 36, 815–836. doi: 10.1111/j.1574-6976.2011.00311.x

Hunter, P. R., and Gaston, M. A. (1988). Numerical index of the discriminatory ability of typing systems: an application of Simpson's index of diversity. *J. Clin. Microbiol.* 26, 2465–2466.

Kadariya, J., Smith, T. C., and Thapaliya, D. (2014). *Staphylococcus aureus* and staphylococcal food-borne disease: an ongoing challenge in public health. *Biomed. Res. Int.* 2014, 827965. doi: 10.1155/2014/827965

Kellermann, M. G., Sobral, L. M., Da Silva, S. D., Zecchin, K. G., Graner, E., Lopes, M. A., et al. (2008). Mutual paracrine effects of oral squamous cell carcinoma cells and normal oral fibroblasts: induction of fibroblast to myofibroblast transdifferentiation and modulation of tumor cell proliferation. *Oral Oncol.* 44, 509–517. doi: 10.1016/j.oraloncology.2007.07.001

Kerouanton, A., Hennekinne, J. A., Letertre, C., Petit, L., Chesneau, O., Brisabois, A., et al. (2007). Characterization of *Staphylococcus aureus* strains associated with food poisoning outbreaks in France. *Int. J. Food Microbiol.* 115, 369–375. doi: 10.1016/j.ijfoodmicro.2006.10.050

Koreen, L., Ramaswamy, S. V., Graviss, E. A., Naidich, S., Musser, J. M., and Kreiswirth, B. N. (2004). spa typing method for discriminating among *Staphylococcus aureus* isolates: implications for use of a single marker to detect genetic micro- and macrovariation. *J. Clin. Microbiol.* 42, 792–799. doi: 10.1128/JCM.42.2.792-799.2004

Kuroda, M., Ohta, T., Uchiyama, I., Baba, T., Yuzawa, H., Kobayashi, I., et al. (2001). Whole genome sequencing of meticillin-resistant *Staphylococcus aureus*. *Lancet* 357, 1225–1240. doi: 10.1016/S0140-6736(00)04403-2

Lindstedt, B. A., Torpdahl, M., Vergnaud, G., Le Hello, S., Weill, F. X., Tietze, E., et al. (2013). Use of multilocus variable-number tandem repeat analysis (MLVA) in eight European countries, 2012. *Eurosurveillance* 18, 20385.

Lv, G., Xu, B., Wei, P., Song, J., Zhang, H., Zhao, C., et al. (2014). Molecular characterization of foodborne-associated *Staphylococcus aureus* strains isolated in Shijiazhuang, China, from 2010 to 2012. *Diagn. Microbiol. Infect. Dis.* 78, 462–468. doi: 10.1016/j.diagmicrobio.2013.12.006

Martin, M. C., Fueyo, J. M., Gonzalez-Hevia, M. A., and Mendoza, M. C. (2004). Genetic procedures for identification of enterotoxigenic strains of *Staphylococcus aureus* from three food poisoning outbreaks. *Int. J. Food Microbiol.* 94, 279–286. doi: 10.1016/j.ijfoodmicro.2004.01.011

Maâtallah, M., Bakhrouf, A., Habeeb, M. A., Turlej-Rogacka, A., Iversen, A., Pourcel, C., et al. (2013). Four genotyping schemes for phylogenetic analysis of *Pseudomonas aeruginosa*: comparison of their congruence with multi-locus sequence typing. *PLoS ONE* 8:e82069. doi: 10.1371/journal.pone. 0082069

Mehrotra, M., Wang, G., and Johnson, W. M. (2000). Multiplex PCR for detection of genes for *Staphylococcus aureus* enterotoxins, exfoliative toxins, toxic shock syndrome toxin 1, and methicillin resistance. *J. Clin. Microbiol.* 38, 1032–1035.

Morandi, S., Brasca, M., Lodi, R., Cremonesi, P., and Castiglioni, B. (2007). Detection of classical enterotoxins and identification of enterotoxin genes in *Staphylococcus aureus* from milk and dairy products. *Vet. Microbiol.* 124, 66–72. doi: 10.1016/j.vetmic.2007.03.014

Murchan, S., Kaufmann, M. E., Deplano, A., De Ryck, R., Struelens, M., Zinn, C. E., et al. (2003). Harmonization of pulsed-field gel electrophoresis protocols for epidemiological typing of strains of methicillin-resistant *Staphylococcus aureus*:

a single approach developed by consensus in 10 European laboratories and its application for tracing the spread of related strains. *J. Clin. Microbiol.* 41, 1574–1585.

Omoe, K., Hu, D. L., Takahashi-Omoe, H., Nakane, A., and Shinagawa, K. (2003). Identification and characterization of a new staphylococcal enterotoxin-related putative toxin encoded by two kinds of plasmids. *Infect. Immun.* 71, 6088–6094. doi: 10.1128/IAI.71.10.6088-6094.2003

Ono, H. K., Omoe, K., Imanishi, K., Iwakabe, Y., Hu, D. L., Kato, H., et al. (2008). Identification and characterization of two novel staphylococcal enterotoxins, types S and T. *Infect. Immun.* 76, 4999–5005. doi: 10.1128/IAI.00045-08

Ostyn, A., De Buyser, M. L., Guillier, F., Groult, J., Felix, B., Salah, S., et al. (2010). First evidence of a food poisoning outbreak due to staphylococcal enterotoxin type E, France, 2009. *Eurosurveillance* 15, 19528.

Pourcel, C., Hormigos, K., Onteniente, L., Sakwinska, O., Deurenberg, R. H., and Vergnaud, G. (2009). Improved multiple-locus variable-number tandem-repeat assay for *Staphylococcus aureus* genotyping, providing a highly informative technique together with strong phylogenetic value. *J. Clin. Microbiol.* 47, 3121–3128. doi: 10.1128/JCM.00267-09

Reskova, Z., Korenova, J., and Kuchta, T. (2014). Effective application of multiple locus variable number of tandem repeats analysis to tracing *Staphylococcus aureus* in food-processing environment. *Lett. Appl. Microbiol.* 58, 376–383. doi: 10.1111/lam.12200

Ruzickova, V., Karpiskova, R., Pantucek, R., Pospisilova, M., Cernikova, P., and Doskar, J. (2008). Genotype analysis of enterotoxin H-positive *Staphylococcus aureus* strains isolated from food samples in the Czech Republic. *Int. J. Food Microbiol.* 121, 60–65. doi: 10.1016/j.ijfoodmicro.2007.10.006

Sato'o, Y., Omoe, K., Naito, I., Ono, H. K., Nakane, A., Sugai, M., et al. (2014). Molecular epidemiology and identification of a *Staphylococcus aureus* clone causing food poisoning outbreaks in Japan. *J. Clin. Microbiol.* 52, 2637–2640. doi: 10.1128/JCM.00661-14

Schouls, L. M., Spalburg, E. C., Van Luit, M., Huijsdens, X. W., Pluister, G. N., Van Santen-Verheuvel, M. G., et al. (2009). Multiple-locus variable number tandem repeat analysis of *Staphylococcus aureus*: comparison with pulsed-field gel electrophoresis and spa-typing. *PLoS ONE* 4:e5082. doi: 10.1371/journal.pone.0005082

Severiano, A., Pinto, F. R., Ramirez, M., and Carriço, J. A. (2011). Adjusted wallace coefficient as a measure of congruence between typing methods. *J. Clin. Microbiol.* 49, 3997–4000. doi: 10.1128/JCM.00624-11

Sharma, N. K., Rees, C. E., and Dodd, C. E. (2000). Development of a single-reaction multiplex PCR toxin typing assay for *Staphylococcus aureus* strains. *Appl. Environ. Microbiol.* 66, 1347–1353. doi: 10.1128/AEM.66.4.1347-1353.2000

Shimizu, A., Fujita, M., Igarashi, H., Takagi, M., Nagase, N., Sasaki, A., et al. (2000). Characterization of *Staphylococcus aureus* coagulase type VII isolates from staphylococcal food poisoning outbreaks (1980-1995) in Tokyo, Japan, by pulsed-field gel electrophoresis. *J. Clin. Microbiol.* 38, 3746–3749.

Shopsin, B., Gomez, M., Montgomery, S. O., Smith, D. H., Waddington, M., Dodge, D. E., et al. (1999). Evaluation of protein a gene polymorphic region DNA sequencing for typing of *Staphylococcus aureus* strains. *J. Clin. Microbiol.* 37, 3556–3563.

Sobral, A. P., De Oliveira Lima, D. N., Cazal, C., Santiago, T., Das Gracas Granja Mattos, M., Melo, B., et al. (2007). Myxoid liposarcoma of the lip: correlation of histological and cytological features and review of the literature. *J. Oral Maxillofac. Surg.* 65, 1660–1664. doi: 10.1016/j.joms.2006.06.264

Sobral, D., Schwarz, S., Bergonier, D., Brisabois, A., Fessler, A. T., Gilbert, F. B., et al. (2012). High throughput multiple locus variable number of tandem repeat analysis (MLVA) of *Staphylococcus aureus* from human, animal and food sources. *PLoS ONE* 7:e33967. doi: 10.1371/journal.pone.0033967

Strommenger, B., Kettlitz, C., Weniger, T., Harmsen, D., Friedrich, A. W., and Witte, W. (2006). Assignment of *Staphylococcus* isolates to groups by spa typing, SmaI macrorestriction analysis, and multilocus sequence typing. *J. Clin. Microbiol.* 44, 2533–2540.

Wattinger, L., Stephan, R., Layer, F., and Johler, S. (2012). Comparison of *Staphylococcus aureus* isolates associated with food intoxication with isolates from human nasal carriers and human infections. *Eur. J. Clin. Microbiol. Infect. Dis.* 31, 455–464. doi: 10.1007/s10096-011-1330-y

Wei, H. L., and Chiou, C. S. (2002). Molecular subtyping of *Staphylococcus aureus* from an outbreak associated with a food handler. *Epidemiol. Infect.* 128, 15–20. doi: 10.1017/S0950268801006355

Yu, G. X., Snyder, E. E., Boyle, S. M., Crasta, O. R., Czar, M., Mane, S. P., et al. (2007). A versatile computational pipeline for bacterial genome annotation improvement and comparative analysis, with Brucella as a use case. *Nucleic Acids Res.* 35, 3953–3962. doi: 10.1093/nar/gkm377

Conflict of Interest Statement: The authors declare that the research was conducted in the absence of any commercial or financial relationships that could be construed as a potential conflict of interest.

8

Intrinsic plasmids influence MicF-mediated translational repression of *ompF* in *Yersinia pestis*

Zizhong Liu[1], Haili Wang[1], Hongduo Wang[1], Jing Wang[1,2], Yujing Bi[1], Xiaoyi Wang[1], Ruifu Yang[1]* and Yanping Han[1]*

[1] State Key Laboratory of Pathogen and Biosecurity, Beijing Institute of Microbiology and Epidemiology, Beijing, China,
[2] Animal Husbandry Base Teaching and Research Section, College of Animal Science and Technology, Hebei North University, Zhangjiakou, China

Edited by:
*Yi-Cheng Sun,
Chinese Academy of Medical
Sciences and Peking Union Medical
College, China*

Reviewed by:
*Jeff Shannon,
National Institutes of Health, USA
Weili Liang,
Chinese Center for Disease Control
and Prevention, China*

*Correspondence:
*Ruifu Yang and Yanping Han,
20, Dongdajie, Fengtai,
Beijing 100071, China
ruifuyang@gmail.com;
yanpinghan@gmail.com*

Yersinia pestis, which is the causative agent of plague, has acquired exceptional pathogenicity potential during its evolution from *Y. pseudotuberculosis*. Two laterally acquired plasmids, namely, pMT1 and pPCP1, are specific to *Y. pestis* and are critical for pathogenesis and flea transmission. Small regulatory RNAs (sRNAs) commonly function as regulators of gene expression in bacteria. MicF, is a paradigmatic sRNA that acts as a post-transcriptional repressor through imperfect base pairing with the 5′-UTR of its target mRNA, *ompF*, in *Escherichia coli*. The high sequence conservation and minor variation in the RNA duplex of MicF-*ompF* has been reported in *Yersinia*. In this study, we utilized super-folder GFP reporter gene fusion to validate the post-transcriptional MicF-mediated regulation of target mRNA *ompF* in *Y. pestis*. Unexpectedly, upon MicF overexpression, the slightly upregulated expression of OmpF were found in the wild-type strain, which contradicted the previously established model. Interestingly, the translational repression of *ompF* target fusions was restored in the intrinsic plasmids-cured *Y. pestis* strain, suggesting intrinsic plasmids influence the MicF-mediated translational repression of *ompF* in *Y. pestis*. Further examination showed that plasmid pPCP1 is likely the main contributor to the abolishment of MicF-mediated translational repression of endogenous or plasmid-borne *ompF*. It represents that the possible roles of intrinsic plasmids should be considered upon investigating sRNA-mediated gene regulation, at least in *Y. pestis*, even if the exact mechanism is not fully understood.

Keywords: *Yersinia pestis*, sRNA regulation, MicF-ompF, intrinsic plasmid, translational fusion

Introduction

The human pathogenic species of *Yersiniae* include *Yersinia pestis*, *Y. pseudotuberculosis*, and *Y. enterocolitica* (Perry and Fetherston, 1997). *Y. pestis* evolved from its ancestor *Y. pseudotuberculosis* several thousands of years ago (Achtman et al., 1999). Interestingly, the two species induce remarkably different diseases by distinct transmission routes. *Y. pestis* causes pandemics of bubonic or pneumonic plague, which is a fatal disease to rodents and humans, and is transmitted by flea biting, whereas *Y. pseudotuberculosis* only causes a chronic and relatively mild disease as an enteric pathogen (Zhou et al., 2006). Although different plasmid combinations exist among *Y. pestis* strains, typical strains of *Y. pestis* harbor three virulence plasmids (pCD1, pMT1, and pPCP1), which encode various virulence determinants. The plasmid pCD1 is commonly shared in all three pathogenic species of *Yersiniae*. The other two plasmids (pPCP1 and pMT1) are laterally acquired by *Y. pestis* (Hu et al., 1998). These three plasmids in *Y. pestis* possess different replication systems. The replicon of plasmid pCD1 belongs to IncFIIA replication system (Perry et al., 1998). Plasmid pPCP1 has a ColE1-like replicon (Hu et al., 1998). The replication region of the pMT1 consists of a structural gene (*repA*) and accessory elements of replication (Lindler et al., 1998). Although the plasmid-encoded virulence determinants are well documented, the way by which the laterally acquired plasmids have complicated the regulatory networks of *Y. pestis* during its evolution remains ambiguous.

Small RNAs (sRNAs) are involved in regulatory networks as regulators of gene expression to facilitate the quick adjustment of bacterial cells to environmental stresses (Waters and Storz, 2009). *Trans*-encoded sRNAs, which represent a major class of sRNAs in bacteria generally activate or repress mRNA translation by limited base-pairing with mRNA (Desnoyers et al., 2013). The RNA-binding protein Hfq is usually required to help the sRNA-mRNA interaction and RNA stability (Vogel and Luisi, 2011). Hfq has been implicated in stress adaptation and virulence in many bacterial pathogens such as *Y. pestis* (Sittka et al.,

2007; Kulesus et al., 2008; Geng et al., 2009; Chiang et al., 2011). MicF is defined as a canonical *trans*-encoded sRNA that regulates outer membrane protein F (OmpF) synthesis in *Escherichia coli* and other related bacteria (Andersen et al., 1989). Approximately 20 nt of MicF forms a perfect RNA duplex by directly pairing with the Shine-Dalgarno region of the *ompF* mRNA. This process occludes the initiation of the 30S ribosome; thus translation is inhibited and the cleavage of mRNA is possibly induced (Andersen et al., 1989). MicF is a highly conserved sRNA in closely related *Enterobacteriaceae* genomes (Delihas, 1997). A strong phylogenetic relationship is also found in MicF/*ompF* interacting sites and RNA duplex in *Yersiniae* (Delihas, 2003). Thus, the regulatory outcome is expected to fit the generalized model, in which MicF forms a duplex with the 5′ UTR of the *ompF* mRNA thereby inhibiting the translation and promoting degradation of the *ompF* transcript. In this study we utilized super-folder GFP reporter gene fusion to validate the MicF-mediated regulation of the *ompF* 5′ UTR in *Y. pestis*. Unexpectedly, the OmpF translation was slightly induced by overexpressed MicF instead of being inhibited in *Y. pestis* Microtus strain 201. This phenomenon is paradoxical to the previous prediction. Interestingly, the translational repression

TABLE 1 | Bacterial strains, plasmids, and oligonucleiotides used in this study.

Name	Characteristics	Sources
STRAINS		
MG1655	*E. coli* serotype K12, strain MG1655	Wang's Lab
201	*Y. pestis* wild-type strain 201 (pCD1$^+$, pMT1$^+$, pPCP1$^+$, and pPCRY1$^+$)	Song et al., 2004
201-null	*Y. pestis* strain 201 derivative (pCD1$^-$, pMT1$^-$, pPCP1$^-$, and pPCRY1$^-$)	Ni et al., 2008
201-pCD1$^+$	*Y. pestis* strain 201 derivative (pCD1$^+$, pMT1$^-$, pPCP1$^-$, and pPCRY1$^-$)	Ni et al., 2008
201-pMT1$^+$	*Y. pestis* strain 201 derivative (pCD1$^-$, pMT1$^+$, pPCP1$^-$, and pPCRY1$^-$)	Ni et al., 2008
201-pPCP1$^+$	*Y. pestis* strain 201 derivative (pCD1$^-$, pMT1$^-$, pPCP1$^+$, and pPCRY1$^-$)	Ni et al., 2008
201-pCD1$^+$pMT$^+$	*Y. pestis* strain 201 derivative (pCD1$^+$, pMT1$^+$, pPCP1$^-$, and pPCRY1$^-$)	Ni et al., 2008
201-pCD1$^+$pPCP1$^+$	*Y. pestis* strain 201 derivative (pCD1$^+$, pMT1$^-$, pPCP1$^+$, and pPCRY1$^-$)	Ni et al., 2008
201-pMT$^+$pPCP1$^+$	*Y. pestis* strain 201 derivative (pCD1$^-$, pMT1$^+$, pPCP1$^+$, and pPCRY1$^-$)	Ni et al., 2008
PLASMIDS		
pXG10-SF	A low-copy translational fusion vector with pSC101 origin	Corcoran et al., 2012
pXG-1	Modified pXG10-SF in which sfGFP expression is P$_{LtetO}$-controlled	This study
pXG-OmpF::gfp	OmpF::GFP fusion plasmid by inserting a DNA fragment amplified by primer ompF/R into pXG10-SF	This study
pBAD/HisA	A high-copy expression vector	Invitrogen
pBAD-TF	An inducible transcriptional fusion vector modified from pBAD/HisA	This study
pBAD-MicF	MicF expressing plasmid by inserting a DNA fragment amplified by primer micF/R into pBAD-TF	This study

Name	Sequence (5′-3′)	
OLIGONUCLEIOTIDES		
ompF-F	TGGATGCATACACAGACGACACCAAACTC	
ompF-R	CTTGCTAGCGGCTAACAGAGCTGGGATTAC	
pBAD-F	TCTGCAGAGCTCGGTACCAAGCTTGCCTGGCGGCAGTAGCGCGGTGGTCCCAC	
pBAD-R	TTGGTACCGAGCTCTGCAGAATTCTATGGAGAAACAGTAGAGAGTTGCGATAAAAAGCG	
pXG-1-F	GAGGGGAAATCTGATGGCTAGCGGATCCGCTGGCTCCGCTGCTGG	
pXG-1-R	CATCAGATTTCCCCTCATGCATGTGCTCAGTATCTCTATCACTGATAG	
micF-F	GTGGAATTCGCTATCATCATTATTTTCCTATCATTGTGG	
micF-R	CATGGTACCTATTCAACTTGAAGTATGACGGGTATAAC	

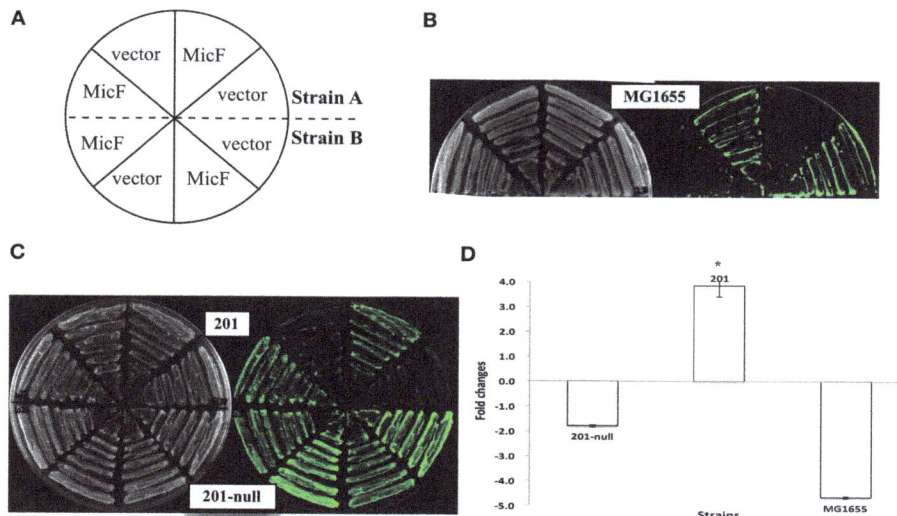

FIGURE 1 | Expression of the *ompF::gfp* fusions upon MicF overexpression in *Escherichia coli* and *Yersinia pestis* strains.
(A) Layout map of bacterial strains carrying the control plasmid without sRNA overexpression (denoted as "vector") or MicF-overexpression plasmid (denoted as "MicF"). Duplicate images of representative strains are shown in parallel on plates. **(B)** Representative fluorescence images of *E. coli* strain MG1655. The image obtained under visible light mode is shown at the left panel and that of the same plate under the fluorescence mode at the right panel. **(C)** Representative fluorescence images of *Y. pestis* WT strain (201) and its plasmid-cured derivative strain (201-null). **(D)** Quantitative measurements of fluorescence produced by the tested strains. Fold changes are provided as the ratio of fluorescence values detected in MicF-overexpressed bacterial strains divided by those detected in strains carrying pBAD blank vectors. Values presented are means ± standard deviations of two independent experiments. The asterisks indicate statistically significant differences compared to the values detected in strain MG1655.

was restored in the intrinsic plasmids-cured strain. Further investigations were conducted to determine which plasmid(s) are responsible for the abrogation of MicF-mediated *ompF* regulation.

Materials and Methods

Bacterial Strains and Growth
The bacterial strains in this study are listed in **Table 1**. *Y. pestis* strain 201 (F1$^+$, VW$^+$, Pst$^+$, and Pgm$^+$) isolated from *Microtus brandti* in Inner Mongolia, China, belongs to biovar *Microtus*. This strain is avirulent to humans but highly lethal to mice (Zhou et al., 2004). Strain 201 has gene content that is almost identical to that of *Y. pestis* strain 91001, which possesses four plasmids, namely, pCD1, pMT1, pPCP1, and pCRY1 (Song et al., 2004). All the combinations of plasmid(s)-cured strains derived from strain 201 were constructed by Bin et al. based on the plasmid incompatibility in our laboratory (Ni et al., 2008). *E. coli* and *Y. pestis* were grown to exponential phase on LB (Luria–Bertani) agar with 0.1% arabinose at 37°C or 26°C. Approximate concentrations of antibiotics (100 μg/mL ampicillin and 34 μg/mL chloramphenicol) were added to LB agar throughout this study.

Construction of *gfp* Reporter Fusions Plasmid
Plasmid pXG10-SF, which is an improved gfp-based translational fusion vector with the constitutive PLtetO promoter (Corcoran et al., 2012), was used to construct the GFP reporter fusion to *ompF*. An amplicon containing the 5′ UTR of the AUG start codon (92 nt) and 16 amino acids of OmpF was obtained by using the primer pair ompF-F/R. The resulting fragment was fused to sfGFP by inserting it into the *Nsi*I/*Nhe*I-digested pXG10-SF, which yielded the plasmid pXG-*ompF::gfp*.

Construction of sRNA Overexpression Plasmid and pXG-1
QuikChange® Lightning Site-Directed Mutagenesis Kit (Stratagene) was used to construct an inducible transcriptional fusion vector by modifying an expression vector, pBAD/HisA. According to the manufacturer's instructions, the pBAD/HisA plasmid was amplified with pBAD-F/R primer pair by PCR followed by *Dpn*I treatment. The resulting plasmid pBAD-TF removed the 327-nt fragment containing several elements (RBS, AUG, and polyclonal sites) and introduced *Eco*RI and *Hind*III restriction sites upstream of the *rrnB* terminator sequence. The modified vector pBAD-TF was used as pBAD control. To construct the MicF overexpression vector, a DNA fragment spanning full-length (85 nt) and 58 nt downstream of MicF was ligated to modified pBAD vector and pBAD-MicF was obtained.

The similar protocol was also used to construct a control plasmid designated as pXG-1 with sfGFP expression controlled by PtetO by modifying pXG10-SF. The pXG10-SF plasmid was amplified with primer pair pXG-1-F/R. The resulting plasmid removed the 725-nt *lacZ*-containing fragment between *Aat*II and *Bhe*I and introduced RBS and AUG sequence (GAGGGGAAAUCUGAUG) upstream of *lacZ* from the transcriptional fusion vector pRW50 into the same site.

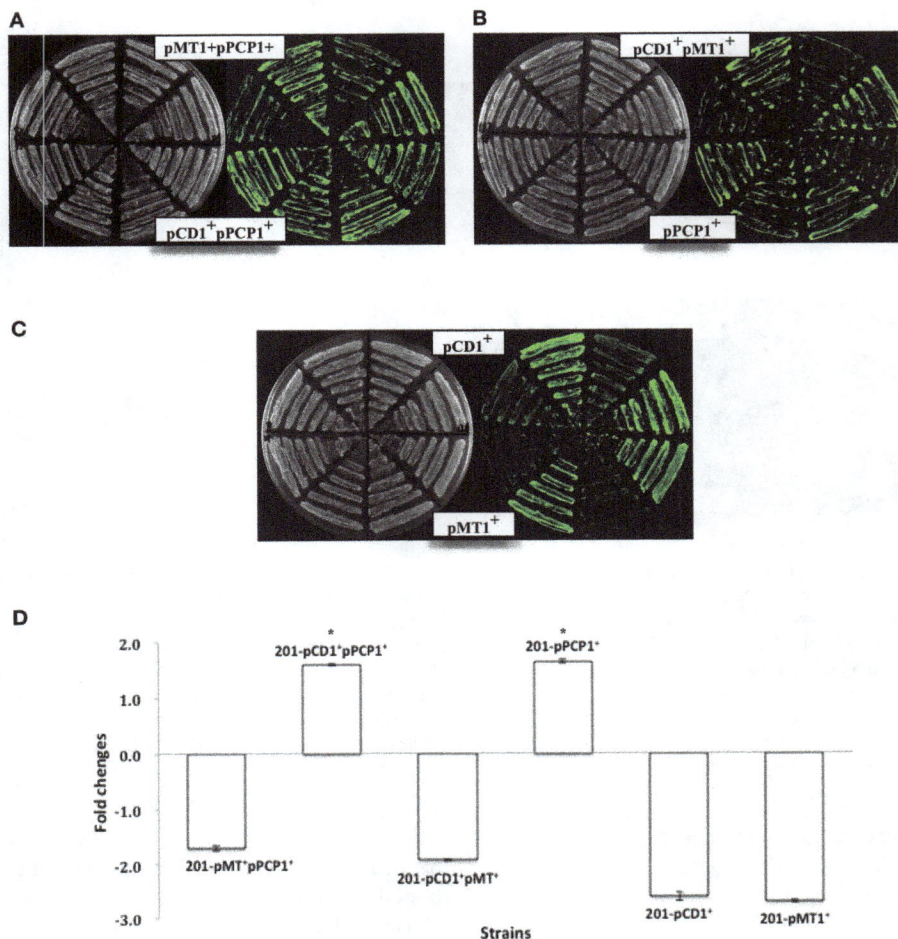

FIGURE 2 | Expression of the *ompF::gfp* fusions upon MicF overexpression in the *Yersinia pestis* strains with different plasmid combinations. Representative fluorescence images of *Y. pestis* derivative strains are shown **(A–C)**. The corresponding quantitative results are also shown **(D)**, in which the asterisks indicate statistically significant differences compared to the values detected in strain MG1655 shown in **Figure 1**.

Imaging and Quantitative Measurements of GFP Fluorescence

Images of bacteria expressing plasmid-borne *gfp* fusions and grown overnight on LB plates were taken using a CCD camera in Gel Doc XR+ image analyzer (Bio-Rad) under the SYBR Green mode. The aliquots of each strain were scraped from the plates and resuspended into PBS buffer. Cell densities were adjusted to $OD_{620nm} \approx 1.5$ and 200 µL of bacterial suspensions were placed into 96-well microtiter plates. Two independent cultures as biological replicates and three aliquots as technical replicates were used throughout the study for each strain. Green fluorescence images were captured by SpectraMax M2 MicroplateReaders (Molecular Devices) with an excitation/emission wavelength of 485/525 nm. Fold changes in the MicF-mediated OmpF expression were calculated by dividing the specific fluorescence of strains with MicF overexpression by that of strains with negative control plasmid. The data were analyzed using One-Way analysis of variance (ANOVA) and

Sidak's multiple comparisons test, where P-values of <0.01 were considered significant.

OmpF Detection by Using Western Blot

Y. pestis were grown in LB agar plate at 26°C for 36 h. Equal amounts of bacterial aliquots were collected and lysed by ultrasonication. Total cellular proteins were separated on SDS-PAGE and immunoblotted with anti-OmpF multiclonal antibody and DyLight 680–labeled goat anti-rabbits antibody followed by detection using the Odyssey Infrared Imaging System. The abundance values were calculated as the expression level of the derivatives divided by that of strain 201 using the Quantity One software. The GroEL protein was detected in parallel as control.

RNA Detection by Using Northern Blot

Total RNA was then extracted from various bacterial strains grown in LB agar plate in the presence of 0.1% arabinose at 26°C for 36 h using the TRIzol Reagent (Invitrogen). Total RNA

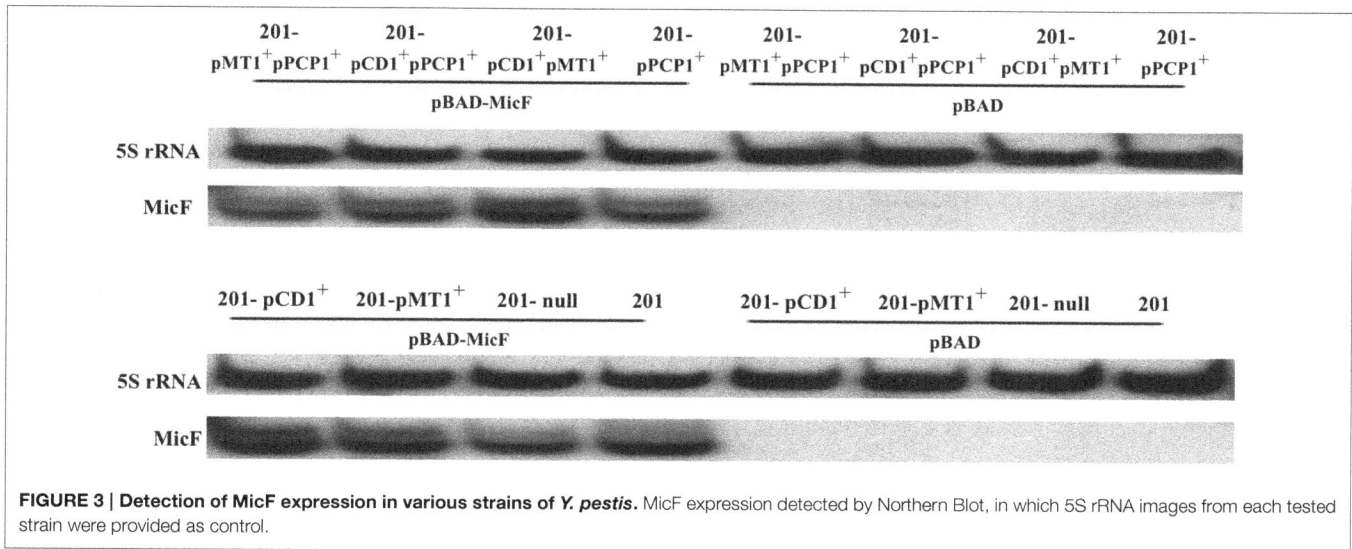

FIGURE 3 | Detection of MicF expression in various strains of Y. pestis. MicF expression detected by Northern Blot, in which 5S rRNA images from each tested strain were provided as control.

samples (1 μg) were denatured at 70°C for 5 min, separated on 6% polyacrylamide 7 M urea gel, and transferred onto Hybond N+ membranes (GE) via electroblot. The membranes were UV-crosslinked and pre-hybridized for 1 h. Northern hybridization was performed by adding the DIG-labeled MicF-specific RNA probe synthesized by *in vitro* transcription using T7 RNA polymerase. The RNAs were immunologically detected according to the instructions on the DIG Northern Starter Kit (Roche). Band intensities on the Northern blots were quantified by Quantity One software. The 5s rRNA species was monitored in parallel as control.

Results and Discussion

In this study, the MicF-mediated *ompF* regulation was validated using a two-compatible-plasmid reporter system established by (Corcoran et al., 2012). The sRNA plasmid is a high-copy vector (pBAD), which ensures the high level of expressed MicF and minimizes the inference of chromosome-encoded MicF. The target plasmid is a low-copy vector (pXG10), where transcription of the *ompF::gfp gene* is driven by the constitutive promoter P_{LtetO}, thereby uncouples the translation from *ompF* transcription. In this system MicF is overexpressed and *ompF-gfp* fusion represents the OmpF abundance. We found that the intrinsic plasmid(s) have an effect on sRNA-mediated regulation, at least on the MicF-mediated *ompF* translation.

Intrinsic Plasmids Accounted for the Abolishment of the MicF-mediated Translational Repression of *ompF* in *Y. pestis*

Y. pestis-specific *ompF-gfp* fusion and sRNA MicF were cloned into low- and high-copy vectors, respectively. Both plasmids were transformed into *E. coli* and *Y. pestis*. All the plasmids were checked throughout all the tested strains by PCR. The results showed that all the plasmids (pMT1, pCD1, pPCP1,

and MicF-expressing plasmid pBAD-MicF) were present as expected (Figure S1). The reporter GFP fluorescence activity was monitored to evaluate the regulatory roles in various bacterial strains.

The inhibitory translation of *ompF-gfp* fusion upon MicF overexpression (approximately five-fold repression) was observed in *E. coli* strain MG1655. Only the 1.8-fold repression was observed in the *Y. pestis* strain 201 that is cured of all the endogenous plasmids (**Figure 1**). This finding is consistent with the results previously reported on *E. coli* (Mizuno et al., 1984) and also confirms that MicF-mediated *ompF* repression occurs in the 5′-UTR. Interestingly, the inhibitory effect was not found in the *Y. pestis* WT strain 201. Instead, more than three-fold upregulation of OmpF-GFP expression was observed under the same conditions (**Figure 1**). The relative quantification of fluorescence value was also measured in *Y. pestis* strains with different combinatorial plasmids grown to exponential phase in LB medium followed by arabinose induction for 1 h. Similar tendency was found in *Y. pestis* strains grown in liquid medium as that found in solid medium (**Figures 1, 2** and Figure S2). To exclude the possibility that the effect was caused by the translational reporter system, we construct plasmid pXG-1 which sfGFP expression is constitutively P_{LtetO}-controlled. No changes in fluorescence intensity were found between pBAD-MicF and pBAD-vector groups in *Y. pestis* strain 201 and 201-null carrying pXG-1 (Figure S3). The different regulatory consequences in *Y. pestis* strains carrying or cured of plasmids indicated that the virulence-associated plasmids might have been recruited to sRNA-mediated regulatory networks in the chromosome during evolution.

Plasmid pPCP1 Alone Abolished the MicF-mediated Translational Repression of *OmpF*

Y. pestis strains carrying different combinations of the three plasmids (pMT1, pPCP1, and pCD1) were used to determine

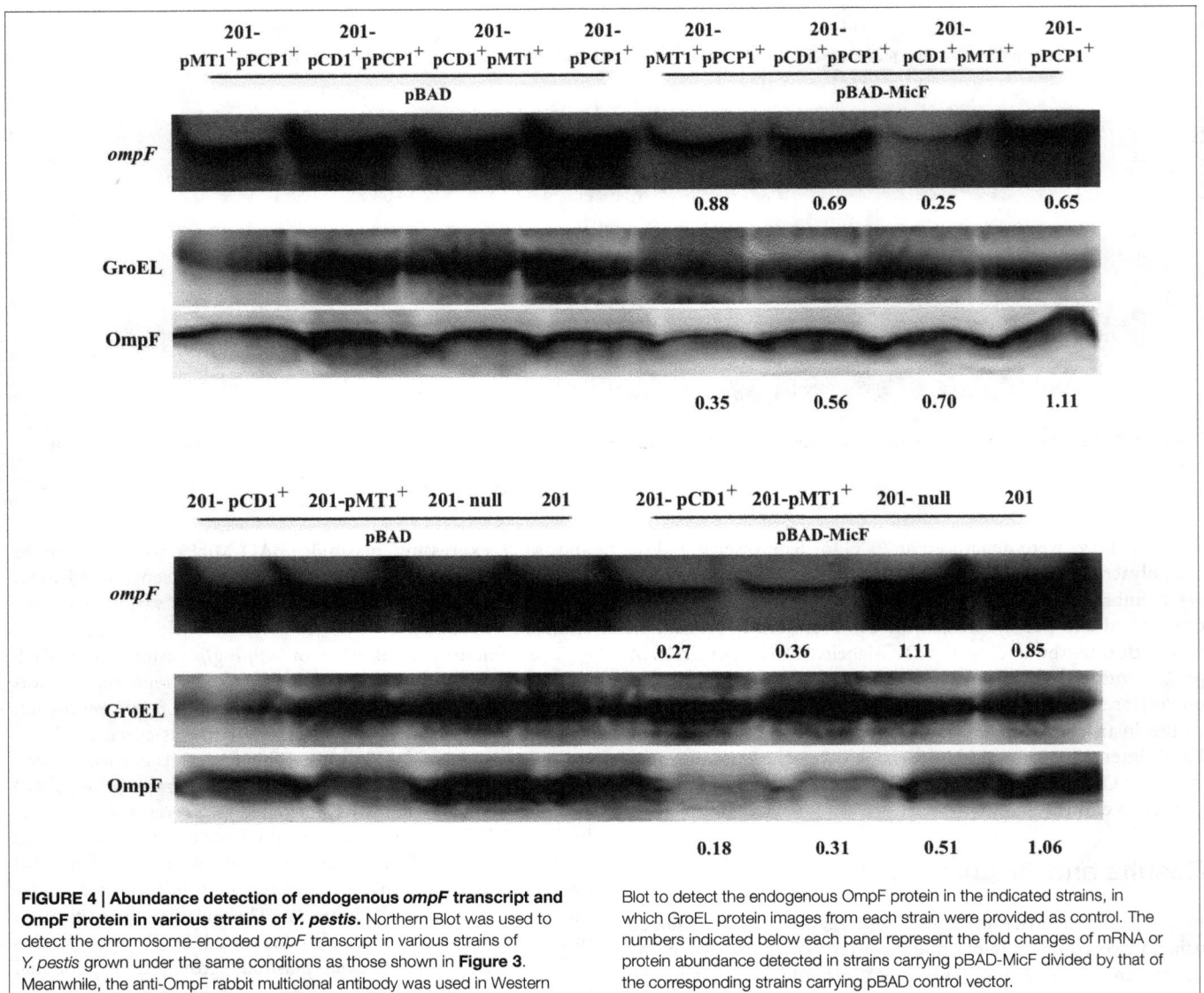

FIGURE 4 | Abundance detection of endogenous *ompF* transcript and OmpF protein in various strains of *Y. pestis*. Northern Blot was used to detect the chromosome-encoded *ompF* transcript in various strains of *Y. pestis* grown under the same conditions as those shown in **Figure 3**. Meanwhile, the anti-OmpF rabbit multiclonal antibody was used in Western Blot to detect the endogenous OmpF protein in the indicated strains, in which GroEL protein images from each strain were provided as control. The numbers indicated below each panel represent the fold changes of mRNA or protein abundance detected in strains carrying pBAD-MicF divided by that of the corresponding strains carrying pBAD control vector.

the probable plasmid(s) responsible for the abolished MicF-mediated regulation of *ompF*. Upon MicF overexpression, *ompF* translation remained repressed in the pCD1- or pMT1-containing strains, as observed in the plasmid-null strain. However, the expression level of OmpF-GFP fusion protein was even slightly upregulated about 1.6-fold in the *Y. pestis* strain carrying plasmid pPCP1. Such activation was also observed in the strain carrying both plasmids pPCP1 and pCD1 (**Figure 2**). Plasmid pPCP1 alone could overwhelm the translational repression in the plasmid-cured strain, thereby suggesting that pPCP1 may mainly contribute to the abolishment of sRNA-mediated regulation. Paradoxically, translational repression was still found in the strain with pPCP1 and pMT1. We speculated that addition of pMT1 might interfere with the effect of pPCP1 on sRNA-mediated regulation because of the potential interactions between these two plasmids.

We also monitored MicF in the different bacterial strains under the same conditions as those shown in **Figures 1**, **2**. No MicF were found expressed in the pBAD control of various *Y. pestis* strains by using Northern Blot, which might be mainly due to huge disparity in copy number of MicF between plasmid pBAD and chromosome and/or low expression levels of endogeneous MicF under our experimental conditions. Additionally, no significant differences in MicF overexpression were found among *Y. pestis* strains (**Figure 3**).

Validation of MicF-mediated Regulation of Endogenous *OmpF* Elicited by Intrinsic Plasmids

To test whether the intrinsic plasmids interfere with the MicF-mediated regulation of the chromosome-encoded target gene *ompF*, the abundances of endogenous *ompF* transcript and OmpF protein were validated in various strains of *Y. pestis* by Northern Blot and Western Blot, respectively (**Figure 4**). In agreement with the results of gene fusion reporter systems, the abundance of *ompF* transcript or OmpF protein were found decreased 2.0–5.0 fold in four pPCP1-cured strains (122-null, 122-pCD1$^+$pMT1$^+$, 201-pCD1$^+$, and 201-pMT1$^+$) upon MicF overexpression relative to that in the control strains. Strikingly, the *ompF* transcript was stable, but approximate three-fold decrease was found in the expression level of OmpF in the strain 122-pMT1$^+$pPCP1$^+$, which is also consistent with the findings presented in **Figure 2**. However, only the comparable levels of *ompF* and OmpF were found among three groups of pPCP-containing strains (122, 122-pPCP1, and 122-pCD1$^+$pPCP1$^+$). The downregulation phenomenon is roughly consistent with that of plasmid-borne *gfp* fusion. The discrepancy is likely due to the different sensitivity between translational fusion assay and Western blot. However, no obvious upregulation was found in strain 201. Maybe the actual activation was magnified by translational fusion assay or veiled by detection threshold of Western blot or Northern Blot. Taken together, this observation further confirmed the conclusion that intrinsic plasmids have the potential impacts on abolishment of MicF-mediated *ompF* regulation in *Y. pestis*.

Our study demonstrated that the mobile elements affect sRNA-mediated regulation in *Y. pestis*. Distinct conclusions may be drawn if various strains carrying different plasmids are used to investigate the sRNA-mediated regulation. For example, Hfq reportedly represses the biofilm formation in *Y. pestis* KIM6+ (an avirulent derivative of the fully virulent strain KIM, which was cured of the pCD1 plasmid) grown in BHI medium (Bellows et al., 2012). Opposite effects on biofilm formation were observed in the *hfq* mutant of *Y. pestis* wild-type strain 201 and its derivative strain lacked the plasmid pCD1 (unpublished data). Therefore, the background bacteria to be used as control strain should be carefully selected. The possible roles of plasmids in gene regulation should be considered even if the exact mechanism is not fully understood. Although a relationship exists between the intrinsic plasmids and sRNA-mediated regulation, the presence or absence of any plasmid did not cause the clear-cut effects on MicF-mediated *ompF* regulation. This phenomenon may indicate that mutual interactions exist among intrinsic plasmids in *Y. pestis*, and such interactions further influence sRNA-mediated regulation.

Acknowledgments

We greatly thank Jörg Vogel's lab for kindly providing the pXG superfolder GFP series of plasmids and Prof. Hengliang Wang from Beijing Institute of Biotechnology for offering *E. coli* K12 strain MG1655. This study was funded by the National Basic Research Program of China (2014CB744405), the National Natural Science Foundation of China (31171248 and 31430006), and the State Key of Pathogen and Biosecurity (Academy of Military Medical Science, SKLPBS1418).

References

Achtman, M., Zurth, K., Morelli, G., Torrea, G., Guiyoule, A., and Carniel, E. (1999). *Yersinia pestis*, the cause of plague, is a recently emerged clone of *Yersinia pseudotuberculosis*. *Proc. Natl. Acad. Sci. U.S.A.* 96, 14043–14048. doi: 10.1073/pnas.96.24.14043

Andersen, J., Forst, S. A., Zhao, K., Inouye, M., and Delihas, N. (1989). The function of micF RNA. micF RNA is a major factor in the thermal regulation of OmpF protein in *Escherichia coli*. *J. Biol. Chem.* 264, 17961–17970.

Bellows, L. E., Koestler, B. J., Karaba, S. M., Waters, C. M., and Lathem, W. W. (2012). Hfq-dependent, co-ordinate control of cyclic diguanylate synthesis and catabolism in the plague pathogen *Yersinia pestis*. *Mol. Microbiol.* 86, 661–674. doi: 10.1111/mmi.12011

Chiang, M. K., Lu, M. C., Liu, L. C., Lin, C. T., and Lai, Y. C. (2011). Impact of Hfq on global gene expression and virulence in *Klebsiella pneumoniae*. *PLoS ONE* 6:e22248. doi: 10.1371/journal.pone.0022248

Corcoran, C. P., Podkaminski, D., Papenfort, K., Urban, J. H., Hinton, J. C., and Vogel, J. (2012). Superfolder GFP reporters validate diverse new mRNA targets of the classic porin regulator, MicF RNA. *Mol. Microbiol.* 84, 428–445. doi: 10.1111/j.1365-2958.2012.08031.x

Delihas, N. (1997). Antisense micF RNA and 5′-UTR of the target ompF RNA: phylogenetic conservation of primary and secondary structures. *Nucleic Acids Symp. Ser.* 36, 33–35.

Delihas, N. (2003). Annotation and evolutionary relationships of a small regulatory RNA gene *micF* and its target *ompF* in *Yersinia* species. *BMC Microbiol.* 3:13. doi: 10.1186/1471-2180-3-13

Desnoyers, G., Bouchard, M. P., and Massé, E. (2013). New insights into small RNA-dependent translational regulation in prokaryotes. *Trends Genet.* 29, 92–98. doi: 10.1016/j.tig.2012.10.004

Geng, J., Song, Y., Yang, L., Feng, Y., Qiu, Y., Li, G., et al. (2009). Involvement of the post-transcriptional regulator Hfq in *Yersinia pestis* virulence. *PLoS ONE* 4:e6213. doi: 10.1371/journal.pone.0006213

Hu, P., Elliott, J., McCready, P., Skowronski, E., Garnes, J., Kobayashi, A., et al. (1998). Structural organization of virulence-associated plasmids of *Yersinia pestis*. *J. Bacteriol.* 180, 5192–5202.

Kulesus, R. R., Diaz-Perez, K., Slechta, E. S., Eto, D. S., and Mulvey, M. A. (2008). Impact of the RNA chaperone Hfq on the fitness and virulence potential of uropathogenic *Escherichia coli*. *Infect. Immun.* 76, 3019–3026. doi: 10.1128/IAI.00022-08

Lindler, L. E., Plano, G. V., Burland, V., Mayhew, G. F., and Blattner, F. R. (1998). Complete DNA sequence and detailed analysis of the *Yersinia pestis* KIM5 plasmid encoding murine toxin and capsular antigen. *Infect. Immun.* 66, 5731–5742.

Mizuno, T., Chou, M. Y., and Inouye, M. (1984). A unique mechanism regulating gene expression: translational inhibition by a complementary RNA transcript (micRNA). *Proc. Natl. Acad. Sci. U.S.A.* 81, 1966–1970. doi: 10.1073/pnas.81.7.1966

Ni, B., Du, Z., Guo, Z., Zhang, Y., and Yang, R. (2008). Curing of four different plasmids in *Yersinia pestis* using plasmid incompatibility. *Lett. Appl. Microbiol.* 47, 235–240. doi: 10.1111/j.1472-765X.2008.02426.x

Perry, R. D., and Fetherston, J. D. (1997). *Yersinia pestis*—etiologic agent of plague. *Clin. Microbiol. Rev.* 10, 35–66.

Perry, R. D., Straley, S. C., Fetherston, J. D., Rose, D. J., Gregor, J., and Blattner, F. R. (1998). DNA sequencing and analysis of the low-Ca2+-response plasmid pCD1 of *Yersinia pestis* KIM5. *Infect. Immun.* 66, 4611–4623.

Sittka, A., Pfeiffer, V., Tedin, K., and Vogel, J. (2007). The RNA chaperone Hfq is essential for the virulence of Salmonella typhimurium. *Mol. Microbiol.* 63, 193–217. doi: 10.1111/j.1365-2958.2006.05489.x

Song, Y., Tong, Z., Wang, J., Wang, L., Guo, Z., Han, Y., et al. (2004). Complete genome sequence of *Yersinia pestis* strain 91001, an isolate avirulent to humans. *DNA Res.* 11, 179–197. doi: 10.1093/dnares/11.3.179

Vogel, J., and Luisi, B. F. (2011). Hfq and its constellation of RNA. *Nat. Rev. Microbiol.* 9, 578–589. doi: 10.1038/nrmicro2615

Waters, L. S., and Storz, G. (2009). Regulatory RNAs in bacteria. *Cell* 136, 615–628. doi: 10.1016/j.cell.2009.01.043

Zhou, D., Han, Y., and Yang, R. (2006). Molecular and physiological insights into plague transmission, virulence and etiology. *Microbes Infect.* 8, 273–284. doi: 10.1016/j.micinf.2005.06.006

Zhou, D., Tong, Z., Song, Y., Han, Y., Pei, D., Pang, X., et al. (2004). Genetics of metabolic variations between *Yersinia pestis* biovars and the proposal of a new biovar, microtus. *J. Bacteriol.* 186, 5147–5152. doi: 10.1128/JB.186.15.5147-5152.2004

Conflict of Interest Statement: The authors declare that the research was conducted in the absence of any commercial or financial relationships that could be construed as a potential conflict of interest.

9

The allosteric behavior of Fur mediates oxidative stress signal transduction in *Helicobacter pylori*

*Simone Pelliciari, Andrea Vannini, Davide Roncarati and Alberto Danielli**

Department of Pharmacy and Biotechnology (FaBiT), University of Bologna, Bologna, Italy

Edited by:
Beiyan Nan,
Texas A&M University, USA

Reviewed by:
Paul S. Hoffman,
University of Virginia, USA
Héctor Toledo,
Universidad de Chile, Chile

*Correspondence:
Alberto Danielli,
Department of Pharmacy and
Biotechnology (FaBiT), University of
Bologna, Via Selmi 3, 40126 Bologna,
Italy
alberto.danielli@unibo.it

The microaerophilic gastric pathogen Helicobacter pylori is exposed to oxidative stress originating from the aerobic environment, the oxidative burst of phagocytes and the formation of reactive oxygen species, catalyzed by iron excess. Accordingly, the expression of genes involved in oxidative stress defense have been repeatedly linked to the ferric uptake regulator Fur. Moreover, mutations in the Fur protein affect the resistance to metronidazole, likely due to loss-of-function in the regulation of genes involved in redox control. Although many advances in the molecular understanding of HpFur function were made, little is known about the mechanisms that enable Fur to mediate the responses to oxidative stress. Here we show that iron-inducible, apo-Fur repressed genes, such as *pfr* and *hydA*, are induced shortly after oxidative stress, while their oxidative induction is lost in a *fur* knockout strain. On the contrary, holo-Fur repressed genes, such as *frpB1* and *fecA1*, vary modestly in response to oxidative stress. This indicates that the oxidative stress signal specifically targets apo-Fur repressed genes, rather than impairing indiscriminately the regulatory function of Fur. Footprinting analyses showed that the oxidative signal strongly impairs the binding affinity of Fur toward apo-operators, while the binding toward holo-operators is less affected. Further evidence is presented that a reduced state of Fur is needed to maintain apo-repression, while oxidative conditions shift the preferred binding architecture of Fur toward the holo-operator binding conformation, even in the absence of iron. Together the results demonstrate that the allosteric regulation of Fur enables transduction of oxidative stress signals in *H. pylori*, supporting the concept that apo-Fur repressed genes can be considered oxidation inducible Fur regulatory targets. These findings may have important implications in the study of *H. pylori* treatment and resistance to antibiotics.

Keywords: ferric uptake regulator, oxidative stress, antibiotic resistance, allosteric regulation, redox regulation, metalloproteins, metal homeostasis, transcriptional regulation

Introduction

Helicobacter pylori is an obligate microaerophilic human pathobiont that colonizes the gastric mucosa of half of the world's population. Infections can persist asymptomatically for the lifespan of the host. However, in many cases the colonization of *H. pylori* constitutes a major cause of acute and chronic gastritis, gastric and duodenal ulcer diseases, as well as gastric cancer (Salama et al., 2013). The microaerophilic nature renders the bacterium highly vulnerable to oxygen toxicity originating from the aerobic environment and endogenous sources of reactive oxygen species (ROS; Hazell et al., 2001). In addition, *H. pylori* has to counter exogenous sources of ROS derived by the inflammatory response of the gastric epithelium (Bagchi et al., 1996), eventually leading to the oxidative burst of infiltrating macrophages and neutrophils (Ramarao et al., 2000). As such, the bacterium has adapted to neutralize noxious oxigen species by expressing a rich repertoire of antioxidant factors and enzymes (Wang et al., 2006). Even though many advances have been made in understanding their molecular function and regulation, it is less clear how the oxidative signal is transduced by the bacterium to provide the coordinated responses to counteract the oxidative damage. In fact, the *H. pylori* genome lacks annotated orthologs of potential oxidative stress regulators involved in the transcriptional control of antioxidant proteins, such as OxyR, PerR, OhrR, or SoxRS (Dubbs and Mongkolsuk, 2012). Beside the post-transcriptional regulator CsrA (Barnard et al., 2004), the DNA binding (HP0119) and repair (MutS) proteins (Wang et al., 2005; Wang and Maier, 2015), and the essential orphan response regulator HsrA (HP1043), for which a role in the oxidative stress response to low levels of metronidazole (MTZ) or oxygen was recently proposed (Olekhnovich et al., 2014), the regulation of oxidative stress defenses in *H. pylori* has been repeatedly linked to the ferric uptake regulator Fur (Cooksley et al., 2003; van Vliet et al., 2004; Ernst et al., 2005; Danielli and Scarlato, 2010). Point mutations in the apo-repressed *sodB* superoxide dismutase promoter affect its Fur-dependent regulation (Carpenter et al., 2009a), while several amino acid substitutions in the coding sequence of HpFur were shown to affect resistance to MTZ (Albert et al., 2005; Choi et al., 2011), likely through the derepression of SodB (Tsugawa et al., 2011). This involvement of Fur in redox homeostasis is not surprising since the formation of hydroxyl radical species, the most reactive ROS, is intimately linked to the availability of free intracellular iron ions through the Fenton reaction (Touati, 2000). Moreover, the Fur superfamily of regulators comprises PerR and other orthologs whose roles in the transduction of oxidative stress signals have been well characterized also in other bacteria (Lee and Helmann, 2006a,b, 2007; Belzer et al., 2011; Dubbs and Mongkolsuk, 2012; Troxell and Hassan, 2013).

The Fur protein of *H. pylori* is peculiar, because of its allosteric regulation mechanism which confers the function of a transcriptional commutator switch: holo-Fur and apo-Fur each bind different regulatory elements, repressing oppositely either iron-repressible (FeOFF) or iron-inducible (FeON) gene targets, according to the intracellular iron concentration. Thereby, transcription of iron uptake genes such as *frpB* or *fecA* is repressed by holo-Fur when iron is abundant (Delany et al., 2001a; Danielli et al., 2009), while the transcription of genes encoding iron storage protein like the Pfr ferritin is induced (Delany et al., 2001b), along with genes coding for iron-cofactor proteins such as SodB and HydA (Ernst et al., 2005). On the contrary, when iron is limiting, the transcription of iron uptake genes is derepressed, together with the apo-Fur mediated repression of iron storage genes (*pfr*) and genes encoding the iron-cofactored enzymes, including *sodB* and *hydA* (Carpenter et al., 2013). Recently it has been demonstrated that this mechanism is based on the allosteric behavior of Fur which adopts different conformations and binding architectures to DNA when complexed to the iron co-factor (Agriesti et al., 2014). The peculiar apo-repression mechanism appears to rely on the presence of an additional α-helix at the N-terminus of the

Fur protein, conserved in the ε-proteobacteria clade including *H. pylori* and *C. jejuni* (Carpenter et al., 2009b, 2013; Butcher et al., 2012; Agriesti et al., 2014). Interestingly, the point mutations in the same α-helix have been shown to strongly influence the resistance to MTZ (Choi et al., 2011), suggesting a direct link between apo-Fur repression and oxidative stress response in *H. pylori*.

It is with this background, and with the bystanding interest in the molecular basis of *H. pylori* persistence and antibiotic resistance, that we sought to further investigate the contribution of apo- and holo-Fur to the redox homeostasis of the bacterium, trying to characterize the mechanisms that enable Fur to mediate the responses to oxidative stress.

Materials and Methods

Bacterial Strains and Culture Conditions

Helicobacter pylori strains (listed in **Table 1**), were revitalized from glycerol stocks on Brucella broth agar plates added with 5% fetal calf serum and Skirrow's antibiotic supplement under microaerophilic conditions in jars (Oxoid gas packs). After re-streaking on fresh plates, bacteria were cultured in a 9% CO_2, 91% air atmosphere at 37°C and 95% humidity in a water-jacketed incubator (Thermo Forma Scientific). Liquid cultures were grown in modified Brucella broth medium supplemented with 5% fetal bovine serum in glass flasks or 25 cm^3 sterile plastic flasks with vented cap (Corning). For transcriptional analysis, *H. pylori* planktonic cultures ($OD_{600} \sim 0.8$) were treated for 10 min either with 1 mM $(NH_4)_2Fe(SO_4)_2$, 150 μM 2,2 Dipyridyl (Dipy) or 10 mM H_2O_2 before RNA extraction.

DNA manipulations

DNA manipulations were performed with standard techniques. Restriction and modification enzymes were purchased from New England Biolabs. Preparations of plasmid DNA were carried out with a NucleoBond Xtra Midi plasmid purification kit (Macherey-Nagel).

DNAse I Footprinting

Promoter probes were prepared as previously described (Agriesti et al., 2014). Briefly, pGEM K-F and pGEMpfr plasmids were digested with HindIII or BamHI (NEB) respectively, dephosphorylated by treatment with calf intestine phosphatase (NEB) and subsequently 5′-end labeled with [γ-^{32}P]-ATP and T4 polynucleotide kinase (NEB). After a second digestion at the 3′-end of the probe, the DNA fragments were recovered by gel extraction. Recombinant His$_6$-Fur was overexpressed and purified under native conditions (Delany et al., 2001b), treated with thrombin protease (10 U/mg) to remove the N-terminal histidine tag and dialyzed against Fur Footprinting buffer (10 mM Tris-Cl, pH 7.85, 50 mM NaCl, 10 mM KCl, 0.02% Igepal CA-630, 10% glycerol, 0.1 mM DTT).

The DNase I footprinting reactions were performed in 1X Fur footprinting buffer incubating approximatively 15 fmol of radiolabeled probe with different amounts of Fur protein for 20 min at room temperature, with 300 ng of sonicated salmon

TABLE 1 | Bacterial strains, plasmids, and oligonucleotides.

Helicobacter pylori strain	Genotype	References
G27	Clinical isolate, wild type	Xiang et al. (1995)
G27(fur::km)	Derived from G27 strain; bp 25 to 434 of the CDS of fur (G27_401) were substituted with a kanamycin resistance cassette, Km^R	Delany et al. (2005)

Plasmid	Description	
pGEMK-F	Derivative of pGEM3Z containing a 447 bp EcoRI–PstI fragment comprising the intergenic region between katA and frpB and the 5′ end of each gene. Amp^r	Delany et al. (2001b)
pGEMpfr	Derivative of pGemT containing the pfr–G27_615 intergenic region as a 390 bp from base 699220 to base 699609 of the G27 genome, Amp^r	Delany et al. (2001b)

Oligos	Sequence	
FrpB RT FW	TGTGAGAGGCATTGAAGACAGGCT	Agriesti et al. (2014)
FrpB RT RV	CGCCTTTGGTAACTTCCACGCTTT	Agriesti et al. (2014)
Pfr RT FW	TGCTGTTCAGCCACATACCATTGC	Agriesti et al. (2014)
Pfr RT RV	GCGCCTGAGCATAAGTTTGAAGGT	Agriesti et al. (2014)
FecA1 RT FW	AGCGTGCATGGTGTCAAAAC	This work
FecA1 RT RV	AACTTCCTTGCTCCTCCAGC	This work
HydA RT FW	GAAAGCCGCTCAATACGCAG	This work
HydA RT RV	TTGCGCGTTAGAGGGGTTAG	This work
16s RT FW	GGAGTACGGTCGCAAGATTAAA	Loh et al. (2011)
16s RT RV	CTAGCGGATTCTCTCAATGTCAA	Loh et al. (2011)

sperm DNA as non-specific competitor, 150 μM $(NH_4)_2Fe(SO_4)_2$ or 150 μM Dipy, in a final volume of 50 μl. To perform the assay in different redox conditions, DTT (1 mM or 5 mM) or H_2O_2 (5 mM) were added to the binding reaction.

DNAse I (Novagen) was diluted in 1X Fur FPB added with 10 mM $CaCl_2$ and 5 mM $MgCl_2$. Samples containing iron and DTT were digested with 0,03 U of DNase for 90 s; for all the other conditions the concentration of DNase was raised to 0,15 U.

Reactions were stopped with the addition of 140 μl of STOP buffer (192 mM NaOAc pH 5.2, 32 mM EDTA, 0.14% SDS, 64 μg/μL salmon sperm DNA), then purified and extracted. Samples were resuspended in 10 μL of formamide loading buffer, denatured at 100°C for 3 min, separated on 8 M urea-6% polyacrylamide sequencing gels and autoradiographed.

RNA Extraction and cDNA Synthesis

Total RNA was extracted with TRI-Reagent (Sigma-Aldrich) following the manufacturer's protocol. To remove contaminating genomic DNA, 5 μg of total RNA were treated with 1U DNAseI in 1X DNAse buffer (80 mM Hepes pH 7.5, 10 mM NaCl, 5 mM $MgCl_2$, 10 mM DTT) at 37°C for 30 min in a final volume of 50 μl; then the samples were phenol/chloroform extracted, ethanol precipitated and resuspended in RNAse free mQH$_2$O.

For cDNA synthesis, 1 μg of DNA-free RNA was incubated with 50 ng of random primers in a final volume of 10 μl, denatured for 5 min at 70°C and immediately chilled on ice; then 5 U of reverse transcriptase (RT-AMV, Promega), dNTPs (final concentration 1 mM each) and RT-AMV Buffer were added and the reaction was incubated for 1 h at 37°C.

Real Time PCR

Two μL of the diluted (1:5) cDNA samples were mixed with 5 μL of 2X iTaq Universal SYBR Green Supermix (Bio-Rad) and oligonucleotides specific for the gene of interest (**Table 1**) in a final volume of 10 μL. Real time PCR was performed using the following cycling protocol: 50°C for 2 min, 95°C for 2 min, then 40 cycles consisting of a denaturation for 15 s at 95°C followed by 1 min at 60°C (annealing and extension step). For each real time experiment, the specificity of the reaction was checked by including a melting profile at the end of the run. Data were analyzed using the ΔΔCt method. The levels of expression of the genes of interest were normalized against the measured level of the RNA coding for the housekeeping 16S rRNA gene.

Results

Hydrogen Peroxide Induces Apo- Fur Repressed Genes

In order to investigate the role of Fur on the transduction of oxidative stress signals, we analyzed the transcriptional responses of Fur-regulated genes in H. pylori cultures exposed to hydrogen peroxide. We selected holo-Fur repressed (frpB and fecA1) and apo-Fur repressed genes (pfr and hydA) since they are oppositely regulated by Fur in virtue of an extensively described allosteric regulation mechanism responsive to iron (Delany et al., 2001b; Carpenter et al., 2013; Agriesti et al., 2014). To find the optimal conditions for the analysis, several preliminary assays were carried out (data not shown): bacterial cultures were treated with 10 to 100 mM H_2O_2 for 5 to 20 min, and the mRNA levels of pfr and frpB genes were assayed by RT-qPCR and compared to the untreated samples. The more reproducible results were obtained in the samples treated with 10 mM of H_2O_2 for 10 min. Higher concentrations of hydrogen peroxide led to erratic variations of mRNA levels, while prolonged treatments showed a return to nearly non-stressed levels after 20 min, with a bell-shaped trend of the response (data not shown).

Thus, planktonic cultures of wild type and Δfur strains were treated with 10 mM of H_2O_2 to induce the oxidative stress, or with

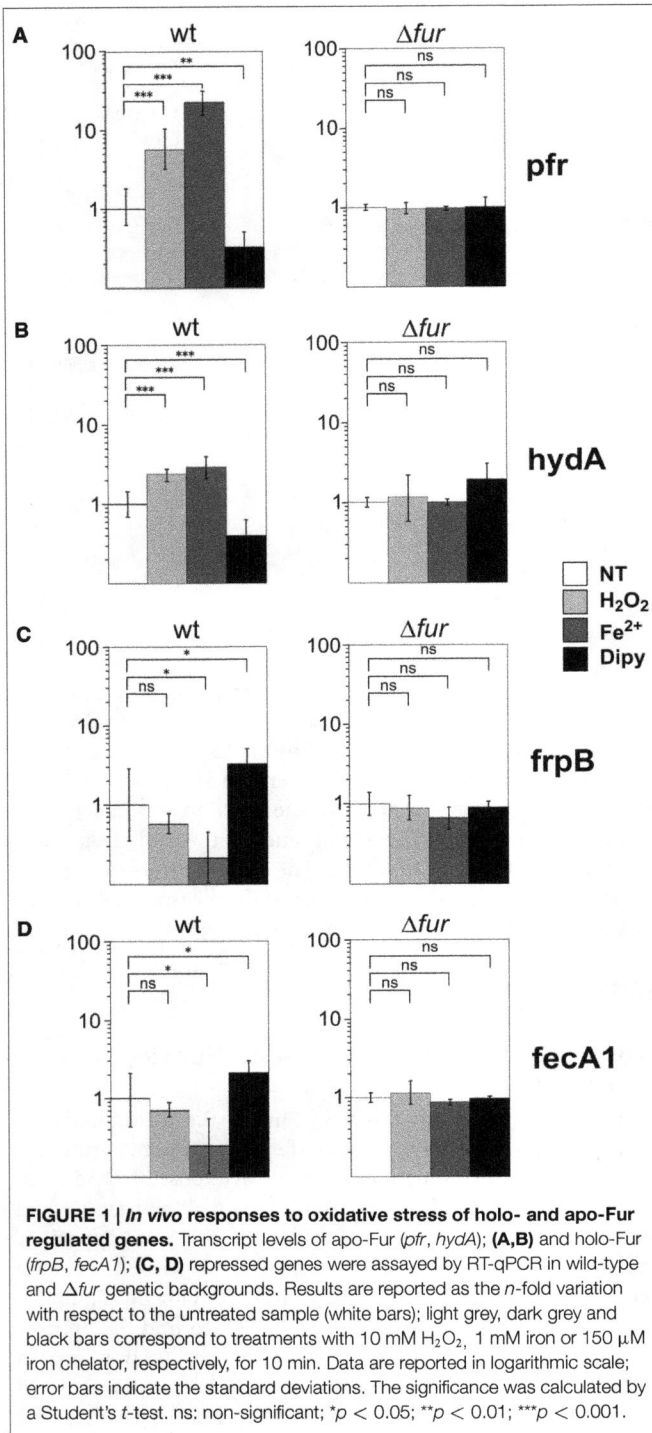

FIGURE 1 | *In vivo* responses to oxidative stress of holo- and apo-Fur regulated genes. Transcript levels of apo-Fur (*pfr*, *hydA*); **(A,B)** and holo-Fur (*frpB*, *fecA1*); **(C, D)** repressed genes were assayed by RT-qPCR in wild-type and Δ*fur* genetic backgrounds. Results are reported as the *n*-fold variation with respect to the untreated sample (white bars); light grey, dark grey and black bars correspond to treatments with 10 mM H_2O_2, 1 mM iron or 150 μM iron chelator, respectively, for 10 min. Data are reported in logarithmic scale; error bars indicate the standard deviations. The significance was calculated by a Student's *t*-test. ns: non-significant; *$p < 0.05$; **$p < 0.01$; ***$p < 0.001$.

1 mM iron sulfate (Fe^{2+}) and 150 μM iron chelator (2,2-dipyridyl; Dipy) to elicit the well-described responses occurring after iron repletion or chelation. Messenger RNA levels of genes subjected to either holo- or apo-repression were followed by RT-qPCR with results reported in **Figure 1**.

In the wild type strain, we observed an increase of *pfr* transcript level in response to iron and a decrease of mRNA levels upon iron chelation, as expected for an apo-Fur regulated gene. Interestingly, a sixfold increase of *pfr* mRNA level was also observed in response to the H_2O_2 treatment (**Figure 1A**). This response was lost in a Δ*fur* strain, suggesting that the observed regulation is Fur-dependent. To ascertain whether inducibility upon oxidative stress could be a conserved feature of apo-repressed Fur targets, the analysis was repeated on *hydA*, that codes for a subunit of a quinone-reactive Ni/Fe-hydrogenase, reported previously to be apo-Fur repressed as *pfr* (Carpenter et al., 2013). Consistently, the transcript levels of *hydA* were induced upon iron repletion and hydrogen peroxide treatment, and proved to be lost in a *fur* knockout background, paralleling the responses observed for *pfr* (**Figure 1B**).

On the other hand, holo-Fur repressed genes, such as *frpB* and *fecA1*, which are induced by the withdrawal of iron and repressed in a Fur- and iron-dependent fashion, exhibited a different, weak response to oxidative stress. In fact, the slight down-regulation upon hydrogen peroxide treatment resulted in both cases statistically non-significant (**Figures 1C,D**).

We conclude that apo-Fur but not holo-Fur repressed genes are responsive to hydrogen peroxide treatment, and that the oxidative stress signal mimics the effects of iron repletion on the transcription of these genes. Notably, the responses are lost in a Δ*fur* strain, suggesting that the transduction of the oxidative signal is directly or indirectly mediated by Fur.

Hydrogen Peroxide Selectively Impairs Fur Binding to DNA

To establish if the Fur protein mediates the oxidative stress signal response through a direct regulation, DNaseI footprinting assays were performed with the purified recombinant Fur protein and the radiolabeled P*pfr* promoter region as probe. Binding experiments were conducted under iron repletion (**Figure 2A**) or chelation (**Figure 2B**), both in reducing (5 mM DTT, 1 mM DTT) or oxidizing conditions (5 mM H_2O_2).

Under reducing conditions, Fur protects the P*pfr* probe in correspondence of three regions when iron is chelated, pOPI, pOPII, and pOPIII, each encasing a hypersensitive band (**Figure 2B**; 1 mM DTT, 5 mM DTT). These regions correspond precisely, to the three previously characterized *bona-fide* apo-operators of the P*pfr* promoter (Delany et al., 2001b). Under the same redox conditions, but in the presence of iron ions, Fur loses affinity for pOPI, pOPII, and pOPIII (**Figure 2B**; center and right panels), in agreement with the allosteric behavior reported recently for Fur on this promoter (Agriesti et al., 2014). Interestingly, a fourth, low-affinity, distal and iron-dependent region of protection appears (*holo*-pOPIV; **Figure 2A**), reported also originally by Delany and co-workers (Delany et al., 2001b).

When hydrogen peroxide is added to the reaction, the protection pattern conferred by Fur changes dramatically: the binding of apo-Fur to pOPII and pOPIII is strongly impaired, while a protection on the low-affinity holo-pOPIV operator appears, even in the absence of iron ions (**Figure 2B**, left panel). Similarly, hydrogen peroxide has a negative effect on Fur binding even in the presence of iron, with a general loss of affinity for all the apo-operators. Seemingly, also Fur binding to pOPIV is

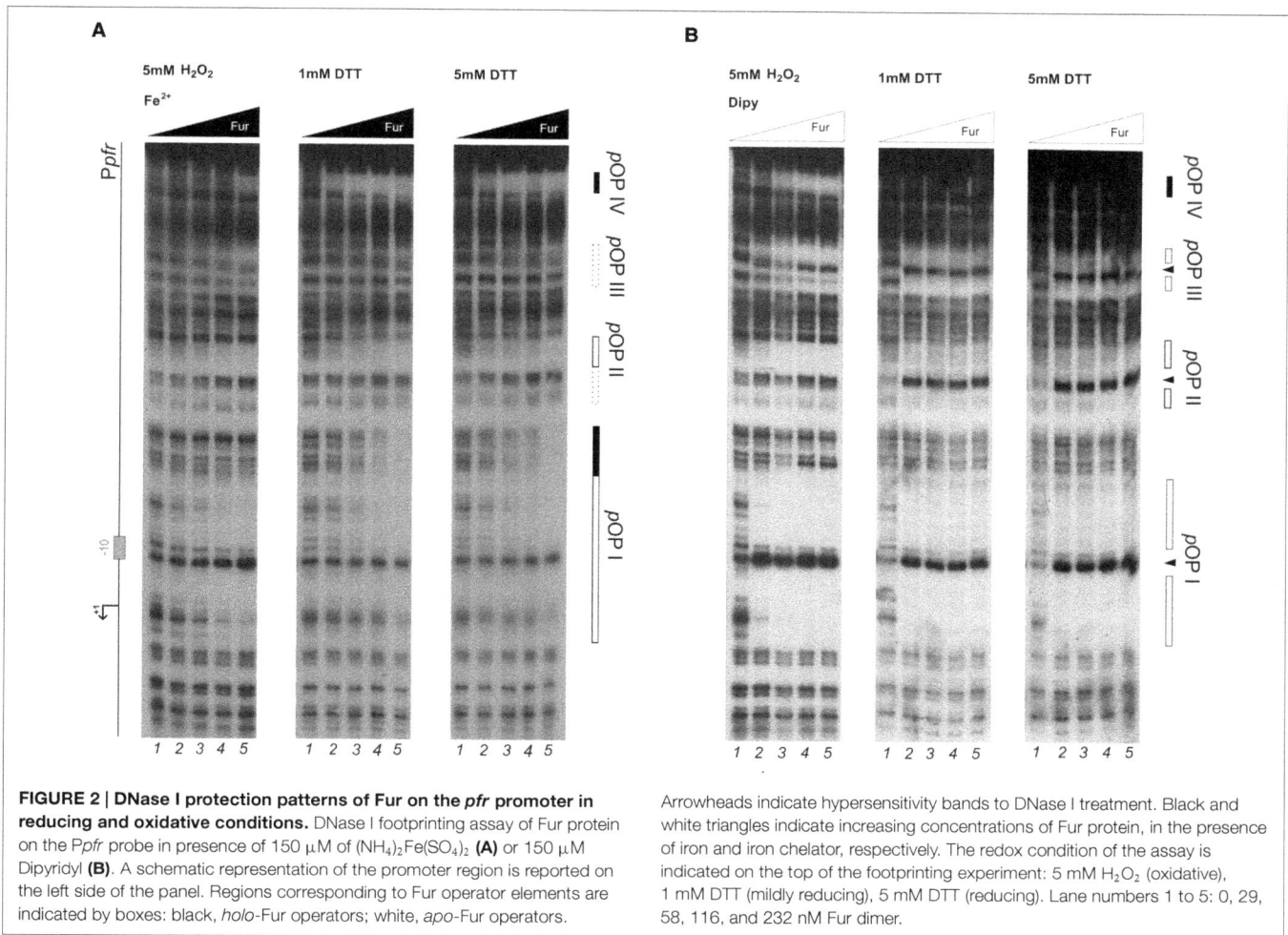

FIGURE 2 | DNase I protection patterns of Fur on the *pfr* promoter in reducing and oxidative conditions. DNase I footprinting assay of Fur protein on the P*pfr* probe in presence of 150 μM of $(NH_4)_2Fe(SO_4)_2$ **(A)** or 150 μM Dipyridyl **(B)**. A schematic representation of the promoter region is reported on the left side of the panel. Regions corresponding to Fur operator elements are indicated by boxes: black, *holo*-Fur operators; white, *apo*-Fur operators.

Arrowheads indicate hypersensitivity bands to DNase I treatment. Black and white triangles indicate increasing concentrations of Fur protein, in the presence of iron and iron chelator, respectively. The redox condition of the assay is indicated on the top of the footprinting experiment: 5 mM H_2O_2 (oxidative), 1 mM DTT (mildly reducing), 5 mM DTT (reducing). Lane numbers 1 to 5: 0, 29, 58, 116, and 232 nM Fur dimer.

affected, resulting in the absence of a clear protected region on the P*pfr* probe at the highest protein concentration (**Figure 2A**; left panel).

Oxidative Stress Signal Transduction on Apo-repressed Fur Targets

Since the most prominent change of Fur binding to P*pfr* in response to the oxidative signal was observed especially on the distal pOPIII and pOPIV elements, additional experiments with lowered Fur protein concentration were carried out to investigate whether also the binding to the high-affinity proximal operator elements pOPI and pOPII could be affected (**Figure 3**). In the footprinting experiment, an equivalent Fur protection pattern in pOPI and pOPII was observed at a fivefold higher protein concentration when hydrogen peroxide was added to the binding reaction, confirming the loss of apo-Fur binding also to the proximal high-affinity pOPI operator overlapping the core promoter region (**Figure 3A**). The same results where observed when the binding to the whole promoter region was assayed by EMSA (**Figure 3B**). These results demonstrate that oxidative conditions can impair the binding affinity of Fur to apo-operators, and promote the binding of the protein to a low affinity

holo-operator (pOPIV) even in the absence of the iron-cofactor. Recalling the robust transcriptional derepression of *pfr* observed *in vivo* after hydrogen peroxide treatment, we conclude that the oxidative stress signal can be transduced in a transcriptional response on apo-repressed (FeON) Fur targets, as the direct result of a decreased binding of the Fur repressor.

Oxidative Stress Imposes Different Binding Architectures of Fur

To investigate whether a similar effect could pertain also holo-Fur targets, the footprinting analysis was extended to the P*frpB* probe, encompassing the promoter region of the iron-repressed (FeOFF) *frpB* gene (**Figure 4**). In agreement with the allosteric behavior of Fur (Agriesti et al., 2014), iron influenced the binding affinity of the protein on this promoter oppositely with respect to P*pfr*. Under reducing conditions, the highest affinity was observed in the presence of the iron co-factor for the proximal fOPI holo-operator, overlapping the core promoter, with a second region of protection (fOPII) appearing only at higher protein concentration upstream of fOPI (**Figure 4A**, center and right panels). When iron was chelated the affinity of Fur for these two elements swapped, resulting in the strong

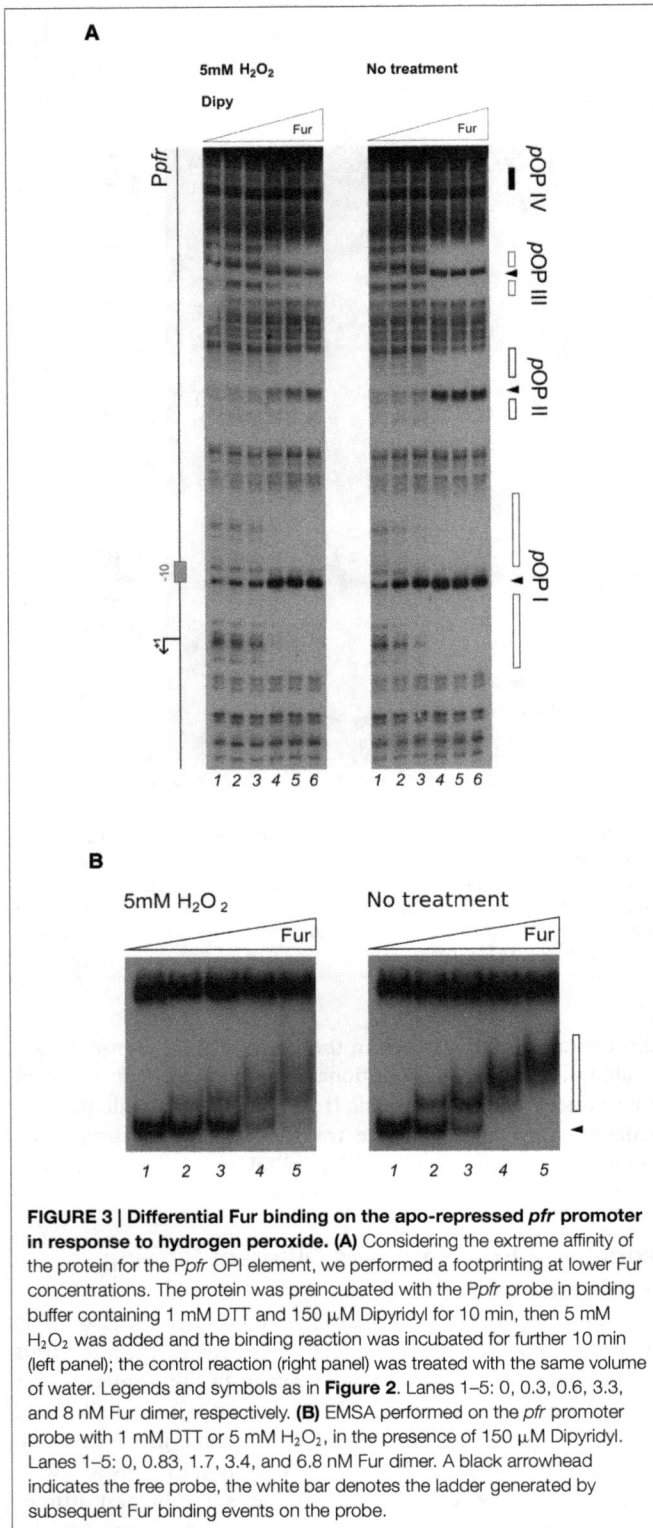

FIGURE 3 | Differential Fur binding on the apo-repressed *pfr* promoter in response to hydrogen peroxide. (A) Considering the extreme affinity of the protein for the P*pfr* OPI element, we performed a footprinting at lower Fur concentrations. The protein was preincubated with the P*pfr* probe in binding buffer containing 1 mM DTT and 150 μM Dipyridyl for 10 min, then 5 mM H_2O_2 was added and the binding reaction was incubated for further 10 min (left panel); the control reaction (right panel) was treated with the same volume of water. Legends and symbols as in **Figure 2**. Lanes 1–5: 0, 0.3, 0.6, 3.3, and 8 nM Fur dimer, respectively. **(B)** EMSA performed on the *pfr* promoter probe with 1 mM DTT or 5 mM H_2O_2, in the presence of 150 μM Dipyridyl. Lanes 1–5: 0, 0.83, 1.7, 3.4, and 6.8 nM Fur dimer. A black arrowhead indicates the free probe, the white bar denotes the ladder generated by subsequent Fur binding events on the probe.

protection of the fOPII apo-operator even at the lowest Fur concentration, while binding to the fOPI holo-operator was impaired (**Figure 4B**; center and right panels). In addition, a third distal apo-operator appeared, fOPIII, immediately upstream

of fOPII. Note also the formation of two hypersensitive bands between the three operators, indicative of modifications affecting the DNA structure.

Strikingly, the treatment with hydrogen peroxide induced distinct modifications in the protection patterns of Fur. The high-affinity binding to the fOPII and fOPIII apo-operators was abolished (**Figure 4B**, left panel), while the iron-dependent binding to the fOPI holo-operator was only modestly affected (**Figure 4A**, left panel). In addition, hydrogen peroxide provoked a Fur-dependent protection of fOPI which resembled that of the holo-protein even though the iron co-factor was chelated (**Figure 4B**, left panel), paralleling the effect observed for pOPIV on the P*pfr* promoter (**Figure 2B**; left panel). Thus, the oxidative stress signal promotes a swap in the binding architecture of Fur, impairing the binding to apo-operators and favoring to a certain extent the binding of holo-operators which require a different binding architecture of the regulator. In other words, oxidative stress induces holo-mimetic binding architectures, which appear to disfavor the binding to the apo-operators.

Cumulative Effects of Fur Binding Affinity and Fur Binding Conformation to the *frpB* Promoter

The allosteric behavior of Fur induced by hydrogen peroxide suggests that the oxidative stress signal specifically targets apo-Fur repressed genes, rather than impairing indiscriminately the regulatory function of Fur. To ascertain this hypothesis and better estimate the affinity loss on holo-operators, an additional set of footprinting analyses on the P*frpB* promoter was conducted, with lowered Fur protein concentrations (**Figure 5**). Under reducing conditions, the protection of the fOPI holo-operator is partial at 4 nM and results in a complete protection at 8 nM Fur dimer (**Figure 5**; right panel, lanes 2–3). After hydrogen peroxide treatment similar protections were elicited only at 34 nM Fur dimer (left panel, lane 5). Thus we can estimate a three- to fourfold loss of holo-Fur affinity for fOPI provoked by the oxidative stress signal. However, this effect can be compensated by the gain in apo-Fur binding affinity for the same operator (**Figure 4B**). This evidence explains the observed responses to oxidative stress under physiological growth conditions, in which holo-Fur repressed targets such as *frpB* and *fecA1* are not significantly deregulated by oxidative stress (**Figures 1C,D**), while apo-Fur repressed targets (*pfr* and *hydA*) are induced by the same stimulus (**Figures 1A,B**).

Discussion

Helicobacter pylori is highly adapted to persist in the human gastric niche and establish lifespan infections. Several lines of evidence suggest that in this environment not only *H. pylori* is exposed to low pH (Scott et al., 2007), but also to oxidative stress induced by the host inflammatory response (Ramarao et al., 2000; Ding et al., 2007). To counter these conditions, which pose a threat to the integrity of proteins and genomic DNA, the bacterium adopts a rich repertoire of antioxidant detoxification factors, including DNA binding and repairing systems (Wang et al., 2005, 2006, 2012; Wang and Maier, 2015). Clearly, these systems need to be coordinately expressed in response to an oxidative stress signal. Some regulatory mechanisms were proposed in the recent past,

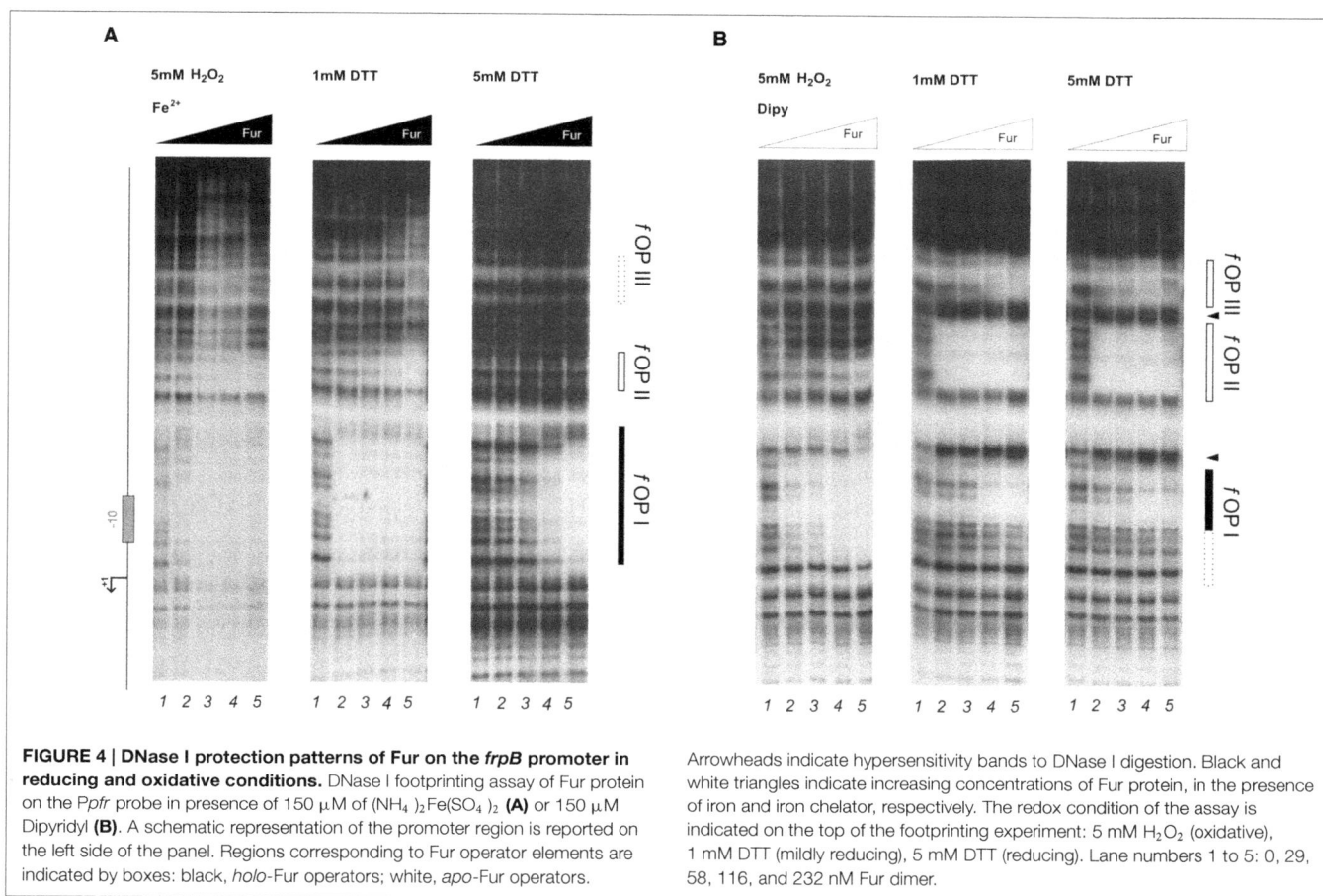

FIGURE 4 | DNase I protection patterns of Fur on the *frpB* promoter in reducing and oxidative conditions. DNase I footprinting assay of Fur protein on the P*pfr* probe in presence of 150 μM of (NH$_4$)$_2$Fe(SO$_4$)$_2$ **(A)** or 150 μM Dipyridyl **(B)**. A schematic representation of the promoter region is reported on the left side of the panel. Regions corresponding to Fur operator elements are indicated by boxes: black, *holo*-Fur operators; white, *apo*-Fur operators.

Arrowheads indicate hypersensitivity bands to DNase I digestion. Black and white triangles indicate increasing concentrations of Fur protein, in the presence of iron and iron chelator, respectively. The redox condition of the assay is indicated on the top of the footprinting experiment: 5 mM H$_2$O$_2$ (oxidative), 1 mM DTT (mildly reducing), 5 mM DTT (reducing). Lane numbers 1 to 5: 0, 29, 58, 116, and 232 nM Fur dimer.

however, our current knowledge on how the oxidative signal is transduced to provoke a transcriptional response to ROS is still limited, mainly because *H. pylori* lacks the dedicated oxygen response regulators described in other microorganisms (Danielli et al., 2010).

Here we demonstrate that HpFur can mediate the response to an oxidative signal into a specific transcriptional output. Interestingly, this transcriptional response appears to pertain mostly the apo-Fur repressed genes, i.e. genes that are inducible by free intracellular iron (FeON). Accordingly, the treatment with hydrogen peroxide mimics *in vivo* the transcriptional effect exerted by Fur in response to iron repletion on these genes (**Figure 1**). These results are also paralleled in the DNA binding assays *in vitro*, in which the oxidative signal confers a holo-mimetic DNA binding behavior to the transcriptional regulator. In fact, while the affinity of Fur for apo-elements as well as holo-elements is impaired in response to H$_2$O$_2$, the binding architecture of the regulator is switched to a conformation favoring the binding of holo-operators even in the absence of the iron co-factor (see pOPIV and fOPI; **Figures 2** and **4**). The evidence that Fur adopts holo-mimetic binding conformations in response to hydrogen peroxide strongly suggests that the allosteric behavior of HpFur, responsible for its function as transcriptional commutator switch (Agriesti et al., 2014), also allows for a specific transduction of the oxidative stress signal

on the apo-repressed gene targets. Therefore we propose a model supporting the hypothesis that apo-Fur repressed genes can be considered oxidation inducible Fur regulatory targets (**Figure 6**). Consistently, many genes that are responsive to aerobic oxygen tension (Park and Lee, 2013) or ROS have also been independently listed as Fur targets in transcriptomic and ChIP-Chip analyses (Ernst et al., 2005; Danielli et al., 2006; Gancz et al., 2006)

The concept that iron-inducible apo-Fur repressed genes can be considered oxidation-inducible Fur regulatory targets (FeON = OxON) is of particular interest, especially recalling the prominent role of this metallo-regulator in the transcriptional regulatory network of the bacterium (Danielli and Scarlato, 2010). This is coherent with the intimate association of redox control and iron homeostasis, and in general with the physiology of *H. pylori*. As such, apo-Fur repressed genes such as *sodB*, which protect the cell from ROS excess, are induced by iron (FeON) and by oxygen stress (OxON), likely through the same allosteric behavior of Fur described for *pfr* and *hydA*. Similarly, the *oor* operon, which codes for an essential but oxygen-labile oxidoreductase, is induced by iron through the regulatory activity of Fur (Gilbreath et al., 2012), and is likely regulated by the allosteric behavior of Fur after an oxidative stress to compensate for oxidative inactivation. Another gene that appears to be under a resembling control is the *nifS* gene encoding a Fe-S cluster synthesis protein (Alamuri et al., 2006).

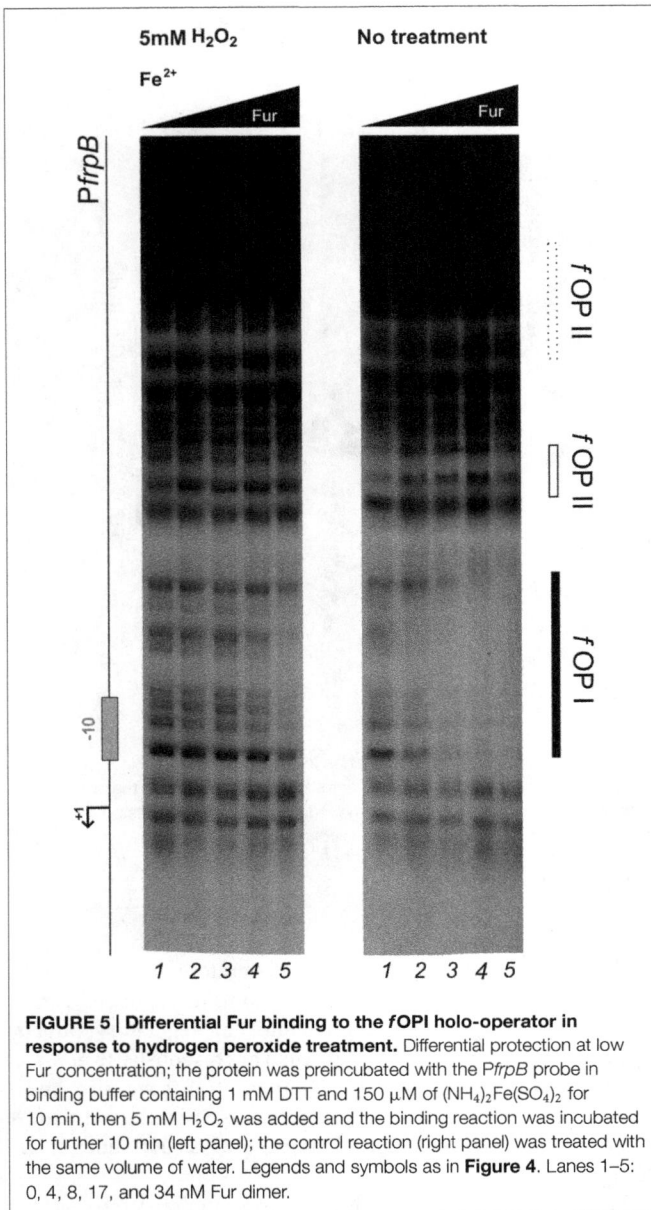

FIGURE 5 | Differential Fur binding to the fOPI holo-operator in response to hydrogen peroxide treatment. Differential protection at low Fur concentration; the protein was preincubated with the *PfrpB* probe in binding buffer containing 1 mM DTT and 150 μM of $(NH_4)_2Fe(SO_4)_2$ for 10 min, then 5 mM H_2O_2 was added and the binding reaction was incubated for further 10 min (left panel); the control reaction (right panel) was treated with the same volume of water. Legends and symbols as in **Figure 4**. Lanes 1–5: 0, 4, 8, 17, and 34 nM Fur dimer.

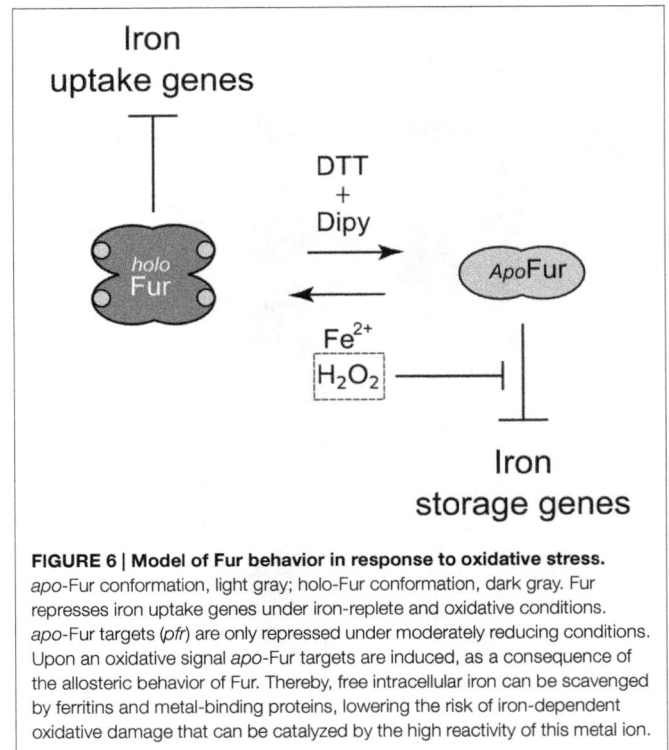

FIGURE 6 | Model of Fur behavior in response to oxidative stress. *apo*-Fur conformation, light gray; holo-Fur conformation, dark gray. Fur represses iron uptake genes under iron-replete and oxidative conditions. *apo*-Fur targets (*pfr*) are only repressed under moderately reducing conditions. Upon an oxidative signal *apo*-Fur targets are induced, as a consequence of the allosteric behavior of Fur. Thereby, free intracellular iron can be scavenged by ferritins and metal-binding proteins, lowering the risk of iron-dependent oxidative damage that can be catalyzed by the high reactivity of this metal ion.

In a similar fashion, *H. hepaticus* PerR is able to control both peroxide- and iron-responsive transcription of oxidative stress defenses genes (Belzer et al., 2011).

Thus, in addition to the transcriptional responses to free iron, Fur appears to cover in *H. pylori* the functions regulated by PerR in other bacteria. Interestingly, in terms of metal binding, the dimerization domain of HpFur, including the structural S1 metal-binding site, is more similar to *B. subtilis* PerR than to other Fur orthologs (Dian et al., 2011). Moreover, in HpFur two additional regulatory binding sites are found: S2 which seems to be conserved in all Fur-like proteins with some variability in the metal coordination geometry, and S3 which is present in several Fur proteins but absent in PerR orthologs. S2 is predicted to be the regulatory site responsible for the conformational changes

that activate Fur for DNA binding (Dian et al., 2011), and it is interesting to recall that point mutations affecting the formation of this site impair the repression of *sodB* by HpFur (Tsugawa et al., 2011). On the contrary, the S3 site is dispensable for DNA binding but its disruption reduces the HpFur DNA-binding affinity. It was suggested that S3 may amplify the DNA-binding affinity of Fur under metal repletion (Dian et al., 2011). Thus, HpFur seems to combine the features of iron-sensing (Fur) and oxidation-sensing (PerR) Fur-superfamily members in one molecule. As such it will be worth exploring whether the additional regulatory metal binding site encompassed in the HpFur structure may be involved the oxidant sensing properties of Fur reported in this work.

Another important structural feature of HpFur is its unique N-terminal extension. It has been proposed that the N-terminal α-helix may participate in the DNA-binding activity of apo-Fur, since apo-repression could not be complemented by Fur orthologs from other species that do not contain the N-terminal extension (Miles et al., 2010). The latter is supposed to favor a V-shape conformation of HpFur in the absence of metal ion at the regulatory S2 site, as also supported by recent crystallographic studies on *Campylobacter jejuni* apo-Fur (Butcher et al., 2012). Strikingly, the mutations in this N-terminal extension greatly affect *H. pylori* resistance to MTZ, which needs to be activated by chemical reduction. Since MTZ antibiotic activity depends on the intracellular redox condition of the cell, the association of mutations in the N-terminal extension of Fur and MTZ resistance suggest a direct link between apo-repression and the Fur-dependent regulation of redox homeostasis and/or antioxidant factors in *H. pylori*.

De facto, although many oxidative defense genes may be regulated by other transcription factors or ncRNAs, our results support for the first time the assumption that the allosteric behavior of Fur, and apo-Fur regulation in particular, is closely associated with the transduction of oxidative stress signals, with important implications in the study of *H. pylori* treatment and resistance to antibiotics.

References

Agriesti, F., Roncarati, D., Musiani, F., Del Campo, C., Iurlaro, M., Sparla, F., et al. (2014). FeON-FeOFF: the *Helicobacter pylori* Fur regulator commutates iron-responsive transcription by discriminative readout of opposed DNA grooves. *Nucleic Acids Res.* 42, 3138–3151. doi: 10.1093/nar/gkt1258

Alamuri, P., Mehta, N., Burk, A., and Maier, R. J. (2006). Regulation of the *Helicobacter pylori* Fe-S cluster synthesis protein NifS by iron, oxidative stress conditions, and fur. *J. Bacteriol.* 188, 5325–5330. doi: 10.1128/JB.00104-06

Albert, T. J., Dailidiene, D., Dailide, G., Norton, J. E., Kalia, A., Richmond, T. A., et al. (2005). Mutation discovery in bacterial genomes: metronidazole resistance in *Helicobacter pylori*. *Nat. Methods* 2, 951–953. doi: 10.1038/nmeth805

Bagchi, D., Bhattacharya, G., and Stohs, S. J. (1996). Production of reactive oxygen species by gastric cells in association with *Helicobacter pylori*. *Free Radic. Res.* 24, 439–450. doi: 10.3109/10715769609088043

Barnard, F. M., Loughlin, M. F., Fainberg, H. P., Messenger, M. P., Ussery, D. W., Williams, P., et al. (2004). Global regulation of virulence and the stress response by CsrA in the highly adapted human gastric pathogen *Helicobacter pylori*. *Mol. Microbiol.* 51, 15–32. doi: 10.1046/j.1365-2958.2003.03788.x

Belzer, C., van Schendel, B. A., Hoogenboezem, T., Kusters, J. G., Hermans, P. W. M., van Vliet, A. H. M., et al. (2011). PerR controls peroxide- and iron-responsive expression of oxidative stress defense genes in Helicobacter hepaticus. *Eur. J. Microbiol. Immunol.* 1, 215–222. doi: 10.1556/EuJMI.1.2011.3.5

Butcher, J., Sarvan, S., Brunzelle, J. S., Couture, J.-F., and Stintzi, A. (2012). Structure and regulon of Campylobacter jejuni ferric uptake regulator Fur define apo-Fur regulation. *Proc. Natl. Acad. Sci. U.S.A.* 109, 10047–10052. doi: 10.1073/pnas.1118321109

Carpenter, B. M., Gancz, H., Gonzalez-Nieves, R. P., West, A. L., Whitmire, J. M., Michel, S. L., et al. (2009a). A single nucleotide change affects Fur-dependent regulation of sodB in H. pylori. *PLoS ONE* 4:e5369. doi: 10.1371/journal.pone.0005369

Carpenter, B. M., Whitmire, J. M., and Merrell, D. S. (2009b). This is not your mother's repressor: the complex role of Fur in pathogenesis. *Infect. Immun.* 77, 2590–2601. doi: 10.1128/IAI.00116-09

Carpenter, B. M., Gilbreath, J. J., Pich, O. Q., McKelvey, A. M., Maynard, E. L., Li, Z.-Z., et al. (2013). Identification and characterization of novel *Helicobacter pylori* apo-Fur-regulated target genes. *J. Bacteriol.* 195, 5526–5539. doi: 10.1128/JB.01026-13

Choi, S. S., Chivers, P. T., and Berg, D. E. (2011). Point mutations in *Helicobacter pylori*'s fur regulatory gene that alter resistance to metronidazole, a prodrug activated by chemical reduction. *PLoS ONE* 6:e18236. doi: 10.1371/journal.pone.0018236

Cooksley, C., Jenks, P. J., Green, A., Cockayne, A., Logan, R. P. H., and Hardie, K. R. (2003). NapA protects *Helicobacter pylori* from oxidative stress damage, and its production is influenced by the ferric uptake regulator. *J. Med. Microbiol.* 52, 461–469. doi: 10.1099/jmm.0.05070-0

Danielli, A., Amore, G., and Scarlato, V. (2010). Built shallow to maintain homeostasis and persistent infection: insight into the transcriptional regulatory network of the gastric human pathogen *Helicobacter pylori*. *PLoS Pathog.* 6:e1000938. doi: 10.1371/journal.ppat.1000938

Danielli, A., Romagnoli, S., Roncarati, D., Costantino, L., Delany, I., and Scarlato, V. (2009). Growth phase and metal-dependent transcriptional regulation of the fecA genes in *Helicobacter pylori*. *J. Bacteriol.* 191, 3717–3725. doi: 10.1128/JB.01741-08

Danielli, A., Roncarati, D., Delany, I., Chiarini, V., Rappuoli, R., and Scarlato, V. (2006). In vivo dissection of the *Helicobacter pylori* Fur regulatory circuit by genome-wide location analysis. *J. Bacteriol.* 188, 4654–4662. doi: 10.1128/JB.00120-06

Danielli, A., and Scarlato, V. (2010). Regulatory circuits in *Helicobacter pylori*: network motifs and regulators involved in metal-dependent responses. *FEMS Microbiol. Rev.* 34, 738–752. doi: 10.1111/j.1574-6976.2010.00233.x

Delany, I., Ieva, R., Soragni, A., Hilleringmann, M., Rappuoli, R., and Scarlato, V. (2005). In vitro analysis of protein-operator interactions of the NikR and fur metal-responsive regulators of coregulated genes in *Helicobacter pylori*. *J. Bacteriol.* 187, 7703–7715. doi: 10.1128/JB.187.22.7703-7715.2005

Delany, I., Pacheco, A. B., Spohn, G., Rappuoli, R., and Scarlato, V. (2001a). Iron-dependent transcription of the frpB gene of *Helicobacter pylori* is controlled by the Fur repressor protein. *J. Bacteriol.* 183, 4932–4937. doi: 10.1128/JB.183.16.4932-4937.2001

Delany, I., Spohn, G., Rappuoli, R., and Scarlato, V. (2001b). The Fur repressor controls transcription of iron-activated and -repressed genes in *Helicobacter pylori*. *Mol. Microbiol.* 42, 1297–1309. doi: 10.1046/j.1365-2958.2001.02696.x

Dian, C., Vitale, S., Leonard, G. A., Bahlawane, C., Fauquant, C., Leduc, D., et al. (2011). The structure of the *Helicobacter pylori* ferric uptake regulator Fur reveals three functional metal binding sites. *Mol. Microbiol.* 79, 1260–1275. doi: 10.1111/j.1365-2958.2010.07517.x

Ding, S.-Z., Minohara, Y., Fan, X. J., Wang, J., Reyes, V. E., Patel, J., et al. (2007). *Helicobacter pylori* infection induces oxidative stress and programmed cell death in human gastric epithelial cells. *Infect. Immun.* 75, 4030–4039. doi: 10.1128/IAI.00172-07

Dubbs, J. M., and Mongkolsuk, S. (2012). Peroxide-sensing transcriptional regulators in bacteria. *J. Bacteriol.* 194, 5495–5503. doi: 10.1128/JB.00304-12

Ernst, F. D., Bereswill, S., Waidner, B., Stoof, J., Mäder, U., Kusters, J. G., et al. (2005). Transcriptional profiling of *Helicobacter pylori* Fur- and iron-regulated gene expression. *Microbiol. Read. Engl.* 151, 533–546. doi: 10.1099/mic.0.27404-0

Gancz, H., Censini, S., and Merrell, D. S. (2006). Iron and pH homeostasis intersect at the level of Fur regulation in the gastric pathogen *Helicobacter pylori*. *Infect. Immun.* 74, 602–614. doi: 10.1128/IAI.74.1.602-614.2006

Gilbreath, J. J., West, A. L., Pich, O. Q., Carpenter, B. M., Michel, S., and Merrell, D. S. (2012). Fur activates expression of the 2-oxoglutarate oxidoreductase genes (oorDABC) in *Helicobacter pylori*. *J. Bacteriol.* 194, 6490–6497. doi: 10.1128/JB.01226-12

Hazell, S. L., Harris, A. G., and Trend, M. A. (2001). "Evasion of the toxic effects of oxygen," in *Helicobacter pylori: Physiology and Genetics*, eds H. L. Mobley, G. L. Mendz, and S. L. Hazell (Washington, DC: ASM Press), 167–175.

Lee, J.-W., and Helmann, J. D. (2006a). Biochemical characterization of the structural Zn^{2+} site in the *Bacillus subtilis* peroxide sensor PerR. *J. Biol. Chem.* 281, 23567–23578. doi: 10.1074/jbc.M603968200

Lee, J.-W., and Helmann, J. D. (2006b). The PerR transcription factor senses H2O2 by metal-catalysed histidine oxidation. *Nature* 440, 363–367. doi: 10.1038/nature04537

Lee, J.-W., and Helmann, J. D. (2007). Functional specialization within the Fur family of metalloregulators. *Biometals Int. J. Role Met. Ions Biol. Biochem. Med.* 20, 485–499. doi: 10.1007/s10534-006-9070-7

Loh, J. T., Shaffer, C. L., Piazuelo, M. B., Bravo, L. E., McClain, M. S., Correa, P., et al. (2011). Analysis of cagA in *Helicobacter pylori* strains from Colombian populations with contrasting gastric cancer risk reveals a biomarker for disease severity. *Cancer Epidemiol. Biomarks Prev.* 20, 2237–2249. doi: 10.1158/1055-9965.EPI-11-0548

Miles, S., Carpenter, B. M., Gancz, H., and Merrell, D. S. (2010). *Helicobacter pylori* apo-Fur regulation appears unconserved across species. *J. Microbiol.* 48, 378–386. doi: 10.1007/s12275-010-0022-0

Olekhnovich, I. N., Vitko, S., Valliere, M., and Hoffman, P. S. (2014). Response to metronidazole and oxidative stress is mediated through homeostatic regulator HsrA (HP1043) in *Helicobacter pylori*. *J. Bacteriol.* 196, 729–739. doi: 10.1128/JB.01047-13

Acknowledgments

The authors wish to thank Vincenzo Scarlato for fruitful discussions and advice. This work was supported by a grant from the Italian Ministry of Education and University (2010P3S8BR_003) and by the University of Bologna.

Park, S. A., and Lee, N. G. (2013). Global regulation of gene expression in the human gastric pathogen *Helicobacter pylori* in response to aerobic oxygen tension under a high carbon dioxide level. *J. Microbiol. Biotechnol.* 23, 451–458. doi: 10.4014/jmb.1209.09064

Ramarao, N., Gray-Owen, S. D., and Meyer, T. F. (2000). *Helicobacter pylori* induces but survives the extracellular release of oxygen radicals from professional phagocytes using its catalase activity. *Mol. Microbiol.* 38, 103–113. doi: 10.1046/j.1365-2958.2000.02114.x

Salama, N. R., Hartung, M. L., and Müller, A. (2013). Life in the human stomach: persistence strategies of the bacterial pathogen *Helicobacter pylori*. *Nat. Rev. Microbiol.* 11, 385–399. doi: 10.1038/nrmicro3016

Scott, D. R., Marcus, E. A., Wen, Y., Oh, J., and Sachs, G. (2007). Gene expression in vivo shows that *Helicobacter pylori* colonizes an acidic niche on the gastric surface. *Proc. Natl. Acad. Sci. U.S.A.* 104, 7235–7240. doi: 10.1073/pnas.0702300104

Touati, D. (2000). Iron and oxidative stress in bacteria. *Arch. Biochem. Biophys.* 373, 1–6. doi: 10.1006/abbi.1999.1518

Troxell, B., and Hassan, H. M. (2013). Transcriptional regulation by Ferric Uptake Regulator (Fur) in pathogenic bacteria. *Front. Cell. Infect. Microbiol.* 3:59. doi: 10.3389/fcimb.2013.00059

Tsugawa, H., Suzuki, H., Satoh, K., Hirata, K., Matsuzaki, J., Saito, Y., et al. (2011). Two amino acids mutation of ferric uptake regulator determines *Helicobacter pylori* resistance to metronidazole. *Antioxid. Redox Signal.* 14, 15–23. doi: 10.1089/ars.2010.3146

van Vliet, A. H., Kuipers, E. J., Stoof, J., Poppelaars, S. W., and Kusters, J. G. (2004). Acid-responsive gene induction of ammonia-producing enzymes in *Helicobacter pylori* is mediated via a metal-responsive repressor cascade. *Infect. Immun.* 72, 766–773. doi: 10.1128/IAI.72.2.766-773.2004

Wang, G., Alamuri, P., Humayun, M. Z., Taylor, D. E., and Maier, R. J. (2005). The *Helicobacter pylori* MutS protein confers protection from oxidative DNA damage. *Mol. Microbiol.* 58, 166–176. doi: 10.1111/j.1365-2958.2005.04833.x

Wang, G., Alamuri, P., and Maier, R. J. (2006). The diverse antioxidant systems of *Helicobacter pylori*. *Mol. Microbiol.* 61, 847–860. doi: 10.1111/j.1365-2958.2006.05302.x

Wang, G., Lo, L. F., and Maier, R. J. (2012). A histone-like protein of *Helicobacter pylori* protects DNA from stress damage and aids host colonization. *DNA Repair.* 11, 733–740. doi: 10.1016/j.dnarep.2012.06.006

Wang, G., and Maier, R. J. (2015). A novel DNA-binding protein plays an important role in *Helicobacter pylori* stress tolerance and survival in the host. *J. Bacteriol.* 197, 973–982. doi: 10.1128/JB.02489-14

Xiang, Z., Censini, S., Bayeli, P. F., Telford, J. L., Figura, N., Rappuoli, R., et al. (1995). Analysis of expression of CagA and VacA virulence factors in 43 strains of *Helicobacter pylori* reveals that clinical isolates can be divided into two major types and that CagA is not necessary for expression of the vacuolating cytotoxin. *Infect. Immun.* 63, 94–98.

Conflict of Interest Statement: The authors declare that the research was conducted in the absence of any commercial or financial relationships that could be construed as a potential conflict of interest.

10

Insights into the bacterial community and its temporal succession during the fermentation of wine grapes

Hailan Piao[1], Erik Hawley[2], Scott Kopf[3], Richard DeScenzo[4], Steven Sealock[3], Thomas Henick-Kling[1] and Matthias Hess[5,6]*

[1] Department of Viticulture and Enology, Washington State University, Richland, WA, USA, [2] ZeaChem Inc., Boardman, OR, USA, [3] Pacific Rim Winemakers, West Richland, WA, USA, [4] ETS Laboratories, Saint Helena, CA, USA, [5] Functional Systems Microbiology Laboratory, University of California, Davis, Davis, CA, USA, [6] Department of Energy Joint Genome Institute, Walnut Creek, CA, USA

Edited by:
Giuseppe Spano,
University of Foggia, Italy

Reviewed by:
Vittorio Capozzi,
University of Foggia, Italy
Angela Capece,
University of Basilicata, Italy

*Correspondence:
Matthias Hess,
Functional Systems Microbiology
Laboratory, Department of Animal
Science, University of California,
Davis, 2251 Meyer Hall, One Shields
Avenue, Davis, CA 95616, USA
mhess@ucdavis.edu

Grapes harbor complex microbial communities. It is well known that yeasts, typically *Saccharomyces cerevisiae*, and bacteria, commonly the lactic acid fermenting *Oenococcus oeni*, work sequentially during primary and secondary wine fermentation. In addition to these main players, several microbes, often with undesirable effects on wine quality, have been found in grapes and during wine fermentation. However, still little is known about the dynamics of the microbial community during the fermentation process. In previous studies culture dependent methods were applied to detect and identify microbial organisms associated with grapes and grape products, which resulted in a picture that neglected the non-culturable fraction of the microbes. To obtain a more complete picture of how microbial communities change during grape fermentation and how different fermentation techniques might affect the microbial community composition, we employed next-generation sequencing (NGS)—a culture-independent method. A better understanding of the microbial dynamics and their effect on the final product is of great importance to help winemakers produce wine styles of consistent and high quality. In this study, we focused on the bacterial community dynamics during wine vinification by amplifying and sequencing the hypervariable V1–V3 region of the 16S rRNA gene—a phylogenetic marker gene that is ubiquitous within prokaryotes. Bacterial communities and their temporal succession was observed for communities associated with organically and conventionally produced wines. In addition, we analyzed the chemical characteristics of the grape musts during the organic and conventional fermentation process. These analyses revealed distinct bacterial population with specific temporal changes as well as different chemical profiles for the organically and conventionally produced wines. In summary these results suggest a possible correlation between the temporal succession of the bacterial population and the chemical wine profiles.

Keywords: wine bacteria, wine fermentation, temporal succession, organic grape products, 16S rRNA gene profile, next-generation sequencing

Introduction

Wine is an alcoholic beverage that is produced by fermenting grapes and represents a heterogeneous mixture of complex compounds. Many of the wines' compounds contribute to their characteristic color, aroma, and flavor (Styger et al., 2011; González-Barreiro et al., 2015), and are released during the fermentation process. The metabolic conversion of grape juice into wine is a complex process of alcoholic fermentation and malolactic fermentation (MLF) and involves a mixture of different microorganisms (Fugelsang and Edwards, 2007). Yeasts play important roles during the alcoholic fermentation step and have significant impact on wine quality. Although bacteria are not the main driving force behind wine characteristics and quality, they do have a significant effect on the final product. For example, lactic acid bacteria are known to convert L-malic acid to lactic acid through MLF and to impart flavor complexity, while acetic acid bacteria (AAB) produce acetic acid, which is a key factor in wine spoilage. MLF is important in winemaking by regulating deacidification and microbial stability. MLF usually occurs after the alcoholic fermentation but it may occur during the alcoholic fermentation process. It is possible that monitoring bacterial community profiles during alcoholic fermentation might allow predicting and controlling wine quality more efficiently. Microorganisms that are present during the various stages of vinification have significant impact on the wine quality both positively and negatively (Fleet, 1993; Fugelsang and Edwards, 2007). To ensure consistent high quality wines and allow reliable risk management, it is essential to monitor the microbial populations throughout the vinification process. NGS represents a fast and precise approach to obtain high-resolution insights into the population dynamics.

In past years, several microorganisms have been found in association with wine grapes and wine musts using culture-dependent techniques (Cappello et al., 2004). These conventional microbiology methods facilitated the isolation of a number of yeasts (e.g., *Brettanomyces/Dekkera, Issatchenkia, Zygoascus,* and *Zygosaccharomyces*) (Curtin et al., 2007; Barata et al., 2012; Di Toro et al., 2015), AAB (e.g., *Acetobacter and Gluconacetobacter*) (Barata et al., 2012), and lactic acid bacteria (e.g., *Enterococcus, Lactobacillus, Lactococcus, Oenococcus,* and *Pediococcus*) (Beneduce et al., 2004; Bae et al., 2006; Capozzi et al., 2010; Garofalo et al., 2015). Due to the viable but non-culturable nature of many wine microorganisms or the dominance of a few organisms that grow very well under laboratory conditions, these conventional microbiology approaches resulted in a rather incomplete and biased picture of the microbial community that is involved in the fermentation process (Millet and Lonvaud-Funel, 2000; Oliver, 2005; Cocolin et al., 2013). In more recent years, a culture-independent method called PCR-DGGE, which combines polymerase chain reaction (PCR) with denaturing gradient gel electrophoresis (DGGE), has been frequently used for detecting specific microorganisms during different stages of the wine fermentation process (Renouf et al., 2007; Spano et al., 2007; Andorrá et al., 2008; Laforgue et al., 2009; Pérez-Martín et al., 2014). Although PCR-DGGE remains a useful tool to detect and discriminate microbial organisms potentially

present in wine grapes and musts without cultivation, it has its limitation due to the challenge of distinguishing co-migrating bands from multiplexed PCR products and requirement of intensive bands (Laforgue et al., 2009; Cocolin et al., 2013). With next-generation sequencing (NGS) technologies being a commodity now, powerful tools for high-throughput analysis of complex microbial communities via amplification and subsequent sequencing of the 16S ribosomal RNA (rRNA) hypervariable regions are now available (Sinclair et al., 2015). NGS have been applied widely and resulted in new insights into microbial community dynamics from diverse environmental samples (Piao et al., 2014; Trexler et al., 2014; Nguyen and Landfald, 2015; Pessoa-Filho et al., 2015) including grape and botrytized wine (Bokulich et al., 2012, 2014), but it is still not well known how the microbial communities associated with different grapes change over time and how these changes affect the final quality of the fermentation products.

There has been a fast growing demand for organic foods and beverages and the market for organically produced wines has experienced a significant boost. To obtain an enhanced understanding of how the different winemaking techniques affect bacterial community dynamics and further find out the bacterial community dynamics affect wine fermentation, we analyzed the temporal succession of the bacterial community and its effects on the changes of chemical characteristics during organic and conventional wine fermentation using 16S rRNA amplicon sequencing. The obtained results revealed a broad bacterial diversity in wine including known wine bacteria. Many of the identified organisms have to our knowledge not been reported to date. By analyzing the dynamics of the bacterial population during the fermentation process, it was possible to detect bacteria that were previously not associated with wine fermentation. The chemical characteristics of the wines, combined with the results of bacterial community profiles, indicated that there might be a possible link between specific bacteria, their succession and some wine characteristics.

Materials and Methods

Sample Collection

Both organic and conventional pied-de-cuve (PDC) were obtained by stomping and fermenting hand-harvested organically grown Riesling grapes in a 200 gallon tote. No sulfur dioxide (SO_2) was added to the organic PDC fermentation, whereas SO_2 (55.8 mg/L) was added during the conventional PDC fermentation process. For organic and conventional bulk fermentation, the organically grown Riesling grapes were machine pressed and transferred to a 15,000 gallon fermentation tank. Juice was allowed to settle for 36 h before heavy solids were removed. When sugar content of the organic or conventional PDC reached approximately 10 Brix, the PDCs were transferred to bulk fermentation tanks. Fermentation temperature was maintained between 10 and 13°C. Neither SO_2 nor fining agents were added to the organic musts during primary fermentation, while SO_2 (38.5 mg/L) and bentonite were added to the conventional musts. Yeast assimilable nitrogen was added in the form of autolyzed yeast product and diammonium phosphate (DAP) to the organic and conventional wine respectively. Brix and ethanol measurements were taken to monitor fermentation progress and fermentation was terminated when a Brix of 2.5 and 6.9 was reached for organic and conventional wine, respectively.

DNA Extraction and 16S rRNA Gene Amplification

Total microbial DNA was extracted from 500 mg of the organic and conventional wine samples using a FastDNA SPIN Kit for Soil (MP Biomedical, Solon, OH) according to the manufacturer's instructions. Extracted DNA was quantified with a spectrophotometer (Nanodrop ND1000; Thermo Scientific, USA). The hypervariable V1–V3 region of the 16S rRNA gene was amplified from the environmental DNA using the primer set 28F/519R (28F: 5′-ccatctcatccctgcgtgtctccgactcagxxxxxxxxGAG TTTGATCNTGGCTCAG-3′ and 519R: 5′-cctatccctgtgtgccttg gcagtctcagGTNTTACNGCGGCKGCTG-3′). Primer sequences were modified by the addition of 454 A or B adapter sequences (lower case) and ended with the sequencing key "TCAG" (underlined). In addition, the forward primer included a 8 bp barcode, indicated by xxxxxxxx in the forward primer sequence above, for multiplexing of samples during sequencing. The barcode sequence for each sample is listed in Table S1.

The V1–V3 region of the 16S rRNA genes was amplified with primer pair 28F/519R by emulsion PCR. Subsequent PCR reactions were performed using the Roche Live amplification mix (according to the Roche protocol) with the following PCR conditions: initial denaturation for 1 min at 94°C, followed by 50 amplification cycles of (30 s at 94°C, 4.5 min at 58°C, and 30 s at 68°C), and hold at 10°C. Emulsion PCR and sequencing of the PCR amplicons were performed following the Roche 454 GS FLX Titanium technology instructions provided by the manufacturer.

Data Analysis

Raw pyrosequencing data were demultiplexed and processed using QIIME version 1.7.0 (Caporaso et al., 2010b). Sequencing primers and barcodes were removed from the raw sequence reads by allowing 1.5 mismatches to the barcode and 2 mismatches to the primer sequence. Sequences were removed if they had homopolymeric regions of more than 6 nt, were smaller than 200 nt, had quality scores lower than 25, or if they were identified as being chimeric. This resulted in a total of 16,142 and 28,490 high quality 16S rRNA gene sequences from organic and conventional wine samples, respectively.

Quality filtered sequences were clustered into operational taxonomic units (OTUs) at a 97% sequence identity cut-off using UCLUST (Edgar, 2010). The most abundant sequence of each OTU was picked as representative sequence. Singleton and

doubleton abundance, Shannon, Simpson, and Chao1 estimators were calculated using the QIIME software. Representative sequences were aligned using the PyNAST algorithm (Caporaso et al., 2010a) and the alignment was filtered to remove common gaps. Following the quality filtering and grouping steps, 1340 unique sequences (representing 44,632 total sequences) were aligned and taxonomically classified using the RDP classifier program (Wang et al., 2007) with 80% confidence rating against the Greengenes database (McDonald et al., 2012).

Chemical Analysis

Chemical analyses of the wine samples were performed at ETS Laboratories (Saint Helena, CA) using an Agilent 7700 inductively coupled plasma-mass spectrometer according to manufacturer's instructions and as described by Hopfer et al. (2013).

Results

Bacterial Community Profile of Organically and Conventionally Produced Wine

To determine bacterial community dynamics and their effects on wine components, we compared the profiles of the bacterial community in wines that were produced using organic and conventional fermentation protocols. Grape juice was inoculated with indigenous yeasts from the grape skins by adding PDC. This traditional wine making technique reduces the needs for commercial yeast and usually increases wine complexity. Samples for bacterial community profiling were collected from the PDC (0 day) and must at different fermentation stages after PDC was added to the grape juice. Environmental DNA was extracted from PDC and must followed by pyrosequencing of the hypervariable V1–V3 region of the 16S rRNA genes. The quality-filtered pyrotag reads were clustered into OTUs at a 97% of sequence identity level, which resulted in 529 and 1099 distinct OTUs, representing 16,142 and 28,490 sequences from organic and conventional wine, respectively (**Table 1**). Analysis of OTUs profiles suggests that community richness within organic wine was stable at early stage of fermentation (0, 2, and 3 days; **Table 1**; Table S2). Continuing the fermentation process, increased community richness at 10 days was measured, whereas decreased community richness was observed afterwards (**Table 1**; Table S2). Compared to organically producing wine, bacterial community richness increased significantly at 6 days of fermentation (**Table 1**; Table S2) then decreased rapidly within 24 h (**Table 1**; Table S2) during conventional wine production. These findings are supported by the calculated rarefaction curves (Figure S1). Shannon's diversity and Simpson indices are higher

TABLE 1 | Summary of generated reads and OTUs observed.

Duration of fermentation [days]	Organic						Conventional					
	0	2	3	10	16	Total	0	2	6	7	12	Total
Quality filtered reads	5,420	3,569	4,188	1,583	1,382	16,142	16,001	1,531	7,588	2,127	1,243	28,490
OTUs observed	173	165	176	202	146	529	268	201	612	220	160	1099

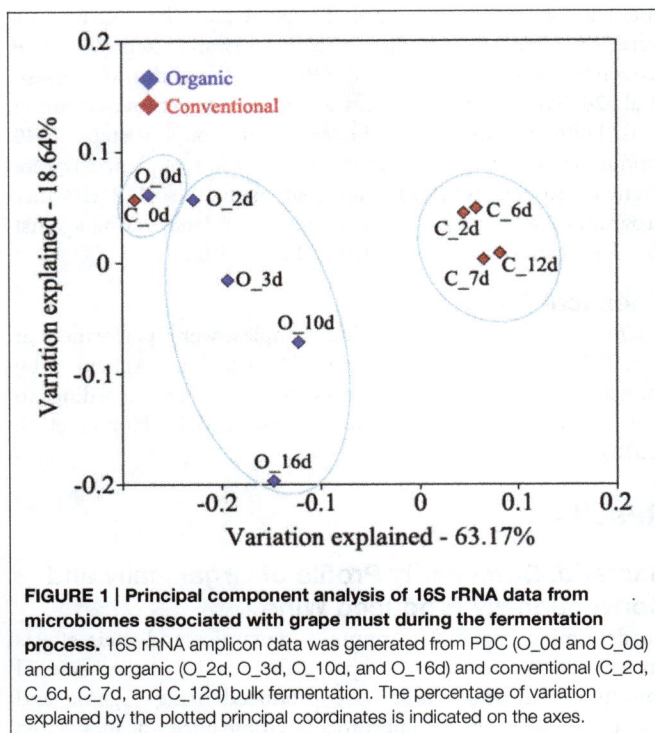

FIGURE 1 | Principal component analysis of 16S rRNA data from microbiomes associated with grape must during the fermentation process. 16S rRNA amplicon data was generated from PDC (O_0d and C_0d) and during organic (O_2d, O_3d, O_10d, and O_16d) and conventional (C_2d, C_6d, C_7d, and C_12d) bulk fermentation. The percentage of variation explained by the plotted principal coordinates is indicated on the axes.

in conventionally fermented wine (Table S2), suggesting that the bacterial community in conventionally produced wine became more diverse than in organically produced wine. Principal component analysis suggests that the wine microbiome profiles associated with grape must during conventional fermentation were distinct from the microbiome profiles associated with grape must from organic fermentation (**Figure 1**).

Phylogenetic Profiles of the Bacterial Communities during the Fermentation Processes

Clustering of the obtained 16S rRNA gene sequences based on a 97% sequence identity cut-off and assigning phylogeny to each of the obtained OTUs suggest that a total of 15 phyla (contributing ≥1 of the reads) were present during the fermentation process of the two grape musts under observation (**Figure 2** and Table S3). Nine of the observed 15 phyla were found in musts from both fermentation techniques (i.e., *Proteobacteria, Cyanobacteria, Bacteroidetes, Firmicutes, Actinobacteria, Acidobacteria, Spirochaetes, Verrucomicrobia,* and *Fusobacteria*), while the presence of some phyla depended on the applied fermentation technique. Specifically, *Nitrospirae, Planctomycetes,* and *Tenericutes* were detected solely in the samples from organically fermented must, whereas *Fibrobacteres* and members of the candidate phylum WYO were detected only in the conventionally produced wine musts (**Figure 2** and Table S3). It is possible that members of these specific phyla might contribute to the distinct chemical characteristics of the produced wines. *Proteobacteria* is the predominant phylum in both wine musts (**Figure 2** and Table S3), which was represented primarily by the *Gammaproteobacteria* within the PDC (0 day). During fermentation the relative abundance of

Gammaproteobacteria decreased significantly in both wine musts (6–8 fold), which was partially complemented by an increase of other members of the *Proteobacteria*, i.e., *Alphaproteobacteria, Betaproteobacteria,* and *Deltaproteobacteria* (**Table 2**). During organic fermentation, the abundance of *Alphaproteobacteria* increased and this phylogenetic group became the dominant class (57% at 15 days). During conventional fermentation, population of *Alphaproteobacteria* increased as well (~4.5 fold) but did not dominate the community (21.72–27.63%). Abundance of *Betaproteobacteria* increased 250–380 fold to a relative abundance between 18.15 and 27.10% (**Table 2**). Overall population changes suggest a notable reduction of *Proteobacteria* (**Figure 2** and Table S3), which is similar to what has been observed previously during botrytized wine fermentation (Bokulich et al., 2012). This decrease in *Proteobacteria*, specifically of the *Gammaproteobacteria*, was accompanied by an increase of the *Bacteroidetes, Firmicutes,* and *Actinobateria*. The increase was in particular notable within the microbiome from the conventionally fermented wine, while the increase was less notable within the microbiome from organically fermented wine (**Table 2** and Table S3). Within the conventionally fermented wine, the increase of abundance of *Bacteroidetes* was caused through a significant increase in *Spingobacteriia* and a moderate increase in *Bacteroidia* (**Figure 2**; **Table 2** and Table S3). The increase of *Firmicutes* was due largely to an increase of the *Bacilli* and a moderate increase of the *Clostridia* (**Table 2**). Further analysis of the bacterial community resulted in the detection of 96 genera across all samples, of which 33 genera were found both in organically and conventionally fermented must. Twenty-one of the 96 genera were detected only within the bacterial communities associated with organically fermented must, whereas 42 genera were found only within the bacterial communities associated with conventionally fermented grapes (**Table 3**). Increased genus diversity was observed for the microbiome from conventionally fermented must (75 genera total) when compared to the microbiome from organically fermented must (54 genera total). Representatives of the genus *Gluconobacter*, an acetic acid bacterium commonly found associated with grape skin (Joyeux et al., 1984), was detected in the microbiome of both wine types, however discrete changes within the *Gluconobacter* population were observed between organically and conventionally fermented wines. Comparison between organically and conventionally produced wines revealed that the population of *Gluconobacter* was highly abundant in organic PDC fermentation (8.67% at 0 day), while it possessed very low abundance in conventional PDC fermentation (0.47% at 0 day; **Table 3**). During the fermentation process, the *Gluconobacter* population increased in both musts and eventually represented the predominant genus from organically produced wine at late stage (49%; 16 day), while it was relatively stable, accounting for 5–7% of population, throughout the conventional fermentation process (5–7%; **Table 3**). Beside the dominant genus *Gluconobacter*, a number of other genera (total sequences detected >1% in data from at least one of the time points) were also detected during both fermentation procedures (i.e., *Clavibacter, Propionibacterium, Hymenobacter, Pedobacter, Bacillus, Staphylococcus, Acetobacter,*

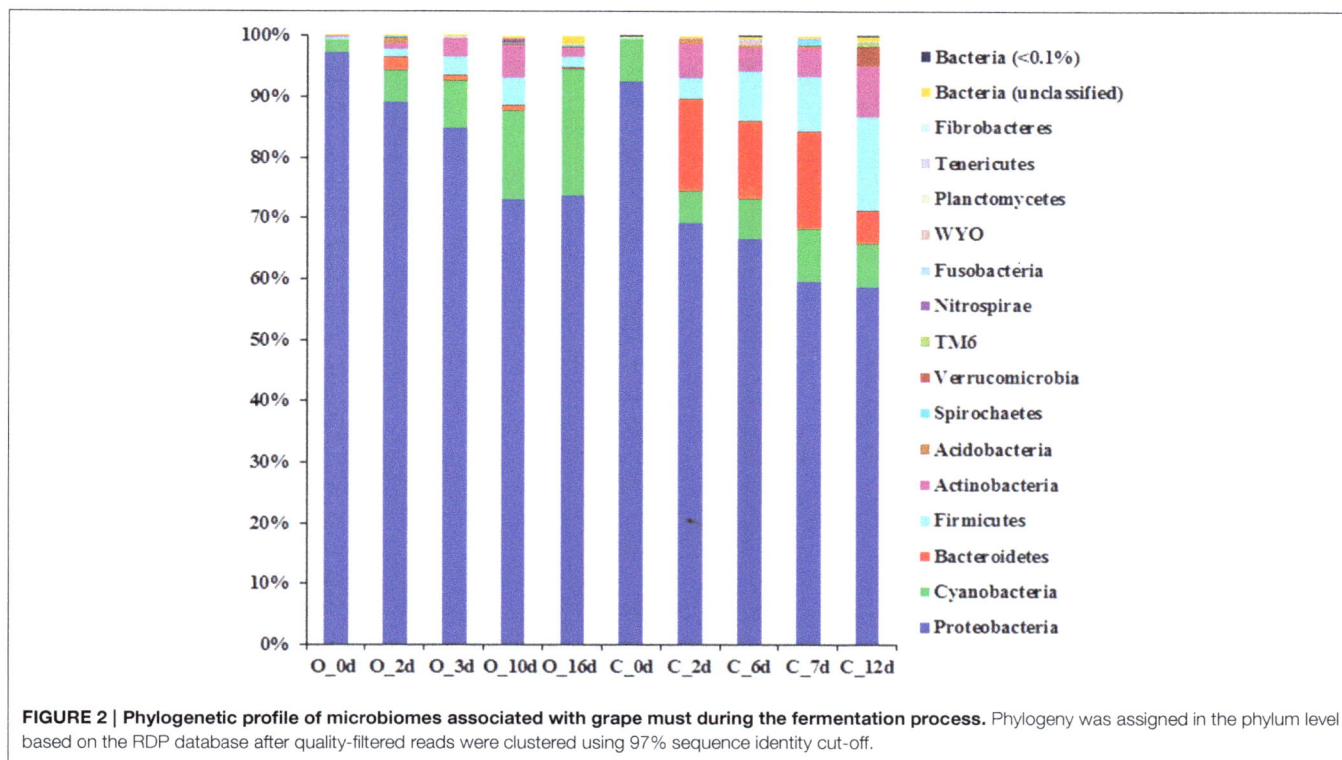

FIGURE 2 | Phylogenetic profile of microbiomes associated with grape must during the fermentation process. Phylogeny was assigned in the phylum level based on the RDP database after quality-filtered reads were clustered using 97% sequence identity cut-off.

Spingomonas, Diaphorobacter, Janthinobacterium, Ralstonia, Neisseria, Acinetobacter, Pseudomonas, and *Leptospira*), with *Pedobacter, Spingomonas, Janthinobacterium,* and *Pseudomonas* exhibiting dominance only during the conventional fermentation process (**Table 3**). In addition, other less abundant phylogenetic groups (total sequences detected between 0.1 and 1%) were observed during the two distinct fermentation processes (i.e., *Corynebacterium, Micrococcus, Sediminibacterium, Dyadobacter, Exiguobacterium, Lactobacillus, Clostridium, Roseburia, Faecalibacterium, Fusobacterium, Bradyrhizobium, Methylobacterium, Roseomonas, Salinispora, Curvibacter, Pelomonas, Trabulsiella,* and *Haemophilus*) (**Table 3**). Interestingly, *Oenococcus,* a genus containing known lactic acid bacteria, was detected only in the microbiome of conventionally fermented wine (**Table 3**).

Chemical Component Analysis from Organic and Conventional Wine

Several parameters, such as sugar concentration, temperature, pH value, ethanol concentration and a variety of chemical characteristics, of the grape must were monitored during the fermentation process (**Figure 3** and **Table 4**). Sugar concentrations were stable until 3 days into the fermentation process, after this period sugar concentration decreased linearly in both wine fermentations (**Figure 3A**). Overall pH values were slightly lower from organically produced wine than conventionally produced wine, while ethanol reached a higher concentration during the organic fermentation process (**Figures 3C,D**). Lactic acid concentration at the end of the

organic PDC fermentation was higher, while it was same in both wine fermentation processes, suggesting that wine fermentation was terminated before secondary fermentation was initiated. Malic acid content increased during both fermentation processes, however overall malic acid content was higher in conventionally fermented wine. Volatile acidity (VA) content changed irregularly, at early stage of fermentation (2–3 days) lower VA contents were measured for both types of wine samples, afterwards it increased to about three-fold in conventionally fermented wine, while it returned to first day level in organically fermented wine. Overall tartaric acid concentration was higher in organically fermented wine compare to conventionally fermented wine. A summary of the chemical characteristics of the grape musts is provided in **Table 4**. Initial nitrogen concentration was similar in both juices at the first day of fermentation and additional nitrogen was provided during the fermentation process to support continuous growth of yeast. Nitrogen concentrations are summarized in **Table 4**. More detailed and controlled studies will help to enhance our understanding of the molecular processes and microbe-microbe and microbe-must interaction would be of great value.

Discussion

Culture-independent NGS is a cost-effective approach to study composition and the spatial and temporal changes of microbial communities and it has been applied to various environment samples (Piao et al., 2014; Nguyen and Landfald, 2015). However, to our knowledge, as of today only a few studies have been

TABLE 2 | Relative abundance of prokaryotes associated with grape musts during organic and conventional fermentation at the class level.

Duration of fermentation [days]	Organic					Conventional				
	0	2	3	10	16	0	2	6	7	12
Acidobacteria;c_Acidobacteria-2	0.00	0.90	0.00	0.38	0.29	0.00	0.72	0.46	0.33	0.00
Actinobacteria;c_Actinobacteria	0.06	0.98	2.96	5.31	1.52	0.05	5.68	3.58	4.84	8.21
Actinobacteria;c_Thermoleophilia	0.00	0.00	0.00	0.00	0.00	0.00	0.00	0.13	0.00	0.00
Bacteroidetes;c_Bacteroidia	0.04	0.03	0.05	0.32	0.00	0.01	2.55	1.74	2.77	0.24
Bacteroidetes;c_Flavobacteriia	0.00	0.73	0.00	0.13	0.00	0.00	0.26	0.58	0.47	0.00
Bacteroidetes;c_Sphingobacteriia	0.02	1.37	0.72	0.38	0.14	0.02	12.48	10.46	12.60	5.15
Cyanobacteria;c_4C0d-2	0.04	0.03	0.00	0.00	0.00	0.00	0.13	0.91	0.00	0.00
Cyanobacteria;c_S15B-MN24	0.02	0.11	0.31	0.63	0.43	0.00	0.98	0.61	1.13	2.65
Cyanobacteria;c_Synechococcophycideae	0.00	0.00	0.00	1.01	0.00	0.00	0.00	0.00	0.00	0.00
Fibrobacteres;c_Fibrobacteria	0.00	0.00	0.00	0.00	0.00	0.00	0.00	0.18	0.00	0.00
Firmicutes;c_Bacilli	0.44	0.98	2.65	3.79	1.52	0.19	1.44	6.80	5.88	10.62
Firmicutes;c_Clostridia	0.02	0.17	0.43	0.76	0.22	0.06	1.96	1.32	3.10	4.83
Fusobacteria;c_Fusobacteria	0.00	0.00	0.00	0.13	0.00	0.01	0.00	0.00	0.00	0.16
Nitrospirae;c_Nitrospira	0.07	0.00	0.00	0.32	0.00	0.00	0.00	0.00	0.00	0.00
Planctomycetes;c_Planctomycetia	0.00	0.00	0.19	0.00	0.00	0.00	0.00	0.00	0.00	0.00
Proteobacteria;c_Alphaproteobacteria	12.49	9.78	24.52	30.32	57.16	5.82	27.63	22.26	23.93	21.72
Proteobacteria;c_Betaproteobacteria	0.06	4.82	3.15	7.14	2.53	0.07	22.73	27.10	18.15	23.65
Proteobacteria;c_Deltaproteobacteria	0.11	0.31	0.00	0.44	0.43	0.02	0.00	0.29	1.50	1.93
Proteobacteria;c_Gammaproteobacteria	84.59	74.05	57.07	34.87	13.46	86.56	18.68	16.75	15.84	11.34
Spirochaetes;c_Leptospirae	0.00	0.22	0.05	0.19	0.14	0.00	0.07	0.00	1.18	0.00
Tenericute;c_Mollicutes	0.00	0.00	0.14	0.00	0.00	0.00	0.00	0.03	0.00	0.00
TM6;c_SJA-4	0.00	0.00	0.05	0.06	0.07	0.00	0.00	0.00	0.00	0.72
Verrucomicrobia;c_Opitutae	0.00	0.00	0.00	0.00	0.00	0.00	0.00	0.00	0.00	3.06
Verrucomicrobia;c_Verruco-5	0.00	0.00	0.00	0.13	0.00	0.00	0.00	0.00	0.00	0.00

published that employed NGS to study the dynamics of the microbial wine ecosystem (Bokulich et al., 2012, 2014, 2015). To enhance our understanding of the microbial dynamics, specifically of bacterial dynamics, during grape fermentation, we employed culture-independent 16S rRNA amplicon sequencing to determine changes in the bacterial population of grape must during the fermentation process. Currently, the most commonly used culture-independent method within the wine industry for comparing microbial populations associated with different grape products is PCR-DGGE (Cocolin et al., 2000; Lopez et al., 2003). PCR-DGGE possesses only a limited ability to provide detailed information about biodiversity within a sample as bands associated with different phylogenetic groups might be visible as a single band resulting in underestimation of microbial community diversity.

In this study we identified 96 genera and discriminated over 30 species that were present during wine fermentation. Importantly, most of the species we detected have not been reported previously during wine fermentation (Table S4), with the exception of a few species (i.e., *Propionibacterium acnes*, *Bacillus thermoamylovorans*, *Pseudomonas stutzeri*) that were isolated from grapevine, palm wine, and wine corks (Combet-Blanc et al., 1995; Bañeras et al., 2013; Yousaf et al., 2014). The genus *Gluconobacter* increased significantly during organic fermentation (from 3.28 to 49.42%), while it exhibited less notable changes during the conventional fermentation (from

5.63 to 7.57%) process (**Table 3**). A major difference of the organic and conventional wine making processes employed in this study was the addition of SO_2 to the conventional wine prior to PDC fermentation (50 mg/L) and bulk fermentation (38.5 mg/L), while no SO_2 added to the organic wine until completion of primary fermentation. The availability of SO_2 during primary fermentation might represent a selective effect on the *Gluconobacter* population. Bokulich and colleagues showed that *Gluconobacter* population was significantly suppressed by SO_2 at concentrations ≥ 25 mg/L (Bokulich et al., 2015). At higher taxonomic resolution the genus *Gluconobacter* was dominated by one distinct OTU (i.e., OTU denovo952) during the fermentation process (Table S5). To further define this specific OTU, its representing nucleotide sequence was compared to sequences deposited in NCBI database. Results revealed a 99.6% sequence identity with *Gluconobacter oxydans*, the main representative of AAB on grapes (Joyeux et al., 1984). *Gluconobacter oxydans* is known as spoilage acetic acid bacterium together with *Acetobacter* during winemaking; *Gluconobacter oxydans* is often detected in grapes, while *Acetobacter* is found in wine (Bartowsky and Henschke, 2008). Although AAB have been identified as wine spoilage bacteria previously, the population of AAB are often underestimated with culture-dependent method due to the lack of appropriate cultivation techniques (Millet and Lonvaud-Funel, 2000). Amplicon sequencing data allowed us to observe significant population changes of *Gluconobacter*

TABLE 3 | Relative abundance of prokaryotes associated with grape musts during organic and conventional fermentation at the genus level.

Duration of fermentation [days]	Organic						Conventional				
	0	0	2	3	10	16	0	2	6	7	12
Actinobacteria;c_Actinobacteria;o_Actinomycetales;f_Corynebacteriaceae;g_Corynebacterium	0.00	0.00	0.00	0.79	0.69	0.29	0.01	0.39	0.18	0.24	0.48
Actinobacteria;c_Actinobacteria;o_Actinomycetales;f_Dermabacteraceae;g_Brachybacterium	0.00	0.00	0.00	0.00	0.00	0.00	0.00	0.20	0.00	0.00	0.00
Actinobacteria;c_Actinobacteria;o_Actinomycetales;f_Dietziaceae;g_Dietzia	0.00	0.00	0.00	0.00	0.00	0.14	0.00	0.00	0.00	0.00	0.00
Actinobacteria;c_Actinobacteria;o_Actinomycetales;f_Gordoniaceae;g_Gordonia	0.00	0.00	0.00	0.00	1.45	0.00	0.00	0.00	0.00	0.00	0.24
Actinobacteria;c_Actinobacteria;o_Actinomycetales;f_Microbacteriaceae;g_Clavibacter	0.00	0.00	0.00	0.19	0.32	0.00	0.00	0.20	0.00	0.00	1.53
Actinobacteria;c_Actinobacteria;o_Actinomycetales;f_Microbacteriaceae;g_Curtobacterium	0.00	0.00	0.08	0.00	0.00	0.00	0.00	0.20	0.67	0.47	0.00
Actinobacteria;c_Actinobacteria;o_Actinomycetales;f_Microbacteriaceae;g_Frigoribacterium	0.00	0.00	0.00	0.00	0.00	0.00	0.00	0.26	0.00	0.00	0.00
Actinobacteria;c_Actinobacteria;o_Actinomycetales;f_Microbacteriaceae;g_Leucobacter	0.00	0.00	0.00	0.00	0.00	0.00	0.00	0.00	0.00	1.03	0.00
Actinobacteria;c_Actinobacteria;o_Actinomycetales;f_Microbacteriaceae;g_Microbacterium	0.00	0.00	0.34	0.05	0.00	0.00	0.00	0.00	0.05	0.00	0.00
Actinobacteria;c_Actinobacteria;o_Actinomycetales;f_Microbacteriaceae;g_Rathayibacter	0.00	0.00	0.00	0.00	0.00	0.00	0.00	0.26	0.18	0.00	0.00
Actinobacteria;c_Actinobacteria;o_Actinomycetales;f_Microbacteriaceae;g_Salinibacterium	0.00	0.00	0.00	0.00	0.00	0.00	0.00	0.46	0.12	0.00	0.00
Actinobacteria;c_Actinobacteria;o_Actinomycetales;f_Micrococcaceae;g_Kocuria	0.00	0.00	0.00	0.00	0.00	0.14	0.00	0.00	0.00	0.00	0.00
Actinobacteria;c_Actinobacteria;o_Actinomycetales;f_Micrococcaceae;g_Micrococcus	0.00	0.00	0.00	0.00	0.76	0.00	0.00	0.33	0.11	0.28	0.08
Actinobacteria;c_Actinobacteria;o_Actinomycetales;f_Mycobacteriaceae;g_Mycobacterium	0.00	0.00	0.08	0.00	0.06	0.00	0.00	0.00	0.03	0.00	0.56
Actinobacteria;c_Actinobacteria;o_Actinomycetales;f_Nocardiaceae;g_Rhodococcus	0.00	0.00	0.00	0.26	0.00	0.00	0.00	0.07	0.00	0.00	0.00
Actinobacteria;c_Actinobacteria;o_Actinomycetales;f_Propionibacteriaceae;g_Propionibacterium	0.06	0.06	0.48	0.55	1.52	0.43	0.03	0.65	1.75	1.97	4.51
Bacteroidetes;c_Bacteroidia;o_Bacteroidales;f_Bacteroidaceae;g_Bacteroides	0.04	0.00	0.00	0.00	0.00	0.00	0.01	0.07	0.03	0.47	0.00
Bacteroidetes;c_Bacteroidia;o_Bacteroidales;f_Prevotellaceae;g_Prevotella	0.00	0.00	0.03	0.00	0.00	0.00	0.00	0.52	0.37	0.71	0.16
Bacteroidetes;c_Bacteroidia;o_Bacteroidales;f_[Paraprevotellaceae];g_[Prevotella]	0.00	0.00	0.00	0.00	0.00	0.00	0.00	0.20	0.01	0.05	0.00
Bacteroidetes;c_Flavobacteria;o_Flavobacteriales;f_Flavobacteriaceae;g_Chryseobacterium	0.00	0.00	0.00	0.06	0.00	0.00	0.00	0.26	0.46	0.47	0.00
Bacteroidetes;c_Sphingobacteria;o_Sphingobacteriales;f_Chitinophagaceae;g_Sediminibacterium	0.00	0.00	0.00	0.13	0.00	0.00	0.00	0.00	0.21	0.09	0.80
Bacteroidetes;c_Sphingobacteria;o_Sphingobacteriales;f_Chitinophagaceae;g_Segetibacter	0.00	0.00	0.00	0.00	0.00	0.00	0.00	0.00	0.04	0.47	0.00
Bacteroidetes;c_Sphingobacteria;o_Sphingobacteriales;f_Flexibacteraceae;g_Dyadobacter	0.00	0.00	0.28	0.00	0.19	0.00	0.00	0.00	0.17	0.00	0.00
Bacteroidetes;c_Sphingobacteria;o_Sphingobacteriales;f_Flexibacteraceae;g_Hymenobacter	0.00	0.00	0.56	0.10	0.00	0.00	0.00	1.05	1.79	1.74	1.53
Bacteroidetes;c_Sphingobacteria;o_Sphingobacteriales;f_Flexibacteraceae;g_Spirosoma	0.00	0.00	0.00	0.00	0.00	0.00	0.00	0.00	0.18	0.00	0.00
Bacteroidetes;c_Sphingobacteria;o_Sphingobacteriales;f_Sphingobacteriaceae;g_Pedobacter	0.00	0.00	0.50	0.62	0.00	0.14	0.00	9.67	7.39	8.56	2.33
Cyanobacteria;c_Synechococcophycideae;o_Synechococcales;f_Synechococcaceae;g_Prochlorococcus	0.00	0.00	0.00	0.00	1.01	0.00	0.00	0.00	0.00	0.00	0.00
Fibrobacteres;c_Fibrobacteria;o_Fibrobacterales;f_Fibrobacteraceae;g_Fibrobacter	0.00	0.00	0.00	0.00	0.00	0.00	0.00	0.00	0.18	0.00	0.00
Firmicutes;c_Bacilli;o_Bacillales;f_Bacillaceae;g_Anoxybacillus	0.00	0.00	0.00	0.06	0.00	0.00	0.00	0.00	0.03	0.33	0.00
Firmicutes;c_Bacilli;o_Bacillales;f_Bacillaceae;g_Bacillus	0.02	0.00	0.87	1.27	0.95	0.43	0.08	0.33	0.12	0.00	1.77
Firmicutes;c_Bacilli;o_Bacillales;f_Bacillaceae;g_Geobacillus	0.00	0.00	0.00	0.25	0.00	0.00	0.00	0.00	0.00	0.00	0.00
Firmicutes;c_Bacilli;o_Bacillales;f_Bacillaceae;g_Terribacillus	0.00	0.00	0.00	0.00	0.00	0.51	0.00	0.00	0.00	0.00	0.00
Firmicutes;c_Bacilli;o_Bacillales;f_Paenibacillaceae;g_Paenibacillus	0.00	0.00	0.00	0.06	0.14	0.00	0.00	0.00	0.09	0.00	0.00
Firmicutes;c_Bacilli;o_Bacillales;f_Staphylococcaceae;g_Staphylococcus	0.07	0.00	0.00	0.91	0.69	0.00	0.04	0.33	1.67	0.52	2.98
Firmicutes;c_Bacilli;o_Bacillales;f_Thermoactinomycetaceae;g_Planifilum	0.00	0.00	0.11	0.00	0.00	0.00	0.00	0.00	0.00	0.00	0.00
Firmicutes;c_Bacilli;o_Exiguobacterales;f_Exiguobacteraceae;g_Exiguobacterium	0.00	0.00	0.00	0.00	0.00	0.22	0.00	0.00	0.00	0.00	0.64

(Continued)

TABLE 3 | Continued

Duration of fermentation [days]	Organic					Conventional				
	0	2	3	10	16	0	2	6	7	12
Firmicutes;c_Bacilli;o_Lactobacillales;f_Aerococcaceae;g_Aerococcus	0.00	0.00	0.00	0.19	0.00	0.00	0.00	0.00	0.00	0.00
Firmicutes;c_Bacilli;o_Lactobacillales;f_Lactobacillaceae;g_Lactobacillus	0.07	0.00	0.24	0.00	0.00	0.00	0.13	0.00	0.00	0.00
Firmicutes;c_Bacilli;o_Lactobacillales;f_Leuconostocaceae;g_Leuconostoc	0.04	0.00	0.00	0.00	0.00	0.00	0.00	0.00	0.00	0.00
Firmicutes;c_Bacilli;o_Lactobacillales;f_Leuconostocaceae;g_Oenococcus	0.00	0.00	0.00	0.00	0.00	0.00	0.00	1.83	4.00	0.88
Firmicutes;c_Bacilli;o_Lactobacillales;f_Leuconostocaceae;g_Weissella	0.00	0.00	0.00	0.13	0.00	0.00	0.00	0.00	0.00	0.00
Firmicutes;c_Bacilli;o_Lactobacillales;f_Streptococcaceae;g_Streptococcus	0.00	0.00	0.00	0.00	0.00	0.01	0.00	0.05	0.05	0.16
Firmicutes;c_Clostridia;o_Clostridiales;f_Clostridiaceae;g_Anaerococcus	0.00	0.00	0.00	0.00	0.00	0.00	0.00	0.00	0.00	3.30
Firmicutes;c_Clostridia;o_Clostridiales;f_Clostridiaceae;g_Clostridium	0.00	0.00	0.00	0.13	0.00	0.00	0.00	0.28	0.00	0.00
Firmicutes;c_Clostridia;o_Clostridiales;f_Lachnospiraceae;g_Blautia	0.00	0.00	0.00	0.00	0.00	0.01	0.00	0.01	0.28	0.00
Firmicutes;c_Clostridia;o_Clostridiales;f_Lachnospiraceae;g_Butyrivibrio	0.00	0.00	0.00	0.00	0.00	0.00	0.26	0.00	0.00	0.00
Firmicutes;c_Clostridia;o_Clostridiales;f_Lachnospiraceae;g_Moryella	0.00	0.00	0.00	0.00	0.00	0.00	0.00	0.00	0.00	0.16
Firmicutes;c_Clostridia;o_Clostridiales;f_Lachnospiraceae;g_Oribacterium	0.00	0.00	0.00	0.00	0.00	0.00	0.00	0.22	0.00	0.00
Firmicutes;c_Clostridia;o_Clostridiales;f_Lachnospiraceae;g_Roseburia	0.00	0.00	0.00	0.19	0.00	0.00	0.00	0.01	0.00	1.37
Firmicutes;c_Clostridia;o_Clostridiales;f_Peptococcaceae;g_Desulfosporosinus	0.00	0.00	0.00	0.00	0.14	0.00	0.00	0.00	0.00	0.00
Firmicutes;c_Clostridia;o_Clostridiales;f_Ruminococcaceae;g_Faecalibacterium	0.02	0.11	0.00	0.00	0.07	0.00	0.13	0.11	0.61	0.00
Firmicutes;c_Clostridia;o_Clostridiales;f_Ruminococcaceae;g_Oscillospira	0.00	0.00	0.02	0.00	0.00	0.00	0.00	0.16	0.33	0.00
Firmicutes;c_Clostridia;o_Clostridiales;f_Veillonellaceae;g_Megamonas	0.00	0.00	0.00	0.00	0.00	0.00	0.00	0.03	0.14	0.00
Firmicutes;c_Clostridia;o_Clostridiales;f_Veillonellaceae;g_Veillonella	0.00	0.00	0.00	0.00	0.00	0.00	0.33	0.00	0.09	0.16
Fusobacteria;c_Fusobacteria;o_Fusobacteriales;f_Fusobacteriaceae;g_Fusobacterium	0.00	0.00	0.00	0.13	0.00	0.00	0.00	0.00	0.00	0.00
Nitrospirae;c_Nitrospira;o_Nitrospirales;f_Nitrospiraceae;g_Nitrospira	0.07	0.00	0.00	0.32	0.00	0.00	0.00	0.00	0.00	0.00
Proteobacteria;c_Alphaproteobacteria;o_Rhizobiales;f_Bradyrhizobiaceae;g_Balneimonas	0.00	0.00	0.00	0.00	0.14	0.00	0.00	0.00	0.00	0.00
Proteobacteria;c_Alphaproteobacteria;o_Rhizobiales;f_Bradyrhizobiaceae;g_Bosea	0.00	0.00	0.00	0.00	0.00	0.00	0.00	0.36	0.00	0.00
Proteobacteria;c_Alphaproteobacteria;o_Rhizobiales;f_Bradyrhizobiaceae;g_Bradyrhizobium	0.00	0.00	0.00	0.88	0.14	0.00	0.00	0.16	0.00	0.00
Proteobacteria;c_Alphaproteobacteria;o_Rhizobiales;f_Methylobacteriaceae;g_Methylobacterium	0.00	0.28	0.00	0.76	0.07	0.00	0.33	0.82	0.00	0.40
Proteobacteria;c_Alphaproteobacteria;o_Rhizobiales;f_Phyllobacteriaceae;g_Phyllobacterium	0.00	0.00	0.29	0.00	0.00	0.00	0.00	0.00	0.00	0.00
Proteobacteria;c_Alphaproteobacteria;o_Rhizobiales;f_Rhizobiaceae;g_Agrobacterium	0.00	0.00	0.00	0.00	0.00	0.00	0.13	0.00	0.00	0.00
Proteobacteria;c_Alphaproteobacteria;o_Rhodospirillales;f_Acetobacteraceae;g_Acetobacter	0.00	0.00	0.29	0.00	0.29	0.00	0.65	0.32	2.07	0.24
Proteobacteria;c_Alphaproteobacteria;o_Rhodospirillales;f_Acetobacteraceae;g_Acidocella	0.02	1.01	0.00	0.95	0.80	0.00	0.26	0.00	0.00	0.00
Proteobacteria;c_Alphaproteobacteria;o_Rhodospirillales;f_Acetobacteraceae;g_Gluconobacter	8.67	3.28	19.56	17.50	49.42	0.47	7.32	6.10	7.57	5.63
Proteobacteria;c_Alphaproteobacteria;o_Rhodospirillales;f_Acetobacteraceae;g_Roseomonas	0.00	0.31	0.19	0.19	0.00	0.00	0.00	0.18	0.47	0.32
Proteobacteria;c_Alphaproteobacteria;o_Rhodospirillales;f_Rhodospirillaceae;g_Telmatospirillum	0.00	0.00	0.00	0.00	0.00	0.00	0.00	0.00	0.00	0.32
Proteobacteria;c_Alphaproteobacteria;o_Rickettsiales;f_Rickettsiaceae;g_Rickettsia	0.00	0.00	0.00	0.00	0.00	0.00	0.00	0.00	1.13	0.00
Proteobacteria;c_Alphaproteobacteria;o_Rickettsiales;f_Rickettsiaceae;g_Wolbachia	0.02	0.00	0.00	0.00	0.22	0.00	0.07	0.00	0.00	0.00
Proteobacteria;c_Alphaproteobacteria;o_Sphingomonadales;f_Sphingomonadaceae;g_Sphingomonas	0.02	1.01	0.41	0.95	0.80	0.00	8.43	9.83	6.58	6.60
Proteobacteria;c_Betaproteobacteria;o_Burkholderiales;f_Alcaligenaceae;g_Pigmentiphaga	0.00	0.00	0.00	0.00	0.00	0.00	0.07	0.08	0.42	0.00
Proteobacteria;c_Betaproteobacteria;o_Burkholderiales;f_Burkholderiaceae;g_Burkholderia	0.00	0.00	0.00	0.00	0.00	0.00	0.00	0.00	0.00	0.64
Proteobacteria;c_Betaproteobacteria;o_Burkholderiales;f_Burkholderiaceae;g_Salinispora	0.00	0.00	0.00	0.13	0.07	0.00	0.13	0.07	0.00	0.48

(Continued)

TABLE 3 | Continued

Duration of fermentation [days]	Organic						Conventional			
	0	2	3	10	16	0	2	6	7	12
Proteobacteria;c_Betaproteobacteria;o_Burkholderiales;f_Comamonadaceae;g_Curvibacter	0.00	0.00	0.17	0.06	0.36	0.00	0.00	0.24	0.00	0.08
Proteobacteria;c_Betaproteobacteria;o_Burkholderiales;f_Comamonadaceae;g_Diaphorobacter	0.00	0.17	0.00	0.13	0.00	0.00	0.00	0.00	0.00	1.05
Proteobacteria;c_Betaproteobacteria;o_Burkholderiales;f_Comamonadaceae;g_Methylibium	0.00	0.00	0.00	0.00	0.00	0.00	0.07	0.30	0.56	0.00
Proteobacteria;c_Betaproteobacteria;o_Burkholderiales;f_Comamonadaceae;g_Pelomonas	0.00	0.00	0.14	0.00	0.00	0.00	0.59	0.00	0.00	0.00
Proteobacteria;c_Betaproteobacteria;o_Burkholderiales;f_Comamonadaceae;g_Ramlibacter	0.00	0.00	0.00	0.00	0.00	0.00	0.00	0.04	0.47	0.00
Proteobacteria;c_Betaproteobacteria;o_Burkholderiales;f_Comamonadaceae;g_Rubrivivax	0.00	0.00	0.00	0.00	0.00	0.00	0.13	0.25	0.00	0.00
Proteobacteria;c_Betaproteobacteria;o_Burkholderiales;f_Comamonadaceae;g_Schlegelella	0.00	0.00	0.00	0.00	0.00	0.00	0.00	0.13	0.00	0.00
Proteobacteria;c_Betaproteobacteria;o_Burkholderiales;f_Comamonadaceae;g_Variovorax	0.00	0.03	0.00	0.00	0.00	0.00	0.20	0.16	0.00	0.00
Proteobacteria;c_Betaproteobacteria;o_Burkholderiales;f_Oxalobacteraceae;g_Janthinobacterium	0.00	0.84	0.38	0.63	0.14	0.00	7.12	8.91	5.59	1.37
Proteobacteria;c_Betaproteobacteria;o_Burkholderiales;f_Oxalobacteraceae;g_Ralstonia	0.00	0.00	0.05	1.83	0.36	0.03	0.59	0.18	0.09	2.98
Proteobacteria;c_Betaproteobacteria;o_Neisseriales;f_Neisseriaceae;g_Neisseria	0.00	0.00	0.10	0.25	0.00	0.00	0.13	0.00	0.00	1.05
Proteobacteria;c_Betaproteobacteria;o_Rhodocyclales;f_Rhodocyclaceae;g_KD1-23	0.00	0.00	0.00	0.00	0.00	0.00	0.26	0.00	0.00	0.00
Proteobacteria;c_Gammaproteobacteria;o_Alteromonadales;f_Shewanellaceae;g_Shewanella	0.00	0.00	0.00	0.13	0.00	0.00	0.00	0.00	0.00	0.00
Proteobacteria;c_Gammaproteobacteria;o_Enterobacteriales;f_Enterobacteriaceae;g_Candidatus Hamiltonella	0.00	0.00	0.00	0.00	0.00	0.00	0.26	0.00	0.00	0.00
Proteobacteria;c_Gammaproteobacteria;o_Enterobacteriales;f_Enterobacteriaceae;g_Citrobacter	0.00	0.28	0.00	0.00	0.00	0.00	0.00	0.04	0.05	0.00
Proteobacteria;c_Gammaproteobacteria;o_Enterobacteriales;f_Enterobacteriaceae;g_Erwinia	0.02	0.00	0.00	0.06	0.07	0.01	0.52	0.30	0.33	0.00
Proteobacteria;c_Gammaproteobacteria;o_Enterobacteriales;f_Enterobacteriaceae;g_Escherichia	0.00	0.00	0.00	0.00	0.00	0.00	0.00	0.01	0.24	0.00
Proteobacteria;c_Gammaproteobacteria;o_Enterobacteriales;f_Enterobacteriaceae;g_Trabulsiella	0.20	0.22	0.17	0.25	0.14	0.36	0.00	0.00	0.00	0.00
Proteobacteria;c_Gammaproteobacteria;o_Pasteurellales;f_Pasteurellaceae;g_Haemophilus	0.00	0.00	0.29	0.32	0.07	0.00	0.07	0.00	0.00	0.40
Proteobacteria;c_Gammaproteobacteria;o_Pseudomonadales;f_Moraxellaceae;g_Acinetobacter	0.00	0.20	0.14	0.00	0.14	0.00	0.33	1.16	3.06	4.99
Proteobacteria;c_Gammaproteobacteria;o_Pseudomonadales;f_Moraxellaceae;g_Enhydrobacter	0.00	0.00	0.00	0.00	0.00	0.00	0.20	0.12	0.00	0.00
Proteobacteria;c_Gammaproteobacteria;o_Pseudomonadales;f_Pseudomonadaceae;g_Pseudomonas	0.02	0.45	1.03	0.63	0.14	0.12	10.39	8.95	6.54	2.57
Spirochaetes;c_[Leptospirae];o_[Leptospirales];f_Leptospiraceae;g_Leptospira	0.00	0.22	0.05	0.19	0.14	0.00	0.07	0.00	1.18	0.00

oxydans during wine fermentation and less abundant changes of *Acetobacter* from both organically and conventionally fermented wine (**Table 3** and Table S5). The increased abundance of *G. oxydans* during the organic fermentation process might explain the increased susceptibility to wine spoilage in wines that are produced using organic fermentation techniques. Overall, these results demonstrate that 16S rRNA gene sequencing technique can be used efficiently to obtain a detailed description of the bacterial population associated with grape juice and must and to discover novel microorganisms that might lead to wine spoilage. This ability will allow wine makers to prevent losing revenues and investing in NGS technologies pose a promising avenue for wine makers, in particular as NGS has become a commodity and software for NGS data analysis is freely available. By comparing community dynamics of organically and conventionally fermented grape musts, we also observed that the population of *Pedobacter*, *Sphingomonas*, *Janthinobacterium*, and *Pseudomonas* were significantly higher in musts subjected to

conventional than organic fermentation practices. It also appears that the bacterial population associated with the conventionally produced wine, experiences more significant community changes during the vinification process. This finding can be explained by the fact that commonly additives such as DAP have a significant effect on the indigenous bacterial population (**Figure 1**) and affect the community profile almost instantly. On the other hand, the increased community complexity of conventionally fermented must is less expected although it can also be explained by the affect of the additives that are employed in the conventional fermentation process. These additives appear to affect primarily phylogenetic groups that are undesired during the fermentation process and that dominate the prokaryotic community prior to their addition. Additionally, decreased community complexity and diversity in the organically fermented grape juice might be caused by the presence of indigenous yeasts on the skin of grapes that are not subjected to fungicide (i.e., SO_2) treatments during the organic PDC fermentation. This antimicrobial affect by indigenous yeasts in bacteria during the fermentation process was reported previously (Lonvaudfunel et al., 1988; Henick-Kling and Park, 1994) and it is possible that a defined mixture of naturally occurring yeast strains might represent a highly sustainable approach for controlling the composition and temporal succession of the bacterial population during the fermentation process. In order to make such yeast mixtures effective they would need to include additional strains that are efficient against the wine spoilage bacteria (e.g., *Gluconobacter oxydans*) that appear to be little affected by currently known indigenous grape skin yeasts.

Previously it was reported that winery surfaces were dominated by non-fermentation-related bacteria (i.e., *Pseudomonas*, *Comamonadaceae*, *Flavobaterium*, *Enterbacteraceae*, *Brevundimonas*, and *Bacillus*). Accordingly, we detected *Pseudomonas*, *Comamonadaceae*, *Enterbacteraceae*, and *Bacillus* during both organic and conventional fermentation (Table S6). The population of *Pseudomonas* and *Comamonadaceae* are larger at the early stage of conventional fermentation (2 days), which suggests that *Pseudomonas* and some members of *Comamonadaceae* originated from conventionally vinification process or their growth was not instantly inhibited by addition of SO_2 prior to conventional

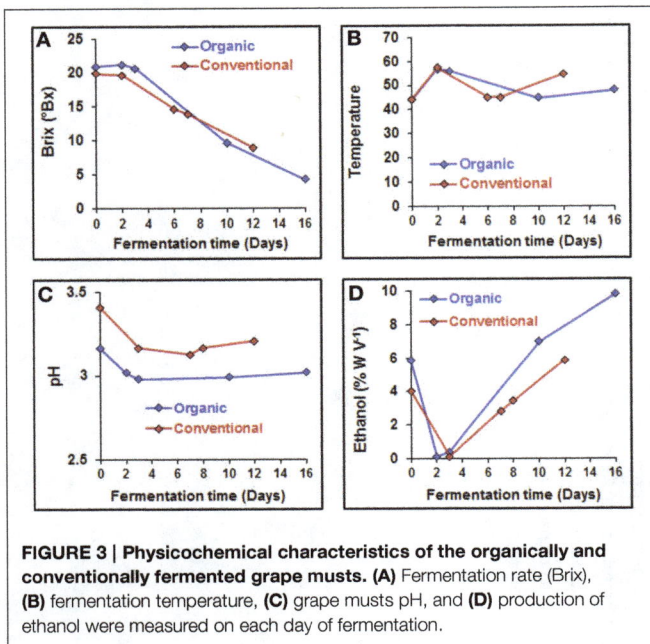

FIGURE 3 | Physicochemical characteristics of the organically and conventionally fermented grape musts. (A) Fermentation rate (Brix), **(B)** fermentation temperature, **(C)** grape musts pH, and **(D)** production of ethanol were measured on each day of fermentation.

TABLE 4 | Chemical profile of grape musts during organic and conventional fermentation.

Duration of fermentation [days]	Organic					Conventional				
	0	2	3	10	16	0	2	6	7	12
Ethanol at 20°C (% Vol)	5.9	0.1	0.4	7	9.8	4	0.1	2.8	3.4	5.9
L-lactic acid (g/L)	0.97	0.05	0.05	0.05	0.05	0.29	0.05	0.05	0.05	0.05
L-malic acid (g/L)	1.97	3.08	3.04	2.45	2.31	2.7	4.56	4.02	4.09	3.79
Volatile acidity(acetic) (g/L)	0.16	0.05	0.09	0.12	0.2	0.14	0.07	0.34	0.42	0.46
Tartaric acid (g/L)	2.1	4.2	4	3.2	3	1.7	2.5	2.1	1.9	1.9
Titratable acidity (g/L)	7.3	7.6	7.3	7.3	7.5	5.8	7.6	7.3	7.5	7.2
Yeast assimilable nitrogen (mg/L)	18	137	103	18	18	18	219	101	155	143
Alpha-amino compounds (as N) (mg/L)	10	91	70	14	12	14	112	54	56	59
Ammonia (mg/L)	10	56	40	10	10	10	130	57	120	102

vinification. The other possibility might be that the growth of *Pseudomonas* and some members of *Comamonadaceae* was suppressed by antimicrobial components produced by indigenous yeasts associated with organically fermented wine. *Enterbacteraceae*, a dominant family from grapevine (Pinto et al., 2014), is extremely abundant during PDC fermentation (about 85% in both samples), with a rapid population decrease during conventional fermentation (5% at day 2), this might be caused by addition of SO_2. A less significant decrease was observed during organic fermentation [73% (day 2), 34% (day 10), 13% (day 13)], which might also be explained by the antimicrobial activity of an indigenous yeast that might have been associated with the grapes.

In this study, we obtained a more detailed understanding of the temporal succession of the bacterial population and associated changes of the wine chemistry during conventionally and organically fermented grapes using NGS technologies, which could not be studied with less sensitive molecular approaches (i.e., PCR-DGGE). The sequences generated during this study were deposited in NCBI's short read archive using the study accession number SRP058864. In summary, these results suggest that there are temporal changes in the bacterial population that is associated with the fermentation process and that these populations might contain microorganisms that have until today not been linked with the fermentation process. Further comprehensive study of how the bacterial species of wine interact and how the microbial community dynamics correlated with grape must and wine components during the fermentation process will be of great value for developing improved methods to control wine quality.

Author Contributions

Conceived and designed the experiments: MH. Performed the experiments: MH, HP, EH, SK, and SS. Generated and analyzed the data: MH, HP, EH, SK, RD, and SS. Wrote the paper: MH, HP, EH, and TH.

References

Andorrá, I., Landi, S., Mas, A., Guillamón, J. M., and Esteve-Zarzoso, B. (2008). Effect of oenological practices on microbial populations using culture-independent techniques. *Food Microbiol.* 25, 849–856. doi: 10.1016/j.fm.2008.05.005

Bae, S., Fleet, G. H., and Heard, G. M. (2006). Lactic acid bacteria associated with wine grapes from several Australian vineyards. *J. Appl. Microbiol.* 100, 712–727. doi: 10.1111/j.1365-2672.2006.02890.x

Bañeras, L., Trias, R., Godayol, A., Cerdán, L., Nawrath, T., Schulz, S., et al. (2013). Mass spectrometry identification of alkyl-substituted pyrazines produced by *Pseudomonas* spp. isolates obtained from wine corks. *Food Chem.* 138, 2382–2389. doi: 10.1016/j.foodchem.2012.12.030

Barata, A., Malfeito-Ferreira, M., and Loureiro, V. (2012). Changes in sour rotten grape berry microbiota during ripening and wine fermentation. *Int. J. Food Microbiol.* 154, 152–161. doi: 10.1016/j.ijfoodmicro.2011.12.029

Bartowsky, E. J., and Henschke, P. A. (2008). Acetic acid bacteria spoilage of bottled red wine - A review. *Int. J. Food Microbiol.* 125, 60–70. doi: 10.1016/j.ijfoodmicro.2007.10.016

Beneduce, L., Spano, G., Vernile, A., Tarantino, D., and Massa, S. (2004). Molecular characterization of lactic acid populations associated with wine spoilage. *J. Basic Microbiol.* 44, 10–16. doi: 10.1002/jobm.200310281

Bokulich, N. A., Joseph, C. M., Allen, G., Benson, A. K., and Mills, D. A. (2012). Next-generation sequencing reveals significant bacterial diversity of botrytized wine. *PLoS ONE* 7:e36357. doi: 10.1371/journal.pone.0036357

Bokulich, N. A., Swadener, M., Sakamoto, K., Mills, D. A., and Bisson, L. F. (2015). Sulfur dioxide treatment alters wine microbial diversity and fermentation progression in a dose-dependent fashion. *Am. J. Enol. Vitic.* 66, 73–79. doi: 10.5344/ajev.2014.14096

Bokulich, N. A., Thorngate, J. H., Richardson, P. M., and Mills, D. A. (2014). Microbial biogeography of wine grapes is conditioned by cultivar, vintage, and climate. *Proc. Natl. Acad. Sci. U.S.A.* 111, E139–E148. doi: 10.1073/pnas.1317377110

Caporaso, J. G., Bittinger, K., Bushman, F. D., Desantis, T. Z., Andersen, G. L., and Knight, R. (2010a). PyNAST: a flexible tool for aligning sequences to a template alignment. *Bioinformatics* 26, 266–267. doi: 10.1093/bioinformatics/btp636

Caporaso, J. G., Kuczynski, J., Stombaugh, J., Bittinger, K., Bushman, F. D., Costello, E. K., et al. (2010b). QIIME allows analysis of high-throughput community sequencing data. *Nat. Methods* 7, 335–336. doi: 10.1038/nmeth.f.303

Capozzi, V., Russo, P., Beneduce, L., Weidmann, S., Grieco, F., Guzzo, J., et al. (2010). Technological properties of *Oenococcus oeni* strains isolated from typical southern Italian wines. *Lett. Appl. Microbiol.* 50, 327–334. doi: 10.1111/j.1472-765X.2010.02795.x

Cappello, M. S., Bleve, G., Grieco, F., Dellaglio, F., and Zacheo, G. (2004). Characterization of *Saccharomyces cerevisiae* strains isolated from must of grape grown in experimental vineyard. *J. Appl. Microbiol.* 97, 1274–1280. doi: 10.1111/j.1365-2672.2004.02412.x

Cocolin, L., Alessandria, V., Dolci, P., Gorra, R., and Rantsiou, K. (2013). Culture independent methods to assess the diversity and dynamics of microbiota during food fermentation. *Int. J. Food Microbiol.* 167, 29–43. doi: 10.1016/j.ijfoodmicro.2013.05.008

Cocolin, L., Bisson, L. F., and Mills, D. A. (2000). Direct profiling of the yeast dynamics in wine fermentations. *FEMS Microbiol. Lett.* 189, 81–87. doi: 10.1111/j.1574-6968.2000.tb09210.x

Combet-Blanc, Y., Ollivier, B., Streicher, C., Patel, B. K., Dwivedi, P. P., Pot, B., et al. (1995). *Bacillus thermoamylovorans* sp. nov., a moderately thermophilic and amylolytic bacterium. *Int. J. Syst. Bacteriol.* 45, 9–16. doi: 10.1099/00207713-45-1-9

Curtin, C. D., Bellon, J. R., Henschke, P. A., Godden, P. W., and Lopes, M. A. D. B. (2007). Genetic diversity of *Dekkera bruxellensis* yeasts isolated from Australian wineries. *FEMS Yeast Res.* 7, 471–481. doi: 10.1111/j.1567-1364.2006.00183.x

Di Toro, M. R., Capozzi, V., Beneduce, L., Alexandre, H., Tristezza, M., Durante, M., et al. (2015). Intraspecific biodiversity and 'spoilage potential' of *Brettanomyces bruxellensis* in Apulian wines. *Lwt-Food Sci. Technol.* 60, 102–108. doi: 10.1016/j.lwt.2014.06.059

Edgar, R. C. (2010). Search and clustering orders of magnitude faster than BLAST. *Bioinformatics* 26, 2460–2461. doi: 10.1093/bioinformatics/btq461

Fleet, G. H. (1993). *Wine Microbiology and Biotechnology.* New York, NY: Taylor & Francis Group.

Fugelsang, K. C., and Edwards, C. G. (2007). *Wine Microbiolog Practical Application and Procedures.* New York, NY: Springer.

Garofalo, C., El Khoury, M., Lucas, P., Bely, M., Russo, P., Spano, G., et al. (2015). Autochthonous starter cultures and indigenous grape variety for regional wine production. *J. Appl. Microbiol.* 118, 1395–1408. doi: 10.1111/jam.12789

González-Barreiro, C., Rial-Otero, R., Cancho-Grande, B., and Simal-Gándara, J. (2015). Wine aroma compounds in grapes: a critical review. *Crit. Rev. Food Sci. Nutr.* 55, 202–218. doi: 10.1080/10408398.2011.650336

Henick-Kling, T., and Park, Y. H. (1994). Considerations for the use of yeast and bacterial starter cultures: SO₂ and timing of inoculation. *Am. J. Enol. Vitic.* 45, 464–469.

Hopfer, H., Nelson, J., Mitchell, A. E., Heymann, H., and Ebeler, S. E. (2013). Profiling the trace metal composition of wine as a function of storage temperature and packaging type. *J. Anal. At. Spectrom.* 28, 1288–1291. doi: 10.1039/c3ja50098e

Joyeux, A., Lafonla-fourcade, S., and Ribéreaugayon, P. (1984). Evolution of acetic-acid bacteria during fermentation and storage of wine. *Appl. Environ. Microbiol.* 48, 153–156.

Laforgue, R., Guérin, L., Pernelle, J. J., Monnet, C., Dupont, J., and Bouix, M. (2009). Evaluation of PCR-DGGE methodology to monitor fungal communities on grapes. *J. Appl. Microbiol.* 107, 1208–1218. doi: 10.1111/j.1365-2672.2009.04309.x

Lonvaudfunel, A., Joyeux, A., and Desens, C. (1988). Inhibition of malolactic fermentation of wines by products of yeast metabolism. *J. Sci. Food Agric.* 44, 183–191. doi: 10.1002/jsfa.2740440209

Lopez, I., Ruiz-Larrea, F., Cocolin, L., Orr, E., Phister, T., Marshall, M., et al. (2003). Design and evaluation of PCR primers for analysis of bacterial populations in wine by denaturing gradient gel electrophoresis. *Appl. Environ. Microbiol.* 69, 6801–6807. doi: 10.1128/AEM.69.11.6801-6807.2003

McDonald, D., Price, M. N., Goodrich, J., Nawrocki, E. P., Desantis, T. Z., Probst, A., et al. (2012). An improved Greengenes taxonomy with explicit ranks for ecological and evolutionary analyses of bacteria and archaea. *ISME J.* 6, 610–618. doi: 10.1038/ismej.2011.139

Millet, V., and Lonvaud-Funel, A. (2000). The viable but non-culturable state of wine micro-organisms during storage. *Lett. Appl. Microbiol.* 30, 136–141. doi: 10.1046/j.1472-765x.2000.00684.x

Nguyen, T. T., and Landfald, B. (2015). Polar front associated variation in prokaryotic community structure in Arctic shelf seafloor. *Front. Microbiol.* 6:17. doi: 10.3389/fmicb.2015.00017

Oliver, J. D. (2005). The viable but nonculturable state in bacteria. *J. Microbiol.* 43, 93–100.

Pérez-Martín, F., Seseña, S., Fernández-González, M., Arévalo, M., and Palop, M. L. (2014). Microbial communities in air and wine of a winery at two consecutive vintages. *Int. J. Food Microbiol.* 190, 44–53. doi: 10.1016/j.ijfoodmicro.2014.08.020

Pessoa-Filho, M., Barreto, C. C., Dos Reis Junior, F. B., Fragoso, R. R., Costa, F. S., De Carvalho Mendes, I., et al. (2015). Microbiological functioning, diversity, and structure of bacterial communities in ultramafic soils from a tropical savanna. *Antonie Van Leeuwenhoek* 107, 935–949. doi: 10.1007/s10482-015-0386-6

Piao, H., Lachman, M., Malfatti, S., Sczyrba, A., Knierim, B., Auer, M., et al. (2014). Temporal dynamics of fibrolytic and methanogenic rumen microorganisms during *in situ* incubation of switchgrass determined by 16S rRNA gene profiling. *Front. Microbiol.* 5:307. doi: 10.3389/fmicb.2014.00307

Pinto, C., Pinho, D., Sousa, S., Pinheiro, M., Egas, C., and Gomes, A. C. (2014). Unravelling the diversity of grapevine microbiome. *PLoS ONE* 9:e85622. doi: 10.1371/journal.pone.0085622

Renouf, V., Claisse, O., and Lonvaud-Funel, A. (2007). Inventory and monitoring of wine microbial consortia. *Appl. Microbiol. Biotechnol.* 75, 149–164. doi: 10.1007/s00253-006-0798-3

Sinclair, L., Osman, O. A., Bertilsson, S., and Eiler, A. (2015). Microbial community composition and diversity via 16S rRNA gene amplicons: evaluating the illumina platform. *PLoS ONE* 10:e0116955. doi: 10.1371/journal.pone.0116955

Spano, G., Lonvaud-Funel, A., Claisse, O., and Massa, S. (2007). *In vivo* PCR-DGGE analysis of *Lactobacillus plantarum* and *Oenococcus oeni* populations in red wine. *Curr. Microbiol.* 54, 9–13. doi: 10.1007/s00284-006-0136-0

Styger, G., Prior, B., and Bauer, F. F. (2011). Wine flavor and aroma. *J. Ind. Microbiol. Biotechnol.* 38, 1145–1159. doi: 10.1007/s10295-011-1018-4

Trexler, R., Solomon, C., Brislawn, C. J., Wright, J. R., Rosenberger, A., McClure, E. E., et al. (2014). Assessing impacts of unconventional natural gas extraction on microbial communities in headwater stream ecosystems in Northwestern Pennsylvania. *Front. Microbiol.* 5:522. doi: 10.3389/fmicb.2014.00522

Wang, Q., Garrity, G. M., Tiedje, J. M., and Cole, J. R. (2007). Naive Bayesian classifier for rapid assignment of rRNA sequences into the new bacterial taxonomy. *Appl. Environ. Microbiol.* 73, 5261–5267. doi: 10.1128/AEM.00062-07

Yousaf, S., Bulgari, D., Bergna, A., Pancher, M., Quaglino, F., Casati, P., et al. (2014). Pyrosequencing detects human and animal pathogenic taxa in the grapevine endosphere. *Front. Microbiol.* 5:327. doi: 10.3389/fmicb.2014.00327

Conflict of Interest Statement: The authors declare that the research was conducted in the absence of any commercial or financial relationships that could be construed as a potential conflict of interest.

11

A comparative study of infrared and microwave heating for microbial decontamination of paprika powder

Lovisa Eliasson*, Sven Isaksson, Maria Lövenklev[†] and Lilia Ahrné

Food and Bioscience, SP Technical Research Institute of Sweden, Gothenburg, Sweden

Edited by:
Kai Reineke,
GNT Europa GmbH, Germany

Reviewed by:
Petros Taoukis,
National Technical University
of Athens, Greece
Nicolás Meneses,
Bühler AG, Switzerland

*Correspondence:
Lovisa Eliasson,
Food and Bioscience, SP Technical
Research Institute of Sweden, Box
5401, SE-402 29 Gothenburg,
Sweden
lovisa.eliasson@sp.se

[†] Present address:
Maria Lövenklev,
Swedish National Food Agency,
Uppsala, Sweden

There is currently a need in developing new decontamination technologies for spices due to limitations of existing technologies, mainly regarding their effects on spices' sensory quality. In the search of new decontamination solutions, it is of interest to compare different technologies, to provide the industry with knowledge for taking decisions concerning appropriate decontamination technologies for spices. The present study compares infrared (IR) and microwave decontamination of naturally contaminated paprika powder after adjustment of water activity to 0.88. IR respectively microwave heating was applied to quickly heat up paprika powder to 98°C, after which the paprika sample was transferred to a conventional oven set at 98°C to keep the temperature constant during a holding time up to 20 min. In the present experimental set-up microwave treatment at 98°C for 20 min resulted in a reduction of 4.8 log units of the total number of mesophilic bacteria, while the IR treatment showed a 1 log unit lower reduction for the corresponding temperature and treatment time. Microwave and IR heating created different temperature profiles and moisture distribution within the paprika sample during the heating up part of the process, which is likely to have influenced the decontamination efficiency. The results of this study are used to discuss the difficulties in comparing two thermal technologies on equal conditions due to differences in their heating mechanisms.

Keywords: infrared heating, microwave heating, microbial decontamination, paprika powder, quality

Introduction

Herbs and spices are important food ingredients. However, while they contribute positively to the sensorial properties of the foods, they also often constitute a microbial hazard due to poor sanitation during growth, harvest, drying, and storage. Various bacteria, such as *Salmonella*, *Bacillus cereus*, *Clostridium perfringens*, and *Escherichia coli* are often found on herbs and spices (Banerjee and Sarkar, 2003; Sagoo et al., 2009). The low water activity in dried herbs and spices creates an unfavorable environment for survival of many kinds of vegetative bacteria, which often results in high numbers of bacterial spores such as *Clostridium* and *Bacillus* species (Spruthi, 1980). High amounts of spores in dried spices form a potential microbial hazard since, e.g., *Bacillus* species are known to have high resistance to heat treatments, which may result in germination of surviving spores when spices are added to high water activity foodstuffs (Daelman et al., 2013).

A few non-thermal decontamination technologies have been applied on dried herbs and spices on industrial scale. Unfortunately, these technologies have so far either had disadvantages from a health perspective (ethylene dioxide) or found poor consumer acceptance (γ-irradiation). This has resulted in steam treatment as the most accepted decontamination technology of today in Europe (Schweiggert et al., 2007). However, since steam treatment has important impact on sensorial properties, there is a continuous interest to investigate other methods. Alternative thermal decontamination methods could be microwave heating (Dehne et al., 1990; Dehne and Bögl, 1993; Emam et al., 1995; Legnani et al., 2001; Dababneh, 2012) or infrared (IR) heating (Staack et al., 2008a; Erdoğdu and Ekiz, 2011, 2013; Eliasson et al., 2014). Both IR and microwave heating show attractive properties, above all fast direct heating of the material, which probably could be used in order to reduce processing time, compared to conventional technologies, and thereby potentially also preserve quality characteristics in a better way.

The heating mechanism of IR radiation, corresponding to the wavelengths 0.76 μm–1 mm of the electromagnetic spectrum, is based on the changes of rotations and vibration of molecules, which leads to an energy absorption that further is transferred to heat when the molecules return to their normal state. IR heating is divided into near (0.76–2 μm), medium (2–4 μm), and far (4–1000 μm) IR. The application of different IR wavelength regions affects the amount of transmitted energy, rate of temperature increase and penetration depth. IR heating has a limited penetration depth, about 0.31–4.76 mm dependent on food product and applied wavelengths, and therefore IR is mainly considered a surface heating technology (Skjöldebrand, 2001).

Microwaves are a portion of the electromagnetic spectrum with wavelengths ranging from 1 mm to 1 m and frequencies between 300 and 300 GHz. Microwave energy has been used in food processing applications mainly due to its ability to cause fast volumetric heating that penetrates considerably into the bulk of the material. The heating mechanism of microwaves is based on the interaction between their electric fields and matter, which causes movement of dipoles and ions, a movement that finally translates into heat. The resulting heating depends on many factors. Not only food-related properties, such as moisture content, density, dielectric properties and temperature, are matters of important influence, but also aspects related to the actual design of the microwave heating device. Microwaves penetrate into the heated material—hence they have the potential to cause a more homogenous heating than many other heating methods. Nonetheless, variations of temperature within the food mostly do occur to some extent (Meredith, 1998); these variations are often referred to as hot- and cold-spots and are seldom easily foreseen without thorough investigations.

While a number of studies already investigated IR or microwave decontamination of herbs and spices (Dehne et al., 1990; Dehne and Bögl, 1993; Emam et al., 1995; Legnani et al., 2001; Staack et al., 2008a; Erdoğdu and Ekiz, 2011, 2013; Dababneh, 2012; Eliasson et al., 2014) only a few made comparative studies where one applied technology is compared to another one. IR heating has been compared to the combined effect of IR heating followed by non-thermal ultraviolet treatment (Erdoğdu and Ekiz, 2011). Microwave heating has been compared to non-thermal methods like γ-radiation (Vajdi and Pereira, 1975; Emam et al., 1995; Legnani et al., 2001), but to our knowledge none so far compared the results of two different thermal technologies such as IR and microwave heating for decontamination of spices. Therefore the objective of the present study is to investigate the differences in decontamination efficiency between IR and microwave heating, applied on paprika powder. Since both technologies have been mentioned in the literature as alternative decontamination methods, there is an interest to make a comparison and examine differences in heating mechanism and its influence on temperature profiles, water activity and decontamination. Such studies are important since they can help providing the industry with knowledge for taking decisions concerning appropriate decontamination technologies for spices.

Materials and Methods

Preparation of the Raw Material

Paprika powder with a moisture content of 5.6% and water activity of 0.38 was provided by Juan José Albarracin (Murcia, Spain). Prior to the IR and microwave treatments the water activity of the paprika powder was increased to 0.88. This water activity was considered favorable in spices with regards to decontamination efficiency, based on data from previous publications (Staack et al., 2008a; Eliasson et al., 2014). Also pre-trials with dry unconditioned paprika powder showed difficulties in creating a fast and homogenous heating by both IR and microwave heating. The adjustment of the water activity was done in a climate chamber (Vötsch VCL 7010, Balingen-Frommern, Germany) of relative humidity 88% and temperature 20°C. Three trays, of size 21 × 30 cm, containing 150 g of paprika powder each was placed in the climate chamber for 1.5 days. The water activity was controlled by removing samples from the top and bottom of each tray. The samples from the three trays were mixed prior to the IR and microwave treatment.

Infrared and Microwave Treatment of the Paprika Powder

Infrared respectively microwave heating was used for heating up the paprika sample to 98°C. After the heating up, the sample was moved to a conventional oven with constant temperature, to keep the desired temperature of the paprika sample for holding times of 10 and 20 min. The IR and microwave heating was thus limited to the heating up period. Nonetheless, this constitutes a logical process, provided the sample has been homogenously heated during the heating up period. After that time, only a low power of IR or microwaves would be needed to keep up temperature. In addition, if the sample remains in a cold place, continuous application of microwaves or IR radiation would most likely result in a process that is difficult to control, since cooling happens on the surface of the sample due to the fact that the surrounding air is not heated by IR (Olsson, 2005) or microwaves (Ohlsson, 1999); this results from both methods being radiative in their nature, thus causing direct heating of the food. On the contrary, application of hot-air circulation means adding heat continuously where it is needed during this part of the process.

Effort was made to design the experiments in such a way that IR and microwave treatments would be as comparable as possible. Usage of the same amount of paprika powder (66 g), the same geometry of the heating unit (Petri dish of diameter

11 cm and height 1.5 cm, with cover) and thereby a density of 0.46 g/cm^3, combined with an identical design of the holding time (10 and 20 min of constant temperature), using an conventional oven (Garomat Electrolux, Stockholm, Sweden), ensured that the samples experienced same conditions during the holding time regardless of the technology used for the heating up. The heating up part of the process was designed in such a way that the paprika sample was heated as fast and homogenously as possible.

The IR heating of paprika powder was performed according to the procedure developed by Staack et al. (2008a) and modified by Eliasson et al. (2014). In brief, 66 g paprika powder was placed in a closed glass Petri dish. The IR transparency of the Petri dish glass cover was measured by the use of a black body. The heat flux received by the black body was measured both without and with the Petri glass cover on the top, and from this the IR transparency of the glass was calculated to be 87%. During IR treatment, the temperature was measured with type-T thermocouples (Pentronic AB, Gunnebo, Sweden) fixed at three different depths of the paprika bed. By the support of a Teflon device, one thermocouple was placed close to the top surface (1.5 mm depth), a second in the middle of the sample (7.5 mm depth) and the third close to the bottom (13.5 mm depth).

The IR oven (Ircon Drying Systems AB, Vänersborg, Sweden) was semi-continuous and equipped with six near infrared (NIR) emitters 20 cm above as well as six NIR emitters 20 cm below the sample tray (**Figure 1**). The sample tray was moving back and forth during the IR treatment. The NIR emitters were of quartz tube type, with tungsten filament and halogen gas, resulting in a maximum IR emission of 1.2 μm (2100°C) at full power level. To reach a homogenous heating up of the paprika sample the IR heat flux was regulated stepwise: 22.6 kW/m^2 in step 1 for 40s, 11 kW/m^2 in step 2 for 90 s, and 0 kW/m^2 in step 3 for 90 s. This

recipe was developed by stepwise making small changes in heat flux and time in each experiment. When the desired temperature was reached the settings were duplicated to confirm the result. The indicated heat flux is the value applied from each side. The stepwise regulation was necessary to avoid burning of the surface of the sample and to let the interior part of the sample equilibrate by conduction. The heating unit was moved to the conventional oven during step 3 (0 heat flux), where it then was kept for the desired holding time of 10 or 20 min.

The microwave treatment of paprika powder was performed in the same type of Petri dish as used in the IR treatment. The temperature during the microwave treatments was monitored by four fiber optic probes (Neoptix Inc, QC, Canada) placed at four different radial locations in the Petri dish. The Petri dish with paprika powder was placed in the middle of the microwave oven (Panasonic 1670, Stockholm, Sweden) and a power level of 650 W, 50% of the maximum output, was applied for 60 s. After that the sample was moved to the conventional oven during a step of 0 power, lasting for 30 s. The heating-up experiments with microwaves were repeated five times.

Visualization of Heat Patterns by Thermal Images

Thermal images of the paprika samples were taken during the IR and microwave heating with a Thermal Tracer TH7700 IR camera (NEC Avio Technologies, Tokyo, Japan). The images were taken at room temperature by removal of samples after 20 and 40 s of heating, as a means to illustrate and evaluate the different nature of the two heating technologies during the heating up part of the process. Thermal images were taken both of the cross-section of the paprika sample and from above.

Analysis of Water Activity

The water activity of the paprika powder was analyzed in an Aqua-Lab (Decagon Device, Pullman, WA, USA). Different layers of the paprika sample were analyzed after the heating up time (i.e., right after IR or microwave heating), after 10 min holding time and after 20 min holding time in the conventional oven. The Petri dish cover was removed directly after treatment and paprika powder was collected from the top, center and bottom of the paprika bed. Samples were analyzed in duplicates from each layer.

Microbial Quantification

The naturally occurring microflora in the paprika powder, i.e., aerobic mesophilic bacteria and bacterial spores, was determined before and after IR respectively microwave treatment. As previously mentioned the water activity of the paprika sample was adjusted to 0.88 in a climate chamber. The naturally occurring microflora was also quantified before and after the placement in the climate chamber to ensure the microbial level was not influenced. After each treatment, the Petri dish with 66 g paprika powder was cooled on ice with the cover of the Petri dish kept on. The entire paprika sample was then mixed and 10 g gram sample from this was further transferred and mixed with 90 mL of 0.1% peptone water (0.85% NaCl and 1% peptone; Difco Laboratories/Becton Dickinson, Stockholm, Sweden) and processed in a stomacher for 30 s. The total number of aerobic mesophilic bacteria was determined using viable count

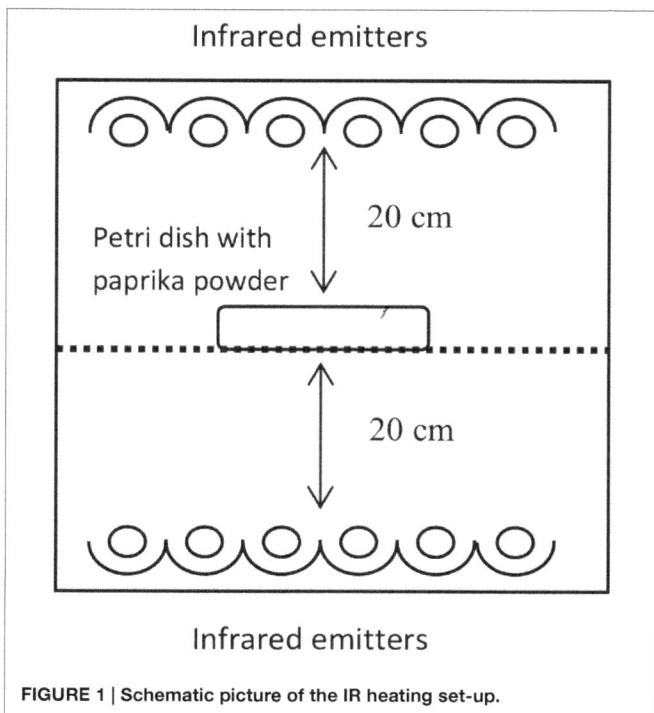

Infrared emitters

Petri dish with paprika powder

20 cm

20 cm

Infrared emitters

FIGURE 1 | Schematic picture of the IR heating set-up.

on Tryptone Soy Agar (TSA; Oxoid Ltd.) and incubation at 30°C for 72 h (NMKL 2013). Aerobic spores were grown on blood agar base supplemented with 5% horse blood (Sahlgrenska University Hospital, Gothenburg, Sweden) and incubated at 37°C for 48 h. Analysis was performed in duplicates.

Results and Discussion

Temperature Profiles

The IR heat flux was regulated stepwise in order to avoid overheating of the paprika sample's surface, and thereby reach the treatment temperature as fast and homogenously as possible for the entire sample. The first step of 22.6 kW/m^2 enabled a fast heating up, while the second step of 11 kW/m^2 reduced the heating rate at the surface and allowed the interior of the sample to be heated by conduction. After the second step, the difference between the highest and lowest temperature in the sample was 24°C as shown in **Figure 2A**. Therefore a third step of 0 heat flux was applied to let the temperature equilibrate by

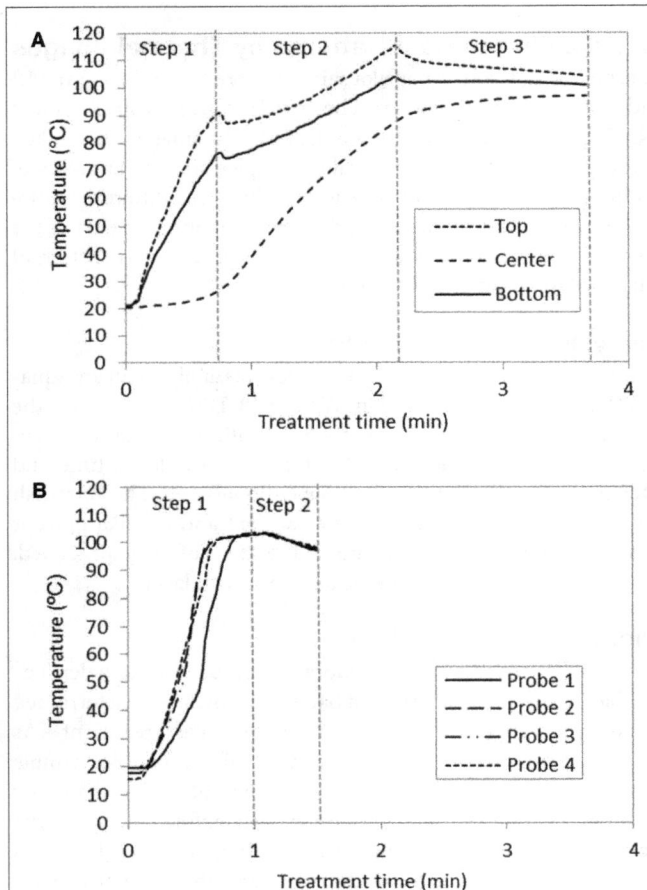

FIGURE 2 | Temperature profile to reach 98°C in the paprika powder, with an initial water activity of 0.88, treated by (A) infrared heating; step (1) 22.6 kW/m^2, step (2) 11 kW/m^2, and step (3) 0 kW/m^2, respectively (B) microwave heating; step (1) 650 W and step (2) 0 W. The Petri dish was moved to the conventional oven during step 3 and step 2 for the infrared and microwave treatment respectively.

conduction, making sure that the temperature in the center of the sample reached the target temperature 98°C. During this step the sample was moved to the conventional oven for holding times of 10 respectively 20 min. The necessary heating up time with IR heating was about 3.7 min to reach minimum 98°C in the entire paprika sample.

The microwave treatment was done by applying a power level of 650 W for 60 s, after which the transfer of the heating unit to the conventional oven was done during the 30 s with 0 power level. The probes reached 98°C after about 45 s; however, taking into account the reduction in temperature when moving the paprika sample to the conventional oven, which can be seen in **Figure 2**, it was chosen to let the temperature go beyond 98°C during the first step of the heating up with microwaves. To judge from the readings of the four probes, the entire paprika sample appeared to be heated homogenously with the current set-up (**Figure 2B**). After having applied 650 W for 60 s (in triplicates) the temperature was 103°C with standard variation ± 0.9°C, while, during the transfer time (no applied power) to the conventional oven, the standard variation was 4.8°C around the mean value 98°C.

Both IR and microwave heating are considered as rapid heating technologies. The longer heating up time required for the IR treatment (3.7 min) compared to the microwave treatment (1.5 min), is likely related to the sample thickness 1.5 cm in the present study, combined with the fact that IR heats the surface of the sample, with a penetration depth of a few mm. Erdoğdu et al. (2015) showed experimentally an IR penetration depth of 3.7 mm into paprika powder of a water activity 0.88 when applying near-IR radiation. Such a penetration depth into the paprika powder explains the necessity to reduce the IR heat flux in a stepwise manner to not overheat the surface of the sample and to let the interior part of the sample be heated by conduction, thus affecting the heating up time. In comparison, Staack et al. (2008b) needed a heating up time of about 4.5 min to reach 100°C in paprika powder of water activity of 0.88. In the study by Staack et al. (2008b), the thickness of the sample bed was 1 cm and the IR heat flux was applied from one single side, compared to the two side heating applied in the present study. This means that dependent on the choice of IR set up and sample thickness the heating up time by IR is affected.

After the heating up part of the process, the IR and microwave treated samples were placed inside the conventional oven, keeping 98°C, during the same experimental run. When the door was opened to remove or place samples inside the oven, an air temperature drop of 20–30°C for a few seconds was noted; however, pre-trials had showed that this did not significantly affect the temperature inside the paprika sample itself.

Thermal Images

Thermal images were taken during the IR and microwave treatment after interruption of the process and removal of the samples from the IR or microwave oven. This was done to evaluate the heat distribution at two times of the heating-up phase. Based on the total 60 s heating-up time of microwave heating, a time step of 60/3 = 20 s was chosen, resulting in images at times 20 and 40 s, with reference images of IR heating at the same times. The resulting images are shown in **Figure 3**. The thermal images

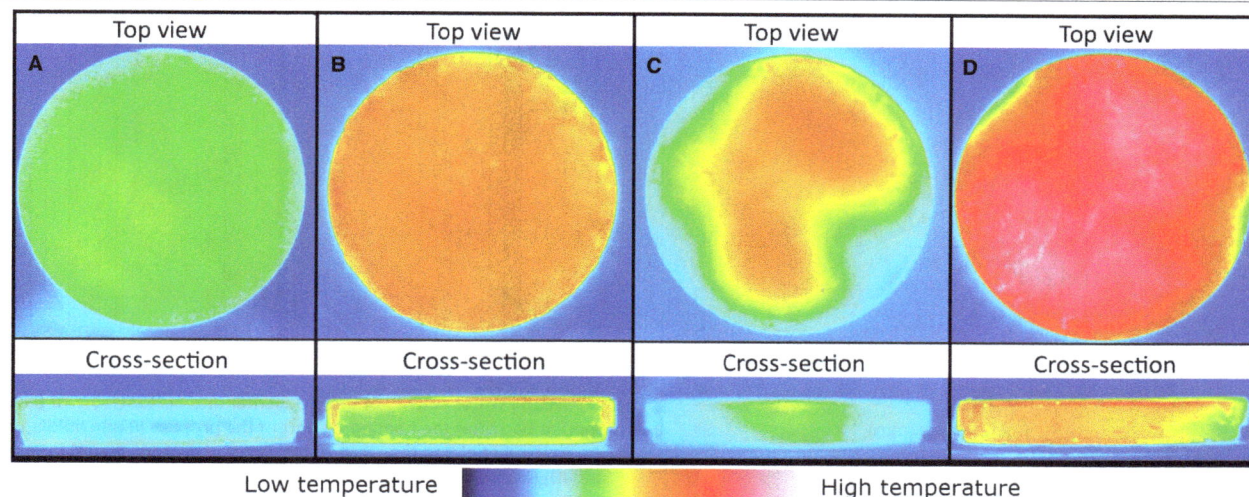

FIGURE 3 | Thermal images of the paprika powder after IR heating for (A) 20 s, (B) 40 s, and corresponding images of microwave heating after (C) 20 s and (D) 40 s. Upper row of images are taken from above, and lower row are images taken of the cross-section of the paprika sample. In the first case the petri dish cover was removed after heating; in the latter case the petri dish was split with bottom and cover still in place.

in the upper row of **Figure 3** were taken from above the sample a few seconds after removal of the petri dish cover. The images of the cross-section (the lower row of thermal images) were taken a few seconds after splitting the petri dishes.

As expected **Figures 3A,B** show that IR generates the highest temperature at the surface of the sample. Furthermore it can be seen that a homogenous heating is achieved over the entire surface (**Figure 3B**). On the other hand, **Figures 3C,D** reveal the internal heating nature of microwaves. Seen from above, the microwave treated samples (**Figures 3C,D**), exhibit a less homogenous surface heating, compared to the IR heating, which can be explained by the formation of hot and cold spots during the microwave heating.

Reduction of Natural Flora and Water Activity

The naturally occurring microflora in the paprika batch was 6.80 ± 0.10 log cfu/g for aerobic mesophilic bacteria and 6.87 ± 0.13 log cfu/g for the bacterial spores. The similar level detected on the plates for aerobic mesophilic bacteria and bacterial spores, in combination with visual observations of the colonies, indicated that the natural flora of the paprika batch mainly consisted of bacterial spores. As presented in **Figure 4**, the best microbial reduction was achieved for the microwave treated paprika powder after a holding time of 20 min, with a 4.8 log units reduction for aerobic mesophilic bacteria and a 3.2 log units reduction for the bacterial spores. The corresponding values for the IR treated samples were 3.8 log units and 2.3 log units. This can be compared to the study by Staack et al. (2008b) where a 4 log units reduction of inoculated *B. cereus* spores was found in paprika powder (water activity 0.88) after 10 min of IR treatment. The less efficient reduction found in the present study could possibly be explained by that a natural flora is more resistant than inoculated *B. cereus* spores, or may possibly be due to differences in the IR heating set up between the two studies.

The differences in the level of reduction between the aerobic mesophilic bacteria and bacterial spores were in the range of 0.13–0.17 log units and 1.37–1.65 log units after 10 respectively 20 min treatment. The natural flora was found to mainly consist of bacterial spores, and thereby the reduction of aerobic mesophilic bacteria and bacterial spores was expected to be similar. The deviation from this expectation for the 20 min treatment, both for IR and microwaves, is likely explained by a better recovery of damaged cells on the blood agar based media used for bacterial spores. This could also explain why the difference between aerobic mesophilic bacteria and bacterial spores are larger after the longer treatment time, when the cells probably are more damaged.

Care was taken to design experiments where the IR and microwave heated samples received a comparable heat-load, by trying to reach 98°C as fast and homogenously as possible. Pre-trials showed that heat treatment at 90°C for 10 min resulted in less than 0.4 log unit reductions of aerobic mesophilic bacteria. Hence it can be assumed that only the last 85 s in **Figure 2A**, and 42 s in **Figure 2B** (i.e., when the temperature is above 90°C) result in significant decontamination. The difference between these numbers, 85 and 42 s, in relation to the total holding time is quite small while the resulting decontamination differed by approximately 1 log between the two technologies. This difference is hence likely explained by other factors, such as different variations in distribution of water activity within the samples during treatment and/or differences in the temperature profiles, generated from different heating mechanisms during the heating up part of the process.

As shown by the temperature profiles in **Figure 2A** and the thermal images in **Figures 3A,B**, IR is heating the sample from the top and bottom surfaces, while heating of the interior predominantly results from heat-transfer by conduction. The microwave heating seems, in this study, to result in an overall more homogenous profile, compared to the IR heating, due to volumetric heating. However, it should be remembered that the

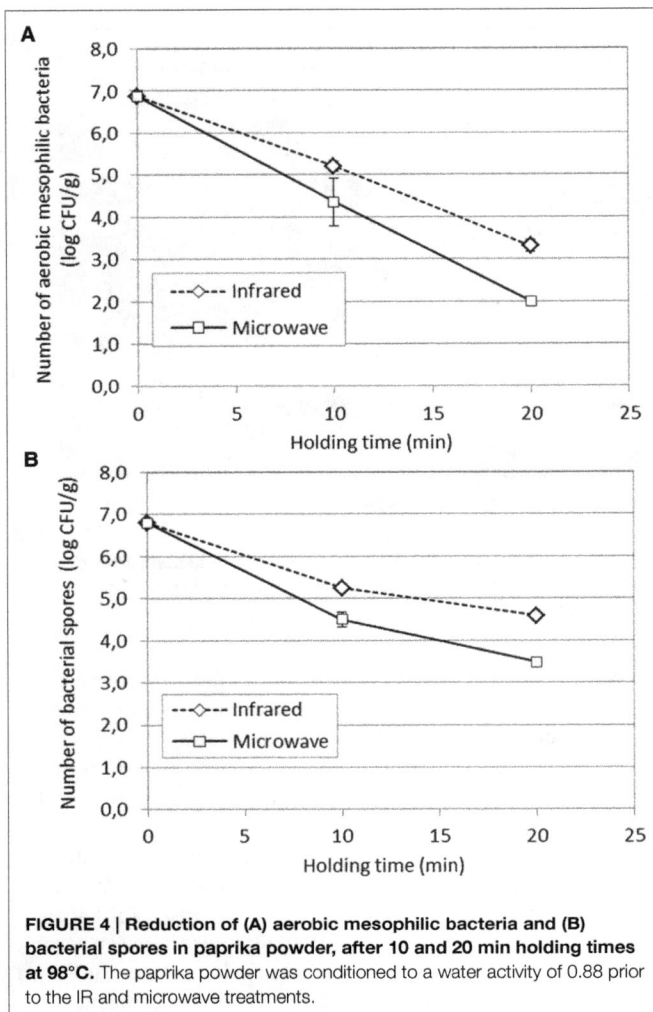

FIGURE 4 | Reduction of (A) aerobic mesophilic bacteria and (B) bacterial spores in paprika powder, after 10 and 20 min holding times at 98°C. The paprika powder was conditioned to a water activity of 0.88 prior to the IR and microwave treatments.

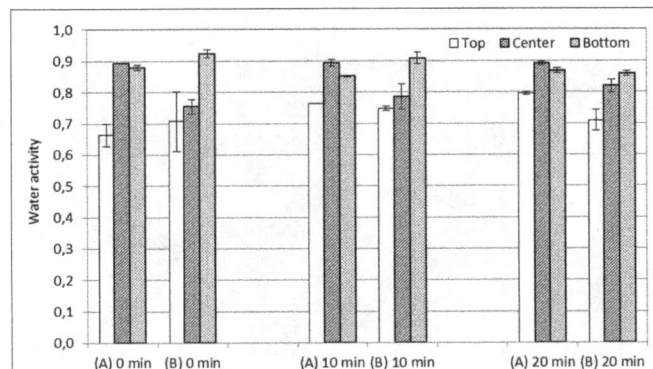

FIGURE 5 | Water activity of the top, center and bottom layer of the paprika bed after the heating up part of the process (0 min holding time), 10 min holding time and 20 min holding time for the (A) infrared and (B) microwave treated samples. The overall water activity of the paprika sample was 0.88 before the treatment.

combination of the inherent complexity of microwave heating, including the formation of hot and cold spots, which can be seen in **Figures 3C,D**, combined with a limited number of measurement points puts a limit to certainty in statements about the overall sample temperature. Another factor that restricted temperature control inside the paprika samples resulted from the fact that, due to technical restrictions, temperature probes had to be removed during movement of the samples from the IR/microwave unit to the conventional oven. Hence temperatures inside the paprika bed could not be recorded during the holding time in the convection oven, even though the air temperature was monitored.

The water activity of the paprika powder was adjusted to 0.88 before IR and microwave treatment. After the heating up part of the process, as well as after 10 and 20 min holding time, the water activity was analyzed in the top, center and bottom layer of the paprika bed. As shown in **Figure 5**, the heating up with microwaves showed a reduced water activity of 0.75 in the center layer of the sample, while the water activity of the IR treated sample was retained at 0.88 for the corresponding location. The rapid internal heating generated by microwaves creates a pressure-driven flow of liquid water and vapor to the surface of the material (Datta and Ni, 2002; Datta and Rakesh, 2012), thus pushing the

water against the wall of the Petri dish. This is a characteristic of the microwave heating, in contrast to IR heating, the latter having more of surface heating properties that do not generate the same pressure-driven flow of moisture to the surface (Datta and Ni, 2002; Datta and Rakesh, 2012). This probably explains both the lower water activity in the center of the sample as well as the increased water activity of 0.92 in the bottom layer, after the heating up with microwaves. However, this pressure-driven flow of moisture is expected to also result in an increased water activity on the top surface of the microwave treated sample. The probable explanation for why this was not confirmed experimentally would be water vapor escaping through the gap between the Petri dish and the cover during treatment. This hypothesis is confirmed by the reduced water activity observed at the top surface also of the IR treated sample.

After holding times of 10 and 20 min, the pattern with a lower water activity in the center of the sample is retained for the microwave treated sample, while the IR treated sample keep the highest water activity in the center of the sample. The differences in the distribution of the water inside the IR respectively microwave treated sample could possibly have influenced the decontamination efficiency. This since previous studies show that the heat resistance of spores is related to the water activity, and its preservation, during heat treatment (Hamanaka et al., 2005; Tiburski et al., 2014).

Another remark can be made on the choice of convection heating during the holding time. As motivated earlier, this is a logical design, although it limits the IR or microwave heating to the heating up part of the process. On the contrary, if one would assume exclusive usage of IR or microwave heating, only small amounts of IR or microwave power would be needed to keep up temperature, unless the samples were intentionally cooled down during the holding time. However, the latter would obviously add a few down-sides to the process, such as poor energy-efficiency.

At a first glance at the results of the present study, one may want to conclude that microwave heating is more effective than IR heating due to the 1 log unit higher reduction of bacteria by microwave heating. However, when comparing the results of two different technologies, one needs to bear in mind that there are technology specific factors that might cause apparent difference in efficiency between two thermal treatments. Applied on the present study, it is likely that such factors are the facts that microwaves generate a volumetric heating while IR heating is better described by surface heating. This creates different heat patterns (**Figures 2** and **3**) as well as variations in the moisture distribution, and thereby water activity (**Figure 5**), within the sample. It is likely that these differences have an impact on the resulting decontamination, which thereby could explain the differences in the observed efficiency of the two technologies. Another choice of experimental parameters, for instance a smaller sample thickness or different level of sample compression, could hence possibly have resulted in a more favorable result for the IR heating. In summary, this illustrates the inherent difficulty in designing a representative experimental set-up for comparative studies between two thermal technologies.

From a practical point of view, the difference of 1 log unit between the two heating technologies compared in this study could be set in relation to potential quality gain due to reduced treatment time. To judge from **Figures 2A,B**, the difference in treatment time would be about 5 min for reaching the same level of decontamination. Further studies are needed to elucidate if this can result in a quality benefit of the paprika powder in terms of sensorial properties.

To our knowledge, IR and microwave heating has not previously been compared in the same study for decontamination of spices. Nonetheless IR and microwave heating have been compared in studies of other applications, such as drying (Chua and Chou, 2005; Sumnu et al., 2005). There are also numerous examples of processes where the technologies have been combined, not the least in baking, with the goal to combine various heating modes for the creation of different temperature and moisture profiles, and thereby tailor the product quality and process efficiency (Datta and Ni, 2002; Datta and Rakesh, 2012).

Conclusion

As a step in providing knowledge for the industry in their work to look for alternative decontamination methods, the decontamination efficiency of IR and microwave heating, applied on paprika powder has been done. With an adjusted water activity in the product of 0.88, the natural flora of aerobic mesophilic bacteria was reduced by 3.8 log units for IR heating and 4.8 log units for microwave heating, under the conditions of the present experimental set-up. The samples had then been subjected to one heating up phase to reach the desired temperature of 98°C which lasted for 1.5 min (microwaves) respectively 3.7 min (IR), and a subsequent holding time of 20 min at 98°C convection heating.

The observed differences in decontamination efficiency are difficult to explain through a discussion purely based on time and temperature. On the other hand, the present study shows that the two technologies used for the heating-up step cause spatial differences in temperature distribution and consequently in water activity. The theoretically complex relation between total decontamination and spatial variations in water activity and time-temperature makes it difficult to link quantitatively the observed difference in level of total decontamination to the measured values of water activity and temperature. Yet it is not unlikely that the difference in decontamination originates from real variations of water activity and time-temperature. This fact illustrates the difficulties in comparing two thermal technologies on equal conditions due to their inherent differences in heating mechanism.

Author Contributions

LE performed the IR and microwave treatments and wrote part of the manuscript. SI planned the work, took the thermal images and wrote part of the manuscript. ML designed and interpreted the microbial analysis and reviewed the manuscript. LA helped planning the work and reviewed the manuscript.

Acknowledgments

The research leading to these results has received funding from the European Union's Seventh Framework Programme (FP7-SME-2011-2) under grant agreement n°285838. The authors gratefully acknowledge I. Karlsson for microbial laboratory work.

References

Banerjee, M., and Sarkar, P. K. (2003). Microbiological quality of some retail spices in India. *Food Res. Int.* 36, 469–474. doi: 10.1016/S0963-9969(02)00194-1

Chua, K. J., and Chou, S. K. (2005). A comparative study between intermittent microwave and infrared drying of bioproducts. *Int. J. Food Sci. Technol.* 40, 23–39. doi: 10.1111/j.1365-2621.2004.00903.x

Dababneh, B. F. (2012). An innovative microwave process for microbial decontamination of spices and herbs. *Afr. J. Microbiol. Res.* 7, 636–645.

Daelman, J., Jacxsens, L., Lahou, E., Devlieghere, F., and Uyttendaele, M. (2013). Assessment of the microbial safety and quality of cooked chilled foods and their production process. *Int. J. Food Microbiol.* 160, 193–200. doi: 10.1016/j.ijfoodmicro.2012.10.010

Datta, A. K., and Ni, H. (2002). Infrared and hot-air-assisted microwave heating of foods for control of surface moisture. *J. Food Eng.* 51, 355–364. doi: 10.1016/S0260-8774(01)00079-6

Datta, A. K., and Rakesh, V. (2012). Principles of microwave combination heating. *Compr. Rev. Food Sci. Food Saf.* 12, 24–39. doi: 10.1111/j.1541-4337.2012.00211.x

Dehne, K. I., and Bögl, K. W. (1993). Pasteurization of spices by microwave and high frequency. *Food Mark. Technol.* 7, 35–38.

Dehne, L. I., Reich, E., Bögl, K. W., Frey, W., Klingler, R. W., Mohr, E., et al. (1990). Zum Stand der Entkeimung von Gewürzen mittels Mikrowellen und Hochfrequenz. *SozEp Hefte* 4, 4–89. [In German].

Eliasson, L., Libander, P., Lövenklev, M., Isaksson, S., and Ahrné, L. (2014). Infrared decontamination of oregano: effects on *Bacillus cereus* spores, water activity, color, and volatile compounds. *J. Food Sci.* 79, E2447–E2455. doi: 10.1111/1750-3841.12694

Emam, O. A., Farag, S. A., and Aziz, N. H. (1995). Comparative effect of gamma and microwave irradiation on the quality of black pepper. *Z. Lebensm. Unters. Forsch.* 201, 557–561. doi: 10.1007/BF01201585

Erdoğdu, S. B., and Ekiz, H. İ. (2011). Effect of ultraviolet and far infrared radiation on microbial decontamination and quality of cumin seeds. *J. Food Sci.* 76, M284–M292. doi: 10.1111/j.1750-3841.2011.02192.x

Erdoğdu, S. B., and Ekiz, H. İ. (2013). Far infrared and ultraviolet radiation as a combined method for surface pasteurization of black pepper seeds. *J. Food Eng.* 116, 310–314. doi: 10.1016/j.jfoodeng.2012.12.026

Erdoğdu, S. B., Eliasson, L., Erdoğdu, F., Isaksson, S., and Ahrné, L. (2015). Experimental determination of penetration depths of various spice commodities (black pepper seeds, paprika powder and oregano leaves) under infrared radiation. *J. Food Eng.* 161, 75–81. doi: 10.1016/j.jfoodeng.2015.03.036

Hamanaka, D., Uchino, T., Furuse, N., and Tanaka, S. I. (2005). Inactivation effect of infrared radiation heating on bacterial spores pretreated at various water activities. *Biocontrol Sci.* 10, 61–65. doi: 10.4265/bio.10.61

Legnani, P. P., Leoni, E., Righi, F., and Zarabini, L. A. (2001). Effect of microwave heating and gamma irradiation on microbiological quality of spices and herbs. *Ital. J. Food Sci.* 13, 337–345.

Meredith, R. J. (1998). *Engineers' Handbook of Industrial Microwave Heating.* London: Institution of Electrical Engineers.

Ohlsson, T. (1999). "Minimal processing of foods with electric heating methods," in *Processing Foods: Quality Optimization and Process Assessment*, eds F. A. R. Oliveira and J. C. Oliveira (Boca Raton, FL: CRC Press), 98–99.

Olsson, E. (2005). *Jet Impingement and Infrared Heating of Cylindrical Foods.* Ph.D. thesis, Chalmers Reproservice, Göteborg, p. 14.

Sagoo, S. K., Little, C. L., Greenwood, M., Mithani, V., Grant, K. A., Mclauchlin, J., et al. (2009). Assessment of the microbiological safety of dried spices and herbs from production and retail premises in the United Kingdom. *Food Microbiol.* 26, 39–43. doi: 10.1016/j.fm.2008.07.005

Schweiggert, U., Reinhold, C., and Schieber, A. (2007). Conventional and alternative processes for spice production—a review. *Trends Food Sci. Technol.* 18, 260–268. doi: 10.1016/j.tifs.2007.01.005

Skjöldebrand, C. (2001). "Infrared heating," in *Thermal Technologies in Food Processing*, ed. P. Richardson (Cambridge: Woodhead Publishing Limited), 208–228.

Spruthi, J. S. (1980). *Spices and Condiments: Chemistry, Microbiology, Technology.* New York: Academic Press.

Staack, N., Ahrné, L., Borch, E., and Knorr, D. (2008a). Effect of infrared heating on quality and microbial decontamination in paprika powder. *J. Food Eng.* 86, 17–24. doi: 10.1016/j.jfoodeng.2007.09.004

Staack, N., Ahrné, L., Borch, E., and Knorr, D. (2008b). Effects of temperature, pH, and controlled water activity on inactivation of spores of *Bacillus cereus* in paprika powder by near-IR radiation. *J. Food Eng.* 89, 319–324. doi: 10.1016/j.jfoodeng.2008.05.010

Sumnu, G., Turabi, E., and Oztop, M. (2005). Drying of carrots in microwave and halogen lamp-microwave combination ovens. *LWT* 38, 549–553. doi: 10.1016/j.lwt.2004.07.006

Tiburski, J. H., Rosenthal, A., Guyot, S., Perrier-Cornet, J. M., and Gervais, P. (2014). Water distribution in bacterial spores: a key factor in heat resistance. *Food Biophys.* 9, 10–19. doi: 10.1007/s11483-013-9312-5

Vajdi, M., and Pereira, R. R. (1975). Comparative effects of ethylene oxide, gamma irradiation and microwave treatments on selected spices. *J. Food Sci.* 38, 893–895. doi: 10.1111/j.1365-2621.1973.tb02102.x

Conflict of Interest Statement: The authors declare that the research was conducted in the absence of any commercial or financial relationships that could be construed as a potential conflict of interest.

12

Ces locus embedded proteins control the non-ribosomal synthesis of the cereulide toxin in emetic *Bacillus cereus* on multiple levels

Genia Lücking[1], Elrike Frenzel[2†‡], Andrea Rütschle[1†], Sandra Marxen[3], Timo D. Stark[3], Thomas Hofmann[3], Siegfried Scherer[1,4] and Monika Ehling-Schulz[2]*

[1] Department of Microbiology, Central Institute for Food and Nutrition Research (Zentralinstitut für Ernährungs- und Lebensmittelforschung), Technische Universität München, Freising, Germany, [2] Functional Microbiology, Institute of Microbiology, Department of Pathobiology, University of Veterinary Medicine Vienna, Vienna, Austria, [3] Chair of Food Chemistry and Molecular Sensory Science, Technische Universität München, Freising, Germany, [4] Lehrstuhl für Mikrobielle Ökologie, Wissenschaftszentrum Weihenstephan, Technische Universität München, Freising, Germany

Edited by:
Michael Gänzle,
University of Alberta, Canada

Reviewed by:
Jinshui Zheng,
Huazhong Agricultural University,
China
Anne-Brit Kolstø,
University of Oslo, Norway

*Correspondence:
Monika Ehling-Schulz,
Functional Microbiology, Institute
of Microbiology, Department
of Pathobiology, University
of Veterinary Medicine Vienna,
Veterinärplatz 1, A-1210 Vienna,
Austria
monika.ehling-
schulz@vetmeduni.ac.at

†Present address:
Elrike Frenzel,
Molecular Genetics Group, Groningen
Biomolecular Sciences
and Biotechnology Institute, Centre
for Synthetic Biology, University
of Groningen, Groningen, Netherlands

‡These authors have contributed
equally to this work.

The emetic toxin cereulide produced by *Bacillus cereus* is synthesized by the modular enzyme complex Ces that is encoded on a pXO1-like megaplasmid. To decipher the role of the genes adjacent to the structural genes *cesA/cesB*, coding for the non-ribosomal peptide synthetase (NRPS), gene inactivation- and overexpression mutants of the emetic strain F4810/72 were constructed and their impact on cereulide biosynthesis was assessed. The hydrolase CesH turned out to be a part of the complex regulatory network controlling cereulide synthesis on a transcriptional level, while the ABC transporter CesCD was found to be essential for post-translational control of cereulide synthesis. Using a gene inactivation approach, we show that the NRPS activating function of the phosphopantetheinyl transferase (PPtase) embedded in the *ces* locus was complemented by a chromosomally encoded Sfp-like PPtase, representing an interesting example for the functional interaction between a plasmid encoded NRPS and a chromosomally encoded activation enzyme. In summary, our results highlight the complexity of cereulide biosynthesis and reveal multiple levels of toxin formation control. *ces* operon internal genes were shown to play a pivotal role by acting at different levels of toxin production, thus complementing the action of the chromosomal key transcriptional regulators AbrB and CodY.

Keywords: *Bacillus cereus*, *ces* gene cluster, regulation, *cesH*, *cesP*, *cesC*, *cesD*, cereulide synthetase

Introduction

The cyclic dodecadepsipeptide cereulide, a heat-, acid-, and proteolytically stable toxin, is responsible for the emetic type of food borne illness caused by a specific subgroup of *Bacillus cereus*. Intoxication with cereulide, which is preformed during vegetative growth of *B. cereus* in foods, causes nausea and heavy vomiting around 0.5–6 h after consumption of contaminated food (Ehling-Schulz et al., 2004). These symptoms are presumably induced by the interaction of cereulide with 5-HT$_3$ serotonin receptors leading to the stimulation of the afferent vagus nerve (Agata et al., 1995). Usually, these symptoms decline after 24 h, but more severe foodborne intoxications requiring hospitalization or even including fatalities are reported increasingly (Dierick et al., 2005; Naranjo et al., 2011; Messelhäusser et al., 2014; Tschiedel et al., 2015).

In agreement with its chemical structure [D-O-Leu-D-Ala-L-O-Val-L-Val]$_3$, cereulide is produced enzymatically by the non-ribosomal cereulide peptide synthetase Ces (Ces-NRPS; Ehling-Schulz et al., 2005b). NRPSs are large multifunctional enzyme complexes consisting of repetitive modules which selectively incorporate amino acid, α-hydroxy acid or carboxylic acid monomers in the peptide product (Marahiel et al., 1997). The order of the modules usually corresponds directly to that of the monomers in the assembled peptide chain, although strict co-linearity is not always reinforced in nature (Wenzel and Müller, 2005; Lane and Moore, 2011). Very recently we could show that the enzymatic activity of the Ces-NRPS does not follow the canonical NRPS biosynthesis logic, but represents a novel mechanism of non-ribosomal peptide assembly, by using dipeptides rather than monomers as basic units (Marxen et al., 2015a,b).

The cereulide biosynthetic genes were found to be located on a pXO1-like megaplasmid, organized in a 24-kb cluster comprising the seven ces genes $cesH$, $cesP$, $cesT$, $cesA$, $cesB$, $cesC$, and $cesD$ (Ehling-Schulz et al., 2006). The $cesPTABCD$ genes represent an operon being transcribed as a single 23-kb polycistronic mRNA, whereas the adjacently located $cesH$ is transcribed from its own promoter (Dommel et al., 2010; **Figure 1**). The structural NRPS genes $cesA$ and $cesB$ encode the modules that are responsible for the activation and incorporation of each two monomers: The CesA2 and CesB2 submodules install D-alanine and L-valine, respectively, whereas CesA1 and CesB1 bind α-keto acids of leucine and valine (α-ketoisocaproic acid and α-ketoisovaleric acid; Magarvey et al., 2006; Alonzo et al., 2015). A central condensation domain in the C-terminal part of CesA

together with a type I thioesterase (TE) domain located in the C-terminal part of CesB is suggested to act as esterification and elongation center before the final dodecadepsipeptide cereulide is released from the TE domain by macrocyclization (Marxen et al., 2015b).

The adjacent to $cesAB$ located gene $cesT$ encodes a putative type II thioesterase (TEII), an enzyme which is often found in association with NRPSs. In contrast to the type I thioesterases, which are an integrated part of the NRPS catalyzing the product release, external TEIIs reactivate the catalytic NRPS domains by removing misprimed monomers (Schwarzer et al., 2002). The $cesP$ gene, located upstream of $cesT$, encodes a putative 4'-phosphopantetheinyl transferase (PPTase). Such enzymes are detected frequently in association with NRPS enzymes, since they are crucial for their activation. PPTases catalyze the transfer of a 4'-phosphopantheteine (Ppant) moiety of coenzyme A to a conserved serine residue of the PCP domain, thereby converting the apo-carrier protein to its active $holo$-form (Lambalot et al., 1996).

Besides these typical NRPS-associated genes, three additional open reading frames with so far unknown functions are present in the ces gene cluster: $cesH$, at the 5' end, encodes a putative hydrolase or acyltransferase, whose role regarding cereulide synthesis is unclear. $cesC$ and $cesD$, encoding a putative ABC transporter, are located at the 3' terminus of the ces operon (Ehling-Schulz et al., 2006).

The aim of this work was to assess the role of the ces genes flanking the structural cereulide synthetase genes in order to gain further insights into the unusual biosynthetic pathway, which catalyzes the production of the highly potent depsipeptide toxin cereulide. An essential prerequisite for translational studies was the availability of a Ces-NRPS specific antibody, which was raised successfully against the CesB1 module. Targeted gene

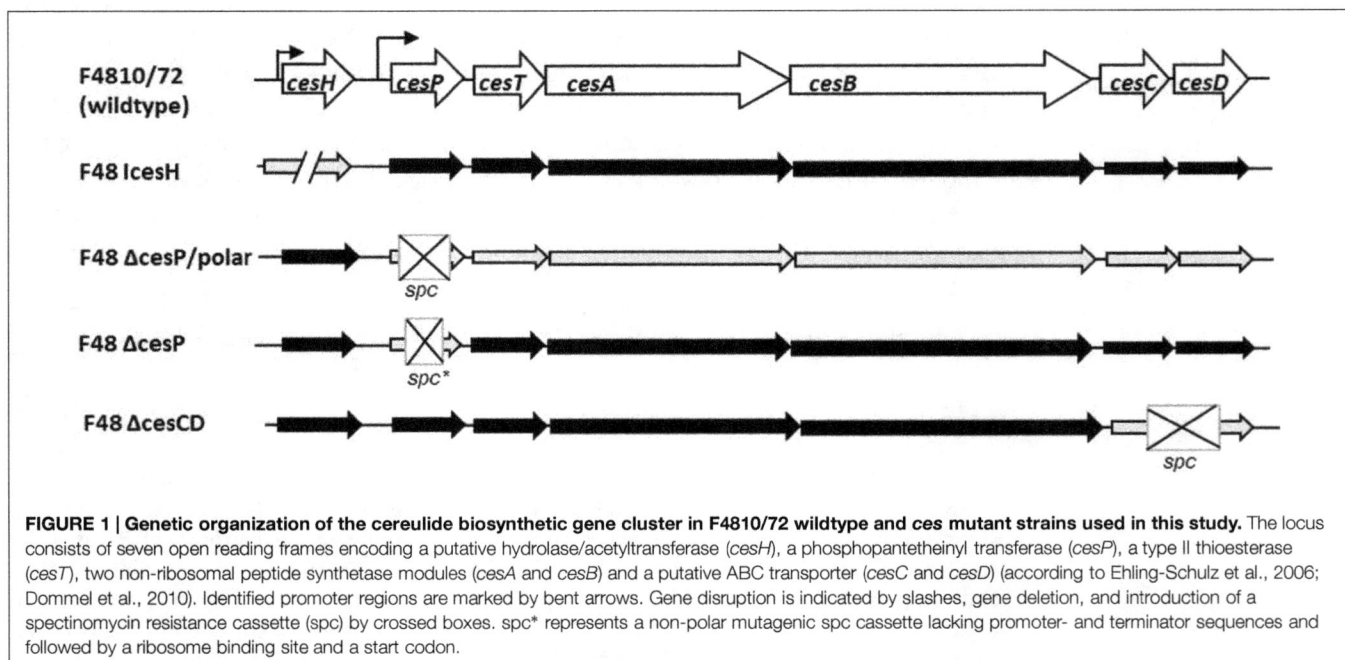

FIGURE 1 | Genetic organization of the cereulide biosynthetic gene cluster in F4810/72 wildtype and *ces* mutant strains used in this study. The locus consists of seven open reading frames encoding a putative hydrolase/acetyltransferase (*cesH*), a phosphopantetheinyl transferase (*cesP*), a type II thioesterase (*cesT*), two non-ribosomal peptide synthetase modules (*cesA* and *cesB*) and a putative ABC transporter (*cesC* and *cesD*) (according to Ehling-Schulz et al., 2006; Dommel et al., 2010). Identified promoter regions are marked by bent arrows. Gene disruption is indicated by slashes, gene deletion, and introduction of a spectinomycin resistance cassette (spc) by crossed boxes. spc* represents a non-polar mutagenic spc cassette lacking promoter- and terminator sequences and followed by a ribosome binding site and a start codon.

inactivation- and overexpression mutants of the emetic reference strain F4810/72 were constructed and characterized regarding the different steps of toxin synthesis. The influence of the accessory *ces* genes on the different cereulide process levels is demonstrated and resulting functional predictions are discussed.

Materials and Methods

Bacterial Strains, Plasmids and Growth Conditions

Details on bacterial strains and plasmids used in this study are provided in **Table 1** and Supplementary Table S1, respectively. Unless otherwise specified, all *B. cereus*, *Bacillus subtilis* and *Bacillus megaterium* were grown in LB broth (10 g/L tryptone, 5 g/L yeast extract, 10 g/L NaCl) or on LB agar plates at 30°C. For liquid cultures, 100 ml LB media was inoculated with approximately 10^3 cfu/ml from a 14–16 h pre-culture and cultures were incubated in 500 ml baffled flasks with 150 rpm. *Escherichia coli* strains used for subcloning were cultured in LB at 37°C. Growth was monitored by optical density at 600 nm (OD_{600}) using a GeneQuant pro spectrophotometer (Biochrom). Concentrations of antibiotics applied were 100 μg/ml for ampicillin, spectinomycin, and polymyxin B; 50 μg/ml for kanamycin; 5 μg/ml for erythromycin; 5 μg/ml for chloramphenicol; and 10 μg/ml for tetracycline.

Sequence Analysis

The genome information of the emetic reference strain F4810/72 (AH187), retrieved from the NCBI website (GenBank accession no. CP001177 and CP001179), was used for sequence analysis of the *ces* locus and homology searches. Amino acid sequences were retrieved from the NCBI database and protein homology search was performed using Blastp available at NCBI[1]. Multiple sequence alignments were carried out with Clustal W using default parameters (Thompson et al., 1994). Maximum-likelihood trees were constructed with MEGA 5 (Tamura et al., 2011) using the Jones–Thornton–Taylor (JTT) model. All positions containing gaps were eliminated. Numbers at nodes (≥50%) represent bootstrap support of 500 resamplings. Annotation of characteristic protein domains and membrane spanning motifs was achieved with the SMART database (Letunic et al., 2012) and the programs TMHMM[2] and TMpred[3].

Construction of *B. cereus* ΔcesP and ΔcesCD Deletion Mutants

Deletion and complementation mutants were constructed as described previously (Lücking et al., 2009). In brief, flanking regions of *cesP* and *cesCD* were amplified by PCR using the primer pairs listed in Supplementary Table S2. Digested PCR products, together with an excised spectinomycin resistance

cassette (spc) from pUC1318spc (Murphy, 1985), were ligated into the TOPO pCR 2.1 vector (Invitrogen). For *cesP*, an additional construct with a non-polar mutagenic spectinomycin cassette from pSPCH+2 (Mesnage et al., 2000) was produced to avoid polar effects of the spectinomycin cassette on the *ces* operon

TABLE 1 | Bacterial strains used in this study.

Strain	Relevant characteristics	Reference
E. coli		
TOP10	General cloning host	Invitrogen
INV110	Methylase-deficient general cloning host	Invitrogen
JM83/pRK24	Donor strain for conjugation; Tra+, Mob+, Amp^r, Tc^r	Trieu-Cuot et al., 1987
B. subtilis		
168		Spizien, 1958
B. cereus		
ATCC 14579	Type strain (non-emetic)	Ivanova et al., 2003
ATCC 10987	Emetic-like strain	Rasko et al., 2004
A529	Emetic food-borne outbreak isolate	Stark et al., 2013
F4810/72	Emetic reference strain (also AH 187)	Ehling-Schulz et al., 2005a
F48ΔcesP/polar	F4810/72 ΔcesP::spc; Spc^r	This study
F48ΔcesP	F4810/72 ΔcesP::spcH+2 (non-polar Spc resistance cassette); Spc^r	This study
F48ΔcesCD	F4810/72 ΔcesCD::spc; Spc^r	This study
F48ΔcesCD/com_cesCD	F48ΔcesCD complemented with pAD/Pro-ces/cesCD; Spc^r, Cm^r	This study
F48IcesH	F4810/72 with disrupted cesH (cesH::pMAD); Ery^r	This study
F48Ippt	F4810/72 with disrupted ppt (ppt::pMAD); Ery^r	This study
F48ΔcesP/Ippt	F48ΔcesP with disrupted ppt; Spc^r, Ery^r	This study
F48ΔcesP/Ippt/com_cesP	F48ΔcesP/Ippt complemented with pAD/Pro-ces/cesP; Spc^r Ery^r, Cm^r	This study
F48ΔcesP/Ippt/com_ppt	F48ΔcesP/Ippt complemented with pMM/ppt; Tc^r	This study
F48pMM/cesH	F4810/12 containing pMM/cesH for CesH overexpression; Tc^r	This study
B. megaterium		
WH320	Protein overexpression strain lacking alkaline proteases	MoBiTec
WH320pWHCesB1His	WH320 containing pWHCesB1His for CesB overexpression; Amp^r, Tc^r	This study

---

[1] http://blast.ncbi.nlm.nih.gov

[2] http://www.cbs.dtu.dk/services/TMHMM/

[3] http://www.ch.embnet.org/software/TMPRED_form.html

genes located downstream of *cesP*. Constructs were excised from TOPO and inserted into the conjugative suicide vector pAT113, giving rise to pAT113ΔcesP/spc, pAT113ΔcesCD/spc, and pAT113ΔcesP/spcH+2 (see Supplementary Table S1 for details). Plasmids were then transformed into *E. coli* JM83/pRK24 by heat shock and the resulting strains were used for transconjugal transfer into *B. cereus* F4810/72 using a mating procedure previously described (Pezard et al., 1991). For complementation of F48ΔcesCD and F48ΔcesP/Ippt, the genes *cesCD* and *cesP* were amplified using the primers cesCD_F_Xba/cesCD_R_Pae and cesP_Pro_F2/cesP_Pro_R (Supplementary Table S2), respectively, and cloned into the shuttle vector pAD123 together with the ~500 bp *ces* promoter region. For complementation of F48ΔcesP/Ippt with *ppt* (NCBI accession no. ACJ79141), the respective gene was amplified with ppt_chr_F and ppt_chr_R and ligated into the shuttle vector pMM1522, containing a xylose-inducible promoter. Constructs were propagated in the non-methylating *E. coli* strain INV110 cells and then introduced into the according *B. cereus* mutant by electroporation, giving rise to F48ΔcesCD/com_cesCD, F48ΔcesP/Ippt/com_cesP, and F48ΔcesP/Ippt/com_ppt.

Construction of *B. cereus cesH* and *ppt* Insertion- and Overexpression Mutants

Gene inactivation by pMAD integration was performed as described previously (Dommel et al., 2011). In brief, internal fragments (~300 bp) of *cesH* and *ppt* (NCBI accession no. ACJ79141) were amplified by PCR using primers listed in Table S2 and ligated into the thermosensitive shuttle vector pMAD. Constructs were introduced into *B. cereus* F4810/72 or F48ΔcesP by electroporation and transformants were obtained after 2 days at 30°C on LB plates with erythromycin and X-Gal (20 μg/ml). Plasmid integration was enforced by re-cultivation of a blue colony in LB medium overnight at 42°C for several times. For further experiments, the resulting strains F48IcesH, F48Ippt, and F48ΔcesP/Ippt were cultured at high temperatures (37 or 42°C), which is essential to avoid the loss of the integrated plasmid.

For *cesH* overexpression, the promoterless *cesH* gene was amplified using the respective primers listed in Supplementary Table S2. The digested fragment was ligated into pMM1522, harboring a xylose-inducible promoter, and then introduced in *B. cereus* F4810/72 by electroporation. To induce gene overexpression, 0.1 % D-xylose (v/v) was added to the culture medium.

Production of a Monoclonal anti-CesB Antibody

For generation of a monoclonal antibody targeting the cereulide NRPS, the DNA fragment (3999 bp) encoding the first CesB submodule (CesB1: consisting of an adenylation-, ketoreductase-, and peptidyl carrier protein domain) including the native *cesB* ribosome binding site was amplified with Pfu polymerase (Promega) from genomic DNA of *B. cereus* F4810/72 using the primer pair cesB135Nco_for/cesB135Xho_rev (Supplementary Table S2). Cloning *via* the XhoI/NcoI sites

into the pET28b(+) vector resulted in a transcriptional C-terminal His$_6$ tag fusion. CesB1-His$_6$ was amplified from pET28-*cesB1* by PCR with cesBpETSpe_for2 and cesBpETSph_rev2 and cloned into the SpeI and SphI restriction sites of pWH1520, giving rise to the xylose-inducible overexpression plasmid pWHCesB1His. This construct was transferred into *B. megaterium* WH320 protoplasts by PEG-fusion according to the manufacturer's protocol (MoBiTec, Göttingen). The CesB1His$_6$ protein was overexpressed in recovered *B. megaterium* cells and purified using Ni-NTA affinity columns (Qiagen) according to the manufacturer's instruction. The monoclonal antibody (mAB) CesB5-6 was raised by BioGenes (Berlin) using the purified CesB1His$_6$ protein, NMRI mice (obtained from Janvier, France), and the myeloma cell line SP2/0-Ag14 (obtained from DSMZ, Germany). Purified CesB1 mAB with a concentration of 1.68 μg ml^{-1} in PBS solution was tested for its specificity by enzyme-linked immunosorbent assays (ELISA; data not shown) and by immunoblotting.

Reverse Transcription-qPCR

To analyze *cesA* transcript levels, 1 ml culture samples of *B. cereus* strains were harvested at OD$_{600}$ 8 by centrifugation (10000 *g*, 4°C, 2 min). Total RNA isolation, cDNA synthesis and RT-qPCR were carried out as described previously (Dommel et al., 2011). Transcript levels of the 16S *rrn* gene served as the reference control for data normalization and relative gene expression ratios were calculated with the REST software (Pfaffl et al., 2002).

Western- and Slot Blot Analysis of CesB

Twenty milliliter of *B. cereus* cultures, grown to different optical densities, were centrifuged (9500 rpm, 6 min, Sigma 3–18K with angle rotor 19776) and cell pellets resuspended in 2 ml lysis buffer [50 mM Tris-HCl pH 7.6, 2 mM EDTA pH 7.5, 1 mM Pefabloc (Merck)]. Cells were disrupted by two passages through a French Pressure cell press (1000 psi) and the soluble protein fraction was collected by centrifugation of the lysates (13000 rpm, 2x 30 min, 4°C). Protein concentration was determined according to Bradford using the Roti-Quant solution (Carl Roth GmbH). For Western blot analysis, 30 μg of total protein was separated by SDS-PAGE on 8% gels and the proteins were immobilized onto a PVDF membrane (Millipore) by semi-dry blotting for 2 h at 120 mA. A 1:2000 dilution of the monoclonal anti-CesB antibody was used to detect the CesB1 module. The second incubation was done with a 1:20000-dilution of HRP-conjugated goat anti-mouse IgG (Dianova) and blots were developed with the SuperSignal West Pico Chemiluminescent Substrate (Thermo Scientific). For slot blot analysis, 25 μg of total protein was blotted directly onto a PVDF membrane by vacuum filtration. Again, the 1:2000 dilution of the monoclonal anti-CesB antibody was used to detect the CesB module of the cereulide synthetase. In addition, blots of the same samples were performed with a 1:5000 dilution of a polyclonal rabbit anti-AtpB serum (Agrisera) detecting the beta subunit of the ATP synthase. As second antibodies served an alkaline phosphatase-conjugated goat anti-mouse IgG (1: 10000 dilution, Dianova) and an alkaline phosphatase-conjugated goat anti-rabbit IgG (1:3000, Dianova), respectively. Chromogenic detection of alkaline phosphatase activity was accomplished with BICP-p-toluidine salt- and NBT solutions (Carl Roth GmbH) according to the manufacturer's instructions.

Sample Preparation and Cereulide Quantification by Means of UPLC-MS/MS

The biosynthetic production of cereulide in LB broth supplemented with 0.2% glucose and of ^{13}C-labeled cereulide in MOD medium supplemented with ^{13}C$_1$-L-valine was carried out as described previously (Bauer et al., 2010). Extraction of cereulide from autoclaved cells of different *B. cereus* strains grown in 100 ml LB broth was carried out as described (Stark et al., 2013), using 1 ng of ^{13}C$_6$-cereulide as internal standard. Cereulide quantification via stable isotope dilution analysis (SIDA) was performed using a Xevo TQ-S Acquity i-class UPLC-MS/MS system and an UPLC BEH C18 column (2.1 mm × 50 mm, 1.7 μm; Waters, UK and USA). One microliter aliquots of ethanolic extracts (of *B. cereus* samples) were injected directly into the UPLC/MS-MS system. Operated with a flow rate of 1.0 ml/min at a temperature of 50°C, the following gradient was used for chromatography: starting with a mixture (90/10, v/v) of methanol and 10 mM ammonium formate (0.1% formic acid), the methanol content was increased to 100% within 0.5 min, kept constant for 0.3 min and decreased within 0.1 min to 90%. Measurements were performed using electrospray with positive ionization and the quantitative calibration mode consisting of the following ion source parameters: capillary voltage +3.5 kV, sampling cone 30 V, source offset 30 V, source temperature 150°C, desolvation temperature 600°C, cone gas 150 L/h, desolvation gas 1000 L/h, collision gas flow 0.15 ml/min and nebuliser gas flow 6.5 bar. Calibration of the TQ-S in the range from m/z 40-1963 was performed using a solution of phosphoric acid (0.1% in acetonitrile). The UPLC and Xevo TQ-S systems were operated with MassLnyx^{TM} 4.1 SCN 813 software, data processing and analysis were performed using TargetLynx. By means of the multiple reaction monitoring (MRM) mode, the ammonium adducts of cereulide (qualifier: m/z 1170.7→172.2; 1170.7→314.2; quantifier: m/z 1170.7→357.2) and ^{13}C$_6$-cereulide (m/z 1176.7→173.2; 1176.7→316.3; 1176.7→358.3) were analyzed using the mass transitions (given in brackets) monitored for a duration of 25 ms. ESI$^+$ mass and product ion spectra were acquired with direct flow infusion using IntelliStart. The MS/MS parameters were tuned for each individual compound, detecting the fragmentation of the [M+NH$_4$]$^+$ molecular ions into specific product ions after collision with argon. For quantitation, 10 ethanolic standard solutions of the analyte cereulide (0.05–10.0 ng/ml) and the internal standard ^{13}C$_6$-cereulide (1 ng/ml) were mixed and analyzed in triplicates by means of UPLC–MS/MS using the MRM mode. Calibration curve was prepared by plotting peak area ratios of analyte to internal standard against concentration ratios of each analyte to the internal standard using linear regression.

Cytotoxicity Test

Aliquots of *B. cereus* cultures were taken after 24 h of incubation and autoclaved (15 min at 121°C) to lyse cells and denature heat-labile toxins. Cereulide amounts of these samples were determined using the HEp-2 cell based cell culture assay as previously described (Lücking et al., 2009). Cereulide titres of the mutant strains were normalized to the value of the parental strain *B. cereus* F4810/72 (grown for 24 h at 30 or 37°C, respectively), which was defined as being 100%.

Results

Comparative Sequence Analysis of the Cereulide Synthetase Flanking Genes

Based on the genome information (NCBI accession no. CP001177 and CP001179) available for the emetic reference strain F4810/72 (AH187), a BLAST database search and extensive sequence analysis was carried out. This *in silico* approach reconfirmed the *ces* gene cluster boundaries suggested previously (Ehling-Schulz et al., 2006; Dommel et al., 2010). The *ces* gene locus comprises seven genes (*cesHPTABCD*), which are present in all emetic *B. cereus* and emetic *Bacillus weihenstephanensis* sequenced so far (data not shown). *cesH* represents the 5′ prime end of the *ces* locus while *cesD*, which is followed by a strong terminator sequence, marks the 3′ prime end.

CesH is a putative 31 kDa hydrolase belonging to the α/β-hydrolase fold superfamily of proteins. Besides the plasmid-borne *cesH*, F4810/72 carries 13 chromosomal genes annotated to code for members of this hydrolase subfamily, with none of them showing notable sequence similarity toward CesH. Comparison of the CesH with amino acid sequences retrieved from annotated genomes of the *B. cereus* group members revealed a large number of orthologs showing significant similarity. However, proteins with 100% identity (amino acids) were only found in emetic *B. cereus* strains possessing the *ces* gene cluster. The CesH sequences retrieved from the *ces* gene cluster in emetic *B. weihenstephanensis* strains showed a homology of 91–93% (**Figure 2**).

cesP codes for a putative 4′-phosphopantetheinyl transferase (PPtase) belonging to the Sfp-like subgroup of the PPTase superfamily. *In silico* analysis revealed the presence of an additional gene encoding a 4′-PPtase (ACJ79141; herein after referred to as *ppt*) in the chromosome of *B. cereus* F4810/72. A homolog of this protein with similarity above 80% is present in almost all *B. cereus* group members, while CesP homologs can only be found in emetic *B. cereus* and emetic *B. weihenstephanensis* strains. An alignment of the amino acid sequences of CesP and Ppt resulted in a rather low degree of homology (27% identity at 56% coverage), nevertheless, according to their size and three conserved motifs, both PPTases can be classified as typical members of the "Sfp-like" family (Copp and Neilan, 2006), showing 32% (CesP) and 26% (Ppt) identity to Sfp (P39135), the prototype PPTase of *B. subtilis,* respectively (**Figure 3**).

The two open reading frames *cesC* and *cesD* at the 3′ terminus of the *ces* gene locus encode a putative ABC

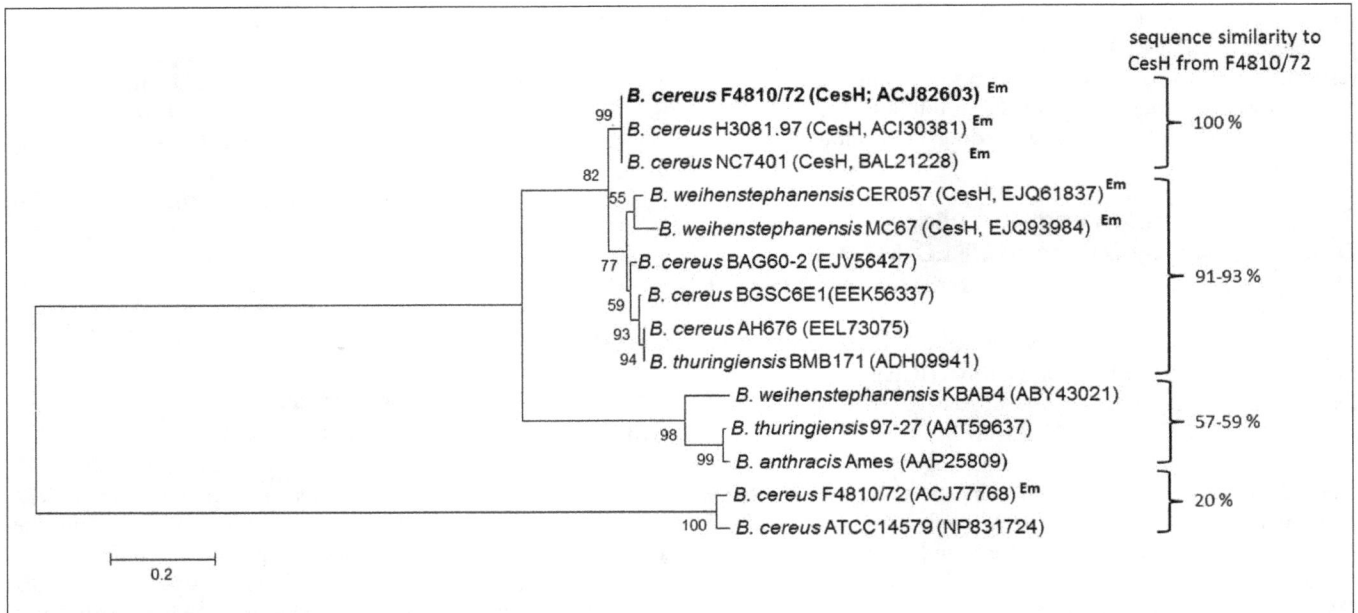

FIGURE 2 | Maximum-likelihood protein similarity tree based on amino acid sequences of CesH from *Bacillus cereus* F4810/72 (AH187) and selected homologs of other *B. cereus* group members. NCBI accession numbers of proteins are given in brackets. Em indicates an emetic strain containing the *ces* gene cluster.

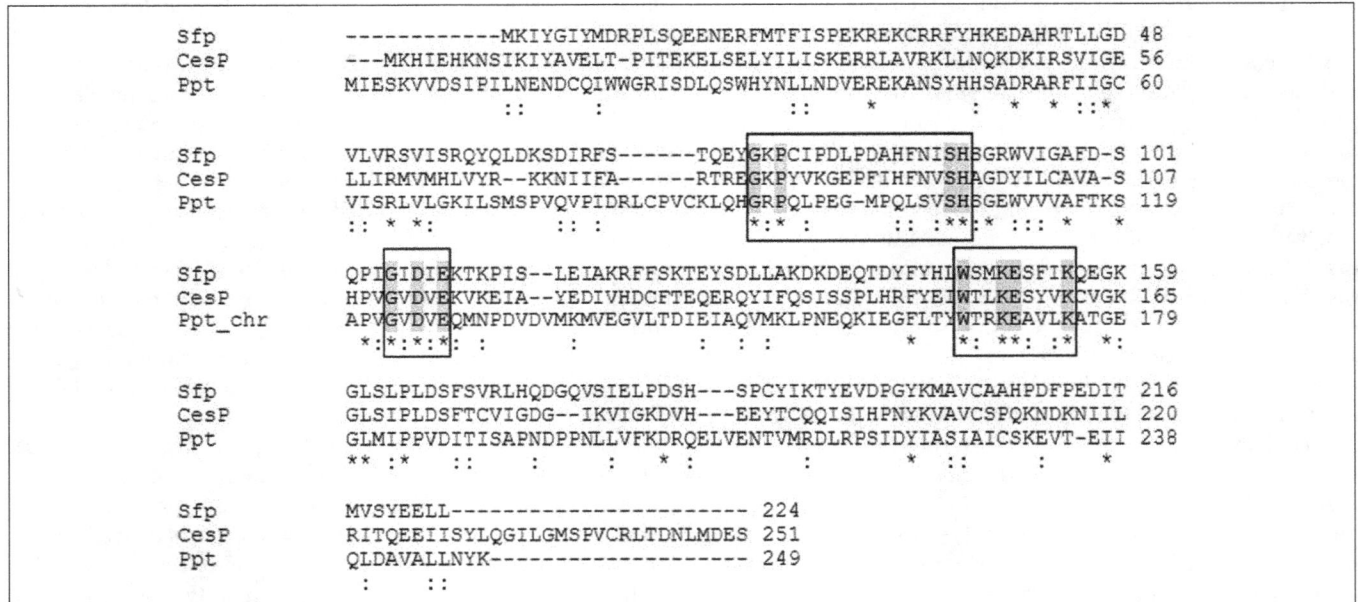

FIGURE 3 | Amino acid sequence alignment of the PPTases CesP (ACJ82723), Ppt (ACJ79141) of *B. cereus* F4810/72, and Sfp (P39135) of *B. subtilis*. Three conserved motifs, which are typical for members of the "Sfp-like" PPTase subfamily, are shown in boxes. Identical residues are highlighted with asterisk, similar residues with colon.

transporter. Protein domain annotation using the SMART database (Letunic et al., 2012) revealed the presence of characteristic motifs involved in ATP- binding and hydrolysis for CesC. The amino acid sequence of CesC was found to be homologous to several known ATP-binding proteins, e.g., BerA of *B. thuringiensis* required for β-exotoxin I production (Espinasse et al., 2002), or TnrB2 of *Streptomyces longisporoflavus*

and BcrA of *B. licheniformis*, which have been shown to confer resistance to tetronasin and bacitracin, respectively (Linton et al., 1994; Podlesek et al., 1995; Supplementary Figure S1). For CesD, five hydrophobic transmembrane domains were predicted, which are typical for membrane-spanning proteins. CesD showed homology (33–34% identity) only toward the putative permease BerB of *B. thuringiensis* and its orthologs present in many *B. cereus* group members. Sequence alignment of CesCD with known ABC transporters of antimicrobial peptides confirmed that CesCD most closely resembles members of the BcrAB subgroup, which generally consist of one ATPase and one permease with six transmembrane helices (Gebhard, 2012; **Figure 4**).

Construction of Mutants and Determination of their Cereulide Production Capacities

To investigate the potential role of *cesH*, *cesP*, *cesC*, and *cesD* on cereulide production, various deletion-, insertion-, and overexpression mutants of *B. cereus* F4810/72 were constructed as described in the materials and methods section. An overview of mutants generated is provided in **Figure 1** and the major genetic characteristics of mutant strains are summarized in **Table 1**, respectively. After growth in LB medium for 24 h at 30 or 37°C (insertion mutants), culture samples were taken and cells disrupted by autoclaving in order to obtain total cereulide amounts including extra- and intracellularly accumulated toxin. Cereulide concentrations of ethanolic extracts were measured by means of SIDA-UPLC-MS/MS. Furthermore, cytotoxicity was determined by the HEp-2 cell culture assay.

The cereulide production of the *cesH* inactivation mutant (F48IcesH) turned out to be almost twice as high compared to the wildtype strain. Interestingly, when *cesH* was overexpressed in the wildtype strain (F48pMM/cesH) with a xylose-inducible shuttle vector, almost no cereulide was detectable, while in the absence of the inducer xylose,

cereulide production was comparable to the wildtype (**Table 2**). These results demonstrate that CesH inhibits cereulide synthesis.

For the analysis of *cesP*, two knockout strains were generated, since *cesP* is the first gene of the *ces* operon and disruption of this gene may affect the *ces* genes located downstream. Thus,

TABLE 2 | Cereulide production of wildtype emetic *B. cereus* and *ces* gene mutants determined by SIDA-UPLC-MS/MS and HEp-2 bioassay.

Strain	Cereulide concentration in μg/ml	Cytotoxicity in %[c]
Wildtype (F4810/72)[a]	4.28 ± 1.06	100
Wildtype (F4810/72)[b]	3.67 ± 0.70	100
F48IcesH[b]	5.76 ± 0.99	106.6 ± 23.2
F48pMM/cesH (+0.1% Xyl)[a]	0.04 ± 0.01	1.3 ± 0.6
F48pMM/cesH (w/o Xyl)[a]	3.47 ± 1.59	97.6 ± 37.9
F48ΔcesP/polar[a]	0.0 ± 0.0	0.0 ± 0.0
F48ΔcesP[a]	3.54 ± 1.07	83.2 ± 23.8
F48Ippt[b]	3.19 ± 1.62	75.4 ± 26.4
F48ΔcesP/Ippt[b]	0.0 ± 0.0	0.9 ± 1.9
F48ΔcesP/Ippt/com_cesP[b]	2.62 ± 1.51	99.9 ± 44.6
F48ΔcesP/Ippt/com_ppt[b]	0.49 ± 0.02	32.6 ± 15.3
F48ΔcesCD[a]	0.0 ± 0.0	0.0 ± 0.0
F48ΔcesCD/com_cesCD[a]	0.87 ± 0.20	43.7 ± 2.4

[a]*Strain grown at 30°C.*
[b]*Strain grown at 37°C.*
[c]*Values normalized to cytotoxicity of wildtype at corresponding temperature.*

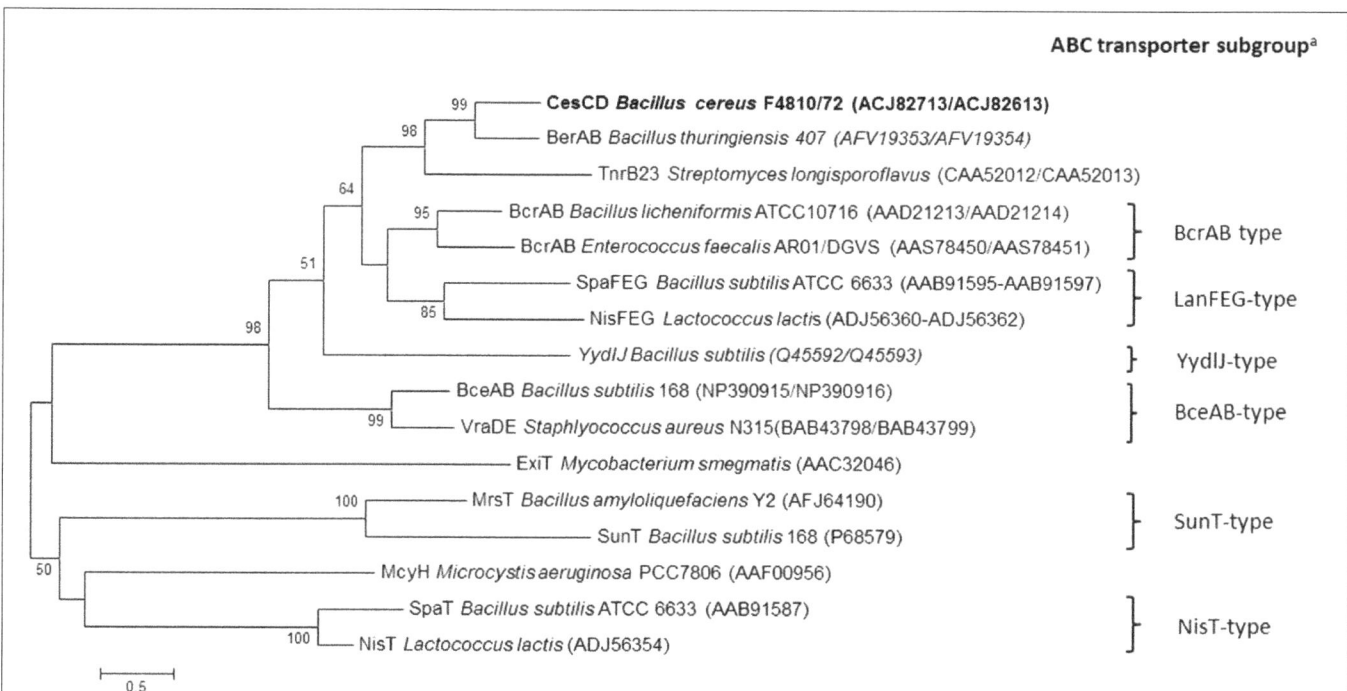

FIGURE 4 | Protein similarity tree of the amino acid sequences of CesCD and previously characterized ABC transporters. [a]Classification of ABC transporters of antimicrobial peptides according to Gebhard (Gebhard, 2012).

in one strain a non-polar spectinomycin resistance cassette (spc) lacking promoter and terminator sequences (Mesnage et al., 2000) was used for mutagenesis (F48ΔcesP), while the other harbors a normal spc potentially leading to polar effects on downstream located genes (F48ΔcesP/polar; see also **Figure 1**). Indeed, SIDA-UPLC-MS/MS of the latter revealed no cereulide production and no toxicity could be detected toward HEp-2 cells. In contrast, F48ΔcesP produced toxin amounts similar to that in the wildtype, demonstrating *cesP*-independent cereulide synthesis (**Table 2**). Since *in silico* analysis exposed the presence of a second sfp-like PPTase in the chromosome of *B. cereus* F4810/72 (*ppt*), a respective gene inactivation mutant was constructed in the wildtype (F48Ippt) to test its possible involvement in toxin formation. UPLC-MS/MS showed that this mutant produces cereulide levels comparable to the wild type. However, when *ppt* was disrupted in F48ΔcesP, leading to a double knock out mutant in *cesP* and *ppt* (F48ΔcesP/Ippt), no cereulide or cytotoxicity was detected (**Table 2**). This effect was reversible by complementation of the mutant with *cesP* and – to a lesser extend – with *ppt*, indicating a redundant role of CesP and Ppt in cereulide production.

For the *cesCD* deletion mutant (F48ΔcesCD) neither cereulide by means of MS nor cytotoxicity in the HEp-2 bioassay could be determined. Both parameters could be restored to some extend by *in trans* complementation of F48ΔcesCD with a shuttle vector containing *cesC* and *cesD* under the control of the *ces* locus promoter (**Table 2**). These results demonstrate that the putative ABC transporter genes *cesC* and *cesD* are essential for cereulide formation in *B. cereus* F4810/72.

Generation of a Monoclonal Antibody Targeting the Cereulide NRPS

Next, we were interested to further dissect the level of regulation leading to the huge differences in cereulide quantities found in the *ces* mutants. To study the impact of the *ces* mutations on the expression of the Ces-NRPS, a specific antibody against the cereulide synthetase was generated. The CesB1 submodule of the cereulide synthetase, which was shown to present a ketoacid-binding module unique among *Bacillus* sp. NRPS (Magarvey et al., 2006), served as antigen for the generation of monoclonal antibodies (mAB) as outlined in the materials and methods section. The specificity of the resulting monoclonal anti-CesB1 antibody was assessed using two emetic-, two non-emetic *B. cereus* strains and one *B. subtilis* strain (possessing the surfactin NRPS genes) by ELISA (data not shown) and immunoblotting. Western blot analysis revealed the specificity of the antibody toward the two cereulide-positive strains *B. cereus* F4810/72 and A529, while no signal was detected for the cereulide-negative strains ATCC 10987 and ATCC 14579 or for *B. subtilis* 168 (Supplementary Figure S2). The CesB1 mAB reacted specifically with a large protein band migrating way above the range of the protein ladder used, which is consistent with the predicted molecular mass of CesB1 at 304 kDa. The appearance of an additional weak band at 170 kDa indicates fragmentation of the large CesB1 complex, which was repeatedly observed after denaturing PAGE.

Transcriptional and Translational Analysis of *ces* Mutant Strains

Quantitative RT-PCR detecting *cesA* mRNA levels as well as immunoblotting using the CesB mAB were carried out to investigate, whether the striking differences of cereulide synthesis of the *B. cereus* F4810/72 wildtype and the *ces* mutants (**Table 2** and **Figure 5C**) were caused by variations in NRPS gene transcription or by translational regulation of the synthetase complex. **Figure 5** combines the results of *ces* transcription-, translation-, and cereulide production analyses for the wildtype and seven mutant strains. For the cereulide-negative mutant F48ΔcesP/polar, no *cesA* transcripts or CesB expression was detectable (**Figures 5A,B**, lane 4), confirming the assumption that insertion of spc in *cesP* led to a transcriptional stop of the downstream located *ces* genes in the operon. In contrast, transcript levels and CesB protein signals for the other two cereulide-deficient mutants, F48ΔcesP/Ippt and F48ΔcesCD, resembled those of the parental strain (**Figures 5A,B**, lanes 6 and 7). These data suggest that *cesP*, *ppt*, *cesCD* and their respective gene products do not interfere with *ces* transcription or translation of the cereulide synthetase complex, but seem to play a role in the post-translational formation of active cereulide. Transcript- and CesB protein levels of the *cesH* insertion mutant (F48IcesH), which revealed higher toxin levels than the wildtype in the MS assay (**Figure 5C**, lane 2), were comparable to those of the wildtype. However, for the *cesH* overexpression mutant (F48pMM/cesH), *cesA* transcription was down-regulated significantly and only a very weak CesB band was detected by slot blot analysis (**Figures 5A,B**, lane 3). Therefore, the toxin deficiency of the latter strain may be due to *cesH* and its gene product, acting as a possible transcriptional repressor of the cereulide synthetase genes.

Discussion

Hitherto, research on the cereulide biosynthetic *ces* operon focused on the molecular and biochemical characterization of the non-ribosomal cereulide peptide synthetase encoded by *cesA* and *cesB* (Ehling-Schulz et al., 2005b; Magarvey et al., 2006; Alonzo et al., 2015; Marxen et al., 2015b). However, little is known about the adjacent genes *cesH*, *cesP*, *cesC*, and *cesD*, which code for a putative hydrolase, a PPtase and a putative transport system of the ABC-type, respectively (Ehling-Schulz et al., 2006). So far it is unclear, why different strains of the emetic *B. cereus* group display highly variable toxicity (Carlin et al., 2006; Stark et al., 2013). Thus, it is important to dissect the role of *ces* locus genes and their gene products on cereulide formation. Since it was recently shown that the Ces-NRPS represents a novel mechanism for non-ribosomal depsipeptide assembly (Marxen et al., 2015b), information on the functional architecture of the *ces* gene locus would also improve our general understanding of the complex biochemical

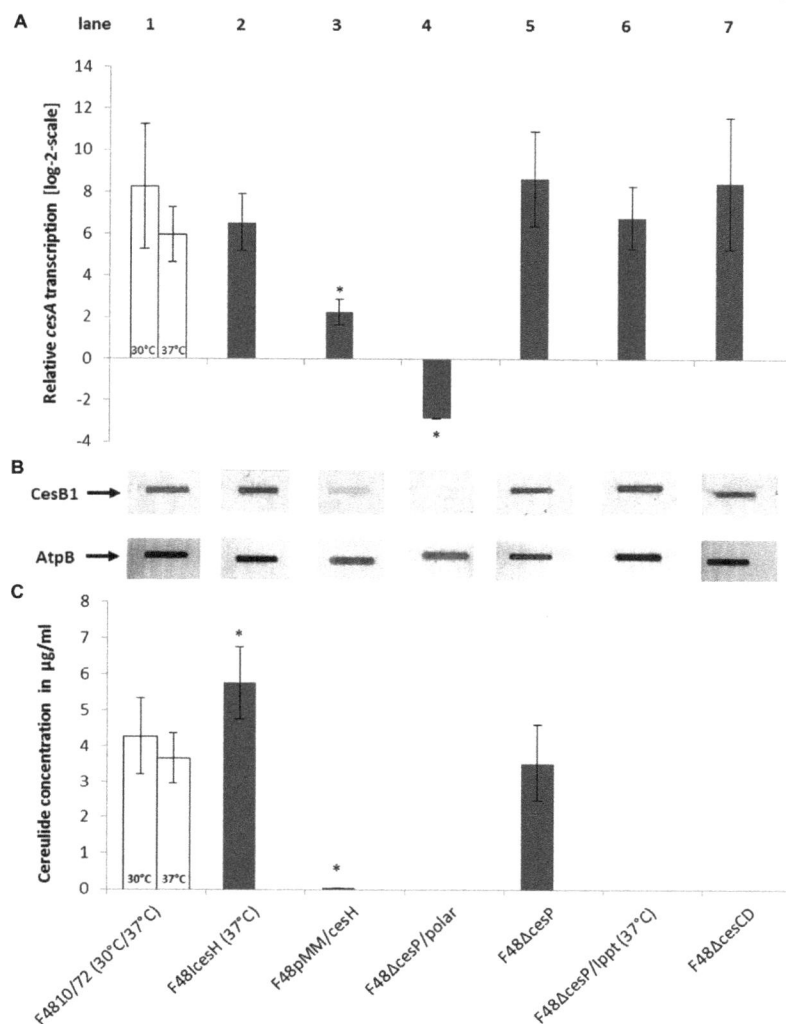

FIGURE 5 | Comparison of *ces* gene transcription, cereulide synthetase expression, and cereulide production of *B. cereus* F4810/72 wildtype (WT; 1) and the mutant strains F48ΔcesP (5), F48ΔcesP/lppt (6), F48ΔcesP/polar (4), F48ΔcesCD (7), F48IcesH (2), and F48pMM/cesH (3). (A) Relative *cesA* transcription of *B. cereus* strains grown in LB medium at 30 or 37°C (insertion mutants) and harvested at exponential growth phase (OD$_{600}$ = 8) for RNA isolation. The asterisks denote statistically significant ($P < 0.05$) differences in *cesA* mRNA levels between wildtype and mutant strains grown under the same conditions. **(B)** Slot blot analysis of CesB and AtpB expression using 25 μg of total protein per slot from *B. cereus* strains harvested at OD$_{600}$ = 8. The CesB1 module of the cereulide synthetase was detected with the CesB1 mAB and AtpB (beta subunit of ATP synthase) with a polyclonal anti-AtpB antibody, serving as a protein loading control. **(C)** Cereulide production of *B. cereus* strains determined by UPLC-MS/MS-SIDA. Strains were incubated in LB medium for 24 h (stationary growth phase) before cereulide was extracted for quantification. Statistically significant ($P < 0.05$) differences in cereulide concentration between wildtype and mutant strains are marked by asterisks.

pathways involved in the production of natural, non-ribosomally synthesized peptides.

CesH, a Putative Repressor Involved in Timing of *ces* Gene Transcription

The first gene in the 5′ prime proximity of the *ces* cluster, *cesH*, codes for a putative hydrolase of the α/β-hydrolase fold superfamily. According to the ESTER classification database, *cesH* belongs to the 6_AlphaBeta_hydrolase subgroup, which to date consists of over 3800 members using various substrates (Hotelier et al., 2004). Although *cesH* is transcribed from its own promoter – while the remaining *ces* genes are co-transcribed – it is considered to be an integral part of the *ces* gene locus, since *in silico* analysis revealed a copy in all emetic *B. cereus* strains as well as in cereulide-producing *B. weihenstephanensis* strains. In general, the presence of hydrolase genes in NRPS gene clusters seems to be rare and their function is unknown. We found one gene coding for a CesH – homolog (presenting 59% identity) in close proximity of a 30 kb polyketide-NRPS locus of the non-emetic *B. weihenstephanensis* KBAB4. Surprisingly, our data revealed an inhibitory effect of CesH on cereulide formation, as CesH overexpression led to a non-toxic

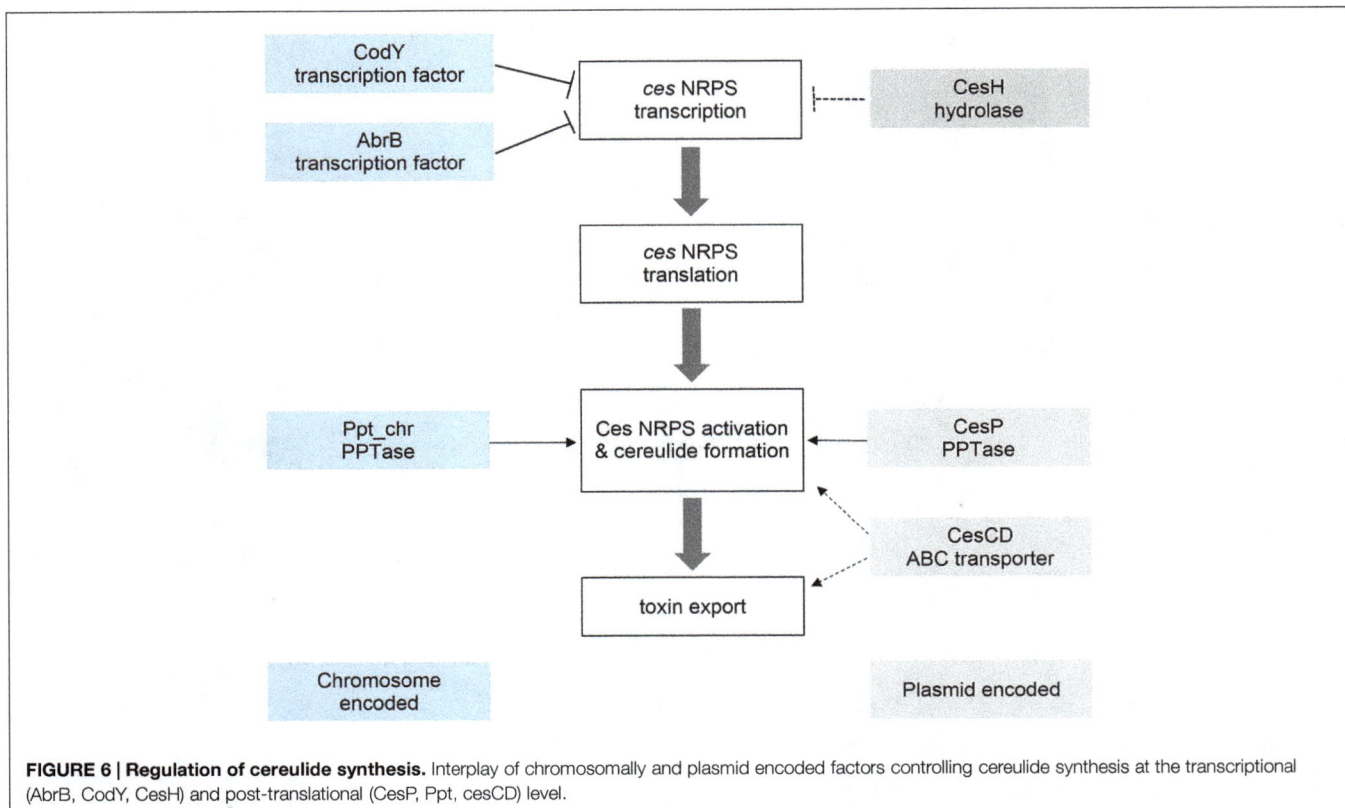

FIGURE 6 | Regulation of cereulide synthesis. Interplay of chromosomally and plasmid encoded factors controlling cereulide synthesis at the transcriptional (AbrB, CodY, CesH) and post-translational (CesP, Ppt, cesCD) level.

phenotype, while the *cesH* insertion mutant produced more cereulide than the wildtype (**Table 2**). Transcriptional analysis of the CesH overexpression mutant resulted in strongly down-regulated *cesA* mRNA levels, which is in line with the very weak CesB slot blot signal obtained for this mutant. These data indicate that CesH is an additional regulatory member of the cereulide pathway by acting directly or indirectly as a transcriptional repressor of the *ces* gene operon. Our previous work demonstrated that transcription of the polycistronic *cesPTABCD* genes is co-regulated in a complex manner by several key transcription factors of the chromosome (Ehling-Schulz et al., 2015), e.g., AbrB and CodY, leading to a tightly regulated transcription peak in late exponential phase (Lücking et al., 2009; Frenzel et al., 2012). While AbrB and CodY are responsible for the onset and strong increase of *cesPTABCD* transcription, CesH may function as a closing signal shutting down mRNA synthesis. Since a direct action of a hydrolase as a transcriptional regulator seems to be unlikely, another possibility may be that CesH acts indirectly by degrading metabolites or quorum sensing signaling molecules that influence *ces* gene transcription in later growth phases. This hypothesis is in line with transcriptional kinetic studies of *cesH* in *B. cereus* F4810/72, demonstrating highest *cesH* expression in stationary growth phase (Supplementary Figure S3B), while the other *ces* operon genes are known to be transcribed earlier in the growth cycle (Dommel et al., 2011). Further experiments including activity tests with purified CesH protein, which are clearly beyond the scope of the current study, are necessary to

elucidate the regulatory role of *cesH* in cereulide production in detail.

Plasmid Encoded CesP and Chromosomally Encoded Ppt are Two Sfp Type- PPTases with Functionally Redundant Enzymatic Activities

In contrast to the frequency of proteins from the α/β-hydrolase fold superfamily, only very few members of the PPtase superfamily are present in *Bacillus* species. By post-translational modification and thereby activation of acyl-, aryl-, or peptidyl-carrier proteins, PPTases are essential enzymes for the synthesis of fatty acids, polyketides and non-ribosomal peptides (Walsh et al., 1997). Based on size, conserved sequence motifs and substrate selectivity, bacterial PPTases can be classified into two major groups, the Sfp-like PPTases and the acyl carrier protein synthases (AcpS; Finking and Marahiel, 2004). While PPTases of the Sfp-type are mostly found in association with NRPS, the enzymes of the AcpS type are linked to the activation of fatty acid and polyketide synthesis. Genome data analysis of *B. cereus* F4810/72 revealed the presence of two chromosomally encoded PPTases: an AcpS-type PPtase (ACJ81272) and a Sfp-like PPtase (ACJ79141, here named Ppt). The latter is encoded in the vicinity of the *dhb* operon, which has been shown to be responsible for the non-ribosomal synthesis of the siderophore bacillibactin in *B. subtilis* (May et al., 2001). The plasmid-encoded *cesP* gene codes for an additional Sfp-like PPtase and presents an integral part of the *ces* gene cluster, being co-transcribed with the NRPS genes (Dommel et al., 2010). Surprisingly, CesP was not essential for cereulide synthesis, as deletion of *cesP* did not significantly alter cereulide production or toxicity of the strain. Only an additionally disrupted *ppt* led to a toxin-negative phenotype, indicating that Ppt can function as a redundant CesP-PPTase in cereulide biosynthesis. Likewise, CesP seems to be able to functionally complement Ppt, as siderophore production (measured by a colorimetric assay) was detectable in low amounts in mutant strains concerning solely *ppt* or *cesP*, but not in the double knockout strain (data not shown). Transcriptional analysis of *ppt* in *B. cereus* F4810/72 revealed a weak, but constitutive expression throughout growth with a slight peak in stationary phase (Supplementary Figure S3A), similar to *sfp* transcription in *B. subtilis*, which was shown to be weak according to low promoter activity (Nakano et al., 1992). In contrast, transcription of *cesP*, which is linked to the *ces* operon, is strongly growth phase dependent, peaking highly in the late logarithmic phase (Dommel et al., 2011). Thus, we propose that the two PPtases encoded by *cesP* and *ppt* present functionally redundant enzymes with temporal different expression profiles in the growth cycle of *B. cereus* F4810/72.

ABC Transporter CesCD is Essential for Post-translational Cereulide Formation

Genes coding for transport systems of the ABC-type are frequently found part of, or in close proximity to, peptide synthetase operons of NRPS products, such as siderophores, antibiotics or lipopeptides. For example the production of lichenysin A of *B. licheniformis*, tyrocidine of *B. brevis*, syringomycin *of Pseudomonas syringae* or pyoverdine *of P. aeruginosa* are linked to different ABC transporters, which are thought to be involved in product secretion or self-resistance (Quigley et al., 1993; McMorran et al., 1996; Mootz and Marahiel, 1997; Yakimov et al., 1998). Gene disruption of the transporter genes *cesC* and *cesD* of *B. cereus* F4810/72 resulted in a complete cereulide deficient phenotype (**Table 2**), which could be restored partly by *in trans* complementation with *cesCD*, indicating that the putative ABC transporter is essential for toxin production. Sequence analysis of CesCD revealed homology to members of the BcrAB- or DRI- (drug resistance and immunity) subfamily of ABC systems (Dassa and Bouige, 2001; Davidson et al., 2008; Gebhard, 2012) with highest similarity toward the BerAB transporter of *B. thuringiensis*, which was shown to be essential for β-Exotoxin I production and suggested to cause toxin immunity by efflux of the molecule; but no experimental proof has been presented so far (Espinasse et al., 2002). Thus, it is tempting to speculate that CesCD is involved in the transport of the emetic toxin, maybe mediating its efflux and thereby conferring resistance. However, if CesCD were only responsible for cereulide export, an intracellular accumulation of toxin would be expected by inactivation of the transporter. Since no accumulation, but complete abolition of cereulide was observed in the CesCD knockout mutant, we propose that CesCD, besides having a potential transport function, plays a more direct role in the cereulide biosynthesis pathway without influencing *ces* gene transcription or NRPS translation. For the lantibiotics nisin and subtilin it was shown that the multimeric synthetase complexes,

which are required for pre-peptide maturation, are associated to the membrane by interaction with ABC transporters linked to the enzyme gene clusters (Siegers et al., 1996; Kiesau et al., 1997). More recently, the NRPS/PKS enzymes responsible for the biosynthesis of the antibiotic bacillaene and the siderophore pyoverdine were found to form membrane-associated mega complexes (Straight et al., 2007; Imperi and Visca, 2013). Presuming that CesCD is a membrane-bound protein, it could be involved in membrane-anchoring of the cereulide synthetase complex or other cereulide-processing enzymes, which otherwise may be non-functional. So far, immunoblot analysis using our anti-CesB antibody detected the cereulide synthetase only in cytosolic but not in membrane cell fractions (Rütschle, unpublished data). Therefore, further experiments targeting the localisation of cereulide and its synthetase complex are in progress to define the exact function of CesCD in cereulide synthesis.

Conclusion

Taken together, our data demonstrate the importance of the *cesAB*- adjacent genes *cesH*, *cesP*, and *cesCD* in the regulation of cereulide biosynthesis. Interestingly, these genes exert their regulatory functions on very different levels of toxin production, ranging from the transcriptional to the post-translational level (**Figure 6**). To our knowledge this is the first report on the impact of genes located within an NRPS encoding operon, and their gene products, on the transcription, translation and synthesis of the NRPS product and its biosynthetic machinery. The fact that not only chromosomal encoded master transcriptional regulators, such as AbrB or CodY, but also genes embedded in the *ces* gene cluster itself, influence cereulide production, highlights the complexity of regulation of synthesis of NRPS products.

Acknowledgments

We would like to thank Romy Wecko and Christine Braig for excellent technical support and Dietmar Hillman for advices concerning *B. megaterium* protoplast transformation procedures. Thanks to Daniel Zeigler (Bacillus Genetic Stock Center) for the gift of pAD123 and pMAD, and Wolfgang Hillen (University of Erlangen-Nürnberg) posthumously for the gift of pWH1520. We also thank Patrice Courvalin (Institut Pasteur) for kindly supplying us with *E. coli* JM83/pRK24, pAT113 and pSPCH+2. This research project was supported by the German Ministry of Economics and Technology (via AiF) and the FEI (Forschungskreis der Ernährungsindustrie e.V., Bonn), through projects AiF 15186 N and AiF 16845 N.

References

Agata, N., Ohta, M., Mori, M., and Isobe, M. (1995). A novel dodecadepsipeptide, cereulide, is an emetic toxin of *Bacillus cereus*. *FEMS Microbiol. Lett.* 129, 17–20. doi: 10.1111/j.1574-6968.1995.tb07550.x

Alonzo, D. A., Magarvey, N. A., and Schmeing, T. M. (2015). Characterization of cereulide synthetase, a toxin-producing macromolecular machine. *PLoS ONE* 10:e0128569. doi: 10.1371/journal.pone.0128569

Bauer, T., Stark, T., Hofmann, T., and Ehling-Schulz, M. (2010). Development of a stable isotope dilution analysis for the quantification of the *Bacillus cereus* toxin cereulide in foods. *J. Agric. Food Chem.* 58, 1420–1428. doi: 10.1021/jf9033046

Carlin, F., Fricker, M., Pielaat, A., Heisterkamp, S., Shaheen, R., Salonen, M. S., et al. (2006). Emetic toxin-producing strains of *Bacillus cereus* show distinct characteristics within the *Bacillus cereus* group. *Int. J. Food Microbiol.* 109, 132–138. doi: 10.1016/j.ijfoodmicro.2006.01.022

Copp, J. N., and Neilan, B. A. (2006). The phosphopantetheinyl transferase superfamily: phylogenetic analysis and functional implications in cyanobacteria. *Appl. Environ. Microbiol.* 72, 2298–2305. doi: 10.1128/AEM.72.4.2298-2305.2006

Dassa, E., and Bouige, P. (2001). The ABC of ABCS: a phylogenetic and functional classification of ABC systems in living organisms. *Res. Microbiol.* 152, 211–229. doi: 10.1016/S0923-2508(01)01194-9

Davidson, A. L., Dassa, E., Orelle, C., and Chen, J. (2008). Structure, function, and evolution of bacterial ATP-binding cassette systems. *Microbiol. Mol. Biol. Rev.* 72, 317–364. doi: 10.1128/MMBR.00031-07

Dierick, K., Van Coillie, E., Swiecicka, I., Meyfroidt, G., Devlieger, H., Meulemans, A., et al. (2005). Fatal family outbreak of *Bacillus cereus*-associated food poisoning. *J. Clin. Microbiol.* 43, 4277–4279. doi: 10.1128/JCM.43.8.4277-4279.2005

Dommel, M. K., Frenzel, E., Strasser, B., Blöchinger, C., Scherer, S., and Ehling-Schulz, M. (2010). Identification of the main promoter directing cereulide biosynthesis in emetic *Bacillus cereus* and its application for real-time monitoring of ces gene expression in foods. *Appl. Environ. Microbiol.* 76, 1232–1240. doi: 10.1128/AEM.02317-09

Dommel, M. K., Lücking, G., Scherer, S., and Ehling-Schulz, M. (2011). Transcriptional kinetic analyses of cereulide synthetase genes with respect to growth, sporulation and emetic toxin production in *Bacillus cereus*. *Food Microbiol.* 28, 284–290. doi: 10.1016/j.fm.2010.07.001

Ehling-Schulz, M., Frenzel, E., and Gohar, M. (2015). Food – bacteria interplay: pathometabolism of emetic *Bacillus cereus*. *Front. Microbiol.* 6:704. doi: 10.3389/fmicb.2015.00704

Ehling-Schulz, M., Fricker, M., Grallert, H., Rieck, P., Wagner, M., and Scherer, S. (2006). Cereulide synthetase gene cluster from emetic *Bacillus cereus*: structure and location on a mega virulence plasmid related to *Bacillus anthracis* toxin plasmid pXO1. *BMC Microbiol.* 6:20. doi: 10.1186/1471-2180-6-20

Ehling-Schulz, M., Fricker, M., and Scherer, S. (2004). *Bacillus cereus*, the causative agent of an emetic type of food-borne illness. *Mol. Nutr. Food Res.* 48, 479–487. doi: 10.1002/mnfr.200400055

Ehling-Schulz, M., Svensson, B., Guinebretiere, M. H., Lindback, T., Andersson, M., Schulz, A., et al. (2005a). Emetic toxin formation of *Bacillus cereus* is restricted to a single evolutionary lineage of closely related strains. *Microbiology* 151, 183–197. doi: 10.1099/mic.0.27607-0

Ehling-Schulz, M., Vukov, N., Schulz, A., Shaheen, R., Andersson, M., Martlbauer, E., et al. (2005b). Identification and partial characterization of the nonribosomal peptide synthetase gene responsible for cereulide production in emetic *Bacillus cereus*. *Appl. Environ. Microbiol.* 71, 105–113. doi: 10.1128/AEM.71.1.105-113.2005

Espinasse, S., Gohar, M., Lereclus, D., and Sanchis, V. (2002). An ABC transporter from *Bacillus thuringiensis* is essential for beta-exotoxin I production. *J. Bacteriol.* 184, 5848–5854. doi: 10.1128/JB.184.21.5848-5854.2002

Finking, R., and Marahiel, M. A. (2004). Biosynthesis of nonribosomal peptides1. *Annu. Rev. Microbiol.* 58, 453–488. doi: 10.1146/annurev.micro.58.030603.123615

Frenzel, E., Doll, V., Pauthner, M., Lücking, G., Scherer, S., and Ehling-Schulz, M. (2012). CodY orchestrates the expression of virulence determinants in emetic *Bacillus cereus* by impacting key regulatory circuits. *Mol. Microbiol.* 85, 67–88. doi: 10.1111/j.1365-2958.2012.08090.x

Gebhard, S. (2012). ABC transporters of antimicrobial peptides in Firmicutes bacteria – phylogeny, function and regulation. *Mol. Microbiol.* 86, 1295–1317. doi: 10.1111/mmi.12078

Hotelier, T., Renault, L., Cousin, X., Negre, V., Marchot, P., and Chatonnet, A. (2004). ESTHER, the database of the alpha/beta-hydrolase fold superfamily of proteins. *Nucleic Acids Res.* 32, D145–D147. doi: 10.1093/nar/gkh141

Imperi, F., and Visca, P. (2013). Subcellular localization of the pyoverdine biogenesis machinery of *Pseudomonas aeruginosa*: a membrane-associated "siderosome". *FEBS Lett.* 587, 3387–3391. doi: 10.1016/j.febslet.2013.08.039

Ivanova, N., Sorokin, A., Anderson, I., Galleron, N., Candelon, B., Kapatral, V., et al. (2003). Genome sequence of *Bacillus cereus* and comparative analysis with *Bacillus anthracis*. *Nature* 423, 87–91. doi: 10.1038/nature01582

Kiesau, P., Eikmanns, U., Gutowski-Eckel, Z., Weber, S., Hammelmann, M., and Entian, K. D. (1997). Evidence for a multimeric subtilin synthetase complex. *J. Bacteriol.* 179, 1475–1481.

Lambalot, R. H., Gehring, A. M., Flugel, R. S., Zuber, P., Lacelle, M., Marahiel, M. A., et al. (1996). A new enzyme superfamily – the phosphopantetheinyl transferases. *Chem. Biol.* 3, 923–936. doi: 10.1016/S1074-5521(96)90181-7

Lane, A. L., and Moore, B. S. (2011). A sea of biosynthesis: marine natural products meet the molecular age. *Nat. Prod. Rep.* 28, 411–428. doi: 10.1039/c0np90032j

Letunic, I., Doerks, T., and Bork, P. (2012). SMART 7: recent updates to the protein domain annotation resource. *Nucleic Acids Res.* 40, D302–D305. doi: 10.1093/nar/gkr931

Linton, K. J., Cooper, H. N., Hunter, I. S., and Leadlay, P. F. (1994). An ABC-transporter from *Streptomyces longisporoflavus* confers resistance to the polyether-ionophore antibiotic tetronasin. *Mol. Microbiol.* 11, 777–785. doi: 10.1111/j.1365-2958.1994.tb00355.x

Lücking, G., Dommel, M. K., Scherer, S., Fouet, A., and Ehling-Schulz, M. (2009). Cereulide synthesis in emetic *Bacillus cereus* is controlled by the transition state regulator AbrB, but not by the virulence regulator PlcR. *Microbiology* 155, 922–931. doi: 10.1099/mic.0.024125-0

Magarvey, N. A., Ehling-Schulz, M., and Walsh, C. T. (2006). Characterization of the cereulide NRPS alpha-hydroxy acid specifying modules: activation of alpha-keto acids and chiral reduction on the assembly line. *J. Am. Chem. Soc.* 128, 10698–10699. doi: 10.1021/ja0640187

Marahiel, M. A., Stachelhaus, T., and Mootz, H. D. (1997). Modular peptide synthetases involved in nonribosomal peptide synthesis. *Chem. Rev.* 97, 2651–2674. doi: 10.1021/cr960029e

Marxen, S., Stark, T. D., Frenzel, E., Rütschle, A., Lücking, G., Purstinger, G., et al. (2015a). Chemodiversity of cereulide, the emetic toxin of *Bacillus cereus*. *Anal. Bioanal. Chem.* 407, 2439–2453. doi: 10.1007/s00216-015-8511-y

Marxen, S., Stark, T. D., Rütschle, A., Lücking, G., Frenzel, E., Scherer, S., et al. (2015b). Depsipeptide Intermediates Interrogate Proposed Biosynthesis of Cereulide, the Emetic Toxin of *Bacillus cereus*. *Sci. Rep.* 5, 10637. doi: 10.1038/srep10637

May, J. J., Wendrich, T. M., and Marahiel, M. A. (2001). The dhb operon of *Bacillus subtilis* encodes the biosynthetic template for the catecholic siderophore 2,3-dihydroxybenzoate-glycine-threonine trimeric ester bacillibactin. *J. Biol. Chem.* 276, 7209–7217. doi: 10.1074/jbc.M009140200

McMorran, B. J., Merriman, M. E., Rombel, I. T., and Lamont, I. L. (1996). Characterisation of the pvdE gene which is required for pyoverdine synthesis in *Pseudomonas aeruginosa*. *Gene* 176, 55–59. doi: 10.1016/0378-1119(96)00209-0

Mesnage, S., Fontaine, T., Mignot, T., Delepierre, M., Mock, M., and Fouet, A. (2000). Bacterial SLH domain proteins are non-covalently anchored to the cell surface via a conserved mechanism involving wall polysaccharide pyruvylation. *EMBO J.* 19, 4473–4484. doi: 10.1093/emboj/19.17.4473

Messelhäusser, U., Frenzel, E., Blöchinger, C., Zucker, R., Kämpf, P., and Ehling-Schulz, M. (2014). Emetic *Bacillus cereus* are more volatile than thought: recent foodborne outbreaks and prevalence studies in Bavaria (2007-2013). *Biomed Res. Int.* 2014, 465603. doi: 10.1155/2014/465603

Mootz, H. D., and Marahiel, M. A. (1997). The tyrocidine biosynthesis operon of *Bacillus brevis*: complete nucleotide sequence and biochemical characterization of functional internal adenylation domains. *J. Bacteriol.* 179, 6843–6850.

Murphy, E. (1985). Nucleotide sequence of a spectinomycin adenyltransferase AAD(9) determinant from *Staphylococcus aureus* and its relationship to AAD(3") (9). *Mol. Gen. Genet.* 200, 33–39. doi: 10.1007/BF00383309

Nakano, M. M., Corbell, N., Besson, J., and Zuber, P. (1992). Isolation and characterization of sfp: a gene that functions in the production of the lipopeptide biosurfactant, surfactin, in *Bacillus subtilis. Mol. Gen. Genet.* 232, 313–321.

Naranjo, M., Denayer, S., Botteldoorn, N., Delbrassinne, L., Veys, J., Waegenaere, J., et al. (2011). Sudden death of a young adult associated with *Bacillus cereus* food poisoning. *J. Clin. Microbiol.* 49, 4379–4381. doi: 10.1128/JCM.05129-11

Pezard, C., Berche, P., and Mock, M. (1991). Contribution of individual toxin components to virulence of *Bacillus anthracis. Infect. Immun.* 59, 3472–3477.

Pfaffl, M. W., Horgan, G. W., and Dempfle, L. (2002). Relative expression software tool (REST) for group-wise comparison and statistical analysis of relative expression results in real-time PCR. *Nucleic Acids Res.* 30, e36. doi: 10.1093/nar/30.9.e36

Podlesek, Z., Comino, A., Herzog-Velikonja, B., Zgur-Bertok, D., Komel, R., and Grabnar, M. (1995). *Bacillus licheniformis* bacitracin-resistance ABC transporter: relationship to mammalian multidrug resistance. *Mol. Microbiol.* 16, 969–976. doi: 10.1111/j.1365-2958.1995.tb02322.x

Quigley, N. B., Mo, Y. Y., and Gross, D. C. (1993). SyrD is required for syringomycin production by *Pseudomonas* syringae pathovar syringae and is related to a family of ATP-binding secretion proteins. *Mol. Microbiol.* 9, 787–801. doi: 10.1111/j.1365-2958.1993.tb01738.x

Rasko, D. A., Ravel, J., Okstad, O. A., Helgason, E., Cer, R. Z., Jiang, L., et al. (2004). The genome sequence of *Bacillus cereus* ATCC 10987 reveals metabolic adaptations and a large plasmid related to *Bacillus anthracis* pXO1. *Nucleic Acids Res.* 32, 977–988. doi: 10.1093/nar/gkh258

Schwarzer, D., Mootz, H. D., Linne, U., and Marahiel, M. A. (2002). Regeneration of misprimed nonribosomal peptide synthetases by type II thioesterases. *Proc. Natl. Acad. Sci. U.S.A.* 99, 14083–14088. doi: 10.1073/pnas.212382199

Siegers, K., Heinzmann, S., and Entian, K. D. (1996). Biosynthesis of lantibiotic nisin. Posttranslational modification of its prepeptide occurs at a multimeric membrane-associated lanthionine synthetase complex. *J. Biol. Chem.* 271, 12294–12301. doi: 10.1074/jbc.271.21.12294

Spizizen, J. (1958). Transformation of biochemically deficient strains of *Bacillus Subtilis* by deoxyribonucleate. *Proc. Natl. Acad. Sci. U.S.A.* 44, 1072–1078. doi: 10.1073/pnas.44.10.1072

Stark, T., Marxen, S., Ruetschle, A., Luecking, G., Scherer, S., Ehling-Schulz, M., et al. (2013). Mass spectrometric profiling of *Bacillus cereus* strains and quantitation of the emetic toxin cereulide by means of stable isotope dilution analysis and HEp-2 bioassay. *Anal. Bioanal. Chem.* 405, 191–201. doi: 10.1007/s00216-012-6485-6

Straight, P. D., Fischbach, M. A., Walsh, C. T., Rudner, D. Z., and Kolter, R. (2007). A singular enzymatic megacomplex from *Bacillus subtilis. Proc. Natl. Acad. Sci. U.S.A.* 104, 305–310. doi: 10.1073/pnas.0609073103

Tamura, K., Peterson, D., Peterson, N., Stecher, G., Nei, M., and Kumar, S. (2011). MEGA5: molecular evolutionary genetics analysis using maximum likelihood, evolutionary distance, and maximum parsimony methods. *Mol. Biol. Evol.* 28, 2731–2739. doi: 10.1093/molbev/msr121

Thompson, J. D., Higgins, D. G., and Gibson, T. J. (1994). CLUSTAL W: improving the sensitivity of progressive multiple sequence alignment through sequence weighting, position-specific gap penalties and weight matrix choice. *Nucleic Acids Res.* 22, 4673–4680. doi: 10.1093/nar/22.22.4673

Trieu-Cuot, P., Carlier, C., Martin, P., and Courvalin, P. (1987). Plasmid transfer by conjugation from *Escherichia coli* to Gram-positive bacteria. *FEMS Microbiol. Lett.* 48, 289–294. doi: 10.1111/j.1574-6968.1987.tb0 2558.x

Tschiedel, E., Rath, P. M., Steinmann, J., Becker, H., Dietrich, R., Paul, A., et al. (2015). Lifesaving liver transplantation for multi-organ failure caused by *Bacillus cereus* food poisoning. *Pediatr. Transplant.* 19, E11–E14. doi: 10.1111/petr.12378

Walsh, C. T., Gehring, A. M., Weinreb, P. H., Quadri, L. E., and Flugel, R. S. (1997). Post-translational modification of polyketide and nonribosomal peptide synthases. *Curr. Opin. Chem. Biol.* 1, 309–315. doi: 10.1016/S1367-5931(97)80067-1

Wenzel, S. C., and Müller, R. (2005). Formation of novel secondary metabolites by bacterial multimodular assembly lines: deviations from textbook biosynthetic logic. *Curr. Opin. Chem. Biol.* 9, 447–458. doi: 10.1016/j.cbpa.2005. 08.001

Yakimov, M. M., Kroger, A., Slepak, T. N., Giuliano, L., Timmis, K. N., and Golyshin, P. N. (1998). A putative lichenysin A synthetase operon in *Bacillus licheniformis*: initial characterization. *Biochim. Biophys. Acta* 1399, 141–153. doi: 10.1016/S0167-4781(98) 00096-7

Conflict of Interest Statement: The authors declare that the research was conducted in the absence of any commercial or financial relationships that could be construed as a potential conflict of interest.

13

Food safety concerns deriving from the use of silver based food packaging materials

Alessandra Pezzuto[1], Carmen Losasso[2]*, Marzia Mancin[2], Federica Gallocchio[2], Alessia Piovesana[1], Giovanni Binato[3], Albino Gallina[3], Alberto Marangon[4], Renzo Mioni[1], Michela Favretti[1] and Antonia Ricci[2]

[1] *Optimization and Control of Food Production Laboratory, Istituto Zooprofilattico Sperimentale delle Venezie, San Donà di Piave, Italy,* [2] *Department of Food Safety, Istituto Zooprofilattico Sperimentale delle Venezie, Legnaro, Italy,* [3] *Laboratory of Chemistry, Istituto Zooprofilattico Sperimentale delle Venezie, Legnaro, Italy,* [4] *Sensory Analysis Laboratory, Veneto Agricoltura, Istituto per la Qualità e le Tecnologie Agroalimentari, Thiene, Italy*

The formulation of innovative packaging solutions, exerting a functional antimicrobial role in slowing down food spoilage, is expected to have a significant impact on the food industry, allowing both the maintenance of food safety criteria for longer periods and the reduction of food waste. Different materials are considered able to exert the required antimicrobial activity, among which are materials containing silver. However, challenges exist in the application of silver to food contact materials due to knowledge gaps in the production of ingredients, stability of delivery systems in food matrices and health risks caused by the same properties which also offer the benefits. Aims of the present study were to test the effectiveness and suitability of two packaging systems, one of which contained silver, for packaging and storing *Stracchino* cheese, a typical Italian fresh cheese, and to investigate if there was any potential for consumers to be exposed to silver, *via* migration from the packaging to the cheese. Results did not show any significant difference in the effectiveness of the packaging systems on packaged *Stracchino* cheese, excluding that the active packaging systems exerted an inhibitory effect on the growth of spoilage microorganisms. Moreover, silver migrated into the cheese matrix throughout the storage time (24 days). Silver levels in cheese finally exceeded the maximum established level for the migration of a non-authorised substance through a functional barrier (Commission of the European Communities, 2009). This result poses safety concerns and strongly suggests the need for more research aimed at better characterizing the new packaging materials in terms of their potential impacts on human health and the environment.

Keywords: food safety, food packaging, antimicrobial, silver migration

Edited by:
Amit Kumar Tyagi,
The University of Texas MD Anderson
Cancer Center, USA

Reviewed by:
Zhao Chen,
Clemson University, USA
Shinjini Singh,
University of Texas MD Anderson
Cancer Center, USA

*Correspondence:
Carmen Losasso
closasso@izsvenezie.it

INTRODUCTION

Food packaging technology is continuously evolving in response to growing consumer demand for minimally processed, more natural, fresh, and longer storable food.

Moreover, every year an increasing amount of edible food is lost along the entire food supply chain. The European Commission estimated that the annual food waste amounts to 89 million tons, or 179 kg per capita, varying considerably between individual countries and the various sectors, without even considering agricultural food waste or fish catches returned to the sea; furthermore, total food waste is expected to rise to approximately 126 million tons (a 40% increase) by 2020, unless additional preventive actions or measures are taken (European Parliament, 2012; European Parliament – STOA, 2013). Thus, packaging optimisation strategies have been proposed to guarantee the maintenance of food safety criteria and reduce food waste.

This scenario has strongly inspired the packaging industry to go beyond the traditional functions of the package and to offer innovative solutions addressing the changing demands of the food industry and consumers as well as the increasing regulatory and legal requirements (Realini and Marcos, 2014). Therefore, the function of food packaging has evolved from being a simple physical barrier, aimed at avoiding food contact with the external environment (passive packaging), to exerting a functional role in slowing down food spoilage (active packaging) by means of specific action on the chemical, enzymatic and mechanical phenomena, thus extending the shelf life of food (Marsh and Bugusu, 2007).

Different active packaging products are considered able to exert antimicrobial activity, especially when the materials contain silver (Quintavalla and Vicini, 2002; Coma, 2008). However, even though silver's antibacterial properties have long been proved (Silver et al., 2006), the lack of standardization in terms of particle characterization and test conditions makes it difficult to define its range of effectiveness and specificity against different bacterial species (Bae et al., 2010).

Moreover, there is an expanding body of scientific studies demonstrating that silver, especially in its nanosize, could introduce new risks to human health (Ahamed et al., 2010; Gaillet and Rouanet, 2015). The use of silver in food contact materials could potentially increase the probability of consumers' exposure: silver could migrate from packaging into foods, even though preliminary results indicate that migration is expected to be minimal (Chaudhry et al., 2008); although these studies seem to give some reassurances about safety, the few migration studies published to date have been targeted at food simulants, so further investigation needs to be performed especially in the case of complex food matrices.

In the European Union, the main regulatory framework related to the use of food contact materials is still Regulation (EC) No (1935/2004); it states that *"materials and articles, including active and intelligent materials and articles do not have to transfer their constituents to food in quantities which could endanger human health or bring about an unacceptable change in the composition of the food or bring about a deterioration in the organoleptic characteristics thereof."* However, the scientific literature has a complete lack of any data quantifying rates of migration of food package components into food.

The European Food Safety Authority stipulated, with reference to article 10 of Regulation (EC) No (1935/2004), an opinion of the panel on food contact materials, enzymes, flavourings and processing aids (CEF) of the risks originating from the migration of substances from food contact materials into food is required [European Food Safety Authority (EFSA), 2009]. According to Commission of the European Communities

(2009), Article 14, a maximum level of migration of 0.001 mg/kg should be observed for the migration of a non-authorized substance through a functional barrier. For some novel substances intended for inclusion in food packaging materials, such as silver, adequate toxicological data is not yet available and so safety assessments are still in progress. Thus, silver must undergo an appropriate authorisation process and safety evaluation before it is introduced.

Aims of the present study were to test the effectiveness of two active packaging systems, one of which contained silver, for the packaging and storage at 4°C of *Stracchino* cheese, a typical Italian fresh cheese, by assessing microbial, chemical, and sensorial parameters. Moreover, the migration of silver from the packaging into the cheese was monitored during the chill storage period.

MATERIALS AND METHODS

Materials

Two active food packaging systems were studied in the present research: the active packaging *Food-touch®* by Microbeguard Corp. (USA), containing silver zeolite for the purpose of exerting antimicrobial properties, and an innovative packaging, *Ovtene®* by Arcadia Spa (Italy), containing calcium carbonate, talc and titanium dioxide, for the purpose of acting as an absorber with the intent to extend food shelf life and supplied by a local producer. Both of these products were used in their original liner formats, as provided by suppliers. As control, a traditional passive packaging system was also used.

Stracchino cheese, a commercially available typical Italian fresh cheese with a declared shelf life of 20 days, preferentially consumed by children and the elderly, was selected as a perishable and high value food suitable for use in these packaging systems.

Stracchino pieces (250 g) were hand-wrapped according to manufacturers' instructions in the selected packaging systems and then analyzed for a range of microbial, chemical and sensorial parameters. Packaged *Stracchino* cheese was stored at 4°C for 25 days.

Microbial, Water Activity, and pH Analyses

Microbial analyses were conducted to monitor the numbers of spoilage and indicator microorganisms in the cheese during storage in the three selected packaging systems as detailed in **Table 1**.

The following microbial determinations were performed:

Total Viable Count at 30°C (ISO 4833, 2003a; Plate Count Agar at 30 ± 1°C for 72 ± 3 h in aerobic conditions), *Pseudomonas* spp. (Cetrimide-Fusidic acid-Cephalosporin Agar at 25 ± 1°C for 44 ± 4 h), *Enterobacteriaceae* (ISO 21528-2, 2004a; Violet Red Bile Glucose Agar at 37 ± 1°C for 24 ± 2 h), Lactic Acid Bacteria at 30°C (MRS Agar at 30 ± 1°C for 72 ± 3 h in aerobic conditions), Moulds and Yeasts (Rose Bengal Chloramphenicol Agar at 25 ± 1°C for 3–5 days). Additionally, water activity was measured (a_w, according to ISO 21807, 2004b), as was pH (ISO 2917, 1999).

TABLE 1 | Sampling design for microbial, chemical, and sensorial determination.

Experimental time (days)	Storage temperature (°C)	Number of samples		
		Traditional packaging	Innovative packaging (*Ovtene®*)	Active packaging (*Food-touch®*)
0	0/+4	20	–	–
8	0/+4	20	20	20
15	0/+4	20	20	20
21	0/+4	20	20	20
25	0/+4	20	20	20
Total number of samples		100	80	80

Chemical Analyses

Chemical analyses were conducted on days 0, 7, 14, 20, and 24 of storage. In total, 20 replicate cheese samples were examined each day. The following chemical parameters were assessed:

(1) *Total Volatile Basic Nitrogen* (TVB-N), determined by the method prescribed in Commission of the European Communities (2005) for the evaluation of the TVN-B in fish;
(2) *Sulfides*, determined by Lead Acetate Test Strips (Sigma-Aldrich, Switzerland);
(3) *Peroxides* via iodometric titration (Biffoli, 1979);
(4) *Stamm test* (Hamm et al., 1965);
(5) *Thiobarbituric acid test* (TBA; Fernandez et al., 1997).

Sensory Evaluation

In order to detect any sensory difference during the storage period, sensory evaluation was conducted on *Stracchino* cheese packaged using only the *Ovtene®* system and the traditional packaging as control. Sensory evaluation on *Stracchino* packaged in the *Food-touch®* system was not conducted, due to the claimed presence of silver and the possibility of its migration into the cheese matrix. Two sensory evaluations were performed after 0 and 7 days of storage using fresh *Stracchino* as a standard reference. The sensory evaluation was performed as described by ISO 13299 (2003b). A panel of 15 judges, experts in sensory evaluation of cheese, was previously trained on the quantitative evaluation of the following selected descriptors: (i) *Odor intensity*; (ii) *Aroma intensity*; (iii) *Saltiness*; (iv) *Acidity*; (v) *Bitterness*; (vi) *Homogeneity*; (vii) *Consistency*; (viii) *Adherence*.

Silver Migration Test

Silver migration from the *Food-touch®* system into the cheese was determined on days 0, 7, 14, 20, and 24 of storage. In total, two independent determinations of 20 cheese samples were examined each day.

Atomic Absorption Instrumentation

Silver concentrations were determined using Electrothermal Atomic Absorption Spectrometry (ETAAS) on an M6 mkII Atomic Absorption Spectrometer (Thermo Electron, Cambridge,

UK) with D_2 background correction, equipped with a GF95 Graphite Furnace atomiser.

Analytical Methods

Cheese samples (2 g) were homogenized, then 8 ml concentrated HNO_3 and 2 ml H_2O_2 were added and the mixture digested in Teflon liners using a CEM (Mattews, NC, USA) Mars Xpress microwave oven. Digested cheese samples were then diluted to up to 25 ml in class A volumetric flasks with deionised water and analyzed. All calibration solutions for metal determination were made by dilution from Certified Standard Solutions (ULTRAgrade® ICP Standards, 1000 mg/mL) provided by Ultra Scientific (North Kingstown, RI, USA).

Trueness of analytical data was verified by means of Certified Reference Materials (NRCC DORM2) analyzed concurrently with samples in each analytical batch. LOQ (6 s) value were found to be 0.0015 mg/kg. Operating conditions are reported in Supplementary Table S1.

Statistical Analysis

The microbial counts were logarithm transformed, and data distributions over time were represented by box and whiskers plots. To evaluate the effect of type of packaging, time and the interaction between the two independent variables on the dynamics of the microbial parameters, analysis of variance (ANOVA) was applied. The observation taken at point 0 (day 1), presenting the same values distribution for each type of packaging and the outliers, identified using Grubbs test, were not considered in the analysis (Grubbs, 1969; Agresti, 1990; Dohoo et al., 2003).

The assumption of homoscedasticity was verified using the Breusch-Pagan & Cook-Weisberg test and the residual plot (Breusch and Pagan, 1979; Cook and Weisberg, 1983).

The normality condition of residuals was verified using the Shapiro-Francia test (Shapiro and Francia, 1972), the graphical analysis of the residuals plotted against the normal probability distribution (q-q plot), and the histogram of residuals versus normal curve (Dohoo et al., 2003).

A post-estimation analysis was performed to evaluate the significant paired contrasts taking into account the multi test correction in the evaluation of statistical significance (Dohoo et al., 2003).

The software STATA 12.0 (Release 12) was used to conduct the statistical analysis of microbial data.

Analysis of variance was also performed to analyze the sensory data by SYSTAT software, in order to assess any significant sensory difference between cheese treated with the active or the traditional packaging systems, after checking the reliability and homogeneity of the obtained data.

P-value < 0.05 was considered significant in the statistical analysis.

RESULTS

Microbial Analyses

The population dynamics of spoilage-related microorganisms (Total Viable Counts, lactic acid bacteria, moulds, yeasts, and

Pseudomonas spp.) and *Enterobacteriaceae* in the *Stracchino* cheese are described in **Figure 1**.

The assumptions of homoscedasticity of data and the normality of residuals were satisfied for each analyzed microbial group (data not shown).

Analysis of variance analysis showed that populations of each of the investigated groups of microorganisms were affected both by storage time (as expected) and by type of packaging ($P < 0.001$).

In particular, Total Viable Counts increased initially (between 0 and 15 days of storage), maintained a constant value during days 15–21, and then increased further during days 21–25 of storage (**Figure 1A**). In detail, in traditionally packaged cheese, the Total Viable Counts were significantly higher than in

Ovtene®-packaged cheese on days 8 and 25. Total Viable Counts in cheese packaged in the *Food-Touch*® system were slightly, but significantly, lower on each sampling day than in traditionally packaged cheese. Total Viable Counts in cheese packaged in the *Food-Touch*® system were slightly but significantly lower than in cheese packaged in the *Ovtene*® system on each given day, except for days 8 and 25, when the cheese in the two packaging systems contained similar numbers of Total Viable Count bacteria.

Lactic acid bacteria showed a similar growth trend in cheese in the examined packaging systems (**Figure 1B**). Lactic acid bacteria numbers in cheese remained constant until day 8, and increased from day 15 until the end of storage. This trend was more evident in the case of the traditional packaging. At the end of storage, lactic acid bacteria numbers in cheese in *Ovtene*® packaging were

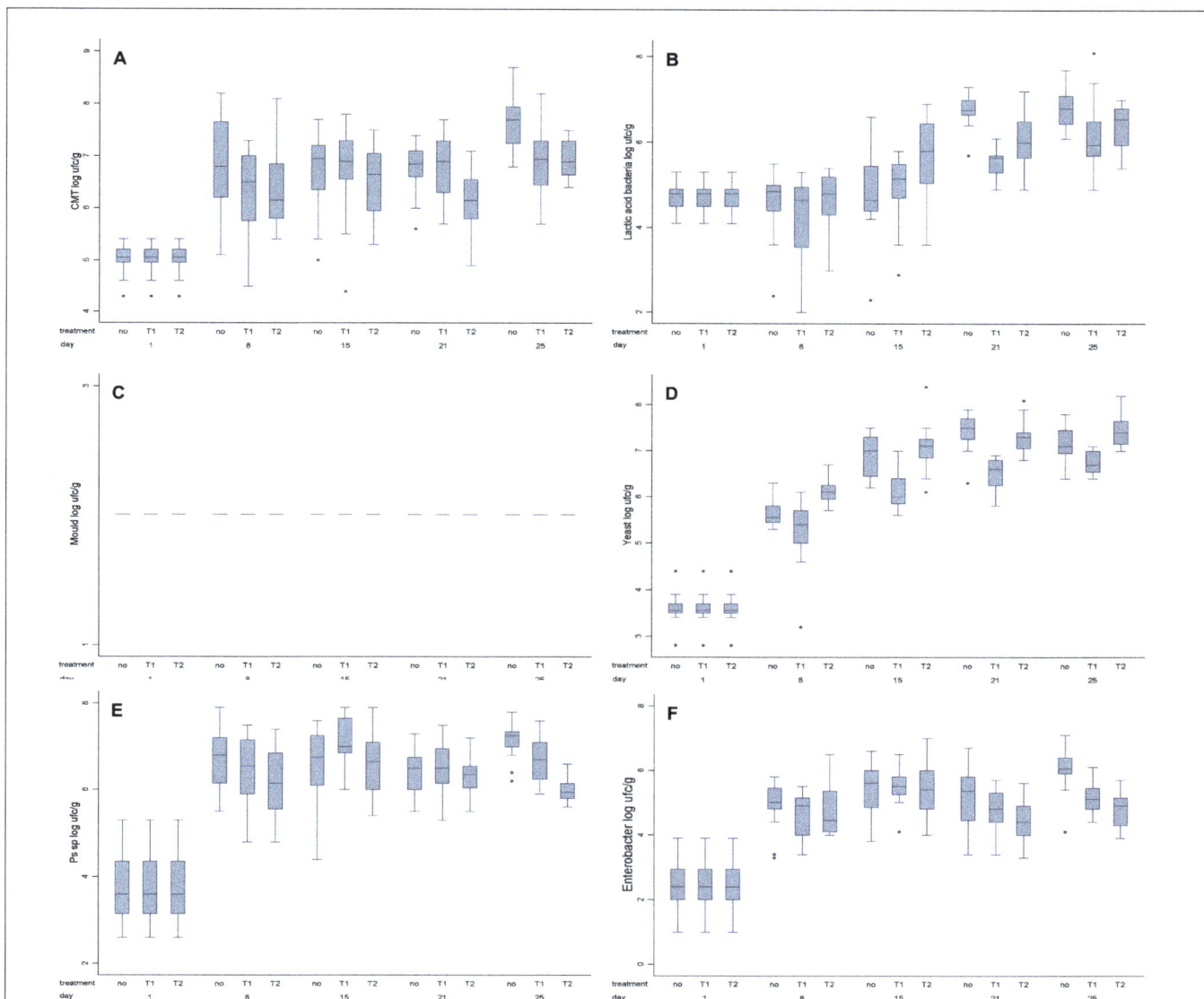

FIGURE 1 | Box plot of the microbiological data per observation time and type of packaging: traditional (no), *Ovtene*® packaging (T1) and *Food-Touch*® system (T2). (A) Total Microbial Counts; **(B)** Lactic Acid Bacteria; **(C)** Moulds; **(D)** Yeasts; **(E)** *Pseudomonas* sp.; **(F)** *Enterobacteriaceae*.

lower than in traditionally packaged cheese and in *Food-Touch®*-packaged cheese.

No moulds were detected in *Stracchino* cheese during the study, so these data are not discussed further.

Ovtene®-packaged cheese had noticeably lower yeast counts than cheese in traditional packaging and *Food-Touch®*-packaged cheese (**Figure 1D**).

Pseudomonas spp. increased in numbers in the cheese in all three packaging systems in the first days of storage (up to around day 8), and remained constant thereafter, except for slight fluctuations (**Figure 1E**).

Finally in the case of *Enterobacteriaceae*, a marked increase in numbers was measured in the first 15 days of storage, then slight fluctuations were observed at the end of the storage time (**Figure 1F**). *Enterobacteriaceae* numbers were lower in cheese packaged in the *Ovtene®* system than in the traditional system after 25 days of storage. *Enterobacteriaceae* numbers were lower in *Food-Touch®*-packaged cheese between days 21 and 25 than in cheese packed in the other two systems.

Chemical Analysis

Table 2 shows the results of chemical analyses on cheese during storage. The only difference between the three examined packaging systems was the level of detected TVB-N. When the *Stracchino* cheese was wrapped with either one of the two active packaging options, higher levels of TVB-N were detected, compared with the traditional system. No differences were noticed for all the other investigated parameters.

TABLE 2 | Chemical determinations obtained from analyses of cheese packaged in traditional, innovative and active packaging.

Experimental Time (days)	TVB-N (mgN/100 g)	Sulfides	Peroxides (MEQ O$_2$/Kg of fat)	Stamm test	TBA test
Traditional packaging					
0	2.00 ± 1.02	Negative	<5	Negative	Negative
7	3.00 ± 1.27	Negative	<5	Negative	Negative
14	4.30 ± 6.30	Negative	<5	Negative	Negative
20	0.10 ± 0.31	Negative	<5	Negative	Negative
24	13.00 ± 8.97	Negative	<5	Negative	Negative
***Ovtene®* packaging**					
0	2.00 ± 1.02	Negative	<5	Negative	Negative
7	3.00 ± 1.06	Negative	<5	Negative	Negative
14	15.70 ± 11.26	Negative	<5	Negative	Negative
20	14.05 ± 6.51	Negative	<5	Negative	Negative
24	1.65 ± 1.42	Negative	<5	Negative	Negative
***Food-Touch®* packaging**					
0	2.00 ± 1.02	Negative	<5	Negative	Negative
7	1.00 ± 1.15	Negative	<5	Negative	Negative
14	14.80 ± 8.54	Negative	<5	Negative	Negative
20	<0.10	Negative	<5	Negative	Negative
24	2.55 ± 1.93	Negative	<5	Negative	Negative

Mean and standard deviation for each observation time are indicated.

Sensory Evaluation

No significant difference due to the packaging characteristics, was displayed by the sensory evaluation tests for the identified parameters except for homogeneity and adherence, with *Ovtene®* packaging displaying the best performance ($P > 0.05$; data not shown).

Silver Migration Test

The extent of Ag migration from the *Food-touch®* composite film into the cheese over time is shown in **Table 3**. The migration of Ag increased gradually from 0.053 mg/kg after 7 days of incubation to 0.103 mg/kg after 24 days (**Table 3**).

DISCUSSION

In this work, the effectiveness of two new active packaging systems on microbial, chemical, and sensorial qualities of *Stracchino* cheese was evaluated. Moreover, the possibility that the cheese could contain chemicals deriving from the active food packaging systems (*Ovtene®* and *Food-Touch®*) was assessed.

Despite the Food-Touch® system resulting in lower bacterial growth at some given times throughout the cheese storage, the final results did not show any significant difference in the cheese microbiota examined, of any packaged *Stracchino* cheese samples, excluding that the investigated packaging systems exerted a different inhibitory effect on the growth of spoilage microorganisms.

On the contrary, a putative effect exerted by the *Ovtene* system, which maintained two of the examined sensory characteristics, homogeneity and adherence, was observed. This effect may have been a consequence of the preservation of the functional cheese microbiota, known to be involved in the typical organoleptic properties of cheeses (Hanniffy et al., 2009; Sgarbi et al., 2013).

These results are coherent with previously published research (Incoronato et al., 2010, 2011; Morsy et al., 2014), suggesting that although application of silver based antimicrobial systems in the food industry is a widespread phenomenon, appraisal

TABLE 3 | Assessment of silver migration to cheese from traditional and active packaging.

Time (days)	Silver (mg/kg)
Traditional packaging	
0	<0.0015
7	<0.0015
14	<0.0015
20	<0.0015
24	<0.0015
***Food-Touch®* packaging**	
0	<0.0015
7	0.053 ± 0.025
14	0.245 ± 0.053
20	0.047 ± 0.023
24	0.103 ± 0.038

Mean and standard deviation for each observation time are indicated.

of the full potential of silver as an antimicrobial and its possible implementation in food packaging technologies is still a challenging task (Berton et al., 2014; Losasso et al., 2014).

However, since health and safety properties of many food contact materials are not fully understood, food safety should be the main concern when formulating materials for food packaging applications. Thus, according to the European Regulation (EC) No (1935/2004), efforts have to be devoted to investigate the overall migration of compounds from new packaging materials to the food, in order to elucidate the risks to humans consuming such packaged foods (de Kruijf et al., 2002; Chaudhry et al., 2008).

In this context, our results pose some safety concerns, as the level of silver migration from the active packaging system containing silver greatly exceeded the maximum established level for the migration of a non-authorised substance through a functional barrier (Commission of the European Communities, 2009).

Despite the relevance of the topic, to date, only a limited number of reports have studied the potential for silver migrating from plastic food containers, with most reports being focused on silver nanoparticles (Echegoyen and Nerín, 2013; von Goetz et al., 2013; Artiaga et al., 2015). In these studies, food containers were exposed to a number of food-simulating solutions (not real foods) under a variety of experimental conditions in an attempt to determine the possible risks for human health. Conversely, our data investigated silver migration using a real food matrix as the acceptor, and clearly showed that silver levels in cheese reached unacceptable levels, up to around 250 times higher than the 0.001 mg/kg level prescribed by EU regulation.

Even though the published reports have revealed that silver has a low tendency to migrate from the investigated materials into solutions which mimic food, under regular use conditions, several discrepancies were found in these studies, particularly with regard both to the obtained results and to the analytical methodologies used. However, unambiguous methodologies to detect and quantify the chemicals migrated from packaging are currently lacking, making it difficult for us to produce an overall assessment of results published to date.

As far as the *Ovtene* system is concerned, although this product did not display any effect in reducing the proliferation of all spoilage microorganisms, the preservation of some typical features of the cheese and the absence of any measured chemical migration into cheese could make this product interesting to the food industry.

CONCLUSION

The development of innovative and active packaging systems could provide important instruments to overcome existing challenges that are associated with packaging materials, positively affecting the shelf life and the quality of foods, which will ultimately benefit both the producers and consumers. However, more in-depth research is needed in order to characterize their potential impacts on consumer health and the environment.

ACKNOWLEDGMENT

The present research was funded by Italian Ministry of Health, Grant code: RC IZSVe 12/08.

REFERENCES

Agresti, A. (1990). *Categorical Data Analysis*. New York, NY: John Wiley & Sons, Inc.

Ahamed, M., Alsalhi, M. S., and Siddiqui, M. K. J. (2010). Silver nanoparticle application and human health. *Clin. Chim. Acta* 411, 1841–1848. doi: 10.1016/j.cca.2010.08.016

Artiaga, G., Ramos, K., Ramos, L., Cámara, C., and Gómez-Gómez, M. (2015). Migration and characterisation of nanosilver from food containers by AF4-ICP-MS. *Food Chem.* 166, 76–85. doi: 10.1016/j.foodchem.2014.05.139

Bae E., Park, H. J., Lee, J., Kim, Y., Yoon, J., Park, K., et al. (2010). Bacterial cytotoxicity of the silver nanoparticle related to physicochemical metrics and agglomeration properties. *Environ. Toxicol. Chem.* 29, 2154–2160. doi: 10.1002/etc.278

Berton, V., Montesi, F., Losasso, C., Facco, D. R., Toffan, A., and Terregino, C. (2014). Study of the interaction between silver nanoparticles and salmonella as revealed by transmission electron microscopy. *J. Prob. Health* 3:123.

Biffoli, R. (1979). *Chimica Degli Alimenti*, Vol. 2, Florence: Uncini Pierucci.

Breusch, T. S., and Pagan, A. R. (1979). A simple test for heteroscedasticity and random coefficient variation. *Econometrica* 47, 1287–1294. doi: 10.2307/1911963

Chaudhry, Q., Scotter, M., Blackburn, J., Ross, B., Boxall, A., Castle, L., et al. (2008). Anals control expo. *Risk Assess.* 25, 241–258.

Coma, V. (2008). Bioactive packaging technologies for extended shelf life of meat-based products. *Meat Sci.* 78, 90–103. doi: 10.1016/j.meatsci.2007.07.035

Commission of the European Communities (2005). No 2074/2005 of 5 December 2005 laying down implementing measures for certain products under Regulation (EC) No 853/2004 of the European Parliament and of the Council and for the organisation of official controls under Regulation (EC) No 854/2004 of the European Parliament and of the Council and Regulation (EC) No 882/2004 of the European Parliament and of the Council, derogating from Regulation (EC) No 852/2004 of the European Parliament and of the Council and amending Regulations (EC) No 853/2004 and (EC) No 854/2004. *Off. J. Eur. Union* 27–59. Available at: http://faolex.fao.org/docs/pdf/eur87031.pdf

Commission of the European Communities (2009). No. 450/2009 of 29 May on active and intelligent materials and articles intended to come into contact with food. *Off. J. Eur. Union* 3–11.

Cook, R. D., and Weisberg, S. (1983). Diagnostics for heteroscedasticity in regression. *Biometrika* 70, 1–10.

de Kruijf, N., van Beest, M., Rijk, R., Sipilainen-Malm, T., Paseiro Losada, P., and De Meulenaer, B. (2002), Active and intelligent packaging: applications and regulatory aspects. *Food Addit. Contam.* 19(Suppl.), 144–162. doi: 10.1080/02652030110072722

Dohoo, I., Wayne, M., and Stryhn, H. (2003). *Veterinary Epidemiologic Research*, 2nd Edn. Moorpark, CA: AVC Inc.

Echegoyen, Y., and Nerín, C. (2013). Nanoparticle release from nano-silver antimicrobial food containers. *Food Chem. Toxicol.* 62, 16–22. doi: 10.1016/j.fct.2013.08.014

European Food Safety Authority (EFSA) (2009). Guidelines on submission of a dossier for safety evaluation by the EFSA of active or intelligent substances present in active and intelligent materials and articles intended to come into contact with food. *EFSA J.* 1208.

European Parliament (2012). *European Parliament Resolution of 19 January on How to Vvoid Food Wastage: Strategies for a More Efficient Food Chain in the EU (2011/2175(INI)).* Available at: http://eur-lex.europa.eu/legal-content/EN/TXT/?uri=uriserv:OJ.CE.2013.227.01.0025.01.ENG

European Parliament – STOA (2013). *Technology Options for Feeding 10 Billion People Recycling Agricultural, Forestry & Food Wastes and Residues for Sustainable Bioenergy and Biomaterials.* Available at: http://www.europarl.europa.eu/RegData/etudes/etudes/join/2013/513513/IPOL.JOIN_ET(2013)513513_EN.pdf

Fernandez, J., Pérez-Álvarez, J., and Fernández-López, J. (1997). Thiobarbituric acid test for monitoring lipid oxidation in meat. *Food Chem.* 59, 345–353. doi: 10.1016/S0308-8146(96)00114-8

Gaillet, S., and Rouanet, J. M. (2015). Silver nanoparticles: their potential toxic effects after oral exposure and underlying mechanisms - a review. *Food Chem. Toxicol.* 77C, 58–63. doi: 10.1016/j.fct.2014.12.019

Grubbs, F. (1969). Procedures for detecting outlying observations in samples. *Technometrics* 11, 1–21. doi: 10.1080/00401706.1969.10490657

Hamm, D. L., Hammond, E. G., Parvanah, V., and Snyder, H. E. (1965). The determination of peroxides by the Stamm method. *J. Am. Oil Chem. Soc.* 42, 920–922. doi: 10.1007/BF02632445

Hanniffy, S. B., Peláez, C., Martínez-Bartolomé, M. A., Requena, T., and Martínez-Cuesta, M. C. (2009). Key enzymes involved in methionine catabolism by cheese lactic acid bacteria. *Int. J. Food Microbiol.* 135, 223–230. doi: 10.1016/j.ijfoodmicro.2009.08.009

Incoronato, A. L., Buonocore, G. G., Conte, A., Lavorgna, M., and Nobile, M. A. (2010). Active systems based on silver-montmorillonite nanoparticles embedded into bio-based polymer matrices for packaging applications. *J. Food Prot.* 73, 2256–2262.

Incoronato, A. L., Conte, A., Buonocore, G. G., and Del Nobile, M. A. (2011). Agar hydrogel with silver nanoparticles to prolong the shelf life of Fior di Latte cheese. *J. Dairy Sci.* 94, 1697–1704. doi: 10.3168/jds.2010-3823

ISO 21528-2 (2004a). *Microbiology of Food and Animal Feeding Stuffs – Horizontal Methods for the Detection and Enumeration of Enterobacteriaceae – Colony-Count Method.* Part 2. Geneva: International Organization for Standardization.

ISO 21807 (2004b). *Microbiology of Food and Animal Feeding Stuffs – Determination of Water Activity.* International standard, 1st Edn. Geneva: International Organization for Standardization.

ISO 2917 (1999). *Meat and Meat Products – Measurement of pH – Reference Method.* 2nd Edn. Geneva: International Organization for Standardization.

ISO 4833 (2003a). Microbiology of Food and Animal Feeding stuffs – Horizontal Method for the Enumeration of Microorganisms -Colony-Count Technique at 30 Degrees C. By the Pour Plate Technique. International standard, 3rd Edn. Geneva: International Organization for Standardization.

ISO 13299 (2003b). *Sensory Analysis – Methodology – General Guidance for Establishing a Sensory Profile.* Geneva: International Organization for Standardization.

Losasso, C., Belluco, S., Cibin, V., Zavagnin, P., Mičetić, I., Gallocchio, F., et al. (2014). Antibacterial activity of silver nanoparticles: sensitivity of different *Salmonella serovars. Front. Microbiol.* 5:227. doi: 10.3389/fmicb.2014.00227

Marsh, K., Bugusu, B. (2007). Food packaging—roles, materials, and environmental issues. *J. Food Sci.* 72, R40–R55. doi: 10.1111/j.1750-3841.2007.00301.x

Morsy, M. K., Khalaf, H. H., Sharoba, A. M., El-Tanahi, H. H., and Cutter, C. N. (2014). Incorporation of essential oils and nanoparticles in pullulan films to control foodborne pathogens on meat and poultry products. *J. Food Sci.* 79, M675–M684. doi: 10.1111/1750-3841.12400

Quintavalla, S., and Vicini, L. (2002). Antimicrobial food packaging in meat industry. *Meat Sci.* 62, 373–380. doi: 10.1016/S0309-1740(02)00121-3

Realini, C., Marcos, B. (2014). Active, and intelligent packaging systems for a modern society. *Meat Sci.* 98, 404–419. doi: 10.1016/j.meatsci.2014.06.031

Regulation (EC) No (1935/2004). Regulation (EC) No 1935/2004 of the european parliament and of the council of 27 October on materials and articles intended to come into contact with food and repealing Directives 80/590/EEC and 89/109/EEC.

Sgarbi, E., Lazzi, C., Iacopino, L., Bottesini, C., Lambertini, F., and Sforza, S. (2013). Microbial origin of non proteolytic aminoacyl derivatives in long ripened cheeses. *Food Microbiol.* 35, 116–120. doi: 10.1016/j.fm.2013.02.013

Shapiro, S. S., and Francia, R. S. (1972). An approximate analysis of variance test for normality. *J. Am. Statist. Assoc.* 67, 215–216. doi: 10.1080/01621459.1972.10481232

Silver, S., Phung, L. T., and Silver, G. (2006). Silver as biocides in burn and wound dressings and bacterial resistance to silver compounds. *J. Ind. Microbiol. Biotechnol.* 33, 627–634. doi: 10.1007/s10295-006-0139-7

von Goetz, N., Lorenz, C., Windler, L., Nowack, B., Heuberger, M., and Hungerbühler, K. (2013). Migration of Ag- and TiO2-(Nano)particles from textiles into artificial sweat under physical stress: experiments and exposure modeling. *Environ. Sci. Technol.* 47, 9979–9987. doi: 10.1021/es304329w

Conflict of Interest Statement: The authors declare that the research was conducted in the absence of any commercial or financial relationships that could be construed as a potential conflict of interest.

14

Survey on antimicrobial resistance patterns in *Vibrio vulnificus* and *Vibrio cholerae* non-O1/non-O139 in Germany reveals carbapenemase-producing *Vibrio cholerae* in coastal waters

Nadja Bier, Keike Schwartz, Beatriz Guerra and Eckhard Strauch *

Department of Biological Safety, Federal Institute for Risk Assessment, Berlin, Germany

Edited by:
Learn-Han Lee,
Monash University Malaysia, Malaysia

Reviewed by:
Weili Liang,
Chinese Center for Disease Control
and Prevention, China
Adrian Canizalez-Roman,
Autonomous University of Sinaloa,
Mexico
Vengadesh Letchumanan,
University of Malaya, Malaysia

*Correspondence:
Eckhard Strauch
eckhard.strauch@bfr.bund.de

An increase in the occurrence of potentially pathogenic *Vibrio* species is expected for waters in Northern Europe as a consequence of global warming. In this context, a higher incidence of *Vibrio* infections is predicted for the future and forecasts suggest that people visiting and living at the Baltic Sea are at particular risk. This study aimed to investigate antimicrobial resistance patterns among *Vibrio vulnificus* and *Vibrio cholerae* non-O1/non-O139 isolates that could pose a public health risk. Antimicrobial susceptibility of 141 *V. vulnificus* and 184 *V. cholerae* non-O1/non-O139 strains isolated from German coastal waters (Baltic Sea and North Sea) as well as from patients and retail seafood was assessed by broth microdilution and disk diffusion. Both species were susceptible to most of the agents tested (12 subclasses) and no multidrug-resistance was observed. Among *V. vulnificus* isolates, non-susceptibility was exclusively found toward aminoglycosides. In case of *V. cholerae*, a noticeable proportion of strains was non-susceptible to aminopenicillins and aminoglycosides. In addition, resistance toward carbapenems, quinolones, and folate pathway inhibitors was sporadically observed. Biochemical testing indicated the production of carbapenemases with unusual substrate specificity in four environmental *V. cholerae* strains. Most antimicrobial agents recommended for treatment of *V. vulnificus* and *V. cholerae* non-O1/non-O139 infections were found to be effective *in vitro*. However, the occurrence of putative carbapenemase producing *V. cholerae* in German coastal waters is of concern and highlights the need for systematic monitoring of antimicrobial susceptibility in potentially pathogenic *Vibrio* spp. in Europe.

Keywords: antimicrobial resistance pattern, Baltic Sea, North Sea, carbapenemase, disk diffusion, broth microdilution

INTRODUCTION

The family *Vibrionaceae* within the class of Gammaproteobacteria comprises eight genera of Gram-negative, facultative anaerobic, straight, or curved rods that are mostly oxidase-positive, halophilic, and motile (Farmer and Janda, 2004). Members of this family are ubiquitously distributed in aquatic ecosystems worldwide. They can be found as free-living bacteria and as commensals of aquatic organisms and play an important role in nutrient cycling of natural aquatic habitats. Due to their metabolic diversity and their adaptive abilities to changing environmental conditions, a seasonal, and geographical variability of total *Vibrio* populations is observed in response to climatic influences and seawater circulations (Mansergh and Zehr, 2014).

Among the *Vibrionaceae*, a number of important human pathogenic bacteria have been identified that can cause gastrointestinal infections, wound infections or septicemia. *Vibrio cholerae*, *Vibrio vulnificus*, and *Vibrio parahaemolyticus* are considered as the most clinically relevant human pathogens within the genus *Vibrio* (Daniels and Shafaie, 2000). The latter species is widely disseminated in estuarine, marine, and coastal so surroundings and the leading cause of human intestinal infections after consumption of raw and undercooked seafood (Letchumanan et al., 2014). *V. cholerae* and *V. vulnificus* are also part of the microbial community in coastal or estuarine aquatic ecosystems with moderate salinities (Thompson and Polz, 2006).

V. cholerae is a well-known human pathogen consisting of more than 200 serogroups (Kaper et al., 1995; Lutz et al., 2013). Toxigenic *V. cholerae* of the O1 or O139 serogroup are the causative agents of cholera, an endemic disease in many Asian and African countries with symptoms of severe watery diarrhea, vomiting, and dehydration. All other serogroups designated as *V. cholerae* non-O1/non-O139 have also been linked to sporadically occurring human infections ranging from extraintestinal wound or ear infections (Huhulescu et al., 2007) to relatively mild or sometimes severe gastroenteritis (Tobin-D'Angelo et al., 2008), whereby smaller diarrheal outbreaks were also reported (Luo et al., 2013). Additionally, rarely occurring bacteremia has been described with mortality rates up to 61.5% (Petsaris et al., 2010). Since 2000, around 40 cases of *V. cholerae* non-O1/non-O139 infections in the United States have been reported to the CDC annually[1]. *V. cholerae* infections contracted in Germany were mainly ear or wound infections caused by non-toxigenic non-O1/non-O139 strains that were acquired through contact to seawater (Huehn et al., 2014). Due to the rare occurrence of *V. cholerae* non-O1/non-O139 infections, there are no official recommendations on antibiotic therapy (Petsaris et al., 2010). However, in case of bacteremia an early administration of antibiotic therapy can prevent a fatal outcome. Several case studies on *V. cholerae* non-O1/non-O139 bacteremia and wound infection exist, where fluoroquinolones and third-generation cephalosporins have been used (Huhulescu et al., 2007; Petsaris et al., 2010). But also treatment with ampicillin or last-line carbapenems has been described (Feghali and Adib, 2011; Lu et al., 2014).

V. vulnificus is known as a highly virulent pathogen. Although infections occur only sporadically, they can rapidly progress to septicemia, especially in persons with predisposing risk factors (e.g., immunocompromising conditions or chronic liver diseases resulting in elevated serum iron levels; Oliver, 2006). Foodborne infections can either result in a relatively mild gastroenteritis

or in primary septicemia with mortality rates of 61% (Shapiro et al., 1998; Oliver, 2006). A second infection route for *V. vulnificus* is through open wounds exposed to seawater. Due to the high multiplication rate of the pathogen, wound infections may quickly progress to necrotizing fasciitis, which often makes surgical debridement or amputation necessary (Daniels and Shafaie, 2000). Delayed treatment promotes progression to secondary septicemia with mortality rates about seventeen per cent (Shapiro et al., 1998; Daniels and Shafaie, 2000). Surgical interventions should be considered early to prevent a fatal outcome as poor blood perfusion in necrotic tissue can impede the achievement of effective concentrations of antimicrobial agents (Chen et al., 2012). However, to avoid septicemia and a further distribution of the pathogen additional antibiotic therapy is indispensable and should be administered as early as possible. Due to the fast progression of *V. vulnificus* infections, the presence of antimicrobial resistance preventing an effective therapy can be fatal for the patient. A combination of a tetracycline with a third-generation cephalosporin or single-agent therapy with fluoroquinolones is recommended by the CDC[2], while trimethoprim-sulfamethoxazole in combination with an aminoglycoside is proposed for the treatment of pregnant women and children.

In the U.S., 95% of all seafood-related deaths can be attributed to *V. vulnificus*, whereas infections in Germany were almost exclusively wound infections occurring after contact to seawater (Oliver, 2006; Huehn et al., 2014). So far, *Vibrio* infections in Germany occur only sporadically but incidences peaked after extreme heatwaves (Huehn et al., 2014). Due to impacts of climate change, a rise in the occurrence of *V. vulnificus* and *V. cholerae* is predicted for European waters (Baker-Austin et al., 2012). Changing demography is expected to further contribute to higher incidences of *Vibrio* infections (Baker-Austin et al., 2012). In view of these forecasts and the potential severity of infections, an investigation on antimicrobial susceptibility of *Vibrio* spp. is demanded to provide guidance for medical intervention, but also for epidemiological purposes. For this reason, our study aimed to assess antimicrobial resistance prevalence among *V. vulnificus* and *V. cholerae* non-O1/non-O139 posing a public health risk for the population. Environmental isolates were obtained from German coastal and estuarine waters of the open North Sea and the intracontinental Baltic Sea. In addition, we also included isolates from clinical sources and retail seafood for comparison and to give a more comprehensive overview of antimicrobial resistance patterns of these two species in Germany. To our knowledge, this is the first study examining antimicrobial susceptibility of *V. vulnificus* and *V. cholerae* non-O1/non-O139 in Northern Europe on a large scale.

MATERIALS AND METHODS

Bacterial Strains

The strains used in this study are summarized in **Table 1** and listed in detail in Supplementary Tables S1, S2. Antimicrobial susceptibilities were determined for a total of 325 bacterial

---

[1] http://www.cdc.gov/cholera/non-01-0139-infections.html

[2] http://emergency.cdc.gov/disasters/vibriovulnificus.asp

TABLE 1 | Origin and source of *V. cholerae* non-O1/non-O139 (*n* = 184) and *V. vulnificus* (*n* = 141) strains included in the study.

Origin	Geographical origin	Source	Source code	No. of strains
V. cholerae				
Environmental (E) (2009–2014; *n* = 131)	Baltic Sea (BS) (*n* = 79)	Seawater (sw)	E-BS-sw	54
		Sediment (sd)	E-BS-sd	4
		Seawater/sediment (sw/sd)	E-BS-sw/sd	21
	North Sea (NS) (*n* = 52)	Bivalve mollusks (bm)	E-NS-bm	26
		Seawater (sw)	E-NS-sw	12
		Seawater/sediment (sw/sd)	E-NS-sw/sd	14
Clinical (C) (1995–2012; *n* = 18)	Travel-associated (ta) (*n* = 7)	Extraintestinal (ext)	C-ta-ext	1
		Intestinal (int)	C-ta-int	6
	Germany/Austria (G/A) (*n* = 11)	Extraintestinal (ext)	C-G/A-ext	9
		Intestinal (int)	C-G/A-int	2
Retail (R) (2008–2014; *n* = 35)	Germany (G) (*n* = 35)	Bivalve mollusks (bm)	R-G-bm	2
		Crustacean (cr)	R-G-cr	26
		Fish (fi)	R-G-fi	7
V. vulnificus				
Environmental (E) (2004–2012; *n* = 122)	Baltic Sea (BS) (*n* = 70)	Seawater (sw)	E-BS-sw	46
		Sediment (sd)	E-BS-sd	24
	North Sea (NS) (*n* = 52)	Seawater (sw)	E-NS-sw	29
		Sediment (sd)	E-NS-sd	21
		Bivalve mollusks (bm)	E-NS-bm	2
Clinical (C) (1994–2011; *n* = 19)	Denmark (D)	Extraintestinal (ext)	C-D-ext	14
	Germany (G)	Extraintestinal (ext)	C-G-ext	5

strains, including 141 isolates of *V. vulnificus* (19 clinical, 122 environmental) and 184 isolates of *V. cholerae* non-O1/non-O139 (18 clinical, 131 environmental, 35 retail). The majority of environmental strains were isolated by health authorities during the German research programs KLIWAS[3] and VibrioNet[4] between 2004 and 2014. Water and sediment samples were mostly collected at bathing sites along the Baltic Sea and North Sea coastline as well as within the estuaries of the rivers Ems and Weser (Böer et al., 2012). Environmental isolates from bivalve mollusks were obtained from coastal areas of the North Sea. Isolates from retail samples were collected by health authorities of Germany and sent to the National Reference Laboratory for Monitoring Bacteriological Contamination of Bivalve Mollusks of the Federal Institute for Risk Assessment (BfR), Germany. Clinical *V. vulnificus* and *V. cholerae* non-O1/non-O139 isolates were characterized in previous studies (Bier et al., 2013; Schirmeister et al., 2014).

DNA Extraction

DNA extraction was performed with two methods that are equally applicable for *Vibrio* species. Genomic DNA of *V. vulnificus* isolates was extracted from 1 ml of an overnight culture using the RTP Bacteria DNA Kit according to the manufacturer's protocol (STRATEC Biomedical AG, Birkenfeld, Germany).

Genomic DNA of *V. cholerae* strains was extracted using a boiling method: 1.5 ml of an overnight culture were centrifuged

[3]http://www.kliwas.de/KLIWAS/EN/03_ResearchTasks/03_vh3/04_304/304_node.html
[4]http://www.vibrionet.de/

at 14,000 g for 4 min. The cell pellet was suspended in 300 μl TE buffer (10 mM Tris-HCl, 0.5 mM EDTA, pH 8), boiled for 10 min at 95°C, and subsequently cooled on ice. After centrifugation at 14,000 g for 2 min, a 200 μl aliquot of the supernatant was transferred to a new sterile tube. DNA preparations were stored at −20°C.

Species Confirmation

Species confirmation of all *V. vulnificus* and *V. cholerae* strains was carried out by species-specific *toxR* PCR amplification as previously described (Bauer and Rørvik, 2007) and in parallel by MALDI-TOF MS analysis. MALDI-TOF MS analysis was performed using a Microflex LT system mass spectrometer (Bruker Daltonik, Bremen, Germany) following the manufacturer's settings. MALDI spectra were obtained by the direct transfer method according to the manufacturer's protocol as previously described (Schirmeister et al., 2014).

Characterization of *V. cholerae* Isolates

V. cholerae isolates were characterized and subtyped via multiplex PCR targeting *rfb* sequences specific for O1 and O139 serogroups, *toxR*, and *ctxA*. PCR amplification was performed in a final volume of 25 μl with 1x PCR buffer (3 mM MgCl$_2$), 0.2 mM of each deoxynucleoside triphosphate, 0.5 μM of O1 *rfb* primers, 0.125 μM of O139 *rfb*, *toxR*, and *ctxA* primers, 1.5 U DreamTaq DNA polymerase (Thermo Fisher Scientific Biosciences GmbH, St. Leon-Rot, Germany), and 2 μl of genomic DNA. After an initial denaturation step at 94°C for 4 min, the cycling conditions were the following: 30 cycles at 94°C for 30 s, 59°C for 30 s, 72°C for 30 s, followed by a final extension step at 72°C for 5 min.

Antimicrobial Susceptibility Testing

Antimicrobial susceptibility to the following 13 antimicrobial agents was determined by broth microdilution according to the guidelines of the Clinical and Laboratory Standards Institute (CLSI, 2012a) using custom-defined microtiter plates (EUMVS2, Trek Diagnostic Systems, East Grinstead, United Kingdom): ampicillin, ceftazidime, cefotaxime, chloramphenicol, ciprofloxacin, colistin, florfenicol, gentamicin, kanamycin, nalidixic acid, streptomycin, tetracycline, and trimethoprim. Test ranges are shown in Supplementary Table S3. Additionally, all isolates were tested for their susceptibility to amoxicillin-clavulanic acid (20/10 μg), cefepime (30 μg), imipenem (10 μg), levofloxacin (5 μg), meropenem (10 μg), and sulfamethoxazole-trimethoprim (23.75/1.25 μg) by the disk diffusion method, according to the guidelines of the CLSI using commercially available disks (Oxoid GmbH, Wesel, Germany; CLSI, 2012b).

Strains showing non-susceptibility to imipenem (zone diameter $\leq$ 19) were tested against an additional panel of β-lactams by broth microdilution (imipenem, ertapenem, cefepime, cefoxitin, temocillin; EUVSEC2, Trek Diagnostic Systems) or disk diffusion (aztreonam, 30 μg).

Following the guidelines of the CLSI, tests were performed with Mueller-Hinton agar and cation-adjusted Mueller-Hinton broth without supplementation of additional sodium chloride (CLSI, 2010a). *Escherichia coli* ATCC 25922 was used for quality assurance. Minimal inhibitory concentration (MIC) values and inhibition zone diameters of all strains are listed in Supplementary Tables S1, S2. Results were interpreted using the criteria summarized in Supplementary Table S3. In general, results were interpreted according to CLSI clinical breakpoints specific for *Vibrio* spp. (CLSI, 2010a), which derived from breakpoints for *Enterobacteriaceae* (CLSI, 2010b). In cases where CLSI breakpoints for *Vibrio* spp. were obsolete or not available, the latest CLSI breakpoints for *Enterobacteriaceae* were used: document M100-S25 for aztreonam, cefepime, ertapenem, gentamicin, kanamycin, imipenem, meropenem, nalidixic acid, and trimethoprim (CLSI, 2015); document Vet01-S2 for florfenicol (CLSI, 2013). Other interpretive criteria were used for colistin (EUCAST clinical breakpoints for *Enterobacteriaceae*[5]; EUCAST, 2015), temocillin (BSAC interpretive criteria for systemic infections; Andrews, 2009), and streptomycin (based on different studies of *Vibrio* spp. and *E. coli*; National Food Institute, 2013; Shaw et al., 2014), as no CLSI breakpoints were available.

Molecular Investigation of Resistance Determinants

PCR amplification was conducted to detect specific antimicrobial resistance determinants depending on the phenotype found. Non-susceptible isolates were screened for genes mediating resistance to streptomycin (*aadA1*, *aadA2*, and *strA/B*) and β-lactams (*bla*$_{PSE-1}$, *bla*$_{OXA-1-like}$, *bla*$_{TEM-1-like}$) that are

---

[5]http://www.eucast.org

widespread in *Enterobacteriaceae* and other Gram-negative bacteria. Specifically imipenem-resistant strains (zone diameter $\leq$ 19) were tested for the presence of several carbapenemase and AmpC β-lactamase encoding genes. Presence of class 1 integrons was investigated by PCR amplification of the corresponding integrase gene *intI1* in all β-lactam and streptomycin non-susceptible strains.

Standard PCR reactions were performed using a Mastercycler EP gradient (Eppendorf, Hamburg, Germany) in a volume of 25 μl with 1x PCR buffer (2 mM MgCl$_2$), 0.2 mM of each deoxynucleoside triphosphate (dNTP), 0.2 μM of each primer, 1.5 U DreamTaq DNA polymerase, and 2 μl of genomic DNA. After an initial denaturation step at 94°C for 4 min, the cycling conditions were the following: 30 cycles of denaturation at 94°C for 30 s, primer annealing for 30 s, and extension at 72 °C for 1 min per kb, followed by a final extension step at 72°C for 10 min.

All primer pairs, target genes, corresponding annealing temperatures, and amplicon sizes are listed in Supplementary Table S4. Enterobacterial strains carrying class 1 integrons and investigated resistance determinants (*aadA1*, *aadA2*, *bla*$_{PSE-1}$, *bla*$_{OXA-1}$, *bla*$_{TEM-1}$, *bla*$_{NDM-1}$, *bla*$_{IMP-1}$, *bla*$_{VIM-2}$, *bla*$_{KPC-3}$, *bla*$_{OXA-48}$, *bla*$_{ACC-1}$, *bla*$_{CMY-1}$, *bla*$_{CMY-2}$, *bla*$_{DHA-1}$, *bla*$_{ACT-1}$, *bla*$_{FOX-1}$, *strA*, and *strB*), as well as one *V. cholerae* non-O1/non-O139 isolate carrying a class 1 integron with the *aadA1* gene were used as positive controls. Susceptible *V. cholerae* and *V. vulnificus* strains were included as negative controls.

All streptomycin resistant *V. cholerae* (VN-3469, VN-5095, VN-10191, and VN-10192) and *V. vulnificus* (VN-0098, VN-0100, VN-0125, VN-0129) isolates were further examined for mutations in the *rpsL* gene encoding ribosomal protein S12. Streptomycin susceptible (*V. cholerae*: VN-0298, VN-2997, VN-3955, VN-4226, and *V. vulnificus*: VN-0096, VN-0274, VN-3368) and intermediate resistant strains (*V. cholerae*: VN-3944, VN-4261, and *V. vulnificus*: VN-3418, VN-3981, VN-10121) were included as controls. For specific amplification and sequencing of the whole *rpsL* gene in *V. cholerae* and *V. vulnificus*, two primer pairs Vc-rpsL-F/Vc-rpsL-R and Vv-rpsL-F/Vv-rpsL-R were designed based on published genome sequences (*V. cholerae* strains NIH41, N16961, O395, and *V. vulnificus* strains CMCP6, YJ016, MO6-24/O). Purification of PCR products was performed using the MSB® Spin PCRapace Kit (STRATEC Biomedical AG, Berlin, Germany). Sequencing was conducted on both strands through sequencing service (Eurofins MWG GmbH, Ebersberg, Germany). Electropherograms were assembled and trimmed using SeqMan Pro (v12; DNASTAR Lasergene, Madison, Wisconsin). Sequences were analyzed and compared to the sequences of reference and control strains using Accelrys Gene (v2.5, Accelrys Inc., San Diego, California).

Test for Carbapenemase Activity: Carba NP Test II/Blue-Carba Test

Strains non-susceptible to imipenem (zone diameter $\leq$ 19) were grown overnight at 37°C on chromID™ CARBA (bioMérieux, Nürtingen, Germany). Bacterial colonies were subsequently tested for carbapenemase activity with the improved Carba NP test II (Dortet et al., 2014) using 12 g/L imipenem/cilastatin

(Zienam®, MSD SHARP, and DOHME GMBH, Haar, Germany) and two calibrated loops (10 μl) as bacterial inoculum to increase enzyme quantity. The Blue-CARBA test was performed as described (Pires et al., 2013) and in addition analogously to the Carba NP test II with two loops of bacterial colonies in 200 μl of the test solution in microcentrifuge tubes and supplementation of tazobactam or EDTA to inhibit class A or metallo-carbapenemases, respectively. Two *E. coli* strains positive for NDM-1 and KPC-2, respectively, as well as a KPC-3-positive *Klebsiella pneumoniae* strain served as positive controls.

Statistical Analyses

Descriptive statistics were used to analyze resistance prevalence to different antimicrobial agents (**Table 2**). Chi-square test for independence was applied with 2×2 contingency tables to test if observed differences displayed in **Table 3** were statistically significant (P-values ≤ 0.05). MIC$_{50}$ and MIC$_{90}$ were defined as the concentration at which growth of 50 and 90% of the isolates was inhibited, respectively.

RESULTS

Antimicrobial Susceptibility of Clinical and Environmental *V. vulnificus* Isolates

Fifty-seven per cent of all examined *V. vulnificus* isolates showed susceptibility to all antimicrobial agents tested (with the exception of colistin). All 141 *V. vulnificus* isolates, regardless of their origin were susceptible to quinolones, fluoroquinolones, phenicols, tetracyclines, folate pathway inhibitors, aminopenicillins with or without β-lactamase inhibitors, carbapenems, and third- and fourth- generation cephalosporins. The clinically relevant agents cefotaxime, ceftazidime, and tetracycline were among the most effective antimicrobial agents *in vitro* as they showed MIC$_{90}$ values identical to the lowest concentration tested (**Table 4**). Non-susceptibility was exclusively observed toward aminoglycosides with 40 and 3% of all strains showing intermediate resistance (MIC 32 mg/L) and resistance (MIC 64 mg/L) to streptomycin, respectively. One clinical isolate was intermediate resistant to kanamycin (MIC 32 mg/L) while all strains were susceptible

to gentamicin. The percentage of non-susceptible strains with respect to different origins is shown in **Table 3**. No significant difference was observed between clinical ($n = 19$) and environmental ($n = 122$) isolates, nor was there a significant difference between isolates from the Baltic Sea and the North Sea ($p > 0.05$, χ^2). None of the examined gene determinants encoding streptomycin resistance (*aadA1, aadA2, strA/B*), nor class 1 integrons were detected in streptomycin non-susceptible *V. vulnificus* isolates. Sequence analysis of the *rpsL* gene revealed that all four streptomycin resistant *V. vulnificus* isolates (VN-0098, VN-0100, VN-0125, VN-0129) as well as two susceptible (VN-0274, VN-3368) and two intermediate resistant isolates (VN-3918, VN-10121) carried one silent point mutation A-291-T compared to the three reference strains (CMCP6, YJ016, and MO6-24/O). An additional silent mutation C-351-T within the *rpsL* gene was observed in strain VN-0100.

Antimicrobial Susceptibility of *V. cholerae* non-O1/non-O139 Isolated from Clinical, Environmental and Seafood Samples

All 184 isolates investigated in this study were confirmed to be non-toxigenic *V. cholerae* non-O1/non-O139 isolates. The

TABLE 3 | Overall resistance occurrence in *V. cholerae* and *V. vulnificus* isolates with respect to different origins.

	Retail	Clinical	Environmental	North Sea	Baltic Sea
V. cholerae	($n = 35$)	($n = 18$)	($n = 131$)	($n = 52$)	($n = 79$)
Strains susceptible to all antimicrobial agents	27 (77%)	10 (56%)	86 (66%)	28 (54%)	58 (73%)
Non-susceptible strains	8 (23%)	8 (44%)	45 (34%)	24(46%)	21 (27%)
V. vulnificus		($n = 19$)	($n = 122$)	($n = 52$)	($n = 70$)
Strains susceptible to all antimicrobial agents		8 (42%)	72 (59%)	28 (54%)	44 (63%)
Non-susceptible strains		11 (58%)	50 (41 %)	24 (46 %)	26 (37 %)

TABLE 2 | Susceptibility vs. resistance occurrence (%) found among *V. cholerae* non-O1/non-O139 isolates from different origins[a].

Antimicrobial agent	Total ($n = 184$)			Retail ($n = 35$)			Clinical ($n = 18$)			Environmental ($n = 131$)			North Sea ($n = 52$)			Baltic Sea ($n = 79$)		
	S	I	R	S	I	R	S	I	R	S	I	R	S	I	R	S	I	R
Amoxicillin/clavulanic acid	98	2	0	100	0	0	100	0	0	97	3	0	94	6	0	99	1	0
Ampicillin	89	0	11	89	0	11	83	0	17	90	0	10	85	0	15	94	0	6
Imipenem	97	1	2	100	0	0	100	0	0	95	2	3	94	0	6	96	3	1
Meropenem	98	2	<1	100	0	0	100	0	0	97	2	1	94	4	2	99	1	0
Nalidixic acid	99	0	1	100	0	0	89	0	11	100	0	0	100	0	0	100	0	0
Streptomycin	78	20	2	86	11	3	83	17	0	75	23	2	65	29	6	81	19	0
Trimethoprim	99	0	1	97	0	3	100	0	0	100	0	0	100	0	0	100	0	0

S, susceptible; I, intermediate resistant; R, resistant.

[a]*All strains were susceptible to ceftazidime, chloramphenicol, ciprofloxacin, cefotaxime, cefepime, florfenicol, gentamicin, kanamycin, levofloxacin, trimethoprim/sulfamethoxazole and tetracycline.*

TABLE 4 | Antimicrobial MIC distributions for the *V. cholerae* and *V. vulnificus* isolates tested.

Antimicrobial agent	Test range (mg/L)	Breakpoints[a] (mg/L)			MIC (mg/L) distribution for *V. vulnificus* (*n* = 141)			MIC (mg/L) distribution for *V. cholerae* (*n* = 184)		
		S	I	R	MIC_{50}	MIC_{90}	Range	MIC_{50}	MIC_{90}	Range
Ampicillin	0.5–32	≤8	16	≥32	1	2	≤0.5–2	2	8	≤0.5–>32
Cefotaxime	0.06–4	≤1	2	≥4	≤0.06	≤0.06	≤0.06–0.12	≤0.06	≤0.06	≤0.06–0.12
Ceftazidime	0.25–16	≤4	8	≥16	≤0.25	≤0.25	≤0.25–0.5	≤0.25	≤0.25	≤0.25–0.5
Chloramphenicol	2–64	≤8	16	≥32	≤2	≤2	≤2	≤2	≤2	≤2
Ciprofloxacin	0.008–8	≤1	2	≥4	0.015	0.03	≤0.008–0.06	≤0.008	≤0.008	≤0.008–0.5
Florfenicol	2–64	≤4	8	≥16	≤2	≤2	≤2	≤2	≤2	≤2
Gentamicin	0.25–32	≤4	8	≥16	2	2	0.5–4	1	2	≤0.25–4
Kanamycin	4–128	≤16	32	≥64	8	16	≤4–32	≤4	8	≤4–16
Nalidixic acid	4–64	≤16		≥32	≤4	≤4	≤4–8	≤4	≤4	≤4–>64
Streptomycin	2–128	≤16	32	≥64	16	32	4–64	16	32	8–64
Tetracycline	1–64	≤4	8	≥16	≤1	≤1	≤1–2	≤1	≤1	≤1
Trimethoprim	0.5–32	≤8		≥16	1	1	≤0.5–4	≤0.5	1	≤0.5–>32

S, susceptible; I, intermediate resistant; R, resistant.
[a]Criteria used for interpretation and corresponding references are given in Supplementary Table S3.

majority of isolates (67%) were susceptible to all antimicrobial agents tested (with the exception of colistin). Eighteen per cent of the strains showed intermediate resistance to one or two antimicrobial agents (mostly to streptomycin) and the remaining strains (15%) showed full resistance to at least one antimicrobial agent. None of the *V. cholerae* isolates showed multidrug-resistance, defined as resistance to three or more classes of antimicrobial agents (Chen et al., 2010). Resistance profiles are shown in Supplementary Table S2, while resistance occurrence is given in **Table 2**.

As observed among *V. vulnificus*, all *V. cholerae* strains were susceptible to ciprofloxacin, chloramphenicol, florfenicol, cefotaxime, sulfamethoxazole-trimethoprim, levofloxacin, ceftazidime, cefepime, gentamicin, kanamycin, and tetracycline. Additionally, 98% of the isolates were susceptible to amoxicillin/clavulanic acid. The most effective clinically relevant agents *in vitro* were ciprofloxacin, cefotaxime, ceftazidime, and tetracycline as they showed MIC_{90} values identical to the lowest concentration tested (**Table 4**).

Similar to the *V. vulnificus* isolates, a small proportion of *V. cholerae* strains showed resistance to streptomycin (2%), while 20% of the strains were intermediate resistant. In contrast to *V. vulnificus*, the most frequent antimicrobial resistance found among all *V. cholerae* isolates was resistance to ampicillin (11%). Resistance to nalidixic acid and trimethoprim was rarely observed in two clinical isolates and in one isolate from seafood. Non-susceptibility to the carbapenems imipenem and meropenem was observed in 5 and 3% of the environmental isolates, respectively.

Clinical strains showed the highest percentage of non-susceptible strains (44%), followed by environmental strains (34%) and by strains isolated from retail seafood (23%) (**Table 3**). However, statistical analysis revealed that the observed differences to environmental isolates are not significant ($p > 0.05$, χ^2). Comparison between the geographical

origin of environmental strains revealed that the percentages of strains non-susceptible to streptomycin, ampicillin, amoxicillin/clavulanic acid, meropenem, and imipenem were higher in the North Sea compared to the Baltic Sea (**Table 2**). The higher occurrence of non-susceptible *V. cholerae* strains in the North Sea (**Table 3**) was statistically significant ($\chi 2 = 5.327$, d.f. = 1, $p < 0.05$).

Analysis of MIC distributions revealed that susceptibilities of the two *Vibrio* species were rather similar (**Table 4**). Differences were exclusively observed for kanamycin and ciprofloxacin, where MIC_{90} values of *V. cholerae* were two and four times lower and in case of ampicillin four times higher than those for *V. vulnificus*.

Susceptibility to colistin was excluded from any statistical analysis and tables, since *V. vulnificus* and *V. cholerae* possess an intrinsic resistance to colistin, which is used for selective growth on cellobiose-polymyxin B-colistin agar (Massad and Oliver, 1987). However, five *V. cholerae* strains were highly susceptible to colistin (MICs ≤ 2 mg/L) and would therefore fail to grow on this selective agar.

Neither class 1 integrons, nor gene determinants encoding streptomycin resistance (*aadA1, aadA2, strA/B*) were detected among non-susceptible *V. cholerae* isolates. Sequence analysis of the *rpsL* gene revealed three silent point mutations C-198-T, A-251-T, and G-360-A in two streptomycin resistant *V. cholerae* isolates (VN-10191, VN-10192) as well as within a susceptible isolate (VN-4226). In addition, PCR amplifications to detect β-lactamase genes (bla_{PSE-1}, $bla_{OXA-1-like}$, $bla_{TEM-1-like}$) were negative in all tested *V. cholerae* isolates.

Examination of Carbapenem Non-susceptible *V. cholerae* Isolates

Among the 131 environmental *V. cholerae* non-O1/non-O139 isolates analyzed, resistance to the carbapenem imipenem (zone diameter ≤ 19) was observed in four strains (VN-2808, VN-2825,

VN-2923, and VN-2997). These strains additionally showed resistance to ampicillin, intermediate resistance to amoxicillin-clavulanic acid, as well as intermediate or full resistance to meropenem (**Table 5**). In contrast, they were susceptible to the third- and fourth-generation cephalosporins ceftazidime, cefotaxime, and cefepime. Further characterization revealed resistance to aztreonam and ertapenem, while the strains were susceptible to temocillin and intermediate resistant to cefoxitin. The four strains grew on chromID™ CARBA agar, while growth was inhibited on chromID™ OXA-48 (bioMérieux GmbH, Nürtingen, Germany), indicating the expression of carbapenem-hydrolyzing enzymes other than OXA-48 type carbapenemases. To further investigate the presence of carbapenemases, Blue-CARBA and Carba NP II tests were conducted on these strains. Imipenem-hydrolyzing activity was detected in intact cells and crude cell extracts of each of the four strains. This activity was inhibited by tazobactam but not by EDTA, suggesting the presence of Ambler class A carbapenemases rather than class B metallo-carbapenemases or class D OXA-carbapenemases (Dortet et al., 2012).

PCRs to identify genes encoding Ambler class A carbapenemases were performed. However, no products with expected sizes were observed using primers for detection of NMC-A, SME 1-3, IMI 1-3, or KPC 1-5. In the case of NMC-A, SME, and IMI, a general failure of PCR amplification cannot be excluded as no positive control strains were available.

In addition, the strains were negative for PCR-amplification of specific genes encoding Ambler class B metallo-carbapenemases (VIM 1-2, IMP, NDM-1), class D carbapenemase (OXA-48), and AmpC-β-lactamases (MOX 1-2, CMY 1-11, LAT 1-4, BIL-1, DHA 1-2, ACC, MIR-1T, ACT-1, FOX 1-5b) as well as of class 1 integrons.

DISCUSSION

Prevalence of Antimicrobial Resistance in *V. vulnificus* and *V. cholerae* non-O1/non-O139

In this study, *V. vulnificus* isolates from German coastal waters as well as of clinical origin were susceptible to quinolones, fluoroquinolones, phenicols, tetracyclines, folate pathway inhibitors, aminopenicillins with or without β-lactamase inhibitors, carbapenems, and third- and fourth-generation

TABLE 5 | β-lactam MIC values and inhibition zone diameters found in the putative carbapenemase-producers and in carbapenem susceptible isolates selected as negative controls.

| | Antimicrobial agent | Breakpoints[d] | | | Strain ID of *V. cholerae* isolates | | | | | | |
| | | | | | Putative carbapenemase-producers | | | | Carbapenem susceptible controls | | |
		S	I	R	VN-02997	VN-02825	VN-02923	VN-02808	VN-10145	VN-00301	VN-00161
MIC values (mg/L)[a]	Ampicillin	≤8	16	≥32	>32	>32	>32	>32	2	2	2
	Cefepime	≤8	16	≥32	0.5	0.25	0.5	0.25	0.12	≤0.06	≤0.06
	Cefoxitin	≤8	16	≥32	16	16	16	16	4	4	4
	Cefotaxime	≤1	2	≥4	0.12	≤0.06	≤0.06	≤0.06	≤0.06	≤0.06	≤0.06
	Ceftazidime	≤4	8	≥16	0.5	0.5	≤0.25	≤0.25	≤0.25	≤0.25	≤0.25
	Ertapenem	≤0.5	1	≥2	2	>2	>2	2	0.12	0.03	0.06
	Imipenem	≤1	2	≥4	16	>16	>16	16	1	0.5	1
	Temocillin	≤8	-	>8	4	4	4	4	4	1	1
Inhibition zone diameter (mm)[b]	Amoxicillin/ clavulanic acid	≥18	14–17	≤13	14	13.5	14	15	23	28	18
	Aztreonam	≥21	18–20	≤17	12	15	17	16	NA	NA	29
	Cefepime	≥25	19–24	≤18	28	26	28	28	34	40	30
	Ceftazidime	≥21	18–20	≤17	28	26	28	28	NA	NA	31
	Imipenem	≥23	20–22	≤19	14	15	16	15	30	36	26
	Meropenem	≥23	20–22	≤19	20	20	20	19	34	36	28
β-lactam resistance phenotype[c]					(AMC)-AMP-ATM-ETP-(FOX)-IPM-(MEM)	(AMC)-AMP-ATM-ETP-(FOX)-IPM-(MEM)	(AMC)-AMP-ATM-ETP-(FOX)-IPM-(MEM)	(AMC)-AMP-ATM-ETP-(FOX)-IPM-MEM	–	–	–

MIC, minimal inhibitory concentration; S, susceptible; I, intermediate resistant; R, resistant; NA, not assessed; AMC, amoxicillin/clavulanic acid; AMP, ampicillin; ATM, aztreonam; ETP, ertapenem; FOX, cefoxitin; IPM, imipenem; MEM, meropenem.
[a]MIC values obtained by broth microdilution.
[b]Inhibition zone diameter obtained by disk diffusion.
[c]Results against non-β-lactams are not displayed; intermediate resistance is shown in brackets.
[d]Criteria used for interpretation with corresponding references are given in Supplementary Table S3.

cephalosporins. Non-susceptibility was exclusively observed toward aminoglycosides; predominantly streptomycin and sporadically kanamycin. Similar observations were made by Han et al. (2007), who reported total susceptibility with comparable MIC$_{90}$ values to chloramphenicol, ampicillin, ceftazidime, cefotaxime, ciprofloxacin, gentamicin, and tetracycline among *V. vulnificus* isolates from oysters of the Louisiana Gulf coast, USA (Han et al., 2007). In a recent study of *V. vulnificus* isolates from the Chesapeake Bay, USA (Shaw et al., 2014), the highest percentage of resistance was also observed against streptomycin. However, a large percentage of intermediate resistant strains to chloramphenicol (78%) and sporadically non-susceptibility to β-lactams was also reported in that study (Shaw et al., 2014). Compared to our study, higher percentages of non-susceptible strains to ampicillin, tetracycline, nalidixic acid, trimethoprim, and especially to the aminoglycosides streptomycin and gentamicin were observed in a study of 151 *V. vulnificus* isolates from South Carolina, USA, while resistance to chloramphenicol and meropenem was under one per cent (Baker-Austin et al., 2009).

Among *V. cholerae* non-O1/non-O139 isolated from clinical, environmental, and seafood samples in Germany, no multidrug-resistance was observed and the majority (67%) of isolates were susceptible to all antimicrobial agents tested. Full resistance was most frequently found toward ampicillin (11%) and streptomycin (2%). In addition, a considerable proportion of isolates showed intermediate resistance to streptomycin (20%). Resistance to nalidixic acid and trimethoprim was only sporadically found in isolates from clinical and seafood samples, respectively. While numerous studies on antimicrobial resistance of toxigenic *V. cholerae* O1, O139 strains have been published, data on *V. cholerae* non-O1/non-O139 are less frequent. A recent study showed similar antimicrobial resistance patterns among environmental *V. cholerae* non-O1/non-O139 isolates from the Chesapeake Bay, USA (Ceccarelli et al., 2015). No multidrug-resistant isolates were detected and resistance to β-lactams was found in some isolates. In one large-scale study from India on antimicrobial susceptibility of *V. cholerae* non-O1/non-O139 isolates, the highest percentage of resistant strains was also seen for ampicillin (88%) and streptomycin (85%) though with a considerably higher frequency (Kumar et al., 2009). In contrast to our study, Kumar et al. (2009) reported a high prevalence of multidrug resistance and only a small percentage of strains were susceptible to all ten antimicrobial agents tested (12%).

Several PCR analyses were performed to reveal the underlying molecular mechanisms responsible for the observed non-susceptibility to streptomycin, ampicillin, and imipenem.

Resistance to streptomycin is often mediated by enzymatic inactivation through adenylation by aminoglycoside (3″) adenyltransferases (*aadA* genes) or through phosphorylation by aminoglycoside phosphotransferases (*strA/strB* genes; Shaw et al., 1993; Tsai et al., 2014). However, none of the *Vibrio* isolates was positive for amplification of *aadA1*, *aadA2,* or *strA/B* genes that are commonly found in *Enterobacteriaceae* and that have already been identified in *V. cholerae* (Hochhut et al., 2001; Sá et al., 2010; Yu et al., 2012). Ribosomal alterations resulting from mutations in the *rpsL* gene encoding ribosomal protein S12 can be another cause of streptomycin resistance (Shaw et al., 1993; Tsai et al., 2014). However, amino acid sequences of ribosomal protein S12 were identical in all investigated strains, irrespective of the streptomycin resistance phenotype. Observed single nucleotide polymorphisms within the *rpsL* gene of some resistant as well as of some susceptible isolates were silent mutations. This indicates that other streptomycin inactivating enzymes or other resistance mechanisms, such as decreased permeability or other ribosomal alterations (e.g., mutations in the *rrs* gene encoding the 16S ribosomal RNA) may be responsible for the observed phenotype (Shaw et al., 1993; Tsai et al., 2014).

Likewise, all β-lactam non-susceptible *Vibrio* isolates were negative for amplification of bla_{PSE-1}, $bla_{OXA-1-like}$, and $bla_{TEM-1-like}$ genes encoding common β-lactamases in *Enterobacteriaceae* and other Gram-negative bacteria. This suggests that resistance may be encoded by other β-lactamase genes which are known to show a high diversity in Gram-negative bacteria. β-lactam resistance may also be mediated by other mechanisms such as reduced permeability, increased efflux, or target alterations (e.g., reduced affinity or increased amount of penicillin-binding protein (PBP); Foster, 1983).

Carbapenemase Producing *V. cholerae*

Antimicrobial susceptibility patterns as well as growth patterns on different selective media indicated the presence of a β-lactamase with carbapenem hydrolyzing activity in four environmental *V. cholerae* non-O1/non-O139 isolates from the Baltic Sea and the North Sea. The expression of carbapenemases was confirmed by positive Blue-CARBA and positive Carba NP II tests, which specifically detect imipenem hydrolyzing activity (Dortet et al., 2012; Pires et al., 2013). Inhibition of carbapenemase activity by tazobactam but not by EDTA suggested the presence of Ambler class A carbapenemases rather than class B metallo-carbapenemases or class D OXA-carbapenemases (Dortet et al., 2012). However, we found no evidence for the presence of specific genes encoding Ambler class A carbapenemases.

The four strains showed an exceptional resistance profile: They were non-susceptible to aminopenicillins, carbapenems, cefoxitin, and aztreonam and were only slightly inhibited by the β-lactamase inhibitor clavulanic acid. However, they were fully susceptible to third- and fourth-generation cephalosporins as well as to temocillin. The observed resistance to aminopenicillins coupled with susceptibility to extended spectrum cephalosporins may indicate the presence of an OXA-type carbapenemase (Ambler class D) reviewed by Walther-Rasmussen and Høiby (2006). However, with some exceptions, e.g., OXA-23, these enzymes are generally not inhibited by tazobactam, as seen for the four strains in the Carba NP II and Blue-CARBA tests and generally don't mediate resistance to aztreonam (Walther-Rasmussen and Høiby, 2006).

So far, the identity of the enzyme responsible for imipenem hydrolyzing activity in the four strains remains unclear, as none of the examined carbapenemase and AmpC-β-lactamase genes could be detected. It cannot be excluded that in addition to a carbapenem hydrolyzing enzyme other resistance mechanisms, such as reduced affinity of PBPs, porin alterations resulting in decreased membrane permeability or active efflux systems, either alone or in combination may also contribute to the observed phenotype (Walther-Rasmussen and Høiby, 2006; Queenan and Bush, 2007; Nordmann et al., 2012). Carbapenem-resistant *V. cholerae* have already been reported in other studies. NDM-1 carbapenemase was detected in a *V. cholerae* O1 El Tor Ogawa strain isolated from a 2-year old child (Mandal et al., 2012) as well as in *V. cholerae* isolated from seepage water in India (Walsh et al., 2011). Furthermore, increasing resistance to carbapenems was recently described among *V. cholerae* O1 or O139 strains isolated between 1986 and 2012 in southwest China (Gu et al., 2014).

CONCLUDING REMARKS

In this study, antimicrobial agents recommended as first choice agents for the treatment of *V. vulnificus* and *V. cholerae* non-O1/non-O139 infections such as fluoroquinolones, tetracyclines, and extended spectrum cephalosporins were found to be effective *in vitro* against both species. However, the administration of aminopenicillins, carbapenems, or aminoglycosides for treatment of *V. cholerae* non-O1/non-O139 infections, which has been reported in few studies (Daniels and Shafaie, 2000; Feghali and Adib, 2011; Lu et al., 2014) should be considered carefully, as non-susceptibility was most frequently observed against ampicillin and streptomycin and sporadically to carbapenems. For *V. vulnificus*, non-susceptibility was exclusively observed to the aminoglycosides streptomycin and kanamycin. However, gentamicin was effective against both species and could be an aminoglycoside of choice for the treatment of children and pregnant woman, as was also suggested by others (Shaw et al., 2014).

We report the detection of carbapenemase producing *V. cholerae* from different locations of the German coast line (North Sea and Baltic Sea) representing an environmental reservoir of carbapenem resistance. An entry into the sea resulting from sanitary pollution or human recreational activities cannot be excluded, but seems not likely as vibrios are indigenous bacteria of the marine environment and not intestinal commensals of humans or terrestrial animals. The strains displaying carbapenemase activity showed resistance to an unusual pattern of β-lactams. Therefore, characterization of the underlying genetic background is necessary to identify the responsible genes e.g., using whole genome sequencing as the most promising approach. Further investigations on the mobility as well as on the location of encoding genes are also needed, since location on mobile genetic elements would imply a higher risk for interspecies spread. Carbapenems are last line antimicrobial agents for treatment of multidrug-resistant Gram-negative bacteria and are of high therapeutic value (Nordmann et al., 2012). The occurrence of putative carbapenemase producing *V. cholerae* in the North and Baltic Sea is therefore of great concern and highlights the need for systematic monitoring of antimicrobial susceptibility in potentially pathogenic *Vibrio* spp. in Europe.

FUNDING

This work was supported by the Federal Ministry of Education and Research (VibrioNet, BMBF grant 01KI1015A) and the Federal Institute for Risk Assessment (BfR 46-001). The German research program KLIWAS was funded by the Federal Ministry of Transport and Digital Infrastructure.

ACKNOWLEDGMENTS

Dr. G. Hauk and Dr. O. Duty (Governmental Institute of Public Health and Social Affairs of Mecklenburg-Western Pomerania) as well as Dr. N. Brennholt and Dr. S. I. Böer (German Federal Institute of Hydrology), Dr. K. Luden and Dr. E.-A. Heinemeyer (Governmental Institute for Public Health of Lower Saxony) are greatly acknowledged for providing *Vibrio* isolates collected within the German research program KLIWAS. We thank Dr. A. Käsbohrer (Federal Institute for Risk Assessment, BfR) for providing laboratory facilities and A. Schabanowski, S. Schmoger, and T. Skladnikiewicz-Ziemer (BfR) for introduction into techniques of antimicrobial resistance testing. We further thank Dr. B.-A. Tenhagen (BfR) and Dr. B. Malorny (BfR) for advice.

REFERENCES

Andrews, J. M. (2009). BSAC standardized disc susceptibility testing method (version 8). *J. Antimicrob. Chemother.* 64, 454–489. doi: 10.1093/jac/dkp244

Baker-Austin, C., McArthur, J. V., Lindell, A. H., Wright, M. S., Tuckfield, R. C., Gooch, J., et al. (2009). Multi-site analysis reveals widespread antibiotic resistance in the marine pathogen *Vibrio vulnificus*. *Microb. Ecol.* 57, 151–159. doi: 10.1007/s00248-008-9413-8

Baker-Austin, C., Trinanes, J. A., Taylor, N. G. H., Hartnell, R., Siitonen, A., and Martinez-Urtaza, J. (2012). Emerging *Vibrio* risk at high latitudes in response to ocean warming. *Nat. Clim. Change* 3, 73–77. doi: 10.1038/nclimate1628

Bauer, A., and Røervik, L. M. (2007). A novel multiplex PCR for the identification of *Vibrio parahaemolyticus*, *Vibrio cholerae* and *Vibrio vulnificus*. *Lett. Appl. Microbiol.* 45, 371–375. doi: 10.1111/j.1472-765X.2007.02195.x

Bier, N., Bechlars, S., Diescher, S., Klein, F., Hauk, G., Duty, O., et al. (2013). Genotypic diversity and virulence characteristics of clinical and environmental *Vibrio vulnificus* isolates from the baltic sea region. *Appl. Environ. Microbiol.* 79, 3570–3581. doi: 10.1128/AEM.00477-13

Böer, S., Hauk, G., Duty, O., Luden, K., Heinemeyer, E.-A., and Brennholt, N. (2012). "Pathogenic *Vibrio* species in German coastal waters of the North Sea and the Baltic Sea – a comparison," in *Veranstaltungen International Symposium "Pathogenic Vibrio spp. in Northern European Waters"* (Koblenz: German Federal Institute of Hydrology), 36–42. doi: 10.5675/BfG_Veranst_2012.4

Ceccarelli, D., Chen, A., Hasan, N. A., Rashed, S. M., Huq, A., and Colwell, R. R. (2015). Non-O1/non-O139 *Vibrio cholerae* carrying multiple virulence factors and *V. cholerae* O1 in the Chesapeake Bay, Maryland. *Appl. Environ. Microbiol.* 81, 1909–1918. doi: 10.1128/AEM.03540-14

Chen, S. C., Lee, Y. T., Tsai, S. J., Chan, K. S., Chao, W. N., Wang, P. H., et al. (2012). Antibiotic therapy for necrotizing fasciitis caused by *Vibrio vulnificus*: retrospective analysis of an 8 year period. *J. Antimicrob. Chemother.* 67, 488–493. doi: 10.1093/jac/dkr476

Chen, X., Naren, G. W., Wu, C. M., Wang, Y., Dai, L., Xia, L. N., et al. (2010). Prevalence and antimicrobial resistance of *Campylobacter* isolates in broilers from China. *Vet. Microbiol.* 144, 133–139. doi: 10.1016/j.vetmic.2009.12.035

CLSI (2010a). *Methods for Antimicrobial Dilution and Disk Susceptibility Testing of Infrequently Isolated or Fastidious Bacteria; Approved Guideline —Second Edition M45-A2.* Wayne, PA: Clinical and Laboratory Standards Institute, CLSI.

CLSI (2010b). *Performance Standards for Antimicrobial Susceptibility Testing; Twentieth Informational Supplement M100-S20.* Wayne, PA: Clinical and Laboratory Standards Institute, CLSI.

CLSI (2012a). *Methods for Dilution Antimicrobial Susceptibility Tests for Bacteria That Grow Aerobically; Approved Standard—Ninth Edition M7-A9.* Wayne, PA: Clinical and Laboratory Standards Institute, CLSI.

CLSI (2012b). *Performance Standards for Antimicrobial Disk Susceptibility Tests; Approved Standard—Eleventh Edition M2-A11.* Wayne, PA: Clinical and Laboratory Standards Institute, CLSI.

CLSI (2013). *Performance Standards for Antimicrobial Disk and Dilution Susceptibility Tests for Bacteria Isolated From Animals; Second Informational Supplement VET01-S2.* Wayne, PA: Clinical and Laboratory Standards Institute, CLSI.

CLSI (2015). *Performance Standards for Antimicrobial Susceptibility Testing; Twenty-fifth Informational Supplement M100-S25.* Wayne, PA: Clinical and Laboratory Standards Institute, CLSI.

Daniels, N. A., and Shafaie, A. (2000). A review of pathogenic *Vibrio* infections for clinicians. *Infect. Med.* 17, 665–685.

Dortet, L., Bréchard, L., Cuzon, G., Poirel, L., and Nordmann, P. (2014). Strategy for rapid detection of carbapenemase-producing *Enterobacteriaceae*. *Antimicrob. Agents Chemother.* 58, 2441–2445. doi: 10.1128/AAC.01239-13

Dortet, L., Poirel, L., and Nordmann, P. (2012). Rapid identification of carbapenemase types in *Enterobacteriaceae* and *Pseudomonas* spp. By using a biochemical test. *Antimicrob. Agents. Chemother.* 56, 6437–6440. doi: 10.1128/AAC.01395-12

EUCAST (2015). The European Committee on Antimicrobial Susceptibility Testing. *Breakpoint Tables for Interpretation of MICs and Zone Diameters. Version 5.0.* Available online at: http://www.eucast.org.

Farmer, J. J. III, and Janda, J. M. (2004). "Family I. *Vibrionaceae*," in *Bergey's Manual of Systematic Bacteriology, 2nd Edn.*, ed G. M. Garrity (New York, NY: Springer), 491–546.

Feghali, R., and Adib, S. M. (2011). Two cases of *Vibrio cholerae* non-O1/non-O139 septicaemia with favourable outcome in Lebanon. *East. Mediterr. Health J.* 17, 722–724.

Foster, T. J. (1983). Plasmid-determined resistance to antimicrobial drugs and toxic metal ions in bacteria. *Microbiol. Rev.* 47, 361–409.

Gu, W., Yin, J., Yang, J., Li, C., Chen, Y., Yin, J., et al. (2014). Characterization of *Vibrio cholerae* from 1986 to 2012 in Yunnan Province, southwest China bordering Myanmar. *Infect. Genet. Evol.* 21, 1–7. doi: 10.1016/j.meegid.2013.10.015

Han, F., Walker, R. D., Janes, M. E., Prinyawiwatkul, W., and Ge, B. (2007). Antimicrobial susceptibilities of *Vibrio parahaemolyticus* and *Vibrio vulnificus* isolates from Louisiana Gulf and retail raw oysters. *Appl. Environ. Microbiol.* 73, 7096–7098. doi: 10.1128/AEM.01116-07

Hochhut, B., Lotfi, Y., Mazel, D., Faruque, S. M., Woodgate, R., and Waldor, M. K. (2001). Molecular analysis of antibiotic resistance gene clusters in *Vibrio cholerae* O139 and O1 SXT constins. *Antimicrob. Agents Chemother.* 45, 2991–3000. doi: 10.1128/AAC.45.11.2991-3000.2001

Huehn, S., Eichhorn, C., Urmersbach, S., Breidenbach, J., Bechlars, S., Bier, N., et al. (2014). Pathogenic vibrios in environmental, seafood and clinical sources in Germany. *Int. J. Med. Microbiol.* 304, 843–850. doi: 10.1016/j.ijmm.2014.07.010

Huhulescu, S., Indra, A., Feierl, G., Stoeger, A., Ruppitsch, W., Sarkar, B., et al. (2007). Occurrence of *Vibrio cholerae* serogroups other than O1 and O139

in Austria. *Wien. Klin. Wochenschr.* 119, 235–241. doi: 10.1007/s00508-006-0747-2

Kaper, J. B., Morris, J. G. Jr., and Levine, M. M. (1995). Cholera. *Clin. Microbiol. Rev.* 8, 48–86.

Kumar, P. A., Patterson, J., and Karpagam, P. (2009). Multiple antibiotic resistance profiles of *Vibrio cholerae* non-O1 and non-O139. *Jpn. J. Infect. Dis.* 62, 230–232.

Letchumanan, V., Chan, K. G., and Lee, L. H. (2014). *Vibrio parahaemolyticus*: a review on the pathogenesis, prevalence, and advance molecular identification techniques. *Front. Microbiol.* 5:705. doi: 10.3389/fmicb.2014.00705

Lu, B., Zhou, H., Li, D., Li, F., Zhu, F., Cui, Y., et al. (2014). The first case of bacteraemia due to non-O1/non-O139 *Vibrio cholerae* in a type 2 diabetes mellitus patient in mainland China. *Int. J. Infect. Dis.* 25, 116–118. doi: 10.1016/j.ijid.2014.04.015

Luo, Y., Ye, J., Jin, D., Ding, G., Zhang, Z., Mei, L., et al. (2013). Molecular analysis of non-O1/non-O139 *Vibrio cholerae* isolated from hospitalised patients in China. *BMC Microbiol.* 13:52. doi: 10.1186/1471-2180-13-52

Lutz, C., Erken, M., Noorian, P., Sun, S., and McDougald, D. (2013). Environmental reservoirs and mechanisms of persistence of *Vibrio cholerae*. *Front. Microbiol.* 4:375. doi: 10.3389/fmicb.2013.00375

Mandal, J., Sangeetha, V., Ganesan, V., Parveen, M., Preethi, V., Harish, B. N., et al. (2012). Third-generation cephalosporin-resistant *Vibrio cholerae*, India. *Emerg. Infect. Dis.* 18, 1326–1328. doi: 10.3201/eid1808.111686

Mansergh, S., and Zehr, J. P. (2014). *Vibrio* diversity and dynamics in the Monterey Bay upwelling region. *Front. Microbiol.* 5:48. doi: 10.3389/fmicb.2014.00048

Massad, G., and Oliver, J. D. (1987). New selective and differential medium for *Vibrio cholerae* and *Vibrio vulnificus*. *Appl. Environ. Microbiol.* 53, 2262–2264.

National Food Institute, T. U. O. D. (2013). *DANMAP 2013. Use of Antimicrobial Agents and Occurrence of Antimicrobial Resistance in Bacteria from Food Animals, Food and Humans in Denmark.*

Nordmann, P., Dortet, L., and Poirel, L. (2012). Carbapenem resistance in *Enterobacteriaceae*: here is the storm! *Trends Mol. Med.* 18, 263–272. doi: 10.1016/j.molmed.2012.03.003

Oliver, J. D. (2006). "*Vibrio vulnificus*," in *The Biology of Vibrios*, eds F. L. Thompson, B. Austin, and J. Swings, 349–366. doi: 10.1128/9781555815714.ch25

Petsaris, O., Nousbaum, J. B., Quilici, M. L., Le Coadou, G., Payan, C., and Abalain, M. L. (2010). Non-O1, non-O139 *Vibrio cholerae* bacteraemia in a cirrhotic patient. *J. Med. Microbiol.* 59, 1260–1262. doi: 10.1099/jmm.0.02 1014-0

Pires, J., Novais, A., and Peixe, L. (2013). Blue-carba, an easy biochemical test for detection of diverse carbapenemase producers directly from bacterial cultures. *J. Clin. Microbiol.* 51, 4281–4283. doi: 10.1128/JCM.01634-13

Queenan, A. M., and Bush, K. (2007). Carbapenemases: the versatile beta-lactamases. *Clin. Microbiol. Rev.* 20, 440–458. doi: 10.1128/CMR.00001-07

Sá, L. L., Fonseca, E. L., Pellegrini, M., Freitas, F., Loureiro, E. C., and Vicente, A. C. (2010). Occurrence and composition of class 1 and class 2 integrons in clinical and environmental O1 and non-O1/non-O139 *Vibrio cholerae* strains from the Brazilian Amazon. *Mem. Inst. Oswaldo Cruz* 105, 229–232. doi: 10.1590/S0074-02762010000200021

Schirmeister, F., Dieckmann, R., Bechlars, S., Bier, N., Faruque, S. M., and Strauch, E. (2014). Genetic and phenotypic analysis of *Vibrio cholerae* non-O1, non-O139 isolated from German and Austrian patients. *Eur. J. Clin. Microbiol. Infect. Dis.* 33, 767–778. doi: 10.1007/s10096-013-2011-9

Shapiro, R. L., Altekruse, S., Hutwagner, L., Bishop, R., Hammond, R., Wilson, S., et al. (1998). The role of Gulf Coast oysters harvested in warmer months in *Vibrio vulnificus* infections in the United States, 1988-1996. Vibrio Working Group. *J. Infect. Dis.* 178, 752–759. doi: 10.1086/515367

Shaw, K. J., Rather, P. N., Hare, R. S., and Miller, G. H. (1993). Molecular genetics of aminoglycoside resistance genes and familial relationships of the aminoglycoside-modifying enzymes. *Microbiol. Rev.* 57, 138–163.

Shaw, K. S., Rosenberg Goldstein, R. E., He, X., Jacobs, J. M., Crump, B. C., and Sapkota, A. R. (2014). Antimicrobial susceptibility of *Vibrio vulnificus* and *Vibrio parahaemolyticus* recovered from recreational and commercial areas of Chesapeake Bay and Maryland Coastal Bays. *PloS ONE* 9:e89616. doi: 10.1371/journal.pone.0089616

Thompson, J. R., and Polz, M. F. (2006). "Dynamics of vibrio populations and their role in environmental nutrient cycling," in *The Biology of Vibrios*, eds F. L.

Thompson, B. Austin, and J. Swings (Washington, DC: ASM Press), 190–203. doi: 10.1128/9781555815714.ch13

Tobin-D'Angelo, M., Smith, A. R., Bulens, S. N., Thomas, S., Hodel, M., Izumiya, H., et al. (2008). Severe diarrhea caused by cholera toxin-producing *Vibrio cholerae* serogroup O75 infections acquired in the southeastern United States. *Clin. Infect. Dis.* 47, 1035–1040. doi: 10.1086/591973

Tsai, Y. K., Liou, C. H., Lin, J. C., Ma, L., Fung, C. P., Chang, F. Y., et al. (2014). A suitable streptomycin-resistant mutant for constructing unmarked in-frame gene deletions using rpsL as a counter-selection marker. *PLoS ONE* 9:e109258. doi: 10.1371/journal.pone.0109258

Walsh, T. R., Weeks, J., Livermore, D. M., and Toleman, M. A. (2011). Dissemination of NDM-1 positive bacteria in the New Delhi environment and its implications for human health: an environmental point prevalence study. *Lancet Infect. Dis.* 11, 355–362. doi: 10.1016/S1473-3099(11)70059-7

Walther-Rasmussen, J., and Høiby, N. (2006). OXA-type carbapenemases. *J. Antimicrob. Chemother.* 57, 373–383. doi: 10.1093/jac/dki482

Yu, L., Zhou, Y., Wang, R., Lou, J., Zhang, L., Li, J., et al. (2012). Multiple antibiotic resistance of *Vibrio cholerae* serogroup O139 in China from 1993 to 2009. *PLoS ONE* 7:e38633. doi: 10.1371/journal.pone.0038633

Conflict of Interest Statement: The authors declare that the research was conducted in the absence of any commercial or financial relationships that could be construed as a potential conflict of interest.

15

Magnesium ions mitigate biofilm formation of *Bacillus* species via downregulation of matrix genes expression

Hilla Oknin[1,2], Doron Steinberg[2] and Moshe Shemesh[1]*

[1] Department of Food Quality and Safety, Institute for Postharvest Technology and Food Sciences, Agricultural Research Organization, The Volcani Center, Bet-Dagan, Israel, [2] Biofilm Research Laboratory, Faculty of Dental Medicine, Institute of Dental Sciences, Hebrew University-Hadassah, Jerusalem, Israel

Edited by:
Beiyan Nan,
Taxas A&M University, USA

Reviewed by:
Anushree Malik,
Indian Institute of Technology Delhi,
India
Robert Lawrence Brown,
Agricultural Research Service, USA

*Correspondence:
Moshe Shemesh,
Department of Food Quality and
Safety, Institute for Postharvest
Technology and Food Sciences,
Agricultural Research Organization,
The Volcani Center,
50250 Bet-Dagan, Israel
moshesh@agri.gov.il

The objective of this study was to investigate the effect of Mg^{2+} ions on biofilm formation by *Bacillus* species, which are considered as problematic microorganisms in the food industry. We found that magnesium ions are capable to inhibit significantly biofilm formation of *Bacillus* species at 50 mM concentration and higher. We further report that Mg^{2+} ions don't inhibit bacterial growth at elevated concentrations; hence, the mode of action of Mg^{2+} ions is apparently specific to inhibition of biofilm formation. Biofilm formation depends on the synthesis of extracellular matrix, whose production in *Bacillus subtilis* is specified by two major operons: the *epsA-O* and *tapA* operons. We analyzed the effect of Mg^{2+} ions on matrix gene expression using transcriptional fusions of the promoters for *eps* and *tapA* to the gene encoding β galactosidase. The expression of the two matrix operons was reduced drastically in response to Mg^{2+} ions suggesting about their inhibitory effect on expression of the matrix genes in *B. subtilis*. Since the matrix gene expression is tightly controlled by Spo0A dependent pathway, we conclude that Mg^{2+} ions could affect the signal transduction for biofilm formation through this pathway.

Keywords: biofilm formation, magnesium ions, food industry, *Bacillus* species, microbial development

Introduction

The vast majority of bacteria often grow as elaborate multicellular communities, referred to as biofilms (Hall-Stoodley et al., 2004; Kolter and Greenberg, 2006). Biofilm formation represents one of the most successful strategies for survival in natural environments, which protect bacteria and facilitates growth under unfavorable conditions, such as turbulent flow or limited access to nutrients (Stewart and Costerton, 2001; Hall-Stoodley et al., 2004). Biofilm formation is a multistage process in which cells adhere to a surface through production of an extracellular matrix that is typically composed of polysaccharides, proteins, and nucleic acids (Flemming and Wingender, 2010). These exopolymeric substances often surround and protect the bacteria (Shemesh et al., 2010). Thus, biofilm bacteria are more resistant than planktonic cells to various antimicrobials (Costerton, 1999; Mah and O'Toole, 2001).

Biofilms are problematic in a broad range of areas, and specifically in the food, environmental, and biomedical fields (Simoes et al., 2010). Within food industry, biofilm formation in dairy processing plants is a most significant problem. The major source of the contamination of dairy products is often associated with biofilms (Flint et al., 1997), particularly biofilms formed by members of the *Bacillus* genus (Sharma and Anand, 2002; Simoes et al., 2010). As *Bacillus* species are ubiquitously present in nature, they easily spread through food production systems, and contamination with these species is almost inevitable. The biofilm formed by thermo-resistant *Bacillus* species in a milk line can rapidly grow to such an extent that the passing milk is contaminated with cells released from the biofilm (Wirtanen et al., 1996). Thus, biofilms formed by *Bacillus* species is the major type of hygiene problems in dairy industry.

Clearly, preventing biofilm formation would be a much more desirable option than affecting it in the maturation stage, therefore a range of antimicrobial strategies have been proposed to control biofilms. However, conventional cleaning and disinfection regimens or present antimicrobial strategies may contribute to inefficient biofilm control and to the dissemination of resistance (Simoes et al., 2010). Hence, techniques that are able to prevent or control the formation of unwanted biofilms may have adverse side effects. Therefore, it necessitates looking for other methods to prevent and eradicate bacterial biofilms more successfully.

Environmental factors such as electrolyte concentrations and medium composition may have important impacts on biofilm formation (Song and Leff, 2006; Shemesh et al., 2007). Divalent cations, such as Mg^{2+} and Ca^{2+}, can influence biofilm formation directly through their effect on electro-static interactions and indirectly via physiology-dependent attachment processes by acting as important cellular cations and enzyme cofactors (Fletcher, 1988; Malik and Kakii, 2003; Song and Leff, 2006). In spite of the potentially important role, the effect of Mg^{2+} ions on bacterial adhesion and biofilm formation has barely been studied. Mg^{2+} has been shown to influence adherence to surfaces in *Pseudomonas* spp. (Simoni et al., 2000). Moreover, in *Aeromonas hydrophilia*, mutations in Mg^{2+} transport systems result in reduction of swarming and biofilm formation (Merino et al., 2001). Accordingly to Dunne and Burd (1992), 16 mM of Mg^{2+} significantly enhanced *in vitro* adhesion of *Staphylococcus epidermidis* to plastic (Dunne and Burd, 1992). In addition, it was reported that increase in Mg^{2+} concentrations affected positively on biofilm formation by *P. fluorescens* (Song and Leff, 2006). Also, it was hypothesized that low Ca^{2+} or Mg^{2+} concentrations have the potential to inhibit biofilm formation by some *A. flavithermus* and *Geobacillus* spp. strains during the processing of milk formulations (Somerton et al., 2013). Another recent study has showed that biofilm formation decreased with increasing concentration of Mg^{2+} in *Enterobacter cloacae* (Zhou et al., 2014). Although recent studies have shown that magnesium might have diverse effects on biofilms, the effect of Mg^{2+} ions on biofilm formation by sporulation *Bacillus* species remains largely unknown. Therefore, the purpose of this study was to investigate the effect of Mg^{2+} ions on biofilm formation by *Bacillus* species, which are potential importance of biofilm formation in food industrial settings.

Materials and Methods

Strains and Growth Media

The *Bacillus subtilis* wild strain NCIB3610 (Branda et al., 2001) and *Bacillus cereus* ATCC 10987 stain, obtained from Michel Gohar's lab collection (INRA, France), were used in this study. For fluorescent microscopy, we used a strain (YC161 with P_{spank}-gfp) that produced GFP constitutively (Chai et al., 2011) which was obtained from the laboratory collection of Yunrong Chai (Northeastern University, USA). For routine growth, all strains were propagated in Lysogeny broth (LB; 10 g of tryptone, 5 g of yeast extract and 5 g of NaCl per liter) or on solid LB medium supplemented with 1.5% agar. For biofilm generation, bacteria were grown to stationary phase in LB medium at 37°C in shaking culture to around 1×10^8 CFU per ml. Biofilms were generated at 30°C in the biofilm promoting medium LBGM (LB + 1% (v/v) glycerol + 0.1 mM $MnSO_4$) (Shemesh and Chai, 2013). To test the effect of magnesium, sodium or calcium ions on biofilm formation, different concentrations of either $MgCl_2$ (Merck KGaA), NaCl (BIO LAB LTD), or $CaCl_2$ (Merck KGaA) were added directly into the LBGM medium. For colony type biofilm formation, 3 μl of the cells (around 3×10^5 CFU) was spotted onto LBGM medium solidified with 1.5% agar as described previously (Shemesh and Chai, 2013). Plates were incubated at 30°C for 72 h prior to analysis. For pellicle formation, 5 μl of the cells (around 5×10^5 CFU) was mixed within 4 ml of LBGM broth in 12-well plates (Costar). Plates were incubated at 30°C for 24 h. Images were taken using a Zeiss Stemi 2000-C microscope with an axiocam ERc 5s camera.

For experiments performed with *B. cereus*, bacteria were grown to stationary phase in LB medium at 37°C in shaking culture to around 5×10^7 CFU per ml. For pellicle formation, 5 μl of the cells (around 2.5×10^5 CFU) was mixed within 4 ml of LBGM broth in glass tubes in the presence or absence of different concentration of $MgCl_2$. The glass tubes were incubated at 30°C for 24 h.

Assay of β-galactosidase Activity

To analyze the effect of magnesium ions on matrix gene expression we used transcriptional fusions of the promoters for *eps* and *tapA* to the gene encoding β galactosidase (Chai et al., 2008). Samples of generated pellicles as described above were collected and resuspended in phosphate-buffered saline (PBS) buffer. Typical long bundled chains of cells in the biofilm colony were disrupted using mild sonication as described previously (Branda et al., 2006). Optical density of the cell samples was normalized using OD_{600}. One milliliter of cell suspensions was collected and assayed for β-Galactosidase activity as described previously (Chai et al., 2008).

Growth Curve Analysis

Initially, the cells were grown in shaking cultures over night at 23°C/150 rpm in LB to around 2×10^9 CFU per ml. On the next morning, the cultures were diluted 1:100 (to around 2×10^7 CFU) into LBGM with or without addition of different concentration of $MgCl_2$ and incubated at 37°C at 150 rpm. The absorbance of the cultures at 600 nm was measured periodically for each culture for 9 h. Each condition had three replicates, and the growth curve experiments were repeated twice. Representative results are shown.

Fluorescent Microscopy Analysis

For fluorescent microscopy, we used a strain YC161 that produced GFP constitutively. The strain was first grown in shaking culture for 5 h 37°C/150 rpm in LB to around 1×10^8 CFU per ml. Next, $5 \mu l$ (around 5×10^5 CFU) of suspension from the generated culture was introduced into 4 ml of LBGM medium and incubated at 30°C for 24 h statically. Afterwards, one milliliter of suspension from each sample was collected, mildly sonicated (10 s/20% Amp/5) and centrifuged at 5000 rpm for 2 min. Next, the supernatant was removed and the pellet was resuspended by pipetation. For microscopic observation, $3 \mu l$ from the samples were transferred onto a glass slide and visualized in a transmitted light microscope using Nomarski differential interference contrast (DIC), at x40 magnification. A confocal laser scanning microscope was used to visualize GFP expression of strain YC161 using an Olympus IX81 confocal laser scanning microscope (CLSM) (Japan) equipped with 488 nm argon-ion and 543 nm helium neon lasers. For experiments performed with *B. cereus*, the cells were stained with CYTO 9 from the FilmTracer™ LIVE/DEAD Biofilm Viability Kit (Molecular Probes, OR) following instructions of the manufacturer. Fluorescence emission of the stained samples was determined using an Olympus IX81 confocal laser scanning microscope (Japan) equipped with 488 nm argon-ion and 543 nm helium neon lasers.

Statistical Analysis

Statistical analysis was performed using T-test to compare the control and tested samples. Statistical significant was determined at $P < 0.05$.

Results

The starting point of this investigation was the observation that at the elevated concentrations Mg^{2+} ions could inhibit biofilm formation by *B. subtilis*. As seen in **Figure 1**, Mg^{2+} ions inhibited notably pellicle formation by *B. subtilis* in a concentration dependent manner. The inhibitory effect of Mg^{2+} ions was not restricted to $MgCl_2$ compound since we found that other magnesium salts, such as $MgSO_4$ have also inhibited the pellicle formation (data not shown). This indicates that the inhibitory effect of magnesium salts is attributed to Mg^{2+} ions. Moreover, colony type biofilm formation was also inhibited significantly in the presence of high concentrations of Mg^{2+} ions (**Figure 1**).

To confirm that the significant inhibition in biofilm formation by Mg^{2+} ions is not a result of toxicity to bacterial cells, we tested the effect of different concentrations of Mg^{2+} ions on bacterial growth. As shown in Supplementary Figure 1, the growth curve analysis suggests about very little effect of the Mg^{2+} ions on bacterial growth at the tested concentrations; hence, the mode of action of Mg^{2+} ions is apparently specific to inhibition of biofilm formation.

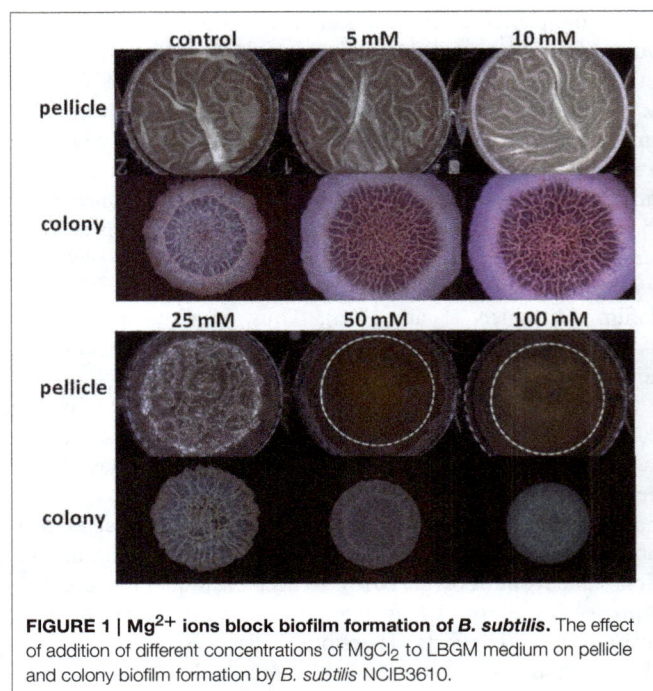

FIGURE 1 | Mg^{2+} ions block biofilm formation of *B. subtilis*. The effect of addition of different concentrations of $MgCl_2$ to LBGM medium on pellicle and colony biofilm formation by *B. subtilis* NCIB3610.

Biofilm formation depends on the synthesis of extracellular matrix, whose production in *B. subtilis* is specified by two major operons: the *epsA-O* and *tapA* operons (Kearns et al., 2005; Branda et al., 2006; Chu et al., 2006). The *epsA-O* operon is responsible for the production of the exopolysaccharides whereas the *tapA* operon is responsible for the production of amyloid-like fibers (Chai et al., 2008; Romero et al., 2010). We hypothesized that the inhibitory effect of Mg^{2+} ions on biofilm formation could be due to down-regulation of the genes involved in matrix synthesis. To test this hypothesis, we analyzed the effect of Mg^{2+} ions on matrix gene expression using transcriptional fusions of the promoters for *epsA-O* and *tapA* to the gene encoding β galactosidase. The expression of the matrix operons was notably reduced in response to the addition of Mg^{2+} ions (**Figure 2A**). The reduction in *eps* expression was relatively small (around four-fold) but significant, while *tapA* expression was decreased almost 14.5-fold at elevated concentrations of Mg^{2+} ions (**Figure 2A**). This result suggests that addition of Mg^{2+} ions down regulates expression of the extracellular matrix genes in *B. subtilis*.

Next, we visualized microscopically the effect of magnesium ions by testing bundling phenotype of fluorescently tagged *B. subtilis* cells (YC161 with P_{spank}-*gfp*), which produce GFP constitutively (Chai et al., 2011). As seen in **Figure 3**, there is significant reduction in bundling ability of *B. subtilis* cells in the presence of 25 mM $MgCl_2$ and higher. This result further confirms the potential of Mg^{2+} ions to inhibit biofilm formation by *B. subtilis*.

Subsequently, we wondered whether other common salts may also affect the biofilm formation at the same concentration as it does $MgCl_2$. Therefore, we tested the effect of NaCl and $CaCl_2$ on biofilm formation by *B. subtilis*. Notably, none of

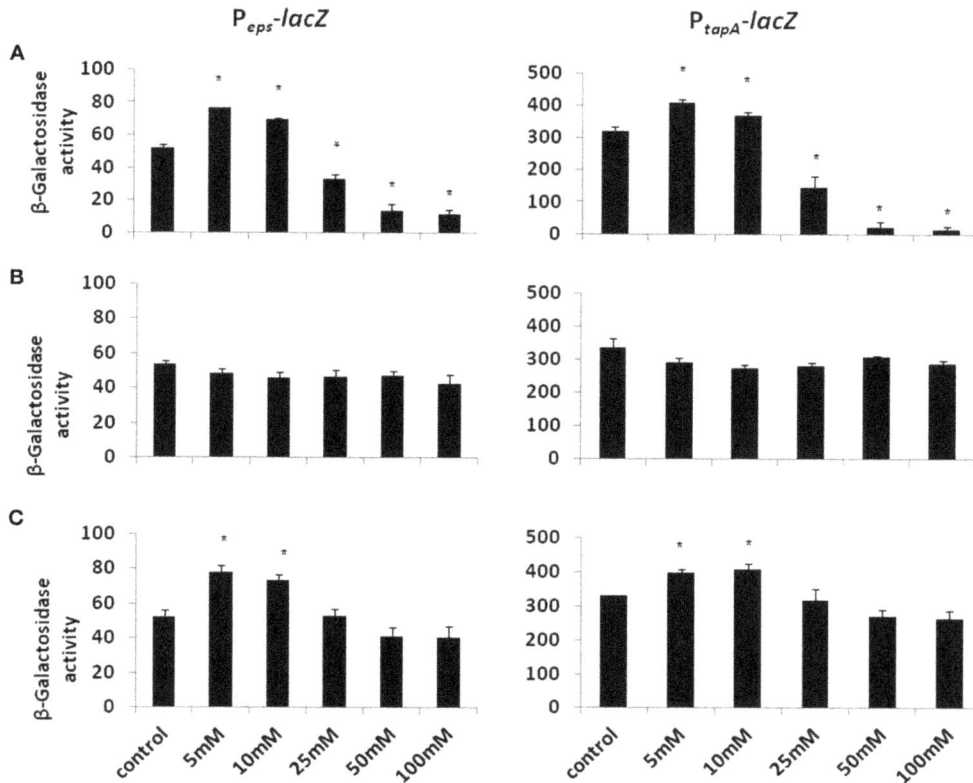

FIGURE 2 | The effect of Mg^{2+}, Ca^{2+}, and Na^+ ions on transcription of the operons responsible for the matrix production. Transcription of the operons responsible for the matrix production is differentially regulated in response to **(A)** Mg^{2+}, **(B)** Na^+, and **(C)** Ca^{2+} ions. The left panel shows results from RL4548 cells that bear the P_{eps}-lacZ transcriptional fusion and the right panel demonstrates results from RL4582 cells that bear the P_{tapA}-lacZ transcriptional fusion;. *P-value < 0.05 compared to control.

those compounds could inhibit the biofilm formation in the same manner as $MgCl_2$. Although, there was not a significant difference in pellicle formation or bundling ability of *B. subtilis* cells in the presence of NaCl at tested concentrations (**Figure 4**), nonetheless, NaCl could slightly affect the expression of *eps* operon at 100 mM concentration (**Figure 2B**). Interestingly, we detected a very slight inhibition in the pellicle formation in the presence of 50 mM or higher concentrations of $CaCl_2$ (**Figure 5**), while expression of the *eps* and *tapA* operons was found to be somewhat downregulated (**Figure 2C**). It should be noted that neither NaCl nor $CaCl_2$ affected notably bacterial growth (Supplementary Figures 2 and 3).

Discussion

It becomes increasingly clear that most of the bacteria in their natural state exist as surface associated matrix enclosed biofilms. Bacteria are much protected from environmental insults as well as various antimicrobial treatments in the biofilm mode of growth. Our results show that Mg^{2+} inhibits biofilm formation by *B. subtilis* at 25 mM and higher concentration, although at low concentrations (5 and 10 mM) Mg^{2+} enhanced biofilm formation of *B. subtilis*. Apparently, the inhibitory effect of ions

is conserved in other *Bacillus* species too. Using CLSM method, we observed a notable inhibition in biofilm formation by *B. cereus* (Supplementary Figure 5), while bacterial growth was not affected in the presence of Mg^{2+} ions (Supplementary Figure 4). Interestingly, the results of our study are in consistence with some of the previous findings regarding the effect of magnesium ions on bacterial adhesion and biofilm formation by different species. Previous studies have shown that Mg^{2+} has varying effects on bacterial adhesion (Marcus et al., 1989; Dunne and Burd, 1992; Tamura et al., 1994), which could be explained due to the difference in bacterial species and Mg^{2+} concentrations used in the various studies. For instance, Tamura et al. (1994) showed that 2 mM magnesium had no significant effect on adherence of *Streptococci*, while higher concentrations enhanced adherence to a small degree (Tamura et al., 1994). Another study showed that at 16 mM magnesium significantly enhanced the *in vitro* adhesion of *S. epidermidis* to plastic surface (Dunne and Burd, 1992). Additional study has showed that Mg^{2+} enhanced adherence of mucoid in one *P. aeruginosa* strain tested and showed no effect on the other (Marcus et al., 1989). It was further found that increase in Mg^{2+} concentrations positively influenced bacterial attachment but the effect changed over time during biofilm formation (Song and Leff, 2006).

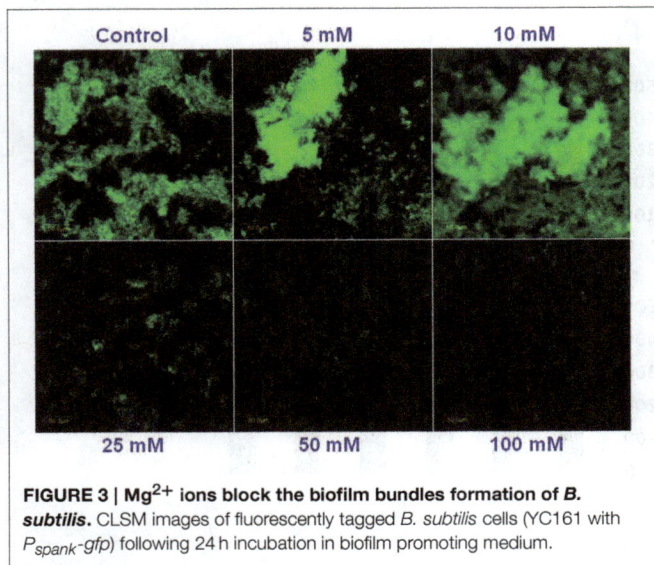

FIGURE 3 | Mg²⁺ ions block the biofilm bundles formation of B. subtilis. CLSM images of fluorescently tagged *B. subtilis* cells (YC161 with P_{spank}-*gfp*) following 24 h incubation in biofilm promoting medium.

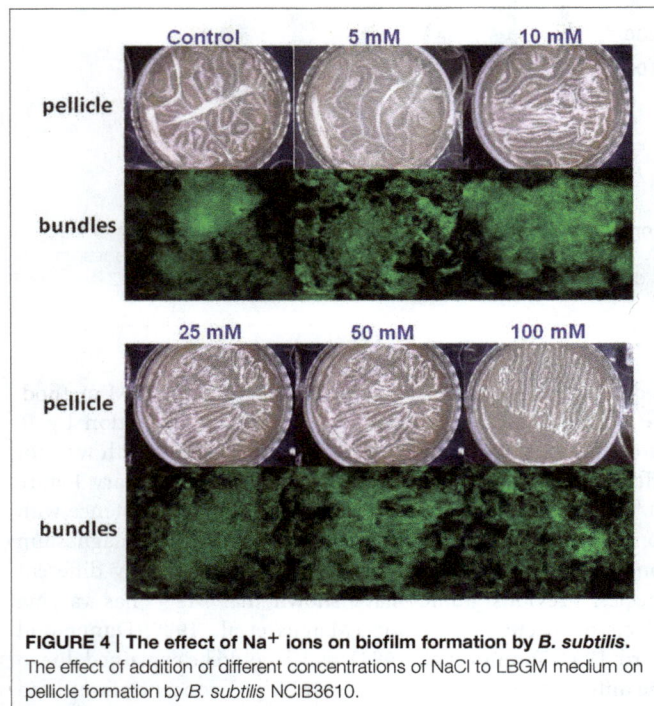

FIGURE 4 | The effect of Na⁺ ions on biofilm formation by B. subtilis. The effect of addition of different concentrations of NaCl to LBGM medium on pellicle formation by *B. subtilis* NCIB3610.

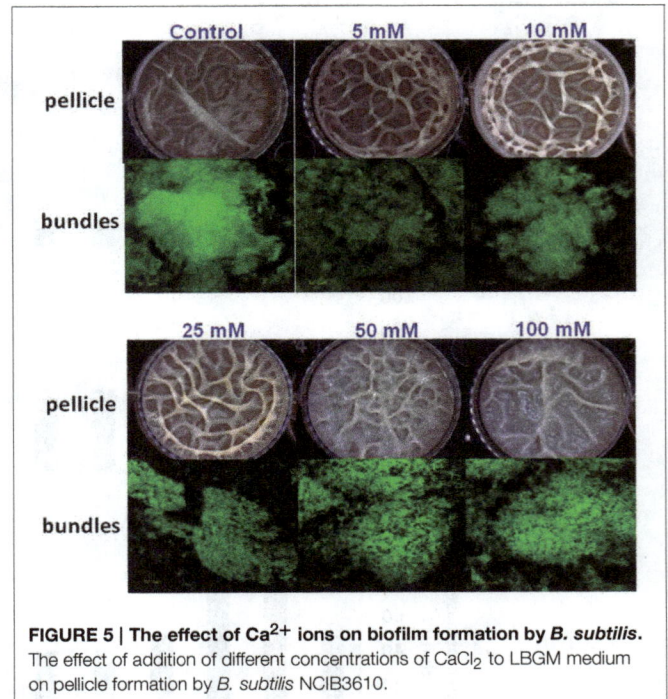

FIGURE 5 | The effect of Ca²⁺ ions on biofilm formation by B. subtilis. The effect of addition of different concentrations of CaCl₂ to LBGM medium on pellicle formation by *B. subtilis* NCIB3610.

It was proposed previously that high Mg^{2+} concentration might contribute to an increase in exopolysaccharide (EPS) production and biofilm stabilization (Costerton et al., 1995). However, it was also found that biofilm formation decreased with increasing concentration of Mg^{2+} in *E. cloacae* (Zhou et al., 2014). In our study, we showed that the expression of the two major operons responsible for biofilm matrix production were reduced notably in response to Mg^{2+} ions, suggesting about an inhibitory effect on expression of the matrix genes in *B. subtilis*. We demonstrated that Mg^{2+} ions are capable to

profoundly inhibit biofilm formation of *B. subtilis* at 25 mM concentration and higher. Since the matrix gene expression is tightly controlled by Spo0A~P dependent pathway (Shemesh and Chai, 2013), it is conceivable that Mg^{2+} ions could affect the signal transduction for biofilm formation through this pathway. Although, it is also possible that matrix gene expression is alternatively turned on by a Spo0A~P independent pathway such as YwcC-SlrA pathway (Chai et al., 2009). It will be interesting to further investigate in future studies how the inhibitory effect of Mg^{2+} ions affect a certain signaling pathway involved in biofilm formation.

In our study we decided to determine whether other divalent metal ions such as Ca^{2+} (CaCl₂) or monovalent metal ions such as Na^+ (NaCl) can inhibit biofilm formation. Our results showed that Ca^{2+} and Na^+ ions did not significantly decrease biofilm formation by *B. subtilis*. It is known that the Ca^{2+} ions have beneficial effect of on the mechanical stability of various biofilms (Rose, 2000). Moreover, it is also established that Ca^{2+} ions is important for bacterial biofilm formation (Geesey et al., 2000). Calcium is thought to promote thicker bacterial biofilms, primarily through ionic bridging of the extracellular matrix material (Rose and Turner, 1998; Körstgens et al., 2001). Previous studies have shown that the addition of Ca^{2+} caused a significant increase in *S. paucimobilis* biofilm formation at different concentration levels (Guvensen et al., 2012). Another study has shown that *Pseudoalteromonas spp.* produces larger amounts of biofilm-associated polysaccharide with increased Ca^{2+} (Patrauchan et al., 2005). Additional study showed that the amount of extracellular polysaccharide material of an alginate-producing of *Pseudomonas aeruginosa is* induced as much as eight-fold in response to Ca^{2+} ions (Sarkisova et al., 2005). It was also recently shown that Na^+ ions could also affect auto-aggregation and biofilm formation in some foodborne pathogens (Xu et al., 2010).

In overall, results of the present study show the inhibitory effect of Mg^{2+} ions on *B. subtilis* biofilm formation as well as reduction in expression of main genes involved in biofilm formation. Findings of this study can open opportunities for development of novel strategies to control biofilm formation in various settings by using small natural molecules. Hence, different magnesium salts can be used to prevent or inhibit bacterial colonization and biofilm formation of *Bacillus* species in industrial and clinical settings.

Acknowledgments

Contribution from the Agricultural Research Organization, the Volcani Center, Beit Dagan, Israel, No. 720/15-E Series is acknowledged. We would like to thank Dr. Y. Chai from the Northeastern University for *B. subtilis* strains and helpful discussions. We thank Dr. Michel Gohar for the *B. cereus* strain. We also acknowledge members of the Shemesh and Steinberg laboratories for helpful discussions and some technical assistance, especially Danielle duanis-assaf and Ievgeniia Ostrov. We are also grateful to Eduard Belausov from ARO for excellent technical assistance with confocal microscopy. This work was partially supported by the COST ACTION FA1202 BacFoodNet. The authors of the paper declare no conflict of interest.

References

Branda, S. S., Chu, F., Kearns, D. B., Losick, R., and Kolter, R. (2006). A major protein component of the *Bacillus subtilis* biofilm matrix. *Mol. Microbiol.* 59, 1229–1238. doi: 10.1111/j.1365-2958.2005.05020.x

Branda, S. S., González-Pastor, J. E., Ben-Yehuda, S., Losick, R., and Kolter, R. (2001). Fruiting body formation by *Bacillus subtilis*. *Proc. Natl. Acad. Sci. U.S.A.* 98, 11621–11626. doi: 10.1073/pnas.191384198

Chai, Y., Chu, F., Kolter, R., and Losick, R. (2008). Bistability and biofilm formation in *Bacillus subtilis*. *Mol. Microbiol.* 67, 254–263. doi: 10.1111/j.1365-2958.2007.06040.x

Chai, Y., Kolter, R., and Losick, R. (2009). Paralogous antirepressors acting on the master regulator for biofilm formation in *Bacillus subtilis*. *Mol. Microbiol.* 74, 876–887. doi: 10.1111/j.1365-2958.2009.06900.x

Chai, Y., Norman, T., Kolter, R., and Losick, R. (2011). Evidence that metabolism and chromosome copy number control mutually exclusive cell fates in *Bacillus subtilis*. *EMBO J.* 30, 1402–1413. doi: 10.1038/emboj.2011.36

Chu, F., Kearns, D. B., Branda, S. S., Kolter, R., and Losick, R. (2006). Targets of the master regulator of biofilm formation in *Bacillus subtilis*. *Mol. Microbiol.* 59, 1216–1228. doi: 10.1111/j.1365-2958.2005.05019.x

Costerton, J. W. (1999). Introduction to biofilm. *Int. J. Antimicrob. Agents* 11, 217–221. discussion: 237–239. doi: 10.1016/S0924-8579(99)00018-7

Costerton, J. W., Lewandowski, Z., Caldwell, D. E., and Korber, D. R., and Lappin-Scott, H. M. (1995). Microbial biofilms. *Annu. Rev. Microbiol.* 49, 711–745. doi: 10.1146/annurev.mi.49.100195.003431

Dunne, W. M. Jr, and Burd, E. M. (1992). The effects of magnesium, calcium, EDTA, and pH on the *in vitro* adhesion of *Staphylococcus epidermidis* to plastic. *Microbiol. Immunol.* 36, 1019–1027. doi: 10.1111/j.1348-0421.1992.tb02106.x

Flemming, H. C., and Wingender, J. (2010). The biofilm matrix. *Nat. Rev. Microbiol.* 8, 623–633. doi: 10.1038/nrmicro2415

Fletcher, M. (1988). Attachment of *Pseudomonas fluorescens* to glass and influence of electrolytes on bacterium-substratum separation distance. *J. Bacteriol.* 170, 2027–2030.

Flint, S. H., Bremer, P. J., and Brooks, J. D. (1997). Biofilms in dairy manufacturing plant - description, current concerns and methods of control. *Biofouling* 11, 81–97. doi: 10.1080/08927019709378321

Geesey, G. G., Wigglesworth-Cooksey, B., and Cooksey, K. E. (2000). Influence of calcium and other cations on surface adhesion of bacteria and diatoms: A review. *Biofouling* 15, 195–205. doi: 10.1080/08927010009386310

Guvensen, N. C., Demir, S., and Ozdemir, G. (2012). Effects of magnesium and calcium cations on biofilm formation by *Sphingomonas Paucimobilis* from an industrial environment. *Fresenius Environ. Bull.* 21, 3685–3692. doi: 10.1016/j.copbio.2013.05.185

Hall-Stoodley, L., Costerton, J. W., and Stoodley, P. (2004). Bacterial biofilms: from the natural environment to infectious diseases. *Nat. Rev. Microbiol.* 2, 95–108. doi: 10.1038/nrmicro821

Kearns, D. B., Chu, F., Branda, S. S., Kolter, R., and Losick, R. (2005). A master regulator for biofilm formation by *Bacillus subtilis*. *Mol. Microbiol.* 55, 739–749. doi: 10.1111/j.1365-2958.2004.04440.x

Kolter, R., and Greenberg, E. P. (2006). Microbial sciences: the superficial life of microbes. *Nature* 441, 300–302. doi: 10.1038/441300a

Körstgens, V., Flemming, H. C., Wingender, J., and Borchard, W. (2001). Influence of calcium ions on the mechanical properties of a model biofilm of mucoid *Pseudomonas aeruginosa*. *Water Sci. Technol.* 43, 49–57.

Mah, T. F., and O'Toole, G. A. (2001). Mechanisms of biofilm resistance to antimicrobial agents. *Trends Microbiol.* 9, 34–39. doi: 10.1016/S0966-842X(00)01913-2

Malik, A., and Kakii, K. (2003). Intergeneric coaggregations among *Oligotropha carboxidovorans* and Acinetobacter species present in activated sludge. *FEMS Microbiol. Lett.* 224, 23–28. doi: 10.1016/S0378-1097(03)00391-4

Marcus, H., Austria, A., and Baker, N. R. (1989). Adherence of *Pseudomonas aeruginosa* to tracheal epithelium. *Infect. Immun.* 57, 1050–1053.

Merino, S., Gavín, R., Altarriba, M., Izquierdo, L., Maguire, M. E., and Tomás, J. M. (2001). The MgtE Mg^{2+} transport protein is involved in *Aeromonas hydrophila* adherence. *FEMS Microbiol. Lett.* 198, 189–195. doi: 10.1111/j.1574-6968.2001.tb10641.x

Patrauchan, M. A., Sarkisova, S., Sauer, K., and Franklin, M. J. (2005). Calcium influences cellular and extracellular product formation during biofilm-associated growth of a marine *Pseudoalteromonas* sp. *Microbiology* 151(Pt 9), 2885–2897. doi: 10.1099/mic.0.28041-0

Romero, D., Aguilar, C., Losick, R., and Kolter, R. (2010). Amyloid fibers provide structural integrity to *Bacillus subtilis* biofilms. *Proc. Natl. Acad. Sci. U.S.A.* 107, 2230–2234. doi: 10.1073/pnas.0910560107

Rose, R. K. (2000). The role of calcium in oral streptococcal aggregation and the implications for biofilm formation and retention. *Biochim. Biophys. Acta* 1475, 76–82. doi: 10.1016/S0304-4165(00)00048-9

Rose, R. K., and Turner, S. J. (1998). Extracellular volume in streptococcal model biofilms: effects of pH, calcium and fluoride. *Biochim. Biophys. Acta* 1379, 185–190. doi: 10.1016/S0304-4165(97)00098-6

Sarkisova, S., Patrauchan, M. A., Berglund, D., Nivens, D. E., and Franklin, M. J. (2005). Calcium-induced virulence factors associated with the extracellular matrix of mucoid *Pseudomonas aeruginosa* biofilms. *J. Bacteriol.* 187, 4327–4337. doi: 10.1128/JB.187.13.4327-4337.2005

Sharma, M., and Anand, S. K. (2002). Biofilms evaluation as an essential component of HACCP for food/dairy industry - a case. *Food Control* 13, 469–477. doi: 10.1016/S0956-7135(01)00068-8

Shemesh, M., and Chai, Y. (2013). A combination of glycerol and manganese promotes biofilm formation in *Bacillus subtilis* via histidine kinase KinD signaling. *J. Bacteriol.* 195, 2747–2754. doi: 10.1128/JB.00028-13

Shemesh, M., Kolter, R., and Losick, R. (2010). The biocide chlorine dioxide stimulates biofilm formation in *Bacillus subtilis* by activation of the histidine kinase KinC. *J. Bacteriol.* 192, 6352–6356. doi: 10.1128/JB.01025-10

Shemesh, M., Tam, A., and Steinberg, D. (2007). Expression of biofilm-associated genes of *Streptococcus mutans* in response to glucose and sucrose. *J Med Microbiol.* 56(Pt 11), 1528–1535. doi: 10.1099/jmm.0.47146-0

Simoes, M., Simoes, L. C., and Vieira, M. J. (2010). A review of current and emerging control strategies. *LWT Food Sci. Technol.* 43, 573–583. doi: 10.1016/j.lwt.2009.12.008

Simoni, S. F., Bosma, T. N. P., Harms, H., and Zehnder, A. J. B. (2000). Bivalent cations increase both the subpopulation of adhering bacteria and their adhesion efficiency in sand columns. *Environ. Sci. Technol.* 34, 1011–1017. doi: 10.1021/es990476m

Somerton, B., Flint, S., Palmer, J., Brooks, J., and Lindsay, D. (2013). Preconditioning with cations increases the attachment of *Anoxybacillus flavithermus* and Geobacillus species to stainless steel. *Appl. Environ. Microbiol.* 79, 4186–4190. doi: 10.1128/AEM.00462-13

Song, B., and Leff, L. G. (2006). Influence of magnesium ions on biofilm formation by *Pseudomonas fluorescens*. *Microbiol. Res.* 161, 355–361. doi: 10.1016/j.micres.2006.01.004

Stewart, P. S., and Costerton, J. W. (2001). Antibiotic resistance of bacteria in biofilms. *Lancet* 358, 135–138. doi: 10.1016/S0140-6736(01)05321-1

Tamura, G. S., Kuypers, J. M., Smith, S., Raff, H., and Rubens, C. E. (1994). Adherence of group B streptococci to cultured epithelial cells: roles of environmental factors and bacterial surface components. *Infect. Immun.* 62, 2450–2458.

Wirtanen, G., Husmark, U., and Mattila-Sandholm, T. (1996). Microbial evaluation of the biotransfer potential from surfaces with *Bacillus* biofilms after rinsing and cleaning procedures in closed food-processing systems. *J. Food Prot.* 59, 727–733.

Xu, H., Zou, Y., Lee, H. Y., and Ahn, J. (2010). Effect of NaCl on the biofilm formation by foodborne pathogens. *J. Food Sci.* 75, M580–M585. doi: 10.1111/j.1750-3841.2010.01865.x

Zhou, G., Li, L. J., Shi, Q. S., Ouyang, Y. S., Chen, Y. B., and Hu, W. F. (2014). Efficacy of metal ions and isothiazolones in inhibiting *Enterobacter cloacae* BF-17 biofilm formation. *Can. J. Microbiol.* 60, 5–14. doi: 10.1139/cjm-2013-0492

Conflict of Interest Statement: The authors declare that the research was conducted in the absence of any commercial or financial relationships that could be construed as a potential conflict of interest.

16

Genetic determinants of heat resistance in *Escherichia coli*

Ryan G. Mercer[1], Jinshui Zheng[2], Rigoberto Garcia-Hernandez[1], Lifang Ruan[2], Michael G. Gänzle[1]* and Lynn M. McMullen[1]

[1] Department of Agricultural, Food and Nutritional Science, University of Alberta, Edmonton, AB, Canada, [2] State Key Laboratory of Agricultural Microbiology, Huazhong Agricultural University, Wuhan, China

Edited by:
Abd El-Latif Hesham,
Assiut University, Egypt

Reviewed by:
Teresa M. Coque,
Hospital Universitario Ramón y Cajal,
Spain
Marc William Allard,
United States Food and Drug
Administration, USA
Susan Bach,
Agriculture and Agri-Food Canada,
Canada

*Correspondence:
Michael G. Gänzle,
Department of Agricultural, Food and
Nutritional Science, University of
Alberta, 4-10 Ag/For Centre,
Edmonton, AB T6G 2P5, Canada
mgaenzle@ualberta.ca

Escherichia coli AW1.7 is a heat resistant food isolate and the occurrence of pathogenic strains with comparable heat resistance may pose a risk to food safety. To identify the genetic determinants of heat resistance, 29 strains of *E. coli* that differed in their of heat resistance were analyzed by comparative genomics. Strains were classified as highly heat resistant strains, exhibiting a D_{60}-value of more than 6 min; moderately heat resistant strains, exhibiting a D_{60}-value of more than 1 min; or as heat sensitive. A ~ 14 kb genomic island containing 16 predicted open reading frames encoding putative heat shock proteins and proteases was identified only in highly heat resistant strains. The genomic island was termed the locus of heat resistance (LHR). This putative operon is flanked by mobile elements and possesses >99% sequence identity to genomic islands contributing to heat resistance in *Cronobacter sakazakii* and *Klebsiella pneumoniae*. An additional 41 LHR sequences with >87% sequence identity were identified in 11 different species of β- and γ-proteobacteria. Cloning of the full length LHR conferred high heat resistance to the heat sensitive *E. coli* AW1.7ΔpHR1 and DH5α. The presence of the LHR correlates perfectly to heat resistance in several species of *Enterobacteriaceae* and occurs at a frequency of 2% of all *E. coli* genomes, including pathogenic strains. This study suggests the LHR has been laterally exchanged among the β- and γ-proteobacteria and is a reliable indicator of high heat resistance in *E. coli*.

Keywords: STEC, VTEC, EHEC, O157, heat resistance, beef, Cronobacter, Klebsiella

Introduction

Escherichia coli are commensals in the human and animal gut but the species also comprises intestinal and extraintestinal pathogens. The ecological versatility of *E. coli* is reflected in its genome plasticity. The average *E. coli* genome is approximately 5.16 Mb, encoding an average of 5190 genes. The core genome of *E. coli* comprises about 1700 genes (Kaas et al., 2012); however, the pan-genome of *E. coli* contains more than 18,000 genes and is still considered to be open (Rasko et al., 2008; Touchon et al., 2009; Kaas et al., 2012).

Lateral gene transfer promotes the evolution and diversity of *E. coli*, and allows acquisition of virulence factors (Dobrindt et al., 2004; Croxen et al., 2013; Gordienko et al., 2013). Genes responsible for colonization, toxin production and antibiotic resistance are encoded on mobile genetic elements and are transmitted between strains of *E. coli* (Croxen et al., 2013). One prominent example is the Shiga toxin (*stx1* or *stx2*), carried on the genomes of lambdoid prophages (O'Brien et al., 1984). The horizontal transfer of large gene clusters, called genomic islands, also provides accessory genes for niche adaptation and pathogenicity (reviewed in Schmidt and Hensel, 2004; Rasko et al., 2008; Croxen et al., 2013). The locus of enterocyte effacement (LEE) is a 35-kb genomic island coding for virulence genes for attachment and effacement of intestinal epithelial cells and other pathogenic traits (McDaniel et al., 1995). Novel combinations of accessory genes present significant challenges to public health. Transduction of an *E. coli* by a Shiga toxin-converting phage resulted in a new pathovar, enteroaggregative hemorrhagic *E. coli*, which caused a large foodborne outbreak in summer 2011 (Bielaszewska et al., 2011).

Pathovars of *E. coli* are characterized by their virulence gene profile, mechanisms for cellular adhesions, and site of colonization, and include enteropathogenic *E. coli* (EPEC), enterohemorrhagic *E. coli* (EHEC), enteroaggregative *E. coli* (EAEC), enteroaggregative hemorrhagic *E. coli* (EAHEC), enterotoxigenic *E. coli* (ETEC), and uropathogenic *E. coli* (UPEC) (Agarwal et al., 2012; Croxen et al., 2013). Due to the severity of infections and the low infectious dose, EHEC and EAHEC are particularly of concern for both public health and the food industry (Bielaszewska et al., 2011; Scallan et al., 2011; Croxen et al., 2013). EHEC carry *stx* genes and are also referred to as Shiga toxin-producing *E. coli* (STEC) (Croxen et al., 2013). STEC causes an estimated 175,000 food-borne infections per year in the United States (Scallan et al., 2011). The most frequent serotype of STEC in North America is O157:H7, but other serotypes have also been implicated in STEC infections and are food adulterants in the U.S. (USDA, 2012).

Ruminants including cattle are the primary reservoir for STEC (Lainhart et al., 2009; Croxen et al., 2013). The beef processing industry applies thermal intervention methods such as steam pasteurization and hot water washes to reduce STEC contamination of meat. However, heat resistance in *E. coli* is highly variable and some strains exhibit a stable thermotolerant phenotype (Rudolph and Gebendorfer, 2010). While most strains of *E. coli* have D_{60} values of less than 1 min, moderately or exceptionally heat resistant strains exhibit D_{60} values of more than 1 and more than 10 min, respectively (Hauben et al., 1997; Dlusskaya et al., 2011; Liu et al., 2015). The beef isolate *E. coli* AW1.7 has a D_{60} value of more than 60 min and survives in beef patties grilled to a core temperature of 71°C or "well done" (Dlusskaya et al., 2011). Heat resistance in *E. coli* AW1.7 has been attributed to the accumulation of compatible solutes (Pleitner et al., 2012) and membrane transport proteins including the outer membrane porin NmpC (Ruan et al., 2011); however, the genetic determinants of heat resistance remain unknown. This study aimed to identify genetic determinants of heat resistance in *E. coli* by comparative genomic analysis of heat sensitive, moderately heat resistant, and extremely heat resistant strains of *E. coli*. Analyses focused on food and clinical isolates of *E. coli* and included Shiga-toxin producing strains.

Materials and Methods

Strain Selection and Heat Treatments
The 29 strains of *E. coli* used in this study included previously characterized heat resistant and sensitive food and clinical isolates (Ruan et al., 2011; Liu et al., 2015). Strains were selected to include isolates differing in their heat resistance, and to include the phylogenetic variability in the species *E. coli*. All strains were grown overnight in Luria-Bertani (LB) broth at 37°C with 200 rpm agitation. To determine the heat resistance, *E. coli* strains were treated at 60°C for 5 min as previously described (Dlusskaya et al., 2011). After heating, the cultures were serially diluted, plated onto LB agar and incubated aerobically overnight at 37°C. Strains were classified into phenotypic groups based on their survival after heating. Strains with a reduction in cell counts of more than 5 log (cfu mL^{-1}) after 5 min at 60°C were classified as heat sensitive. Strains demonstrating a reduction in cell counts of 1 to 5 log (cfu mL^{-1}) were classified as moderately heat resistant while strains with reductions less than 1 log (cfu mL^{-1}) designated as highly heat resistant.

Genomic DNA Isolation, Genome Sequencing, Assembly, and Annotation
DNA for genome sequencing was isolated from overnight cultures of *E. coli* grown in 5 ml of LB broth. Genomic DNA was isolated using the Wizard® Genomic DNA Purification Kit (Promega, Madisson, Wisconsin, USA) following the manufacturer's guidelines. The quality and quantity of each sample was assessed using gel electrophoresis and a NanoDrop® 2000c spectrophotometer (Thermo Scientific, Wilmington, Delaware, USA). DNA samples were sequenced using Illumina HiSeq2000 with an insert size of 300 bp by Axeq Technologies (Seoul, South Korea). The quality of the 100-bp paired-end reads was assessed using the FastQC tool (http://www.bioinformatics.bbsrc.ac.uk/projects/fastqc) and low quality reads were filtered by Quake (Kelley et al., 2010). Assemblies were obtained using ABySS 1.3.4 (Assembly By Short Sequence; Simpson et al., 2009) with the most optimal k-mer value for each genome. Genomes were annotated automatically by the RAST server. For O157:H7 strains, the genomes assemblies were improved by scaffolding the contigs using the reference genomes of strains EDL933 (Accession: NC002655) and Sakai (Accession: NC002695). All genomic sequences of the 29 strains used in this current study are deposited to the NCBI wgs database under BioProject PRJNA277539.

Core Genome Phylogenetic Analysis and Identification of Orthologous Genes Unique to Different Phenotypes
To construct a core-genome phylogenetic tree, the 28 sequenced genomes obtained in this study were combined with 48 reference genomes obtained from NCBI Genbank (ftp://ftp.ncbi.nlm.nih.gov/genome) for a total of 76 *E. coli* and *Shigella* genomes. Reference genomes were selected to prioritize closed genomes over whole genome shotgun sequences, and to represent the phylogroups A, B1, B2, D, E, and *Shigella*. Construction of the core genome phylogenetic tree employed the previously described workflow (Touchon et al., 2009; Hazen et al., 2013) including genome alignment to identify the core genome, extraction of nucleotide sequences of the core genome, and calculation of a maximum likelihood phylogenetic tree. The genomes were aligned with Mugsy with default parameters

(Angiuoli and Salzberg, 2011). Homologous blocks present in each genome were extracted and concatenated using an in-house Perl script. The most disordered regions were eliminated using Gblocks with default parameters (Talavera and Castresana, 2007). The core genome size of the 76 genomes was approximately 2.7 Mbp. A maximum likelihood phylogenetic tree was constructed by RaxML with default parameters (Stamatakis, 2014) using bootstrapping for 1000 replicates.

To identify orthologous genes unique to the different phenotypic groups, protein sequences from all 29 genomes were combined and searched using all-against-all BLAST. The protein sequences with identities and coverage above 70% were clustered into families using the program OrthoMCL (Li et al., 2003). The inflation value of 2 was used for the MCL clustering. Sequence identity and comparisons of open reading frames (ORFs) were analyzed in Geneious (Biomatters, Auckland, New Zealand).

For phylogenetic analysis of the locus of heat resistance, genomes containing homologous sequences with >80% coverage of the ~14 kb LHR nucleotide sequence from *E. coli* AW1.7 were retrieved from NCBI. Sequences with homology to the LHR of *E. coli* AW1.7 were manually extracted and aligned with ClustalW implemented in MEGA6 (Tamura et al., 2013). The MEGA6 oftware package was used to construct a maximum-likelihood tree with default parameters. Bootstrap support values were calculated from 100 replicates.

To assess frequency of the locus of heat resistance in *E. coli*, a BLAST search of both the NCBI Genomes and whole-genome shotgun assemblies (wgs) database was performed. For each study, the number of strains containing sequence corresponding to >80% coverage was counted and totaled to give an approximate percentage of strains that were positive for the locus. Bioinformatic characterization of the genetic island was completed using BPROM (Solovyev and Salamov, 2011) and ARNold (Gautheret and Lambert, 2001) to identify putative promoters and rho-independent terminator sequences, respectively.

Genetic Complementation of the LHR

To construct a plasmid-borne copy of the LHR, primers were designed in Geneious to selectively amplify the entire genomic island in 3 separate fragments. All primers and plasmids used in this study are listed in **Table 1**. PCR reactions were carried out using Phusion High-Fidelity DNA polymerase (Thermo Scientific) according to manufacturer guidelines. F1 (~6 kb) was amplified using primer pair HR-R1/HR-R1 and included the native putative promoter sequence. F2 (~3.3 kb) and F3 (~7 kb) were amplified by primer pairs HR-F2.1/HR-R2 and HR-F3/HR-R3, respectively. F1 and F2 were cloned separately into pUC19 as KpnI/XbaI inserts, while F3 was inserted as a XbaI/HindIII fragment, yielding recombinant vectors pUCF1, pUCF2, and pUCF3 (**Table 1**). All 3 fragments were sequenced and subsequently sub-cloned into the low-copy plasmid, pRK767 (Gill and Warren, 1988), generating pRF1, pRF2, and pRF3 (**Table 1**). To construct the entire LHR on a plasmid, F1 was ligated into pUCF2 as a KpnI/SmaI fragment, resulting in pUCF1-2. The 8.3 kb insert, F1-2, was sub-cloned to pRK767 as a KpnI/XbaI fragment. The new recombinant vector, pRF1-2, and

TABLE 1 | Primers and plasmids used in this study.

Primer	Sequence (5′ → 3′)	References
HR-F1	TTAGGTACCGCTGTCCATTGCCTGA	This study
HS-R1	AGACCAATCAGGAAATGCTCTGGACC	This study
HR-R1	TATCTAGAGTCGCGTGCCAATACCAGTTC	This study
HR-F2.1	AGGGTACCAGCGATATCCGTCAATTGACT	This study
HR-F2.2	GAGGTACCTGTCTTGCCTGACAACGTTG	This study
HR-R2	TATCTAGAATGTCATTTCTATGGAGGCATGAATCG	This study
HR-F3	TATCTAGAGATGGTCAGCGCAGCG	This study
HS-F1	GCAATCCTTTGCCGCAGCTATT	This study
HR-R3	GTCAAGCTTCTAGGGCTCGTAGTTCG	This study

Plasmids	Description	References
pUC19	High-copy cloning vector	Sigma
pRK767	Low-copy cloning vector	Gill and Warren, 1988
pUCF1	LHR fragment 1 in pUC19	This study
pUCF2	LHR fragment 2 in pUC19	This study
pUCF3	LHR fragment 3 in pUC19	This study
pUCF1-2	LHR fragments 1-2 in pUC19	This study
pRF1	LHR fragment 1 in pRK767	This study
pRF2	LHR fragment 2 in pRK767	This study
pRF3	LHR fragment 3 in pRK767	This study
pRF1-2	LHR fragments F1-2 in pRK767	This study
pLHR	The entire LHR sequence, F1-F2-F3, in pRK767	This study

F3 were digested with BglII and HindIII and ligated to form pRF1-2-3, or simply, pLHR. The plasmids pRF1, pRF2, pRF3, and pLHR were electroporated into *E. coli* AW1.7ΔpHR1, a heat sensitive derivative of AW1.7 (Pleitner et al., 2012). The resulting strains, as well as the DH5α strains used for cloning, were assayed for heat resistance as previously described (Liu et al., 2015). All transformants carrying either pUC19- or pRK767-based recombinant vectors were plated on LB media containing 50 mg l^{-1} ampicillin or 15 mg l^{-1} tetracycline, respectively.

PCR Screening of Beef Isolates for Heat Resistance

A set of 55 strains of *E. coli* that were previously isolated from a beef processing plant (Dlusskaya et al., 2011) was screened for heat resistance with primers targeting the locus of heat resistance. *E. coli* AW1.7 and its heat sensitive derivative *E. coli* AW1.7ΔpHR1 (Pleitner et al., 2012) were used as positive and negative controls, respectively. Primers (**Table 1**) were designed and used to selectively target 3 separate regions (size ranging 1.8–2.8 kb) of the locus of heat resistance. The primer pairs HR-F1/HS-R1 and HR-F2.2/HR-R2 amplified regions A (~1.8 kb) and B (~2.8 kb), respectively. Primers HS-F1 and HR-R3 were used to amplify region C (~2.8 kb). Recombinant Taq® DNA polymerase (Invitrogen, Burlington, Ontario) was used to amplify products in a standard colony PCR reaction mixture and amplicons were visualized by staining with SYBRsafe (Invitrogen, Burlington, Ontario) after agarose gel electrophoresis. Heat resistance for strains *E. coli* MB1.8, DM19.2, AW1.1, GM12.6,

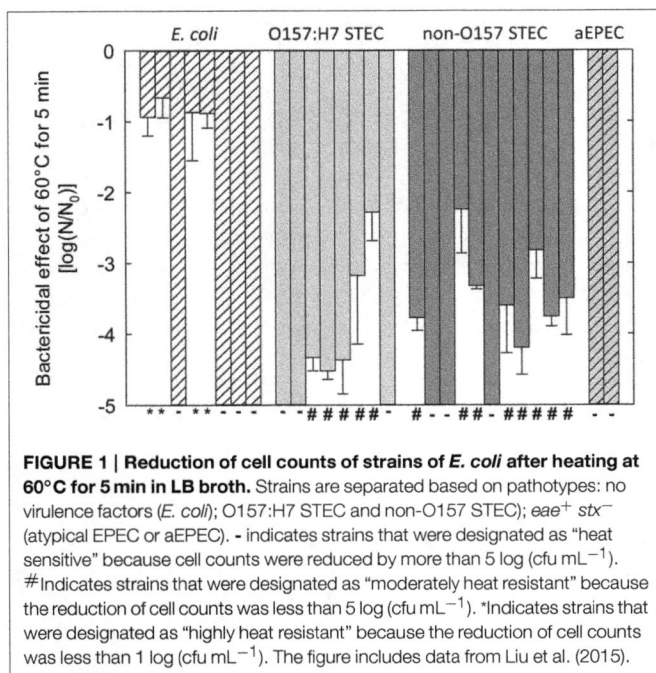

FIGURE 1 | Reduction of cell counts of strains of _E. coli_ after heating at 60°C for 5 min in LB broth. Strains are separated based on pathotypes: no virulence factors (_E. coli_); O157:H7 STEC and non-O157 STEC); _eae+ stx−_ (atypical EPEC or aEPEC). - indicates strains that were designated as "heat sensitive" because cell counts were reduced by more than 5 log (cfu mL−1). #Indicates strains that were designated as "moderately heat resistant" because the reduction of cell counts was less than 5 log (cfu mL−1). *Indicates strains that were designated as "highly heat resistant" because the reduction of cell counts was less than 1 log (cfu mL−1). The figure includes data from Liu et al. (2015).

MB 16.6, MB 3.5, GM9.1, and AW 12.2 (Dlusskaya et al., 2011) was determined by incubation at 60°C for 5 min and enumeration of surviving cells as described above.

Results

Heat Resistance of _E. coli_

Strains of _E. coli_ were selected for genome sequencing to obtain a wide range of heat resistance despite the limited number of strains (**Figure 1**). In **Figure 1**, strains are grouped based on their virulence profiles. O157:H7 STEC and non-O157 STEC were grouped based on serotype and the presence of _stx1_ or _stx2_ coding for Shiga toxins (Liu et al., 2015). Strains from four groups included heat sensitive strains (**Figure 1**), in agreement with the heat sensitivity of a majority of strains of _E. coli_ (Hauben et al., 1997; Dlusskaya et al., 2011; Liu et al., 2015). Both O157:H7 STEC and non-O157:H7 STEC included moderately heat resistant strains (**Figure 1**). Four non-pathogenic strains of _E. coli_ including _E. coli_ AW1.7 were highly heat resistant. All of these strains were obtained from a beef processing facility (Dlusskaya et al., 2011).

Genome Sequences and Characteristics

The 29 _E. coli_ genomes sequences obtained in this study included genomes from 4 highly heat resistant strains, 13 moderately resistant strains, and 12 heat sensitive strains (**Figure 1, Table 2**). Genebank accession numbers of the 29 genomes sequenced in this study are indicated in **Table 2**. The number of contigs larger than 500 bp per genome ranged from 95 to 277, with max sequence lengths ranging from 204263–435416 bp (**Table 2**).

Genome sequence data confirmed the presence or absence of _stx1/stx2_ and _eae_ that was determined earlier by PCR (Liu et al.,

2015, **Table 2**). The atypical EPEC (aEPEC) carried the _eae_ gene, but no pEAF-encoded _bfp_ (bundle-forming pilus) genes (Trabulsi et al., 2002). Other strains of _E. coli_ were negative for _eae_, _stx1/2_ and _bfp_. None of the highly heat resistant strains of _E. coli_ carried any virulence factors (**Table 2**). The genomes of _E. coli_ AW1.7 and its heat-sensitive derivative _E. coli_ AW1.7ΔpHR1 (Pleitner et al., 2012) were virtually identical; however, in addition to the loss of the 4842 bp plasmid pHR1, two additional deletions of 21768 and 16248 bp were identified in the heat sensitive _E. coli_ AW1.7ΔpHR1. Of the 19 STEC, 16 possessed the _eae_ gene/LEE pathogenicity island; the remaining 3 STEC were categorized as LEE negative STECs (**Table 2**), which still have the ability to cause disease (Newton et al., 2009). The 19 STEC included moderately heat resistant and heat sensitive strains (**Table 2**). A moderately heat resistant STEC isolate from the 2011 Germany outbreak, O104:H4 11-3088, carried _stx2_, as well as a gene encoding a β-lactamase from the EHEC plasmid, pHUSEC2011. This plasmid also encodes EAEC virulence factors such as the _aaf_ and _agg_ genes (Estrada-Garcia and Navarro-Garcia, 2012).

Phylogenetic Distribution of Heat Resistant Isolates

To assess the phylogenetic relationships of the heat resistant and sensitive strains, a core genome phylogenetic tree was constructed with the genomes from this study, and 48 obtained from the NCBI database. The _E. coli_ phylogenetic groups A, B1, B2, D and E (Jaureguy et al., 2008; Touchon et al., 2009) were well supported by our core genome tree (**Figure 2**).

Moderately heat resistant strains were found in the phylogenetic groups A, B1, and E (**Figure 2**). Resistant and sensitive strains of the serotype O157H7 and O26:H11, 05-6544 and PARC448, respectively, group together near NCBI strains of similar serotypes. This grouping of heat resistant and sensitive isolates occurs with O145:NM isolates as well, however, these strains are found distinctly separate from other phylogenetic groups (**Figure 2**). Some moderately resistant, non-O157 STEC are located on branches with pathogenic _E. coli_ including O104:H4 11-3088 (**Figure 2**). The overall genomic similarity of sensitive and resistant strains may illustrate the ease of acquiring genetic variations to become moderately heat resistant. Particularly strains within phylogenetic group E, comprising O157:H7 STEC (**Figure 2**), are highly related and therefore the differences in the accessory genes, content or sequence, accounts for differences in heat resistance.

All four highly resistant strains were assigned to group A. The highly heat resistant strains _E. coli_ AW1.7 and GM16.6 are located in divergent branches separate from other _E. coli_ in this group (**Figure 2**). _E. coli_ AW1.3 shares a high degree of sequence similarity to _E. coli_ P12b, a model strain for flagellar studies (Ratiner et al., 2010), while _E. coli_ DM18.3 is closely related to the commensal _E. coli_ strain HS (Levine et al., 1978). The phylogenetic diversity of highly heat resistant strains indicates that these strains do not share a common ancestor (**Figure 2**).

Identification the Locus of Heat Resistance (LHR)

To identify differences in gene content conferring high heat resistance, the genomes were separated into their phenotypic

TABLE 2 | *E. coli* strain used in this study and features of their genome sequences.

Accession Numbers	Strain (references); phylogenetic group	Coverage (X)[a]	Number of contigs assembled	Max contig size (bp)	Number of putative proteins[b]	Heat Resistance	EHEC Virulence Factors	Origin
LDYI00000000	AW1.3 (1); A	579.12	184	317424	5041	High	n.d.[c]	Beef
LDYL00000000	DM18-3 (2); A	469.26	111	353387	4700	High	n.d.	Beef
LDYM00000000	GM16-6 (2); A	442.89	164	209077	4678	High	n.d.	Beef
LDYJ00000000	AW1.7 (1); A	494.63	165	245564	4971	High (1)	n.d.	Beef
LDYK00000000	AW1.7ΔpHR1; A	519.61	152	246187	4952	Sensitive (3)	n.d.	AW1.7 mutant
LECO00000000	O103:H2 PARC444 (2); B1	471.64	98	356993	4864	Sensitive (2)	n.d.	Unknown
LECG00000000	O103:H2 PARC445 (2); B1	584.81	158	327869	5096	Sensitive (2)	n.d.	Unknown
LECL00000000	O44:H18 PARC450 (2); E	458.44	146	343811	4951	Sensitive (2)	n.d.	Unknown
LEAF00000000	O157:H7 CO6CE1353 (2); D	484.64	205	376588	5572	Moderate (2)	*stx1 stx2 eae*	Clinical
LEAG00000000	O157:H7 CO6CE1943 (2); D	477.76	185	374853	5436	Moderate (2)	*stx1 stx2 eae*	Clinical
LEAH00000000	O157:H7 CO6CE2940 (2); D	475.57	197	376618	5537	Moderate (2)	*stx2 eae*	Clinical
LEAE00000000	O157:H7 CO6CE900 (2); D	470.23	225	376513	5554	Moderate (2)	*stx2 eae*	Clinical
LEAJ00000000	O157:H7 E0122 (2); D	480.56	189	399998	5478	Moderate (2)	*stx2 eae*	Cattle
LEAD00000000	O157:H7 1935 (2); D	502.88	194	393069	5523	Sensitive (2)	*stx1 stx2 eae*	Human
LEAI00000000	O157:H7 CO283 (2); D	531.16	184	376583	5296	Sensitive (2)	*stx1 stx2 eae*	Cattle
LEAK00000000	O157:H7 LCDC7236 (2); D	492.65	181	376583	5461	Sensitive (2)	*stx1 stx2 eae*	Human
LDYN00000000	O26:H11 05-6544 (2)	426.65	280	219684	5691	Moderate (2)	*stx1 eae*	Human
LECF00000000	O103:H25 338 (2); B1	439.31	218	376897	5321	Moderate (2)	*stx1 eae*	Clinical
LECH00000000	O104:H4 11-3088 (2); B1	515.77	173	320350	5254	Moderate (2)	*stx2*[d]	Human
LECI00000000	O111:NM 583 (2); B1	492.35	185	323305	5067	Moderate (2)	*stx1 eae*	Clinical
LECK00000000	O113:H4 09-0525 (2); A	475.86	165	254878	5275	Moderate (2)	*stx1 stx2*	Unknown
LDZZ00000000	O121:H19 03-2832 (2); B1	457.58	213	434838	5272	Moderate (2)	*stx2 eae*	Human
LEAA00000000	O121:NM 03-4064 (2); B1	568.02	221	435416	5298	Moderate (2)	*stx2 eae*	Human
LEAB00000000	O145:NM 03-6430 (2); n.a.	528.20	210	359240	5371	Moderate (2)	*stx1 eae*	Human
LECM00000000	O45:H2 05-6545 (2); B1	508.74	263	261384	5352	Sensitive (2)	*stx1 eae*	Human
LECN00000000	O76:H19 09-0523 (2); B1	456.09	191	404223	5432	Sensitive (2)	*stx1 stx2*	Unknown
LECJ00000000	O111:NM PARC447 (2); B1	544.42	200	376589	5672	Sensitive (2)	*stx1 stx2 eae*	Unknown
LDYO00000000	O26:H11 PARC448; B1	489.45	240	204263	5429	Sensitive (2)	*eae*	Unknown
LEAC00000000	O145:NM PARC449 (2); n.a.	502.50	181	328848	5390	Sensitive (2)	*eae*	Unknown

n.a., not assigned;
[a] *Based on the Lander-Waterman equation using an average size of E. coli genome (5.16 Mb);*
[b] *Based on OrthoMCL analysis of all annotated genes;*
[c] *n.d., not detected;*
[d] *Carries at least the beta lactamase gene present on pHUSEC2011-2 present in EAEC. Other genes on this plasmid includes factors for adhesion.*
References: (1) Dlusskaya et al., 2011; (2), Liu et al., 2015; (3) Pleitner et al., 2012.

groups: highly resistant; moderately resistant; and sensitive. An OrthoMCL analysis found 3147 orthologs shared among all 28 genomes, however, none of these were unique to heat sensitive or moderately heat resistant strains (**Figure 3**). A set of 6 genes was unique to the highly heat resistant strains (**Figure 3**); all of these genes are located on a 14,469 bp genomic island present in *E. coli* AW1.7, AW1.3, DM18.3, and GM16.6 (**Figure 4**). The 6 genes specific to the highly heat resistant group are scattered among an additional 10 ORFs in this genomic island (**Figure 4**). Remarkably, this genomic island was absent in *E. coli* AW1.7ΔpHR1. The plasmid curing protocol used to generate *E. coli* AW1.7ΔpHR1 (Pleitner et al., 2012) thus also resulted in a 16,248 bp deletion encompassing the genomic island and the flanking mobile genetic elements. This operon was previously identified in heat resistant strains of *Cronobacter sakazakii*

(Gajdosova et al., 2011) and *Klebsiella pneumoniae* (Bojer et al., 2010). Due to its presence in highly heat resistant *E. coli,* the genomic island was named the locus of heat resistance (LHR).

LHR Confers High Heat Resistance to Heat Sensitive *E. coli*

To verify that high heat resistance in *E. coli* is mediated by proteins encoded by the LHR, the heat sensitive *E. coli* AW1.7ΔpHR1 and DH5α were complemented with LHR or fragments of LHR. LHR or LHR fragments were introduced in *E. coli* AW1.7ΔpHR1 and DH5α after cloning into the low-copy vector pRK767. Fragments F1, F2 and F3 encompassed about 6, 3.3, and 8 kbp, respectively. Cloning of the empty plasmid pRK767 served as control and the heat resistance of the resulting derivatives of *E. coli* AW1.7ΔpHR1 and DH5α was compared to the wild type strains (**Figure 5**). Cloning the low copy number plasmid pRK767 into *E. coli* AW1.7 did not affect the strain's heat resistance (**Figures 1**, **5**). Strains expressing the full length LHR were as heat resistant as *E. coli* AW1.7 while strains with plasmids containing only a portion of the LHR remained heat sensitive (**Figure 5**). Complementation of *E. coli* AW1.7ΔpHR1 with the plasmid pHR1 did not alter heat resistance of the strain (Bédié et al., 2012 and data not shown), confirming that the loss of the

LHR rather than the loss of the plasmid pHR1 are responsible for the heat sensitive phenotype of this strain.

Genes Encoded by the LHR

The LHR codes for 16 putative ORFs (**Figure 4A**, Table S1): 2 small heat-shock proteins (sHSPs); a Clp protease (Bojer et al., 2013); several hypothetical proteins with predicted transmembrane domains; a putative sodium/hydrogen

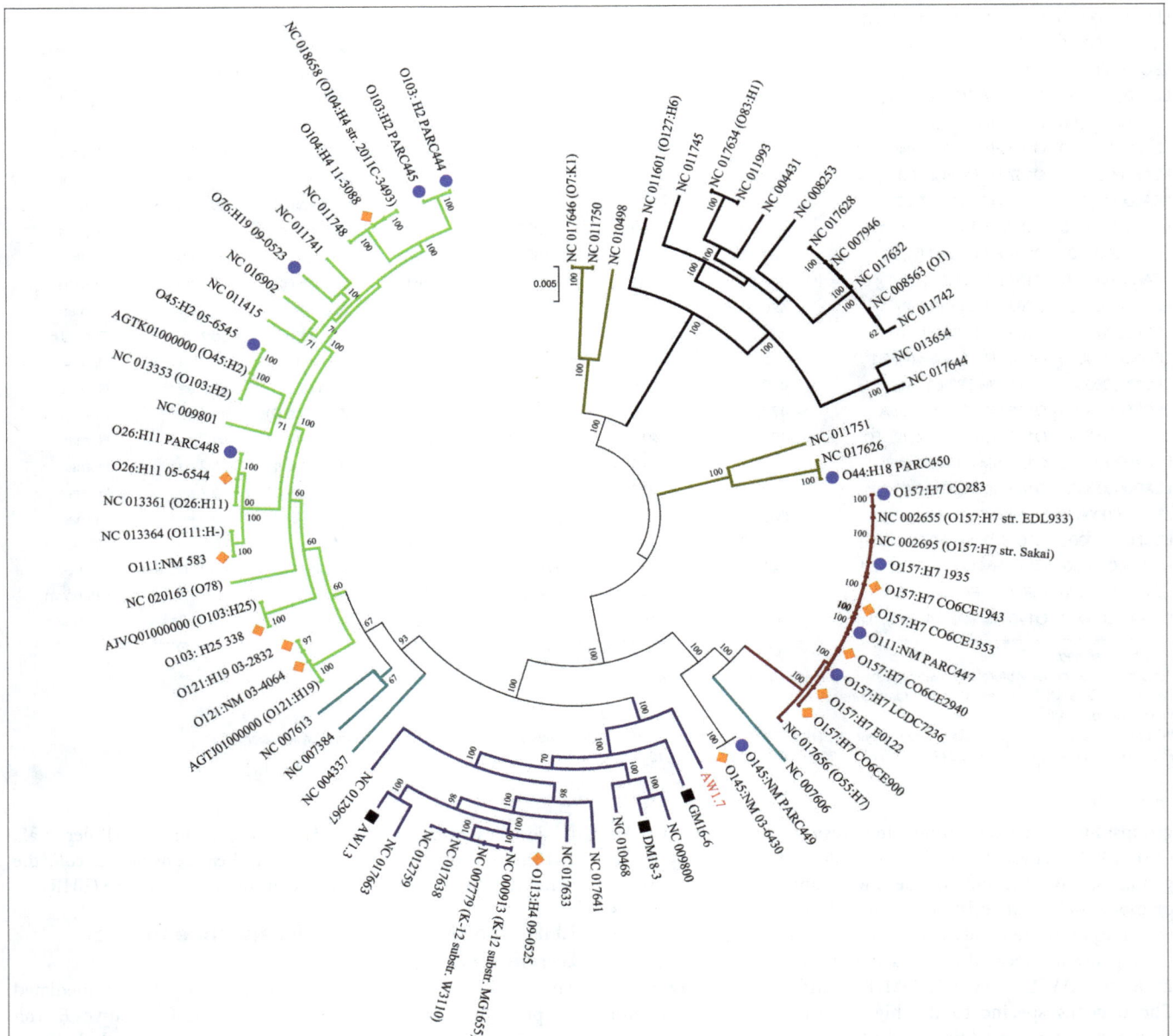

FIGURE 2 | Phylogenomic distribution of strains of *E. coli* and *Shigella* spp. A core genome phylogenetic tree was constructed using the 28 sequenced genomes from this project, indicated by strain numbers and serotype as applicable, and 48 genome sequences from NCBI, indicated by serotype and Accession numbers. The strain numbers of the 48 strains are indicated in Table S2. The phylogenetic groups of *E. coli* are color coded: A, Blue; B1, Green; B2, Black; D, Brown; and E, Maroon; *Shigella* spp. (indicated by teal colored branches) were included in the phylogenetic tree because this genus is considered a host-adapted pathovar of *E. coli*. Bootstrapping values are indicated for each branch. The sequenced genomes from this project are coded by blue circles and orange diamonds indicating heat sensitive and moderately heat resistant strains, respectively. Black squares represent highly heat resistant strains.

exchanger; and several peptidases. **Figure 4** compares the operons in *E. coli*, *C. sakazakii*, and *K. pneumonia*. The predicted function and the conserved functional and transmembrane

domains of the predicted proteins are shown in Table S1. The conservation of the ORFs among *E. coli*, *C. sakazakii*, and *K. pneumonia* is remarkable; most ORFs share more than 99% nucleotide identity to the corresponding genes in *E. coli* AW1.7 (**Figure 4B**). *E. coli* AW1.7 and AW1.3 share 100% nucleotide identity for 10 of the 16 ORFs (**Figure 4B**). In *E. coli* AW1.7, the strongest predicted promoter was located 63 bp upstream from ORF1. BPROM analysis predicted that the transcription factor OmpR interacts with this promoter. Another putative promoter is located 26 bp upstream from ORF 9 (**Figure 4A**). One predicted rho-independent terminator was oriented in the same direction as the ORFs and located 177 bp downstream from ORF 16 (**Figure 4A**).

In all four strains of *E. coli*, the LHR is flanked by mobile elements or putative transposases (**Figure 4B** and data not shown). Accordingly, the GC content of the island is 61.8%, substantially higher than the *E. coli* average of ~50% (**Figure 4A**). In *C. sakazakii* and *K. pneumonia*, the LHR is located on plasmids (Bojer et al., 2010; Gajdosova et al., 2011); however, none of the *E. coli* strains in this study possess plasmids larger than 14 kb (data not shown) and the LHR can thus be assumed to be encoded by the chromosome in the strains of *E. coli* analyzed here. The high degree of sequence identity of the LHR in different species of *Enterobacteriaceae*, the presence of mobile genetic elements adjacent to the LHT, and the divergent GC content suggest that the LHR was acquired by lateral gene transfer.

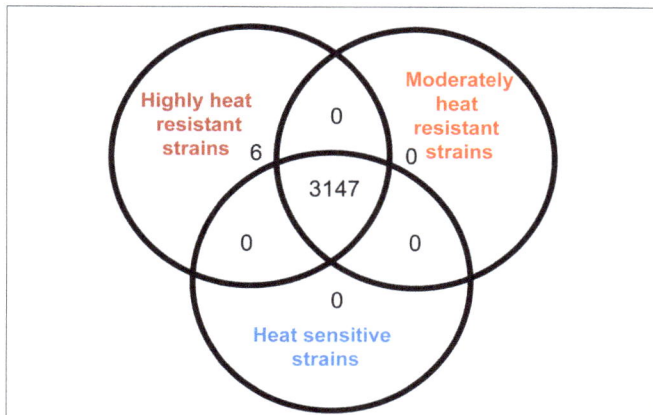

FIGURE 3 | Analysis of orthologous protein coding sequences identified in highly heat resistant, moderately heat resistant and heat sensitive *E. coli* strains by OrthoMCL. The Venn diagram indicates the number of protein coding sequences that are shared by all strains analysed in this study, the number of protein coding sequences that were shared between any two of the phenotypic groups, and the number of protein coding sequences that were found only in one of the three phenotypic groups.

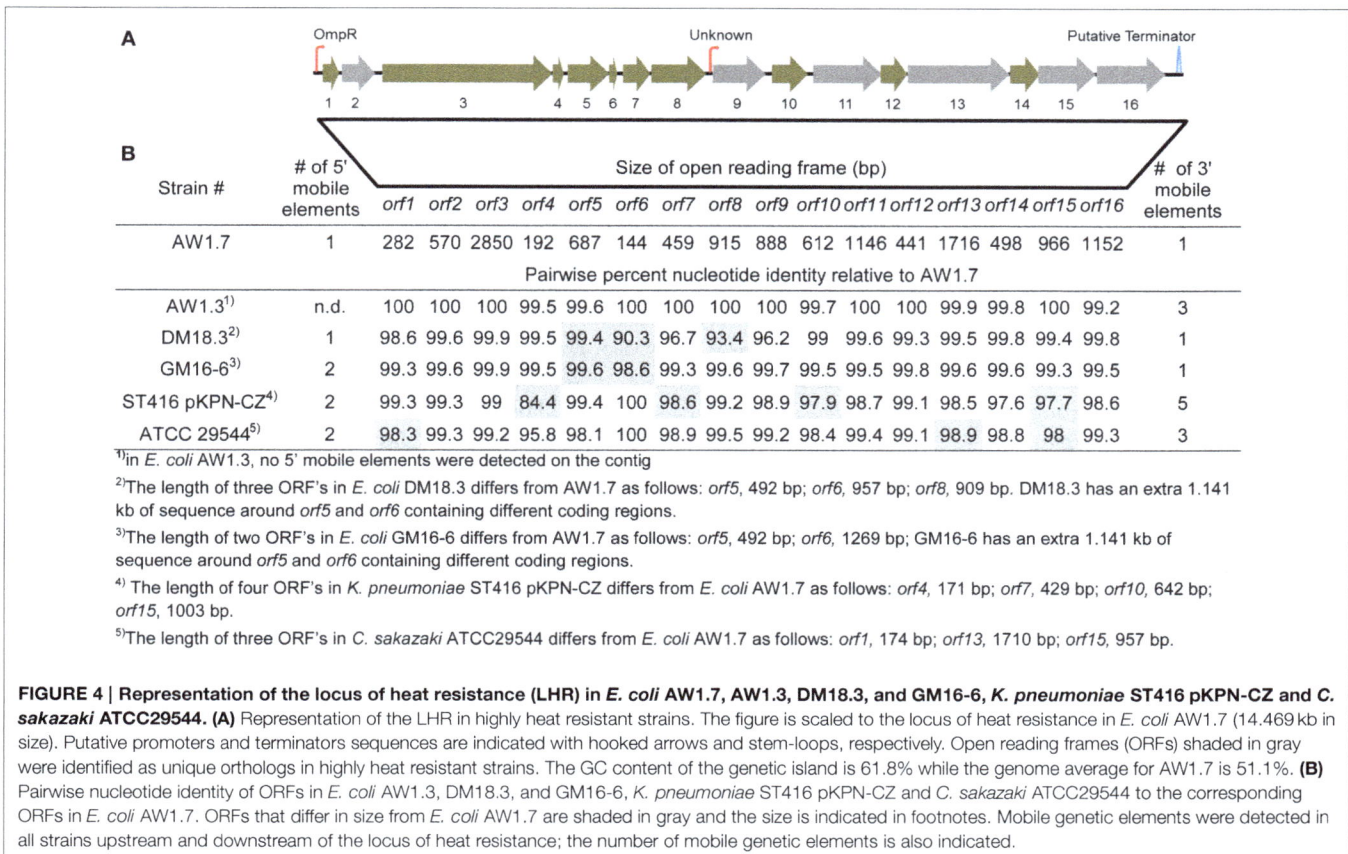

Strain #	# of 5' mobile elements	Size of open reading frame (bp)																# of 3' mobile elements
		orf1	orf2	orf3	orf4	orf5	orf6	orf7	orf8	orf9	orf10	orf11	orf12	orf13	orf14	orf15	orf16	
AW1.7	1	282	570	2850	192	687	144	459	915	888	612	1146	441	1716	498	966	1152	1
		Pairwise percent nucleotide identity relative to AW1.7																
AW1.3[1]	n.d.	100	100	100	99.5	99.6	100	100	100	100	99.7	100	100	99.9	99.8	100	99.2	3
DM18.3[2]	1	98.6	99.6	99.9	99.5	99.4	90.3	96.7	93.4	96.2	99	99.6	99.3	99.5	99.8	99.4	99.8	1
GM16-6[3]	2	99.3	99.6	99.9	99.5	99.6	98.6	99.3	99.6	99.7	99.5	99.5	99.8	99.6	99.6	99.3	99.5	1
ST416 pKPN-CZ[4]	2	99.3	99.3	99	84.4	99.4	100	98.6	99.2	98.9	97.9	98.7	99.1	98.5	97.6	97.7	98.6	5
ATCC 29544[5]	2	98.3	99.3	99.2	95.8	98.1	100	98.9	99.5	99.2	98.4	99.4	99.1	98.9	98.8	98	99.3	3

[1] in *E. coli* AW1.3, no 5' mobile elements were detected on the contig

[2] The length of three ORF's in *E. coli* DM18.3 differs from AW1.7 as follows: *orf5*, 492 bp; *orf6*, 957 bp; *orf8*, 909 bp. DM18.3 has an extra 1.141 kb of sequence around *orf5* and *orf6* containing different coding regions.

[3] The length of two ORF's in *E. coli* GM16-6 differs from AW1.7 as follows: *orf5*, 492 bp; *orf6*, 1269 bp; GM16-6 has an extra 1.141 kb of sequence around *orf5* and *orf6* containing different coding regions.

[4] The length of four ORF's in *K. pneumoniae* ST416 pKPN-CZ differs from *E. coli* AW1.7 as follows: *orf4*, 171 bp; *orf7*, 429 bp; *orf10*, 642 bp; *orf15*, 1003 bp.

[5] The length of three ORF's in *C. sakazaki* ATCC29544 differs from *E. coli* AW1.7 as follows: *orf1*, 174 bp; *orf13*, 1710 bp; *orf15*, 957 bp.

FIGURE 4 | Representation of the locus of heat resistance (LHR) in *E. coli* AW1.7, AW1.3, DM18.3, and GM16-6, *K. pneumoniae* ST416 pKPN-CZ and *C. sakazaki* ATCC29544. (A) Representation of the LHR in highly heat resistant strains. The figure is scaled to the locus of heat resistance in *E. coli* AW1.7 (14.469 kb in size). Putative promoters and terminators sequences are indicated with hooked arrows and stem-loops, respectively. Open reading frames (ORFs) shaded in gray were identified as unique orthologs in highly heat resistant strains. The GC content of the genetic island is 61.8% while the genome average for AW1.7 is 51.1%. **(B)** Pairwise nucleotide identity of ORFs in *E. coli* AW1.3, DM18.3, and GM16-6, *K. pneumoniae* ST416 pKPN-CZ and *C. sakazaki* ATCC29544 to the corresponding ORFs in *E. coli* AW1.7. ORFs that differ in size from *E. coli* AW1.7 are shaded in gray and the size is indicated in footnotes. Mobile genetic elements were detected in all strains upstream and downstream of the locus of heat resistance; the number of mobile genetic elements is also indicated.

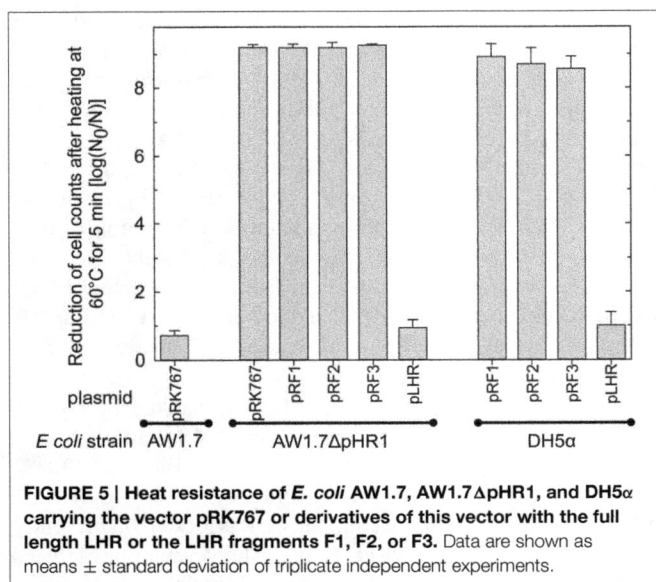

FIGURE 5 | Heat resistance of *E. coli* AW1.7, AW1.7ΔpHR1, and DH5α carrying the vector pRK767 or derivatives of this vector with the full length LHR or the LHR fragments F1, F2, or F3. Data are shown as means ± standard deviation of triplicate independent experiments.

Presence of LHR in *E. coli* and Other Pathogenic Species

Our study and prior studies with *K. pneumonia* and *C. sakazakii* reported a correlation of the presence of the LHR and heat resistance (**Figures 1, 4, 5**, Bojer et al., 2010; Gajdosova et al., 2011). The LHR may thus be a marker for heat resistance in *Enterobacteriaceae* and related organisms. To determine the presence of the LHR in bacterial genomes, we performed a BLAST search using the entire LHR, excluding adjacent transposases, against the NCBI Genomes database. This analysis retrieved 41 sequences with more than 80% coverage from several species in the β- and γ-proteobacteria, including pathogenic strains of *Yersinia entercolitica, Enterobacter cloacae, Citrobacter* sp., *Pseudomonas aeruginosa,* and 16 strains of *E. coli.* The sequences were used to calculate a maximum-likelihood phylogenetic tree (**Figure 6**) that shows remarkable differences from the phylogenetic tree of the bacterial species shown in the tree. The tree is divided into 2 large groups; group A is exclusively composed of sequences γ-proteobacteria (*Enterobacteriaceae* and *Pseudomonas* spp.) while group B includes sequences from β - and γ-proteobacteria (**Figure 6**).

Group A includes sequences from strains of *E. coli* isolated from urinary tract infections (e.g., KTE#) and food isolates of *E. coli.* The conserved sequence identity between the most distantly related sequences from *E. coli,* DM18.3 and KTE233, is 98.9%, suggesting recent lateral transfer of the LHR. LHR sequences from *E. coli* AW1.3 and P12b, two strains that are phylogenetically closely related, cluster in separate branches of group A while LHR sequences from phylogenetically unrelated strains, e.g., *E. coli* AW1.3 and AW1.7, cluster closely together. LHR sequences from *Yersinia enterocolitica, Enterobacter cloacae, Citrobacter* spp., *K. pneumonia,* and *C. sakazakii* are interspersed with LHR sequences from *E. coli* (**Figure 6**). The most divergent LHR sequences in group A belong to 2 *Pseudomonas* spp. (**Figure 6**).

LHR sequences in group B are represented by 13 strains of *Pseudomonas aeruginosa,* including isolates from cystic fibrosis patients. LHR sequences from other pulmonary pathogens include sequences from *Ralstonia pickettii, Burkholderia multivorans,* and *Stenotrophomonas maltophilia* (**Figure 6**). *Dechlorosoma suillum* (now *Azospira suillum;* Byrne-Bailey and Coates, 2012) and *Methylobacillus flagellatus* (Chistoserdova et al., 2007) are found in freshwater and sewage and represent the most divergent LHR sequences in group B. The nucleotide identity between the most distant species from group A (*E. coli* KTE233) and group B (*Pseudomonas aeruginosa* NCAIM B.001380 K260) is 87.2% over >80% of the entire LHR sequence. These data provide evidence that the LHR is highly conserved and has been laterally exchanged within the β- and γ-proteobacteria.

To determine the frequency of the LHR in *E. coli* more accurately, we searched the NCBI whole-genome shotgun assemblies (wgs) database in addition to the NCBI Genome database. This analysis retrieved additional LHR sequences predominantly from clinical isolates including UPEC and ETEC (**Table 3**). Sequences covering >80% of the LHR were identified in 66 out of 3347 strains, with an additional 15 strains found to possess 60–80% of the LHR (**Table 3**). All sequences are more than 99% identical to the LHR sequence of *E. coli* AW1.7. Including genome sequences obtained in this study, the proportion of LHR-positive strains of *E. coli* is approximately 2% (**Table 3**).

PCR Targeting the LHR as a Predictor and Screening Tool for Highly Heat Resistant *E. coli*

To determine whether PCR screening for the LHR reliably identifies highly heat resistant strains of *E. coli,* 55 beef isolates of *E. coli* (Dlusskaya et al., 2011) were screened by PCR using primers targeting 3 different regions of the LHR, spanning several ORF's that are unique to highly heat resistant *E. coli* (Figure S1). Out of the 55 strains of *E. coli,* 13 strains were positive for all 3 LHR amplicons (Figure S1) and 2 strains were positive for 2 of the 3 LHR fragments (Figure S1). We selected 3 LHR positive, 3 LHR negative and the 2 strains containing a partial LHR for evaluation of heat resistance at 5 min at 60°C (**Figure 7**). All LHR positive strains were highly heat resistant but the 2 strains containing a truncated LHR and LHR-negative strains were moderately heat resistant (**Figure 7C**). The results support the hypothesis that the presence of the complete LHR sequence is required for high heat resistance in *E. coli.*

Discussion

The resistance of food-borne pathogens to thermal intervention mechanisms challenges the food industry and public heath sectors, requiring a better understanding of the frequency, distribution and detection of heat resistance. This study employed comparative genomics to identify a genetic island, the LHR, which provides exceptional heat resistance in *E. coli.* Core-genome phylogenetic analysis and phylogenetic analysis of the LHR support the conclusion that the LHR is transmitted via lateral gene transfer. Transfer of the LHR occurred between

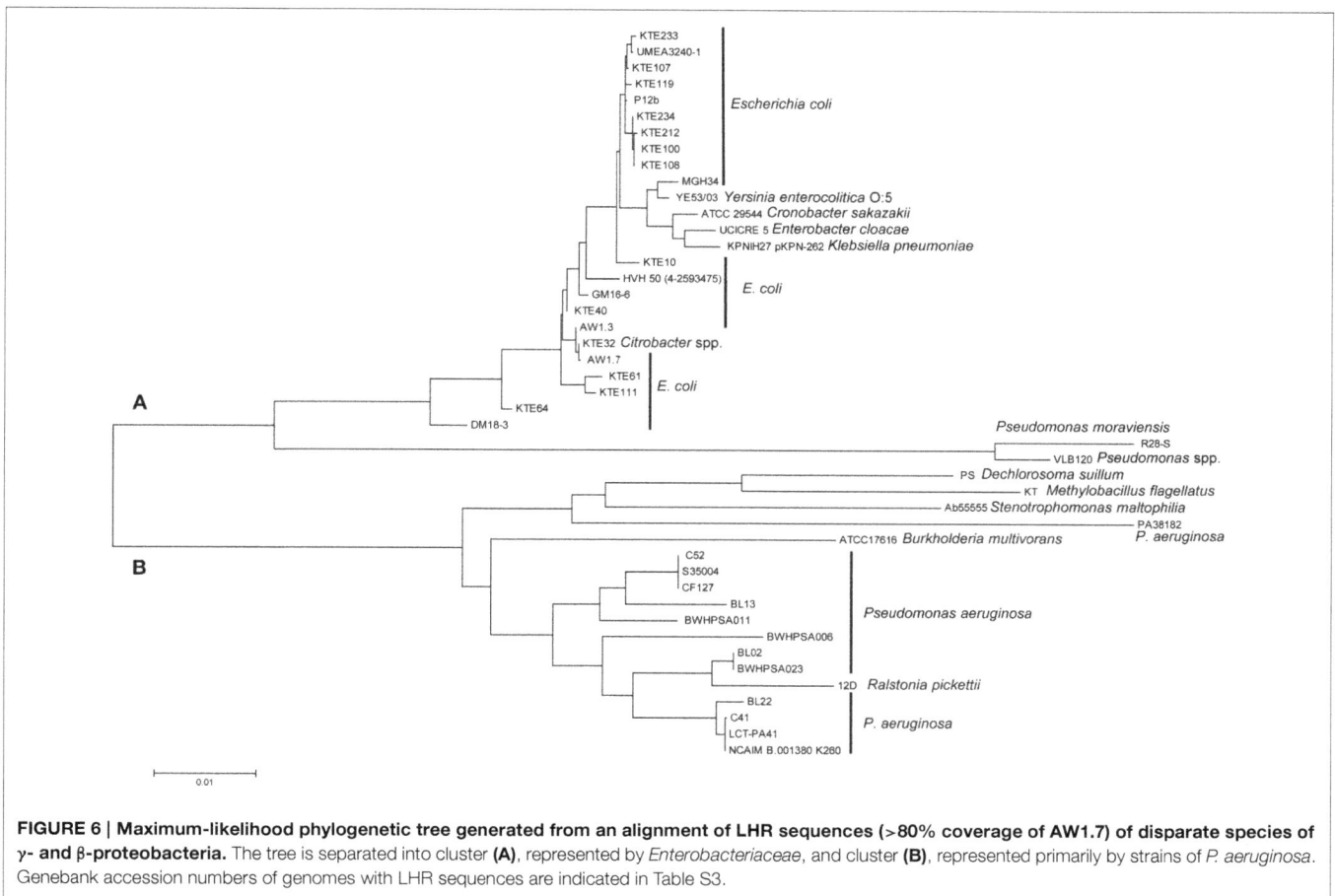

FIGURE 6 | Maximum-likelihood phylogenetic tree generated from an alignment of LHR sequences (>80% coverage of AW1.7) of disparate species of γ- and β-proteobacteria. The tree is separated into cluster **(A)**, represented by *Enterobacteriaceae*, and cluster **(B)**, represented primarily by strains of *P. aeruginosa*. Genebank accession numbers of genomes with LHR sequences are indicated in Table S3.

diverse species in the β- and γ-proteobacteria, including enteric and pulmonary pathogens. Screening of food isolates yielded a number of LHR positive strains, and demonstrated that the LHT is a suitable target for identifying heat resistant *E. coli*.

The LHR Mediates Heat Resistance in *Enterobacteriaceae*

Presence of the LHR in *C. sakazakii* and *K. pneumoniae* correlated to heat resistance of the strains (Bojer et al., 2010; Gajdosova et al., 2011). Of the 36 strains of *E. coli* that were analyzed both with respect to heat resistance and the presence of the LHR, all highly resistant strains carried the LHR and all strains carrying the full length LHR were highly heat resistant. Orthologs of 10 of the 16 ORFs are present in moderately resistant and heat sensitive strains, and a truncated LHR provides only moderate heat resistance. However, presence of the full length locus is unique to highly heat resistant *E. coli*. Complementation with the LHR conferred heat resistance to sensitive strains of *E. coli* only if the entire genomic island was cloned. Heat resistance of *E. coli* is thus dependent on the entire genomic island, and not on the function of a single protein.

The LHR comprises ORFs that are predicted to encode proteins with putative functions in cell envelope maintenance, turnover of misfolded proteins, and heat shock. The predicted products of 5 ORFs possess highly conserved functional domains, including sHSPs (Han et al., 2008) and several proteases. Eight ORFs contain predicted transmembrane domains, including Orfs8-10 and the proteases Orf15 and Orf16. One putative gene, *orf13*, is predicted to encode a sodium/hydrogen antiporter, which corresponds to the interplay of osmotic and heat stress in strains expressing the LHR (Pleitner et al., 2012; Orieskova et al., 2013). Orf16, a predicted membrane protease, possesses a similar domain structure to DegS, a protease involved in the activation of the σ^E stress pathway in *E. coli* (Alba and Gross, 2004). DegS types of proteases are members of the HtrA (high temperature requirement A) family of proteins, which play a role in protein turnover in the periplasm and are induced by heat shock (Kim and Kim, 2005).

The expression of *orf3*, designated as a Clp protease ClpK, increased heat resistance in *E. coli* DH5α; however, transfer of the entire LHR was required for heat resistance in a *clpP* mutant strain (Bojer et al., 2013), suggesting an interplay of ClpP and other proteins encoded within the LHR. Heterologous expression of *orf7-orf10* from *C. sakazakii* in *E. coli* also resulted in an increase in thermotolerance (Gajdosova et al., 2011), but the heat resistance of the resulting transgenic strains was substantially lower than the level of resistance that was observed in *E. coli* AW1.7 carrying the entire LHR (**Figures 1, 5, 7**). Deletion of the

TABLE 3 | Frequency of LHR in _E. coli_. This table lists _E. coli_ genomes or whole genome shotgun sequences containing the locus of heat resistance.

Origin of _E. coli_ strains sequenced (ref)	# of genome sequences	# genomes with LHR 80% (60%) coverage[a]
NCBI genome database[b]	2263	16
Patients with urinary tract infections or bacteremia (1)	317	3 (1)
Clinical isolates of enterotoxigenic _E. coli_ (ETEC) (2)	218	13 (4)
Patients with urinary tract infections (3)	236	15[c]
Clinical isolates after antibiotic treatment (4)	247	21 (9)
Water isolates of O157:H12 (5)	1	1
ETEC (6)	5	1
Woman with recurrent urinary tract infections (7)	27	3 (1)
Intensive care unit patients (8)	5	2 (0)
Clinical and food isolates (this study)	28	4 (0)
	Total # of genomes	**Total LHR**
	3347	**66 (81)**
		% positive
		2.0 (2.4)

[a] > 80% coverage and > 95% pairwise nucleotide identity when compared to _E. coli_ AW1.7; values in brackets indicate BLAST hits with 60–80% coverage and > 95% nucleotide identity when compared to _E. coli_ AW1.7.
[b] Accessed on Aug 11th, 2014.
[c] 13 of these _E. coli_ strains are included in the NCBI genome database.
(1) E.coli UTI Bacteremia initiative, Broad Institute (broadinstitute.org) Accessed Aug 11th, 2014; (2) http://genomesonline.org/project?id=16624 Accessed Aug 11th, 2014; (3) http://www.ncbi.nlm.nih.gov/bioproject/193500 Accessed Aug 11th, 2014; (4) http://www.ncbi.nlm.nih.gov/bioproject/233951 Accessed Aug 11th, 2014; (5) http://www.ncbi.nlm.nih.gov/bioproject/PRJNA51127 Accessed Aug 11th, 2014; (6) Sahl et al., 2010; (7); Chen et al., 2013; (8) Hazen et al., 2014.

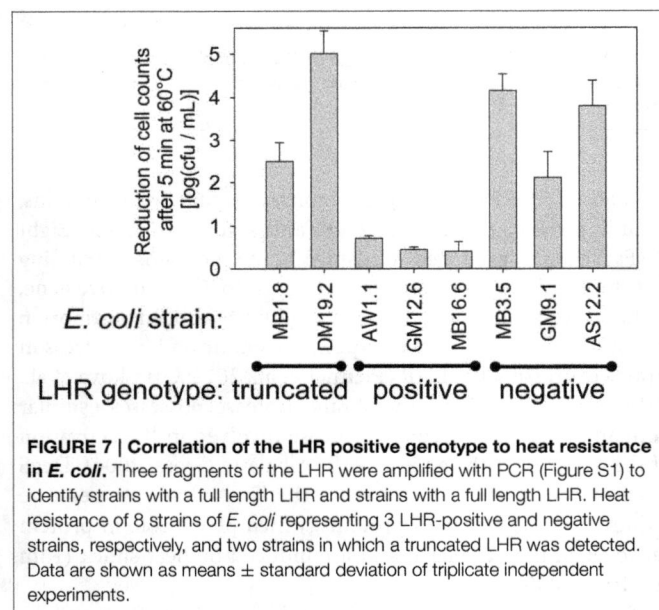

FIGURE 7 | Correlation of the LHR positive genotype to heat resistance in _E. coli_. Three fragments of the LHR were amplified with PCR (Figure S1) to identify strains with a full length LHR and strains with a full length LHR. Heat resistance of 8 strains of _E. coli_ representing 3 LHR-positive and negative strains, respectively, and two strains in which a truncated LHR was detected. Data are shown as means ± standard deviation of triplicate independent experiments.

LHR substantially reduced the resistance of _C. sakazakii_ to heat (Orieskova et al., 2013).

The LHR was suggested to be transcribed as a single polycistronic mRNA in _K. pneumonia_ and _C. sakazakii_ (Gajdosova et al., 2011; Bojer et al., 2013). We identified a strong putative promoter upstream of _orf1_ which is conserved in both _K. pneumoniae_ and _C. sakazakii_. The promoter was predicted to interact with the OmpR, a transcription factor coordinating gene expression in response to osmotic stress (Mizuno and Mizushima, 1990). The LHR is over-expressed in response to osmotic stress (Riedel and Lehner, 2007), which corresponds to the observation that _E. coli_ AW1.7 is resistant to heat only when incubated in growth media containing 1–4% NaCl (Ruan et al., 2011; Pleitner et al., 2012), as well as the observation that deletion of the LHR reduces the tolerance of _C. sakazakii_ to osmotic stress (Orieskova et al., 2013). The LHR may thus function in response to osmotic and heat stress and its function may be partially dependent on the extracellular concentration of compatible solutes.

The LHR is Transmitted by Lateral Gene Transfer between β - and γ-Proteobacteria

The nucleotide identity of the LHR in the _Enterobacteriaceae_ is ~99% and the LHR is consistently flanked by mobile genetic elements. Both imply recent lateral gene transfer. The differences in the phylogenetic relationship between strains _E. coli_ AW1.3 and P12b support this notion. Based on core-genome sequences, _E. coli_ AW1.3 and P12b are highly related and have a recent ancestor. However, their LHR sequences are much more evolutionarily distant; suggesting the strains independently acquired the LHR. Transfer of large genomic elements is well described for genomic islands encoding virulence factors, for example the LEE (Schmidt, 2010). Comparative genomics analysis of the fish pathogen _Edwardsiella tarda_ indicated that the LEE of _E. coli_ is also transmitted to other _Enterobacteriaceae_ (Nakamura et al., 2013). Genomic islands that are transmitted by lateral gene transfer also possess environmental relevance (Juhas et al., 2009) and provide genes for sugar metabolism (Chouikha et al., 2006) or degradation of aromatic compounds (Gaillard et al., 2006). Acquiring multiple genes that require coordinated expression and protein function, e.g., LEE and LHR, can increase the overall fitness of the species.

Genomic islands do not always encode self-transfer capabilities (Shoemaker et al., 2000) and the LHR is located on the chromosome or on plasmids (Bojer et al., 2010; Gajdosova et al., 2011; this study), which may allow exchange through conjugation. Species carrying the LHR occupy similar environmental niches, such as the gastrointestinal tract (*E. coli, Citrobacter* and *Yersinia*), the urinary tract (UPEC and *Yersinia*), and sewage/fresh water (*Enterobacteriaceae*). Remarkably, transfer of the LHR is not restricted to *Enterobacteriaceae* but includes *Pseudomonas* spp. and β-proteobacteria. The GC content and predicted function of the ORFs do suggest a thermophillic origin of the LHR.

The LHR is Present in Approximately 2% of Strains of *E. Coli*, Including Food Isolates and Pathogens

This study, in combination with past studies, has identified 7 LHR-positive and highly heat resistant strains (Dlusskaya et al., 2011; Ruan et al., 2011). None of these strains carry virulence factors; however, bioinformatic analyses revealed that about 2% of all the *E. coli* genome sequences or whole genome shotgun sequences contain the LHR with more than 80% coverage and more than 95% nucleotide identity. All studies on the heat resistance of LHR positive strains of *E. coli, Cronobacter,* and *Klebsiella* confirmed that the full length LHR is a reliable predictor of heat resistance. LHR positive strains of *E. coli* include UPEC and ETEC. Because both the LHR and genes coding for virulence factors are highly mobile, highly heat resistant strains of other pathovars likely also exist. A screening of about 100 strains of STEC has not identified highly heat resistant pathogens (Liu et al., 2015), but screening of 100 strains may not suffice to identify a genetic and physiological trait that is present in about

2% of strains. The identification of the genetic determinants of heat resistance provides a rapid screening tool to identify heat resistant *E. coli* in food or clinical isolates. A broader screening of strains and the assessment of their heat resistance will enable to assess the public health significance of heat resistance in *E. coli*.

This study observed a high frequency of LHR-positive and highly heat resistant strains in beef isolates (Dlusskaya et al., 2011). Beef is an important vector for transmission of STEC (Scallan et al., 2011; USDA, 2012) and highly heat resistant *E. coli* are recovered in high numbers from inoculated beef patties that are cooked medium rare and even survive in burger patties that are cooked "well done," corresponding to an internal temperature of 71°C (Dlusskaya et al., 2011; Liu et al., 2015). To date, the transmission of STEC was attributed to undercooked meat (Schmidt et al., 2002); however, LHR-positive heat resistant pathogens may additionally contribute to foodborne disease. Because these organisms may survive in beef that is cooked to a core temperature of 71°C, cooking meat to a "well done" stage may not always eliminate all pathogenic *E. coli*.

Acknowledgments

We are grateful to Linda Chui, University of Alberta, and Alexander Gill, Health Canada, for providing bacterial strains, and to Andrew Lang, Memorial University, for providing plasmid pRK767. Petr Miller and Gerard Bedié are acknowledged for providing the sequence of plasmid pHR1. Alberta Innovates Bio-Solutions (grant number AI-BIO FSC-12-015) and Alberta Livestock and Meat Agency Ltd (grant number 2013R048R) are acknowledged for funding.

References

Agarwal, J., Srivastava, S., and Singh, M. (2012). Pathogenomics of uropathogenic *Escherichia coli. Indian J. Med. Microbiol.* 30, 141–149. doi: 10.4103/0255-0857.96657

Alba, B. M., and Gross, C. A. (2004). Regulation of the *Escherichia coli* sigma-dependent envelope stress response. *Mol. Microbiol.* 52, 613–619. doi: 10.1111/j.1365-2958.2003.03982.x

Angiuoli, S. V., and Salzberg, S. L. (2011). Mugsy: fast multiple alignment of closely related whole genomes. *Bioinformatics* 27, 334–344. doi: 10.1093/bioinformatics/btq665

Bédié, G., Liu, Y., Miller, P., Ruan, L. F., McMullen, L., and Gänzle, M. G. (2012). "Identification of pHR1 as a major factor in heat and pressure resistance of *E. coli* AW1.7. Book of Abstracts," in *7th International Conference on High Pressure Bioscience and Biotechnology* (Otsu).

Bielaszewska, M., Mellmann, A., Zhang, W., Köck, R., Fruth, A., Bauwens, A., et al. (2011). Characterization of the *Escherichia coli* strain associated with an outbreak of haemolytic uraemic syndrome in Germany, 2011: a microbiological study. *Lancet. Infect. Dis.* 11, 671–676. doi: 10.1016/S1473-3099(11)70165-7

Bojer, M. S., Struve, C., Ingmer, H., Hansen, D. S., and Krogfelt, K. A. (2010). Heat resistance mediated by a new plasmid encoded Clp ATPase, ClpK, as a possible

novel mechanism for nonsocomial persistence of *Klebsiella pneumoniae. PLoS ONE* 5:e15467. doi: 10.1371/journal.pone.0015467

Bojer, M. S., Struve, C., Ingmer, H., and Krogfelt, K. A. (2013). ClpP-dependent and –independent activities encoded by the polycistronic *clpK*-encoding locus contribute to heat shock survival in *Klebsiella pneumoniae. Res. Microbiol.* 164, 205–210. doi: 10.1016/j.resmic.2012.11.005

Byrne-Bailey, K. G., and Coates, J. D. (2012). Complete genome sequence of the anaerobic perchlorate-reducing bacterium *Azospira suillum* strain PS. *J. Bacteriol.* 194, 2767–2768. doi: 10.1128/JB.00124-12

Chen, S. L., Wu, M., Henderson, J. P., Hooton, T. M., Hibbing, M. E., Hultgren, S. J., et al. (2013). Genomic diversity and fitness of *E. coli* strains recovered from the intestinal and urinary tracts of women with recurrent urinary tract infection. *Sci. Transl. Med.* 5, 184ra60 doi: 10.1126/scitranslmed.3005497

Chistoserdova, L., Lapidus, A., Han, C., Goodwin, L., Saunders, L., Brettin, T., et al. (2007). Genome of *Methylobacillus flagellatus*, molecular basis for obligated methylotrophy, and polyphyletic origin of methylotrophy. *J. Bacteriol.* 189, 4020–4027. doi: 10.1128/JB.00045-07

Chouikha, I., Germon, P., Brée, A., Gilot, P., Moulin-Schouleur, M., and Schouler, C. (2006). A *selC*-associated genomic island of the extraintestinal avian pathogenic *Escherichia coli* strain DEN2908 is involved in carbohydrate uptake and virulence. *J. Bacteriol.* 188, 977–987. doi: 10.1128/JB.188.3.977-987.2006

Croxen, M. A., Law, R. J., Scholz, R., Keeney, K. M., Wlodarska, M., and Finlay, B. B. (2013). Recent advances in understanding enteric pathogenic *Escherichia coli. Clin. Microbiol. Rev.* 26, 22–880. doi: 10.1128/cmr.00022-13

Dlusskaya, E. A., McMullen, L. M., and Gänzle, M. G. (2011). Characterization of an extremely heat-resistant *Escherichia coli* obtained from a beef processing facility. *J. Appl. Microbiol.* 110, 840–849. doi: 10.1111/j.1365-2672.2011.04943.x

Dobrindt, U., Hochhut, B., Hentschel, U., and Hacker, J. (2004). Genomic islands in pathogenic and environmental microorganisms. *Nat. Rev. Microbiol.* 2, 414–424. doi: 10.1038/nrmicro884

Estrada-Garcia, T., and Navarro-Garcia, F. (2012). Enteroaggregative *Escherichia coli* pathotype: a genetically heterogeneous emerging foodborne enteropathogen. *FEMS Immunol. Med. Microbiol.* 66, 281–298. doi: 10.1111/j.1574-695X.2012.01008.x

Gaillard, M., Vallaeys, T., Vorhölter, F. J., Minoia, M., Werlen, C., Sentchilo, V., et al. (2006). The clc element of *Pseudomonas* sp. strain B13, a genomic island with various catabolic properties. *J. Bacteriol.* 188, 1999–2013. doi: 10.1128/JB.188.5.1999-2013.2006

Gajdosova, J., Benedikovicova, K., Kamodyova, N., Tothova, L., Kaclikova, E., Stuchlik, S., et al. (2011). Analysis of the DNA region mediating increased thermotolerance at 58°C in *Cronobacter* sp. and other enterobacterial strains. *Antonie Van Leeuwenhoek* 100, 279–289. doi: 10.1007/s10482-011-9585-y

Gautheret, D., and Lambert, A. (2001). Direct RNA motif definition and identification from multiple sequence alignments using secondary structure profiles. *J. Mol. Biol.* 313, 1003–1011. doi: 10.1006/jmbi.2001.5102

Gill, P. R. Jr., and Warren, G. J. (1988). An iron-antagonized fungistatic agent that is not required for iron assimilation from a fluorescent rhizosphere pseudomonad. *J. Bacteriol.* 170, 163–170.

Gordienko, E. N., Kazanov, M. D., and Gelfand, M. S. (2013). Evolution of pan-genomes of *Escherichia coli, Shigella* spp., and *Salmonella enterica. J. Bacteriol.* 195, 2786–2792. doi: 10.1128/JB.02285-12

Han, M. J., Yun, H., and Lee, S. Y. (2008). Microbial small heat shock proteins and their use in biotechnology. *Biotechnol. Adv.* 26, 591–609. doi: 10.1016/j.biotechadv.2008.08.004

Hauben, K. J., Bartlett, D. H., Soontjens, C. C., Cornelis, K., Wuytack, E. Y., and Michiels, W. E. (1997). *Escherichia coli* mutants resistant to inactivation by high hydrostatic pressure. *Appl. Environ. Microbiol.* 63, 945–950.

Hazen, T. H., Sahl, J. W., Fraser, C. M., Donnenberg, M. S., Scheutz, F., and Rasko, D. A. (2013). Refining the pathovar paradigm via phylogenomics of the attaching and effacing *Escherichia coli. Proc. Natl. Acad. Sci. U.S.A.* 110, 12810–12815. doi: 10.1073/pnas.1306836110

Hazen, T. H., Zhao, L., Boutin, M. A., Stancil, A., Robinson, G., Harris, A. D., et al. (2014). Comparative genomics of an IncA/C multidrug resistance plasmid from *Escherichia coli* and *Klebsiella* isolates from intensive care unit patients and the utility of whole-genome sequencing in health care settings. *Antimicrob. Agents Chemother.* 58, 4814–4825. doi: 10.1128/AAC.02573-14

Jaureguy, F., Landraud, L., Passet, V., Diancourt, L., Frapy, E., Guigon, G., et al. (2008). Phylogenetic and genomic diversity of human bacteremic *Escherichia coli* strains. *BMC Genomics* 9:560. doi: 10.1186/1471-2164-9-560

Juhas, M., van der Meer, J. R., Gaillard, M., Harding, R. M., Hood, D. W., and Crook, D. W. (2009). Genomic islands: tools of bacterial horizontal gene transfer and evolution. *FEMS Microbiol. Rev.* 33, 376–393. doi: 10.1111/j.1574-6976.2008.00136.x

Kaas, R. S., Friis, C., Ussery, D. W., and Aarestrup, F. M. (2012). Estimating variation within the genes and inferring the phylogeny of 186 sequenced diverse *Escherichia coli* genomes. *BMC Genomics* 13:577. doi: 10.1186/1471-2164-13-577

Kelley, D. R., Schatz, M. C., and Salzberg, S. L. (2010). Quake: quality-aware detection and correction of sequencing errors. *Genome Biol.* 11:R116. doi: 10.1186/gb-2010-11-11-r116

Kim, D. Y., and Kim, K. K. (2005). Structure and function of HtrA family proteins, the key players in protein quality control. *J. Biochem. Mol. Biol.* 38, 266–274. doi: 10.5483/BMBRep.2005.38.3.266

Lainhart, W., Stolfa, G., and Koudelka, G. B. (2009). Shiga toxin as a bacterial defense against a eukaryotic predator, *Tetrahymena thermophila. J. Bacteriol.* 191, 5116–5122. doi: 10.1128/JB.00508-09

Levine, M. M., Bergquist, E. J., Nalin, D. R., Waterman, D. H., Hornick, R. B., Young, C. R., et al. (1978). *Escherichia coli* strains that cause diarrhoea but do not produce heat-labile or heat-stable enterotoxins and are non-invasive. *Lancet* 311, 1119–1122. doi: 10.1016/S0140-6736(78)90299-4

Li, L., Stoeckert, C. J. Jr., and Roos, D. S. (2003). OrthoMCL: identification of ortholog groups for eukaryotic genomes. *Genome Res.* 13, 2178–2189. doi: 10.1101/gr.1224503

Liu, Y., Gill, A., McMullen, L. M., and Gänzle, M. G. (2015). Variation in heat and pressure resistance of verotoxigenic and non-toxigenic *Escherichia coli. J. Food. Prot.* 78, 111–120. doi: 10.4315/0362-028X.JFP-14-267

McDaniel, T. K., Jarvis, K. G., Donnenberg, M. S., and Kaper, J. B. (1995). A genetic locus of enterocyte effacement conserved among diverse enterobacterial pathogens. *Proc. Natl. Acad. Sci. U.S.A.* 92, 1664–1668. doi: 10.1073/pnas.92.5.1664

Mizuno, T., and Mizushima, S. (1990). Signal transduction and gene regulation through the phosphorylation of two regulatory components: the molecular basis for the osmotic regulation of the porin genes. *Mol. Microbiol.* 4, 1077–1082. doi: 10.1111/j.1365-2958.1990.tb00681.x

Nakamura, Y., Takano, T., Yasuike, M., Sakai, T., Matsuyama, T., and Sano, M. (2013). Comparative genomics reveals that a fish pathogenic bacterium *Edwardsiella tarda* has acquired the locus of enterocyte effacement (LEE) through horizontal gene transfer. *BMC Genomics* 14:642. doi: 10.1186/1471-2164-14-642

Newton, H. J., Sloan, J., Bulach, D. M., Seemann, T., Allison, C. C., Tauschek, M., et al. (2009). Shiga toxin-producing *Escherichia coli* strains negative for locus of enterocyte effacement. *Emerg. Infect. Dis.* 15, 372–380. doi: 10.3201/eid1503.080631

O'Brien, A. D., Newland, J. W., Miller, S. F., Holmes, R. K., Smith, H. W., and Formal, S. B. (1984). Shiga-like toxin-converting phages from *Escherichia coli* strains that cause hemorrhagic colitis or infantile diarrhea. *Science* 226, 694–696.

Orieskova, M., Gajdosova, J., Oslanecova, L., Ondreickova, K., Kaclikova, E., Stuchlik, E., et al. (2013). Function of thermotolerance genomic island in increased stress resistance of *Cronobacter sakazakii. J. Food Nutr. Res.* 52, 37–44.

Pleitner, A., Zhai, Y., Winter, R., Ruan, L., McMullen, L. M., and Gänzle, M. G. (2012). Compatible solutes contribute to heat resistance and ribosome stability in *Escherichia coli* AW1.7. *Biochim. Biophys. Acta* 1824, 1351–1357. doi: 10.1016/j.bbapap.2012.07.007

Rasko, D. A., Rosovitz, M. J., Myers, G. S., Mongodin, E. F., Fricke, W. F., Gajer, P., et al. (2008). The pangenome structure of *Escherichia coli*: comparative genomic analysis of E. coli commensal and pathogenic isolates. *J. Bacteriol.* 190, 6881–6893. doi: 10.1128/JB.00619-08

Ratiner, Y. A., Sihvonen, L. M., Liu, Y., Wang, L., and Siitonen, A. (2010). Alteration of flagellar phenotype of *Escherichia coli* strain P12b, the standard type strain for flagellar antigen H17, possessing a new non-*fliC* flagellin gene *flnA*, and possible loss of original flagellar phenotype and genotype in the course of subculturing through semisolid media. *Arch. Microbiol.* 192, 267–278. doi: 10.1007/s00203-010-0556-x

Riedel, K., and Lehner, A. (2007). Identification of proteins involved in osmotic stress in *Enterobacter sakazakii* by proteomics. *Proteomics* 7, 1217–1231. doi: 10.1002/pmic.200600536

Ruan, L., Pleitner, A., Gänzle, M. G., and McMullen, L. M. (2011). Solute transport proteins and the outer membrane protein NmpC contribute to heat-resistance of *Escherichia coli* AW1.7. *Appl. Environ. Microbiol.* 77, 2961–2967. doi: 10.1128/AEM.01930-10

Rudolph, B., and Gebendorfer, K. M. (2010). Evolution of *Escherichia coli* for growth at high temperatures. *J. Biol. Chem.* 285, 19029–19034. doi: 10.1074/jbc.M110.103374

Sahl, J. W., Steinsland, H., Redman, J. C., Angiuoli, S. V., Nataro, J. P., Sommerfelt, H., et al. (2010). A comparative genomic analysis of diverse clonal types of enterotoxigenic *Escherichia coli* reveals pathovar-specific conservation. *Infect. Immun.* 79, 950–960. doi: 10.1128/IAI.00932-10

Scallan, E., Hoekstra, R. M., Angulo, F. J., Tauxe, R. V., Widdowson, M. A., Roy, S. L., et al. (2011). Foodborne illness acquired in the United States - Major pathogens. *Emerg. Infect. Dis.* 17, 7–14. doi: 10.3201/eid1701.P11101

Schmidt, H., and Hensel, M. (2004). Pathogenicity islands in bacterial pathogenesis. *Clin. Microbiol. Rev.* 17, 14–56. doi: 10.1128/CMR.17.1.14-56.2004

Schmidt, M. A. (2010). LEEways: tales of EPEC, ATEC and EHEC. *Cell Microbiol.* 12, 1544, 1552. doi: 10.1111/j.1462-5822.2010.01518.x

Schmidt, T. B., Keene, M. P., and Lorenzen, C. L. (2002). Improving consumer satisfaction of beef through the use of thermometers and consumer education by wait staff. *J. Food Sci.* 67, 3190–3193. doi: 10.1111/j.1365-2621.2002.tb08880.x

Shoemaker, N. B., Wang, G. R., and Salyers, A. A. (2000). Multiple gene products and sequences required for excision of the mobilizable integrated *Bacteroides* element NBU1. *J. Bacteriol.* 182, 928–936. doi: 10.1128/JB.182.4.928-936.2000

Simpson, J. T., Wong, K., Jackman, S. D., Schein, J. E., Jones, S. J., and Birol, I. (2009). ABySS: a parallel assembler for short read sequence data. *Genome Res.* 19, 1117–1123. doi: 10.1101/gr.089532.108

Solovyev, V., and Salamov, A. (2011). "Automatic annotation of microbial genomes and metagenomic sequences," in *Metagenomics and its Applications in Agriculture, Biomedicine and Environmental Studies*, ed R. W. Li (New York, NY: Nova Science Publishers), 61–78.

Stamatakis, A. (2014). RAxML Version 8: a tool for phylogenetic analysis and post-analysis of large phylogenies. *Bioinformatics* 30, 1312–1313. doi: 10.1093/bioinformatics/btu033

Talavera, G., and Castresana, J. (2007). Improvement of phylogenies after removing divergent and ambiguously aligned blocks from protein sequence alignments. *Systematic Biol.* 56, 564–577. doi: 10.1080/10635150701472164

Tamura, K., Stecher, G., Peterson, D., Filipski, A., and Kumar, S. (2013). MEGA6: molecular evolutionary genetics analysis version 6.0. *Mol. Biol. Evol.* 30, 2725–2729. doi: 10.1093/molbev/mst197

Touchon, M., Hoede, C., Tenaillon, O., Barbe, V., Baeriswyl, S., Bidet, P., et al. (2009). Organised genome dynamics in the *Escherichia coli* species results in highly diverse adaptive paths. *PLoS Genet.* 5:e1000344. doi: 10.1371/journal.pgen.1000344

Trabulsi, L. R., Keller, R., and Tardelli Gomes, T. A. (2002). Typical and atypical enteropathogenic *Escherichia coli*. *Emerg. Infect. Dis.* 8, 508–513. doi: 10.3201/eid0805.010385

USDA. (2012). Shiga-toxin producing *Escherichia coli* in certain raw beef products. *Fed. Regist.* 77, 31975–31981.

Conflict of Interest Statement: The authors declare that the research was conducted in the absence of any commercial or financial relationships that could be construed as a potential conflict of interest.

17

Geriatric respondents and non-respondents to probiotic intervention can be differentiated by inherent gut microbiome composition

Suja Senan[1], Jashbhai B. Prajapati[2]*, Chaitanya G. Joshi[3], Sreeja V.[2], Manisha K. Gohel[4], Sunil Trivedi[5], Rupal M. Patel[5], Himanshu Pandya[6], Uday Shankar Singh[4], Ajay Phatak[7] and Hasmukh A. Patel[1]

[1] Department of Dairy Science, South Dakota State University, Brookings, SD, USA, [2] Department of Dairy Microbiology, Anand Agricultural University, Anand, India, [3] Department of Animal Biotechnology, Anand Agricultural University, Anand, India, [4] Department of Community Medicine, H. M Patel Center for Medical Care and Education, Karamsad, India, [5] Department of Microbiology, H. M Patel Center for Medical Care and Education, Karamsad, India, [6] Department of Medicine, H. M Patel Center for Medical Care and Education, Karamsad, India, [7] Central Research Services, Charutar Arogya Mandal, Karamsad, India

Edited by:
Kate Howell,
University of Melbourne, Australia

Reviewed by:
Stella Maris Reginensi Rivera,
Universidad de la República Oriental del Uruguay, Uruguay
Amit Kumar Tyagi,
The University of Texas MD Anderson Cancer Center, USA

*Correspondence:
Jashbhai B. Prajapati,
Department of Dairy Microbiology,
SMC College of Dairy Science, Anand Agricultural University, Anand,
Gujarat 388 110, India
prajapatijashbhai@yahoo.com,
jbprajapati@aau.in

Scope: Probiotic interventions are known to have been shown to influence the composition of the intestinal microbiota in geriatrics. The growing concern is the apparent variation in response to identical strain dosage among human volunteers. One factor that governs this variation is the host gut microbiome. In this study, we attempted to define a core gut metagenome, which could act as a predisposition signature marker of inherent bacterial community that can help predict the success of a probiotic intervention.

Methods and results: To characterize the geriatric gut microbiome, we designed primers targeting the 16S rRNA hypervariable region V2–V3 followed by semiconductor sequencing using Ion Torrent PGM. Among respondents and non-respondents, the chief genera of phylum Firmicutes that showed significant differences are *Lactobacillus*, *Clostridium*, *Eubacterium*, and *Blautia* ($q < 0.002$), while in the genera of phylum Proteobacteria included *Shigella*, *Escherichia*, *Burkholderia* and *Camphylobacter* ($q < 0.002$).

Conclusion: We have identified potential microbial biomarkers and taxonomic patterns that correlate with a positive response to probiotic intervention in geriatric volunteers. Future work with larger cohorts of geriatrics with diverse dietary influences could reveal the potential of the signature patterns of microbiota for personalized nutrition.

Keywords: geriatric, gut, metagenome, probiotics, MTCC 5463

Abbreviations: µl, microliter; AAU, Anand Agricultural University; CFU, colony forming unit; dL, deciliter; dNTP, deoxynucleotide; ICMR, Indian Council of Medical Research; ISAPP, International Scientific Association for Probiotics and Prebiotics; mg, milligram; ml, milliliter; mM, millimolar; MTCC, microbial type culture collection; nt, nucleotide; PGM, Personal Genome Machine; pmol, picomol; QIIME, Quantitative Insights into Microbial Ecology; rRNA, ribosomal RNA; s, seconds; S, Svedberg unit; U, unit.

Introduction

An integrative study of the host and its surrounding environment is imperative to comprehend the complex biological system of the human body. As a part of the environment, the human host more than 100 trillion bacteria forming the "in-vironment" (de Wouters et al., 2000) made up of millions of microbial genes in the intestine (Lederberg, 2000). The indigenous microbial community plays an integral role in regulating the host's physiological, nutritional, and immunological processes (Hooper et al., 2001). The study of the diversity of the indigenous microbial community can explain the host microbe interaction (Gerritsen et al., 2011). The gut microbiota composition changes with age due to physiological reasons and increased use of medications (Bartosch et al., 2004; Mueller et al., 2006; Mariat et al., 2009; Zwielehner et al., 2009). The microbiota of elderly people showed a higher *Bacteroidetes/Firmicutes* ratio along with a high inter-individual variation in microbiota composition at the phylum level when compared with young adults (Claesson et al., 2011). Studies of the geriatric gut have shown a decline in the count and diversity among Bacteroidetes (Hopkins and Macfarlane, 2002; Woodmansey et al., 2004; Guigoz et al., 2008). Proteolytic bacteria increase on aging in the large bowel, leading to putrefaction (Hopkins et al., 2001). Such changes in intestinal microbiome cause prolonged intestinal transit time and fecal retention among geriatrics (Tiihonen et al., 2009). An understanding of the changes in the microbiome of the elderly has led to the possibility of correcting the dysbiosis by administering probiotics. Probiotics have been successful in increasing the levels of health-promoting bacteria in the fecal microbiota of elderly (Ahmed et al., 2007; Lahtinen et al., 2009; Matsumoto et al., 2009), improving the frequency of bowel movements (Pitkala et al., 2007), *Clostridium difficile*-associated diarrhea incidence (Ouwehand et al., 2009), and frequency of defecation (An et al., 2010).

The translation of the above-mentioned benefits of probiotics to the host cannot be guaranteed. This could be due to the individual differences with respect to diet, the structure and operations of the gut microbiota, nutrient and energy harvest, variations in human environmental exposures, microbial ecology, and genotype (Turnbaugh et al., 2009). In order to assure uniform outcomes of therapy among subjects, the International Scientific Association for Probiotics and Prebiotics (ISAPP) had come up with recommendations for conducting a well-defined trial (Reid et al., 2010). Briefly, they include (1) clearly define the end goal, (2) design the study by identifying precise parameters and defining the level of response that will be tested, (3) base the selection of the intervention on scientific investigations, and (4) carefully select the study cohort. Inter-individual diversity in responses toward probiotics could also be due to core gut microbiome patterns. Recently, role of microbial biomarkers for determining dietary responsiveness were identified in obese individuals (Korpela et al., 2014) and metabolic diseases (McOrist et al., 2011; Walker et al., 2011; Louis, 2012; Lampe et al., 2013), paving the way for personalized nutrition. This study takes up the challenge to identify the factors that differentiate a respondent from a non-respondent and utilize the findings to define the precise dose and response prognosis. This finding can help design probiotic supplements catering to a niche market defined by age, location, or disease state.

From an Indian perspective, gut metagenomics have been studied in malnourished children (Gupta et al., 2011), obese individuals (Patil et al., 2012), and children of varying nutritional status (Ghosh et al., 2014). It was for the first time in India that the present study was conducted to investigate the elderly gut metagenome to identify microbial biomarkers determining responsiveness of the host to a probiotic therapy. We hypothesized that by studying the baseline gut microbiota diversity of elderly subjects, we could identify a core gut microbiome signature pattern that is likely to positively influence the response of an individual to the probiotic strain. The strain under study, *Lactobacillus helveticus* MTCC 5463 is an indigenous potential probiotic with *in vitro*, *in vivo*, and *in silico* studies providing suggestive evidences of the strain's robustness in the gut and transit, adhesion, autoaggregation, colonization, antibacterial property, hypocholesterolemic, and immunomodulatory properties (Senan et al., 2015). The outcome of this study paves the way forward for tailored probiotic therapy.

Materials and Methods

Origin and Maintenance of Bacterial Strains

The indigenous probiotic strain *L. helveticus* MTCC 5463 (Prajapati et al., 2011) and starter culture *Streptococcus thermophilus* MTCC 5460 (Prajapati et al., 2013) were maintained by the Department of Dairy Microbiology, Anand Agricultural University, India at −80°C as 15% glycerol stocks and were routinely cultured in de Man, Rogosa, Sharpe (MRS) and M 17 medium, respectively (HiMedia India Ltd., India).

Product Preparation

The test product was a fermented probiotic drink (*Lassi*) with double toned milk fermented with culture containing *S. thermophilus* MTCC 5460 and *L. helveticus* MTCC 5463. The cultures were added at 0.1% each and incubated aerobically till an acidity of 0.8–0.9% lactic acid was obtained. Both the test and placebo products contained sugar and prebiotic honey in a standardized ratio. The control product was made in a similar manner without the addition of MTCC 5463. The shelf life of the fermented drink was 28 days at 4°C, corresponding to the lower level (10^9 CFU/ml) of strain MTCC 5463.

Participant Selection

Individuals ranging from 64 to 74 years were recruited. Initially, 112 subjects were enrolled in the trial, 36 had to withdraw because of antibiotic consumption. Volunteers were asked to sign the consent form before recruitment. Exclusion criteria included lactose intolerance, recent antibiotic treatment, frequent gastrointestinal disorders, or metabolic diseases. Participants included in the trial had no known allergies or intolerance to dairy foods. The trial had 80% power at a 5% 2-sided significance level to detect a >50% change in the primary outcome among subjects. No antibiotics or laxatives were taken 2 months before or during the study.

Intervention

Sixteen participants showing diversity in lactobacilli count and cholesterol levels were involved in the double-blind, crossover, placebo-controlled, and randomized-feeding trial. The trial was divided into five consecutive periods: a pre-feeding period (2 weeks), followed by a feeding period (4 weeks), a washout period (4 weeks), a second-feeding period (4 weeks), and a final washout period (2 weeks).

Collection and Analysis of the Blood Samples

Blood samples were taken from each volunteer immediately before and after each treatment period using EDTA-containing vacutainers. Total cholesterol (TC) was measured using enzyme-spectrophotometry kits and IgG, IgM, TNF-alpha, INF-gamma, and IL-2 by ELISA capture assay (Siemens Medical Solutions Diagnostics Ltd., India). All the tests were done at the Central Diagnostic Laboratory, Shri Krishna Medical College Karamsad, Anand, Gujarat, an NABL accredited and ISO 15189:2003 laboratory.

Selection of Respondents and Non-Respondents

A respondent was defined as a subject having an improvement in the levels of *L. helveticus* MTCC 5463 strain count in feces and a reduction in cholesterol levels. Similarly, a non-responder was defined as a subject who displayed an absence of decrease in cholesterol levels and increase in viability of the bacterial strain. Based on these criteria, we identified eight subjects each in respondents and non-respondents category.

Fecal Sample Collection

Single fecal samples were collected at the end of every 2 weeks. Participants were given 60 ml sterile stool container with a sterile plastic spoon (Polylab Plasticware, India) and were asked to fill the tube to the 30 ml mark with feces from the midstream defecation period. During the second-feeding period, there was a crossover of the feeding design. For every collection, the stool samples were immediately frozen at −20°C.

DNA Extraction from Stool Samples

DNA was extracted from feces using a QIAamp MiniPrep DNA extraction kit following the manufacturer's instructions. The DNA was stored at −20°C. Quality and purity of the isolated genomic DNA were confirmed by agarose gel electrophoresis and spectrophotometry on the NanoDrop 2000 device (Fisher Scientific, Schwerte, Germany). DNA concentration was estimated with the Qubit 2.0 instruments applying the Qubit dsDNA HS Assay (Life Technologies, Invitrogen division, Darmstadt, Germany).

Quantification of Lactobacilli in Stool

The subjects voided their feces into a 60-ml sterile stool container with a sterile plastic spoon (Polylab Plasticware, India) from the midstream defecation period. Within an hour of sample procurement, samples were diluted and homogenized to give a 10-fold dilution (wet weight/volume). Stool Lactobacilli content was determined by plating aliquots (1 ml) of each dilution on freshly prepared de Mann Rogosa Sharp agar (Himedia, India), incubated for 24–48 h at 37°C under anaerobic conditions. Stool

L. helveticus MTCC 5463 content was determined by quantitative PCR (qPCR) using StepOne Real-Time PCR System (ABI/Thermo Fisher Scientific, Bangalore, India). Primers and 3′ minor groove binder (MGB) probes for accurate detection and quantification of *L. helveticus* MTCC 5463 in human fecal samples were developed with Primer Express v3.0 [Thermo Fisher Scientific (earlier Applied Biosystems), Bangalore, India]. The temperature profile of the qPCR consisted 2 min at 50°C, 10 min at 95°C, followed by 45 cycles of 15 s at 95°C, and 1 min at 60°C. Species specific primers and probe targeted on the bile salt hydrolase gene of *L. helveticus* MTCC 5463 (Accession number AEYL01000315; locus tag AAULH_13111 2049 bp). BLAST[1] and EMBL[2] database were used to ensure the specificity of the primers. Genomic DNA standards prepared with six different serial dilutions (2.68×10^6, 2.68×10^5, 2.68×10^4, 2.68×10^3, 2.68×10^2, and 2.68×10) being equivalent to ranges from 10^6 to 10^1 CFU/ml of target genome (MTCC 5463). The cycle threshold (C_T) was evaluated to create the standard curve. The amplification efficiencies were determined using the formula $E = [10^{(-1/\text{slope})} - 1]$.

16S Primers and Amplicon Library Generation

PCR amplification of the 16S rRNA hypervariable region V2–V3 was performed with a pool of 32 degenerated forward and 1 degenerated reverse primer targeting bacteria as described by Schmalenberger et al. (2001) with barcode sequences at the 5′ end of the primers. Primers were assessed for specificity using the SILVA 108 SSU Reference 16S rRNA gene database and BLASTN matches with corresponding 16S rRNA gene sequences. The addition of the barcodes to the primers resulted in an amplicon approximately 430 nt in length. The primers were diluted and pooled to equimolar quantities. For amplicon library preparation, 4 ng of each genomic DNA, 5 mM dNTPs, 2 mM MgCl$_2$ (Roche Diagnostics, USA), 1 U Platinum Taq DNA polymerase High Fidelity, and 10 pmol primer-mix were used per 25 µl amplification reaction. The PCR conditions were as follows: 95°C for 5 min, followed by 30 cycles of 94°C for 15 s, 60°C for 45 s, 70°C for 30 s, and a final elongation step of 72°C for 10 min. Amplicon product purification was performed via gel electrophoresis on a 1.5% Tris Borat EDTA agarose gel-stained with ethidium bromide (EtBr) (Life Technologies). All positive PCR reactions were electrophoresized in agarose gels and products with the expected size were cut and purified with Qiagen Gel extraction kit (Qiagen, Düsseldorf, Germany). The exact fragment sizes were determined using HT DNA High Sensitivity LabChip Kit (Caliper Life Sciences GmbH, Mainz, Germany). Amplicon library concentration was estimated with the Qubit 2.0 instrument using the Qubit dsDNA HS assay (Life Technologies).

Emulsion PCR and Sequencing

The emulsion PCR was carried out applying the Ion XPress Template kit V2.0 (Life Technologies) as described in the appropriate user Guide (Part No. 4469004 Rev. B 07/2011) provided by the manufacturer. Quality and quantity of the enriched spheres were checked on the Guava easyCyte5 system (Millipore GmbH,

---

[1] http://www.ncbi.nlm.nih.gov/blast/
[2] http://www.ebi.ac.uk/embl/

Schwalbach am Taunus, Germany) as described in the appendix of the Ion Xpress Template Kit User Guide (Part Number 4467389 Rev. B, 05/2011). Sequencing of the amplicon libraries was carried out on the Ion Torrent Personal Genome Machine (PGM) system using the Ion Sequencing 200 kit (all Life Technologies) following the corresponding protocol (Part No. 4471999 Rev B, October 13, 2011). Quality check passed libraries were subjected to emulsion PCR using the Ion PGM 200 Xpress Template Kit (Life Technologies). After bead enrichment, beads were loaded onto Ion 316 chips and sequenced using an Ion Torrent PGM.

Sequence Analysis

The sequence data sets obtained were uploaded to the Metagenome Rapid Annotation using Subsystem Technology (MG-RAST) server[3] and subsequently checked for low-quality reads. The sequence reads that passed the quality filtering step were then subjected to further analysis of the taxonomic annotation of the fecal DNA sequences using QIIME pipeline (Caporaso et al., 2010). To investigate the species diversity, we used rarefaction curves, and richness estimators like Chao1 in QIIME. Statistic comparison of samples organized into respondents and non-respondents for the deferentially abundant microbial diversity was studied using an alignment platform, STAMP.[4]

Results

We conducted a comparative analysis of respondents and non-respondents fecal microbiome to reveal differences and identified biomarkers that differentiate them.

Quantification of Lactobacilli Count Using Traditional Plating and qPCR

Traditional plate counts of lactobacilli at genus level on selective medium ranged from a baseline reading of 8.6 log CFU/g of wet fecal matter, which rose to 9.3 log CFU/g at the end of feeding period and a gradual decrease to 8.7 log CFU/g at the end of the placebo feeding. The qPCR primers targeted the bile salt hydrolase gene of MTCC 5463, which made the gene copy count a fraction of the plate count. On the other hand, the precise *L. helveticus* MTCC 5463 strain count from real-time PCR showed a complete absence of the strain before feeding. At the end of 30 days, the strain appeared in the feces of all subjects in the treated group, reaching a level as high as 8.32 to the lowest amount of 6.17 log gene copies/g fecal matter at end of feeding period.

Summary of Sequence Processing Data

Primers targeting the 16S rRNA gene V2–V3 region (Schmalenberger et al., 2001) precisely generated amplicons from members of the domain bacteria and did not hybridize to sequences of the domains Archaea and Eucarya. By using the ARB SILVA 108 SSU database, a high match of 84.5% at a maximum number of four mismatches was observed. The primer pair theoretically targeted all 16S rRNA gene sequences of the gut microbiota bacterial orders and generated a single amplicon. All 32 amplicons

from the pool of test and placebo groups were mixed together at an equimolar ratio. After pooling and elution of the amplicons from the gel, we got one band for the probiotic fed group of 520 bp having a concentration of 1912.09 pg/µl and molarity of 5562.5 pmol/l. The placebo fed group gave an amplicon of 513 bp having a concentration of 1435.87 pg/µl and 4237.7 pmol/l molarity. The data sets for 16 subjects before probiotic feeding had reads ranging from 13,061 to 980,628 with a read length ranging from 201 to 251 bp and a total amount of 42,52,62,470 bases.

Inter-Individual Differences in Shifts in Phyla Abundance (%) Before and After Probiotic Feeding

The relative abundance of major genera in the elderly gut metagenome and high-level of inter-individual variation is shown in **Figure 1**. We presume that the inter-individual differences are indicative of a highly personal fecal microbiota profile, which determines the response of the host to probiotics. Host factors probably play a major effect in shaping the intestinal microbial ecosystem during an intervention. In the present study, we attempted to understand the core microbiome of respondents and non-respondents to probiotics.

Probiotic Feeding and Effect on Lipid Profile and Immunologic Parameters

In addition to TC (primary outcome), in this study, we also investigated the effect of probiotic intervention on lipid profile, beta glucouronidase activity, and immunological parameters. The mean β-glucuronidase activity was reduced in test group from 1.40 to 0.73 (microgram/min/mg of protein), while in case of placebo group, no effect on enzyme activity was observed. A significant immunomodulatory effect on the TNF-α and IL-2 levels in subjects among probiotic group compared to placebo group was observed. There was however no significant beneficiary effect found on IFN-γ, IgG, or IgM levels. Paired *t*-test showed that there were statistically significant differences in serum cholesterol, VLDL, TC/HDL, LDL/HDL in placebo group and in LDL, TC/HDL, and LDL/HDL in probiotic group. A significant ($p = 0.01$) decrease in the LDL value was seen in the probiotic group at the end of 30 days of feeding (**Table 1**).

Participant Diversification into Respondents and Non-Respondents

The primary outcome of this trial was a reduction in TC after 4 weeks of feeding probiotic MTCC 5463. We defined non-respondents as those subjects who experienced elevations in TC of ≥2.509 mg/dL, whereas respondents were the ones who showed no change in TC or <1.72 mg/dL TC in response to the probiotic intervention of 4 weeks. Among the 59 subjects who could complete the study, we classified a total of 16 subjects into respondents ($n = 8$) and non-respondents ($n = 8$) based on cholesterol levels and lactobacilli counts (**Figure 2**). There were no significant differences in the baseline characteristics of the two groups. This eliminates the influence of gender, weight, and age in influencing the response toward probiotic intervention. The abundance of *L. helveticus* MTCC 5463 was significant

---

[3] http://metagenomics.nmpdr.org/
[4] http://kiwi.cs.dal.ca/Software/STAMP

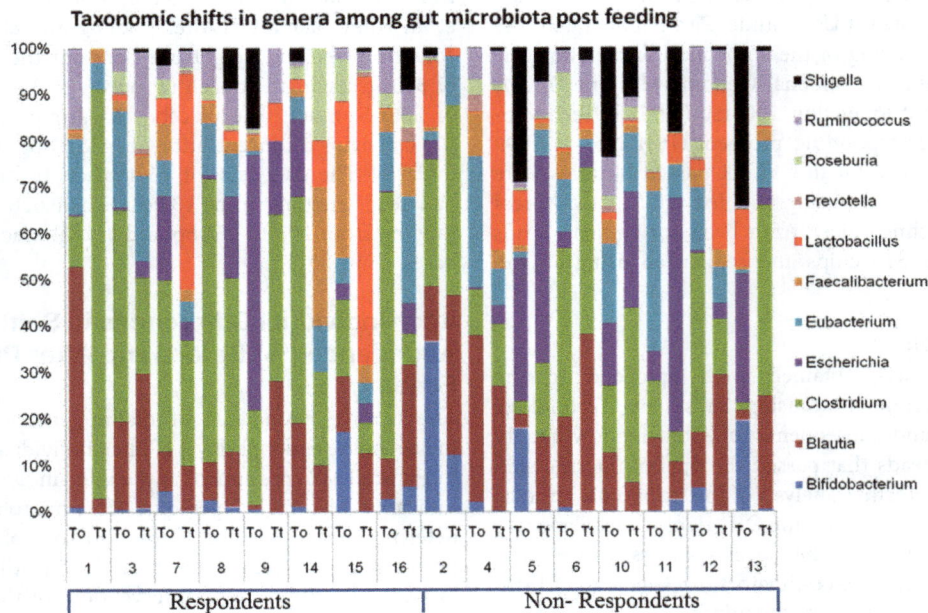

FIGURE 1 | Inter-individual variation in the geriatric gut microbiome pre- and post-probiotic feeding.

TABLE 1 | Effect of probiotic and placebo interventions on lipid profile in geriatric subjects.

Variables	Probiotic (means ± SD)	Placebo (means ± SD)
Total cholesterol (TC) (mg/dL)		
Baseline	161.67 ± 41.05	174.32 ± 49.99
Post-intervention	158.09 ± 42.63	167.09 ± 43.11
p-Value	0.12	<0.001
Triglyceride (mg/dL)		
Baseline	103.77 ± 49.84	116.38 ± 71.01
Post-intervention	104.00 ± 56.43	108.58 ± 70.74
p-Value	0.96	0.03
High density lipoprotein (HDL) (mg/dL)		
Baseline	46.21 ± 12.46	49.67 ± 15.97
Post-intervention	47.08 ± 13.97	48.77 ± 12.98
p-Value	0.24	0.34
Low density lipoprotein (LDL) (mg/dL)		
Baseline	98.48 ± 37.12	88.93 ± 38.37
Post-intervention	92.93 ± 35.79	84.56 ± 31.13
p-Value	0.01	0.09
Very low density lipoprotein (mg/dL)		
Baseline	21.63 ± 12.04	23.28 ± 14.20
Post-intervention	21.34 ± 11.97	21.71 ± 14.12
p-Value	0.74	0.03
Total cholesterol/HDL (mg/dL)		
Baseline	3.91 ± 1.22	3.77 ± 1.33
Post-intervention	3.74 ± 1.20	3.65 ± 1.23
p-Value	<0.001	0.04
LDL/HDL (mg/dL)		
Baseline	2.37 ± 0.96	2.23 ± 1.01
Post-intervention	2.21 ± 0.91	2.13 ± 0.96
p-Value	<0.001	0.04

($p < 0.05$) higher in the individuals with an increase in cholesterol levels, as compared to those with a decrease. The decrease in cholesterol levels among respondents was a maximum 14.19% with a 23.66% increase in lactobacilli count in feces. Among non-respondents, a maximum increase of 34.13% in cholesterol with a 9.31% decrease in lactobacilli count was observed. The increase in lactobacilli counts with a decrease in cholesterol in case of respondents indicated that the observed hypocholesterolemic effect of the strain was dependent on the number of lactobacilli in the gut.

Microbiome Diversity Estimates Associated with Respondents and Non-Respondents

The alpha diversity of the respondent group before probiotic feeding (30.8 ± 4.8) and after feeding (26.6 ± 3.6) was higher than non-respondents' measures for before (25.6 ± 4.5) and after probiotic feeding (24.6 ± 5.8). This indicates that bacterial richness is a factor that promotes responsiveness toward beneficial strains in the gut. To investigate differences in rarefaction measures, we rarified each sample at 33,000 reads and performed the two-sample t-test on the two groups. Respondents had significantly greater alpha diversity indices like phylogenetic distance ($p = 0.022$), Chao1 ($p = 0.019$), and Shannon index ($p = 0.00058$) than the non-respondents (**Figure 3**). A non-significant increase in observed species in case of non-respondents ($p = 0.27$) could be due to presence of distinct commensals that reflect the host's dietary and geographical differences. There was a non-significant abundance of *Clostridium*, *Shigella*, and *Listeria* among rural respondents. Poor sanitation and hygiene maintained in the rural households could have led to the distinct differences in gut microbiota.

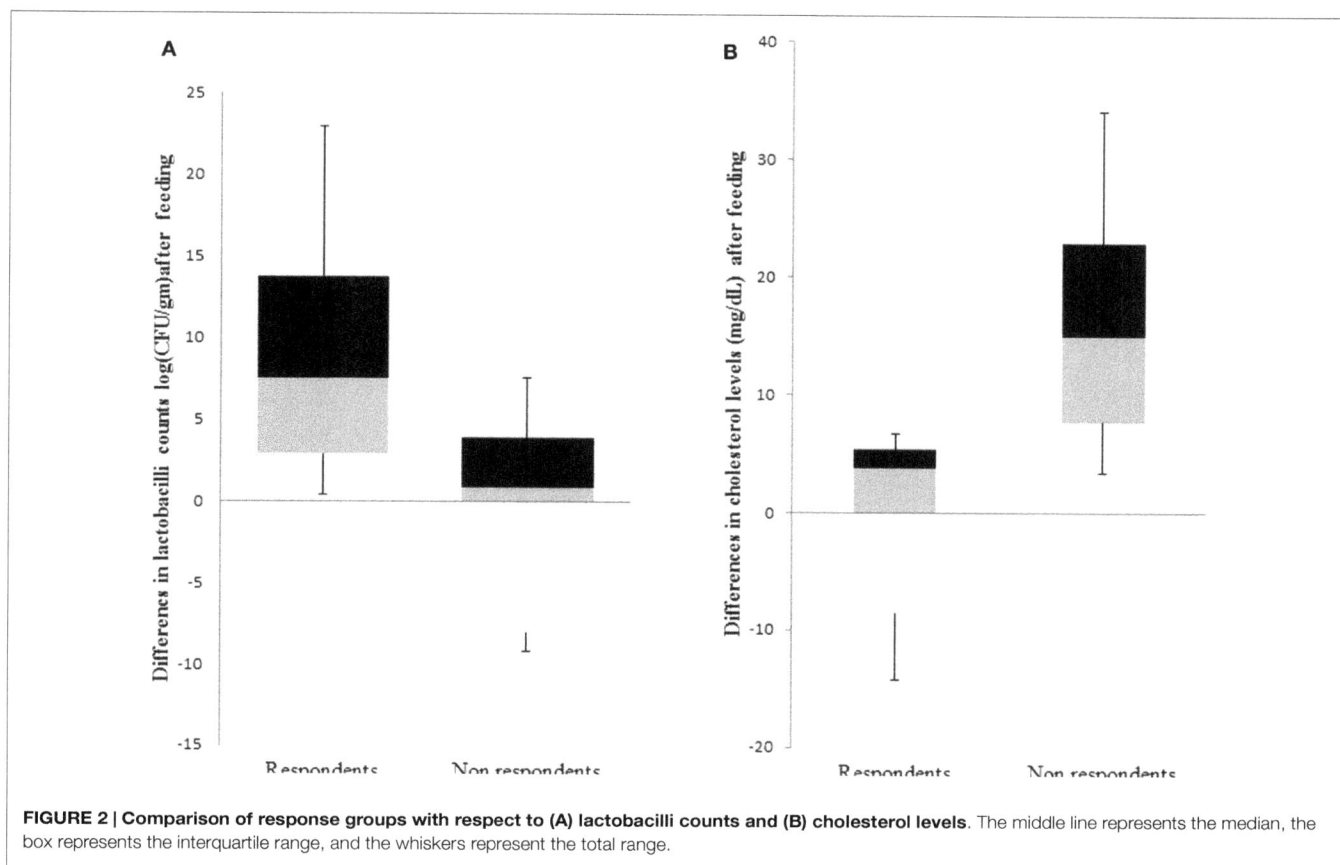

FIGURE 2 | Comparison of response groups with respect to (A) lactobacilli counts and (B) cholesterol levels. The middle line represents the median, the box represents the interquartile range, and the whiskers represent the total range.

Bacterial Taxa Populations Associated with Respondents and Non-Respondents

We performed a comparison of the microbiota between respondents and non-respondents to find specialized bacterial members within the abundant phyla *Firmicutes* and *Proteobacteria*. Respondents carried a lower proportion of *Clostridium* and a higher proportion of *Eubacterium* compared to the non-respondents (**Figure 4**). Surprisingly, although the non-respondents had a higher proportion of gut lactobacilli (31%) compared to 28% in respondents, a favorable reduction in cholesterol corresponding to the increase in strain MTCC 5463 was not observed. This could be due to competitive exclusion by a higher proportion of *Clostridium* (24%) in the gut on non-respondents compared to respondents (6%). The presence of *Listeria* in the non-respondents further emphasizes the need to investigate the association of gut microbiota, especially pathobionts with probiotic strain. Comparing the abundance in the genera of Proteobacteria group, it can be observed (**Figure 5**) that respondents carried a higher amount of *Burkholderia* (63%) and a lower amount of *Shigella* (7%) compared to non-respondents, who harbored lower count of *Burkholderia* (36%) and a higher amount of *Shigella* (31%), which must have affected the colonization of the probiotic strain. Non-respondents carried a higher amount of *Escherichia* and *Brucella* in the gut. *Shigella* seemed to have a symbiont asymptomatic existence in the host, showing no discomfort to the subjects. The higher amount of

Escherichia and *Camphylobacter* could be the deciding biomarkers of non-responsiveness toward probiotic intervention.

Statistical Analysis of Metagenomic Data

A remarkable significant difference among the chief genera of Proteobacteria including *Shigella*, *Escherichia*, *Burkholderia*, and *Camphylobacter* ($q < 0.002$) was observed. The chief genera of Firmicutes that showed remarkable significant difference were *Lactobacillus*, *Clostridium*, *Eubacterium*, and *Blautia* ($q < 0.002$) (**Figure 6**). Although non-respondents carried a higher proportion of Lactobacilli, a favorable physiological function may not be translated to the host possibly due to an increase in *Clostidum*, *Shigella*, and *Escherichia* with a decrease in *Blautia* and *Burkholderia*. We would like to add that the results of population wide samples taken at one time point for a study might not be able to display the entire variation that exists in that population over time and place.

Discussion

A primary beneficial effect of consuming a bile-salt-hydrolyzing *L. helveticus* MTCC 5463 strain is a reduction in serum cholesterol levels. In the clinical trials carried out to prove the hypocholesterolemic effect of the strain (Ashar and Prajapati, 2001; Prajapati et al., 2012) we could observe participants responding differently to the same treatment. Similar cases of inter-individual variability

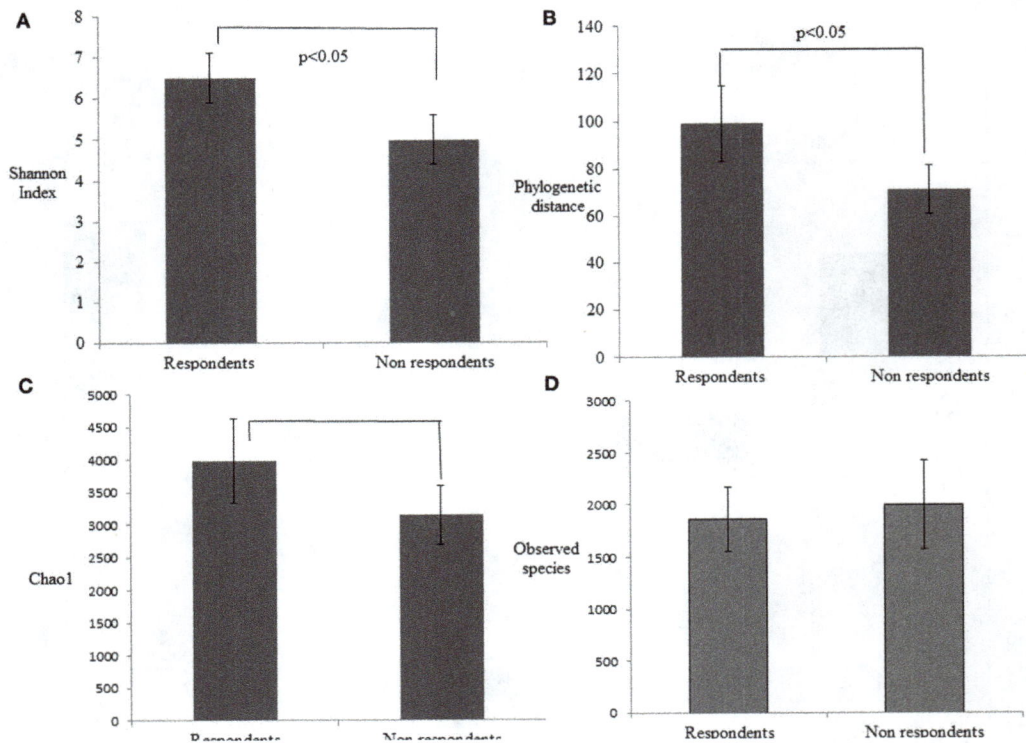

FIGURE 3 | Estimation of the phylogenetic diversity of the gut microbiota in the respondent and non-respondent groups using the (A) Shannon index, (B) phylogenetic distance, (C) Chao1, and (D) observed species. The values are means, and error bars indicate the 95% confidence intervals.

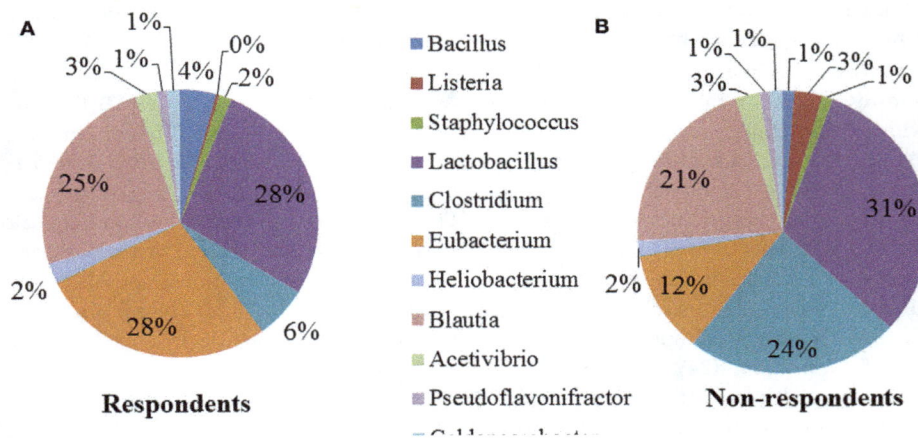

FIGURE 4 | Relative abundances of the dominant genera (Firmicutes) in (A) respondents and (B) non-respondents.

in response to probiotics (van Baarlen et al., 2011; Grzeskowiak et al., 2012; Arboleya et al., 2013) have been published. Previously, subjects had been classified as respondents and non-respondents based on a greater than or less than 10% change in cholesterol (Cox et al., 2014) but this classification was criticized as being impractical (Ding and Schloss, 2014). Classification using a fecal biomarker (Coen et al., 2009) or biomarkers of the host's basic metabolism (Naruszewicz et al., 2002; Herron et al., 2003; Ibrahim

et al., 2010) has been suggested. We conducted a comparative analysis of fecal microbiomes of respondents and non-respondents to identify bacterial biomarkers. Results of such microbial profiling may serve as a clinically useful biomarker in geriatric care (Kostic et al., 2013).

Primer usage is one of the most critical factors affecting 16S rDNA analysis (Armougom and Raoult, 2009). There exists a possibility that amplification efficacy of the primers could have

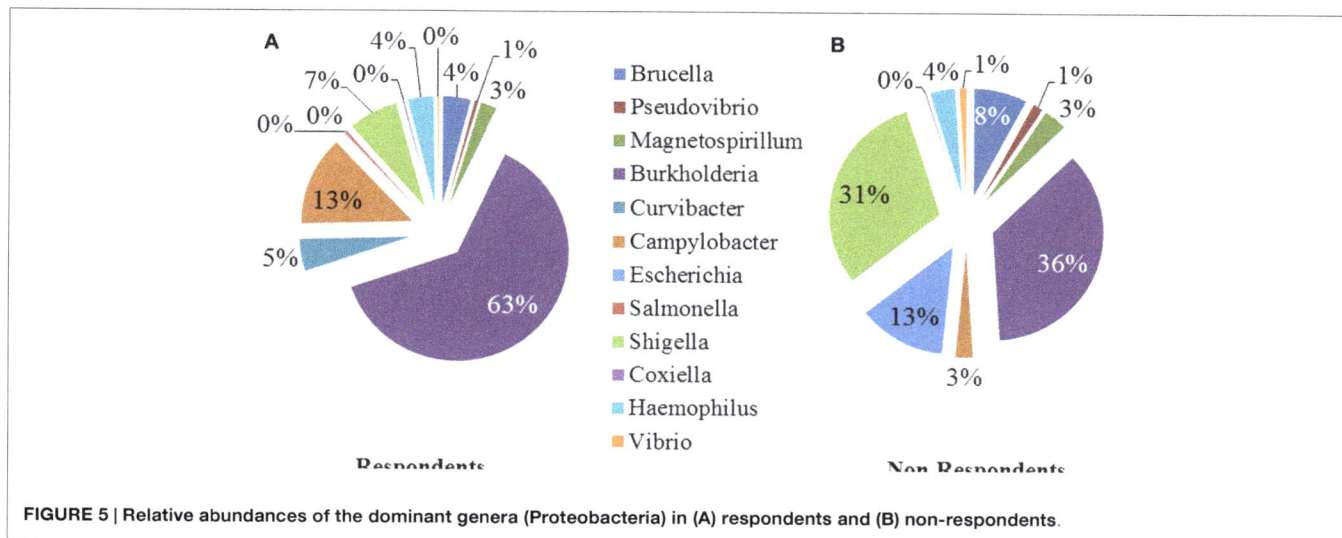

FIGURE 5 | Relative abundances of the dominant genera (Proteobacteria) in (A) respondents and (B) non-respondents.

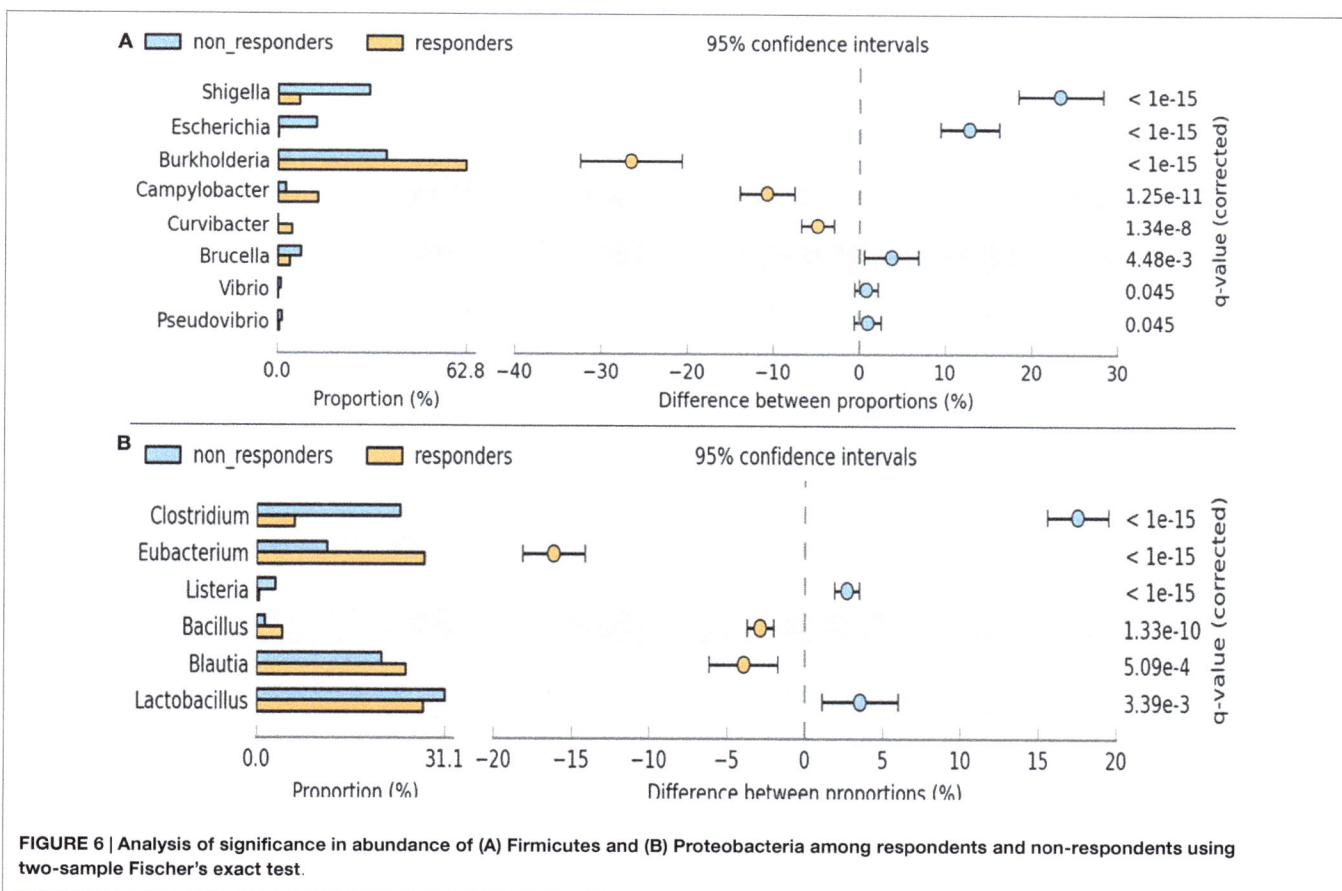

FIGURE 6 | Analysis of significance in abundance of (A) Firmicutes and (B) Proteobacteria among respondents and non-respondents using two-sample Fischer's exact test.

led to the underestimation of bacterial richness, in this study. We chose the Ion Torrent PGM platform due to its inherent low-cost per sequencing run that allowed us to perform the next-generation sequencing on site at the Veterinary faculty at AAU.

There were no significant differences in the age or weight of respondents and non-respondents, though respondents had a tendency to be older. An effect size of 0.8 among the groups further eliminated the influence of gender, weight, and age toward the expected response. Grouping of similar sequences as operational taxonomic units (OTUs) and quantifying the number of OTUs gave an approximation of species diversity in a sample (Sun et al., 2009; Schloss et al., 2014). Species diversity has been reported as a characteristic feature determining state of health or disease. Reduced colonic microbial diversity of dysbiosis is reported in

Crohn disease (Ott et al., 2004), ulcerative colitis (McLaughlin et al., 2010), antibiotic-associated diarrhea (Chang et al., 2008), and *Clostridium difficile* infection (Seto et al., 2014). A higher alpha diversity of gut bacteria among respondents compared to non-respondents seemed to support microbial integrity. Bacterial richness and evenness play an integral role in the success of a probiotic therapy as earlier observed in colitis (Kennedy et al., 2000).

A higher Chao's estimator and the Shannon index among respondents indicate higher saturation and unevenness in taxa abundance within samples. A non-significant increase in observed species in case of non-respondents suggested the presence of species that may have prevented the probiotic to establish and exert the functionality. We could identify distinct microbial diversity in rural subjects compared to the urban dwellers especially in the presence of *Clostridium*, *Shigella*, and *Listeria*, similar to Russian rural communities (Tyakht et al., 2013). Gram-negative bacteria were more abundant than Gram-positive bacteria in the rural population. *Shigella* and *Escherichia* were significantly under-represented in rural African children than urban European children (De Filippo et al., 2010). In the present study, we observed *Shigella* and *Escherichia* higher in geriatric rural dwellers, which reflected the age and geography-induced diversities in human gut microbiome.

Phyla associated with a healthy state in Indian geriatrics as suggested in this study are Firmicutes accounting for at least 50%, followed by Actinobacteria (20%) and Proteobacteria (10%). At the phylum level, the majority of the intestinal bacteria are known to belong to *Bacteroidetes* and *Firmicutes* (Eckburg et al., 2005). Surprisingly, none of the sequences from the present study were assigned to the Bacteroidetes. An under representation of Bacteroidetes could be due to inter-subject variability (Ley et al., 2006), variation due to adiposity (Frank et al., 2007; Wu et al., 2011), or suppression due to inflammatory bowel disease (Lazarevic et al., 2009). We could not ignore the possibility of loss of this phylum in fecal samples when stored for longer periods and MG-RAST-based classifications' sensitivity for Proteobacteria (Korpela et al., 2014). Antibiotic usage in elderly can again cause a decline in commensal anaerobes like *Bacteroides*, *Lactobacillus*, and *Bifidobacterium* (Macfarlane, 2014).

Responses of the host to individual bacterial strains are influenced by the baseline composition of the gut microbiota (Rajilić-Stojanović et al., 2015). The respondents showed a lower percentage of Firmicutes and non-respondents showed a comparative lower amount of Proteobacteria. Among Firmicutes, Clostridia were higher in case of non-respondents (24% compared to 6% in respondents). Although the volunteers were not consuming antibiotics during the trial, prior usage of antibiotics could have diminished the population of total and commensal bacteria (Biagi et al., 2010) leading to an overgrowth of *Clostridium* (Round and Mazmanian, 2009) in non-respondents. Antibiotics are readily available over the counter at pharmacies in India and inconsistent hospital standards toward antibiotic usage could have led to higher proportion of Clostridia in the gut. Respondents carried a higher proportion of *Eubacterium* (28% compared to 12% in non-respondents), which reflected a healthy state. An increased diversity of Eubacteria has been observed in the elderly

(Hopkins and Macfarlane, 2002). Decrease in *Eubacterium* lead to decreased levels of SCFA, facilitating easier entry of *Enterobacteriaceae* into the intestinal mucosa due to an impaired secretion of mucins by the intestinal epithelial cells (Garrett et al., 2010). The outgrowth of anaerobic *Enterobacteriaceae* must have led to a competitive exclusion of aerotolerant MTCC 55463 strains in the host intestine.

Proteobacteria, recently defined as "pathobionts" (Morgan et al., 2004) are considered to be minor and opportunistic components of the human gut ecosystem. The majority of the sequences assigned to the Proteobacteria were Burkholderiales. A higher proportion of *Burkholderia* is a signature of good health as earlier observed in the healthy Indian child data set (Schultz et al., 2004). Non-respondents had a higher proportion of Proteobacteria, especially *Escherichia/Shigella* (indistinguishable as a 16S-based phylotype), previously implicated in intestinal inflammation (Guslandi et al., 2004). Although the volunteers were seemingly healthy with no complaints of gastrointestinal disturbances, a core structural and functional dysbiosis caused by an overgrowth of *Escherichia/Shigella* (Malchow, 1997) could have led to a lack of translation of functionality of MTCC 5463 to the host in spite of being present at a higher proportion in the non-respondents gut.

In this study, it was shown that consumption of probiotic yogurt did not significantly reduce TC levels, the intervention significantly reduced serum levels of LDL, TC/HDL ratio, and LDL/HDL ratio in geriatric volunteers. Many studies in literature support the beneficiary effect of probiotics on lipid profiles of subjects. There also exist some contradictory reports where eating probiotic yogurt did not change lipid profiles (Hatakka et al., 2008; Sadrzadeh-Yeganeh et al., 2010). This indicates that apart from the probiotic strain, the host gut microbiome has a big role to play on the response of the host to the probiotic strain.

STAMP analysis revealed proportions of distinct microbial biomarkers like *Shigella*, *Escherichia*, *Burkholderia*, *Camphylobacter*, *Lactobacillus*, *Clostridium*, *Eubacterium*, and *Blautia* that can help tailor a probiotic therapy to a niche population. The authors would like to strike an analogy to feeding probiotics to a host with imbalanced consortia in the gut to the likes of pouring water to a filled pitcher. Like the water flows out, the probiotic strains are lost in the feces and fail to colonize and translate the functionality to the host. Metatranscriptomic studies could furnish further information for comprehending the molecular basis of responsiveness toward a probiotic therapy because gene expression profiles are more individualized than DNA-level profiles and less variable than microbial composition. Geriatric care is critical in the aging global population and the role of gut metagenomics cannot be overstated in understanding its role in health and disease for the future development of personalized nutrition.

Conclusion

Globally today, the elderly populations are looking for natural means of sustaining digestive health. Compared to the growing awareness and market penetration of probiotics, there is a dearth of scientific evidence on how probiotics affect the composition of gut microbiota. The well-documented probiotic *L. helveticus* MTCC 5463 was administered to geriatrics in a clinical trial, and a deep sequencing technology was employed to study the changes in the resident microbes over the duration of probiotic consumption. We could find that chiefly *Shigella, Escherichia, Burkholderia, Camphylobacter, Lactobacillus, Clostridium, Eubacterium,* and *Blautia* define the response of the host to the probiotic strain. Moreover, we observed a shift in the gut profile of the non-respondents towards a respondent's signature gut profile after consuming the probiotic, which proves the importance of precise personalized selection of dosage for an effective tailored probiotic therapy.

Ethics

Approval for the study design was obtained from the Institutional Ethics Committee (IEC) of Shri Krishna Medical College Karamsad, Anand, Gujarat (HMPCMCE:HREC/FCT/41/01) and Anand Agricultural University (AAU), Anand (AAU/DR/RES/DM/IEC/659/2011). The trial was registered at ICMR Clinical Trial Registry (REF/2012/10/004135).

References

Ahmed, S., Macfarlane, G. T., Fite, A., McBain, A. J., Gilbert, P., Macfarlane, S., et al. (2007). Mucosa-associated bacterial diversity in relation to human terminal ileum and colonic biopsy samples. *Appl. Environ. Microbiol.* 73, 7435–7442. doi:10.1128/AEM.01143-07

An, H. M., Baek, E. H., Jang, S., Lee, D. K., Kim, M. J., Kim, J. R., et al. (2010). Efficacy of lactic acid bacteria (LAB) supplement in management of constipation among nursing home residents. *Nutr. J.* 9, 5. doi:10.1186/1475-2891-9-5

Arboleya, S., Salazar, N., Solís, G., Fernández, N., Gueimonde, M., and de los Reyes-Gavilán, C. G. (2013). *In vitro* evaluation of the impact of human background microbiota on the response to *Bifidobacterium* strains and fructo-oligosaccharides. *Br. J. Nutr.* 110, 2030–2036. doi:10.1017/S0007114513001487

Armougom, F., and Raoult, D. (2009). Exploring microbial diversity using 16S rRNA high-throughput methods. *J. Comput. Sci. Syst. Biol.* 2, 74–92. doi:10.4172/jcsb.1000019

Ashar, M. N., and Prajapati, J. B. (2001). Serum cholesterol level in humans fed acidophilus milk. *Indian J. Microbiol.* 41, 257–263.

Bartosch, S., Fite, A., Macfarlane, G. T., and McMurdo, M. E. (2004). Characterization of bacterial communities in feces from healthy elderly volunteers and hospitalized elderly patients by using real-time PCR and effects of antibiotic treatment on the fecal microbiota. *Appl. Environ. Microbiol.* 70, 3575–3581. doi:10.1128/AEM.70.6.3575-3581.2004

Biagi, E., Nylund, L., Candela, M., Ostan, R., Bucci, L., Pini, E., et al. (2010). Through ageing, and beyond: gut microbiota and inflammatory status in seniors and centenarians. *PLoS ONE* 5:e10667. doi:10.1371/journal.pone.0010667

Caporaso, J. G., Kuczynski, J., Stombaugh, J., Bittinger, K., Bushman, F. D., Costello, E. K., et al. (2010). QIIME allows analysis of high-throughput community sequencing data. *Nat. Methods* 7, 335–336. doi:10.1038/nmeth.f.303

Chang, J. Y., Antonopoulos, D. A., Kalra, A., Tonelli, A., Khalife, W. T., Schmidt, T. M., et al. (2008). Decreased diversity of the fecal Microbiome in recurrent *Clostridium* difficile-associated diarrhea. *J. Infect. Dis.* 197, 435–438. doi:10.1086/525047

Claesson, M. J., Cusack, S., O'Sullivan, O., Greene-Diniz, R., de Weerd, H., Flannery, E., et al. (2011). Composition, variability and temporal stability of the intestinal microbiota of the elderly. *Proc. Natl Acad. Sci. U.S.A.* 108, 4586–4591. doi:10.1073/pnas.1000097107

Coen, M., Want, E. J., Clayton, T. A., Rhode, C. M., Hong, Y. S., Keun, H. C., et al. (2009). Mechanistic aspects and novel biomarkers of responder and non-responder phenotypes in galactosamine-induced hepatitis. *J. Proteome Res.* 8, 5175–5187. doi:10.1021/pr9005266

Cox, L. M., Yamanishi, S., Sohn, J., Alekseyenko, A. V., Leung, J. M., Cho, I., et al. (2014). Altering the intestinal microbiota during a critical developmental window has lasting metabolic consequences. *Cell* 158, 705–721. doi:10.1016/j.cell.2014.05.052

De Filippo, C., Cavalieri, D., Di Paola, M., Ramazzotti, M., Poullet, J. B., Massart, S., et al. (2010). Impact of diet in shaping gut microbiota revealed by a comparative study in children from Europe and rural Africa. *Proc. Natl Acad. Sci. U.S.A.* 107, 14691–14696. doi:10.1073/pnas.1005963107

de Wouters, T., Doré, J., and Lepage, P. (2000). Does our food (environment) change our gut microbiome ('In-Vironment'): a potential role for inflammatory bowel disease. *Dig. Dis.* 30, 33–39. doi:10.1159/000342595

Ding, T., and Schloss, P. D. (2014). Dynamics and associations of microbial community types across the human body. *Nature* 509, 357–360. doi:10.1038/nature13178

Eckburg, P. B., Bik, E. M., Bernstein, C. N., Purdom, E., Dethlefsen, L., Sargent, M., et al. (2005). Diversity of the human intestinal microbial flora. *Science* 308, 1635–1638. doi:10.1126/science.1110591

Frank, D. N., St Amand, A. L., Feldman, R. A., Boedeker, E. C., Harpaz, N., and Pace, N. R. (2007). Molecular-phylogenetic characterization of microbial community imbalances in human inflammatory bowel diseases. *Proc. Natl Acad. Sci. U.S.A.* 104, 13780–13785. doi:10.1073/pnas.0706625104

Garrett, W. S., Gallini, C. A., Yatsunenko, T., Michaud, M., DuBois, A., Delaney, M. L., et al. (2010). Enterobacteriaceae act in concert with the gut microbiota to induce spontaneous and maternally transmitted colitis. *Cell Host Microbe* 8, 292–300. doi:10.1016/j.chom.2010.08.004

Gerritsen, J., Smidt, H., Rijkers, G. T., and de Vos, W. M. (2011). Intestinal microbiota in human health and disease: the impact of probiotics. *Genes Nutr.* 6, 209–240. doi:10.1007/s12263-011-0229-7

Ghosh, T. S., Sen Gupta, S., Bhattacharya, T., Yadav, D., Barik, A., Chowdhury, A., et al. (2014). Gut microbiomes of Indian children of varying nutritional status. *PLoS ONE* 9:e95547. doi:10.1371/journal.pone.0095547

Grzeskowiak, L., Collado, M. C., Mangani, C., Maleta, K., Laitinen, K., Ashorn, P., et al. (2012). Distinct gut microbiota in southeastern African and northern European infants. *J. Pediatr. Gastroenterol. Nutr.* 54, 812–816. doi:10.1097/MPG.0b013e318249039c

Guigoz, Y., Doré, J., and Schiffrin, E. J. (2008). The inflammatory status of old age can be nurtured from the intestinal environment. *Curr. Opin. Clin. Nutr. Metab. Care* 11, 13–20. doi:10.1097/MCO.0b013e3282f2bfdf

Gupta, S. S., Mohammed, M. H., Ghosh, T. S., Kanungo, S., Nair, G. B., and Mande, S. S. (2011). Metagenome of the gut of a malnourished child. *Gut Pathog.* 3, 7. doi:10.1186/1757-4749-3-7

Author Contributions

Conceived and designed the experiments: JP, CJ, and HP; performed metagenomic analysis and manuscript writing: SS; product development: SV; clinical recruitment of participants: MG and US; clinical investigations and data interpretation: ST and RP; literature search and critical review of the manuscript: HAP; statistical data analysis: AP.

Acknowledgments

We are indebted to the technical staff of Ome Reserch Facility, Department of Animal Biotechnology, AAU, India whose names were not included in the author list, but who contributed to sequencing and data analysis of this work. The authors kindly acknowledge the funding support received from DBT, India, through Project BT/PR-14950/FNS/20/490/2010. Partial results of the study were presented in the Symposium on "Gut Microbiota, Probiotics and Their Impact throughout the Lifespan," conducted at Harvard Medical School, Boston, MA, USA. September 22–23, 2014.

Guslandi, M., Mezzi, G., Sorghi, M., and Testoni, P. A. (2004). *Saccharomyces boulardii* in maintenance treatment of Crohn's disease. *Dig. Dis. Sci.* 45, 1462–1466. doi:10.1023/A:1005588911207

Hatakka, K., Mutanen, M., Holma, R., Saxelin, M., and Korpela, R. (2008). *Lactobacillus rhamnosus* LC705 together with *Propionibacterium freudenreichii* ssp shermanii JS administered in capsules is ineffective in lowering serum lipids. *J. Am. Coll. Nutr.* 27, 441–447. doi:10.1080/07315724.2008.10719723

Herron, K. L., Vega-Lopez, S., Conde, K., Ramjiganesh, T., Shachter, N. S., and Fernandez, M. L. (2003). Men classified as hypo- or hyper respondents to dietary cholesterol feeding exhibit differences in lipoprotein metabolism. *J. Nutr.* 133, 1036–1042.

Hooper, L. V., Wong, M. H., Thelin, A., Hansson, L., Falk, P. G., and Gordon, J. I. (2001). Molecular analysis of commensal host-microbial relationships in the intestine. *Science* 291, 881–884. doi:10.1126/science.291.5505.881

Hopkins, M. J., and Macfarlane, G. T. (2002). Changes in predominant bacterial populations in human faeces with age and with *Clostridium* difficile infection. *J. Med. Microbiol.* 51, 448–454. doi:10.1099/0022-1317-51-5-448

Hopkins, M. J., Sharp, R., and Macfarlane, G. T. (2001). Age and disease related changes in intestinal bacterial populations assessed by cell culture, 16S rRNA abundance and community cellular fatty acid profiles. *Gut* 48, 198–205. doi:10.1136/gut.48.2.198

Ibrahim, F., Ruvio, S., Granlund, L., Salminen, S., Viitanen, M., and Ouwehand, A. C. (2010). Probiotics and immunosenescence: cheese as a carrier. *FEMS Immunol. Med. Microbiol.* 59, 53–59. doi:10.1111/j.1574-695X.2010.00658.x

Kennedy, R. J., Hoper, M., Deodhar, K., Kirk, S. J., and Gardiner, K. R. (2000). Probiotic therapy fails to improve gut permeability in a hapten model of colitis. *Scand. J. Gastroenterol.* 35, 1266–1271. doi:10.1080/003655200453601

Korpela, K., Flint, H. J., Johnstone, A. M., Lappi, J., Poutanen, K., Dewulf, E., et al. (2014). Gut microbiota signatures predict host and microbiota responses to dietary interventions in obese Individuals. *PLoS ONE* 9:e90702. doi:10.1371/journal.pone.0090702

Kostic, A. D., Chun, E., Meyerson, M., and Garrett, W. S. (2013). Microbes and inflammation in colorectal cancer. *Cancer Immunol. Res.* 3, 150–157. doi:10.1158/2326-6066.CIR-13-0101

Lahtinen, S. J., Tammela, L., Korpela, J., Parhiala, R., Ahokoski, H., Mykkänen, H., et al. (2009). Probiotics modulate the *Bifidobacterium* microbiota of elderly nursing home residents. *Age (Dordr)* 31, 59–66. doi:10.1007/s11357-008-9081-0

Lampe, J. W., Navarro, S. L., Hullar, M. A. J., and Shojaie, A. (2013). Inter-individual differences in response to dietary intervention: integrating omics platforms towards personalised dietary recommendations. *Proc. Nutr. Soc.* 72, 207–218. doi:10.1017/S0029665113000025

Lazarevic, V., Whiteson, K., Huse, S., Hernandez, D., Farinelli, L., Osterås, M., et al. (2009). Metagenomic study of the oral microbiota by Illumina high-throughput sequencing. *J. Microbiol. Methods* 79, 266–271. doi:10.1016/j.mimet.2009.09.012

Lederberg, J. (2000). Infectious history. *Science* 288, 287–293. doi:10.1126/science.288.5464.287

Ley, R. E., Turnbaugh, P. J., Klein, S., and Gordon, J. I. (2006). Microbial ecology: human gut microbes associated with obesity. *Nature* 444, 1022–1023. doi:10.1038/4441022a

Louis, P. (2012). Dietary modulation of the human gut microbiota. *Agro Food Industry Hi-Tech* 23, 26–28.

Macfarlane, S. (2014). Antibiotic treatments and microbes in the gut. *Environ. Microbiol.* 16, 919–924. doi:10.1111/1462-2920.12399

Malchow, H. A. (1997). Crohn's disease and *Escherichia coli*. A new approach in therapy to maintain remission of colonic Crohn's disease. *J. Clin. Gastroenterol.* 25, 653–658. doi:10.1097/00004836-199712000-00021

Mariat, D., Firmesse, O., Levenez, F., Guimaraes, V., Sokol, H., Doré, J., et al. (2009). The Firmicutes/Bacteroidetes ratio of the human microbiota changes with age. *BMC Microbiol.* 9:123. doi:10.1186/1471-2180-9-123

Matsumoto, M., Sakamoto, M., and Benno, Y. (2009). Dynamics of fecal microbiota in hospitalized elderly fed probiotic LKM512 yogurt. *Microbiol. Immunol.* 53, 421–432. doi:10.1111/j.1348-0421.2009.00140.x

McLaughlin, S. D., Clark, S. K., Tekkis, P. P., Nicholls, R. J., and Ciclitira, P. J. (2010). The bacterial pathogenesis and treatment of pouchitis. *Therap. Adv. Gastroenterol.* 3, 335–348. doi:10.1177/1756283X10370611

McOrist, A. L., Miller, R. B., Bird, A. R., Keogh, J. B., Noakes, M., Topping, D. L., et al. (2011). Fecal butyrate levels vary widely among individuals but are usually increased by a diet high in resistant starch. *J. Nutr.* 141, 883–889. doi:10.3945/jn.110.128504

Morgan, X. C., Tickle, T. L., Sokol, H., Gevers, D., Devaney, K. L., Ward, D. V., et al. (2004). Dysfunction of the intestinal microbiome in inflammatory bowel disease and treatment. *Genome Biol.* 13, R79. doi:10.1186/gb-2012-13-9-r79

Mueller, S., Saunier, K., Hanisch, C., Norin, E., Alm, L., Midtvedt, T., et al. (2006). Differences in fecal microbiota in different European study populations in relation to age, gender, and country: a cross-sectional study. *Appl. Environ. Microbiol.* 72, 1027–1033. doi:10.1128/AEM.72.2.1027-1033.2006

Naruszewicz, M., Johansson, M. L., Zapolska-Downar, D., and Bukowska, H. (2002). Effect of *Lactobacillus plantarum* 299v on cardiovascular disease risk factors in smokers. *Am. J. Clin. Nutr.* 76, 1249–1255.

Ott, S. J., Musfeldt, M., Wenderoth, D. F., Hampe, J., Brant, O., Fölsch, U. R., et al. (2004). Reduction in diversity of the colonic mucosa associated bacterial microflora in patients with active inflammatory bowel disease. *Gut* 53, 685–693. doi:10.1136/gut.2003.025403

Ouwehand, A. C., Tiihonen, K., Saarinen, M., Putaala, H., and Rautonen, N. (2009). Influence of a combination of *Lactobacillus acidophilus* NCFM and lactitol on healthy elderly: intestinal and immune parameters. *Br. J. Nutr.* 101, 367–375. doi:10.1017/S0007114508003097

Patil, D. P., Dhotre, D. P., Chavan, Ş. G., Sultan, A., Jain, D. S., Lanjekar, V. B., et al. (2012). Molecular analysis of gut microbiota in obesity among Indian individuals. *J. Biosci.* 37, 647–657. doi:10.1007/s12038-012-9244-0

Pitkala, K. H., Strandberg, T. E., FinneSoveri, U. H., Ouwehand, A. C., Poussa, T., and Salminen, S. (2007). Fermented cereal with specific bifidobacteria normalizes bowel movements in elderly nursing home residents. A randomized controlled trial. *J. Nutr. Health Aging* 2007, 305–311.

Prajapati, J. B., Khedkar, C. D., Chitra, J., Senan, S., Mishra, V., Sreeja, V., et al. (2011). Whole genome shotgun sequencing of an Indian-origin *Lactobacillus helveticus* strain MTCC 5463 with probiotic potential. *J. Bacteriol.* 193, 4282–4283. doi:10.1128/JB.05449-11

Prajapati, J. B., Nathani, N., Patel, A. K., Senan, S., and Joshi, C. G. (2013). Genomic analysis of dairy starter culture *Streptococcus thermophilus* MTCC 5461. *J. Microbiol. Biotechnol.* 23, 459–466. doi:10.4014/jmb.1210.10030

Prajapati, J. B., Senan, S., Momin, J. K., Damor, R., and Kamalia, K. B. (2012). A randomised double blind placebo controlled trial of potential probiotic strain *Lactobacillus helveticus* MTCC 5463: assessment of its safety, tolerance and influence on intestinal wellbeing and humoral immune response in healthy human volunteers. *Int. J. Health Pharm. Sci.* 1, 92–99.

Rajilić-Stojanović, M., Jonkers, D. M., Salonen, A., Hanevik, K., Raes, J., Jalanka, J., et al. (2015). Intestinal microbiota and diet in IBS: causes, consequences, or epiphenomena. *Am. J. Gastroenterol.* 110, 278–287. doi:10.1038/ajg.2014.427

Reid, G., Gaudier, E., Guarner, F., Huffnagle, G. B., Macklaim, J. M., Munoz, A. M., et al. (2010). Respondents and non-respondents to probiotic interventions: how can we improve the odds? *Gut Microbes* 1, 200–204. doi:10.4161/gmic.1.3.12013

Round, J. L., and Mazmanian, S. K. (2009). The gut microbiota shapes intestinal immune responses during health and disease. *Nat. Rev. Immunol.* 9, 313–323. doi:10.1038/nri2515

Sadrzadeh-Yeganeh, H., Elmadfa, I., Djazayery, A., Jalali, M., Heshmat, R., and Chamary, M. (2010). The effects of probiotic and conventional yoghurt on lipid profile in women. *Br. J. Nutr.* 103, 1778–1783. doi:10.1017/S0007114509993801

Schloss, P. D., Iverson, K. D., Petrosino, J. F., and Schloss, S. J. (2014). The dynamics of a family's gut microbiota reveal variations on a theme. *Microbiome* 2, 25. doi:10.1186/2049-2618-2-25

Schmalenberger, A., Schwieger, F., and Tebbe, C. C. (2001). Effect of primers hybridizing to different evolutionarily conserved regions of the small-subunit rRNA gene in PCR-based microbial community analyses and genetic profiling. *Appl. Environ. Microbiol.* 67, 3557–3563. doi:10.1128/AEM.67.8.3557-3563.2001

Schultz, M., Timmer, A., Herfarth, H. H., Sartor, R. B., Vanderhoof, J. A., and Rath, H. C. (2004). *Lactobacillus* GG in inducing and maintaining remission of Crohn's disease. *BMC Gastroenterol.* 4:5. doi:10.1186/1471-230X-4-5

Senan, S., Prajapati, J. B., and Joshi, C. G. (2015). Whole-genome based validation of the adaptive properties of Indian origin probiotic *Lactobacillus helveticus* MTCC 5463. *J. Sci. Food Agric.* 95, 321–328. doi:10.1002/jsfa.6721

Seto, C. T., Jeraldo, P., Orenstein, R., Chia, N., and DiBaise, J. K. (2014). Prolonged use of a proton pump inhibitor reduces microbial diversity: implications for *Clostridium* difficile susceptibility. *Microbiome* 2, 42. doi:10.1186/2049-2618-2-42

Sun, Y., Cai, Y., Liu, L., Yu, F., Farrell, M. L., McKendree, W., et al. (2009). ESPRIT: estimating species richness using large collections of 16S rRNA pyrosequences. *Nucleic Acids Res.* 37, e76. doi:10.1093/nar/gkp285

Tiihonen, J., Lonnqvist, J., Wahlbeck, K., Klaukka, T., Niskanen, L., Tanskanen, A., et al. (2009). 11-year follow-up of mortality in patients with schizophrenia: a population-based cohort study (FIN11 study). *Lancet* 374, 620–627. doi:10. 1016/S0140-6736(09)60742-X

Turnbaugh, P. J., Ridaura, K., Faith, J. J., Rey, F. E., Knight, R., and Gordon, J. I. (2009). The effect of diet on the human gut microbiome: a metagenomic analysis in humanized gnotobiotic mice. *Sci. Transl. Med.* 1, 6ra14. doi:10.1126/ scitranslmed.3000322

Tyakht, A. V., Kostryukova, E. S., Popenko, A. S., Belenikin, M. S., Pavlenko, A. V., Larin, A. K., et al. (2013). Human gut microbiota community structures in urban and rural populations in Russia. *Nat. Commun.* 4, 2469. doi:10.1038/ ncomms3469

van Baarlen, P., Troost, F., van der Meer, C., Hooiveld, G., Boekschoten, M., Brummer, R. J., et al. (2011). Human mucosal in vivo transcriptome responses to three lactobacilli indicate how probiotics may modulate human cellular pathways. *Proc. Natl Acad. Sci. U.S.A.* 108, 4562–4569. doi:10.1073/pnas.1000079107

Walker, A. W., Ince, J., Duncan, S. H., Webster, L. M., Holtrop, G., Ze, X., et al. (2011). Dominant and diet-responsive groups of bacteria within the human colonic microbiota. *ISME J.* 5, 220–230. doi:10.1038/ismej.2010.118

Woodmansey, E. J., McMurdo, M. E., Macfarlane, G. T., and Macfarlane, S. (2004). Comparison of compositions and metabolic activities of fecal microbiotas in young adults and in antibiotic-treated and non-antibiotic-treated elderly subjects. *Appl. Environ. Microbiol.* 70, 6113–6122. doi:10.1128/AEM.70.10.6113-6122.2004

Wu, G. D., Chen, J., Hoffmann, C., Bittinger, K., Chen, Y. Y., Keilbaugh, S. A., et al. (2011). Linking long-term dietary patterns with gut microbial enterotypes. *Science* 334, 105–108. doi:10.1126/science.1208344

Zwielehner, J., Liszt, K., Handschur, M., Lassl, C., Lapin, A., and Haslberger, A. G. (2009). Combined PCR-DGGE fingerprinting and quantitative-PCR indicates shifts in fecal population sizes and diversity of *Bacteroides*, Bifidobacteria and *Clostridium* cluster IV in institutionalized elderly. *Exp. Gerontol.* 44, 440–446. doi:10.1016/j.exger.2009.04.002

Conflict of Interest Statement: The authors declare that the research was conducted in the absence of any commercial or financial relationships that could be construed as a potential conflict of interest.

18

Diversity and antibiotic susceptibility of autochthonous dairy enterococci isolates: are they safe candidates for autochthonous starter cultures?

Amarela Terzić-Vidojević [1]*, Katarina Veljović [1], Jelena Begović [1], Brankica Filipić [1, 2], Dušanka Popović [1], Maja Tolinački [1], Marija Miljković [1], Milan Kojić [1] and Nataša Golić [1]

[1] Laboratory for Molecular Microbiology, Institute of Molecular Genetics and Genetic Engineering, University of Belgrade, Belgrade, Serbia, [2] Faculty of Pharmacy, University of Belgrade, Belgrade, Serbia

Edited by:
Fausto Gardini,
University of Bologna, Italy

Reviewed by:
Stella Maris Reginensi Rivera,
Universidad de la República, Uruguay
Chiara Montanari,
CIRI Agroalimentare, Italy

*Correspondence:
Amarela Terzić-Vidojević,
Laboratory for Molecular Microbiology,
Institute of Molecular Genetics and
Genetic Engineering, University of
Belgrade, Vojvode Stepe 444a, PO
Box 23, 11010 Belgrade, Serbia
amarela@imgge.bg.ac.rs;
lab6@imgge.bg.ac.rs
website: http://www.imgge.bg.ac.rs

Enterococci represent the most controversial group of dairy bacteria. They are found to be the main constituent of many traditional Mediterranean dairy products and contribute to their characteristic taste and flavor. On the other hand, during the last 50 years antibiotic-resistant enterococci have emerged as leading causes of nosocomial infections worldwide. The aim of this study was to determine the diversity, technological properties, antibiotic susceptibility and virulence traits of 636 enterococci previously isolated from 55 artisan dairy products from 12 locations in the Western Balkan countries (WBC) of Serbia, Croatia and Bosnia and Herzegovina. All strains were identified both by microbiological and molecular methods. The predominant species was *Enterococcus durans*, followed by *Enterococcus faecalis* and *Enterococcus faecium*. Over 44% of the isolates were resistant to ciprofloxacin and erythromycin, while 26.2% of the isolates were multi-resistant to three or more antibiotics belonging to different families. 185 isolates (29.1%) were susceptible to all 13 of the antibiotics tested. The antibiotic-susceptible isolates were further tested for possible virulence genes and the production of biogenic amines. Finally, five enterococci isolates were found to be antibiotic susceptible with good technological characteristics and without virulence traits or the ability to produce biogenic amines, making them possible candidates for biotechnological application as starter cultures in the dairy industry.

Keywords: *Enterococcus* sp., artisan dairy products, diversity, antibiotic susceptibility, virulence

Introduction

Bacteria of the genus *Enterococcus*, or enterococci, are considered lactic acid bacteria (LAB) (Schleifer and Ludwig, 1995). The usual ecological niche for *Enterococcus* species is the gastrointestinal tracts of humans and animals (Garg and Mital, 1991). However, enterococci are widely distributed in large numbers in foods, especially those of animal origin, such as various dairy and meat products, and some are also found in soil, water and on plants.

Enterococcus faecium, *Enterococcus faecalis*, and *Enterococcus durans* are the most prevalent species of enterococci in raw milk cheeses (Terzic-Vidojevic et al., 2007, 2009a,b, 2013, 2014a,b; Golić et al., 2013). They break down lactose and citrate during cheese ripening, which leads to the production of various volatile compounds, such as acetaldehyde, acetoin, diacetyl, and ethanol, which are responsible for the formation of the unique aroma and flavor of the final product (Andrighetto et al., 2001; Sarantinopoulos et al., 2001; Giraffa, 2002; Abeijón et al., 2006; Foulquié-Moreno et al., 2006). Due to their interesting metabolic and biotechnological traits (the ability to metabolize citrate, proteolytic and esterolytic activities, bacteriocin production, and their probiotic characteristics), enterococci may be utilized in food fermentation as commercial starter cultures (Tsakalidou et al., 1993; Centeno et al., 1999; Giraffa, 2003; Menéndez et al., 2004). On the other hand, enterococci have recently been found to be extremely important in clinical microbiology. Food-derived enterococci isolates are known to be resistant to a wide variety of antibiotics and to possess virulence genes (Giraffa, 2002; Saavedra et al., 2003; Veljović et al., 2014). These features contribute to the pathogenicity of enterococci, making them opportunistic pathogens. A major concern is the emergence of vancomycin-resistant enterococci (VRE), since this antibiotic is considered the last alternative for the treatment of multiple resistant infections (Teuber et al., 1999). Ideally, each strain that is intended to be used in a starter culture should be tested individually before any use in food or medicine, must not possess a single virulence factor and should be susceptible to relevant clinical antibiotics (Domig et al., 2003; Franz et al., 2003).

Due to their controversial status enterococci have been studied extensively (Galgano et al., 2001; Gelsomino et al., 2001; Morandi et al., 2006; Psoni et al., 2006; Gomes et al., 2008; Franz et al., 2011; Macedo et al., 2011; Ducková et al., 2014). Preliminary research on autochthonous LAB isolated from artisan dairy products of the Western Balkan region has indicated that about one third of all isolated LAB are *Enterococcus* species (Terzic-Vidojevic et al., 2007, 2009a,b, 2013, 2014a,b; Golić et al., 2013). However, a comprehensive study regarding their safe use as starter cultures in this region is lacking.

This paper evaluates the genetic diversity and antibiotic susceptibility of 636 *Enterococcus* spp. isolates from a laboratory collection including strains previously isolated from various artisan dairy products such as cheese, sour cream and kajmak from Serbia, Croatia and Bosnia and Herzegovina. The genetic diversity was explored by rep-PCR using the $(GTG)_5$ primer, combined with 16S rDNA sequencing. In antibiotic susceptible strains the presence of virulence factors and production of biogenic amines were also determined. We aimed to provide the most complete information about the phenotypic and genotypic diversity, as well as the safety status, of natural enterococcal isolates from traditional dairy products of the Western Balkan region for their potential biotechnological application as starter cultures in the dairy industry.

Materials and Methods

Bacterial Strains, Media, and Growth Conditions
All enterococci strains used in this study were isolated from 55 samples of different artisan dairy products collected from specific rural locations in Serbia, Bosnia and Herzegovina and Croatia in the period from 2003 to 2011. The samples of dairy products,

TABLE 1 | The list of dairy samples used as source of enterococci.

No.	Sample of dairy product	Type of dairy product and number of enterococcal isolates	Region of dairy product sampling and references
1	BGGO1	Cheese (8)	Serbia, Golija mountain;
2	BGGO2	Cheese (6)	(Terzic-Vidojevic et al., 2014a)
3	BGGO5	Cheese (6)	Serbia, Golija mountain; (Golić et al., 2013)
4	BGGO6	Cheese (6)	Serbia, Golija mountain; (Terzic-Vidojevic et al., 2014a)
5	BGGO7	Cheese (6)	Serbia, Golija mountain; (Golić et al., 2013)
6	BGGO8	Cheese (6)	Serbia, Golija mountain;
7	BGGO9	Cheese (8)	(Terzic-Vidojevic et al., 2014a)
8	BGGO10	Cheese (8)	
9	BGGO11	Cheese (6)	Serbia, Golija mountain; (Golić et al., 2013)
10	BGVLJ1	Yogurt (1)	Serbia, Vlasina mountain lake;
11	BGVL1	Cheese (21)	unpublished data
12	BGVL2	Cheese (2)	Serbia, Vlasina mountain lake;
13	BGVL2a	Cheese (15)	(Terzic-Vidojevic et al., 2013)
14	BGPT1	Cheese (11)	Serbia, Stara Planina mountain;
15	BGPT2	Cheese (4)	(Begovic et al., 2011)
16	BGPT3	Cheese (3)	
17	BGPT4	Cheese (5)	
18	BGPT5	Cheese (34)	
19	BGPT6	Cheese (6)	Serbia, Stara Planina mountain;
20	BGPT7	Cheese (4)	unpublished data
21	BGPT9	Cheese (9)	Serbia, Stara Planina mountain;
22	BGPT10	Cheese (11)	(Terzic-Vidojevic et al., 2009b)
23	BGAL1	Cheese (2)	Serbia, surrounding Aleksinac city;
24	BGAL2	Cheese (9)	(Golić et al., 2013)
25	BGAL3	Cheese (12)	
26	BGLE1	Cheese (6)	Serbia, surrounding Leskovac city; (Golić et al., 2013)
27	BGBU1	Cheese (19)	Serbia, Beljanica mountain;
28	BGRE2	Cheese (15)	(Golić et al., 2013)
29	BGZLM1	Milk (10)	Serbia, Zlatar mountain;
30	BGZLS1	Cheese (3)	(Terzic-Vidojevic et al., 2007)
31	BGZLS10	Cheese (12)	
32	BGZLS20	Cheese (13)	
33	BGZLS30	Cheese (12)	
34	BGZLS45	Cheese (20)	
35	BGZLS60	Cheese (24)	
36	BGNV1	Cheese (23)	Serbia, Zlatar mountain; (Terzic-Vidojevic et al., 2009a)

(Continued)

TABLE 1 | Continued

No.	Sample of dairy product	Type of dairy product and number of enterococcal isolates	Region of dairy product sampling and references
37	BGTRS1	Cheese (33)	Bosnia and Herzegovina, Vlašić mountain; (Terzic-Vidojevic et al., 2014b)
38	BGTRS7	Cheese (5)	
39	BGTRS10	Cheese (10)	
40	BGTRM1	Cream (25)	
41	BGTRM7	Cream (11)	
42	BGTRM10	Cream (14)	
43	BGTRK1	Kajmak (2)	
44	BGTRK4	Kajmak(2)	
45	BGTRK7	Kajmak(2)	
46	BGTRK10	Kajmak(11)	
47	BGPAS1	Cheese (26)	Bosnia and Herzegovina, surrounding Pale mountain city; unpublished data
48	ZGPR1	Cheese (15)	Croatia, Prigorje region; (Golić et al., 2013)
49	ZGPR2	Cheese (19)	
50	ZGPR3	Cheese (11)	
51	ZGBP4	Cheese (6)	Croatia, Bilogorsko-Podravska region; (Golić et al., 2013)
52	ZGBP5	Cheese (19)	
53	ZGBP6	Cheese (18)	
54	ZGZA7	Cheese (19)	Croatia, Zagorje region; (Golić et al., 2013)
55	ZGZA9	Cheese (22)	

together with their sources of isolation, are listed in **Table 1**. In total, 636 natural isolates of enterococci from the laboratory collection, initially identified by physiological tests, were used for the investigation of genetic and phenotypic diversity and antibiotic susceptibility.

Enterococci and lactococci used as indicator strains for the analysis of antimicrobial activity of the enterococci strains were grown in M17 broth (Merck, GmbH, Darmstadt, Germany) supplemented with glucose (0.5% w/v) (GM17) at 30°C, while lactobacilli, used to screen the antimicrobial activity of *Enterococcus* sp. were incubated in MRS broth (Merck) at 30°C. Preliminary screening of 636 enterococci isolates for production of antimicrobial compounds was done by the deferred antagonism method using overnight cultures of isolates and various indicator strains. Briefly, soft GM17 and MRS agars (0.7% w/v) containing lactococci or lactobacilli indicator strains were overlaid onto GM17 and MRS plates, respectively. The plates were incubated overnight at the appropriate temperature (30 or 37°C) depending on the indicator strain. A clear zone of inhibition of indicator strain growth around the well was taken as a positive signal for production of antimicrobial compound.

Escherichia coli and *Staphylococcus aureus* were cultivated in Luria broth (LB), containing 0.5% NaCl, 0.5% yeast extract (Torlak, Belgrade, Serbia), and 1% bacto tryptone (Torlak) at

TABLE 2 | The list of strains used in this study for bacteriocin-activity detection.

Bacterial strains	Source
Lactobacillus plantarum A112	Laboratory collection
Lactobacillus casei BGHN14	Laboratory collection
Lactococcus lactis subsp. *lactis* BGMN1-596	Laboratory collection
Lactococcus lactis subsp. *cremoris* NS1	Laboratory collection
Lactococcus lactis subsp. *lactis* biovar. *diacetylactis* S50	Laboratory collection
Enterococcus faecalis BG221	Laboratory collection
Listeria innocua ATCC 33090	ATCC[a]
Escherichia coli ATCC 25922	ATCC
Staphylococcus aureus ATCC 25923	ATCC

[a] ATCC–American Type culture Collection, Manassas, VA, USA.

37°C while *Listeria innocua* was cultivated on BHI medium (Difco, Detroit, MI, USA) at 37°C. Corresponding agar plates were prepared by adding agar (1.7% w/v, Torlak) into each broth. The indicator strains used in this study for screening enterococci antimicrobial activity are listed in **Table 2**. The isolates were stored at $-80°C$ in GM17 broth (Merck) supplemented with glycerol (15% v/v) and revitalized in the same medium by overnight growth at 30°C.

Bacterial Identification

After microscopic examination, enterococci were identified to the genus level by Gram staining, catalase testing, arginine and bile-esculin hydrolysis, growth at 45°C and growth in NaCl 6.5% broth. All 636 enterococci isolates were subjected to the following tests: growth at 15°C, growth in broth with NaCl (8% w/v), production of CO_2 from glucose, citrate utilization, acetoin and diacetyl production, time required for the formation of curd in reconstituted skim milk, exopolysaccharides (EPS) production, aggregation ability, and proteolytic and antimicrobial activity as described previously (Terzic-Vidojevic et al., 2007, 2009a,b, 2013, 2014a,b; Golić et al., 2013).

Identification of enterococci to the species level was performed according to Versalovic et al. (1994) using repetitive element palindromic-polymerase chain reaction (rep-PCR) analysis with (GTG)5 oligonucleotide (50-GTGGTGGTGGTGGTG-30). For this purpose the complete DNA from each of the 636 enterococci isolates was extracted as described by Hopwood et al. (1985) and details of the procedure has been reported by Terzic-Vidojevic et al. (2007). For sequencing of the 16S rRNA region, the complete DNA from certain enterococci isolates was used as a template for PCR amplification with UNI16SF (50-GAGAGTTTGATCCTGGC-30) and UNI16SR (50-AGG AGGTGATCCAGCCG-30) oligonucleotides (Jovcic et al., 2009). The PCR product obtained was purified by Qiagen (GmbH, Hilden, Germany) and sequenced (Macrogen, Amsterdam, the Netherlands and Seoul, South Korea). The BLAST algorithm was used to determine the most related sequences in the NCBI nucleotide sequence database (http://www.ncbi.nlm.nih.gov/BLAST).

Antibiotic Susceptibility Testing

The antibiotic resistance of enterococci isolates was determined by the disc diffusion method recommended by the Clinical and Laboratory Standards Institute (CLSI, 2012). The following antimicrobial drugs (Bio-Rad, Marnes-la-Coquette, France) were used: vancomycin (30 μg), teicoplanin (30 μg), ampicillin, (10 μg), erythromycin (15 μg), tetracycline (30 μg), minocycline (30 μg), quinupristin–dalfopristin (15 μg), ciprofloxacin (5 μg), chloramphenicol (30 μg), nitrofurantoin (300 μg), and linezolid (30 μg), and for high-level resistance (HLR) gentamicin (120 μg) and streptomycin (300 μg).

PCR Detection of Virulence Determinants

The complete DNA of 11 antibiotic-susceptible enterococci strains with the best technological characteristics was used in PCR reactions to detect the presence or absence of genes for the following virulence determinants: cytolysin (*cylA, cylB, and cylM*), aggregation factor (*agg*), gelatinase (*gelE*), enterococcal surface protein (*esp*), cell wall adhesions (*efaA$_{fs}$*, and *efaA$_{fm}$*), and sex-pheromones (*cpd, cob, and ccf*) according to Eaton and Gasson (2001), and collagen adhesin (*ace*) and hyaluronidase (*hyl*) as described by Vankerckhoven et al. (2004).

Biogenic Amines Determination

The ability of the 11 chosen enterococci strains to produce biogenic amines was qualitatively determined on an improved screening medium as described by Bover-Cid and Holzapfel (1999) using four precursor amino acids: histidine, lysine, ornithine, and tyrosine.

Statistical Analysis

Classical ecology indexes were used to obtain species richness (S), with the Shannon–Wiener index (H′) indicating general biodiversity and Simpson's index (D) evaluating dominance of the species in each cheese sample, as follows:

$$S = \Sigma^N; H' = -\Sigma^N p_i \log_2(p_i); D = 1 - \Sigma^N(p_i)^2$$

Where N is the number of species and p_i is the number of isolates belonging to one species in the sample. Clustering was carried out in Statistica 7.0 for Windows (StatSoft Inc. USA) and in BioNumerics 6.5 using the algorithm "Unweighted Pair-Group Average Linkage Analysis." Distances between the clusters were assessed using "Percent of disagreement."

Results

The Diversity of Enterococci Isolates

The general index of enterococci species diversity (H′) was calculated on the basis of the number of different enterococci species among the 636 isolates. The results showed that the dairy samples analyzed contained five *Enterococcus* species: *E. durans, E. faecalis, E. faecium, E. italicus,* and *E. avium*. The highest

diversity and the lowest index of dominance of enterococci species were scored by samples from cheeses from the Prigorje region, Croatia (H′ = 1.61; D = 0.36) and Pirot region, Serbia (H′ = 1.5; D = 0.34), where four of five enterococci species were identified (*E. durans, E. faecalis, E. faecium, E. italicus*). In contrast, the lowest diversity (H′ = 0.99) and the highest dominance index (D = 0.62) were scored by samples from cheese manufactured in Pale, Bosnia and Herzegovina, where almost all isolates (20 out of 26) belonged to *E. faecium*. *E. durans* was the most abundant species in dairy products from Travnik, Bosnia and Herzegovina (84 out of 115), while *E. faecalis* was the most abundant in dairy samples from Zlatar Mountain, Serbia (84 out of 117). Interestingly, only 12 out of 636 isolates belonged to *E. italicus* and they were isolated from five out of 12 regions, while only one out of 636 isolates was identified as *E. avium* and was found in the Zlatar region, Serbia (**Table 3**).

Genotypic Characterization of Enterococci

All 636 enterococci isolates were subjected to (GTG)$_5$-fingerprint analysis. The results revealed that 278 out of 636 isolates (43.71%) differ among each other and have unique (GTG)$_5$-fingerprint profiles, indicating great diversity of the isolates. Among them 124 isolates belonged to *E. durans*, 74 to *E. faecium*, 68 to *E. faecalis*, 11 to *E. italicus* and 1 to *E. avium* species. The fingerprint profiles obtained by (GTG)$_5$-PCR of the enterococci isolates are presented in **Figure S1**. Interestingly, four main clusters have been determined. According to (GTG)$_5$-fingerprint patterns, Cluster 1, comprising 46 isolates, includes the isolates with less genetic distance, mostly isolated from dairy products sampled in close geographical regions situated in the western part of Serbia. Cluster 2, comprising 124 isolates, includes more heterogeneous isolates, mostly isolated from dairy products sampled from southern Serbia, but also from Bosnia and Herzegovina. Cluster 3, comprising 10 isolates, includes only enterococci isolated from Bosnia and Herzegovina, specifically from milk and cheese sampled in the same household. Cluster 4 is the most heterogeneous according to (GTG)$_5$-fingerprint profiles and encompasses the isolates originating from various geographical locations. It can be further divided into groups 4a and 4b. Group 4a is mostly comprised of the isolates originating from Golija Mountain, while group 4b is further divided to subgroups comprising isolates originating from western Serbia and Bosnia and Herzegovina. Hence, the results show that the corresponding groups of strains are partly correlated with the sources of isolation. Interestingly, a subdivision on the basis of households in the same locality was seen within the groups of isolates (e.g., the isolates from dairy products sampled in Croatia are scattered throughout the dendrogram), and isolates from the same households are divided among separate clusters. Finally, the isolates that were undistinguished by rep-PCR were classified on the basis of phenotypic characteristics.

Phenotypic Characterization of Enterococci

Apart from general and genotypic diversity, the ultimate goal in exploring the diversity of natural isolates from various ecological niches is to determine their technological and functional potential. Hence, detailed phenotypic characterization

TABLE 3 | The diversity of *Enterococcus* isolates from autochthonous dairy products collected at various geographic locations of the Western Balkan Countries.

	E. durans	E. faecalis	E. faecium	E. italicus	E. avium	Sum	H′	D
Golija mountain	43	4	13	0	0	60	1.08	0.56
Vlasina lake	28	7	4	0	0	39	1.12	0.56
Pirot	41	21	21	4	0	87	1.5	0.34
Aleksinac	12	8	3	0	0	23	1.4	0.41
Leskovac	3	3	0	0	0	6	1	0.5
Beljanica	9	25	0	0	0	34	0.51	0.61
Zlatar mountain	15	84	15	2	1	117	1.26	0.55
Prigorje	15	21	8	1	0	45	1.61	0.36
Bilogorsko-Podravska	22	12	9	0	0	43	1.48	0.38
Zagorje	6	24	10	1	0	41	1.48	0.42
Travnik	84	16	11	4	0	115	1.22	0.56
Pale	2	4	20	0	0	26	0.99	0.62

of the enterococci strains with regard to acetoin (VP^+) and diacetyl (D^+) production, citrate utilization (C^+), time of milk curdling (TMC), antimicrobial properties (Bac^+), production of proteinases (Prt^+), aggregation ability (Agg^+), and EPS production was performed (**Table 4**). A large number of enterococci produced acetoin and diacetyl, 54.6 and 22.3% respectively. In addition, 40.9% of enterococci could utilize citrate as their only carbon source. On the other hand, only 4.2% of all the enterococci curdled milk within 6 h at 37°C (**Table 4**). Examination of the proteolytic activity of all 636 strains revealed that 17.5% of enterococci exhibited proteolytic activity as evaluated from β-casein hydrolysis by whole cells after 3 h of incubation. Almost 25% of all enterococci isolates showed the ability to produce antimicrobial compounds and most of those were enterococci that belonged to the *E. durans* species (29.3% of all 280 *E. durans* strains). Experiments with pronase E revealed the proteinaceous nature of the antimicrobial compounds, indicating that they could be bacteriocin-like substances (BLIS). Thirteen of 280 *E. durans*, three of 114 *E. faecium*, two of 229 *E. faecalis* strains and one *E. avium* strain had aggregation ability. Furthermore, one *E. durans* and one *E. italicus* strain were EPS producers. Some isolates exhibited two, three, or even four of the tested characteristics (**Table 4**).

Antibiotic Susceptibility and Resistance

An important issue in the selection of strains for safe use in food production is the characterisation of their antibiotic susceptibility in order to avoid uncontrolled spreading of the antibiotic resistance genes through horizontal gene transfer. Susceptibility of all 636 enterococci isolates was examined by the agar disc diffusion method using 13 antibiotics: vancomycin, teicoplanin, ampicillin, erythromycin, tetracycline, minocycline, quinupristin–dalfopristin, ciprofloxacin, chloramphenicol, nitrofurantoin, linezolid, gentamicin (HLR), and streptomycin (HLR) (**Table 5**). The results of antibiotic susceptibility revealed that a total of 451 out of 636 isolates (70.9%) were resistant to at least one of the tested antibiotics. One hundred eighty-two (28.6%) isolates were resistant to one antibiotic (151 to two, 77

to three, 26 isolates to four, 7 isolates to five, 4 isolates to six, 3 isolates to seven antibiotics, while 1 isolate was resistant to even eight antibiotics. It is worthwhile to note that the most resistant isolates belonged to *E. faecalis* species. In general, the highest percent of the enterococci isolates shown to have multiple resistances to various antibiotics originated from dairy products sampled from Pirot, Zlatar Mountain and Vlasina Lake in Serbia and the Bilogorsko-Podravski region and Zagorje, Croatia. On the other hand, a high number of the strains were susceptible to all tested antibiotics (185/636). Eighty nine of them belonged to *E. durans* species, originating mainly from Golija Mountain, Serbia and Travnik, Bosnia and Herzegovina. Thirteen of 31 antibiotic sensitive *E. faecalis* strains were isolated from the region of Zlatar Mountain, Serbia, while 11 of 28 and 9 of 28 antibiotic-sensitive *E. faecium* strains were isolated from the Pale and Travnik regions, Bosnia and Herzegovina. In addition, four strains of *E. italicus* species were also sensitive to all tested antibiotics (data not shown).

Virulence Determinants and Biogenic Amines Production

Finally, 136 out of 185 antibiotic-susceptible isolates, with unique $(GTG)_5$-fingerprint profiles, were further analyzed in order to choose the candidates with the best technological properties. Based on the phenotypic characteristics, 10 *E. durans* strains [BGGO6-15 (D^+, C^+, VP^+), BGAL3-19 (Bac^+, C^+, $VP^\pm$), BGTRM1-52 (Bac^+, Prt^+, VP^+, D^+, TMC 6.5 h), BGTRM7-39 (Bac^+, D^+, C^+, $VP^\pm$), BGTRS1-10 (Bac^+, Prt^+, $D^\pm$), BGTRS7-54 (Bac^+, Prt^+, VP^+, TMC 6.5 h), BGTRS10-42 (Bac^+, D^+, $VP^\pm$), BGTRK10-29 (C^+, VP^+, $D^\pm$), BGPAS1-80 (VP^+, TMC 6 h), and ZGPR2-1 (Bac^+, D^+, $C^\pm$)] and one *E. italicus* [BGTRK4-42 (Bac^+, D^+, $VP^\pm$)] were chosen.

The PCR analysis for detection of virulence genes (see PCR Detection of Virulence Determinants for details) revealed that none of the virulence genes was found in any of the analyzed strains. Moreover, the results obtained in this study indicate that five out of the 11 chosen enterococci strains (BGAL3-19, BGTRS7-54, BGTRM7-39, BGTRS10-42, and ZGPR2-1) were

TABLE 4 | Different technological activities in *Enterococcus* isolates from autochthonous dairy products.

Acetoin production (% of ent.)	Citrate utilization (% of ent.)	Diacetyl production (% of ent.)	TMC[a] up to 6h (% of ent.)	Bacteriocin production (% of ent.)	Proteinases production (% of ent.)	Aggregation ability (% of ent.)	EPS[b] production (% of ent.)
F2[c]	I	D	F1	D	I	A[g]	I
157/229 (68.6)	6/12 (50.0)	75/280 (26.8)	8/114 (7.0)	82/280 (29.3)	4/12 (33.3)	1/1 (100.0)	1/12 (8.3)
D[d]	F2	I	F2	F1	F1	D	D
138/280 (49.3)	113/229 (49.3)	3/12 (25.0)	15/29 (6.6)	31/114 (27.2)	21/114 (18.4)	13/280 (4.6)	1/280 (0.35)
F1[e]	D	F1	D	F2	F2	F1	F1
48/114 (42.1)	105/280 (37.5)	22/114 (19.3)	4/280 (1.4)	44/229 (19.2)	42/229 (18.3)	3/114 (2.6)	0/114 (0.0)
I[f]	F1	F2	I	I	D	F2	F2
3/12 (25.0)	36/114 (31.6)	42/229 (18.3)	0/12 (0.0)	1/12 (8.3)	44/280 (15.7)	2/229 (0.9)	0/229 (0.0)
A[g]	A	A	A	A	A	I	A
1/1 (100.0)	0/1 (0.0)	0/1 (0.0)	0/1 (0.0)	0/1 (0.0)	0/1 (0.0)	0/12 (0.0)	0/1 (0.0)
A TOTAL							
347/636 (54.6)	260/636 (40.9)	142/636 (22.3)	27/636 (4.2)	158/636 (24.8)	111/636 (17.5)	19/636 (2.98)	2/636 (0.31)

[a] Time of milk curdling, [b] Exopolysaccharides, [c] E. faecalis, [d] E. durans, [e] E. faecium, [f] E. italicus, [g] E. avium.

not able to produce biogenic amines in the presence of the precursor amino acids histidine, lysine, ornithine, and tyrosine (**Table 6**).

Discussion

Enterococci are found to be a normal part of the microbiota of artisan dairy products, especially in the Mediterranean region (Saavedra et al., 2003; Morandi et al., 2006; Psoni et al., 2006). However, their use as starter cultures in the dairy industry is still controversial since they have traditionally been considered indicators of fecal contamination. Moreover, enterococci are found to be involved in food spoilage (Franz et al., 1999) and food poisoning (Gardin et al., 2001), as well as in the spread of antibiotic resistance (Giraffa, 2002). Also, their role in the etiology of nosocomial infections cannot be neglected (Franz et al., 2003; Giraffa, 2003; Kayser, 2003). Hence, the aim of this paper was to evaluate the technological potential and safety issues for the use of natural dairy isolates of enterococci as starter cultures in the dairy industry. For that reason, in this study we have analyzed 636 natural isolates of *Enterococcus* sp. originating from various artisan dairy products collected in Western Balkan countries (WBC).

The results obtained in this study revealed the huge diversity among dairy enterococci isolates. The highest enterococci species diversity was recorded in fresh soft cheeses (1–10 days old) sampled from the Prigorje region, Croatia and Pirot, Serbia. Our previous results showed that enterococci and *Leuconostoc* sp. were the dominant species in the fresh soft cheeses of the Prigorje region, while lower diversity was found among other LAB species, indicating that enterococci were probably metabolically active and important for ripening of the artisan cheeses in that region (Golić et al., 2013).

Taking into account the diverse origins of the isolates, considerable genotypic heterogeneity was observed. The results

showed that (GTG)₅-fingerprint analysis was very effective in grouping and discriminating among enterococci strains, partly correlating with the sources of isolation. Hence, the results indicate that variable conditions, such as local vegetation or specific climates in the regions where dairy products are manufactured, significantly contribute to the diversity among enterococci strains. Interestingly, isolates belonging to the same species are found to be located in different clusters, indicating that identification by sequencing of 16S rDNA is not completely reliable in the case of natural isolates originating from complex communities such as dairy products. Possibly, horizontal transfer among bacteria living in the same ecological niche lead to their convergent/divergent evolution.

In order to explore the technological and functional potential of the natural enterococci isolates of artisan dairy origin, phenotypic variability was characterized. Due to various environmental pressures, natural isolates are usually shown to exhibit phenotypic variability (Giraffa et al., 2000, 2004). Enterococci play an important role in cheese ripening and contribute to the formation of the distinctive flavor of dairy products. Their ability to synthesize volatile compounds, such as diacetyl and acetoin, their proteolytic activity, and certainly antimicrobial activity are the important technological characteristics which make enterococci good candidates for starter and functional cultures for the dairy industry (Asteri et al., 2009; Nieto-Arribas et al., 2011). Enterococci usually exhibit weak proteolytic activities (Suzzi et al., 2000), and the best ability for casein degradation is shown by *E. faecalis* strains (Sarantinopoulos et al., 2001; Veljovic et al., 2009). Although the strains *E. faecalis* BGPT1-10P and BGPT1-78 were shown previously to have high activity in milk (Veljovic et al., 2009), in this study we found that only 17.5% of the analyzed strains exhibited proteolytic activity and it was equally distributed among *E. durans* (44/280), *E. faecium* (21/114), and *E. faecalis* (42/229) strains. According to the results of Morea et al. (1999) the degree of milk acidification by enterococci strains depends on

TABLE 5 | Regional distribution of antibiotic resistance of enterococci.

Region (number of enterococcal isolates)	Antibiotics (number of resistant isolates toward certain antibiotics)												
	VAN 30 µg	TEC 30 µg	AMP 10 µg	TET 30 µg	MNO 30 µg	ERY 15 µg	QDP 15 µg	CIP 5 µg	CHL 30 µg	HLG 120 µg	STR 300 µg	FTN 300 µg	LZD 30 µg
SERBIA													
Golija (60)	0	0	0	3	0	18	0	12	0	0	0	22	3
Vlasina (39)	3	0	0	17	12	1	7	14	0	0	0	22	1
Pirot (87)	1	1	1	9	4	38	9	69	6	2	2	54	2
Aleksinac (23)	0	0	0	0	0	0	0	8	0	0	0	6	1
Leskovac (6)	1	0	0	4	0	0	0	5	0	0	0	2	1
Beljanica (34)	0	0	0	11	2	8	0	20	0	0	0	1	0
Zlatar (117)	23	0	0	22	9	9	13	73	0	2	2	9	30
Total Serbia (366)	28 (7.7%)	1 (0.3%)	1 (0.3%)	66 (18.0%)	27 (7.4%)	74 (20.2%)	29 (7.9%)	201 (54.6%)	6 (1.6%)	4 (1.1%)	4 (1.1%)	116 (31.7%)	38 (10.4%)
CROATIA													
Prigorje (45)	3	0	1	12	7	3	7	10	2	3	1	14	14
Bilogorsko Podravski region (43)	5	1	0	3	3	1	10	28	0	2	2	4	8
Zagorje (41)	4	0	0	14	2	5	25	21	0	0	0	3	13
Total Croatia (129)	12 (9.3%)	1 (0.8%)	1 (0.8%)	29 (22.5%)	12 (9.3%)	9 (6.97%)	42 (32.6%)	59 (45.7%)	2 (1.6%)	5 (3.8%)	3 (3.9%)	21 (16.3%)	35 (27.1%)
BOSNIA AND HERZEGOVINA													
Travnik (115)	0	0	1	16	2	2	3	12	1	0	2	11	8
Pale (26)	0	0	0	0	0	2	2	10	0	0	0	4	4
Total BH (141)	0	0	1 (0.7%)	16 (11.3%)	2 (1.4%)	4 (2.8%)	5 (3.5%)	22 (15.6%)	1 (0.7%)	0	2 (1.4%)	15 (10.6%)	12 (8.5%)
Total WBC (636)	40 (6.3%)	2 (0.3%)	3 (0.5%)	111 (17.5%)	41 (6.4%)	87 (13.7%)	76 (11.9%)	282 (44.3%)	9 (1.4%)	9 (1.4%)	9 (1.4%)	152 (23/9%)	85 (13.4%)

VAN, vancomycin; TEC, teicoplanin; AMP, ampicillin; TET, tetracycline; MNO, minocycline; ERY, erythromycin; QDP, quinupristin-dalfopristin; CIP, ciprofloxacin; CHL, chloramphenicol; HLG, gentamicin; STR, streptomycin; FNT, nitrofurantoin; LZD, linezolid.

TABLE 6 | Production of biogenic amines in the presence of the precursor amino acids histidine, lysine, ornithine, and tyrosine.

	Histidine	Lysine	Ornithine	Tyrosine
BGGO6-15	+	+	+	–
BGAL3-19	–	–	–	–
BGTRM1-52	–	–	+	–
BGTRM7-39	–	–	–	–
BGTRS1-10	–	–	+	–
BGTRS7-54	–	–	–	–
BGTRS10-42	–	–	–	–
BGTRK10-29	–	–	+	–
BGPAS1-80	–	+	+	–
ZGPR2-1	–	–	–	–
BGTRK4-42	+	–	–	–

the origin of the strain, hence it is highly likely that enterococci isolates of dairy origin are adapted to growth in milk. On the other hand, synthesis of diacetyl, acetoin from glucose, and citrate in the process of metabolic degradation by enterococci are shown to be very important for flavor formation in dairy products (Nieto-Arribas et al., 2011). Our analysis of enterococci dairy isolates from the WBC region showed that 347 out of 636 strains produced acetoin (54.6%), and 260/636 strains utilized citrate (40.9%), while 142/636 strains synthesized diacetyl (22.3%).

In addition, the excellent antibacterial activity shown by natural enterococci isolates make them promising candidates for food preservation and may contribute to the prevention of food spoilage (Viedma et al., 2009; Ananou et al., 2010). Bacteriocins produced by enterococci, enterocins, are very diverse and widely distributed among isolates. Production of enterocins in combination with a wide range of tolerance to high temperatures, dryness and increased salinity enables enterococci to become the dominant microbiota in fermented products (Franz et al., 1999). Our results support previous reports and indicate that the antimicrobial activity in natural dairy isolates from the WBC region showed a great effect on a number of pathogenic and non-pathogenic strains. Antimicrobial activity detected after treatment with protease suggests the proteinaceous nature of the bacteriocin activity. In our previous work it was shown that a number of E. faecalis strains exhibited an antimicrobial effect on L. innocua and Listeria monocytogenes (Veljovic et al., 2009). In particular, strains BGPT1-10P and BGPT1-78 showed antimicrobial activity against the Gram-negative strain Pseudomonas sp. PA17 and therefore could be used for the production of food biopreservatives (Veljovic et al., 2009). Apart from that, a number of E. durans and E. faecium strains exhibiting antimicrobial activity were found in this study.

Finally, with the aim of the safe use of enterococci as starter cultures in functional food, the frequency of virulence determinants and antibiotic resistance, as well as the synthesis of biogenic amines, was analyzed. The results showed that

185 out of 636 isolates (29.1%) were susceptible to the tested antibiotics, and as many as 59.6% of those isolates were resistant to two or more antibiotics. Interestingly, the presence of virulence determinants and antibiotic resistance is strain-dependent and region-specific. A high number of isolates from localities in Serbia (Zlatar Mountain, Pirot, and Vlasina Lake) had multiple antibiotic resistances. The results implicate the uncontrolled use of antibiotics in these regions, leading to antibiotic residues in feed or foods, which contribute to the occurrence of acquired and/or mutational antibiotic resistance mechanisms. More importantly, the further dissemination of antibiotic resistance genes among food-associated bacteria could cause the loss of natural isolates suitable for application in the dairy industry.

In order to eliminate the enterococci with pathogenic potential from among the possible candidates for use in dairy food production, and to avoid further transfer of virulence genes to other bacteria in the environment, identification of virulence factors is essential (Franz et al., 2003; Mannu et al., 2003). Fortunately, these important features are strain-specific, not species-specific. For this reason, each enterococcal strain intended for use in the dairy industry should be thoroughly tested for the absence of any pathogenic property. Ideally, a strain proposed for use in food production should not possess a single virulence factor and must be sensitive to relevant clinical antibiotics (Domig et al., 2003; Franz et al., 2003).

Virulence factors are commonly detected in clinical isolates of enterococci. Studies by Franz et al. (2001) and Eaton and Gasson (2001) have indicated the presence of virulence factors in food isolates, especially among strains of E. faecalis, which generally carry more virulence factors than strains of E. faecium. Our previous results showed that dairy strains E. faecalis BGPT1-10P and BGPT1-78 carry the genE gene and have gelatinase activity (Veljović et al., 2014). Moreover, in the strain BGPT1–10P a set of three genes of the cytolysin operon (cylM, cylB, and cylA) was detected, although two additional genes, cylL1 and cylL2, required for the expression of hemolytic activity, were not identified in this strain (Veljović et al., 2014). The genes for hemolytic activity (esp and efaA) are found in high frequency in all tested E. faecalis strains (Veljović et al., 2014). However, in this study we demonstrated that none of the 11 selected E. durans dairy strains had any of the tested virulence genes, making them good candidates for use in the dairy industry.

Besides the analysis of virulence determinants and antibiotic resistance, an analysis of biogenic amine synthesis was performed. Biogenic amines are organic bases with aliphatic, aromatic, and heterocyclic structures produced through decarboxylation of the corresponding amino acid (Giraffa, 2002). The capacity for amino acid decarboxylation restricts the use of the strains in the dairy industry. The previous results of Valenzuela et al. (2009) showed that E. faecalis BGPT1-10P has the ability of tyrosine and ornithine decarboxylation. The strains tested in this study did not have the ability of amino acid decarboxylation.

Conclusion

The results of this study reveal that autochthonous dairy products in the Western Balkan region are a rich source of new and diverse enterococci strains with considerable genetic, metabolic and technological potential. Since the role of enterococci in food spoilage and opportunistic infections is well known, before recommending the use of a particular strain as a starter culture it is necessary to characterize each strain in detail. This study reveals that out of 636 natural dairy enterococci isolates, only five strains belonging to *E. durans* have good technological potential and meet the safety criteria for use in the dairy industry. This finding points out the necessity for detailed characterization of enterococci isolated from dairy food (as a food of animal origin), since they could be reservoirs of antibiotic resistance and virulence genes, as well as producers of biogenic amines.

Acknowledgments

This work was funded by the Ministry of Education and Science of the Republic of Serbia, grant No. 173019. We are grateful to Nathaniel Aaron Sprinkle, native English editor for the proofreading of the manuscript.

References

Abeijón, M. C., Medina, R. B., Katz, M. B., and González, S. N. (2006). Technological properties of *Enterococcus faecium* isolated from ewe's milk and cheese with importance for flavour development. *Can. J. Microbiol.* 52, 237–245. doi: 10.1139/W05-136

Ananou, S., Garriga, M., Jofré, A., Aymerich, T., Gálvez, A., Maqueda, M., et al. (2010). Combined effect of enterocin AS-48 and high hydrostatic pressure to control food-borne pathogens inoculated in low acid fermented sausages. *Meat Sci.* 84, 594–600. doi: 10.1016/j.meatsci.2009.10.017

Andrighetto, C., Knijff, E., Lombardi, A., Torriani, S., Vancanneyt, M., Kersters, K., et al. (2001). Phenotypic and genetic diversity of enterococci isolated from Italian cheeses. *J. Dairy Res.* 68, 303–316. doi: 10.1017/S0022029901004800

Asteri, I. A., Robertson, N., Kagkli, D. M., Andrewes, P., Nychas, G., Coolbear, T., et al. (2009). Technological and flavour potential of cultures isolated from traditional Greek cheese-A pool of novel species and starters. *Int. Dairy J.* 19, 595–604. doi: 10.1016/j.idairyj.2009.04.006

Begovic, J., Brandsma, J. B., Jovcic, B., Tolinacki, M., Veljovic, K., Meijer, W. C., et al. (2011). Analysis of dominant lactic acid bacteria from artisanal raw milk cheeses produced on the mountain Stara Planina, Serbia. *Arch. Biol. Sci.* 63, 11–20. doi: 10.2298/ABS1101011B

Bover-Cid, S., and Holzapfel, W. H. (1999). Improved screening procedure for biogenic amine production by lactic acid bacteria. *Int. J. Food Microbiol.* 53, 33–41. doi: 10.1016/S0168-1605(99)00152-X

Centeno, J. A., Menéndez, S., Hermida, M. A., and Rodríguez-Otero, J. L. (1999). Effects of the addition of *Enterococcus faecalis* in Cebreiro cheese manufacture. *Int. J. Food Microbiol.* 48, 97–111. doi: 10.1016/S0168-1605(99)00030-6

CLSI. (2012). *Performance Standards for Antimicrobial Susceptibility Testing*, Vol. 32 (Clinical and Laboratory Standards Institute, Wayne, PA), 22nd Informational Supplement M100–S22.

Domig, K. J., Mayer, H. K., and Kneifel, W. (2003). Methods used for the isolation, enumeration, characterisation and identification of *Enterococcus* spp.: 1. Media for isolation and enumeration. *Int. J. Food Microbiol.* 88, 147–164. doi: 10.1016/S0168-1605(03)00177-6

Ducková, V., Èanigová, M., and Kroèko, M. (2014). Enterococci and their resistance to antibiotics and thyme essential oil. *J. Microbiol. Biotechnol. Food Sci.* 3, 1–4. Available online at: http://www.jmbfs.org/wp-content/uploads/2014/01/1_jmbfs_duckova_2014_m.pdf

Eaton, T. J., and Gasson, M. J. (2001). Molecular screening of *Enterococcus* virulence determinants and potential for genetic exchange between food and medical isolates. *Appl. Environ. Microbiol.* 67, 1628–1635. doi: 10.1128/AEM.67.4.1628-1635.2001

Foulquié-Moreno, M. R., Sarantinopoulos, P., Tsakalidou, E., and De Vuyst, L. (2006). The role of application of enterococci in food and health. *Int. J. Food Microbiol.* 106, 1–24. doi: 10.1016/j.ijfoodmicro.2005.06.026

Franz, C. M., Holzapfel, W. H., and Stiles, M. E. (1999). Enterococci at the crossroads of food safety? *Int. J. Food Microbiol.* 47, 1–24. doi: 10.1016/S0168-1605(99)00007-0

Franz, C. M., Huch, M., Abriouel, H., Holzapfel, W., and Gálvez, A. (2011). Enterococci as probiotics and their implications in food safety. *Int. J. Food Microbiol.* 151, 125–140. doi: 10.1016/j.ijfoodmicro.2011.08.014

Franz, C. M. A. P., Muscholl-Silberhorn, N., Yousif, M. K., Vancanneyt, M., Swings, J., and Holzapfel, W. H. (2001). Incidence of virulence factors and antibiotic resistance among enterococci isolated from food. *Appl. Environ. Microbiol.* 67, 4385–4389. doi: 10.1128/AEM.67.9.4385-4389.2001

Franz, C. M. A. P., Stiles, M. E., Schleifer, K. H., and Holzapfel, W. H. (2003). Enterococci in foods-a conundrum for food safety. *Int. J. Food Microbiol.* 88, 105–122. doi: 10.1016/S0168-1605(03)00174-0

Galgano, F., Suzzi, G., Favati, F., Caruso, M., Martuscelli, M., Gardini, F., et al. (2001). Biogenic amines during ripening in "Semicotto Caprino" cheese: role of enterococci. *Int. J. Food Sci. Technol.* 36, 153–160. doi: 10.1046/j.1365-2621.2001.00443.x

Gardin, F., Martuscelli, M., Caruso, M. C., Galgano, F., Crudele, M. A., Favati, F., et al. (2001). Effects of pH, temperature and NaCl concentration on the growth kinetics, proteolytic activity and biogenic amine production of *Enterococcus faecalis*. *Int. J. Food Microbiol.* 64, 105–117. doi: 10.1016/S0168-1605(00)00445-1

Garg, S. K., and Mital, B. K. (1991). Enterococci in milk and milk products. *Crit. Rev. Microbiol.* 18, 15–45. doi: 10.3109/10408419109113508

Gelsomino, R., Vancanneyt, M., Condon, S., Swings, J., and Cogan, T. M. (2001). Enterococcal diversity in the environment of an Irish Cheddar-type cheesemaking factory. *Int. J. Food Microbiol.* 171, 177–188. doi: 10.1016/S0168-1605(01)00620-1

Giraffa, G. (2002). Enterococci from foods. *FEMS Microbiol. Rev.* 26, 163–171. doi: 10.1111/j.1574-6976.2002.tb00608.x

Giraffa, G. (2003). Functionality of enterococci in dairy products. *Int. J. Food Microbiol.* 88, 215–222. doi: 10.1016/S0168-1605(03)00183-1

Giraffa, G., Andrighetto, C., Antonello, C., Gatti, M., Lazzi, C., Marcazzan, G., et al. (2004). Genotypic and phenotypic diversity of *Lactobacillus delbrueckii* subsp. lactis strains of dairy origin. *Int. J. Food Microbiol.* 91, 129–139. doi: 10.1016/S0168-1605(03)00368-4

Giraffa, G., Gatti, M., Rossetti, L., Senini, L., and Neviani, E. (2000). Molecular diversity within *Lactobacillus helveticus* as revealed by genotypic characterization. *Appl. Environ. Microbiol.* 66, 1259–1265. doi: 10.1128/AEM.66.4.1259-1265.2000

Golić, N., Cadež, N., Terzić-Vidojević, A., Suranská, H., Beganović, J., Lozo, J., et al. (2013). Evaluation of lactic acid bacteria and yeast diversity in traditional white pickled and fresh soft cheeses from the mountain regions of Serbia and lowland regions of Croatia. *Int. J. Food Microbiol.* 166, 294–300. doi: 10.1016/j.ijfoodmicro.2013.05.032

Gomes, B. C., Esteves, C. T., Palazzo, I. C., Darini, A. L., Felis, G. E., Sechi, L. A., et al. (2008). Prevalence and characterization of *Enterococcus* spp. isolated from Brazilian foods. *Food Microbiol.* 25, 668–675. doi: 10.1016/j.fm.2008.03.008

Hopwood, D. A., Bibb, M. J., Chater, K. F., Kieser, T., Bruton, C. J., Kieser, H. M., et al. (1985). *Genetic Manipulation of Streptomyces: A Laboratory Manual.* Norwich: John Innes Foundation.

Jovcic, B., Begovic, J., Lozo, J., Topisirovic, L., and Kojic, M. (2009). Dynamic of sodium dodecyl sulfate utilization and antibiotic susceptibility of strain *Pseudomonas* sp. ATCC19151. *Arch. Biol. Sci.* 61, 159–165. doi: 10.2298/ABS0902159J

Kayser, F. H. (2003). Safety aspects of enterococci from the medical point of view. *Int. J. Food Microbiol.* 88, 255–262. doi: 10.1016/S0168-1605(03)00188-0

Macedo, A. S., Freitas, A. R., Abreu, C., Machado, E., Peixe, L., Sousa, J. C., et al. (2011). Characterization of antibiotic resistant enterococci isolated from untreated waters for human consumption in Portugal. *Int. J. Food Microbiol.* 145, 315–319. doi: 10.1016/j.ijfoodmicro.2010.11.024

Mannu, L., Paba, A., Daga, E., Comunian, R., Zanetti, S., Dupré, I., et al. (2003). Comparison of the incidence of virulence determinants and antibiotic resistance between *Enterococcus faecium* strains of dairy, animal and clinical origin. *Int. J. Food Microbiol.* 88, 291–304. doi: 10.1016/S0168-1605(03)00191-0

Menéndez, S., Godinez, R., Hermida, M., Centeno, J. A., and Rodríguez-Otero, J. L. (2004). Characteristics of "Tetilla" pasteurized milk cheese manufactured with the addition of autochthonous cultures. *Food Microbiol.* 21, 97–104. doi: 10.1016/S0740-0020(03)00014-5

Morandi, S., Brasca, M., Andrighetto, C., Lombardi, A., and Lodi, R. (2006). Technological and molecular characterisation of enterococci isolated from north–west Italian dairy products. *Int. Dairy J.* 16, 867–875. doi: 10.1016/j.idairyj.2005.09.005

Morea, M., Baruzzi, F., and Cocconcelli, P. S. (1999). Molecular and physiological characterization of dominant bacterial populations in traditional Mozzarella cheese processing. *J. Appl. Microbiol.* 87, 574–582. doi: 10.1046/j.1365-2672.1999.00855.x

Nieto-Arribas, P., Seseña, S., Poveda, J. M., Chicón, R., Cabezas, L., and Palop, L. (2011). *Enterococcus* populations in artisanal Manchego cheese: Biodiversity, technological and safety aspects. *Food Microbiol.* 28, 891–899. doi: 10.1016/j.fm.2010.12.005

Psoni, L., Kotzamanides, C., Andrighetto, C., Lombardi, A., Tzanetakis, N., and Litopoulou-Tzanetaki, E. (2006). Genotypic and phenotypic heterogeneity in Enterococcus isolates from Batzos, a raw goat milk cheese. *Int. J. Food Microbiol.* 109, 109–120. doi: 10.1016/j.ijfoodmicro.2006.01.027

Saavedra, L., Taranto, M. P., Sesma, F., and Valdez, G. F. (2003). Homemade traditional cheeses for the isolation of probiotic *Enterococcus faecium* strains. *Int. J. Food Microbiol.* 88, 241–245. doi: 10.1016/S0168-1605(03)00186-7

Sarantinopoulos, P., Andrigheto, C., Georgalaki, M. D., Rea, M. C., Lombardi, A., Cogan, T. M., et al. (2001). Biochemical properties of enterococci relevant to their technological performance. *Int. Dairy J.* 11, 621–647. doi: 10.1016/S0958-6946(01)00087-5

Schleifer, K. H., and Ludwig, W. (1995). "Phylogenetic relationships of lactic acid bacteria," in *The Genera of Lactic Acid Bacteria,* Vol. 2, eds. B. J. B. Wood and W. H. Holzapfel (London: Blackie Academic & Professional), 7–18.

Suzzi, G., Caruso, M., Gardini, F., Lombardi, A., Vannini, L., Guerzoni, M. E., et al. (2000). A survey of the enterococci isolated from an artisanal Italian goat's cheese (semicotto caprino). *J. Appl. Microbiol.* 89, 267–274. doi: 10.1046/j.1365-2672.2000.01120.x

Terzic-Vidojevic, A., Lozo, J., and Topisirovic, L. J. (2009b). Dominant lactic acid bacteria in artisanal Pirot cheeses of different ripening period. *Genetika* 41, 341–352. doi: 10.2298/GENSR0903341T

Terzic-Vidojevic, A., Mihajlović, S., Uzelac, G., Golić, N., Fira, Đ., Kojić, M., et al. (2014a). Identification and characterization of lactic acid bacteria isolated from artisanal white brined Golija cows' milk cheeses. *Arch. Biol. Sci.* 66, 179–192. doi: 10.2298/ABS1401179T

Terzic-Vidojevic, A., Mihajlovic, S., Uzelac, G., Veljovic, K., Tolinacki, M., Nikolic, M., et al. (2014b). Characterization of lactic acid bacteria isolated from artisanal Travnik young cheeses, sweet creams and sweet kajmaks over four seasons. *Food Microbiol.* 39, 27–38. doi: 10.1016/j.fm.2013.10.011

Terzic-Vidojevic, A., Tolinacki, M., Nikolic, M., Veljovic, K., Jovanovic, S., Macej, O., et al. (2013). Artisanal Vlasina raw goat's milk cheeses: evaluation and selection of autochthonous lactic acid bacteria as starter cultures. *Food Technol. Biotechnol.* 51, 554–563. Available online at: http://www.ftb.com.hr/images/pdfarticles/2013/October-december/ftb_51-4_554-563.pdf

Terzic-Vidojevic, A., Veljovic, K., Tolinacki, M., Nikolic, M., Ostojic, M., and Topisirovic, L. J. (2009a). Characterization of lactic acid bacteria isolated from artisanal Zlatar cheeses produced at two different geographical location. *Genetika* 41, 117–136. doi: 10.2298/GENSR0901117T

Terzic-Vidojevic, A., Vukasinovic, M., Veljovic, K., Ostojic, M., and Topisirovic, L. (2007). Characterization of microflora in homemade semi-hard white Zlatar cheese. *Int. J. Food Microbiol.* 114, 36–42. doi: 10.1016/j.ijfoodmicro.2006.10.038

Teuber, M., Meile, L., and Schwarz, F. (1999). Acquired antibiotic resistance in lactic acid bacteria from food. *Antonie Van Leeuwenhoek* 76, 115–137. doi: 10.1023/A:1002035622988

Tsakalidou, E., Manolopoulou, E., Tsilibari, V., Georgalaki, M., and Kalantzopoulos, G. (1993). Esterolytic activities of *Enterococcus durans* and *Enterococcus faecium* strains isolated from Greek cheese. *Neth. Milk Dairy J.* 47, 145–150.

Valenzuela, A. S., Ben Omar, N., Abriouel, H., López, R. L., Veljovic, K., Cañamero, M. M., et al. (2009). Virulence factors, antibiotic resistance, and bacteriocins in enterococci from artisan foods of animal origin. *Food Control* 20, 381–385. doi: 10.1016/j.foodcont.2008.06.004

Vankerckhoven, V., Autgaerden, T. V., Vael, C., Lammens, C., Chapelle, S., Rossi, R., et al. (2004). Development of a multiplex PCR for the detection of *asa 1, gelE, cylA, esp,* and *hyl* genes in enterococci and survey for virulence determinants among European hospital isolates of *Enterococcus faecium.* *J. Clin. Microbiol.* 42, 4473–4479. doi: 10.1128/JCM.42.10.4473-4479.2004

Veljovic, K., Fira, D., Terzic-Vidojevic, A., Abriouel, H., Galvez, A., and Topisirovic, L. (2009). Evaluation of antimicrobial and proteolytic activity of enterococci isolated from fermented products. *Eur. Food Res. Technol.* 230, 63–70. doi: 10.1007/s00217-009-1137-6

Veljović, K., Terzić-Vidojević, A., Tolinaèki, M., Mihajlović, S., Vukotić, G., Golić, N., et al. (2014). "Molecular characterization of natural dairy isolates of *Enterococcus faecalis* and evaluation of their antimicrobial potential," in *Enterococcus faecalis Molecular Characteristics, Role in Nosocomial Infections and Antimicrobial Effects. Bacteriology Research Developments,* ed. H. L. Mack (New York, NY: Nova Publishers, Inc), 123–135.

Versalovic, J., Schneider, M., De Bruijn, F. J., and Lupski, J. R. (1994). Genomic fingerprinting of bacteria using repetitive sequence-based polymerase chain reaction. *Method Mol. Cell. Biol.* 5, 25–40.

Viedma, P. M., Abriouel, H., ben Omar, N., López, R. L., and Gálvez, A. (2009). Antistaphylococcal effect of enterocin AS-48 in bakery ingredients of vegetable origin, alone and in combination with selected antimicrobials. *J. Food Sci.* 74, M384–M389. doi: 10.1111/j.1750-3841.2009.01288.x

Conflict of Interest Statement: The authors declare that the research was conducted in the absence of any commercial or financial relationships that could be construed as a potential conflict of interest.

19

The controversial nature of the *Weissella* genus: technological and functional aspects versus whole genome analysis-based pathogenic potential for their application in food and health

Hikmate Abriouel[1]*, Leyre Lavilla Lerma[1], María del Carmen Casado Muñoz[1], Beatriz Pérez Montoro[1], Jan Kabisch[2], Rohtraud Pichner[2], Gyu-Sung Cho[2], Horst Neve[2], Vincenzina Fusco[3], Charles M. A. P. Franz[2], Antonio Gálvez[1] and Nabil Benomar[1]

[1] *Área de Microbiología, Departamento de Ciencias de la Salud, Facultad de Ciencias Experimentales, Universidad de Jaén, Jaén, Spain,* [2] *Department of Microbiology and Biotechnology, Federal Research Institute of Nutrition and Food, Max Rubner-Institut, Kiel, Germany,* [3] *Institute of Sciences of Food Production, National Research Council of Italy, Bari, Italy*

Edited by:
Andrea Gomez-Zavaglia,
Centro de Investigación y Desarrollo
en Criotecnología de Alimentos,
Argentina

Reviewed by:
Zhao Chen,
Clemson University, USA
Paula Carasi,
Universidad Nacional de La Plata,
Argentina

*Correspondence:
Hikmate Abriouel
hikmate@ujaen.es

Despite the use of several *Weissella (W.)* strains for biotechnological and probiotic purposes, certain species of this genus were found to act as opportunistic pathogens, while strains of *W. ceti* were recognized to be pathogenic for farmed rainbow trout. Herein, we investigated the pathogenic potential of weissellas based on *in silico* analyses of the 13 whole genome sequences available to date in the NCBI database. Our screening allowed us to find several virulence determinants such as collagen adhesins, aggregation substances, mucus-binding proteins, and hemolysins in some species. Moreover, we detected several antibiotic resistance-encoding genes, whose presence could increase the potential pathogenicity of some strains, but should not be regarded as an excluding trait for beneficial weissellas, as long as these genes are not present on mobile genetic elements. Thus, selection of weissellas intended to be used as starters or for biotechnological or probiotic purposes should be investigated regarding their safety aspects on a strain to strain basis, preferably also by genome sequencing, since nucleotide sequence heterogeneity in virulence and antibiotic resistance genes makes PCR-based screening unreliable for safety assessments. In this sense, the application of *W. confusa* and *W. cibaria* strains as starter cultures or as probiotics should be approached with caution, by carefully selecting strains that lack pathogenic potential.

Keywords: *Weissella, in silico* analysis, genome, virulence, antibiotic resistance

INTRODUCTION

Weissella species are non-spore forming, catalase-negative and Gram-positive bacteria that are non-motile, with the exception of *Weissella (W.) beninensis*. To date, the *Weissella* genus comprises 19 validly described species (February 2015[1]). Most of these were isolated from and associated with fermented foods, e.g., *W. confusa*, *W. cibaria* (Björkroth et al., 2002), *W. kimchii* (Choi et al., 2002),

---

[1] http://www.bacterio.net

W. koreensis (Lee et al., 2002), and *W. viridescens* (Comi and Iacumin, 2012). *Weissella viridescens* is considered the type species of the *Weissella* genus (Collins et al., 1994). As a member of the lactic acid bacteria (LAB), *Weissella* have complex nutritional requirements. Because of this, they inhabit nutrient-rich environments and can be isolated from a variety of such sources, including vegetables, meat, fish, raw milk, sewage, blood, soil, the gastrointestinal tracts of humans and animals, as well as the oral cavity and uro-genital tract of humans (Fusco et al., 2015).

From a technological point of view, *Weissella* plays an important role in fermentation processes such as the production of silage, as well as in food fermentations based on vegetables or meat as substrate (Björkroth et al., 2002; Santos et al., 2005).

Several weissellas, mainly belonging to the *W. confusa* and *W. cibaria* species, are being extensively studied for their ability to produce significant amounts of non-digestible oligosaccharides and extracellular polysaccharides, which can be used as prebiotics or for other applications in food, feed, clinical, and cosmetics industries.

Furthermore, several *Weissella* strains have been found to act as probiotics, mainly due to their antimicrobial activity, as is the case for certain bacteriocinogenic strains of *W. paramesenteroides*, *W. hellenica*, and *W. cibaria* (Fusco et al., 2015). For example, *W. hellenica* DS-12 isolated from flounder intestine has been used as probiotic in fish, due to its antimicrobial activity against fish pathogens, such as *Edwardsiella*, *Pasteurella*, *Aeromonas*, and *Vibrio* (Cai et al., 1998). Also, strains of *W. cibaria* were proposed as probiotics for oral health, inhibiting *Streptococcus mutans* glucan biofilm formation (Kang et al., 2006). Recently, weissellas were also shown to exhibit chemopreventive and anti-tumor effects (Kwak et al., 2014).

On the detrimental side, some weissellas were reported to be involved in disease outbreaks such as otitis, sepsis, endocarditis, and even fish mortality (Flaherty et al., 2003; Harlan et al., 2011; Lee et al., 2011; Welch and Good, 2013). Human infections caused by *Weissella* spp. are, however, rarely reported, and occur mostly in patients with impaired host defenses (Lee et al., 2011). Curiously, therefore, in the same species both beneficial and detrimental strains can be found. As an example, *W. confusa* causes sepsis and other serious infections in humans and animals, while it has a functional role in food fermentations and has also been suggested as a probiotic (Fusco et al., 2015).

The probiotic and pro-technological potential of weissellas therefore collides with the potential of these bacteria as human pathogens. As for enterococci, whose use for food and health application has been controversial (Franz et al., 2003; Ogier and Serror, 2008), a safety assessment of each strain that is intended to be used as starter culture or as probiotic, should thus be recommended.

Whole genome sequencing and sequence annotation is increasingly being used as valuable tool for assessing microbial food quality and safety aspects (Alkema et al., 2015), allowing the identification of new genes that may have an important impact on cell metabolism, fitness, and virulence.

Herein, we report the investigation of the pathogenic potential of weissellas based on *in silico* analyses of the 13 whole genome

sequences (of 13 strains belonging to nine *Weissella* species) to date available.

MATERIALS AND METHODS

Data Sequences

Data sequences of genomes of the 13 strains belonging to nine *Weissella* species (**Table 1**) were retrieved from the National Centre for Biotechnology Information (NCBI[2]; accessed on February, 2015). All genome sequences available were analyzed for the presence of different virulence determinants (aggregation substances, adhesins, toxins, pili, hemolysins) and of antibiotic resistance genes, such as fosfomycin and methicillin resistance genes. The accession numbers of each target gene sequence are indicated in **Tables 1–3** and **Figures 1–5**.

Phylogenetic Analyses

All selected target genes were subjected to phylogenetic analyses to determine phylogenetic relationships with those of closely related genera. Alignment of sequences was done using the CLUSTAL W module of the Lasergene program, version 5.05 (MegAlign, Inc., Madison, WI, USA). Phylogenetic trees were reconstructed by the maximum parsimony method using MegAlign (Lasergene program, version 5.05).

RESULTS AND DISCUSSION

Virulence Determinants in *Weissella*

The safety of many LAB species have been recognized as GRAS (Generally Regarded As Safe; for the USA; Food and Drug Administration, 1999) or have attained the QPS (*Qualified Presumption of Safety*; for the European Commission; European Food Safety Authority (EFSA), 2004) status. Indeed, many people refer to these bacteria as being innocuous, and even associated with health beneficial properties. However, it is well known that certain species within a genus, or even certain strains within a specific species, may have different health impacts, as has been pointed out for the enterococci (Franz et al., 2003; Ogier and Serror, 2008). So far, no *Weissella* species have QPS status (European Food Safety Authority (EFSA), 2015) Data on virulence factors present in *Weissella* species are quite scarce, and genomic analysis can therefore aid in detecting and describing the occurrence of virulence determinants that may be present at species or strain level within this genus. In this regard, Ladner et al. (2013) found in the genome sequence of *W. ceti* NC36, an emerging pathogen of farmed rainbow trout in the United States, the presence of several putative virulence factor genes, which did not have homologs encoded in any of the other sequenced *Weissella* genomes. In particular, these include five collagen adhesin genes (WCNC_00912, WCNC_00917, WCNC_00922, WCNC_05547, and WCNC_06207), a platelet-associated adhesin gene (WCNC_01820) and a gene for a mucus-binding protein (WCNC_01840; Ladner et al., 2013). The five collagen adhesin genes included those from *W. ceti* WS105 (three collagen adhesins), *W. ceti* WS74 (three collagen adhesins), *W. ceti* WS08

---

[2]http://www.ncbi.nlm.nih.gov/gene

TABLE 1 | Genome characteristics of *Weissella* species isolated from different sources.

Strains	Genome size (bp)	Source	RefSeq/GenbankWGS
Weissella ceti WS08	1.355.850	Fish brain	NZ_CP007588.1
Weissella ceti WS74	1.389.510	Fish brain	NZ_CP009223.1
Weissella ceti WS105	1.390.400	Fish brain	NZ_CP009224.1
Weissella ceti NC36	1.352.640	Fish spleen	NZ_ANCA00000000.1
Weissella cibaria KACC 11862	2.320.000	Kimchi	NZ_AEKT00000000.1
Weissella confusa LBAE C39-2	2.280.000	Wheat sourdough	NZ_CAGH00000000.1
Weissella halotolerans DSM 20190	1.360.000	Sausage	NZ_ATUU00000000.1
Weissella hellenica Wikim14	1.920.000	Kimchi	BBIK00000001.1
Weissella koreensis KACC 15510	1.441.470	Kimchi	NC_015759.1
Weissella koreensis KCTC 3621	1.728.940	Kimchi	NZ_AKGG00000000.1
Weissella oryzae SG25	2.130.000	Fermented rice grains	NZ_BAWR00000000.1
Weissella paramesenteroides ATCC 33313	1.962.173	Human	NZ_ACKU00000000.1
Weissella thailandensis fsh4-2	1.968.992	Jeotkal (Korean fermented fish condiment)	HE575133 to HE575182

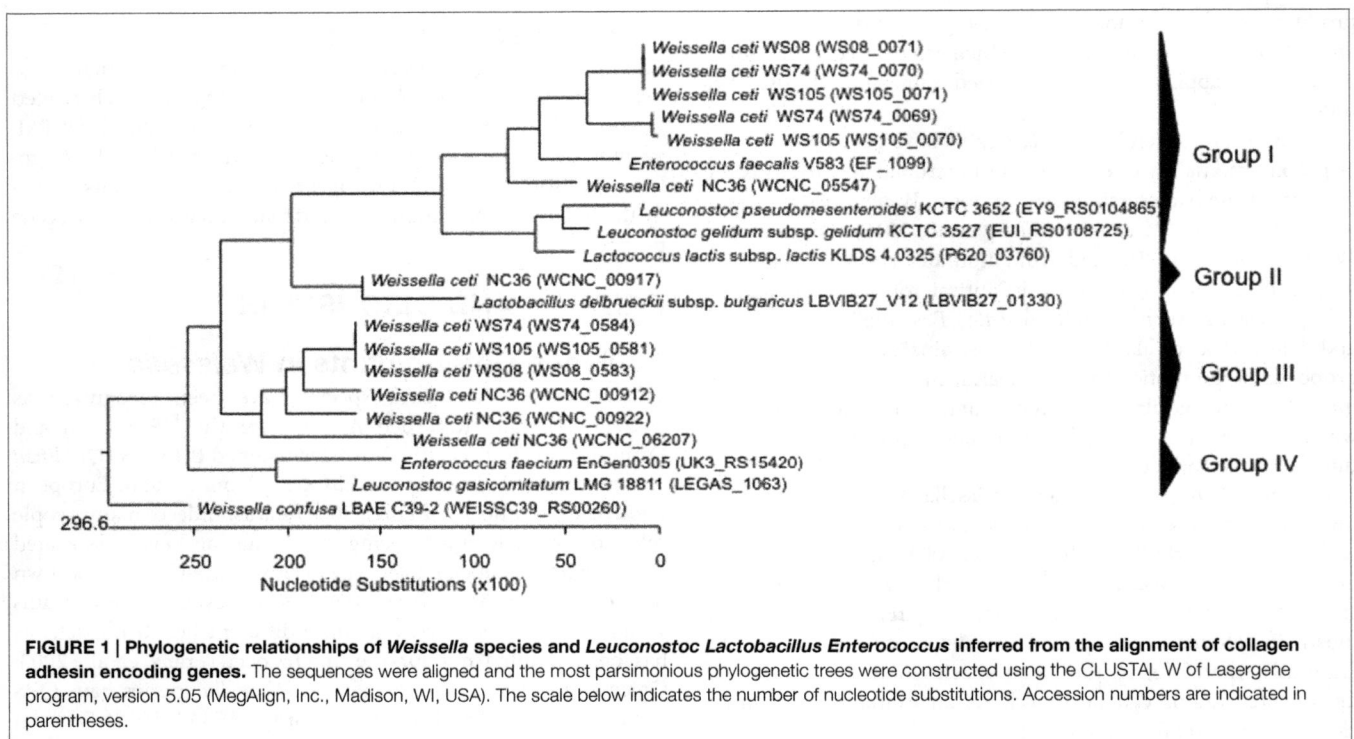

FIGURE 1 | Phylogenetic relationships of *Weissella* species and *Leuconostoc Lactobacillus Enterococcus* inferred from the alignment of collagen adhesin encoding genes. The sequences were aligned and the most parsimonious phylogenetic trees were constructed using the CLUSTAL W of Lasergene program, version 5.05 (MegAlign, Inc., Madison, WI, USA). The scale below indicates the number of nucleotide substitutions. Accession numbers are indicated in parentheses.

(two collagen adhesins), and *W. confusa* LBAE C39-2 (one collagen adhesin; **Table 2**). DNA sequences coding for collagen adhesin proteins in *Weissella* spp. and other LAB (*Leuconostoc* "*Ln.*", *Lactococcus* "*Lc.*", *Enterococcus* "*E.*", and *Lactobacillus* "*Lb.*") were aligned using CLUSTAL W of the Lasergene program, version 5.05 (MegAlign, Inc., Madison, WI, USA) and the most parsimonious phylogenetic tree was reconstructed. The dendrogram generated showed four groups (**Figure 1**). The first group contained two of the three collagen adhesin genes found in *W. ceti* strains, *E. faecalis*, *Leuconostoc* and *Lactococcus*, while genes for group II collagen adhesins were only present in *W. ceti* NC36 and *Lb. delbrueckii*. The third group presented the genes for the third collagen produced by *W. ceti* strains. In the last group (group IV) *E. faecium* and *Leuconostoc* collagen adhesin genes clustered together, however a *W. confusa* LBAE C39-2 collagen

adhesin gene also formed a single lineage (**Figure 1**). In general, the close relatedness of *Weissella*, *Enterococcus*, and *Leuconostoc* collagen adhesion genes at the level of the primary protein sequence may suggest the same evolutionary origin of these proteins in these bacteria. Further genome sequencing of other *Weissella* species would be required to analyze this relatedness with enterococci and leuconostocs in more depth. The presence of virulence determinants commonly present in enterococci also in weissellas and leuconostocs may suggest a horizontal gene transfer event from the *Enterococcus* genus, the latter of which indeed includes strains that are commonly present in many different habitats (Franz et al., 2003).

The role of the genes above in the virulence of *Weissella* is still unknown. Further studies are needed to confirm whether these virulence genes are expressed or not, and if they are located

TABLE 2 | Potential virulence genes of *Weissella* species as revealed by *in silico* screening of the annotated genome sequences.

Strains	Virulence factors detected by analysis *in silico* of genome sequences (locus_tag)				
	Aggregation substance	Collagen adhesion	Hemolysin	Mucus-binding proteins	Staphylococcal surface protein
W. ceti WS08	–	2 CA (WS08_0071, WS08_0583)	1 Hly (WS08_0556), 1 HlyA (WS08_0902)	–	5SPA (WS08_0360, WS08_0450, WS08_0978, WS08_1156, WS08_1190)
W. ceti WS74	–	3 CA (WS74 0069, WS74 0070, WS74_0584)	1 Hly (WS74_0557), 1 HlyA (WS74_0968)	–	4 SPA (WS74_0360, WS74_0451, WS74_1225, WS74_1261)
W. ceti WS105	–	3 CA (WS105 0070, WS105 0071, WS105_0581)	1 Hly (WS105_0554), 1 HlyA (WS105_0965), 1 Hly-like protein (WS105_0227)	–	4 SPA (WS105_0358, WS105_0448, WS105_1219, WS105_1255)
W. ceti NC36	–	5 CA (WCNC 00912, WCNC 00917, WCNC 00922, WCNC 05547, WCNC 06207) 1 PA-ADHE	–	1 MBP (WCNC_01840)	–
W. cibaria KACC 11862	–	–	2 Hly (ESE_RS0108795, ESE_RS11605)	–	–
W. confusa LBAE C39-2	–	1 (WEISSC39_RS00260)	2 Hly (WEISSC39_RS08935, WEISSC39_RS08940)	1 MBP (WEISSC39_RS05980)	–
W. halotolerans DSM 20190	–	–	2 Hly (G414_RS0106810, G414_RS0106815)	–	–
W. hellenica Wikim14	–	–	2 Hly (TY24_RS08990, TY24_RS08995)	–	–
W. koreensis KACC 15510	–	–	–	–	–
W. koreensis KCTC 3621	–	–	2 Hly (JC2156_RS08435, JC2156_10680)	–	–
W. oryzae SG25	2 AGS (WOSG25_200030, WOSG25_200040) 2 Asa1/PrgB (WOSG25_200030, WOSG25_200040)	–	3 hemolysins (WOSG25 150280, WOSG25_RS09625, WOSG25_RS09635)	–	–
W. paramesenteroides ATCC 33313	–	–	2 Hly (HMPREF0877_RS09665, HMPREF0877_RS09670)	–	–
W. thailandensis fsh4-2	–	–	2 Hly (WT2_01519, WT2_01520)	–	–

on mobile genetic elements (e.g., transposons). However, the collagen adhesin protein (Ace) in *Enterococcus* was reported to be involved in adhesion to collagen and to contribute to infective endocarditis and urinary tract infection (Aymerich et al., 2006; Nallapareddy et al., 2011). In addition, the common localization of this gene on plasmids or transposon may pose a risk related with horizontal gene transfer to pathogenic bacteria. On the other hand, adherence is an important pre-requisite for the colonization of probiotics, providing a competitive advantage in different ecosystems. The presence of adhesin genes in weissellas that are not located on mobile genetic elements, and which are not transferable, should therefore not negatively reflect on the strains

probiotic potential, and should not be an excluding factor for its use. It could be regarded as a key factor for the attachment of probiotic bacterial cell in the human host, and thus could be regarded as a colonization factor rather than a virulence factor (Dunne et al., 2004).

In silico analyses of *Weissella* genomes also showed the presence of genes coding for two unnamed aggregation substances (WOSG25_050600 and WOSG25_190240) as well as two aggregation substances named Asa1/PrgB (WOSG25_200030 and WOSG25_200040) in *W. oryzae* SG25 (**Table 2**). The sequences of the genes encoding the unnamed aggregation substances, the aggregation substances Asa1/PrgB from *W. oryzae* SG25 and

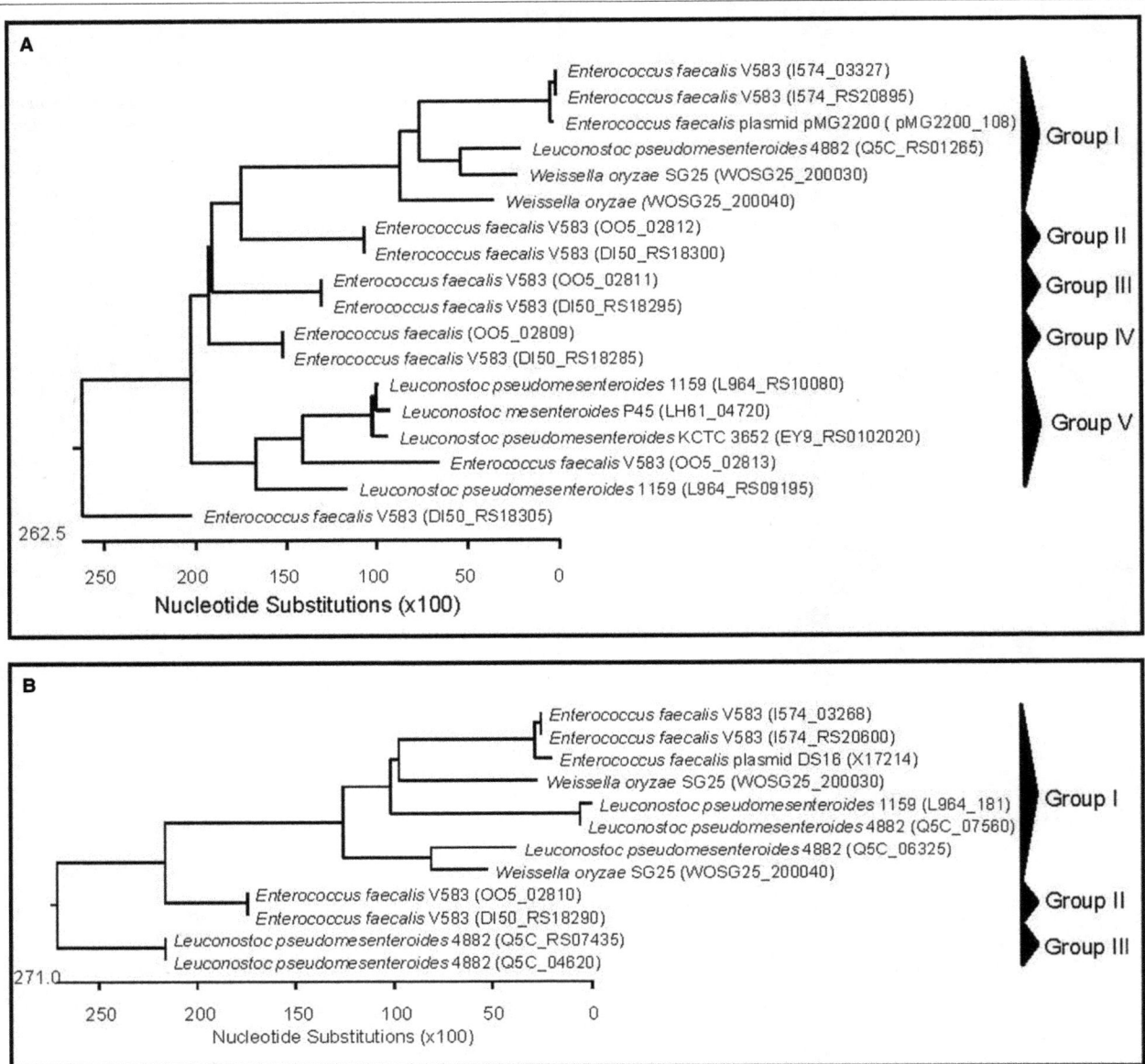

FIGURE 2 | Phylogenetic relationships of *Weissella* species and *Leuconostoc Enterococcus* inferred from the alignment of aggregation substances (A) and aggregation substance Asa1/PrgB (B) encoding genes. The sequences were aligned and the most parsimonious phylogenetic trees were constructed using the CLUSTAL W of Lasergene program, version 5.05 (MegAlign, Inc., Madison, WI, USA). The scale below indicates the number of nucleotide substitutions. Accession numbers are indicated in parentheses.

those of other LAB (*Leuconostoc* and *Enterococcus*) were aligned using the CLUSTAL W of Lasergene program, and the most parsimonious phylogenetic tree was constructed. The results obtained showed that the aggregation substance genes found in *W. oryzae* SG25 clustered with those of *Ln. pseudomesenteroides* and *E. faecalis* (group I in dendrogram), the latter being the aggregation substance gene of *E. faecalis* pMG2200, which is located on a plasmid. This suggests that horizontal gene transfer from *E. faecalis* to other LAB such as *Weissella* and *Leuconostoc* (**Figure 2A**) probably occurred. Other aggregation substance genes of different strains of *E. faecalis*, *Ln. mesenteroides*,

and *Ln. pseudomesenteroides* were very divergent, forming several clusters (**Figure 2A**). When aggregation substance gene sequences of Asa1/PrgB from *W. oryzae* SG25, *E. faecalis*, and *Ln. pseudomesenteroides* were clustered, the dendrogram showed aggregation substances Asa1/PrgB coding genes from *W. oryzae* SG25 and *E. faecalis* or *Leuconostoc* (Group I) to cluster very closely, also suggesting a horizontal gene transfer between these bacteria (**Figure 2B**). The presence of an *asa1* gene was also reported in other potential probiotic LAB, such as *Lb. casei* SJRP35, *Ln. citreum* SJRP44, and *Ln. mesenteroides* subsp. *mesenteroides* SJRP58 isolated from Water-Buffalo Mozzarella

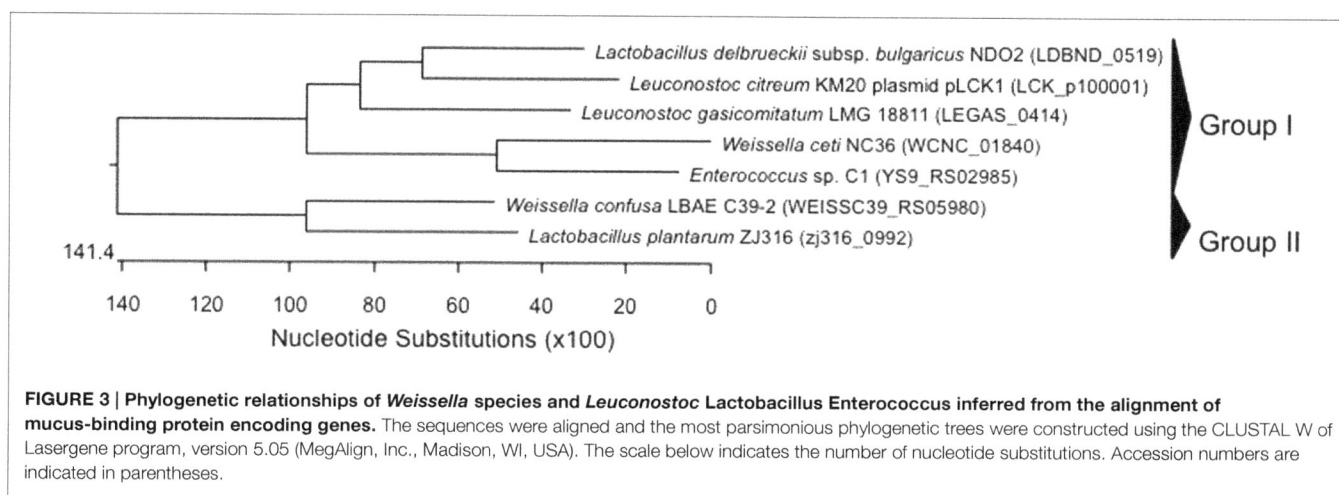

FIGURE 3 | Phylogenetic relationships of *Weissella* species and *Leuconostoc* Lactobacillus Enterococcus inferred from the alignment of mucus-binding protein encoding genes. The sequences were aligned and the most parsimonious phylogenetic trees were constructed using the CLUSTAL W of Lasergene program, version 5.05 (MegAlign, Inc., Madison, WI, USA). The scale below indicates the number of nucleotide substitutions. Accession numbers are indicated in parentheses.

Cheese (Jeronymo-Ceneviva et al., 2014), and also in *Lc. lactis* subsp. *Lactis* KT2W2L (Hwanhlem et al., 2013). As for adhesins, aggregation substances are considered virulence factors. However, the safety assessment of strains harboring adhesins or aggregation substances may highly depend on the localization of such genes on mobile genetic elements and on other safety aspects of the strain. However, the difference in infectivity between weissellas that are either probiotic or involved in infections such as endocarditis, cannot be explained by differences in adherence potential alone, but other factors involved in promoting disease must exist in pathogenic strains, as was similarly postulated by Vankerckhoven et al. (2007) for lactobacilli involved in human infections.

Genes for mucus-binding protein (Mub), which can serve as effector molecules involved in mechanisms of adherence of bacteria to the host, were detected in *W. ceti* NC36 and *W. confusa* LBAE C39-2 (**Table 2**) genome sequences. The phylogenetic tree of nucleotide sequences encoding mucus-binding proteins of weissellas, as well as of the closely related genera *Leuconostoc*, *Lactobacillus*, and *Enterococcus*, showed that the *W. ceti* NC36 *mub* gene clustered closely with *Enterococcus* sp. C1 *mub* in the same group together with sequences from *Lb. delbrueckii* and *Leuconostoc*. However, the *mub* of *W. confusa* LBAE C39-2 clustered with that of *Lb. plantarum* (**Figure 3**). The presence of mucus-binding protein may be a desirable feature in probiotic bacteria, as it may play an important role in the adhesion of the probiotic strain to host surfaces (Mack et al., 1999). However, this property is obviously problematic in potentially pathogenic strains.

Regarding other virulence factors, such as the Serine-rich adhesin for platelets, which is also important in bacterial adhesion and is considered the central event in the pathogenesis of infective endocarditis (Sullam, 1994), this adhesin was only detected in the pathogen *W. ceti* NC36 (WCNC_RS02030; **Table 2**). On the other hand, the gene encoding the Serine-rich adhesin for platelets was detected also in probiotic lactobacilli such as *Lb. johnsonii* NCC533 and *Lb. reuteri* (Zhou and Wu, 2009), but not in leuconostocs. Concerning the staphylococcal surface protein A, which is involved in invasion and infectivity, several genes coding

for a similar protein were detected in *W. ceti* WS08 (WS08_0360, WS08_0450, WS08_0978, WS08_1156, WS08_1190), *W. ceti* WS74 (WS74_0360, WS74_0451, WS74_1225, WS74_1261), and *W. ceti* WS105 (WS105_0358, WS105_0448, WS105_1219, WS105_1255). However, no homologous genes were detected in either leuconostocs or in lactobacilli (**Table 2**). Due to the low homology of virulence determinants in weissellas, it is difficult to detect such genes by PCR, thus genome sequencing would be the only way to detect and identify these determinants.

The presence of genes for hemolysin or hemolysin-like proteins in weissellas was revealed by *in silico* analysis of the annotated genomes sequences (**Table 2**). The ubiquitous occurrence of such genes in probiotic lactobacilli and bifidobacteria, and also in other LAB such as leuconostocs, may raise questions about their function and virulence potential. The role of these genes in LAB virulence is unknown, and their presence in the genome of a strain intended to be used as probiotic should not necessarily be an exclusion factor for targeting strains for probiotic use. As shown in **Table 2**, weissellas harbored in their genome 1–3 hemolysin encoding genes, i.e., genes for hemolysin A and also hemolysin-like proteins.

Antibiotic Resistance

Studies on antibiotic resistance profiles in *Weissella* genus are very limited, and MIC breakpoints have not been defined by EFSA. Thus, to categorize weissellas as sensitive or resistant to different antibiotics of clinical relevance is a difficult task. These bacteria are known for their intrinsic resistance to antibiotics inhibiting cell wall biosynthesis such as vancomycin and fosfomycin, similar to other Gram-positive bacteria (Arca et al., 1997), and also to antibiotics that inhibit tetrahydrofolate biosynthesis such as sulfamethoxazol and trimethoprim (Liu et al., 2009). Furthermore, resistance to gentamicin, kanamycin, and norfloxacin was also reported in food-associated weissellas (Patel et al., 2012).

Reports about molecular detection of antibiotic resistance genes in weissellas are very scarce, possibly due to the high divergence of resistance genes. In this sense, only *mef*(A/E)

TABLE 3 | Antibiotic resistance genes of *Weissella* species as revealed by *in silico* screening of the annotated genome sequences.

Strains	Antibiotic resistance determinants detected by analysis *in silico* of genome sequences (locus_tag)					
	Daunorubicin	Fosfomycin	Methicillin	Glycopeptide	Sulfonamide	Tetracycline
W. ceti WS08	–	1 FosB (WS08_1256)	–	–	Sul (WS08_0966)	–
W. ceti WS74	–	1 FosB (WS74_1327)	–	–	Sul (WS74_1032)	–
W. ceti WS105	–	1 FosB (WS105_1321)	–	–	Sul (WS105_1028)	Tet (WS105_0392)
W. ceti NC36	–	1 MDT-FosB (WCNC_RS02205)	2 MRP (WCNC_02142, WCNC_02627)	–	–	–
W. cibaria KACC 11862	–	1 MDT-FosB (ESE_RS0106205)	3 MRP (ESE_RS0109255, ESE_RS0102540, ESE_RS0105180)	VanZ (ESE_RS0111030)	–	–
W. confusa LBAE C39-2	–	1 MDT-FosB (WEISSC39_RS10580)	1 MRP (WEISSC39_RS07020)	VanZ (WEISSC39_RS04975)	–	–
W. halotolerans DSM 20190	1 DrrC (G414_RS0101040)	1 MDT-FosB (G414_RS0105405)	1 MRP (G414_RS0103120)	–	–	–
W. hellenica Wikim14	1 DrrC (TY24_RS06500)	1 MDT-FosB (TY24_RS09455)	2 MRP (TY24_RS04745, TY24_RS00485)	–	–	–
W. koreensis KACC 15510	–	–	2 MRP (WKK_01735, WKK_02350)	–	–	–
W. koreensis KCTC 3621	–	1 MDT-FosB (JC2156_RS02465)	2 MRP (JC2156_RS06850, JC2156_07490)	–	–	–
W. oryzae SG25	1 DrrC (WOSG25_RS07165)	1 MDT-FosB (WOSG25_RS02065)	2 MRP (WOSG25_091020, WOSG25_RS07655)	–	–	–
W. paramesenteroides ATCC 33313	–	1 MDT-FosB (HMPREF0877_RS05895)	2 MRP (HMPREF0877_RS07670, HMPREF0877_RS03440)	VanZ (HMPREF0877_1234)	–	–
W. thailandensis fsh4-2	–	–	1 MRP (WT2_00144)	–	–	Tet (WT2_00189)
Weissella sp.	ND	ND	ND	ND	ND	sul1, sul2 genes*

*DrrC, daunorubicin resistance protein; FosB, fosfomycin resistance protein; MDT-FosB, multidrug transporter involved in fosfomycin resistance; MRP, methicillin resistance protein; Sul, sulfonamide resistance protein; Tet, tetracycline resistance protein; VanZ, glycopeptide resistance protein. *Byrne-Bailey et al. (2009).*

drug efflux pump genes involved in the active efflux of macrolides (Clancy et al., 1996; Tait-Kamradt et al., 1997) were detected in *W. cibaria* of aquatic origin (Muñoz-Atienza et al., 2013). When we analyzed the *in silico* genome sequences of *Weissella* spp., different antibiotic resistance genes were detected, such as those coding for fosfomycin and methicillin resistance proteins in almost all the genomes of *Weissella* strains sequenced (**Table 3**) to date. For example, the *fosB* gene coding for fosfomycin resistance was detected in *W. ceti* WS08 (WS08_1256), *W. ceti* WS74 (WS74_1327) and *W. ceti* WS105 (WS105_1321), and nucleotide sequences of the genes were identical in all cases. In addition, other weissellas harbored multidrug transporters involved in fosfomycin and deoxycholate resistance, as was the case for *W. ceti* NC36 (WCNC_RS02205), *W. cibaria* KACC 11862 (ESE_RS0106205), *W. confusa* LBAE C39-2 (WEISSC39_RS10580), *W. halotolerans* DSM 20190 (G414_RS0105405), *W. hellenica* Wikim14 (TY24_RS09455), *W. koreensis* KCTC 3621 (JC2156_RS02465), *W. oryzae* SG25 (WOSG25_RS02065), and *W. paramesenteroides* ATCC 33313 (HMPREF0877_RS05895; **Table 3**). An alignment of nucleotide sequences encoding multidrug transporters involved in fosfomycin resistance of weissellas and other LAB by CLUSTAL W alignment showed sequences to cluster into

two groups. High similarities in sequences again suggested an evolutionary relationship of weissellas, leuconostocs, enterococci, and lactobacilli (**Figure 4**) sequences. Fosfomycin is a broad antibacterial agent that targets several pathogens such as *Haemophilus* spp., *Staphylococcus* spp. and most of the enteric gram-negative bacteria. Although no cross resistances are known to occur for other antibiotics, there are no data on whether or not the *fosB* gene is transferable to other bacteria. Thus, weissellas harboring the *fosB* gene should be investigated in terms of the transferability of this gene, especially to pathogens. In the case of enterococci, these bacteria contain a conjugative plasmid that harbors the novel *fosB* transposon (ISL3-like transposon), as well as the Tn1546-like transposon (containing *vanA* and *fosB*, Qu et al. (2014)). Should weissellas harbor no transferable *fosB* gene, a potential probiotic use of these bacteria should be possible, especially when infection treatment with a combination of other antibiotics is possible.

The intrinsic resistance of weissellas toward vancomycin may be attributed to the absence of D-Ala-D-lactate in their cell wall, which is the target of vancomycin. A similar situation exists in other LAB such as *Lactobacillus*, *Pediococcus*, and *Leuconostoc* species (Ammor et al., 2007). Thus, such resistance cannot be attributed to acquisition of resistance genes. Nevertheless, some

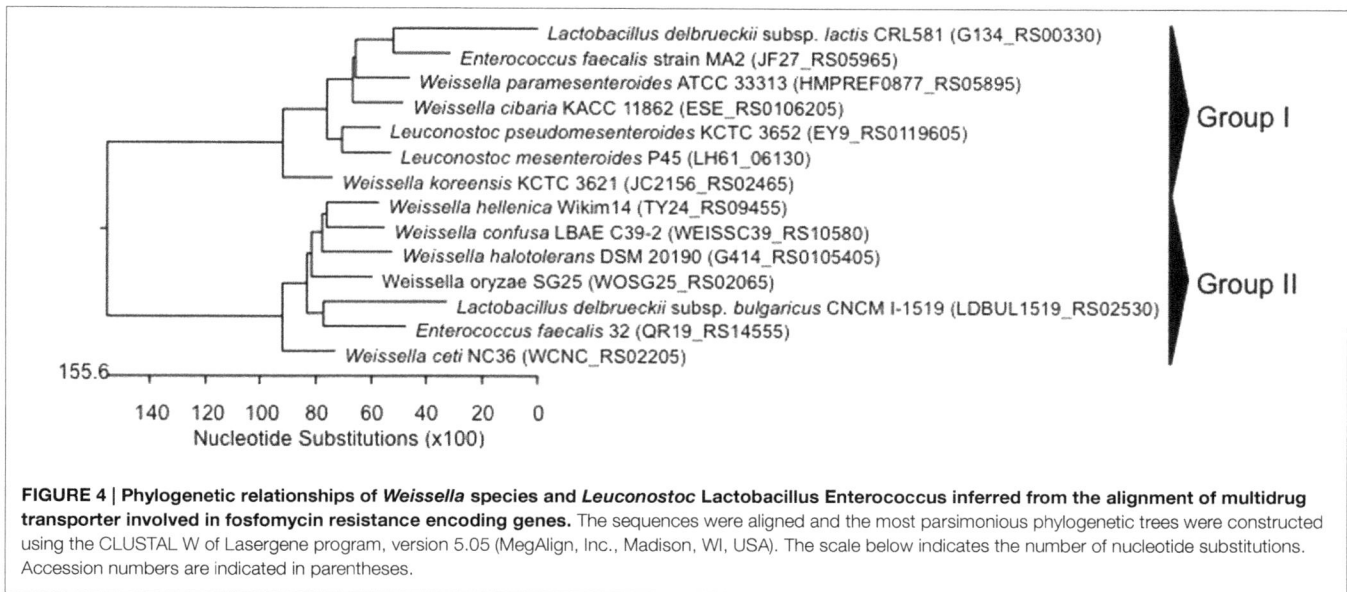

FIGURE 4 | Phylogenetic relationships of *Weissella* species and *Leuconostoc* Lactobacillus Enterococcus inferred from the alignment of multidrug transporter involved in fosfomycin resistance encoding genes. The sequences were aligned and the most parsimonious phylogenetic trees were constructed using the CLUSTAL W of Lasergene program, version 5.05 (MegAlign, Inc., Madison, WI, USA). The scale below indicates the number of nucleotide substitutions. Accession numbers are indicated in parentheses.

weissellas harbored in their genomes the *vanZ* resistance gene, as is the case for *W. cibaria* KACC 11862 (ESE_RS0111030), *W. confusa* LBAE C39-2 (WEISSC39_RS04975), and *W. paramesenteroides* ATCC 33313 (HMPREF0877_1234). This gene confers resistance to teicoplanin and does not involve the incorporation of a substituent of D-alanine into the peptidoglycan precursors (Arthur et al., 1995; **Table 3**).

No data have been reported on methicillin resistance to date. However, analyzing the *in silico* genome sequences of weissellas available in NCBI database, methicillin resistance protein encoding genes were detected in all weissellas (sequenced genomes of nine species; **Table 3**). Methicillin resistance proteins are found in *Staphylococcus* spp. and especially in *S. aureus*, and the presence of genes encoding such proteins in *Weissella* species may suggest a horizontal gene transfer between the genera. The alignment of nucleotide sequences for methicillin resistance protein in weissellas and other Gram-positive bacteria (*S. aureus*, *Leuconostoc*, *E. faecalis*, and *Lactococcus*) clearly showed an evolutionary relationship between the methicillin-resistance sequences of *Weissella* and the other Gram-positive bacteria *S. aureus*, *Leuconostoc*, and *E. faecalis* (**Figure 5**). The dendrogram showed two main groups, in which the genes from *Weissella* spp. were distributed regardless of the species, strain or origin. A divergence of methicillin resistance genes could even be observed in a single strain of *W. ceti* NC36, which carried multiple methicillin resistance genes, varying considerably in nucleotide sequence (**Figure 5**). On the other hand, one of two genes encoding for methicillin resistance in *W. cibaria* KACC 11862 (ESE_RS0105180) formed a single lineage, as such being divergent from the other weissellas and other Gram-positive bacteria (**Figure 5**). Further studies should elucidate the functionality of the methicillin resistance genes of weissellas, and the transferability of these genes to other bacteria.

Sulfonamide resistance could be due either to mutation in the chromosomal gene that mediates dihydropteroate synthesis, which is a folic acid precursor, or to the acquisition of resistance genes coding for resistant forms of the enzyme (Franklin and Snow, 2005). Resistance of weissellas to sulfonamides was reported by Byrne-Bailey et al. (2009) in *Weissella* spp. isolated from un-amended pig slurry and relied on the presence of *sul1* and *sul2* resistance genes. Figueiredo et al. (2012) isolated sulfonamide-resistant *W. ceti* that caused outbreaks characterized by acute haemorrhagic septicaemia and high mortality rates in rainbow trout. However, no genetic elements responsible for this resistance were described. *Weissella* genomes investigated in this study showed the presence of sulfonamide-resistance genes in *W. ceti* strains WS08 (WS08_0966), WS74 (WS74_1032), and WS105 (WS105_1028) isolated from rainbow trout (Figueiredo et al., 2012). Overall, resistance to vancomycin, fosfomycin, and sulfonamides appears to be a common trait in the genera *Weissella* and *Leuconostoc*. On the other hand, it is frequent that bacterial resistance to sulfonamides often co-exists with trimethoprim resistance, since this substance is used in combination with sulfonamides to minimize bacterial resistance. The known *dfr* trimethoprim resistance genes, which are usually associated with integrons (Skold, 2001) could, however, not be detected in any of the *Weissella* genome sequences analyzed here. Thus, the reported resistance of *W. confusa* (Fairfax et al., 2014; Medford et al., 2014) to trimethoprim may be caused by other modifications than those produced in the target enzyme dihydrofolate reductase (Dfr) that is encoded by the *dfr*-genes.

Weissellas are generally susceptible toward tetracycline. Screening of the published genomes revealed the presence of a gene encoding tetracycline (class C) resistance in *W. ceti* WS105 (WS105_0392) and in *W. thailandensis* fsh4-2 (WT2_00189; **Table 3**).

It is frequent to find resistance to the most widely used chemotherapeutic agent daunorubicin in probiotic bacteria such as lactobacilli (*Lb. acidophilus* NCFM, *Lb. casei* BD11) and bifidobacteria (*Bifidobacterium animalis* subsp. *lactis* Bb-12, *Bifidobacterium longum* subsp. *infantis* ATCC 15697). When the annotated genome sequences of weissellas were

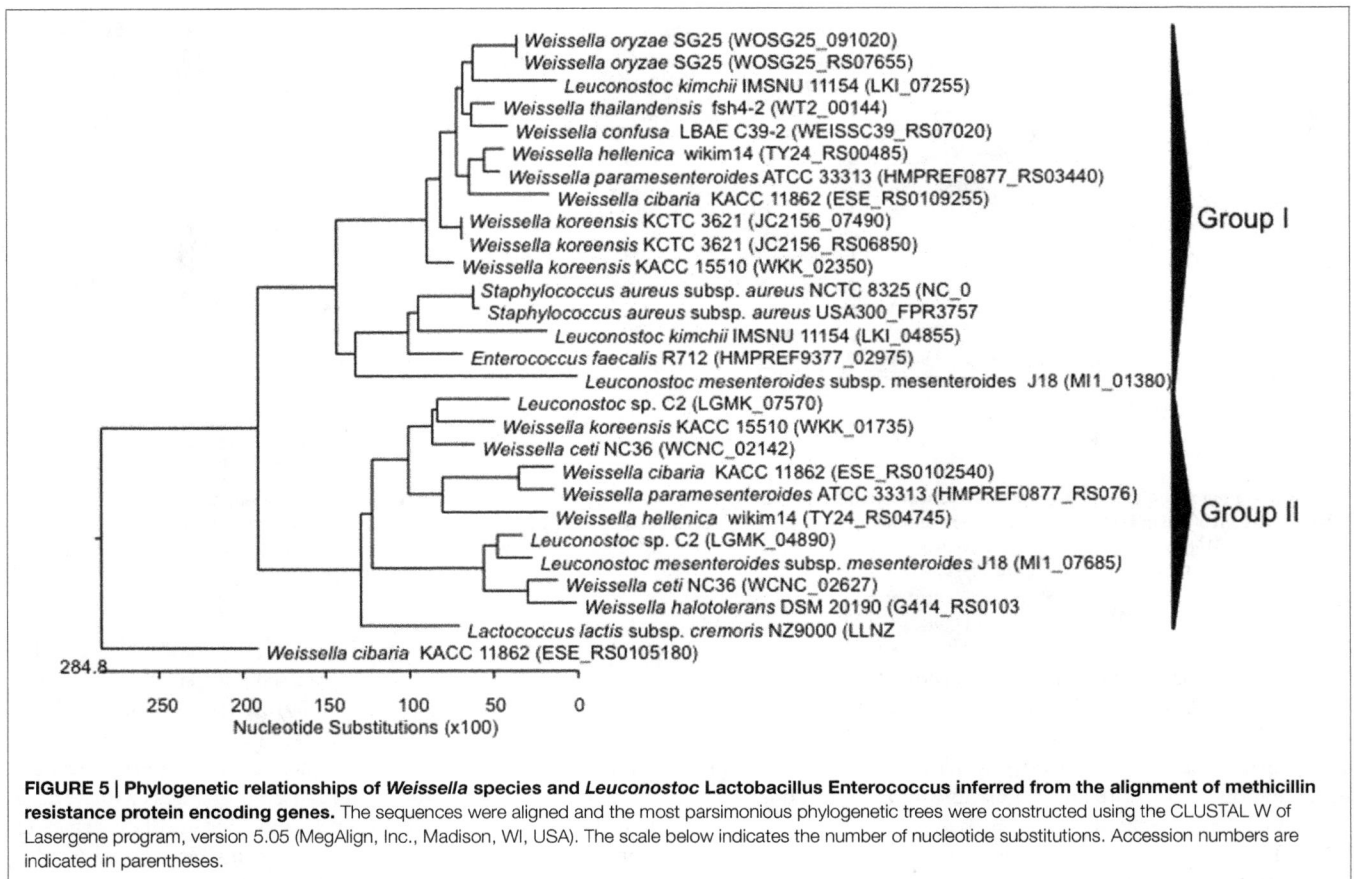

FIGURE 5 | Phylogenetic relationships of *Weissella* species and *Leuconostoc* Lactobacillus Enterococcus inferred from the alignment of methicillin resistance protein encoding genes. The sequences were aligned and the most parsimonious phylogenetic trees were constructed using the CLUSTAL W of Lasergene program, version 5.05 (MegAlign, Inc., Madison, WI, USA). The scale below indicates the number of nucleotide substitutions. Accession numbers are indicated in parentheses.

screened for daunorubicin resistance genes, ABC transporter-based resistance genes could be found in several *Weissella* spp. [*W. halotolerans* DSM 20190 (G414_RS0101040), *W. hellenica* Wikim14 (TY24_RS06500), and *W. oryzae* SG25 (WOSG25_RS07165)] (**Table 3**).

Antibiotic resistance is often based on unspecific mechanisms such as efflux pumps, which are widespread throughout evolution in different bacteria and can pump a variety of drugs out the cells (Piddock, 2006). When we screened the annotated genome sequences of *Weissella* spp., several efflux pumps were detected, such as multiple antibiotic resistance protein MarR, multidrug resistance SMR family protein, MFS multidrug transporter, ABC transporter, and a DedA family protein, which was recently shown to be associated with antibiotic resistance (Kumar and Doerrler, 2014).

Species-Specific Comparison of Biotechnological and Biosafety Issues

Based on our genomic *in silico* investigation, the potential deleterious and beneficial effects for each of the *Weissella* species in view of their application as starter cultures or probiotics are discussed below, highlighting the controversial issue of strains of the same species being industrially important in some cases, while being problematic from a safety point of view in others.

Weissella ceti

Weissella ceti (Vela et al., 2011) was isolated from different organs of beaked whales (*Mesoplodon bidens*) such as the spleen, kidney, muscle, brain, and lymph. Several strains of *W. ceti* were reported as pathogens in fish, such as *W. ceti* WS08, *W. ceti* WS74, and *W. ceti* WS105 from farmed rainbow trout in Brazil (Figueiredo et al., 2014; Costa et al., 2015), and *W. ceti* NC36 from farmed rainbow trout in the United States (Ladner et al., 2013). This suggests that weissellosis is a rapidly emerging disease in farmed rainbow trout in different geographical locations. Recently, Welch and Good (2013) reported that *Weissella* spp. closely related with *W. ceti* strains on the basis of their 16S rRNA gene sequences (99% identity) from Chinese and Brazilian out breaks were involved in mortality of farmed rainbow trout in the USA. On the other hand, no reports were found on the application of *W. ceti* strains as starter cultures or as probiotics. In this study it was found that *W. ceti* genomes harbored several virulence factors and antibiotic resistance genes, which point toward a pathogenic potential. The industrial application of such strains thus seems problematic and should be considered carefully on a strain to strain basis in the background of a detailed safety evaluation for the presence of antibiotic resistances and virulence determinants.

Weissella cibaria

Originally isolated from Thai fermented foods (Björkroth et al., 2002), *W. cibaria* was also isolated from other sources such as sourdough, fermented milk, cheese, fermented vegetables, fermented fish and meat, and also silage (Srionnual et al., 2007; reviewed by Fusco et al., 2015). However, clinical samples have also been a source of *W. cibaria* (Björkroth et al., 2002), with these bacteria being found in human saliva and the vagina, in human and animal feces and milk, and also in human blood and urine (reviewed by Fusco et al., 2015). *Weissella cibaria* has been targeted for use as starter culture in foods for different purposes such as for probiotic effects. In this sense, the dextran produced by dextransucrase from *W. cibaria* JAG8 had potential prebiotic effect for health benefits, stimulating the growth of probiotic bacteria (Tingirikari et al., 2014). Other beneficial effect of *W. cibaria* were suggested to be its capacity to produce higher levels of exopolysaccharides (EPS), which indicates more acid resistance, thus improving its probiotic capacity during passage throughout gastrointestinal tract (Park et al., 2013). Potentially probiotic *W. cibaria* strains were isolated from kimchi and also from goat milk (Elavarasi et al., 2014), and some strains were suggested to have anticancer activity, immune modulating activity, anti-inflammatory activity, antioxidant activity (patent related to the *W. cibaria*'s anticancer activity registered by Cha et al., 2008; Kang et al., 2011; Kwak et al., 2014), antiviral activity (the avian influenza virus; Rho et al., 2009) and anti-obesity effects being suggested to be more effective than the well-known probiotic bacterium *Lb. rhamnosus* GG (LGG; Ahn et al., 2013). It was also reported that the water-soluble polymers produced by *W. cibaria* inhibited biofilm formation by *Streptococcus mutans*, and thus production of such compounds could reduce oral plaque formation by approximately 20.7% *in vivo* and *in vitro* (Kang et al., 2006). Furthermore, the same authors reported that the hydrogen peroxide produced by *W. cibaria* inhibited the growth of the bacterial agent of periodontal disease, *Fusobacterium nucleatum*, and was effective in reducing the production of hydrogen sulfide and methanethiol responsible for the associated foul smell (Kang et al., 2006).

Several reports proposed different strains of *W. cibaria* as starter cultures in different fermentations, and these were based on its antimicrobial capacity. *W. cibaria* strains produce a variety of antagonistic substances, including organic acids and bacteriocins (i.e., weissellicin 110 produced by *W. cibaria* 110; Srionnual et al., 2007), as mentioned above. Wolter et al. (2014) reported that exopolysaccharide-producing *W. cibaria* MG1 might be a suitable starter culture for sourdough fermentation of buckwheat, quinoa and teff flour. *W. cibaria* was successfully tested in a defined semi-liquid sourdough starter (Gaggiano et al., 2007; Ricciardi et al., 2009), and due to its ability to grow at 45°C and produce EPS, which could be used as an alternative to additives for conditioning the textural properties of bread (Di Cagno et al., 2006), it was considered to be a suitable strain for this application. *Weissella cibaria* could also play an important role in meat fermentation. Thongsanit et al. (2009) proposed *W. cibaria* MSS2 as starter culture for the production of *nham* fermented sausage and for kimchi, due to the capacity of these bacteria to produce glutaminase, which is indispensable in food processing. Furthermore, the use of *W. cibaria* as starter for food fermentations promoted

the formation of ornithine from arginine, which in turn may have beneficial health effects, such as anti-obesity effect due to high levels of ornithine in fermented foods (Yu et al., 2009).

On the other hand, *W. cibaria* was reported as an emerging pathogen being associated with bacteremia (Kulwichit et al., 2008) and dog ear otitis (Björkroth et al., 2002), and also as food spoilage organism in sliced vacuum-packed cooked ham (Han et al., 2010). In the present study, we showed that *W. cibaria* KACC 11862 harbored in its genome some virulence determinants (hemolysins) and antibiotic resistance genes (fosfomycin, methicillin, and glycopeptide; **Tables 2–3**). This argues for a detailed investigation on the virulence potential for this species on a strain basis.

Weissella confusa

Due to its previous confusion with *viridans* streptococci, most of infections caused by *W. confusa* were underestimated due to the misidentification of this bacterium by commercial identification systems. When the genus *Weissella* was described by Collins et al. (1993), it was shown that *W. confusa* played an important role in human and animal sepsis and also bacteremia (Green et al., 1990; Olano et al., 2001; Björkroth et al., 2002; Flaherty et al., 2007; Shin et al., 2007; Salimnia et al., 2010, 2011; Harlan et al., 2011; Kumar et al., 2011; Lee et al., 2011; Fairfax and Salimnia, 2012), thumb abscess (Bantar et al., 1991), endocarditis (Shin et al., 2007), osteomyelitis (Kulwichit et al., 2007), and recently it was also shown to be involved in infection of a prosthetic joint (Medford et al., 2014). Here, *in silico* analysis of the *W. confusa* LBAE C39-2 genome showed the presence of virulence determinant genes (encoding collagen adhesion, hemolysin, and mucus-binding proteins) and antibiotic resistance genes (fosfomycin, methicillin, and glycopeptide; **Tables 2–3**).

On the other hand, due to the widespread use of *W. confusa* in fermented foods (sourdough, cereals, vegetables, fermented milk, cheese; reviewed by Fusco et al., 2015), it was proposed as a starter culture, and also as probiotic which provides various beneficial health effects such as the inhibition of *Helicobacter pylori*, a bacterium that causes chronic inflammation and ulcers in the stomach (Nam et al., 2002). Furthermore, *W. confusa* strains isolated from human feces were proposed as potential probiotics (Lee et al., 2012). Other *W. confusa* strains (UI006 and UI007) isolated from traditional dairy foods from Nigeria were proposed as adjunct cultures for the dairy manufacture industry, because of their antagonistic activities against entero- and uro-pathogens (organic acids, ethanol, and hydrogen peroxide), and their lack of toxic compounds (Ayeni et al., 2009, 2011). In this regard Yang (2013) proposed *W. confusa* LK4 isolated from leek kimchi as a functional starter culture for fermentation of leeks. Due to the aforementioned role in human infections, the biotechnological use of *W. confusa* should also be carefully assessed on a strain to strain basis.

Weissella koreensis

These bacteria isolated from a Korean fermented food "kimchi" (Lee et al., 2002) was used as starter culture in kimchi fermentations and production of functional foods, since *W. koreensis* OK1-6 has anti-obesity effects in high-fat diet (HF) induced obese mice (Park et al., 2012). Moon et al. (2012) reported the same anti-obesity effect for *W. koreensis* OK1-6, which produced ornithine from arginine, implying its functional role in reducing obesity. Also, Pi et al. (2014) showed that *W. koreensis* 521 was able to prevent and suppress obesity via the inhibition of pre-adipocyte mitogenesis and differentiation. On the other hand, the application of *W. koreensis* as starter culture together with *Ln. citreum* and baker's yeast in sourdough to make whole wheat bread was also considered as successful, since a good texture and an extended shelf-life of bread could be obtained (Choi et al., 2012). The use of *W. koreensis* as starter culture in broken rice was also found to provide good organoleptic properties to *jeungpyeon*, which is a Korean fermented rice cake (Choi et al., 2013). *Weissella koreensis* was also proposed for use as probiotic in pigs, since dietary supplementation with *W. koreensis* WKG2 in growing pigs could improve the average daily gain (ADG) and could have a beneficial effect on the immune response during an inflammatory challenge (Wang et al., 2014). On the other hand, no negative effects on heath were yet attributed to this species. However, in our *in silico* genome investigation, we showed the presence of potential virulence determinants and antibiotic resistance genes in some strains of *W. koreensis* (**Tables 2–3**), again indicating that a safety assessment for strains targeted for biotechnological use should be done.

Other *Weissella* spp.

Regarding their health-promoting activities, several species of *Weissella* exhibited beneficial effects. *W. kimchii* is a hydrogen peroxide-producing species that has been proposed as probiotic to prevent vaginal infections against *Candida albicans*, *Escherichia coli*, *S. aureus*, and *Streptococcus agalactiae* (Lee, 2005; Lee et al., 2011). *W. hellenica* was isolated from different sources such as fermented vegetables, fermented sausage, fermented fish, cheeses, cow's milk and flounder intestine (Leong et al., 2013; reviewed by Fusco et al., 2015), and is known for its potential probiotic activity due to its bacteriocin-producing capacity (weissellicins D, L, M, and Y). Strains from this species are active against several pathogens (Masuda et al., 2011; Leong et al., 2013; Chen et al., 2014) and have potential also as bio-protective culture to improve safety and shelf-life of foods like tofu (Chen et al., 2014). *W. hellenica* strains isolated from different sources (cheese, fermented sausage, fermented vegetables, cow's milk, flounder intestine; reviewed by Fusco et al., 2015) were also reported as glucan (EPS) producers (Kim et al., 2008), which may have industrial applications. Finally, *W. paramesenteroides* isolated from soil, vegetables, cheese, fermented sausage, fermented sea food, fermented vegetables and also from feces, cow's milk, and gut of rainbow (reviewed by Fusco et al., 2015) can have antibacterial activity by production of bacteriocins such as weissellin A from *W. paramesenteroides* DX (Papagianni and Papamichael, 2011), a bacteriocin from *W. paramesenteroides* DFR-8 (Pal and Ramana, 2010), and also non-proteinaceous antibacterial compounds (Pal and Ramana, 2009) with a broad inhibitory spectrum against

Gram-positive and Gram-negative bacteria. These strains may therefore be interesting for their use as food biopreservatives. On the detrimental side, Han et al. (2010) showed that *W. paramesenteroides* was involved in sliced vacuum-packed cooked ham spoilage.

Due to their heterofermentative metabolism, some weissellas are also involved in food spoilage and lead to sensory defects of, e.g., meat products (Marsden et al., 2009). In this regards, *W. viridescens* plays an important role in meat spoilage, producing a greenish slime on meat surfaces as a result of its production of H_2O_2. In cooked hams, formation of cavities can result from the production of CO_2 by these bacteria. Furthermore, taking into account data presented here from *in silico* analyses of *Weissella* spp. genomes, the presence of virulence determinants, especially hemolysins, were detected in all strains analyzed. Also antibiotic resistance genes, such as those coding for fosfomycin and methicillin resistances (**Tables 2–3**), occur in this species. Also in this case, therefore, the safety of strains intended for industrial use should be investigated for each strain in detail.

CONCLUSION

Considering the large number of health-promoting benefits which could arise from the use of strains of *Weissella* spp., such as antibacterial, anti-viral, anti-tumoral, anti-obesity, anti-inflammatory, and antioxidant activities, several weissellas could be targeted for use as starter cultures or probiotics. By contrast, specific strains of *Weissella* species have also been involved as pathogens in the etiology of different diseases such as bacteremia, endocarditis, sepsis, and may even cause mortality. In fact, the safety of this genus has not been deeply studied, as only some strains are considered as opportunistic pathogens. Thus, the application of strains in foods and feeds or for humans as probiotics should be done with caution. In this report we showed that screening of genome sequences revealed the presence of several virulence and antibiotic resistance genes, which could be the basis of the potential pathogenicity of some strains. However, the presence of single determinants should not be an exclusion criterium for weissellas that may have overall beneficial effects. Thus, selection of weissellas intended to be used as starters or as probiotics should be investigated carefully regarding their safety aspects, preferably by genome sequencing an annotation, since the heterogeneity in nucleotide sequences of well-known virulence and antibiotic resistance-genes make these undetectable by PCR methods. Generally, the application of *W. confusa* and *W. cibaria* strains as starter cultures or as probiotics should be approached with caution, carefully selecting strains which lack pathogenic potential and which do not possess transferable antibiotic resistance genes.

ACKNOWLEDGMENTS

We acknowledge research grants AGL2013-43571-P (Ministerio de Economía y Competitividad, MINECO, FEDER) and UJA2014/07/02 (Plan Propio UJA) and research group AGR230.

REFERENCES

Ahn, S. B., Park, H. E., Lee, S. M., Kim, S. Y., Shon, M. Y., and Lee, W. K. (2013). Characteristics and immuno-modulatory effects of *Weissella cibaria* JW15 isolated from Kimchi, Korea traditional fermented food, for probiotic use. *J. Biomed. Res.* 14, 206–211. doi: 10.12729/jbr.2013.14.4.206

Alkema, W., Boekhorst, J., Wels, M., and van Hijum, S. A. F. T. (2015). Microbial bioinformatics for food safety and production. *Brief. Bioinform.* doi: 10.1093/bib/bbv034 [Epub ahead of print].

Ammor, M. S., Florez, A. B., and Mayo, B. (2007). Antibiotic resistance in non-enterococcal lactic acid bacteria and bifidobacteria. *Food Microbiol.* 24, 559–570. doi: 10.1016/j.fm.2006.11.001

Arca, P., Reguera, G., and Hardisson, C. (1997). Plasmid-encoded fosfomycin resistance in bacteria isolated from the urinary tract in a multicentre Surrey. *J. Antimicrob. Chemother.* 40, 393–399. doi: 10.1093/jac/40.3.393

Arthur, M., Depardieu, F., Molinas, C., Reynolds, P., and Courvalin, P. (1995). The *vanZ* gene of Tn1546 from *Enterococcus faecium* BM4147 confers resistance to teicoplanin. *Gene* 154, 87–92. doi: 10.1016/0378-1119(94)00851-I

Ayeni, F. A., Adeniyi, B. A., Ogunbanwo, S. T., Tabasco, R., Paarup, T., Peláez, C., et al. (2009). Inhibition of uropathogens by lactic acid bacteria isolated from dairy foods and cow's intestine in western Nigeria. *Arch. Microbiol.* 191, 639–648. doi: 10.1007/s00203-009-0492-9

Ayeni, F. A., Adeniyi, B. A., Ogunbanwo, S. T., Nader-Macias, M. E., and Ruas-Madiedo, P. (2011). Survival of *Weissella confusa* and *Lactobacillus paracasei* strains in fermented milks under cold storage and after freeze-drying. *Milchwissenschaft* 66, 61–64.

Aymerich, T., Martin, B., Garriga, M., Vidal-Carou, M. C., Bover-Cid, S., and Hugas, M. (2006). Safety properties and molecular strain typing of lactic acid bacteria from slightly fermented sausages. *J. Appl. Microbiol.* 100, 40–49. doi: 10.1111/j.1365-2672.2005.02772.x

Bantar, C. E., Relloso, S., Castell, F. R., Smayevsky, J., and Bianchini, H. M. (1991). Abscess caused by vancomycin-resistant *Lactobacillus confusus*. *J. Clin. Microbiol.* 29, 2063–2064.

Björkroth, K. J., Schillinger, U., Geisen, R., Weiss, N., Hoste, B., Holzapfel, W. H., et al. (2002). Taxonomic study of *Weissella confusa* and description of *Weissella cibaria* sp. nov., detected in food and clinical samples. *Int. J. Syst. Evol. Microbiol.* 52, 141–148. doi: 10.1099/00207713-52-1-141

Byrne-Bailey, K. G., Gaze, W. H., Kay, P., Boxall, A. B. A., Hawkey, P. M., and Wellington, E. M. H. (2009). Prevalence of sulfonamide resistance genes in bacterial isolates from manured agricultural soils and pig slurry in the United Kingdom. *Antimicrob. Agents Chemother.* 53, 696–702. doi: 10.1128/AAC.00652-07

Cai, Y., Benno, Y., Nakase, T., and Oh, T. K. (1998). Specific probiotic characterization of *Weissella hellenica* DS-12 isolated from flounder intestine. *J. Gen. Appl. Microbiol.* 44, 311–316. doi: 10.2323/jgam.44.311

Cha, S. K., Ahn, B. H., and Kim, J. R. (2008). *Korea Food Research Institute, Assignee Weissella cibaria 148-2 Lactic Bacteria for Functional Healthy Effect and Makgeolli Containing the Same.* Korea patent KR 1020080133488. Inventors Probiotic Properties of Lactic Acid Bacteria Isolated from Water-Buffalo Mozzarella Cheese.

Chen, C., Rui, X., Lu, Z., Li, W., and Dong, M. (2014). Enhanced shelf-life of tofu by using bacteriocinogenic *Weissella hellenica* D1501 as bioprotective cultures. *Food Control* 46, 203–209. doi: 10.1016/j.foodcont.2014.05.004

Choi, H., Kim, Y. W., Hwang, I., Kim, J., and Yoon, S. (2012). Evaluation of *Leuconostoc citreum* HO12 and *Weissella koreensis* HO20 isolated from kimchi as a starter culture for whole wheat sourdough. *Food Chem.* 134, 2208–2216. doi: 10.1016/j.foodchem.2012.04.047

Choi, H., Ok Park, J., and Yoon, S. (2013). Fermentation of broken rice using kimchi-derived *Weissella koreensis* HO20 and its use in Jeungpyeon. *Food Sci. Biotechnol.* 22, 1275–1283. doi: 10.1007/s10068-013-0213-7

Choi, H. J., Cheigh, C. I., Kim, S. B., Lee, J. C., Lee, D. W., Choi, S. W., et al. (2002). *Weissella kimchii* sp. nov., a novel lactic acid bacterium from kimchi. *Int. J. Syst. Evol. Microbiol.* 52, 507–511. doi: 10.1099/00207713-52-2-507

Clancy, J., Petitpas, J., Dib-Hajj, F., Yuan, W., Cronan, M., Kamath, A. V., et al. (1996). Molecular cloning and functional analysis of a novel macrolide-resistance determinant, mefA, from *Streptococcus pyogenes*. *Mol. Microbiol.* 22, 867–879. doi: 10.1046/j.1365-2958.1996.01521.x

Collins, M. D., Samelis, J., Metaxopoulos, J., and Wallbanks, S. (1993). Taxonomic studies on some *Leuconostoc*-like organisms from fermented sausages: description of a new genus *Weissella* for the *Leuconostoc paramesenteroides* group of species. *J. Appl. Bacteriol.* 75, 595–603. doi: 10.1111/j.1365-2672.1993.tb01600.x

Collins, M. D., Samelis, J., Metaxopoulos, J., and Wallbanks, S. (1994). Validation of the publication of new names and new combinations previously effectively published outside the IJSB List No. 49. *Int. J. Syst. Bacteriol.* 44, 370–371. doi: 10.1099/00207713-44-2-370

Comi, G., and Iacumin, L. (2012). Identification and process origin of bacteria responsible for cavities and volatile of flavour compounds in artisan cooked ham. *Int. J. Food Sci. Tech.* 47, 114–121. doi: 10.1111/j.1365-2621.2011.02816.x

Costa, F. A. A., Leal, C. A. G., Schuenker, N. D., Leite, R. C., and Figueiredo, H. C. P. (2015). Characterization of *Weissella ceti* infections in Brazilian rainbow trout, *Oncorhynchus mykiss* (Walbaum), farms and development of an oil-adjuvanted vaccine. *J. Fish Dis.* 38, 295–302. doi: 10.1111/jfd.12236

Di Cagno, R., De Angelis, M., Limitone, A., Minervini, F., Carnevali, P., Corsetti, A., et al. (2006). Glucan and fructan production by sourdough *Weissella cibaria* and *Lactobacillus plantarum*. *J. Agric. Food Chem.* 54, 9873–9881. doi: 10.1021/jf061393+

Dunne, C., Kelly, P., O'Halloran, S., Soden, D., Bennett, M., von Wright, A., et al. (2004). Mechanisms of adherence of a probiotic *Lactobacillus* strain during and after *in vivo* assessment in ulcerative colitis patients. *Microb. Ecol. Health Dis.* 16, 96–104. doi: 10.1080/08910600410032295

Elavarasi, V., Pugazhendhi, A., Poornima Priyadharsani, T. K., Valsala, H., and Thamaraiselvi, K. (2014). Screening and characterization of *Weissella cibaria* isolated from food source for probiotic properties. *Int. J. Comput. Appl.* 1, 29–32.

European Food Safety Authority (EFSA). (2004). *EFSA Scientific Colloquium Summary Report*. Brussels: QPS

European Food Safety Authority (EFSA). (2015). Statement on the update of the list of QPS-recommended biological agents intentionally added to food or feed as notified to EFSA. 2: suitability of (taxonomic) units notified to EFSA until March 2015. *EFSA J.* 13, e4138. doi: 10.2903/j.efsa.2015.4138

Fairfax, M. R., Lephart, P. R., and Salimnia, H. (2014). *Weissella confusa*: problems with identification of an opportunistic pathogen that has been found in fermented foods and proposed as a probiotic. *Front. Microbiol.* 5:254. doi: 10.3389/fmicb.2014.00254

Fairfax, M. R., and Salimnia, H. (2012). "Beware of unusual organisms masquerading as skin contaminants," in *Sepsis–An Ongoing and Significant Challenge*, ed. L. Azevedo (Rijeka: Intech Open), 275–286.

Figueiredo, H. C., Costa, F. A., Leal, C. A., Carvalho-Castro, G. A., and Leite, R. C. (2012). *Weissella* sp. outbreaks in commercial rainbow trout (*Oncorhynchus mykiss*) farms in Brazil. *Vet. Microbiol.* 156, 359–366. doi: 10.1016/j.vetmic.2011.11.008

Figueiredo, H. C., Leal, C. A., Dorella, F. A., Carvalho, A. F., Soares, S. C., Pereira, F. L., et al. (2014). Complete genome sequences of fish pathogenic *Weissella ceti* Strains WS74 and WS105. *Genome Announc.* 2, 1014–1014. doi: 10.1128/genomeA.01014-14

Flaherty, J. D., Levett, P. N., Dewhirst, F. E., Troe, T. E., Warren, J. R., and Johnson, S. (2003). Fatal case of endocarditis due to *Weissella confusa*. *J. Clin. Microbiol.* 41, 2237–2239. doi: 10.1128/JCM.41.5.2237-2239.2003

Flaherty, J. D., Levett, P. N., Dewhirst, F. E., Troe, T. E., Warren, J. R., and Johnson, S. (2007). Fatal case of endocarditis due to *Weissella confusa*. *J. Clin. Microbiol.* 41, 2237–2239. doi: 10.1128/JCM.41.5.2237-2239.2003

Food and Drug Administration. (1999). *Federal Food, Drug, and Cosmetic Act*. Washington, DC: US Food and Drug Administration.

Franklin, T. J., and Snow, G. A. (2005). *Biochemistry and Molecular Biology of Antimicrobial Drug Action*, 6th Edn. New York, NY: Springer Science And Business Media Incorporated.

Franz, C. M. A. P., Stiles, M. E., Schleifer, K.-H., and Holzapfel, W. H. (2003). Enterococci in foods—a conundrum for food safety. *Int. J. Food Microbiol.* 88, 105–122. doi: 10.1016/S0168-1605(03)00174-0

Fusco, V., Quero, G. M., Cho, G., Kabisch, J., Meske, D., Neve, H., et al. (2015). The genus *Weissella*: taxonomy, ecology and biotechnological potential. *Front. Microbiol.* 6:155. doi: 10.3389/fmicb.2015.00155

Gaggiano, M., Di Cagno, R., De Angelis, M., Arnault, P., Tossut, P., Fox, P. F., et al. (2007). Defined multi-species semi-liquid ready-to-use sourdough starter. *Food Microbiol.* 24, 15–24. doi: 10.1016/j.fm.2006.04.003

Green, M., Wadowsky, R. M., and Barbadora, K. (1990). Recovery of vancomycin-resistant gram-positive cocci from children. *J. Clin. Microbiol.* 28, 484–488.

Han, Y., Xu, X., Yun Jiang, Y., Zhou, G., Sun, X., and Xu, B. (2010). Inactivation of food spoilage bacteria by high pressure processing: evaluation with conventional media and PCR–DGGE analysis. *Food Res. Int.* 43, 1719–1724. doi: 10.1016/j.foodres.2010.05.012

Harlan, N. P., Kempker, R. R., Parekh, S. M., Burd, E. M., and Kuhar, D. T. (2011). *Weissella confusa* bacteremia in a liver transplant patient with hepatic artery thrombosis. *Transpl. Infect. Dis.* 13, 290–293. doi: 10.1111/j.1399-3062.2010.00579.x

Hwanhlem, N., Biscola, V., El-Ghaish, S., Jaffrès, E., Dousset, X., Haertlé, T., et al. (2013). Bacteriocin-producing lactic acid bacteria isolated from mangrove forests in Southern Thailand as potential bio-control agents: purification and characterization of bacteriocin produced by *Lactococcus lactis* subsp. *lactis* KT2W2L. *Probiotics Antimicrob. Proteins* 5, 264–278. doi: 10.1007/s12602-013-9150-2

Jeronymo-Ceneviva, A. B., de Paula, A. T., Silva, L. F., Todorov, S. D., Franco, B. D., and Penna, A. L. (2014). Probiotic properties of lactic acid bacteria isolated from water-buffalo mozzarella cheese. *Probiotics Antimicrob Proteins* 6, 141–156. doi: 10.1007/s12602-014-9166-2

Kang, M. S., Chung, J., Kim, S. M., Yang, K. H., and Oh, J. S. (2006). Effect of *Weissella cibaria* isolates on the formation of *Streptococcus mutans* biofilm. *Caries Res.* 40, 418–425. doi: 10.1159/000094288

Kang, M.-S., Lim, H.-S., Kim, S.-M., Lee, H.-C., and Oh, J.-S. (2011). Effect of *Weissella cibaria* on *Fusobacterium nucleatum*-induced interleukin-6 and interleukin-8 production in KB cells. *J. Bacteriol. Virol.* 41, 9–18. doi: 10.4167/jbv.2011.41.1.9

Kim, M. J., Seo, H. N., Hwang, T. S., Lee, S. H., and Park, D. H. (2008). Characterization of exopolysaccharide (EPS) produced by *Weissella hellenica* SKkimchi3 isolated from kimchi. *J. Microbiol.* 46, 535–541. doi: 10.1007/s12275-008-0134-y

Kulwichit, W., Nilgate, S., Chatsuwan, T., Krajiw, S., Unhasuta, C., and Chongthaleong, A. (2007). Accuracies of *Leuconostoc* phenotypic identification: a comparison of API systems and conventional phenotypic assays. *BMC Infect. Dis.* 7:69. doi: 10.1186/1471-2334-7-69

Kulwichit, W., Nilgate, S., Krajiw, S., Unhasuta, C., Chatsuwan, T., and Chongthaleong, A. (2008). "*Weissella* spp.: lactic acid bacteria emerging as a human pathogen," in *Proceedings of the 18th European Congress of Clinical Microbiology and Infectious Diseases*, Barcelona, 19–22.

Kumar, A., Augustine, D., Sudhindran, S., Kurian, A. M., Dinesh, K. R., Karim, S., et al. (2011). *Weissella confusa*: a rare cause of vancomycin-resistant Gram-positive bacteraemia. *J. Med. Microbiol.* 60, 1539–1541. doi: 10.1099/jmm.0.027169-0

Kumar, S., and Doerrler, W. T. (2014). Members of the conserved DedA family are likely membrane transporters and are required for drug resistance in *Escherichia coli*. *Antimicrob. Agents Chemother.* 58, 923–930. doi: 10.1128/AAC.02238-13

Kwak, S.-H., Cho, Y.-M., Noh, G.-M., and Om, A.-S. (2014). Cancer preventive potential of kimchi lactic acid bacteria (*Weissella cibaria*, *Lactobacillus plantarum*). *J. Cancer Prev.* 19, 253–258. doi: 10.15430/JCP.2014.19.4.253

Ladner, J. T., Welch, T. J., Whitehouse, C. A., and Palacios, G. F. (2013). Genome sequence of *Weissella ceti* NC36, an emerging pathogen of farmed rainbow trout in the United States. *Genome Announc.* 1, e00187-12. doi: 10.1128/genomeA.00187-12

Lee, J. S., Lee, K. C., Ahn, J. S., Mheen, T. I., Pyun, Y. R., and Park, Y. H. (2002). *Weissella koreensis* sp. nov., isolated from kimchi. *Int. J. Syst. Evol. Microbiol.* 52, 1257–1261. doi: 10.1099/00207713-52-4-1257

Lee, K. W., Park, J. Y., Jeong, H. R., Heo, H. J., Han, N. S., and Kim, J. H. (2012). Probiotic properties of *Weissella* strains isolated from human faeces. *Anaerobe* 18, 96–102. doi: 10.1016/j.anaerobe.2011.12.015

Lee, M. R., Huang, Y. T., Liao, C. H., Lai, C. C., Lee, P. I., and Hsueh, P. R. (2011). Bacteraemia caused by *Weissella confusa* at a university hospital in Taiwan 1997–2007. *Clin. Microbiol. Infect.* 17, 1226–1231. doi: 10.1111/j.1469-0691.2010.03388.x

Lee, Y. (2005). Characterization of *Weissella kimchii* PL9023 as a potential probiotic for women. *FEMS Microbiol. Lett.* 250, 157–162. doi: 10.1016/j.femsle.2005.07.009

Leong, K. H., Chen, Y. S., Lin, Y. H., Pan, S. F., Yu, B., Wu, H. C., et al. (2013). Weissellicin L, a novel bacteriocin from *sian-sianzih*-isolated *Weissella hellenica* 4-7. *J. Appl. Microbiol.* 115, 70–76. doi: 10.1111/jam.12218

Liu, J. Y., Li, A. H., Ji, C., and Yang, W. M. (2009). First description of a novel *Weissella* species as an opportunistic pathogen for rainbow trout

Oncorhynchus mykiss (Walbaum) in China. *Vet. Microbiol.* 136, 314–320. doi: 10.1016/j.vetmic.2008.11.027

Mack, D. R., Michail, S., Wei, S., McDougall, L., and Hollingsworth, M. A. (1999). Probiotics inhibit enteropathogenic *Escherichia coli* adherence *in vitro* by inducing intestinal mucin gene expression. *Am. J. Physiol.* 276, 941–950.

Marsden, J., Ahmed-Kotrola, N., Saini, J., Fung, D. Y. C., and Phebus, R. (2009). "Cooked meats, poultry and their products," in *Meat Products*, ed. R. Fernandes (Leatherhead: Leatherhead Food International Ltd.), 53–82.

Masuda, Y., Zendo, T., Sawa, N., Perez, R. H., Nakayama, J., and Sonomoto, K. (2011). Characterization and identification of weissellicin Y and weissellicin M, novel bacteriocins produced by *Weissella hellenica* QU 13. *J. Appl. Microbiol.* 112, 99–108. doi: 10.1111/j.1365-2672.2011.05180.x

Medford, R., Patel, S. N., and Evans, G. A. (2014). A confusing case—*Weissella confusa* prosthetic joint infection: a case report and review of the literature. *Can. J. Infect. Dis. Med. Microbiol.* 25, 173–175.

Moon, Y. J., Soh, J. R., Yu, J. J., Sohn, H. S., Cha, Y. S., and Oh, S. H. (2012). Intracellular lipid accumulation inhibitory effect of *Weissella koreensis* OK1-6 isolated from kimchi on differentiating adipocyte. *J. Appl. Microbiol.* 113, 652–658. doi: 10.1111/j.1365-2672.2012.05348.x

Muñoz-Atienza, E., Gómez-Sala, B., Araújo, C., Campanero, C., del Campo, R., Hernández, P. E., et al. (2013). Antimicrobial activity, antibiotic susceptibility and virulence factors of lactic acid bacteria of aquatic origin intended for use as probiotics in aquaculture. *BMC Microbiol.* 13:15. doi: 10.1186/1471-2180-13-15

Nallapareddy, S. R., Singh, K. V., Sillanpaa, J., Zhao, M., and Murray, B. E. (2011). Relative contributions of Ebp Pili and the collagen adhesion ace to host extracellular matrix protein adherence and experimental urinary tract infection by *Enterococcus faecalis* OG1RF. *Infect. Immun.* 79, 2901–2910. doi: 10.1128/IAI.00038-11

Nam, H., Ha, M., Bae, O., and Lee, Y. (2002). Effect of *Weissella confusa* strain PL9001 on the adherence and growth of *Helicobacter pylori*. *Appl. Environ. Microbiol.* 68, 4642–4645. doi: 10.1128/AEM.68.9.4642-4645.2002

Ogier, J.-C., and Serror, P. (2008). Safety assessment of dairy microorganisms: the *Enterococcus* genus. *Int. J. Food Microbiol.* 126, 291–301. doi: 10.1016/j.ijfoodmicro.2007.08.017

Olano, A., Chua, J., Schroeder, S., Minari, A., Lasalvia, M., and Hall, G. (2001). *Weissella confusa* (Basonym: *Lactobacillus confusus*) bacteremia: a case report. *J. Clin. Microbiol.* 39, 1604–1607. doi: 10.1128/JCM.39.4.1604-1607.2001

Pal, A., and Ramana, K. V. (2009). Isolation and preliminary characterization of a non bacteriocin antimicrobial compound from *Weissella* paramesenteroides DFR-8 isolated from cucumber (*Cucumis sativus*). *Process. Biochem.* 44, 499–503. doi: 10.1016/j.procbio.2009.01.006

Pal, A., and Ramana, K. V. (2010). Purification and characterization of bacteriocin from *Weissella paramesenteroides* dfr-8, an isolate from cucumber (*Cucumis sativus*). *J. Food Biochem.* 34, 932–948. doi: 10.1111/j.1745-4514.2010.00340.x

Papagianni, M., and Papamichael, E. M. (2011). Purification, amino acid sequence and characterization of the class IIa bacteriocin weissellin A, produced by *Weissella* paramesenteroides DX. *Bioresour. Technol.* 102, 6730–6734. doi: 10.1016/j.biortech.2011.03.106

Park, J. A., Tirupathi Pichiah, P. B., Yu, J. J., Oh, S. H., Daily, J. W. III, and Cha, Y. S. (2012). Anti-obesity effect of kimchi fermented with *Weissella koreensis* OK1-6 as starter in high-fat diet-induced obese C57BL/6J mice. *J. Appl. Microbiol.* 113, 1507–1516. doi: 10.1111/jam.12017

Park, J. H., Ahn, H. J., Kim, S. G., and Chung, C. H. (2013). Dextran-like exopolysaccharide-producing *Leuconostoc* and *Weissella* from kimchi and its ingredients. *Food Sci. Biotechnol.* 22, 1047–1053. doi: 10.1007/s10068-013-0182-x

Patel, A. R., Lindström, C., Patel, A., Prajapati, J. B., and Holst, O. (2012). Probiotic properties of exopolysaccharide producing lactic acid bacteria isolated from vegetables and traditional Indian fermented foods. *Int. J. Ferment. Foods* 1, 87–101.

Pi, K., Lee, K., Kim, Y., and Lee, E.-J. (2014). The inhibitory effect of *Weissella koreensis* 521 isolated from kimchi on 3T3-L1 adipocyte differentiation. *Int. J. Med. Health Biomed. Pharm. Eng.* 8, 7–10.

Piddock, L. J. (2006). Multidrug-resistance efflux pumps—not just for resistance. *Nat. Rev. Microbiol.* 4, 629–636. doi: 10.1038/nrmicro1464

Qu, T.-T., Shi, K.-R., Ji, J.-S., Yang, Q., Du, X.-X., Wei, Z.-Q., et al. (2014). Fosfomycin resistance among vancomycin-resistant enterococci owing to transfer of a plasmid harbouring the fosB gene. *Int. J. Antimicrob. Agents* 43, 361–365. doi: 10.1016/j.ijantimicag.2013.11.003

Rho, J. B., Poo, H., Choi, Y. K., Kim, C. J., and Sung, M. H. (2009). *New Lactic Acid Bacteria Having its Inhibitory Effect on Avian Influenza Virus Infection and Composition Containing the Same*. United States Patent US 20120156172 A1. Inventors.

Ricciardi, A., Parente, E., and Zotta, T. (2009). Modelling the growth of *Weissella cibaria* as a function of fermentation conditions. *J. Appl. Microbiol.* 107, 1528–1535. doi: 10.1111/j.1365-2672.2009.04335.x

Salimnia, H., Alangaden, G. J., Bharadwaj, R., Painter, T. M., Chandrasekar, P. H., and Fairfax, M. R. (2010). *Weissella confusa*: an unexpected cause of vancomycin-resistant Gram-positive bacteremia in immune compromised hosts. *Transpl. Infect. Dis.* 12, 526–528. doi: 10.1111/j.1399-3062.2010.00539.x

Salimnia, H., Alangaden, G. J., Bharadwaj, R., Painter, T. M., Chandrasekar, P. H., and Fairfax, M. R. (2011). *Weissella confusa*: an unexpected cause of vancomycin-resistant Gram-positive bacteremia in immunocompromised hosts. *Transpl. Infect. Dis.* 13, 294–298. doi: 10.1111/j.1399-3062.2010.00586.x

Santos, E. M., Jaime, I., Rovira, J., Lyhs, U., Korkeala, H., and Björkroth, J. (2005). Characterization and identification of lactic acid bacteria in "morcilla de Burgos". *Int. J. Food Microbiol.* 97, 285–296. doi: 10.1016/j.ijfoodmicro.2004.04.021

Shin, J. H., Kim, D. I., Kim, H. R., Kim, D. S., Kook, J. K., and Lee, J. N. (2007). Severe infective endocarditis of native valves caused by *Weissella confusa* detected incidentally on echocardiography. *J. Infect.* 54, 149–151. doi: 10.1016/j.jinf.2006.09.009

Skold, O. (2001). Resistance to trimethoprim and sulfonamides. *Vet. Res.* 32, 261–273. doi: 10.1051/vetres:2001123

Srionnual, S., Yanagida, F., Lin, L. H., Hsiao, K. N., and Chen, Y. S. (2007). Weissellicin 110, a newly discovered bacteriocin from *Weissella cibaria* 110, isolated from Plaa-Som, a fermented fish product from Thailand. *Appl. Environ. Microbiol.* 73, 2247–2250. doi: 10.1128/AEM.02484-06

Sullam, P. M. (1994). Host-pathogen interactions in the development of bacterial endocarditis. *Curr. Opin. Infect. Dis.* 4, 304–309. doi: 10.1097/00001432-199406000-00004

Tait-Kamradt, A., Clancy, J., Cronan, M., Dib-Hajj, F., Wondrack, L., Yuan, W., et al. (1997). mefE is necessary for the erythromycin-resistant M phenotype in *Streptococcus pneumoniae*. *Antimicrob. Agents Chemother.* 41, 2251–2255.

Thongsanit, J., Tanikawa, M., Yano, S., Tachiki, T., and Wakayama, M. (2009). Identification of glutaminase-producing lactic acid bacteria isolated from Nham, a traditional Thai fermented food and characterisation of glutaminase activity of isolated *Weissella cibaria*. *Ann. Microbiol.* 59, 715–720. doi: 10.1007/BF03179213

Tingirikari, J. M., Kothari, D., Shukla, R., and Goyal, A. (2014). Structural and biocompatibility properties of dextran from *Weissella cibaria* JAG8 as food

additive. *Int. J. Food Sci. Nutr.* 65, 686–691. doi: 10.3109/09637486.2014.917147

Vankerckhoven, V., Moreillon, P., Piu, S., Giddey, M., Huys, G., Vancanneyt, M., et al. (2007). Infectivity of *Lactobacillus rhamnosus* and *Lactobacillus paracasei* isolates in a rat model of experimental endocarditis. *J. Med. Microbiol.* 56, 1017–1024. doi: 10.1099/jmm.0.46929-0

Vela, A. I., Fernández, A., de Quirós, Y. B., Herráez, P., Domínguez, L., and Fernández-Garayzábal, J. F. (2011). *Weissella ceti* sp. nov., isolated from beaked whales (*Mesoplodon bidens*). *Int. J. Syst. Evol. Microbiol.* 61, 2758–2762. doi: 10.1099/ijs.0.028522-0

Wang, J. P., Yoo, J. S., Jang, H. D., Lee, J. H., Cho, J. H., and Kim, I. H. (2014). Effect of dietary fermented garlic by *Weissella koreensis* powder on growth performance, blood characteristics, and immune response of growing pigs challenged with *Escherichia coli* lipopolysaccharide. *J. Anim. Sci.* 89, 2123–2131. doi: 10.2527/jas.2010-3186

Welch, T. J., and Good, C. M. (2013). Mortality associated with weissellosis (*Weissella* sp.) in USA farmed rainbow trout: potential for control by vaccination. *Aquaculture* 388–391, 122–127. doi: 10.1016/j.aquaculture.2013.01.021

Wolter, A., Hager, A. S., Zannini, E., Galle, S., Gänzle, M., Waters, D. M., et al. (2014). Evaluation of exopolysaccharide producing *Weissella cibaria* MG1 strain for the production of sourdough from various flours. *Food Microbiol.* 37, 44–50. doi: 10.1016/j.fm.2013.06.009

Yang, J. S. (2013). Selection of functional lactic acid bacteria as starter cultures for the fermentation of Korean leek (*Allium tuberosum* Rottler). *J. Prob. Health* 1, 4. doi: 10.4172/2329-8901.S1.012

Yu, J. J., Park, H. J., Kim, S. G., and Oh, S. H. (2009). Isolation, identification, and characterization of *Weissella* strains with high ornithine producing capacity from kimchi. *Korean J. Microbiol.* 45, 339–345.

Zhou, M., and Wu, H. (2009). Glycosylation and biogenesis of a family of serine-rich bacterial adhesins. *Microbiology* 155, 317–327. doi: 10.1099/mic.0.025221-0

Conflict of Interest Statement: The authors declare that the research was conducted in the absence of any commercial or financial relationships that could be construed as a potential conflict of interest.

20

Effects of sub-lethal high-pressure homogenization treatment on the outermost cellular structures and the volatile-molecule profiles of two strains of probiotic lactobacilli

Giulia Tabanelli[1], Pamela Vernocchi[1,2], Francesca Patrignani[3], Federica Del Chierico[2], Lorenza Putignani[2,4], Gabriel Vinderola[5], Jorge A. Reinheimer[5], Fausto Gardini[1,3] and Rosalba Lanciotti[1,3]*

[1] Centro Interdipartimentale di Ricerca Industriale Agroalimentare, Università degli Studi di Bologna, Cesena, Italy, [2] Unit of Metagenomics, Bambino Gesù Children's Hospital, IRCCS, Rome, Italy, [3] Dipartimento di Scienze e Tecnologie Agro-alimentari, Università degli Studi di Bologna – Sede di Cesena, Cesena, Italy, [4] Unit of Parasitology, Bambino Gesù Children's Hospital, IRCCS, Rome, Italy, [5] Facultad de Ingeniería Química, Instituto de Lactología Industrial (INLAIN, UNL-CONICET), Universidad Nacional del Litoral, Santa Fe, Argentina

Edited by:
Javier Carballo,
University of Vigo, Spain

Reviewed by:
Maria Guadalupe Vizoso Pinto,
National Univeristy of Tucumán,
Argentina
Qingping Zhong,
South China Agricultural University,
China

*Correspondence:
Rosalba Lanciotti,
Dipartimento di Scienze e Tecnologie
Agro-alimentari, Università degli Studi
di Bologna – Sede di Cesena - Piazza
Goidanich 60, 47521 Cesena, Italy
rosalba.lanciotti@unibo.it

Applying sub-lethal levels of high-pressure homogenization (HPH) to lactic acid bacteria has been proposed as a method of enhancing some of their functional properties. Because the principal targets of HPH are the cell-surface structures, the aim of this study was to examine the effect of sub-lethal HPH treatment on the outermost cellular structures and the proteomic profiles of two known probiotic bacterial strains. Moreover, the effect of HPH treatment on the metabolism of probiotic cells within a dairy product during its refrigerated storage was investigated using SPME-GC-MS. Transmission electron microscopy was used to examine the microstructural changes in the outermost cellular structures due to HPH treatment. These alterations may be involved in the changes in some of the technological and functional properties of the strains that were observed after pressure treatment. Moreover, the proteomic profiles of the probiotic strains treated with HPH and incubated at 37°C for various periods showed different peptide patterns compared with those of the untreated cells. In addition, there were differences in the peaks that were observed in the low-mass spectral region (2000–3000 Da) of the spectral profiles of the control and treated samples. Due to pressure treatment, the volatile-molecule profiles of buttermilk inoculated with treated or control cells and stored at 4°C for 30 days exhibited overall changes in the aroma profile and in the production of molecules that improved its sensory profile, although the two different species imparted specific fingerprints to the product. The results of this study will contribute to understanding the changes that occur in the outermost cellular structures and the metabolism of LAB in response to HPH treatment. The findings of this investigation may contribute to elucidating the relationships between these changes and the alterations of the technological and functional properties of LAB induced by pressure treatment.

Keywords: high-pressure homogenization, probiotic lactobacilli, MALDI-TOF MS, transmission electron microscopy, buttermilk, volatile profile

Introduction

Recently, there has been increasing interest in the potential of applying techniques such as pulsed electric field (PEF), high-hydrostatic pressure (HHP), or high-pressure homogenization (HPH) to enhance the survival rate of probiotic strains or to modify their overall functionality in a positive manner. Among these processes, HPH has been proposed for the treatment of raw materials or the sub-lethal treatment of starters or non-starters and probiotic cells for use in the production of probiotic fermented milks or cheeses with improved sensorial, technological, or functional properties (Lanciotti et al., 2006; Burns et al., 2008a; Patrignani et al., 2009; Tabanelli et al., 2012).

Lanciotti et al. (2007a) demonstrated that a sub-lethal HPH treatment could control the fermentation kinetics of bacterial strains used as starters and modify their metabolic profiles, leading to products with enhanced sensorial properties. Moreover, although the responses varied according to the characteristics of individual strains of lactic acid bacteria (LAB), HPH increased the activity of extracellular or cell wall-associated proteolytic enzymes without having detrimental effects on their viability, confirming their tolerance of moderate pressures.

In addition, sub-lethal HPH treatment improved the acid tolerance and bile tolerance of *L. acidophilus* LA-K (Muramalla and Aryana, 2011) and enhanced some of the biological and functional properties of known probiotic strains both *in vitro* and *in vivo*, and more specifically in mice, trials (Tabanelli et al., 2013, 2014). In particular, the latter authors demonstrated that an HPH treatment applied at 50 MPa modulated the hydrophobicity and auto-aggregation of the treated strains *in vitro* and modified their interaction with the small-intestinal structures of BALB mice. The HPH-treated cells showed a different behavior in the mouse gut and induced a stronger IgA response compared to those of untreated cells, in strain- and feeding-period-dependent manners. These effects were attributed to HPH having modified the outermost cellular structures that play roles in the interactions of probiotic cells and immune cells and are the main targets of sub-lethal pressure (Muramalla and Aryana, 2011; Tabanelli et al., 2014). It is known that when pressure is applied at a sub-lethal level, various cellular responses occur and that the composition of the cellular membrane can change to withstand the exposure to a sub-lethal stress (Russell et al., 1995). Tabanelli et al. (2014) showed that the composition of the membranes and their unsaturation levels affected the response mechanisms adopted by probiotic strains, such as *L. paracasei* A13 and *L. acidophilus* DRU, when they were subjected to sub-lethal HPH treatments. In particular, these authors reported that these treatments reduced the level of cyclic fatty acids and increased the unsaturation level, leading to modification of the membrane fatty-acid profile in a strain-dependent manner. Some authors showed that modulating the membrane fatty-acid composition in response to environmental conditions affected the cell-surface hydrophobicity and adhesive ability of bacterial strains and, consequently, their functional features (Kirjavainen et al., 1998; Kankaanpää et al., 2001, 2004).

Considering that some probiotic properties are associated with the bacterial cell wall, which is also the principal target of HPH, the aim of this study was to evaluate the effect of sub-lethal HPH treatment on the outermost structures of two strains endowed with probiotic features (*Lactobacillus acidophilus* DRU and *Lactobacillus paracasei* A13) using transmission electron microscopy (TEM). These strains were chosen based on the results of previous studies that demonstrated their ability to increase certain functional properties in response to HPH treatment (Tabanelli et al., 2012, 2013). Moreover, bacterial profiling analyses of treated and untreated strains were performed using MALDI-TOF MS (matrix-assisted laser desorption ionization time-of-flight mass spectrometry) to evaluate the effect of HPH on the cellular peptide profiles. MALDI-TOF MS has been applied to microbial detection because it generates characteristic mass spectra that are unique for each species, permitting identification at the genus and species levels, and potentially, at the strain level (Croxatto et al., 2012). This technique has also been employed to evaluate changes in the peptide profiles of microbial cells that were induced by various growth conditions and physico-chemical stresses (Šedo et al., 2013). The high level of versatility and the speed and accuracy of this methodology played key roles in its adoption in many fields, including clinical diagnostics, environmental monitoring, and food-quality control. Although MALDI-TOF MS analysis is an interesting approach to microbial characterization, it has rarely been applied to food-related microorganisms.

Finally, the effect of HPH treatment on the metabolism of LAB in a dairy product, such as buttermilk, was investigated; buttermilk was chosen as the vehicle for the probiotic cells because it was reported to be a suitable medium for maintaining adequate levels of LAB during refrigerated storage (Burns et al., 2008b). Additionally, the volatile profiles of buttermilk inoculated with treated or control cells during 30 days of refrigerated storage were investigated to evaluate the impact of the sub-lethal HPH treatment on the accumulation of molecules that can impart aromas to the product.

Materials and Methods

Strain Culture Conditions and Microbiological Analyses

L. paracasei A13 and *L. acidophilus* DRU are two commercially available probiotic strains that are commonly used in commercial dairy products (Vinderola et al., 2000). The stock cultures were maintained in de Man, Rogosa and Sharpe (MRS) broth (Biokar, Beauvais, France) containing sterile glycerol (20% v/v) at $-70°C$ in the collection of the Instituto de Lactologia Industrial (INLAIN, UNL-CONICET, Santa Fe, Argentina). Fresh cultures of each strain were obtained by two consecutive passages of a 1% (v/v) inoculum of the frozen stocks in MRS broth, with incubation at 37°C for 18 h under aerobic conditions.

The cell counts were performed before and immediately after HPH. Cell counts were obtained by plating the cultures on MRS agar (37°C, 48 h, under aerobic conditions).

High-Pressure Homogenization Treatment

The cells in overnight cultures were harvested by centrifugation (8000 g, 10 min, 4°C). The pellets were washed twice using a solution of 9 g NaCl/l and they were re-suspended in sterile phosphate-buffered saline (PBS) (pH 7.4) at a final concentration of approximately 8 log CFU/ml. The cells were subjected to high-pressure homogenization (HPH) at 50 MPa using a PANDA high-pressure homogenizer (Niro Soavi, Parma, Italy). The inlet temperature of the samples was 20°C and the temperature was increased during the treatment at a rate of 3°C/10 MPa. To prepare control samples, the suspended cells were homogenized at 0.1 MPa. Immediately after the treatment, the samples were rapidly cooled to 10°C in a water bath.

Transmission Electron Microscopy (TEM)

Transmission electron microscopy (TEM) was used to investigate the morphological changes caused by the HPH treatment. Ten milliliters of the control samples and the HPH-treated samples were centrifuged (8000 g, 10 min) and the pelleted cells were fixed by suspending them in 2.5% glutaraldehyde (in 0.1 M K₂HPO₄/KH₂PO₄ buffer, pH 7). These samples were stored at 4°C for 2 h. After aldehyde fixation, the samples were prepared according to Bury et al. (2001). The post-fixed cells were washed using the same buffer and then they were dehydrated for 15 min using the following series of ethanol solutions: 50, 75, 90, and 100%. The dehydrated cells were infiltrated with increasing concentrations of Spurr resin (Agar Scientific, Stansted, Essex, United Kingdom) over 24 h. Polymerization of the resin was achieved by heating the samples in an oven at 65°C for 18 h. Thin sections (approximately 90 nm thick) were placed on carbon-coated Formvar-covered 300-mesh copper grids for approximately 15 min, rinsed using 20 drops of distilled water, negatively stained using 6–7 drops of 2% aqueous uranyl acetate and then examined using a Philips CM 10 transmission electron microscope.

Whole Cell MALDI-TOF MS Fingerprinting Profiles

After being subjected to HPH treatment at 50 MPa, the suspended cells were incubated as follows: (*i*) no incubation (50 MPa T0); (*ii*) incubation at 37°C for 30 min (50 MPa T30); (*iii*) incubation at 37°C for 60 min (50 MPa T60); (*iv*) incubation at 37°C for 120 min (50 MPa T120). To prepare control samples, suspended cells were treated using the homogenizer at 0.1 MPa (0.1 MPa C).

After HPH treatment and incubation, the cells were collected by centrifugation and stored at −80°C until analysis by MALDI-TOF. Then, the cells were washed using H₂O/CH₃CH₂OH (300/600 µl) and were treated according to the method of Putignani et al. (2011). The dried pellets were thoroughly mixed with 50 µl of 70% formic acid (HCOOH) and then with 50 µl of ACN (Sigma-Aldrich, Milan, Italy), and the mixtures were maintained at RT for 10 min at each step. The peptide mixtures derived from acidic hydrolysis were decanted rather than being separated from the insoluble material by centrifugation to avoid possible peptide co-precipitation. These samples were placed (1.5 µl) on an MSP 96 polished steel target (Bruker Daltonics GmbH, Bremen, Germany) and were overlaid with CHCA matrix in 50% ACN/2.5% TFA (1.5 µl) (Sigma-Aldrich) (Putignani et al., 2011). Peptide mass spectra were acquired using

a Microflex MALDI-TOF-MS (Bruker Daltonik GmbH) mass spectrometer that was operated in the linear positive mode at the maximum frequency (20 Hz). Spectral measurements were performed using a Microflex LT mass spectrometer (Bruker Daltonics GmbH), using FlexControl software (version 3.0, Bruker Daltonics GmbH). Eight replicates of each spectrum were collected for each species and were analyzed to evaluate the reproducibility of the results, and 500 laser shots/spots were manually collected using the FlexControl software package. The spectral profiles were visualized using FlexAnalysis 3.0 software (Bruker Daltonics GmbH).

Evaluation of the Effects of HPH Treatment and the Medium pH Value on the Volatile Profiles of Buttermilk

To study the effect of HPH treatment on the aroma-compound production of *L. acidophilus* DRU and *L. paracasei* A13 in a dairy medium, the cells were grown in MRS medium for 18 h at 37°C, harvested by centrifugation (8000 g, 10 min, 4°C) and then re-suspended in buttermilk, previously acidified or not to pH 4.6 using lactic acid (Sigma, Milan, Italy), at a level of approximately 8 log CFU/ml. These preparations were HPH treated as described above (paragraph 2.2) at 0.1 MPa (control samples) or at 50 MPa and were stored for 30 days at 4°C. The buttermilk used was reconstituted (77 g/l) from lyophilized buttermilk purchased from a local dairy and was sterilized at 115°C for 30 min. Buttermilk was chosen as the dairy medium for this study because previous studies had shown that it supported the growth and the survival of adequate numbers of LAB cells during storage (Burns et al., 2008b).

The strain viability rate and the aroma profiles of the LAB-containing buttermilk were determined immediately after inoculation and at 15 and 30 days of refrigerated storage. Solid-phase microextraction and gas-chromatography-mass spectrometry (SPME-GC-MS) were used to detect the aroma compounds as reported by Patrignani et al. (2008). Samples (5 g) were placed in 10 ml sterilized vials, sealed using PTFE/silicon septa and heated for 10 min at 45°C, after which the volatile compounds were allowed to adsorb to a fused silica fiber covered with a 75 µm carboxen polydimethylsiloxane (CAR/PDMS StableFlex) (Supelco, Steiheim, Germany). The adsorbed molecules were desorbed in the gas chromatograph for 10 min. The peaks were detected using an Agilent Hewlett-Packard 6890 GC gas chromatograph equipped with a 5970 MSD MS detector (Hewlett-Packard, Geneva, Switzerland) and a Varian Chrompack CP Wax 52 CB capillary column (50 m × 320 µm × 1.2 µm) (Chrompack, Middelburg, The Netherlands) as the stationary phase. The conditions used were as follows: injection temperature, 250°C; detector temperature, 250°C; carrier gas (He); and flow rate, 1 ml/min. The oven-temperature program used was as follows: 50°C for 1 min; increasing from 50°C to 100°C at 2°C/min; increasing from 100°C to 200°C at 6.5°C/min, and then holding at 200°C for 5 min. Volatile-peak identification was conducted via computerized matching of the mass spectral data with those for the compounds contained in the Agilent Hewlett-Packard NIST 98 and Wiley vers. six mass spectral databases. The SPME-GC-MS results for each sample at each time point (un-inoculated and untreated buttermilk at 0, 15, and 30 days of storage) were expressed as the mean values of six independent analyses.

Statistical Analysis of the Data

The results of volatile profile analysis of each sample were expressed as the mean values of six independent replicate analyses (conducted on different days) and the data were analyzed using Principal Component Analysis (PCA) using Statistica 6.1 software (StatSoft Italy srl, Vigonza, Italy). Eight replicate MALDI TOF MS Biotyper analyses of each sample were conducted and each replica was considered independently. Prior to performing principal component analysis (PCA)-based hierarchical clustering, each spectrum was subjected to mass adjustment, smoothing, baseline subtraction, Normalization and peak picking. The dendrogram for a single organism was created using with distance measurements (correlation), linkage (average) and a 300-score threshold values using MALDI Biotyper 3.1 software (Bruker Daltonics GmbH). The Pearson's correlation coefficients were calculated using spectral row data using R-Bioconductor to establish the reproducibility of the intra- (replicates) and inter-strain conditional data.

Results

Cell Viability Following HPH Treatment

The sub-lethal HPH treatment of *L. paracasei* A13 and *L. acidophilus* DRU cells did not significantly affect their viability, as observed in other studies of LAB strains (Lanciotti et al., 2007a; Tabanelli et al., 2013). The viability rate of the cells of both strains that were HPH treated at 50 MPa was 0.3 log CFU/ml lower than that of the untreated cells.

Transmission Electron Microscopy (TEM)

Figures 1, **2** show TEM images of treated and untreated *L. paracasei* A13 and *L. acidophilus* DRU cells. As shown in **Figures 1A,B**, cell-wall and inner-membrane structures were clearly visible in the control (0.1 MPa) cells of *L. paracasei* A13, as it was an external capsule of proteinaceous material surrounding the cell wall. TEM images of cells of the same strain after HPH treatment at 50 MPa (**Figures 1C,D**) showed changes in the structures of 70–80% of the cells. The external capsule of proteinaceous material surrounding the wall of non-treated *L. paracasei* A13 cells was no longer visible after HPH treatment and the cell surface appeared jagged (indicated using an arrow in **Figure 1C**). Moreover, the cytoplasm appeared to be compressed and it was detached from the outermost cellular structures (indicated using an arrow in **Figure 1D**). The effects of pressure treatment on the outermost cellular structures were also visible in the TEM images of *L. acidophilus*, a species that is characterized by the presence of an S-layer surrounding the cell wall. As shown in **Figures 2A,B**, a continuous, thin, electron-dense layer was visible at the outer edge of the walls of the *L. acidophilus* DRU control cells, whereas this layer appeared discontinuous in the pressure-treated cells (**Figures 2C,D**; indicated using arrows). The HPH-treatment induced morphological changes observed in the TEM images did not significantly impaired cell viability, as demonstrated

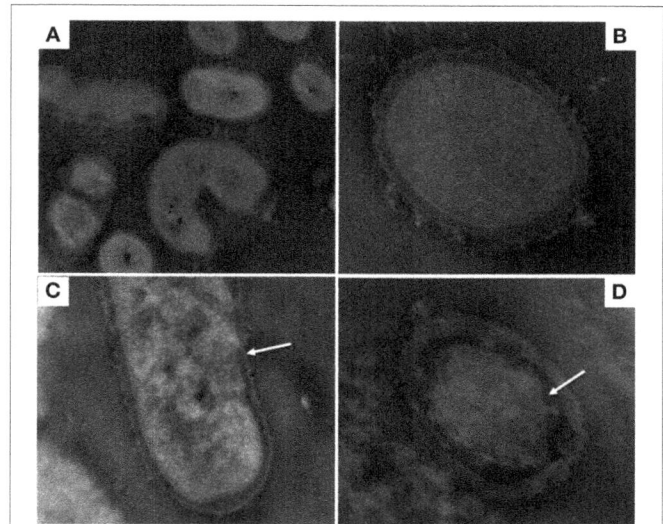

FIGURE 1 | Transmission electron micrographs of *Lactobacillus paracasei* A13: control cells (0.1 MPa) (A,B); 50 MPa HPH treated cells (C,D). Magnification: 28,500x (A) 73,000x (B) and 52,000x (C,D).

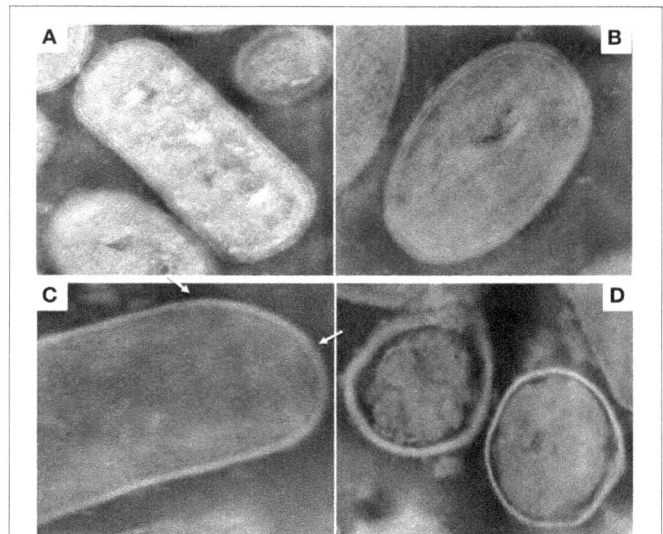

FIGURE 2 | Transmission electron micrographs of *Lactobacillus acidophilus* DRU: control cells (0.1 MPa) (A,B); 50 MPa HPH treated cells (C,D). Magnification: 52,000x (A–C) and 39,000x (D).

by cell counts obtained after HPH treatment. However, these morphological changes could be related to changes in some of the technological and functional properties of the strains observed after pressure treatment that were reported by Tabanelli et al. (2012, 2013).

Whole Cell MALDI-TOF MS Fingerprinting Profile

A proteomic approach using MALDI-TOF MS-based protocols was used to investigate the effect of HPH treatment and the subsequent incubation on the peptide profiles of *L. acidophilus* DRU and *L. paracasei* A13 cells.

Figure 3A shows the peptide spectra of *L. paracasei* A13 cells. Differences in the peptide profiles of the 0.1 MPa C cells compared to that of the treated A13 samples were observed, particularly in the low-mass spectral region (2000–3000 Da). Increasing the incubation period to 60 min after HPH treatment increased the signals in the region between 3500 and 5200 Da. However, the 50 MPa T30 samples showed characteristic peaks at 6367.97, 7818.22, and 7818.22 Da that were absent in cells under the other conditions. Further increasing the incubation period to 120 min decreased the intensity of the peaks in the region between 3000 and 5500 Da.

Figure 3B shows the peptide spectral profiles of *L. acidophilus* DRU cells. The spectral profiles of the 0.1 MPa C cells and the 50 MPa T0 and 50 MPa T30 cells included more peptide peaks, particularly in the 2000–5000 Da region, compared with the spectral profiles of 50 MPa T60 and 50 MPa T120 samples. In contrast, the profiles of treated cells had peaks of higher intensity in the mass range of 7000–10300 Da compared with those in the profile of untreated cells. In addition, the profiles of the treated cells were characterized by the presence of peaks (i.e., at 6247.613, 6446.94, 9161.94, and 9605.68 Da) that were absent in the profiles of the control cells. Moreover, the intensity of the peaks in the 6951.51 and 7641.54 Da region of the spectra of the treated cells differed from that of the control-cell spectra.

The dendrograms that were derived using the MALDI-TOF MS Biotyper profiles showed clusters associated with HPH treatment and with the period of incubation at 37°C. In particular, the dendrogram of the spectra of *L. paracasei* A13 cells showed two major clusters (**Figure 4A**). The first cluster grouped the *L. paracasei* A13 50 MPa T0 and 50 MPa T30 cells. The second major cluster grouped 0.1 MPa C cells and 50 MPa T60, and 50 MPa T120 cells, meaning that the peptide profiles of the untreated cells and the treated cells that were incubated for 60 or 120 min were less than 0.8 apart. However, all of the tested conditions were well differentiated.

The spectral dendrograms of *L. acidophilus* DRU cells showed two major clusters, as follows: the first cluster included 50 MPa T30 and 50 MPa T120 cells and the second cluster included the 0.1 MPa C, 50 MPa T0, and 50 MPa T60 cells (**Figure 4B**). The last cluster could be sub-grouped into two minor clusters in which the control cells and the cells analyzed immediately after the hyperbaric treatment were grouped together, whereas 50 MPa T60 cells were in a separate cluster. In the case of this strain, all of the tested conditions were also well distinguished by their associated spectra.

Pearson's correlation analysis showed a high level of reproducibility of the data for *L. paracasei* A13 cells, ranging from 0.89 to 0.98, whereas the level of reproducibility of the data for *L. acidophilus* cells was slightly lower, ranging from 0.71 to 0.97.

Evaluation of the Effects of HPH Treatment and the Medium pH Value on the Volatile Profiles of Buttermilk

To study the effects of HPH treatment on the volatile profiles of probiotic-containing buttermilk samples (at pH 7 or pH 4.6), buttermilk was inoculated with *L. acidophilus* DRU or *L. paracasei* A13 cells at a concentration of approximately 8 log

CFU/ml and was passed through a high pressure homogenizer at 0.1 MPa (control) or at 50 MPa (HPH treated). All of the samples were stored at 4°C for 30 days. The aroma profiles and cell viability rates were monitored throughout the storage period. The viability results confirmed that buttermilk supported the survival of adequate numbers of probiotic LAB cells during refrigerated storage, independently of the level of HPH treatment, as demonstrated previously (Burns et al., 2008b; Tabanelli et al., 2013). The viability rates of the two strains remained greater than 7.6 log CFU/ml at both pH levels throughout the refrigerated storage period.

The content of the compounds detected in the aroma profile of uninoculated buttermilk and buttermilk samples that were inoculated with *L. paracasei* A13 and *L. acidophilus* DRU cells were expressed as the % of the total peak area, as shown in **Tables 1–3**, respectively.

The most abundant molecules detected in the non-inoculated and untreated buttermilk samples were ketones, such as 2-propanone, 2-butanone, 2-pentanone, and 2-heptanone, aldehydes, such as hexanal, furfural, and benzaldehyde, alcohols, such as hexanol and 2-furanmethanol, and acetic acid (**Table 1**).

As expected, the aroma profiles of the buttermilk samples containing cells of the two strains studied were quite different from the profiles of the control samples due to microbial activity. However, In the case of the samples containing *L. paracasei* A13 cells, the differences between the samples at 15 and 30 days of storage were not pronounced. All the samples at pH 7 were characterized by a higher content of ketones, mainly 2-nonanone, 2-pentanone, and 2-heptanone. Significant levels of 2,3-butanedione and 3 hydroxy-2-butanone were also detected in the HPH-treated samples at pH 7 at 30 days. Alcohols, mainly furanmethanol, were more abundant in buttermilk samples at pH 7, whereas acids (mainly acetic acid) predominated at pH 4.6, reaching relative proportions that were always greater than 45%. The content of esters in the samples containing *L. paracasei* A13 cells was always negligible, unlike that of samples containing *L. acidophilus* DRU cells.

Aldehydes accounted for approximately 15% of the total peak area of the volatile molecule profile of the *L. acidophilus* DRU-containing samples after 15 days of refrigerated storage, with no significant differences due to the pH value or HPH treatment, and the most prominent of these compounds were hexanal, 4-methyl-benzaldehyde, benzaldehyde and furfural. However, the levels of these compounds in the non-treated samples remained stable during 30 days of storage, whereas their levels in the HPH-treated buttermilk samples fell down below 2% during this period. Ketones were the most prominent chemical group present under all of the conditions, but significant differences were observed in relation to HPH treatment and the pH level. At 15 days of storage, the total ketones represented approximately 29 and 45% of the total peak area at pH 4.6 and 7, respectively, with no differences related to HPH treatment. The higher level of total ketones in buttermilk at pH 7 was mainly due to the higher concentration of 2-heptanone compared to the buttermilk at pH 4.6. The content of 2-butanone was higher in the treated samples, whereas the level of 2-propanone was higher in the control samples. 2-pentanone and 2-nonanone were detected in

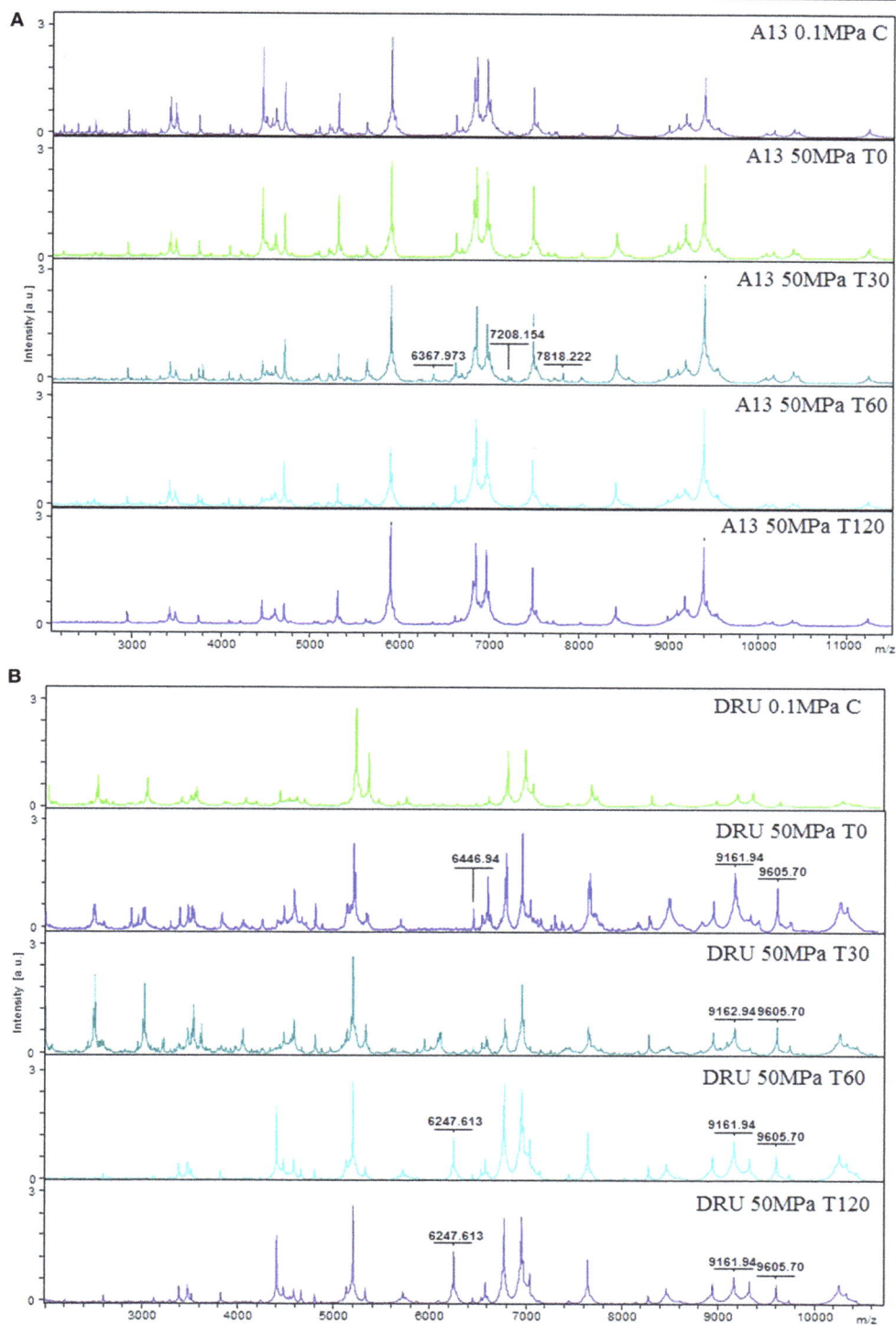

FIGURE 3 | MS proteomic profiling of *L. paracasei* A13 (A) and *L. acidophilus* DRU (B) showing MS fingerprinting of the sample conditions (0.1 MPa C, 50 MPa T0, 50 MPa T30, 50 MPa T60, and 50 MPa T120). The m/z-values are expressed in Da and the amplitudes are reported in a scale of intensity 10^4 arbitrary units (a.u.). Legend: (a) A13 0.1 MPa C, (b) A13 50 MPa T0, (c) A13 50 MPa T30, (d) A13 50 MPa T60, (e) A13 50 MPa T120, (f) DRU 0.1 MPa C, (g) DRU 50 MPa T0, (h) DRU 50 MPa T30, (i) DRU 50 MPa T60, (j) DRU 50 MPa T120.

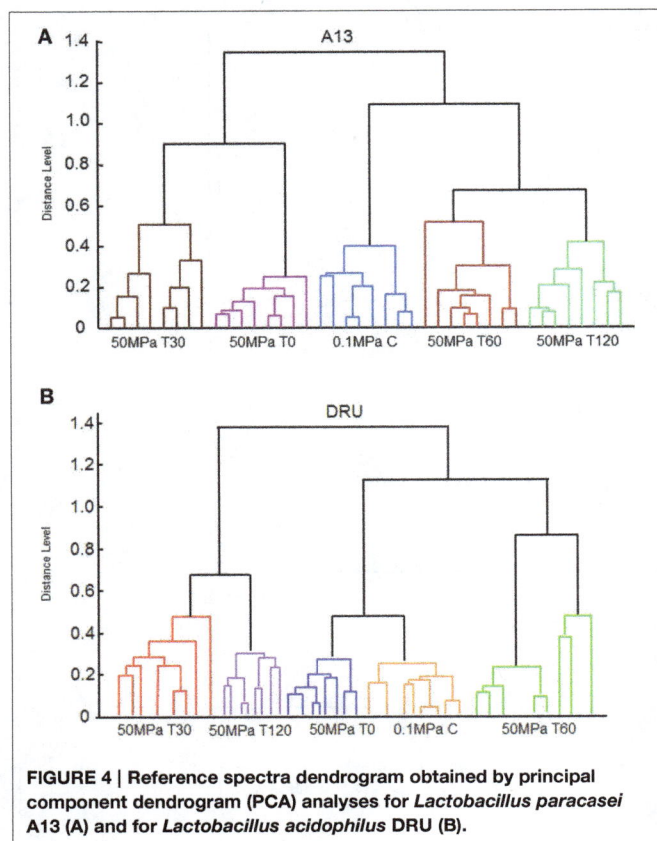

FIGURE 4 | Reference spectra dendrogram obtained by principal component dendrogram (PCA) analyses for *Lactobacillus paracasei* A13 (A) and for *Lactobacillus acidophilus* DRU (B).

TABLE 1 | Volatile compounds (expressed as % peak area) detected in uninoculated buttermilk after 15 and 30 days of refrigerate storage.

Compounds	Uninoculated buttermilk		
	0 days	15 days	30 days
Hexanal	0.61 (±0.07)	4.73 (±0.23)	3.39 (±0.32)
Furfural	2.68 (±0.17)	2.92 (±0.13)	2.67 (±0.21)
Benzaldehyde	2.35 (±0.19)	1.86 (±0.17)	2.05 (±0.13)
Octanal	–*	0.60 (±0.07)	–
Aldehydes	**5.65**	**10.12**	**8.11**
2-propanone	7.40 (±0.39)	8.62 (±0.47)	6.59 (±0.42)
2-butanone	2.01 (±0.11)	2.26 (±0.15)	2.27 (±0.12)
1- hydroxy -2-propanone	1.23 (±0.03)	2.79 (±0.13)	2.37 (±0.21)
2-pentanone	15.69 (±0.78)	13.64 (±0.67)	15.06 (±0.96)
2-heptanone	42.78 (±1.39)	35.55 (±1.56)	37.71 (±1.23)
3,5-octadien-2-one	0.66 (±0.07)	0.30 (±0.02)	0.29 (±0.06)
2-nonanone	6.49 (±0.29)	5.08 (±0.11)	5.68 (±0.38)
2-undecanone	0.86 (±0.02)	0.85 (±0.05)	0.87 (±0.06)
1-phenyl-ethanone	1.88 (±0.05)	1.34 (±0.08)	1.53 (±0.09)
Ketons	**79.01**	**70.43**	**72.38**
2-furanmethanol	11.42 (±0.69)	13.87 (±0.17)	14.84 (±0.88)
Alcohols	**11.42**	**13.87**	**14.84**
Acetic acid	3.34 (±0.28)	5.11 (±0.36)	4.14 (±0.21)
Hexanoic acid	0.59 (±0.02)	0.48 (±0.03)	0.53 (±0.06)
Acids	**3.93**	**5.59**	**4.68**

*Under detection limit. The data are mean of six repetitions. The standard deviation is reported within brackets.

significant and constant concentrations in all of the samples. At 30 days of storage, the ketone concentrations in the non-treated control buttermilk were little changed, but those of the HPH-treated buttermilks had drastically increased, largely due to increases in the relative concentrations of 2-pentanone, 2-heptanone, and 2-nonanone. The alcohols present were mainly represented by furanmethanol, the concentration of which was higher in the non-treated buttermilk sample than in the HPH-treated buttermilk samples at 15 days of storage. This difference was significantly greater at 30 days of storage regardless of the pH value. In contrast, the pH value strongly affected the relative percentage of acids, the levels of which were higher at pH 4.6 than at pH 7, independently of HPH treatment and the storage period. Finally, a significantly higher concentration of ethyl acetate was observed in HPH-treated buttermilk at pH 7 at 15 days of storage. The level of this ester was increased at 30 days of storage and was accompanied by an increased level of ethyl hexanoate.

Principal Component Analysis on the Volatile Profiles of Buttermilk

To better evaluate the effect of HPH treatment on the aroma profile of buttermilk, a Principal Component Analysis (PCA) was conducted using the % of the peak area of the volatile compounds listed in **Tables 1**, **2**. **Figures 5A,C** show PCA loading plots of the aroma profile data for buttermilk that was inoculated with *L. paracasei* A13 cells or *L. acidophilus* DRU cells, respectively,

which demonstrated that the first two principal components (PC1 and PC2) explained more than 85% of the total variability. In both cases, PC1 accounted for the greater part of the variability (approximately 65.38 and 55.99% for the *L. paracasei* A13-containing samples and the *L. acidophilus* DRU-containing samples, respectively), and the samples could be grouped into four clusters according to whether they were HPH treated and the pH of the medium. The control buttermilk samples were grouped in the upper part of the plot, whereas the HPH-treated samples were grouped in the lower part. Moreover, the acidified samples were grouped on the left side of the plot, and the pH 7 buttermilk samples were grouped on the right side.

Figure 5B shows the variable-factor coordinates for the first two principal components for the *L. paracasei* A13 samples. The acidified (pH 4.6) control samples were characterized by the presence of non-anal and acids, whereas the pH 7 control samples were characterized mainly by the presence of ketones, such as 2-heptanone, 2-propanone, and 2-pentanone. The variable factor coordinates for the first two principal components of the buttermilk samples that were inoculated with *L. acidophilus* DRU cells are shown in **Figure 5D**. As was the case for the control samples, the acidified *L. acidophilus* DRU-containing samples were characterized mainly by the presence of acids, whereas the untreated samples and the HPH-treated samples at pH 7 were characterized mainly by the presence of 2-heptanone and

TABLE 2 | Volatile compounds (expressed as % peak area) detected in buttermilk at pH 4.6 and 7, inoculated with *Lactobacillus paracasei* A13 treated at 0.1 and 50 MPa, after 15 and 30 days of refrigerate storage.

Compounds	15 days				30 days			
	pH 4.6		pH 7.0		pH 4.6		pH 7.0	
	0.1 MPa	50 MPa	0.1 MPa	50 MPa	0.1 MPa	50 MPa	0.1 MPa	50 MPa
Hexanal	1.61 (±0.12)	0.58 (±0.03)	1.51 (±0.09)	–*	–	–	1.99 (±0.12)	–
Furfural	2.29 (±0.13)	2.06 (±0.17)	3.30 (±0.14)	4.22 (±0.35)	4.72 (±0.28)	1.25 (±0.78)	2.32 (±0.13)	2.56 (±0.11)
Benzaldehyde	0.89 (±0.03)	1.16 (±0.09)	1.87 (±0.11)	1.96 (±0.12)	1.17 (±0.08)	0.77 (±0.02)	2.52 (±0.18)	1.20 (±0.09)
4-methyl-benzaldehyde	3.25 (±0.28)	3.07 (±0.19)	4.98 (±0.23)	6.65 (±0.47)	3.38 (±0.29)	2.33 (±0.11)	5.91 (±0.27)	4.73 (±0.14)
Non-anal	4.67 (±0.32)	–	–	–	–	2.77 (±0.20)	–	–
Aldehydes	**12.71**	**6.87**	**11.66**	**12.82**	**9.27**	**7.11**	**12.74**	**8.50**
2-propanone	8.44 (±0.38)	5.19 (±0.23)	13.13 (±0.51)	8.81 (±0.41)	4.36 (±0.38)	5.06 (±0.31)	19.69 (±1.07)	6.25 (±0.21)
2-butanone	4.30 (±0.18)	7.63 (±0.47)	5.32 (±0.27)	9.16 (±0.44)	2.14 (±0.11)	5.85 (±0.29)	5.58 (±0.32)	7.97 (±0.35)
2,3-butanedione	–	–	–	2.24 (±0.15)	–	–	–	5.63 (±0.22)
3-hydroxy-2-butanone	–	–	–	–	–	–	–	6.11 (±0.46)
1-hydroxy-2-propanone	1.38 (±0.05)	0.62 (±0.05)	0.97 (±0.03)	1.06 (±0.06)	1.80 (±0.07)	–	0.89 (±0.02)	0.88 (±0.09)
2-pentanone	3.04 (±0.29)	3.36 (±0.13)	6.95 (±0.26)	4.05 (±0.21)	3.59 (±0.28)	1.62 (±0.11)	6.93 (±0.28)	3.11 (±0.18)
2-heptanone	4.11 (±0.28)	7.95 (±0.61)	16.32 (±1.56)	8.32 (±0.78)	9.18 (±0.39)	5.03 (±0.58)	10.67 (±0.43)	11.99 (±0.57)
3,5-octadien-2-one	–	0.37 (±0.03)	0.57 (±0.04)	0.84 (±0.05)	0.96 (±0.07)	0.41 (±0.04)	5.41 (±0.35)	6.06 (±0.37)
2 non-anone	1.55 (±0.08)	2.89 (±0.18)	6.66 (±0.27)	8.24 (±0.59)	3.15 (±0.20)	1.88 (±0.11)	–	–
2-undecanone	0.17 (±0.03)	0.69 (±0.04)	1.38 (±0.07)	1.87 (±0.13)	0.45 (±0.03)	0.49 (±0.01)	0.81 (±0.02)	1.41 (±0.11)
Ketons	**22.99**	**28.70**	**51.29**	**44.59**	**25.64**	**20.34**	**49.98**	**49.41**
Ethyl alcohol	1.49 (±0.10)	2.03 (±0.13)	2.67 (±0.16)	3.51 (±0.28)	–	1.77 (±0.11)	2.84 (±0.21)	1.61 (±0.12)
1-pentanol	–	1.22 (±0.08)	–	1.24 (±0.03)	–	1.14 (±0.07)	1.73 (±0.13)	1.17 (±0.09)
1-hexanol	–	–	–	1.20 (±0.09)	–	0.38 (±0.03)	0.96 (±0.08)	0.72 (±0.07)
2-furanmethanol	13.24 (±0.96)	11.25 (±0.83)	17.79 (±0.91)	19.17 (±1.56)	18.97 (±1.02)	11.83 (±0.96)	17.48 (±1.36)	17.78 (±1.22)
1-nonenol	–	–	3.52 (±0.29)	3.45 (±0.28)	–	–	–	0.79 (±0.08)
Alcohols	**14.73**	**14.50**	**23.98**	**28.57**	**18.97**	**15.12**	**23.02**	**22.08**
Acetic acid	26.20 (±1.30)	24.42 (±1.16)	7.87 (±0.25)	9.20 (±0.37)	27.42 (±1.33)	30.06 (±1.68)	9.83 (±0.53)	14.44 (±1.01)
Hexanoic acid	13.21 (±0.84)	13.12 (±0.91)	2.55 (±0.13)	2.10 (±0.16)	11.39 (±0.59)	14.12 (±0.77)	3.01 (±0.23)	3.39 (±0.26)
Butanoic acid	10.16 (±0.86)	9.84 (±0.65)	2.65 (±0.21)	1.50 (±0.11)	7.32 (±0.47)	11.06 (±0.86)	1.43 (±0.07)	2.18 (±0.18)
Acids	**49.58**	**47.37**	**13.07**	**12.80**	**46.12**	**55.24**	**14.27**	**20.01**
Ethylacetate	–	2.56 (±0.19)	–	4.71 (±0.21)	–	2.19 (±0.18)	–	–
Esters	**0.00**	**2.56**	**0.00**	**4.71**	**0.00**	**2.19**	**0.00**	**0.00**

Under detection limit. The data are mean of six repetitions. The standard deviation is reported within brackets.

2-nonanone and the presence of ethyl acetate, 2-butanone and ethyl alcohol, respectively.

Discussion

In previous studies, it was demonstrated that HPH treatment at 50 MPa enhanced certain probiotic properties of LAB and changed the fatty acid composition of the cell membrane as response to the sub-lethal stress applied (Tabanelli et al., 2013, 2014). Moreover, the HPH treatment significantly reduced the hydrophobicity of the *L. acidophilus* DRU cells, whereas this treatment increased that of the *L. paracasei* A13 cells by five-fold. Tabanelli et al. (2012) correlated these differences with the different gastrointestinal-transit behaviors and gut-epithelial interactions shown by these two probiotic strains in mice. Furthermore, the hydrophobicity of the cells was found to correlate with their adhesive ability (Basson et al., 2007) and changes in the outermost cellular structures, following HPH treatment, could influence the intestinal transit and behavior of the tested strains. TEM micrographs demonstrated the effect of HPH treatment on the cell-wall structures, which was strain-dependent. The TEM images of the control samples (0.1 MPa-treated) of *L. acidophilus*, a species characterized by the presence of an S-layer, showed an intact layer surrounding each cell. Tabanelli et al. (2013) demonstrated that the cells of this strain had a higher level of *in vitro* hydrophobicity compared to that of *L. paracasei* A13 cells and attributed this property to this additional external proteinaceous structure. Furthermore,

TABLE 3 | Volatile compounds (expressed as % peak area) detected in buttermilk at pH 4.6 and 7, inoculated with *Lactobacillus acidophilus* DRU treated at 0.1 and 50 MPa, after 15 and 30 days of refrigerate storage.

Compounds	15 days				30 days			
	pH 4.6		pH 7.0		pH 4.6		pH 7.0	
	0.1 MPa	50 MPa	0.1 MPa	50 MPa	0.1 MPa	50 MPa	0.1 MPa	50 MPa
Hexanal	2.91 (± 0.28)	2.95 (± 0.15)	3.16 (± 0.18)	4.63 (± 0.32)	1.29 (± 0.10)	–*	3.28 (± 0.15)	–
Furfural	2.72 (± 0.19)	2.78 (± 0.11)	3.47 (± 0.22)	2.86 (± 0.14)	2.79 (± 0.13)	0.49 (± 0.04)	3.81 (± 0.29)	0.76 (± 0.04)
Benzaldehyde	1.74 (± 0.16)	2.15 (± 0.16)	2.87 (± 0.26)	2.41 (± 0.16)	2.42 (± 0.15)	0.08 (± 0.01)	3.90 (± 0.35)	–
Octanal	0.46 (± 0.02)	0.91 (± 0.14)	0.51 (± 0.04)	0.48 (± 0.03)	–	–	–	–
4-methyl, benzaldehyde	4.37 (± 0.35)	2.33 (± 0.20)	5.35 (± 0.51)	2.76 (± 0.13)	4.43 (± 0.30)	0.42 (± 0.03)	5.70 (± 0.28)	0.84 (± 0.06)
Non-anal	1.40 (± 0.10)	1.16 (± 0.08)	1.22 (± 0.06)	1.30 (± 0.09)	–	–	–	–
Aldehydes	**13.60**	**12.29**	**16.58**	**14.44**	**10.93**	**0.98**	**16.70**	**1.61**
2-propanone	7.19 (± 0.42)	5.09 (± 0.39)	6.34 (± 0.40)	5.80 (± 0.53)	9.24 (± 0.44)	2.49 (± 0.11)	11.72 (± 0.88)	2.46 (± 0.13)
2-butanone	3.39 (± 0.19)	8.32 (± 0.61)	4.39 (± 0.16)	10.76 (± 0.80)	4.42 (± 0.31)	3.38 (± 0.36)	7.19 (± 0.35)	4.45 (± 0.39)
2-pentanone	3.45 (± 0.23)	3.15 (± 0.20)	4.85 (± 0.38)	5.95 (± 0.42)	3.93 (± 0.25)	23.34 (± 1.86)	6.44 (± 0.28)	6.60 (± 0.47)
2-heptanone	9.72 (± 0.45)	8.77 (± 0.68)	23.03 (± 1.16)	14.87 (± 0.56)	9.01 (± 0.74)	39.15 (± 2.04)	8.31 (± 0.51)	30.25 (± 1.74)
3,5-octadien-2-one	0.47 (± 0.02)	0.34 (± 0.02)	0.56 (± 0.04)	0.20 (± 0.01)	0.29 (± 0.02)	–	0.59 (± 0.03)	–
2-nonanone	3.56 (± 0.27)	3.12 (± 0.15)	4.88 (± 0.40)	4.79 (± 0.27)	3.78 (± 0.28)	10.85 (± 0.84)	6.54 (± 0.44)	20.53 (± 1.33)
2-undecanone	0.90 (± 0.07)	0.68 (± 0.04)	0.96 (± 0.06)	1.67 (± 0.14)	0.52 (± 0.03)	0.68 (± 0.07)	1.23 (± 0.16)	0.76 (± 0.10)
Ketons	**28.69**	**29.47**	**45.00**	**44.03**	**31.19**	**79.88**	**42.01**	**65.05**
Ethyl alcohol	–	1.49 (± 0.13)	0.68 (± 0.04)	1.59 (± 0.11)	1.67 (± 0.09)	0.99 (± 0.06)	1.64 (± 0.15)	1.58 (± 0.11)
2-heptanol	–	–	0.86 (± 0.06)	0.76 (± 0.05)	–	0.98 (± 0.04)	–	1.58 (± 0.13)
2-furanmethanol	15.19 (± 1.13)	12.21 (± 0.71)	16.32 (± 0.86)	13.98 (± 1.06)	13.22 (± 0.83)	4.30 (± 0.18)	16.26 (± 0.95)	3.59 (± 0.20)
Alcohols	**15.19**	**13.69**	**17.85**	**16.33**	**14.89**	**6.27**	**17.89**	**6.75**
Acetic acid	6.81 (± 0.53)	7.23 (± 0.32)	0.94 (± 0.06)	0.77 (± 0.06)	5.23 (± 0.44)	1.58 (± 0.11)	1.04 (± 0.07)	0.28 (± 0.03)
Butanoic acid	10.99 (± 0.56)	11.35 (± 1.06)	–	–	6.26 (± 0.35)	2.75 (± 0.18)	–	–
Hexanoic acid	9.38 (± 0.55)	8.64 (± 0.55)	0.63 (± 0.04)	0.56 (± 0.02)	8.73 (± 0.51)	1.84 (± 0.12)	3.03 (± 0.15)	0.63 (± 0.04)
Acids	**27.18**	**27.22**	**1.57**	**1.32**	**20.22**	**6.17**	**4.07**	**0.91**
Ethylacetate	0.15 (± 0.01)	3.62 (± 0.15)	1.14 (± 0.06)	7.56 (± 0.70)	7.88 (± 0.61)	0.42 (± 0.03)	1.43 (± 0.13)	10.31 (± 0.76)
Ethylhexanoate	–	–	–	–	–	–	–	8.62 (± 0.53)
Esters	**0.15**	**3.62**	**1.14**	**7.56**	**7.88**	**0.42**	**1.43**	**18.93**

Under detection limit. The data are mean of six repetitions. The standard deviation is reported within brackets.

several of the cell-surface proteins of the S-layer, which represent approximately 10% of the total cellular proteins, were reported to have adhesion domains and to be involved in cell adhesion (Åvall-Jääskeläinen and Palva, 2005; Jakava-Viljanen and Palva, 2007). After HPH treatment, the layer surrounding the wall of *L. acidophilus* DRU cells was discontinuous, which could account for the loss of hydrophobicity observed in this strain after pressure treatment by Tabanelli et al. (2013). Additionally, it is evident that factors and treatments that modify the outermost cellular structures, such as HPH, could affect the functional properties of probiotic strains. In several lactobacilli species, such as *L. crispatus* and *L. acidophilus,* the removal or damage of the S-layer proteins resulted in a decreased ability to bind to the epithelium of the host (Buck et al., 2005; Frece et al., 2005).

TEM micrographs of *L. paracasei* A13 cells subjected to HPH treatment showed many changes in the outermost

cellular structures (i.e., proteinaceous material that normally surrounding the cell wall was no longer visible). These changes could be responsible for the increased hydrophobicity of the cells observed by Tabanelli et al. (2013) following HPH treatment.

Although analysis of MALDI-TOF spectra did not permit the identification of the MS/MS peptides, specific peptide fingerprints associated with the strain, the HPH treatment and the incubation period were obtained using this technique. The proteomic profiles of cells of the probiotic strains *L. paracasei* A13 and *L. acidophilus* DRU that were treated using HPH and were incubated at 37°C for different periods showed peptide patterns different from those of untreated cells. These differences can most likely be attributed to the effect of HPH on the cell-surface proteins and the cellular response to the HPH treatment. However, *L. paracasei* A13 and *L. acidophilus* DRU cells showed different behaviors independently of the HPH treatment and the

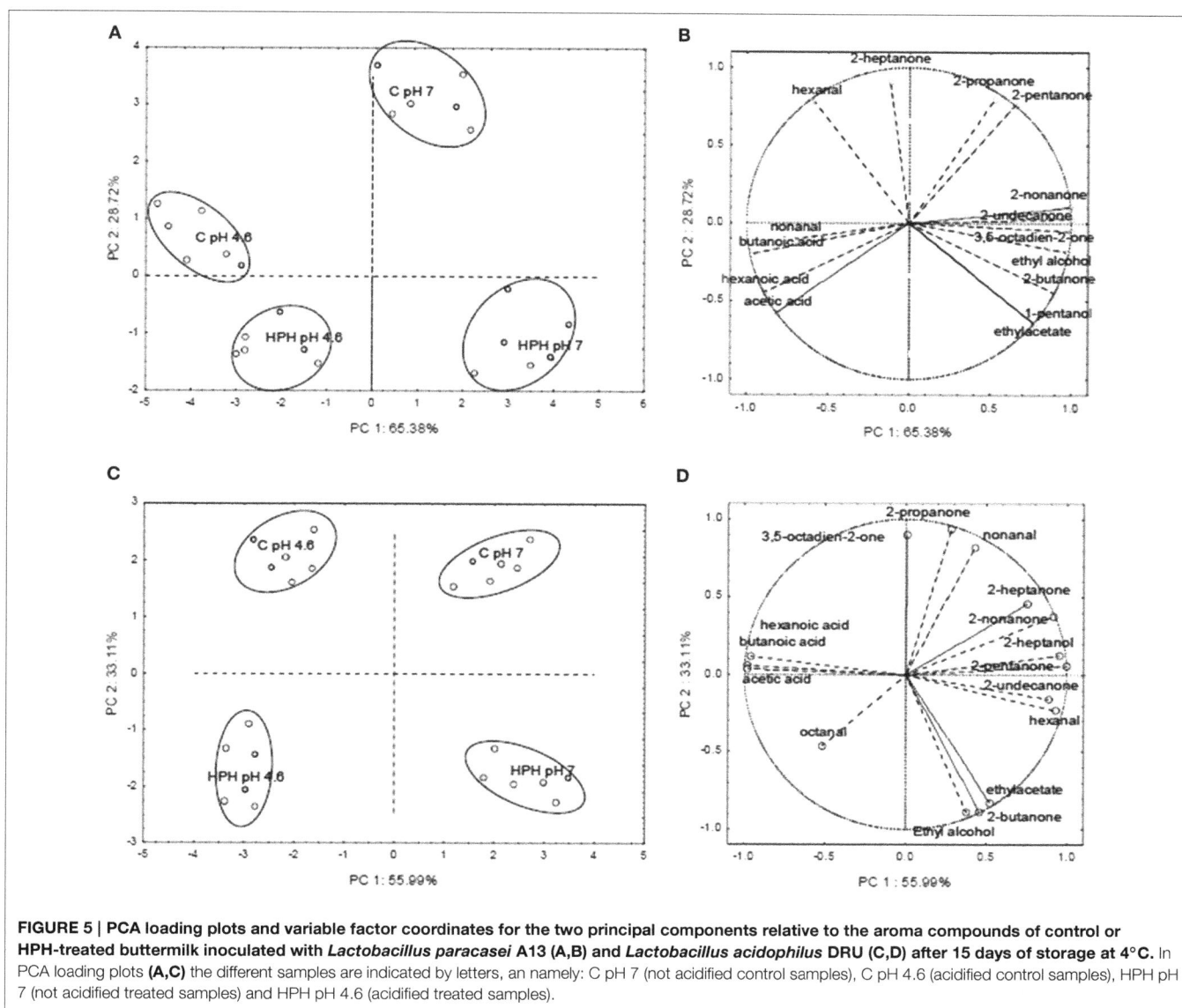

FIGURE 5 | PCA loading plots and variable factor coordinates for the two principal components relative to the aroma compounds of control or HPH-treated buttermilk inoculated with *Lactobacillus paracasei* A13 (A,B) and *Lactobacillus acidophilus* DRU (C,D) after 15 days of storage at 4°C. In PCA loading plots **(A,C)** the different samples are indicated by letters, an namely: C pH 7 (not acidified control samples), C pH 4.6 (acidified control samples), HPH pH 7 (not acidified treated samples) and HPH pH 4.6 (acidified treated samples).

incubation period that appeared to be related to their differential responses to the applied stress.

In particular, the dendrograms associated with *L. paracasei* A13 cells demonstrated that the peptide profiles of treated cells obtained immediately after the hyperbaric treatment were significantly different from those of the control cells, whereas the differences diminished during the incubation period. In fact, the peptide profiles of treated cells incubated for 60 and 120 min were more similar to that of the control cells. The changes in the peptide profiles of treated cells after 30 min of incubation with respect to the peptide profiles of the control cells can be attributed to the presence of specific enzymes and stress proteins that are involved in the restoration of the status *quo ante*.

In contrast, the analysis of the dendrograms associated with *L. acidophilus* DRU cells showed that the peptide profiles of these

cells obtained immediately after pressure treatment were more similar to those of the control cells than were the profiles obtained later, showing that the HPH treatment had less effect on this strain than on *L. paracasei* A13, most likely due to the higher resistance to physical stresses conferred by the S-layer. However, the hyperbaric treatment caused significant modifications of the peptide profile that became evident during the incubation period, indicating that the HPH treatment induced persistent metabolic changes. The lack of peptide identification did not allow distinguishing among the released and novel peptides. However, in the case of *L. paracasei* A13, the similarity of the peptide spectrum of the control cells and that of the treated cells immediately after hyperbaric treatment, as well as the appearance of characteristic peaks during the incubation of the treated cells, suggested that both released and novel peptides contributed to the profile changes that were observed over time. In contrast, in the case of *L. acidophilus* DRU, the similarity of the spectra of the control cells and those of the cells immediately following the HPH treatment suggested that the synthesis of novel proteins was a key mechanism in the stress response of this S-layer-endowed strain.

It is well-known that exposure to physico-chemical stresses results in increased levels of synthesis of stress-response proteins. Jofré et al. (2007) showed that after 2 h of recovery from high hydrostatic-pressure treatment (HHP), several Gram-positive strains expressed transcription factors and proteins related to the synthesis of enzymes involved in energy metabolism. In addition, pressure application (both HHP and HPH) was reported to cause conformational changes in proteins, protein unfolding and the dissociation of oligomeric or aggregated proteins while also affecting enzymatic activities (Fantin et al., 1996; Vannini et al., 2008). In particular, HPH treatment was reported to cause protein conformational changes as well as protein aggregation and to affect the interactions of proteins with other macromolecules, such as polysaccharides and lipids (Floury et al., 2000; Patrignani et al., 2009). The effects of HPH on microorganisms could be attributed to the following: (I) a direct effect of the pressure exerted on the integrity of the cell wall or the outer membranes; (II) the passage of proteins through the damaged cell walls and membranes; and (III) indirect stimulatory effects on the functions of proteins caused by small structural changes that affect their active sites (Diels and Michiels, 2006).

The analysis of the aroma profiles of the two different species showed that they had specific fingerprints (i.e., acetic acid was more pronounced in the profile of the *L. paracasei* samples). Moreover, the pH value of the medium affected the volatile profiles through affecting the metabolic pathways of the bacteria and the level of activity of their enzymes. In addition, the pH value was shown to alter the volatility of compounds such as acids through affecting their interactions with the buttermilk matrix as well as the water-binding capacity of the proteins present (Innocente et al., 2011). In particular, 2-propanone, 2-butanone, furfural, and furanmethanol were detected at both 15 and 30 days of refrigerated storage of buttermilk prepared using any of the samples of both strains under any of the conditions tested. In buttermilk with a pH value of 7 that was inoculated with *L. paracasei* A13 cells and HPH treated at 50 MPa, acetoin was detected at 30 days of storage and diacetyl (2, 3-butanone) was detected at 15 and 30 days of storage. Lanciotti et al. (2007a) showed an increase in the content of several molecules in the aroma profiles of dairy products containing the cells of several *Lactobacillus* species that had been directly treated using a sub-lethal HPH level. Additionally, Patrignani et al. (2007) reported that increasing the level of HPH treatment of probiotic-containing fermented milks increased the diacetyl content.

Several alcohols (ethanol, non-anol, 3-methyl-1-butanol, and 2-ethyl-hexanol), acids (butanoic, heptanoic, and decanoic acids) and ethyl-esters (ethylacetate, ethylbutanoate, ethylhexanoate, ethylheptanoate, and ethyldecanoate) were detected in inoculated samples compared to untreated and un-inoculated buttermilk, although in different amounts in relation to the strain used and treatment applied.

The significant differences in the volatile-molecule profiles of these samples could be due to the effects of the HPH stress on the microbial cells. In particular, the increase in the content of ketones and esters observed after pressure treatment could be associated with higher levels of activity of lipases and esterases. There is much evidence showing that sub-lethal HPH treatment affects the membrane fatty-acid desaturase enzymes that are involved in the active response of cells to high-pressure stress (Somero, 1992; Guerzoni et al., 1997; Tabanelli et al., 2014). In addition, HPH treatment has been reported to alter the activity of several enzymes of microbial origin as well as some of those that naturally occur in food matrices (Vannini et al., 2004; Iucci et al., 2006; Lanciotti et al., 2007b). Moreover, Patrignani et al. (2013) showed increased levels of esters and ketones in yeast cells subjected to HPH treatment in fruit juice. The involvement of ketones in the stress-response mechanisms of microbial cells was documented, whereas esters were regarded as yeast signaling molecules (Isakoff et al., 1996; Kocsis and Weselake, 1996).

The results of the present study demonstrated overall changes in the aroma profile and the production of molecules that positively affected the sensory profile of probiotic cell-containing buttermilk samples that were pressure treated. Therefore, because probiotic products manufactured using only probiotic strains are often characterized by the lack of desirable sensory features or a homogeneous aroma profile, treating probiotic cells with HPH might differentiate the products and enhance their positive sensory properties.

In addition, the results obtained suggested that HPH has several biotechnological applications, including modulating the volatile-molecule profiles of dairy products, improving specific enzymatic activities of cells and enhancing the probiotic properties of bacterial strains. Finally, the proteomic approach used in this study has contributed to add another dowel to the understanding of the mechanisms underlying the stress responses of probiotic strains by demonstrating the involvement of the peptide profile in the response to HPH, which is one of the most promising technologies for application at the industrial level, particularly in the dairy-product sector.

However, the promising aspects of HPH treatment indicated by the results obtained in this study must be further investigated to better understand the relationships among the genomic, volatilomic and peptide-metabolic profiles.

References

Åvall-Jääskeläinen, S., and Palva, A. (2005). *Lactobacillus* surface layers and their applications. *FEMS microbial. Rev.* 29, 511–529. doi: 10.1016/j.femsre.2005.04.003

Basson, A., Flemming, L. A., and Chenia, H. Y. (2007). Evaluation of adherence, hydrophobicity, aggregation, and biofilm development of *Flavobacterium johnsoniae*-like isolates. *Microb. Ecol.* 55, 1–14. doi: 10.1007/s00248-007-9245-y

Buck, B. L., Altermann, E., Svingerud, T., and Klaenhammer, T. R. (2005). Functional analysis of putative adhesion factors in *Lactobacillus acidophilus* NCFM. *Appl. Environ. Microbiol.* 71, 8344–8355. doi: 10.1128/AEM.71.12.8344-8351.2005

Burns, P., Patrignani, F., Serrazanetti, D., Vinderola, G., Reinheimer, J., Lanciotti, R., et al. (2008a). Probiotic Crescenza cheese containing *Lactobacillus paracasei* and *Lactobacillus acidophilus* manufactured with high pressure-homogeneized milk. *J. Dairy Sci.* 91, 500–512. doi: 10.3168/jds.2007-0516

Burns, P., Vinderola, G., Molinari, F., and Reinheimer, J. (2008b). Suitability of whey and buttermilk for the growth and frozen storage of probiotic lactobacilli. *Int. J. Dairy Technol.* 61, 156–164. doi: 10.1111/j.1471-0307.2008.00393.x

Bury, D., Jelen, P., and Kaláb, M. (2001). Disruption of *Lactobacillus delbrueckii* ssp. *bulgaricus* 11842 cells for lactose hydrolysis in dairy products: a comparison of sonication, high-pressure homogenization and bead milling. *Innov. Food Sci. Emerg.* 2, 23–29. doi: 10.1016/S1466-8564(00)00039-4

Croxatto, A., Prod'Hom, G., and Greub, G. (2012). Applications of MALDI-TOF mass spectrometry in clinical diagnostic microbiology. *FEMS Microbial. Rev.* 36, 380–407. doi: 10.1111/j.1574-6976.2011.00298.x

Diels, A. M., and Michiels, C. W. (2006). High-pressure homogenization as a non-thermal technique for the inactivation of microorganisms. *Crit. Rev. Microbiol.* 32, 201–216. doi: 10.1080/10408410601023516

Fantin, G., Fogagnolo, M., Guerzoni, M. E., Lanciotti, R., Medici, A., Pedrini, P., et al. (1996). Effect of high hydrostatic pressure and high pressure homogenization on the enantioselectivity of microbial reduction. *Tetrahedron-Asymmet.* 7, 2879–2887. doi: 10.1016/0957-4166(96)00379-5

Floury, J., Desrumaux, A., and Lardières, J. (2000). Effect of high-pressure homogenization on droplet size distributions and rheological properties of model oil-in-water emulsions. *Innov. Food Sci. Emerg.* 1, 127–134.

Frece, J., Kos, B., Svetec, I. K., Zgaga, Z., Mrša, V., and Šušković, J. (2005). Importance of S-layer proteins in probiotic activity of *Lactobacillus acidophilus* M92. *J. Appl. Microbiol.* 98, 285–292. doi: 10.1111/j.1365-2672.2004.02473.x

Guerzoni, M. E., Ferruzzi, M., Sinigaglia, M., and Criscuoli, G. C. (1997). Increased cellular fatty acid desaturation as a possible key factor in thermotolerance in *Saccharomyces cerevisiae*. *Can. J. Microbiol.* 43, 569–576.

Innocente, N., Marchesini, G., and Biasutti, M. (2011). Feasibility of the SPME method for the determination of the aroma retention capacity of proteose-peptone milk protein fraction at different pH values. *Food Chem.* 124, 1249–1257. doi: 10.1016/j.foodchem.2010.07.056

Isakoff, S. J., Wang, Y., and Skolnik, E. Y. (1996). Finally, some signaling molecules find a home in yeast. *Nat. Biotechnol.* 14, 578–578. doi: 10.1038/nbt0596-578

Iucci, L., Patrignani, F., Vallicelli, M., Guerzoni, M. E., and Lanciotti, R. (2006). Effects of high pressure homogenization on the activity of lysozyme and lactoferrin against *Listeria monocytogenes*. *Food Control* 18, 558–565. doi: 10.1016/j.foodcont.2006.01.005

Jakava-Viljanen, M., and Palva, A. (2007). Isolation of surface (S) layer protein carrying *Lactobacillus* species from porcine intestine and faeces and characterization of their adhesion properties to different host tissues. *Vet. Microbial.* 124, 264–273. doi: 10.1016/j.vetmic.2007.04.029

Jofré, A., Champomier-Vergès, M., Anglade, P., Baraige, F., Martín, B., Garriga, M., et al. (2007). Protein synthesis in lactic acid and pathogenic bacteria during recovery from a high pressure treatment. *Res. Microbiol.* 158, 512–520. doi: 10.1016/j.resmic.2007.05.005

Kankaanpää, P., Salminen, S. J., Isolauri, E., and Lee, Y. K. (2001). The influence of polyunsaturated fatty acids on probiotic growth and adhesion. *FEMS Microbiol. Lett.* 194, 149–153. doi: 10.1016/S0378-1097(00)00519-X

Kankaanpää, P., Yang, B., Kallio, H., Isolauri, E., and Salminen, S. (2004). Effects of polyunsaturated fatty acids in growth medium on lipid composition and on

physicochemical surface properties of lactobacilli. *Appl. Environ. Microbiol.* 70, 129–136. doi: 10.1128/AEM.70.1.129-136.2004

Kirjavainen, P. V., Ouwehand, A. C., Isolauri, E., and Salminen, S. J. (1998). The ability of probiotic bacteria to bind to human intestinal mucus. *FEMS Microbiol. Lett.* 167, 185–189. doi: 10.1016/S0378-1097(98)00387-5

Kocsis, M. G., and Weselake, R. J. (1996). Phosphatidate phosphatases of mammals, yeast, and higher plants. *Lipids* 31, 785–802. doi: 10.1007/BF02522974

Lanciotti, R., Patrignani, F., Iucci, L., Guerzoni, M. E., Suzzi, G., Belletti, N., et al. (2007b). Effects of milk high pressure homogenization on biogenic amine accumulation during ripening of ovine and bovine Italian cheese. *Food Chem.* 104, 693–701. doi: 10.1016/j.foodchem.2006.12.017

Lanciotti, R., Patrignani, F., Iucci, L., Saracino, P., and Guerzoni, M. E. (2007a). Potential of high pressure homogenization in the control and enhancement of proteolytic and fermentative activities of some *Lactobacillus* species. *Food Chem.* 102, 542–550. doi: 10.1016/j.foodchem.2006.06.043

Lanciotti, R., Vannini, L., Patrignani, F., Iucci, L., Vallicelli, M., Ndagijimana, M., et al. (2006). Effect of high pressure homogenisation of milk on cheese yield and microbiology, lipolysis and proteolysis during ripening of Caciotta cheese. *J. Dairy Res.* 73, 216–226. doi: 10.1017/S0022029905001640

Muramalla, T., and Aryana, K. J. (2011). Some low homogenization pressures improve certain probiotic characteristics of yogurt culture bacteria and *Lactobacillus acidophilus* LA-K. *J. Dairy Sci.* 94, 3725–3738. doi: 10.3168/jds.2010-3737

Patrignani, F., Burns, P., Serrazanetti, D., Vinderola, G., Reinheimer, J. A., Lanciotti, R., et al. (2009). Suitability of high pressure-homogenized milk for the production of probiotic fermented milk containing *Lactobacillus paracasei* and *Lactobacillus acidophilus*. *J. Dairy Res.* 76, 74–82. doi: 10.1017/S0022029908003828

Patrignani, F., Iucci, L., Belletti, N., Gardini, F., Guerzoni, M. E., and Lanciotti, R. (2008). Effects of sub-lethal concentrations of hexanal and 2-(E)-hexenal on membrane fatty acid composition and volatile compounds of *Listeria monocytogenes*, *Staphylococcus aureus*, *Salmonella enteritidis* and *Escherichia coli*. *Int. J. Food Microbiol.* 123, 1–8. doi: 10.1016/j.ijfoodmicro.2007.09.009

Patrignani, F., Iucci, L., Lanciotti, R., Vallicelli, M., Mathara, M., Holzapfel, W. H., et al. (2007). Effect of high pressure homogenization, not fat milk solids and milkfat on the technological performances of a functional strain for the production of probiotic fermented milks. *J. Dairy Sci.* 90, 4513–4523. doi: 10.3168/jds.2007-0373

Patrignani, F., Tabanelli, G., Siroli, L., Gardini, F., and Lanciotti, R. (2013). Combined effects of high pressure homogenization treatment and citral on microbiological quality of apricot juice. *Int. J. Food Microbiol.* 160, 273–281. doi: 10.1016/j.ijfoodmicro.2012.10.021

Putignani, L., Del Chierico, F., Onorm, M., Mancinelli, L., Argentierim, M., Bernaschi, P., et al. (2011). MALDI-TOF mass spectrometry proteomic phenotyping of clinically relevant fungi. *Mol. Biosyst.* 7, 620–629. doi: 10.1039/c0mb00138d

Russell, N. J., Evans, R. I., Ter Steeg, P. F., Hellemons, J., Verheul, A., and Abee, T. (1995). Membranes as a target for stress adaptation. *Int. J. Food Microbiol.* 28, 255–261. doi: 10.1016/0168-1605(95)00061-5

Somero, G. N. (1992). Adaptations to hydrostatic pressure. *Anim. Rev. Phys.* 54, 557–577. doi: 10.1146/annurev.physiol.54.1.557

Šedo, O., Vávrová, A., Vaďurová, M., Tvrzová, L., and Zdráhal, Z. (2013). The influence of growth conditions on strain differentiation within the *Lactobacillus acidophilus* group using matrix-assisted laser desorption/ionization time-of-flight mass spectrometry profiling. *Rapid Commun. Mass Spectrom.* 27, 2729–2736. doi: 10.1002/rcm.6741

Tabanelli, G., Burns, P., Patrignani, F., Gardini, F., Lanciotti, R., Reinheimer, J., et al. (2012). Effect of a non-lethal High Pressure Homogenization treatment on the *in vivo* response of probiotic lactobacilli. *Food Microbiol.* 32, 302–307. doi: 10.1016/j.fm.2012.07.004

Tabanelli, G., Patrignani, F., Gardini, F., Vinderola, C. G., Reinheimer, J. A., Grazia, L., et al. (2014). Effect of a sub-lethal high pressure homogenization treatment on the fatty acid membrane composition of probiotic lactobacilli. *Lett. Appl. Microbiol.* 58, 109–117. doi: 10.1111/lam.12164

Tabanelli, G., Patrignani, F., Vinderola, G. C., Reinheimer, J. A., Gardini, F., and Lanciotti, R. (2013). Effect of sub-lethal high pressure homogenization treatments on *in vitro* functional and biological properties of lactic acid bacteria. *Food Sci. Technol.* 53, 580–586. doi: 10.1016/j.lwt.2013.03.013

Vannini, L., Lanciotti, R., Baldi, D., and Guerzoni, M. E. (2004). Interactions between high pressure homogenization and antimicrobial activity of lizozyme and lactoperoxidase. *Int. J Food Microbiol.* 94, 123–135. doi: 10.1016/j.ijfoodmicro.2004.01.005

Vannini, L., Patrignani, F., Iucci, L., Ndagijimana, M., Lanciotti, R., et al. (2008). Effect of a pre-treatment of milk with high pressure homogenization on yield as well as on microbiological, lipolytic and proteolytic patterns of "Pecorino" cheese. *Int. J. Food Microbiol.* 128, 329–335. doi: 10.1016/j.ijfoodmicro.2008.09.018

Vinderola, C. G., Prosello, W., Ghiberto, D., and Reinheimer, J. A. (2000). Viability of probiotic (*Bifidobacterium, Lactobacillus acidophilus* 512) and non probiotic microflora in Argentinean Fresco cheese. *J. Dairy Sci.* 83, 1905–1911.

Conflict of Interest Statement: The authors declare that the research was conducted in the absence of any commercial or financial relationships that could be construed as a potential conflict of interest.

21

Fighting Ebola with novel spore decontamination technologies for the military

Christopher J. Doona[1]*, Florence E. Feeherry[1], Kenneth Kustin[2], Gene G. Olinger[3], Peter Setlow[4], Alexander J. Malkin[5] and Terrance Leighton[6]

[1] U.S. Army Natick – Soldier RD&E Center, Warfighter Directorate, Natick, MA, USA, [2] Department of Chemistry, Emeritus, Brandeis University, Waltham, MA, USA, [3] National Institute of Allergy and Infectious Diseases, Integrated Research Facility – Division of Clinical Research, Fort Detrick, MD, USA, [4] Department of Molecular Biology and Biophysics, University of Connecticut Health Center, Farmington, CT, USA, [5] Biosciences and Biotechnology Division, Physical and Life Sciences Directorate, Lawrence Livermore National Laboratory, Livermore, CA, USA, [6] Children's Hospital – Oakland Research Institute, University of California San Francisco - Benioff, Oakland, CA, USA

Edited by:
*Alexander Mathys,
German Institute of Food
Technologies, Germany*

Reviewed by:
*Bei Li,
Hubei University of Medicine, China
Bastian Dörrbecker,
German Institute of Food
Technologies, Germany*

*Correspondence:
*Christopher J. Doona,
U.S. Army Natick – Soldier RD&E
Center, Warfighter Directorate,
Kansas Street, Natick,
MA 01760-5018, USA
christopher.j.doona.civ@mail.mil*

Recently, global public health organizations such as Doctors without Borders (MSF), the World Health Organization (WHO), Public Health Canada, National Institutes of Health (NIH), and the U.S. government developed and deployed Field Decontamination Kits (FDKs), a novel, lightweight, compact, reusable decontamination technology to sterilize Ebola-contaminated medical devices at remote clinical sites lacking infrastructure in crisis-stricken regions of West Africa (medical waste materials are placed in bags and burned). The basis for effectuating sterilization with FDKs is chlorine dioxide (ClO_2) produced from a patented invention developed by researchers at the US Army Natick Soldier RD&E Center (NSRDEC) and commercialized as a dry mixed-chemical for bacterial spore decontamination. In fact, the NSRDEC research scientists developed an ensemble of ClO_2 technologies designed for different applications in decontaminating fresh produce; food contact and handling surfaces; personal protective equipment; textiles used in clothing, uniforms, tents, and shelters; graywater recycling; airplanes; surgical instruments; and hard surfaces in latrines, laundries, and deployable medical facilities. These examples demonstrate the far-reaching impact, adaptability, and versatility of these innovative technologies. We present herein the unique attributes of NSRDEC's novel decontamination technologies and a Case Study of the development of FDKs that were deployed in West Africa by international public health organizations to sterilize Ebola-contaminated medical equipment. FDKs use bacterial spores as indicators of sterility. We review the properties and structures of spores and the mechanisms of bacterial spore inactivation by ClO_2. We also review mechanisms of bacterial spore inactivation by novel, emerging, and established non-thermal technologies for food preservation, such as high pressure processing, irradiation, cold plasma, and chemical sanitizers, using an array of *Bacillus subtilis* mutants to probe mechanisms of spore germination and inactivation. We employ techniques of high-resolution atomic force microscopy and phase contrast microscopy to examine the effects of γ-irradiation on bacterial spores of *Bacillus anthracis*, *Bacillus thuringiensis*, and *Bacillus atrophaeus* spp. and of ClO_2 on *B. subtilis* spores, and present in detail assays using spore bio-indicators to ensure sterility when decontaminating with ClO_2.

Keywords: Ebola, decontamination technologies, spores, chrloine dioxide, military medicine

1. Introduction

Innovation in Science and Technology comes from myriad sources, such as thinking outside-the-box, applying expertise to new areas, or adapting novel technologies that advance the frontiers of knowledge to fill needs in the commercial marketplace for consumers or to meet critical capability gaps on the battlefield. Researchers at the U.S. Army Natick Soldier Research, Development and Engineering Center (NSRDEC) have invented and patented (Doona et al., 2014) an ensemble of novel decontamination technologies (**Table 1**) involving innovative dry, mixed-chemical technologies designed to be lightweight, compact, portable, easy-to-carry, energy-independent, flameless, almost waterless, inexpensive, safe to end-users and the environment ("green" technologies), and effective in addressing a diverse array of decontamination applications in far-forward military deployments or other high-intensity, austere environments by using the disinfectant chlorine dioxide (ClO_2). These characteristics also make NSRDEC's novel decontamination technologies well-suited for use during large-scale emergencies, natural disasters (Hurricane Katrina, tsunamis, superstorm Sandy) or in humanitarian relief in third-world countries.

The use of chlorine dioxide (ClO_2) to decontaminate *Bacillus anthracis* spores (causative agent of 'Anthrax') following the letter attacks on Washington, DC and other locations was facilitated by data from the author's laboratories and other studies. These attacks were unprecedented in their use of mail processing delivery systems to create large-scale and wide-area *B. anthracis* spore contamination. They highlighted the need for more efficacious, agile and adaptive decontamination modalities that could extinguish primary and secondary nosocomial contact and transmission hazards. Concerns regarding transmission control of existential nosocomial diseases were further highlighted by the SARS and H1N1 pan epidemics. As Joshua Lederberg presciently observed in 1998 (*profiles.nlm.nih.gov/ps/access/BBBDLP.pdf*), with modern transportation and distribution infrastructure no infectious disease is more than 24 away from any location on the earth. Dr. Lederberg's insight became salient with the emergence of the Ebola crisis emanating from West Africa and the concomitant challenges of controlling secondary chains of transmission at their origin and globally. In the summer of 2014, global public health and medical personnel adapted NSRDEC's novel decontamination technologies for field use to fight the spread of Ebola by decontaminating medical equipment at remote clinical sites in West Africa.

Ebola virus disease (EVD) is a severe and often fatal disease in humans that is communicated between humans through contact with infected blood, organs or tissues, bodily fluids (saliva, sweat, vomit, urine, semen, and breast milk), or items they contaminate (clothing, bedding, gauze, needles and syringes, and medical equipment). In March, 2015, WHO estimated 24,842 cases and 10,299 deaths from this outbreak (World Health Organization [WHO], 2015a), and concerns of EVD heightened as EVD cases spread internationally. As an enveloped virus – one with a lipid and protein membrane – Ebola is vulnerable to chemical disinfectants, such as household bleach (OCl^-) and chlorine dioxide (ClO_2), which can be used to sanitize infected surfaces, patient rooms, and to sterilize contaminated medical equipment at remote clinical sites in West Africa (World Health Organization [WHO], 2014). In parts of the world that consume non-traditional foods (bats, monkeys, bush meat) as protein sources, basic food hygiene for preventing the transmission of biological hazards apply equally well to the Ebola virus (Anelich and Moy, 2014; World Health Organization [WHO], 2015b): Keep clean, Separate raw and cooked, Cook thoroughly (specifically, boiling for 5 min or heating for 60 min at $T = 60°C$ inactivates the Ebola virus – Anelich and Moy, 2014), Keep food at safe temperatures, and Use safe water and raw materials.

Natick Soldier RD&E Center's ensemble of patented novel decontamination technologies have the acronyms NCC, PCS, D-FENS, D-FEND ALL, and CoD (**Table 1**) and feature a variety of embodiments designed to produce ClO_2 for killing bacterial spores (*B. anthracis*), vegetative pathogens (*Listeria monocytogenes*, *Escherichia coli*, etc.), viruses (Ebola), and bacteriophage in cross-cutting applications (Setlow et al., 2009), such as sterilizing surgical instruments, decontaminating textiles (uniforms, tents, shelters), sanitizing fresh fruits and vegetables procured in host nations, disinfecting wastewater, providing potable water quality and safety, and promoting hygiene by decontaminating surfaces bathrooms, showers, laundries, Army Field Kitchens, Navy Galleys, and deployable medical units. Deployments of military personnel worldwide generate thousands of tons of wastewater and food waste annually that support disease vectors capable of adversely affecting human health and account for Disease and Non-Battle Injuries (DNBIs) that at an average rate of 1.5% of assigned personnel, would have cost all branches of the military an estimated $32.5M annually (300 personnel per day) during the time of the Balkan conflicts in the 1990s! Finding inexpensive, convenient, and effective decontamination technologies improves hygiene and reduces incidences of DNBIs and other foodborne illnesses, thereby saving all branches of the military millions of dollars in medical costs and promoting health and well-being. For innovative patents such as these, Thomson Reuters in 2012 named the U.S. Army among the Top 100 Global Innovators (Foran, 2013 - available at http://www.army.mil/article/99816/).

TABLE 1 | Acronyms and attributes of NSRDEC decontamination technologies.

Acronym	Technology	Attributes
NCC	**N**ovel **C**hemical **C**ombination	Dry powders mix with water
PCS	**P**ortable **C**hemical **S**terilizer	Plastic suitcase sterilizer
D-FENS	**D**isinfectant-sprayer **F**or **EN**vironmentally friendly **S**anitation	Collapsible handheld sprayer
D-FEND ALL	**D**isinfectant **F**or **EN**vironmentally friendly **D**econtamination, **ALL**-purpose	All purpose decontamination
CoD	**C**ompartment **o**f **D**efense	In-package disinfectant
FDK	**F**ield **D**econtamination **K**it	Ebola disinfectant

Other technologies that decontaminate with gaseous ClO_2 include electrically powered equipment to decontaminate facemasks worn as personal protective equipment (PPE) by emergency first-responders (Stubblefield and Newsome, 2015), and now Field Decontamination Kits (FDKs). FDKs are adapted from the commercial-off-the-shelf (COTS) version of NRDEC's decontamination technologies and are presently being used by global public health organizations [Doctors without Borders (MSF), World Health Organization (WHO), Public Health Canada, National Institutes of Health (NIH)] and the U.S. government to sterilize Ebola-contaminated medical equipment at remote clinical sites in West Africa. While the Ebola virus is classified as Biosafety level 4 (BSL-4) due to the severity of disease in humans, the Ebola virus itself is relatively fragile and presently without a standard test assay under representative conditions even in a high-level containment facility. Bacterial spores therefore provide the standard test assay for sterility and/or decontamination efficacy, primarily because bacterial spores exhibit more resistance to chemical and physical decontamination methods. The author's laboratories have studied the processes of *Bacillus* sporulation, spore germination, spore resistance and persistence, spore decontamination and spore structural biology for many years. Accordingly, we review bacterial spore properties, structures, and resistance mechanisms and focus on the mechanisms through which ClO_2 inactivates bacterial spores as the indicators of efficient bio-decontamination.

We present the ontogeny of NSRDEC's novel ClO_2 decontamination technologies for spores that evolved to field-ready FDKs to meet an urgent need in protecting healthcare workers in West Africa from the spread of EVD during the heights of this international public health crisis. And just as ClO_2 is also a non-thermal food processing technology for sanitizing fresh fruits and vegetables, we explore the characteristics of other Non-thermal technologies (chemical sanitizers, high pressure, irradiation, heat, plasmas, and UV light), particularly high pressure and γ-irradiation, that have also been used in the decontamination of *B. anthracis* (Cléry-Barraud et al., 2004) or other types of spores, and we also consider the mechanisms of bacterial spore inactivation by these agents. The ability of ClO_2 to kill spores used as bio-indicators of sterility is also examined in detail, and we use high-resolution Atomic Force Microscopy (AFM) and phase contrast microscopy to examine the effects of nonthermal technologies on bacterial spores (**Figure 1**).

1.1. Additional Reference Materials

More information relating to this manuscript is available through the following links:

a. http://www.necn.com/news/new-england/Mass-Researchers-Create-Disinfectant-to-Fight-Ebola-280284792.html
b. http://www.metrowestdailynews.com/article/20141029/NEWS/141025763/0/SEARCH
http://www.wcvb.com/news/natick-labs-innovation-could-help-prevent-spread-of-ebola/29370460?utm_source=hootsuite&utm_medium=facebook&utm_campaign=wcvb%2Bchannel%2B5%2Bboston

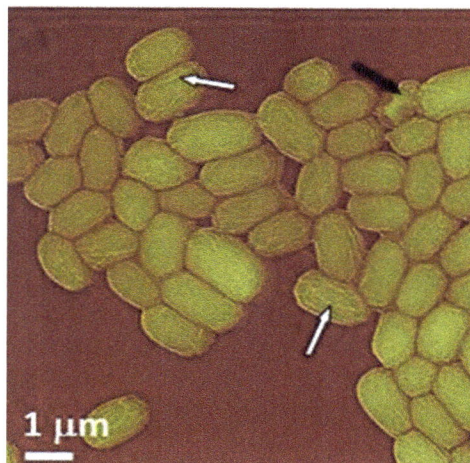

FIGURE 1 | AFM height image of native air-dried anthracis Sterne spores showing ridges (white arrows) and a collapsed spore (black arrow). Further details are discussed with **Figure 9**.

c. http://www.nbcnews.com/watch/nbc-news-channel/researchers-develop-ebola-disinfectant-348828227663
d. http://www.army.mil/article/136641/Natick_plays_key_role_in_helping_to_fight_spread_of_Ebola/
e. http://www.bizjournals.com/boston/blog/techflash/2015/04/why-a-u-s-army-research-facility-in-natick-is.html

2. Materials and Methods

2.1. Novel Redox Chemistry for the Production of ClO_2

Natick Soldier RD&E Center's bacterial spore decontamination technologies comprise a number of different dry chemical oxidation–reduction systems that mix with water in a glass beaker or plastic vessel to produce chlorine dioxide (ClO_2) inside a carryable plastic suitcase, a rigid or a flexible plastic, hand-triggered spray-bottle, a sealable Mylar bag, or inside closable flexible or rigid plastic packaging (such as the plastic clamshell packaging used commercially to store and distribute fresh berries and other fresh or fresh-cut produce).

The NCC (**Table 1**) is the primary source of generating ClO_2 and involves the oxidation–reduction reaction of chlorite (Sodium chlorite, Sigma-Aldrich Cat. No. 244155) and sulfite (Sodium sulfite, Sigma-Aldrich Cat. No. 239312) through the use of a unique chemical effector (Sodium ascorbate, Sigma-Aldrich A7631 – see Curtin et al., 2004) that initiates and controls the rate of the otherwise kinetically inert reaction, such that the production of ClO_2 takes place at near-neutral pH on a practical and relatively short timescale (Curtin et al., 2014). The NCC improves the generation of ClO_2 compared to existing methods, which include (i) the reduction of chlorate [ClO_3^-, Cl(V)] in high acid (HCl or H_2SO_4); (ii) the oxidation of chlorite [ClO_2^-, Cl(III)] by dichlorine [Cl_2, Cl(0)], hypochlorite [OCl^-, Cl(I)], or persulfate [$S_2O_8^{2-}$, S(VII)], or (iii) the acidification of chlorite for the formation and subsequent disproportionation of chlorous acid [$HClO_2$, Cl(III); Horváth et al., 2003]. While the PCS uses the NCC system to generate gaseous ClO_2, D-FENS uses the NCC in conjunction with a novel two-step mixing process (i.e., *pre-concentration* followed by *ii. post-reaction dilution*), to generate aqueous ClO_2 inside a rigid or collapsible plastic spray-bottle. The chemical systems for D-FEND ALL and CoD eliminate the need for the effector and involve only two chemical components and one-step mixing for greater user convenience, but neither has been formally disclosed yet through patent procedures.

2.2. NSRDEC's Novel Decontamination Technologies

As reported previously (Doona et al., 2014) the novel decontamination technologies (NCC, PCS, D-FENS, D-FEND ALL, FDKs, CoD) were validated using laboratory chemical reagents and suitable challenge organisms and substrates to confirm sterility and material compatibility.

2.2.1. The PCS

The PCS (**Figure 2**) is a Modern Field Autoclave, a revolutionary medical device invented to meet a stated Army need and an urgent battlefield demand for a field-portable, non-steam sterilizer technology that can be used by far-forward surgical teams (Doona et al., 2014). The PCS produces gaseous ClO_2 and proceeds where no commercial device existed previously, with a 100% reduction in power usage, 98% reduction in water, 95% reduction in weight, and 96% reduction in cubic footprint compared to conventional steam autoclaves. The PCS used the NCC to sterilize live cultures of *Geobacillus*

stearothermophilus spores in aqueous suspensions (recovered on Antibiotic Assay Medium with 1% soluble starch – see Feeherry et al., 1987; Doona et al., 2014), bio-indicators of *G. stearothermophilus* (BT Sure biological indicator (BI), Thermo Fisher Scientific Cat No. AY759X3) and *Bacillus atrophaeus* [EZ Test (EtO), SGM Biotech Inc., Cat. No. EZG/6] spores, live cultures of *L. monocytogenes* (recovered on Tryptic Soy Agar-Yeast Extract (TSAYE) incubated at $T = 35°C$ for 48–96 h) and *E. coli* (recovered on Nutrient Agar) inoculated on hard surfaces (glass or stainless steel coupons) or on fresh whole tomatoes.

2.2.2. D-FENS

D-FENS (**Figure 3**) generates aqueous chlorine dioxide in a collapsible handheld spray-bottle (Doona et al., 2014) for decontaminating surfaces (fresh produce or medical, food handling, and surfaces in showers and latrines) anywhere large numbers of deployed personnel co-exist in close proximity. The D-FENS system uses the NCC with novel two-step mixing to generate aqueous ClO_2 inside a rigid or collapsible plastic spray-bottle and was validated against a cocktail of *Staphylococcus aureus* (*S. aureus* A-100, produces enterotoxin A; *S. aureus* ATCC 14458, produces enterotoxin B; and *S. aureus* 993, produces enterotoxin D – all strains were recovered on Baird-Parker Agar containing egg yolk tellurite and Yeast Extract). For validation testing, the *S. aureus* aqueous suspension was spread onto on agar surfaces, representative of porous materials with hard-to-reach places, such as those found in real-world decontamination applications, and challenged with ClO_2 solution from D-FENS.

FIGURE 2 | The PCS (top left) consists of a rigid plastic suitcase embellished with reactors, valves (top right), scrubbers, and other design features (bottom left) to effectuate sterilization (bottom right) with ClO_2 while protecting users and the environment.

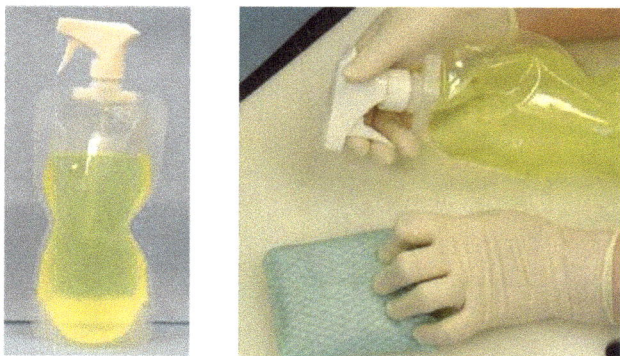

FIGURE 3 | The D-FENS sprayer generates aqueous ClO$_2$ on-site and at point-of-use in a collapsible spray-bottle (left) and easily sprays ClO$_2$ onto contact surfaces (right) to wipe away contaminating pathogens in Army Field Kitchens and Navy galleys.

2.2.3. D-FEND ALL and CoD

D-FEND ALL and CoD are chemical systems that eliminate the need for an effector in generating ClO$_2$ with two-components and one-step mixing, to provide more convenience in treating water, decontaminating textiles, or for in-package anti-microbial treatments of sterile instruments or fresh fruits and vegetables. The D-FEND ALL and CoD systems use separate oxidation–reduction chemical systems that eliminate the need for the added chemical effector. D-FEND ALL (Doona et al., 2014) was validated experimentally in the laboratory to decontaminate live cultures of *B. anthracis* Delta Sterne (recovered on Nutrient Agar and incubated for 16–20 h at $T = 35°C$) and *Bacillus amyloliquefaciens* spores (recovered on ST-1 Nutrient Agar incubated for 18 h at $T = 30°C$) on military textiles, and compared with commercial household bleach (5–6% aqueous hypochlorite, OCl$^-$) and high pressure processing (HPP) for textiles immersed in water using a PT-1 high pressure unit (Avure Technologies, Inc., Kent, WA, USA) with conditions of pressure $P = 550$ MPa, temperature $T = 65°C$, and time $t = 100$ min. CoD was validated for decontamination in rigid plastic packaging using the *G. stearothermophilus* and *B. atrophaeus* spore bio-indicators mentioned above.

2.2.4. FDKs

Field Decontamination Kits (FDKs) are based on adapting commercial versions of the NSRDEC decontamination technologies for use by global public health organizations (MSF, WHO, NIH, etc.) to sterilize Ebola-contaminated medical equipment at remote clinical sites in West Africa. First, the NCC is sold commercially as CHEM-CD (ClorDiSys Solutions, Inc., Lebanon, NJ, USA) as a result of Technology Transfer licensing agreements with private industry. CHEM-CD controllably produces gaseous ClO$_2$ to decontaminate HEPA housings, biosafety cabinets (BSCs) and hoods, and bio-aerosol chambers. CHEM-CD consists of oxidant (Part A), reductant (Part B), and neutralizer (Part C) in separate foil pouches and

wrapped in plastic bag to extend shelf-life to ∼30 months. A video demonstration of CHEM-CD in action is available at http://www.youtube.com/watch?v=EAh_Vz3TNTo. The benefits of using this product relate to the approval of ClO$_2$ as a sterilant by the National Sanitation Foundation (NSF/ANSI 49 Annex G 2009) for advantages in safety, speed, and environmental-friendliness compared to conventional formaldehyde. The CHEM-CD product is the method for producing ClO$_2$ in the deployed FDKs (**Figure 4**). A more detailed description of the materials and operation of the FDKs are discussed in detail in Section 3. Validation of the PCS using the CHEM-CD formulation from the FDK to sterilize *G. stearothermophilus* and *B. atrophaeus* bio-indicators, as reported previously (Doona et al., 2014), are presented in Section 3.1 below.

2.3. Atomic Force Microscopy (AFM)

Detailed experimental procedures for AFM imaging of spores were described previously (Plomp et al., 2005a,b, 2014) and are summarized here. Droplets (∼2.0 μl) of spore suspensions were deposited on plastic cover slips and incubated for 10 min at room temperature, then the sample substrate was carefully rinsed with double-distilled water and allowed to dry. Images were collected using a Nanoscope IV Atomic Force Microscope (Bruker Corporation, Santa Barbara, CA, USA) operated in tapping mode. For rapid, low-resolution analysis of spore samples, fast scanning AFM probes (DMASP Micro-Actuated, Bruker Corporation, Santa Barbara, CA, USA) with resonance frequencies of ∼210 kHz were utilized. For high-resolution imaging, SuperSharpSilicon (SSS) AFM probes (NanoWorld Inc., Neuchâtel, Switzerland) with tip radii <2 nm and resonance frequencies of ∼300 kHz were used. Nanoscope software 5.30r3sr3 was used for data acquisition and subsequent processing of AFM images. In order to assess low-resolution and high-resolution spore features, raw AFM images typically needed to be modified. In particular, the *contrast enhancement* command, which runs a statistical differencing filter on the acquired image, was typically utilized to bring all of the features of an image to the same height and to equalize the contrast among them. This allows all features of an image to be seen simultaneously, and thus a single spore or a group of spores can be imaged at relatively low resolution while spore coat attributes can be visualized at high-resolution. Heights of spore surface features (i.e., folds, coat layers, etc.) were measured from *height* images using the *section* command. Tapping amplitude, phase, and height images were collected simultaneously. Height images allow quantitative height determinations, providing precise measurements of spore surface topography. Amplitude and phase images do not provide height information, but provide similar morphological and structural information as height images, often displaying a greater amount of structural detail and contrast compared to height images and making them a preferred choice for presentation purposes. Prior to AFM characterization, spore preparations were examined for refractility by phase-contrast light microscopy (Nikon Eclipse 50i) to determine the fraction

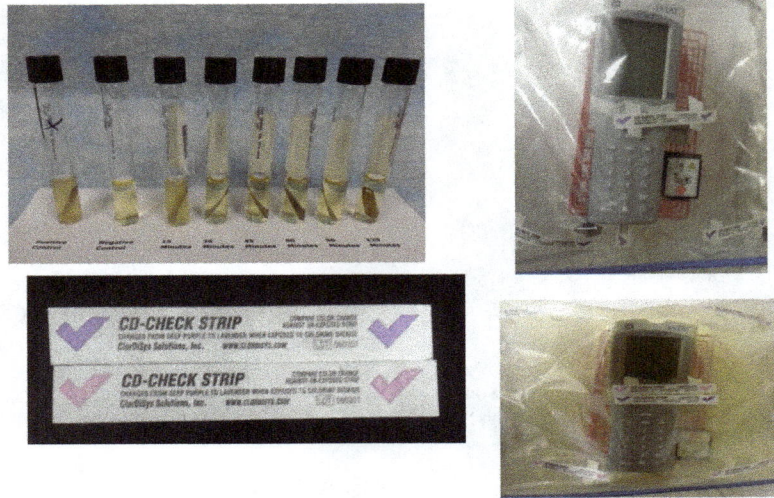

FIGURE 4 | Laboratory testing showed the FDKs were non-destructive to household electronics devices after 15 cycles (top right) and non-destructive to an iStat device after five decontamination cycles (bottom right). The FDK used a CHEM-CD configuration (see Section 3) and generated copious ClO_2 gas **(bottom left)** and $T = 80–120°C$ to kill spore indicators in 30 min **(top left)**.

of ungerminated (phase bright) and germinated (phase dark) spores.

2.4. Bacterial Spore Bio-indicators (BI's)

2.4.1. Spore Strips

Spore strips (Raven Biological Labs, Omaha, NE, USA) are inch-long pieces of cellulose paper inoculated with a known concentration of bacterial spores, and packaged in a barrier material that permits the diffusion of sterilant gas or humidified air, but excludes contaminants such as vegetative bacterial cells or other spores. Spore strips act as bio-indicators (BI's) for various sterilization/decontamination processes (ethylene oxide, abbreviated EtO, ClO_2, autoclaving, irradiation, etc.). Three *Bacillus* spore species were assayed for susceptibility to ClO_2: *B. atrophaeus* (ATCC 9372, formerly *B. subtilis* var. niger and *B. globigii*), *B. thuringiensis* (ATCC 29730), and *G. stearothermophilus* (ATCC 7953, formerly *Bacillus stearothermophilus*). Spore strip populations for all species were determined by the manufacturer, as were D and Z (the slope of a thermal resistance curve) values for two of the three species. D values for ethylene oxide (D_{EtO}) and dry heat (D_{160}) were determined for *B. atrophaeus*, while D values for saturated steam (D_{121} and $D_{132.2}$) were determined for *G. stearothermophilus*. No such values were determined for *B. thuringiensis*. Species lot numbers, populations, and relevant D and Z values are listed in **Table 2**.

2.4.2. BI Packaging Material

The BI packaging material contributes a crucial factor in the efficacy of the ClO_2 sterilizing process. In this report, we investigate BI's packaged either in 1059B medical grade Tyvek (single-sided with a plastic backing; Raven Industries, Sioux Falls, SD, USA) or nothing, as our preliminary data suggested that 1059B was relatively non-reactive and non-attenuating for ClO_2. To comply with current industry standards, we also show that ClO_2 is an efficacious sterilant of *B. atrophaeus* BI's packaged in medical grade glassine (Raven Industries, Sioux Falls, SD, USA).

2.4.3. ClO_2 Generation and Treatment

Chlorine dioxide was generated by the oxidation of technical grade sodium chlorite ($NaClO_2$) by sodium persulfate ($Na_2S_2O_8$; Sigma-Aldrich Co., St. Louis, MO, USA) in aqueous solution:

$$2NaClO_2 + Na_2S_2O_8 \rightarrow 2ClO_2 + 2Na_2SO_4$$

Pure gaseous ClO_2, free of volatile by-products such as Cl_2, was purged from the reaction flask and diluted with ratios of filtered, dehumidified and humidified air to attain the target $ClO_{2(g)}$ concentration and RH conditions. RH was controlled with a series of flow meters (Cole Parmer, Vernon Hills, IL, USA) passing filtered and dehumidified air through a 500 mL gas wash bottle half filled with deionized water. The humidified diluted gas was directed into a 5 L glass test chamber (Thermo Fisher Scientific, Waltham, MA, USA), and ClO_2 concentration ($[ClO_2]$) in parts per million (ppm) was determined by iodometric titration of a 50 or 100 mL volume/sample of gas taken/removed/sampled from the exit port of the reaction chamber in a gas-tight Hamilton sample-lock syringe (Fisher). Experiments were carried out at ambient temperature and atmospheric pressure (measured and monitored with a Traceable® Digital Hygrometer/Thermometer (VWR International, Radnor, PA, USA), with RH ranging from 30 to 90%, and $[ClO_2]$ ranging from approximately 50 to 2000 ppm. Experimentally

TABLE 2 | Spore strip bio-indicator characteristics.

Species	ATCC#	Lot#	Population	D_{160} value	D_{EtO} value	Z-value
Bacillus atrophaeus	9372	1162052	2.0×10^6	2.6	3.0	44.5
B. atrophaeus	9372	1161841	1.3×10^6	2.5	3.3	23.5
B. atrophaeus	9372	1161911	1.3×10^6	2.5	3.8	46.0
Batch 204GB:						
B. atrophaeus	9372	1142042	1.2×10^4	2.8	3.1	39.0
B. atrophaeus	9372	1152041	1.2×10^5	2.8	3.1	39.0
B. atrophaeus	9372	1162043	1.2×10^6	2.8	3.1	39.0
B. atrophaeus	9372	1172041	1.2×10^7	2.8	3.1	39.0
Batch 214GB:						
B. atrophaeus	9372	1142141	3.5×10^4	2.8	5.0	33.3
B. atrophaeus	9372	1152141	2.0×10^5	2.8	5.0	33.3
B. atrophaeus	9372	1162141	1.5×10^6	2.8	5.0	33.3
B. atrophaeus	9372	1172141	1.5×10^7	2.8	5.0	33.3
B. atrophaeus	9372	1182141	1.5×10^8	2.8	5.0	33.3
B. thuringiensis	29730	616022	1.2×10^6	n/a	n/a	n/a
				D_{121} value	$D_{132.2}$ value	Z-value
Geobacillus stearothermophilus	7953	3166031	1.0×10^6	2.0	0.07	7.5

TABLE 3 | Experimental ClO$_2$ target and actual concentrations.

ClO$_2$ concentration (ppm)			ClO$_2$ dose			%RH		
Target	Measurement	SD	Target	Measurement	SD	Target	Measurement	SD
50	50	4	400	414	20	30	30.32	0.006
67	68	4	1000	1037	32	40	38.75	0.011
100	110	17	2000	2020	40	50	49.51	0.015
125	131	6	4000	4046	88	60	59.41	0.013
167	177	21				70	69.36	0.011
200	218	14				80	79.42	0.015
250	256	23				90	89.67	0.01
400	405	31						
500	498	40						
800	784	53						
1000	1027	115						
2000	2027	303						

Room temperature = 21.5°C, SD = 0.770.

TABLE 4 | Tyvek spore strip data summary.

Species	Runs		N	Dose	%RH	D_{EtO}
	Tyvek	No pkg		(ppm ClO$_2$ × t)		
Bacillus atrophaeus	152	144	1.2×10^4–1.0×10^6	110–1991	79	3.1
B. atrophaeus	490	350	3.5×10^4–1.5×10^8	110–1991	79	5.0
B. atrophaeus	500	500	1.3×10^6	438–4106	30–90	3.3
B. atrophaeus	498	495	1.3×10^6	438–4106	30–90	3.8
B. thuringiensis	98	98	1.2×10^6	438–4106	79	–
G. stearothermophilus	80	80	1.0×10^6	438–4106	79	–

measured values for temperature, RH and [ClO$_2$] are listed in **Table 3**.

To ensure exposure and contact of BIs with ClO$_2$, no more than 20 spore strips were placed in the test chamber at one time, and, thus, multiple runs at each reported dose were performed in order to (i) assure repeatability of our results, and (ii) gather enough data to achieve statistically significant results (**Table 4**). One negative control strip (no inoculum and packaged appropriately) was included in the chamber for each set of 10 test strips, and a positive control strip remained outside of the test chamber and away from other potential sterilizing agents for the duration of each experiment.

2.4.4. Microbiological Assays of Sterility

After exposure to the appropriate ClO_2 dose, the strips were placed, using aseptic technique, into tubes of Tryptic Soy Broth containing a Bromocresol Purple pH indicator (Raven Industries, Sioux Falls, SD, USA) and incubated at 37°C (*B. atrophaeus* and *B. thuringiensis*) or 60°C (*G. stearothermophilus*) for 7 days. We monitored the tubes on a daily basis for both turbidity (indicative of bacterial growth) and color change of the pH indicator from purple to yellow (indicative of metabolism). Criteria for a strip being considered "killed" were findings of both *no* turbidity and of *no* color change of the pH indicator, coupled with growth and metabolism for the positive control associated with the sample test set and no growth for the negative control.

2.4.5. Statistical Analysis and Modeling

We fitted a binomial generalized linear model with a complementary log-log link function to the proportion of strips still having live spores after treatment, allowing the dispersion parameter to be greater than one to account for over-dispersion. We adjusted for covariates such as (logarithm of) the number of spores on a strip, the type of packaging used to store the spore strips, RH, and D_{EtO} values. The model in its most general form is expressed as Eq. (1)

$$\log(-\log(1 - \emptyset)) = b_0 + b_1 * Dose + b_2 * \log N + \\ b_3 * Pack + b_4 * RH \qquad (1) \\ + b_5 * D_{EtO}$$

in which $\emptyset$ is the probability of a strip still having living spores after treatment, p_N is the probability of a spore remaining alive after treatment (depends on N), *Dose* is the dose of ClO_2 as calculated by ClO_2 ppm multiplied by exposure time (in units of hours), N is the number of spores on a strip, *Pack* equals 1, if spore strips came in Tyvek packaging, and *Pack* equals 0, if there were no packaging, *RH* is the relative humidity (RH; as a proportion), D_{EtO} is the time (in units of minutes) to reduce the number of living spores to 10% of the original value by ethylene oxide (**Table 5**), and *b*'s are regression coefficients. The relationship between the strip survival probability $\emptyset$ and the spore survival probability p_N is expressed in Eq. (2)

$$1 - \emptyset = (1 - p_N)^N \qquad (2)$$

and Eq. (1) can be re-written as Eq. (3)

$$\log(-\log(1 - p_N)) = b_0 + b_1 * Dose + (b_2 - 1) * \log N + \\ b_3 * Pack + b_4 * RH + b_5 * D_{EtO}$$
$$(3)$$

The range of values of each of the covariates in the various experiments is described in (**Table 4**). Although the independent variable is the strip survival probability $\emptyset$, **Figures 5–7** are reported in terms of our primary measure of interest, which is the probability p_N that any single spore survives.

3. Results – Field Decontamination Kits in the Ebola crisis

Emerging and re-emerging viral infectious diseases have been identified as a major threat to human health, and the current outbreak of EVD has impacted a large part of Western Africa. Conventional procedures for the decontamination of equipment during a filovirus outbreak rely on the use of chemical chlorine (bleach, aqueous hypochlorite, OCl^-) for decontamination when exiting an isolation/treatment center. Electrical equipment that enters an isolation/treatment center and diagnostic facilities can only be surface decontaminated by aqueous rinses with 0.5–5.0% chlorine bleach solution (household bleach is about 6% OCl^-). This procedure is likely *not* adequate for complete decontamination of the devices, particularly for accessing inside the devices.

Devices such as personal cell phones, computers, and most importantly, expensive medical point-of-care (POC) electronic devices used for clinical assessment of patients in isolation/treatment ward are not sufficiently decontaminated by surface rinses. Some medical diagnostic devices have closed systems internal to the device that may have infrequent but probable contamination that cannot be adequately decontaminated by surface rinses with a chemical decontaminating agent. Often these devices contain valuable internal components that may be recycled after being surface decontaminated. This is extremely problematic for items used in a high hazard environment. For low resource, remote sites, the deployment of traditional decontamination tools is nearly impossible. Often there is limited space for equipment, so another desirable aspect of the FDK method is its compact size and simple disposal routine post-decontamination cycle.

Given the wide use of and acceptability of chlorine bleach solutions in laboratories and in isolation/treatment centers, and the historical use of chlorine dioxide in the decontamination of equipment in the U.S., we decided to adapt a chlorine dioxide method to provide equipment decontamination for deployment in the field in areas of active outbreak. This safe and adaptable method uses a commercially available chemical combination kit called CHEM-CD (ClorDiSys Solutions, Inc., Lebanon, NJ, USA) in which the components are packaged separately and isolated from water to prevent chemical reaction from taking place (**Figure 8A**). Once mixed together with water, the chemical reduction of sodium chlorite initiates within minutes, and a well-controlled, exothermic reaction takes place that releases chlorine dioxide gas from solution. This reaction is carefully designed to occur inside a closable plastic bag, such that the ClO_2 released decontaminates thoroughly the entire interior volume of the bag and all items contained therein, including permeating the interior regions of the electronic medical and other devices inside the decontamination bag. Humidity ($\geq$70% RH) and mild heat and mild pressure accumulate inside the bag from the chemical reaction occurring inside a closed

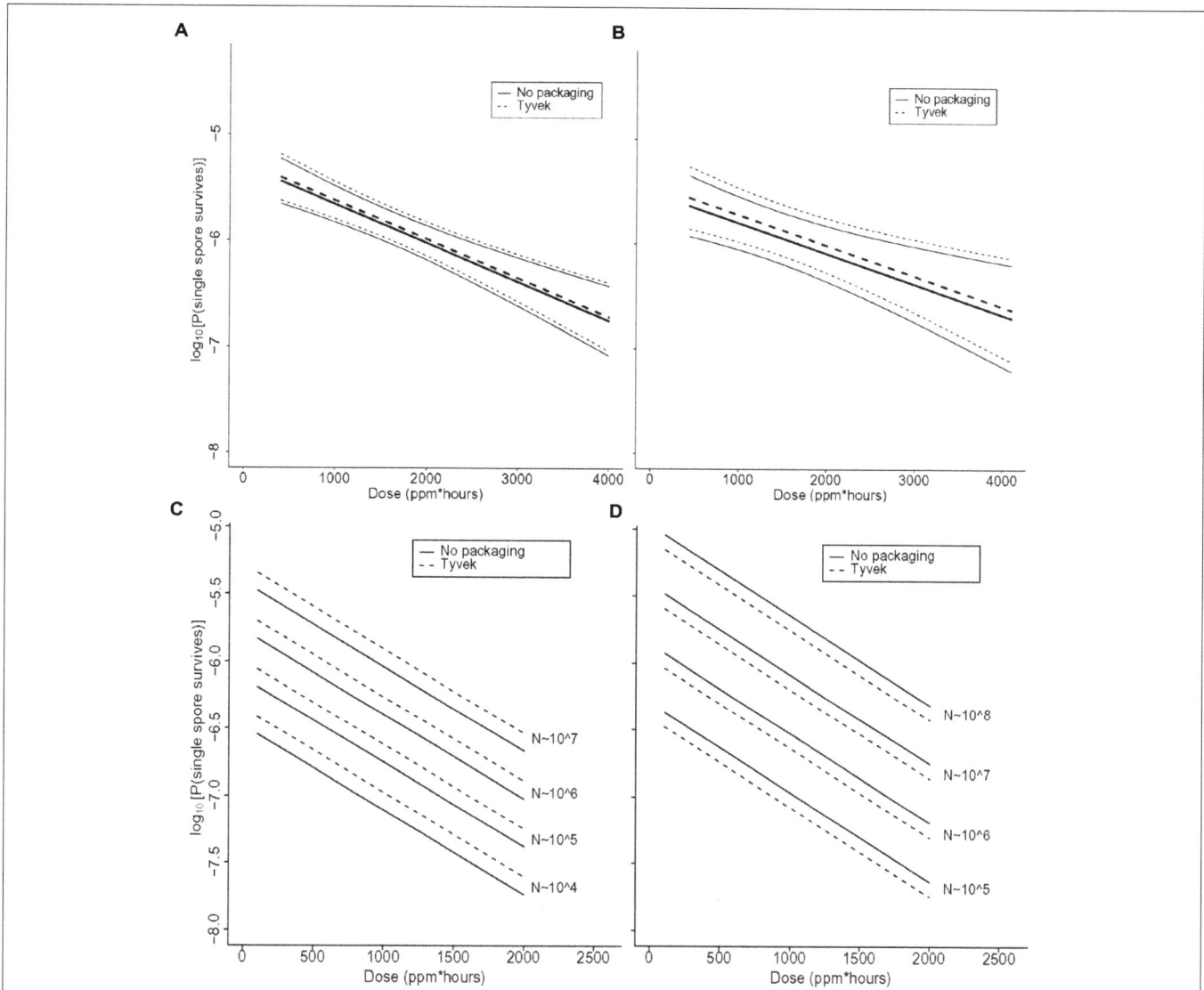

FIGURE 5 | (A) Statistical modeling showing the probability, with 95% confidence, of a single *B. thuringiensis* spore surviving a ClO$_2$ dose range of 438–4106 ppm-h at 79% RH. The spore population of each strip is fixed at 1.2×10^6, and $n = 98$ each for strips contained in Tyvek or with no packaging. In the figure, the lighter set of dashed and solid lines represent the confidence interval for spore strips packaged in Tyvek or no packaging, respectively, while the middle bold lines represent the predicted probabilities. In this model, there is strong evidence of an effect of dose ($p = 0.002$) and no evidence of an effect of packaging ($p = 0.76$). **(B)** Statistical modeling showing the probability, with 95% confidence, of a single *G. stearothermophilus* spore surviving a ClO$_2$ dose range of 438–4106 ppm-h at 79% RH. The spore population of each strip is fixed at 1.0×10^6, and $n = 80$ each for strips contained in Tyvek or no packaging. Lighter dashed and solid lines represent the confidence interval for spore strips packaged in Tyvek or no packaging, respectively, while the middle bold lines represent the predicted probabilities. There is evidence of an effect of dose ($p = 0.02$) and no evidence of an effect of packaging ($p = 0.68$).

(C) Predicted probabilities of a single spore surviving for varying numbers of *B. atrophaeus* spores (1.2×10^4 to 1.2×10^7), with a D$_{EtO}$ value of 3.1, in Tyvek ($n = 152$) and no packaging ($n = 144$) at 79% RH and ClO$_2$ dose ranging from 110 to 1991 ppm-h. There is strong evidence that increased dose decreases the probability of any strips having live spores ($p < 0.01$) and that as the number of spores on the strip increases, so does the probability of survival ($p < 0.01$). There is no evidence of a significant effect of packaging on survival of spores ($p = 0.21$), though there is some difference with a slightly higher rate of survival for spores packaged in Tyvek. **(D)** Predicted probabilities of a single spore surviving for varying numbers of *B. atrophaeus* spores (3.5×10^4 to 1.5×10^8), with a D$_{EtO}$ value of 5.0, in Tyvek ($n = 490$) and no packaging ($n = 350$) at 79% RH and ClO$_2$ dose ranging from 110 to 1991 ppm-h. These results indicate that the probability of a strip having live spores after treatment increases with decreasing dose ($p < 0.01$) and with increasing numbers of spores ($p < 0.01$). There is no evidence of an effect of packaging ($p = 0.21$), although there is a slightly lower rate of survival for spores in Tyvek packagi.

container, but the heat and pressure gently subside over the 30–60 min decontamination period. Human exposure to ClO$_2$ above permissible concentrations and durations is known to cause

irritation of the eyes, skin, nose, throat and lungs, and thus this method is typically used in outdoor environments with exposure to the sun.

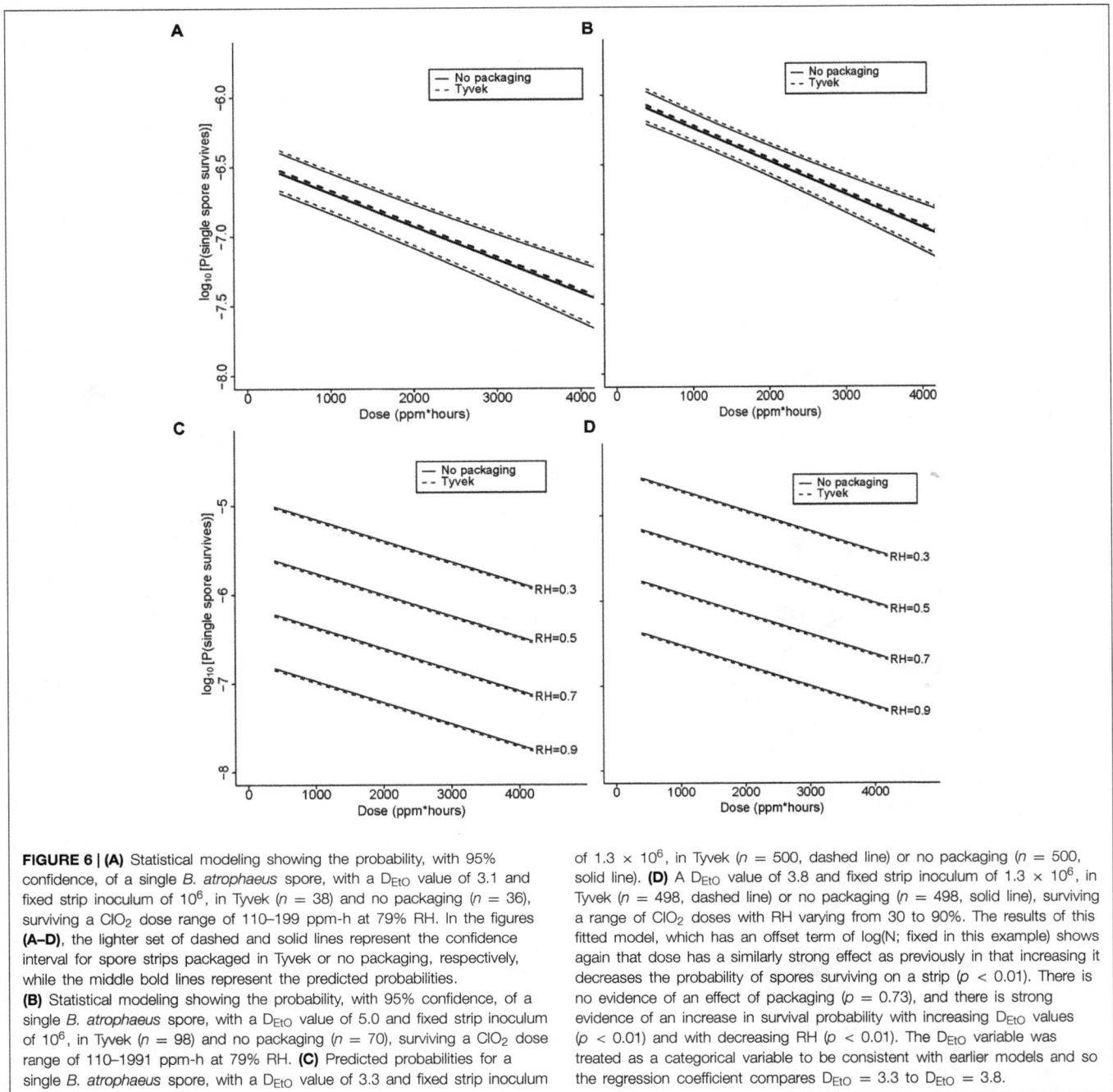

FIGURE 6 | (A) Statistical modeling showing the probability, with 95% confidence, of a single *B. atrophaeus* spore, with a D_{EtO} value of 3.1 and fixed strip inoculum of 10^6, in Tyvek ($n = 38$) and no packaging ($n = 36$), surviving a ClO_2 dose range of 110–199 ppm-h at 79% RH. In the figures **(A–D)**, the lighter set of dashed and solid lines represent the confidence interval for spore strips packaged in Tyvek or no packaging, respectively, while the middle bold lines represent the predicted probabilities. **(B)** Statistical modeling showing the probability, with 95% confidence, of a single *B. atrophaeus* spore, with a D_{EtO} value of 5.0 and fixed strip inoculum of 10^6, in Tyvek ($n = 98$) and no packaging ($n = 70$), surviving a ClO_2 dose range of 110–1991 ppm-h at 79% RH. **(C)** Predicted probabilities for a single *B. atrophaeus* spore, with a D_{EtO} value of 3.3 and fixed strip inoculum of 1.3×10^6, in Tyvek ($n = 500$, dashed line) or no packaging ($n = 500$, solid line). **(D)** A D_{EtO} value of 3.8 and fixed strip inoculum of 1.3×10^6, in Tyvek ($n = 498$, dashed line) or no packaging ($n = 498$, solid line), surviving a range of ClO_2 doses with RH varying from 30 to 90%. The results of this fitted model, which has an offset term of log(N; fixed in this example) shows again that dose has a similarly strong effect as previously in that increasing it decreases the probability of spores surviving on a strip ($p < 0.01$). There is no evidence of an effect of packaging ($p = 0.73$), and there is strong evidence of an increase in survival probability with increasing D_{EtO} values ($p < 0.01$) and with decreasing RH ($p < 0.01$). The D_{EtO} variable was treated as a categorical variable to be consistent with earlier models and so the regression coefficient compares $D_{EtO} = 3.3$ to $D_{EtO} = 3.8$.

Adapting this method for field decontamination was ideal, because it is devoid of electrical power sources, uses small quantities of water from available sources, and the low cost of the FDKs, and with the required supplies simple and easy to obtain, store, and handle for ClO_2 generation. Specifically, this method was used to develop FDKs (**Table 1**; **Figure 8B**) comprising a self-sealable bag (2 or 10 gallon capacity), dry chemical components (Part A and Part B from the commercially available CHEM-CD kit) that react in water to produce chlorine dioxide. The FDK also includes a device to measure water (50 mL tube) and

ClO_2 indicators of sterility (CD Check Strip, ClorDiSys Solutions, Inc., Lebanon, NJ, USA). In addition, laboratory testing with spore BIs (*G. stearothermophilus* for steam sterilization from BT Sure, Thermo Scientific, Marietta, OH, USA, and *B. atrophaeus* for ethylene oxide sterilization from EZ Test, SGM Biotech, Inc., Bozeman, MT, USA as discussed above) showed complete inactivation of spores (sterility) was achieved within 30 min of exposure.

Figure 8A illustrates the procedural steps carried out in utilizing the FDKs. Briefly, the ClO_2 indicators are taped to the

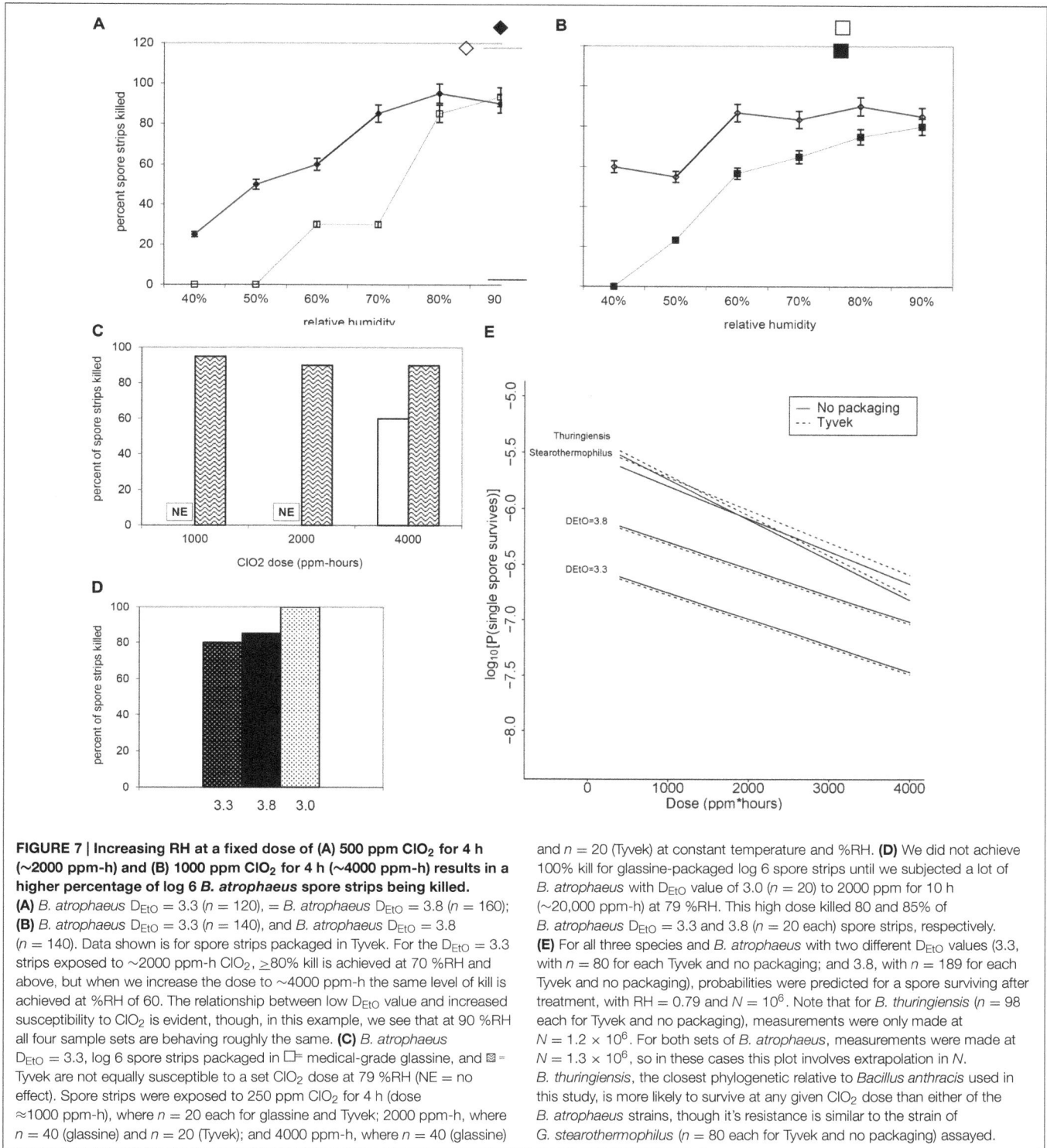

FIGURE 7 | Increasing RH at a fixed dose of (A) 500 ppm ClO$_2$ for 4 h (~2000 ppm-h) and (B) 1000 ppm ClO$_2$ for 4 h (~4000 ppm-h) results in a higher percentage of log 6 *B. atrophaeus* spore strips being killed.
(A) *B. atrophaeus* D$_{EtO}$ = 3.3 (*n* = 120), = *B. atrophaeus* D$_{EtO}$ = 3.8 (*n* = 160);
(B) *B. atrophaeus* D$_{EtO}$ = 3.3 (*n* = 140), and *B. atrophaeus* D$_{EtO}$ = 3.8
(*n* = 140). Data shown is for spore strips packaged in Tyvek. For the D$_{EtO}$ = 3.3 strips exposed to ~2000 ppm-h ClO$_2$, ≥80% kill is achieved at 70 %RH and above, but when we increase the dose to ~4000 ppm-h the same level of kill is achieved at %RH of 60. The relationship between low D$_{EtO}$ value and increased susceptibility to ClO$_2$ is evident, though, in this example, we see that at 90 %RH all four sample sets are behaving roughly the same. **(C)** *B. atrophaeus* D$_{EtO}$ = 3.3, log 6 spore strips packaged in □= medical-grade glassine, and ▨ = Tyvek are not equally susceptible to a set ClO$_2$ dose at 79 %RH (NE = no effect). Spore strips were exposed to 250 ppm ClO$_2$ for 4 h (dose ≈1000 ppm-h), where *n* = 20 each for glassine and Tyvek; 2000 ppm-h, where *n* = 40 (glassine) and *n* = 20 (Tyvek); and 4000 ppm-h, where *n* = 40 (glassine)

and *n* = 20 (Tyvek) at constant temperature and %RH. **(D)** We did not achieve 100% kill for glassine-packaged log 6 spore strips until we subjected a lot of *B. atrophaeus* with D$_{EtO}$ value of 3.0 (*n* = 20) to 2000 ppm for 10 h (~20,000 ppm-h) at 79 %RH. This high dose killed 80 and 85% of *B. atrophaeus* D$_{EtO}$ = 3.3 and 3.8 (*n* = 20 each) spore strips, respectively. **(E)** For all three species and *B. atrophaeus* with two different D$_{EtO}$ values (3.3, with *n* = 80 for each Tyvek and no packaging; and 3.8, with *n* = 189 for each Tyvek and no packaging), probabilities were predicted for a spore surviving after treatment, with RH = 0.79 and *N* = 10^6. Note that for *B. thuringiensis* (*n* = 98 each for Tyvek and no packaging), measurements were only made at *N* = 1.2 × 10^6. For both sets of *B. atrophaeus*, measurements were made at *N* = 1.3 × 10^6, so in these cases this plot involves extrapolation in *N*. *B. thuringiensis*, the closest phylogenetic relative to *Bacillus anthracis* used in this study, is more likely to survive at any given ClO$_2$ dose than either of the *B. atrophaeus* strains, though it's resistance is similar to the strain of *G. stearothermophilus* (*n* = 80 each for Tyvek and no packaging) assayed.

inner bag to be observed during the decontamination process and the chemical reagents Part A and Part B are placed in a container with water. The item to be decontaminated is placed in the bag near or over the mixture and the bag is subsequently sealed. Within a few minutes the reaction produces condensation

(humidity) and yellow ClO$_2$ gas that inflates the bag due to a mild build-up of heat and pressure, although the bag itself does *not* become turgid. After a minimum of 30 min, the color change of the indicator shows adequate concentration and exposure to ClO$_2$ gas to achieve sterility, and the bag can be opened to

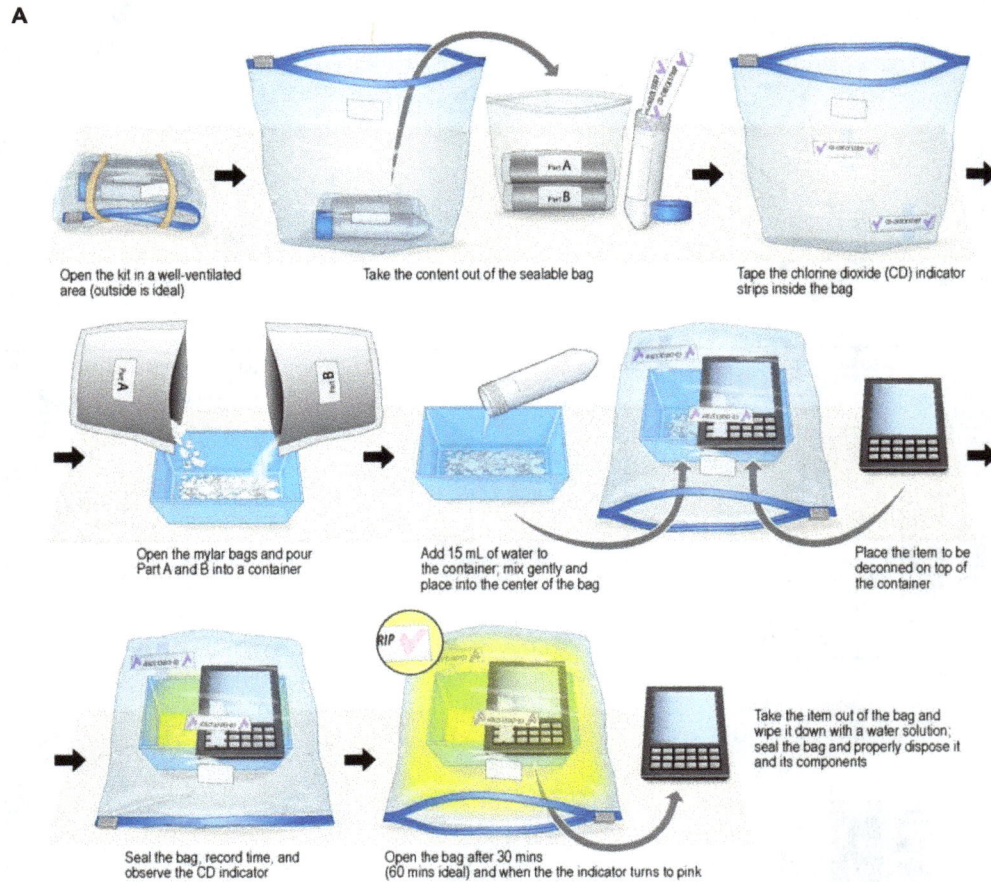

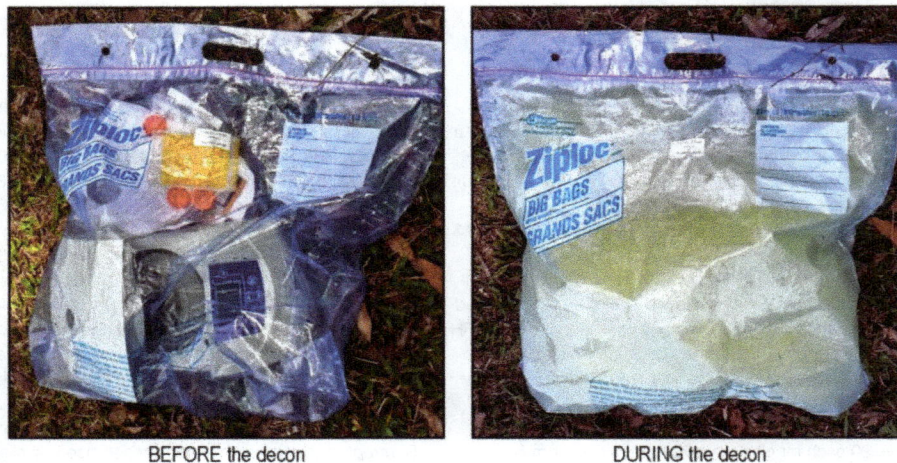

FIGURE 8 | (A) Diagrammatic representation of steps for using FDKs. Top Row. (i) Open FDK in well-ventilated area (outdoors), (ii) remove contents form sealable bag, and (iii) tape ClO_2 indictor strips to inside of bag. Middle Row: (iv) Empty foil pouches of chemical reagents (Part A and Part B) into plastic container, (v) add 15 mL of water with mixing, and (vi) place plastic container inside bag, then place contaminated medical device on top of container. Bottom Row: (vii) Seal bag, run for 30–60 min, and observe purple ⟶ pink color change of ClO_2 indicator strip, then (viii) with gloves on, open bag, remove device from bag, and re-seal bag for proper disposal. Caution: when opening bag, gas has a strong odor – avoid inhaling gas. With gloves on, use water to wipe residue from device/equipment (Illustration provided courtesy of Mr. Jiro Wada of NIH). **(B)** Images of an actual decontamination procedure (before and during) of a microcentrifuge with an FDK on the ground in Liberia (Photos courtesy of Dr. Lisa Hensley and Mr. James Pettitt of NIH/NIAID).

release the trapped gas. The equipment can then be removed and surface-cleaned, since the water, ClO$_2$, and inert salts from the spent reaction can be present on the surface. The remaining solution and materiel can be bagged and disposed of according to local policies.

More than 150 FDKs have been deployed for use by global public health organizations during the Ebola outbreaks in Western Africa, which includes Guinea, Sierra Leone, and Liberia, to protect patients and health care personnel. The FDKs have been used to decontaminate laboratory equipment, POC medical devices and personal cell phones. Overall, the performance has been acceptable and only on occasion have the bags leaked gas during the sterilization cycle, thereby requiring longer exposure times to effectuate the color change of the ClO$_2$ indicator strip and signify exposure sufficient to ensure sterility. For the purpose of field deployment, these kits are safe, compact and easy to ship to remote sites that have limited infrastructure and resources available.

3.1. Results – Laboratory Testing NIH/NIAID Field Decontamination Kits

Instructions for the FDKs involve putting Part A and Part B together into a reaction vessel, then adding water. Independent laboratory testing showed that increasing the volume of water slightly and adding the water to the reaction vessel first, followed by adding Part A (oxidant) with stirring, then Part B (reductant) with stirring helped ensure the reaction runs controllably and smoothly, while evolving the maximum amount of ClO$_2$ gas. Small FDKs consist of a 2.5 gallon sealable Mylar bag, 15 g of Part A, 4 g of Part B, and 30 mL of water mixed inside a 100 mL beaker as the reaction vessel. Large FDKs were scaled proportionately to consist of a 10-gallon bag, 60 g of Part A, 16 g of Part B, and 120 mL of water in a 600 mL beaker as the reactor vessel. In both configurations, the chemical reaction initiates slowly, then releases copious gaseous ClO$_2$ at ~2 min and 15 s after all of the reagents were combined. These reaction compositions with various permutations were run inside the different FDKs (2.5- and 10-gallon bags) and also inside the PCS, using spectrophotometry to monitor [ClO$_2$] at $\lambda = 360$ nm (the absorbance maximum of ClO$_2$) and a combination probe that measures %RH and temperature simultaneously (6621 – Commercial HVAC Temperature/RH Transmitter, Testo, Inc., Sparta, NJ, USA). In addition, the reaction was monitored with ClO$_2$ color-change indicator strips (CD-CHEK, ClorDiSys Solutions, Inc., Lebanon, NJ, USA) and commercially available spore indicators of G. stearothermophilus in Tyvek (ClorDiSys Solutions, Inc., Lebanon, NJ, USA), G. stearothermophilus for steam sterilization (BT Sure, Thermo Scientific, Marietta, OH, USA), and B. atrophaeus for ethylene oxide sterilization (EZ Test, SGM Biotech, Inc., Bozeman, MT, USA). Representative results are summarized in Table 5 using the commercial chemical sets and the three different container units (2.5-gal bag, 10-gal bag, and PCS) run for 30 min. In all instances, the CD-CHEK strips turned color (indicative of sufficient ClO$_2$ exposure for sterilization) and all of the spore bio-indicators confirmed sterility had been achieved.

TABLE 5 | Laboratory tests of FDKs and the PCS.

Test code	Container	Conditions	Observations	Microbiological Results
Test a	2.5-gal bag	- 15 g Part A - 4 g Part B - 30 mL H$_2$O (tap) in a 100 mL beaker	Reaction in 2:20 RH >96.4% $T = 30°C$ [ClO$_2$] >7000 ppm	Sterilized[a,b]
Test h	10-gal bag	- 15g Part A - 4 g Part B - 30 mL H$_2$O (tap) in a 100 mL beaker	Reaction at 2:10 RH >74% $T = 25.4°C$	Sterilized[a,b]
Test i	PCS	- 16 g Part A - 4 g Part B - 30 mL H$_2$O (tap) in a 100 mL beaker	Reaction at 2:30 RH > 90.2% $T = 24.4°C$ [ClO$_2$] >7000 ppm	Sterilized[a,b]
Test n	PCS	- 15 g Part A - 4 g Part B - 30 mL H$_2$O (tap) in a 100 mL beaker	Reaction at 2:12 RH >93.5% [ClO$_2$] >7000 ppm Run time 15 min	Sterilized[a,b]

[a,b]Sterility confirmed with G. stearothermophilus and B. atrophaeus bio-indicators, respectively.

3.2. Results – Decontamination of Textiles

Some textile technologies have certain material properties with the potential to self-decontaminate by inactivating biological hazards (B. anthracis spores) and/or chemical agents on their surfaces, thus making them a form of novel decontamination technology. Other types of textiles require external applications to remove biological and chemical hazards, such as the use of sorbents to remove chemical agents or the use of cold sterilants (bleach, chlorine dioxide, etc.) to safely inactivate biological hazards.

Natick Soldier RD&E Center's novel decontamination technologies (Table 1) have also been applied in the decontamination of textiles using live cultures of B. anthracis Delta Sterne as a surrogate for B. anthracis. Specifically, NSRDEC's "D-FEND ALL" technology was used to create 20–200 ppm ClO$_2$ solutions that completely inactivated 7.5 logs of B. anthracis Sterne inoculated onto two different types of fabrics (a nylon-cotton blend and an experimental weatherproof fabric) within 10 min (minor bleaching of the weatherproof fabric occurred at only the 200 ppm ClO$_2$ solution). Other experimental fabrics were inoculated with B. anthracis Delta Sterne spores, then subjected to either (i) 30 min rinse in bleach (5–6% aqueous OCl$^-$), (ii) gaseous ClO$_2$ using NSRDEC's NCC and PCS technologies (Table 1), or (iii) HPP. In all cases, the decontamination treatments achieved a 100% spore kill on the fabric samples (a 6.59-, 5.73-, and 6.13-log kill for bleach, gaseous ClO$_2$, and HPP, respectively). For HPP experiments, inoculated fabric samples were treated with conditions of pressure = 550 MPa, temperature = 65°C, and time = 100 min.

3.3. Results – AFM Characterization of Morphological and Structural Attributes of *Bacillus* spores

Atomic force microscopy can be used to analyze high-resolution architecture, assembly, structural dynamics, and function of dormant and germinating spores of various wild type and mutant

Bacillus (Plomp et al., 2005a,b,c, 2007a, 2014; Carroll et al., 2008; Ghosh et al., 2008; Plomp and Malkin, 2009; Malkin and Plomp, 2010; Malkin, 2011; Elhadj et al., in preparation) and *Clostridium* species (Plomp et al., 2007b). Specifically, AFM has been used to directly visualize and analyze spore morphological, dimensional, and high-resolution coat structural attributes, and these results demonstrated that spore morphological and coat structures are phylogenetically (Plomp et al., 2005a,b,c, 2014) and growth medium (Malkin, 2011; Elhadj et al., in preparation) dependent. We have found that strikingly different species-dependent spore coat structures are a consequence of nucleation and crystallization mechanisms that regulate the assembly of the outer spore coat (Plomp et al., 2005a,b,c, 2007b; Malkin and Plomp, 2010). Spore morphological, dimensional, and structural attributes could serve as a baseline for assessing the effects of sterilization and/or decontamination treatments on the morphological and structural integrity and the ultra-structural damage of spores treated by irradiation or ClO$_2$.

3.3.1. AFM Characterization of γ-Irradiated *Bacillus* Spores

3.3.1.1. *Native air-dried Bacillus spore*

Size distributions (spore height/width and length) from large populations (several 100s) of air-dried solution- and agar-grown spores were determined for spores of *B. atrophaeus*, *B. thuringiensis*, *B. subtilis* (Plomp et al., 2005a) and *B. anthracis* spores (Elhadj et al., in preparation), with more than 30 spore preparations being utilized for the assessment of *B. anthracis* spore dimensional attributes (Elhadj et al., in preparation). Representative results are compiled in **Table 6**.

Typical AFM images of *B. anthracis* Sterne and *B. thuringiensis* spores are presented in **Figures 9A,B**, respectively. The corresponding optical phase microscopy is shown in (**Figures 9C,D**), demonstrating that these spores are phase bright (refractile, ungerminated). The vast majority of spores (**Figures 9A,B**) are intact with heights within ranges indicated in **Table 6** and exhibiting surface ridges (some indicated with white arrows) extending along the long axis of the spore. These ridges are the characteristic attribute of air-dried *Bacillus* spores (Plomp et al., 2005a,b,c, 2014) and appear due to coat folding caused by changes in spore size upon dehydration (Driks, 2003; Westphal et al., 2003; Plomp et al., 2005a). Occasionally, collapsed spores (as one indicated with a black arrow in **Figure 9A**) with heights in the range of 200–500 nm are observed in the spore preparations.

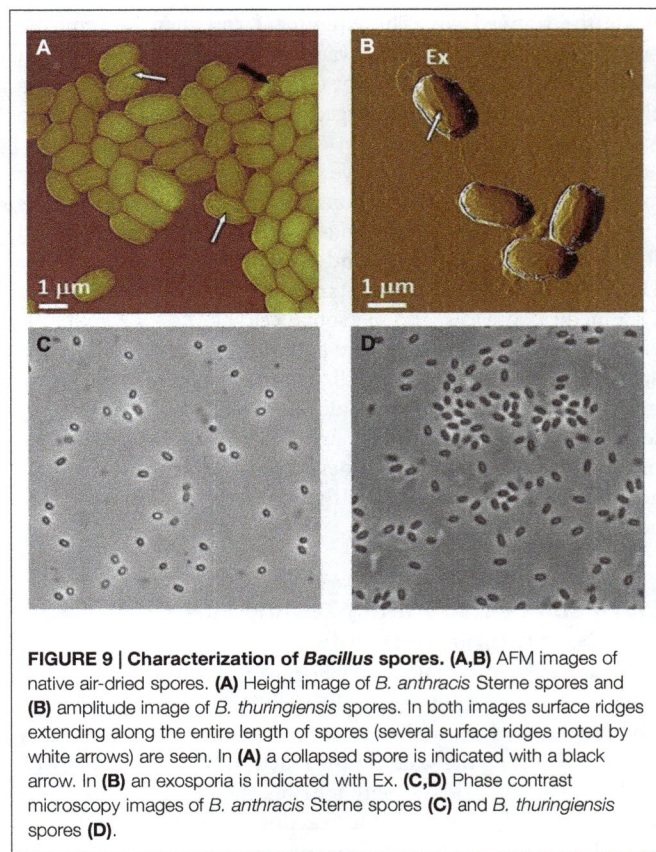

FIGURE 9 | Characterization of *Bacillus* spores. (A,B) AFM images of native air-dried spores. **(A)** Height image of *B. anthracis* Sterne spores and **(B)** amplitude image of *B. thuringiensis* spores. In both images surface ridges extending along the entire length of spores (several surface ridges noted by white arrows) are seen. In **(A)** a collapsed spore is indicated with a black arrow. In **(B)** an exosporia is indicated with Ex. **(C,D)** Phase contrast microscopy images of *B. anthracis* Sterne spores **(C)** and *B. thuringiensis* spores **(D)**.

This phenomenon could be attributable to partial germination, and subsequent partial collapse of the germinated spore upon air drying due to germination-induced internal structural changes.

3.3.1.2. *γ-Irradiated Bacillus spores*

Dauphin et al. (2008) reported that subjecting virulent *B. anthracis* spores at a concentration of 10^7 CFU/ml to a dose of 2.5 × 10^6 rads results in complete spore inactivation (sterilization). The spore images shown below in **Figures 10** and **11** were certified as sterilized for shipping and likely to have received a standard dose higher than 2.5 × 10^6 rads to assure sterility. Specifically, we characterized γ-irradiated *B. anthracis* Ames spore samples produced on nutrient sporulation medium (NSM) agar, Nutrient agar-BBL, Mueller Hinton-BBL, and brain heart infusion (BHI)-BBL using AFM imaging and demonstrated that upon dehydration, as illustrated in **Figure 10**, the architecture of these spores collapsed. AFM examination of irradiated spores prepared as air-dried samples revealed further pronounced morphological and structural differences from native spores (**Figures 9A,B**). As illustrated in **Figure 11**, air-dried samples of γ-irradiated spores comprised partially collapsed spores (PCSs), spore coat remnants (SCRs), and exosporia remnants (ERs). The heights of PCS were in the range of 400–600 nm, which is significantly lower than the height of an air-dried non-irradiated native spores (**Table 6**). The thickness of SCR and ER was in the range of 100–250

TABLE 6 | Spore height determinations.

Spore species	Spore height air-dried (solution-grown)	Spore height air-dried (agar-grown)
*Bacillus thuringiensis**	750–1000 nm avg ≈ 872 nm (AD = 5.4%)	740–1080 nm average ≈ 937 nm (AD = 5.3%)
B. anthracis‡	800–880 nm average ≈ 835 nm (AD = 5.4%)	750–800 nm average ≈ 780 nm (AD = 5.4%)

*Plomp et al., 2005a; ‡Elhadj et al., in preparation.

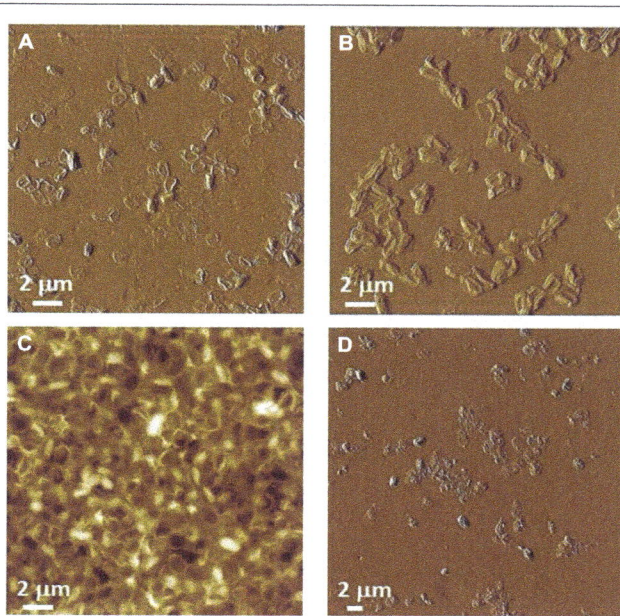

FIGURE 10 | Atomic force microscopy (AFM) images γ-irradiated (standard dose ≥ 2.5 × 10^6 rads to assure sterility) air-dried *B. anthracis* Ames spores. **(A)** Spores produced on Mueller Hinton-BBL; **(B)** spores produced on NSM agar; **(C)** spores produced on nutrient agar-BBL; and **(D)** spores produced spores produced on BHI-BBL.

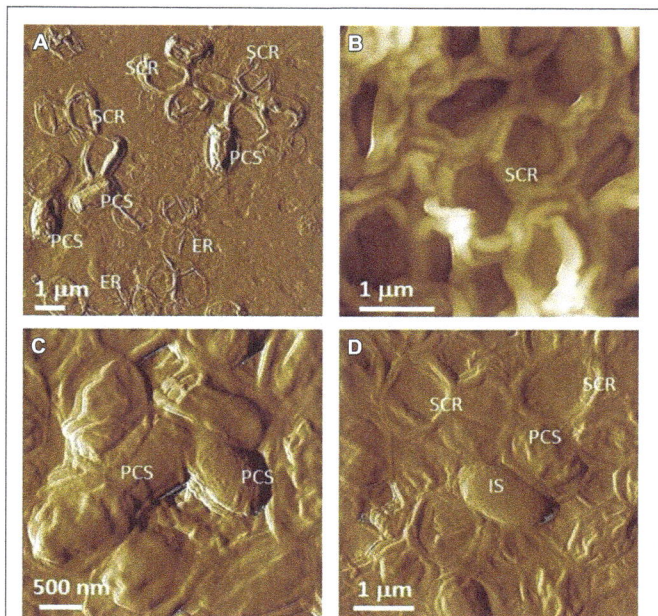

FIGURE 11 | Atomic force microscopy examination of γ-irradiated (≥2.5 × 10^6 rads) *B. anthracis* Ames spore air-died samples. **(A)** Spores produced on Mueller Hinton-BBL; **(B)** spores produced on NSM agar; **(C,D)** spores produced on nutrient agar-BBL. **(A–D)** are amplitude and height AFM images respectively.

and 40–45 nm respectively, which are comparable with the dimensions of native spore coats and exosporia. Only a small proportion of the air-dried samples of γ-irradiated *B. anthracis* Ames spores were present as intact spores (IS, **Figure 11D**) exhibiting heights equal to non-irradiated air-dried *B. anthracis* spores.

These studies demonstrated that exposure to sterilizing γ-irradiation produced profound structural changes in *B. anthracis* spores. Irradiation damages spore internal structural integrity (membranes, cortex etc.) and causes either partial (PCS) or complete (SCR) evacuation of the spore core. It is likely that in the hydrated sample, internal spore components that have sustained damage from the irradiation treatment (protein, DNA, ribosomes, small molecules, etc.) have partially/or completely diffused from the spore core into the bulk liquid phase. Note, that the leakage of spore core contents into the bulk media could also adversely affect biochemical and chemical analysis of the collected irradiated sample. Because of the significant internal structural damage induced by irradiation, the dehydration of irradiated spores suspended in liquid resulted in their collapse (**Figures 10** and **11**).

There was an excellent cross-correlation between phase contrast optical microscopy and AFM in the characterization of the irradiated spore samples. Thus, as seen in **Figure 12**, phase contrast microscopy demonstrated that the vast majority of spores in irradiated samples were either phase dark, which corresponds to the evacuation of the spore coat, or spore ghosts, which corresponds to SCRs.

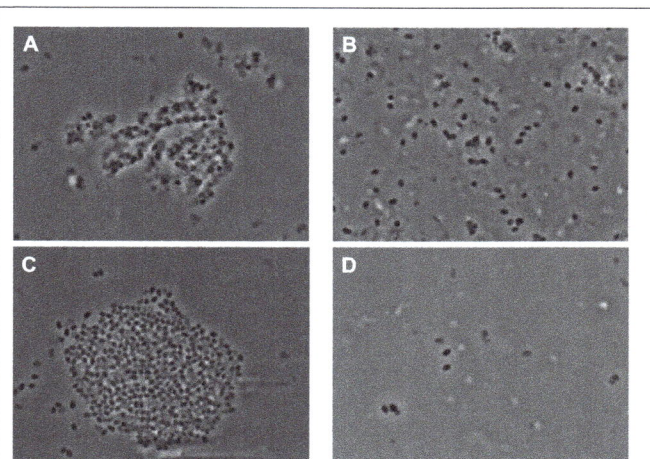

FIGURE 12 | Phase contrast microscopy images of γ-irradiated (≥2.5 × 10^6 rads) *Bacillus anthracis* Ames spores (A) produced on Mueller Hinton-BBL, (B) produced on NSM agar, (C) produced on nutrient agar-BBL, and (D) produced BHI-BBL.

With AFM of air-dried spores for γ-irradiated *B. thuringiensis* spores, similar types of ultrastructural damage and collapse have been observed (**Figure 13**). These samples comprise PCS with heights in the range of 30–40% of the height of non-irradiated native air-dried *B. thuringiensis* spores (**Table 6**, Plomp et al., 2005a).

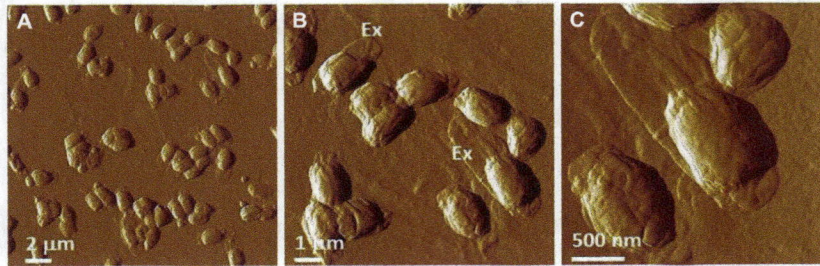

FIGURE 13 | AFM images of γ-irradiated (≥2.5 × 10⁶ rads) _B. thuringiensis_ spores in (A) a low magnification AFM image. In a higher magnification AFM image **(B)**, exosporia (denoted with Ex) are observed. Image **(C)** is a close-up image taken of the lower right area in panel **(B)**.

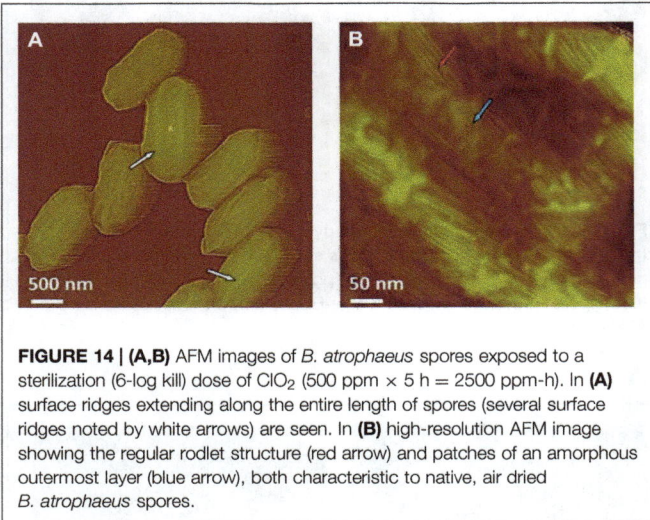

FIGURE 14 | (A,B) AFM images of _B. atrophaeus_ spores exposed to a sterilization (6-log kill) dose of ClO₂ (500 ppm × 5 h = 2500 ppm-h). In **(A)** surface ridges extending along the entire length of spores (several surface ridges noted by white arrows) are seen. In **(B)** high-resolution AFM image showing the regular rodlet structure (red arrow) and patches of an amorphous outermost layer (blue arrow), both characteristic to native, air dried _B. atrophaeus_ spores.

3.3.2. AFM Characterization of Chlorine Dioxide-Treated _Bacillus_ Spores

Chlorine dioxide is sporicidal and a well-known decontamination regime for spores of _B. anthracis_. As illustrated in **Figure 14**, _B. atrophaeus_ spores exposed to sporicidal levels of chlorine dioxide (dose = 2500 ppm-h) remain intact upon air-drying. The height of the chlorine dioxide-treated spores varies in the range of 500–650 nm, which corresponds well with the height of native air-dried _B. atrophaeus_ spores (Plomp et al., 2005a). Similarly, the morphology of ClO₂-treated _B. atrophaeus_ spores is indistinguishable from the morphology of untreated spores with pronounced surface ridges seen in **Figure 9A**. Furthermore, ClO₂ is a selective oxidant affecting cysteine, tryptophan and tyrosine amino acids in proteins and is not expected to grossly alter spore coat ultrastructure. Indeed, the high-resolution spore coat architecture and topology of ClO₂-treated spores is unaltered in character from those of native air-dried spores (Plomp et al., 2005a,b,c, 2014), with rodlet coat structures and patches of an amorphous outermost layer clearly seen (**Figure 14B**).

3.3.3. AFM Characterization of Spore Responses in the Fully Hydrated vs. Air-Dried State

Atomic force microscopy allows a direct comparison of fully hydrated and air-dried native spores visualized under water and in air, respectively. Particularly, AFM studies of fully hydrated _B. atrophaeus_ (Plomp et al., 2005a, 2007a), _B. anthracis_ (Plomp and Malkin, unpublished results) and _Clostridium novyi_ NT spores (Plomp et al., 2007b) demonstrated that high-resolution attributes (i.e., rodlet, honeycomb, and inner coat layer structures), maintained the same patterns, lattice periodicities, and step heights as seen with air-dried spores (Plomp et al., 2005a,b,c, 2007b, 2014; Elhadj et al., in preparation). Furthermore, the ability of the coat to fold and unfold concomitant with changes in spore size was suggested (Driks, 2003; Westphal et al., 2003) based on measurements of _B. thuringiensis_ spore dimensions induced by humidity transients. Images of a fully hydrated _B. atrophaeus_ spores are presented in **Figure 15A**. Surface ridges, the prominent structural features of air-dried spores (**Figure 9**), are typically absent from the surface of fully hydrated spores (**Figure 15A**). The direct visualization of 35 individual spores was performed to probe the dynamic response of their aqueous to aerial phase transition (Plomp et al., 2005a). Spores were visualized under water, then air-dried for ∼40 h and imaged in air (65% RH), then re-imaged after rehydration. Typical examples of hydration/dehydration ultrastructural transitions are presented in **Figures 15A,B**. As illustrated in **Figure 15A**, the coat of a fully hydrated spore appears to be tightly attached to the cortex and upon dehydration (**Figure 15B**) forms an ∼50 nm surface ridge/fold extending along the entire length of the spore.

The changes in spore surface architecture with dehydration were accompanied by a decrease in spore size. As illustrated in **Figure 15C**, the average width/height of 35 individual air-dried spores was reduced to 88% of the size measured for fully hydrated spores. The rehydration of air-dried spores, by placing them in water for 2 h, restored the spores to 97% of their original size, thereby establishing the reversibility of the size transition concurrent with rehydration (**Figure 15C**).

The observed decrease in the width of bacterial spores with dehydration is apparently due to the contraction of

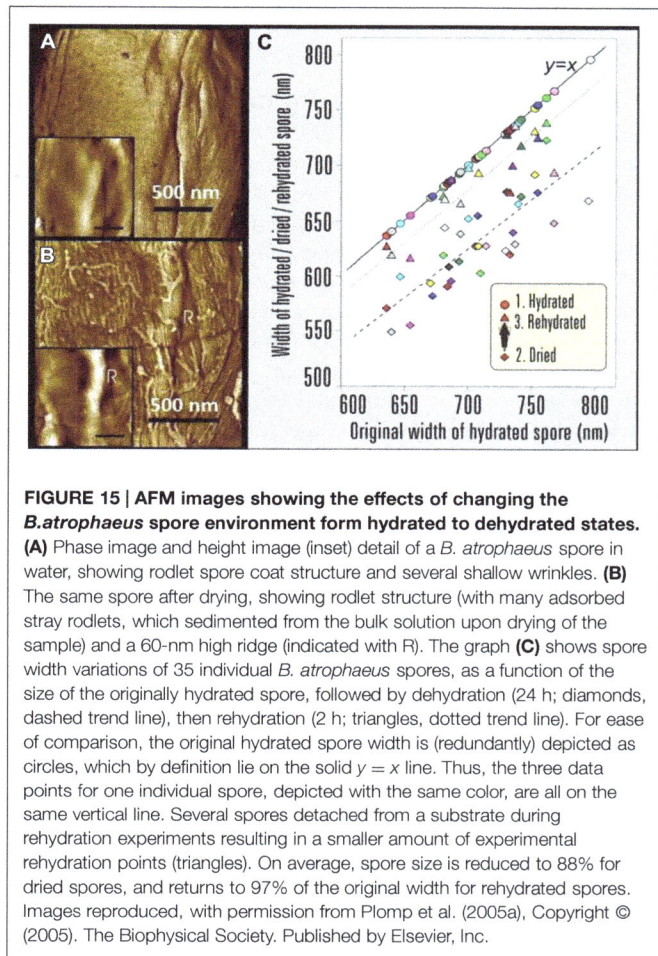

**FIGURE 15 | AFM images showing the effects of changing the
B.atrophaeus spore environment form hydrated to dehydrated states.
(A)** Phase image and height image (inset) detail of a *B. atrophaeus* spore in
water, showing rodlet spore coat structure and several shallow wrinkles. **(B)**
The same spore after drying, showing rodlet structure (with many adsorbed
stray rodlets, which sedimented from the bulk solution upon drying of the
sample) and a 60-nm high ridge (indicated with R). The graph **(C)** shows spore
width variations of 35 individual *B. atrophaeus* spores, as a function of the
size of the originally hydrated spore, followed by dehydration (24 h; diamonds,
dashed trend line), then rehydration (2 h; triangles, dotted trend line). For ease
of comparison, the original hydrated spore width is (redundantly) depicted as
circles, which by definition lie on the solid $y = x$ line. Thus, the three data
points for one individual spore, depicted with the same color, are all on the
same vertical line. Several spores detached from a substrate during
rehydration experiments resulting in a smaller amount of experimental
rehydration points (triangles). On average, spore size is reduced to 88% for
dried spores, and returns to 97% of the original width for rehydrated spores.
Images reproduced, with permission from Plomp et al. (2005a), Copyright ©
(2005). The Biophysical Society. Published by Elsevier, Inc.

the spore core and/or cortex. The spore coat itself does not
shrink/expand but is flexible enough to compensate for the
internal volume decrease of core/cortex compartments by surface
folding and the formation of ridges. These studies establish that
dormant spores are dynamic physical structures, which exhibit
profound morphological and structural responses with changes
in its natural environment. These changes play important
roles in selecting and implementing successful decontamination
strategies.

3.4. Bio-indicators for the Inactivation of *Bacillus spp.* Spores by ClO₂ Gas

As a gas-phase sporicide, ClO_2 has advantages over other gas-
phase sterilizing agents that leave residues or may pose health
hazards (Bruch, 1973; Dolovich et al., 1984; Chapman et al., 1986;
Lettman et al., 1986; Lykins et al., 1994; Muttamara et al., 1995).
The efficacy of ClO_2 decontamination treatments are typically
verified through the use of spore BIs. We present extensive
statistical analysis and modeling results that can be used to
predict survival probabilities for three species of spores at varying
doses of ClO_2 by analyzing its effectiveness to kill 6 logs of
bacterial spores (sterilize). Our analysis includes four lots of
B. atrophaeus spore strips with different Decimal reduction (D)

values, which are the times to reduce surviving spores by 90%, for
ethylene oxide (D_{EtO}) to allow their use as BI's to monitor process
efficacy in large-scale emergency decontamination efforts, as well
as the spores of *G. stearothermophilus* and *B. thuringiensis*, a close
relative of *B. anthracis*. Parameters contributing to the efficacy
of spore sterilization by ClO_2 include, but are not limited to,
ClO_2 concentration, RH, temperature, exposure time, and the BI
packaging materials.

The data in **Figures 5A–D** clearly demonstrate that the
probability, with a confidence level of 95%, of a spore surviving
greatly diminishes as dose increases. In fact, the correlation
between increasing [ClO_2] and sporicidal activity is strong for
all species and lots tested. At a constant spore strip population
(1.2×10^6), exposure time, and %RH (79%), the percent of
spore strips killed increases (probability of a spore surviving
decreases) as a function of increasing dose (dose $\equiv$ ppm
$ClO_2 \times$ exposure time) as the gas concentration increases
(**Figures 5A,B**). Similarly, we have also found that for a constant
[ClO_2], RH, and spore population, increasing the dose by
increasing the exposure time also results in higher levels of kill
(lower probability of survival, **Figures 5C,D**).

A particular point of interest is the dose at which the predicted
spore survival probability is at most 10^{-6} (the EPA target)
with 95% confidence. If we consider species *B. atrophaeus* as
an example, for which there is a D_{EtO} value of 3.1 or 5.0,
an RH of 0.79, and take the number of spores to be 10^6
(not stratifying by packaging), then the dose at which a 6-log
drop is achieved is, with 95% probability, between 268 and
592 or between 628 and 822 ppm-h, respectively (data not
shown), which is significantly less than the dose of 2500 ppm-h
required in some whole building decontaminations. This suggests
that a lower D_{EtO} value correlates with a greater susceptibility
to ClO_2, a phenomenon that we also (see graphically in
Figures 6A–D). Though *B. atrophaeus* spores are accepted as
a standard test organism for sterility (Rosenblatt et al., 1987),
the strains of *G. stearothermophilus* and *B. thuringiensis* spores
assayed required higher ClO_2 doses (at 0.79 RH), between
1235–2415 and 1497–2063 ppm-h, respectively, to achieve the
same spore survival probability ($\leq 10^{-6}$) with 95% confidence
(**Figures 5A,B**).

With [ClO_2] $=$ 500 or 1000 ppm and exposure time fixed
at 4 h, we varied the %RH (40–90%) of the exposure chamber
to assess the bioeffects of humidified ClO_2 on two lots of
B. atrophaeus BI's with D_{EtO} values of 3.3 and 3.8 (**Figures 6C,D**,
respectively). We found, in general, that increasing %RH at a
given dose increases the level of kill (**Figures 7A,B**). Additionally,
to achieve equivalent levels of kill at a fixed ClO_2 dose, higher
RH is required for *B. atrophaeus* strips with a higher D_{EtO} value
(**Figures 7A,B**). In general, we found that RH levels of $\geq 80\%$
are optimal for ClO_2 sporicidal bioeffects over a wide range of
doses.

The humidity dependence of ClO_2 sterilization may be
twofold. Increasing RH results in an increase in localized sorption
of water molecules on the spore surface, causing the spore
to swell and hypothetically resulting in increased pore size
through which ClO_2 can pass (Westphal et al., 2003). ClO_2
partitioning from the gas phase into aqueous solution is about

1:40, and it has been reported that ClO_2, a radical, is a dissolved gas in solution (Aieta and Berg, 1986; Simpson et al., 1993). Therefore, as RH increases, thus does the local concentration of ClO_2, and the spore becomes more swollen, perhaps more readily passing water and its associated dissolved ClO_2 gas through opened channels. Other authors have reported a similar relationship between increased RH during ClO_2 treatment with increased mortality of vegetative bacterial cells (Han et al., 1999, 2001).

We consistently observed no statistically significant difference in spore killing between BI's packaged in 1059B medical grade Tyvek versus no packaging (**Figures 5–7**). Identical lots of *B. atrophaeus* log 6 spore strips, 1161841 and 1161911, with D_{EtO} values of 3.3 and 3.8, respectively, and packaged by the manufacturer in their standard and industry-accepted medical grade glassine, yielded grossly inconsistent results when exposed to a wide range of ClO_2 doses (data not shown) – at 2000 ppm-h and 80% RH, 90% of the spore strips (lot 1161841) were killed when packaged in Tyvek and there was no effect on the strips packaged in glassine (**Figure 7C**). A much greater dose (~20,000 ppm-h) was required to kill 100% of the *B. atrophaeus* strips packaged in glassine (**Figure 7D**). **Figure 7E** shows predicted probabilities of a spore remaining live after treatment, for Tyvek and no packaging, for all three species (two different D_{EtO} values for *B. atrophaeus*) and with RH = 0.79 and the number of spores fixed at $N = 10^6$.

4. Bacterial Spore Properties – Mechanisms of Resistance and Killing

The novel ClO_2 decontamination technologies mentioned above are laboratory inventions that were patented, commercialized, and adapted for actual field use in austere environments (e.g., far-forward military deployments or global humanitarian relief in third-world countries, etc.). The FDKs were adapted from the COTS version of NSRDEC's decontamination technologies and became an important contributor for global public health organizations (MSF, WHO, Public Health Canada, NIH) in sterilizing Ebola-contaminated medical equipment and preventing the spread of disease at remote clinical sites in West Africa. All decontamination technologies have to be validated for efficacy in the laboratory and during their actual field use. Lethal chemical agents typically inactivate viruses by damaging the protein capsid or DNA (Kingsley et al., 2014). While the Ebola virus itself is an enveloped virus and relatively fragile, the virus is classified as Biosafety level 4 (BSL-4) due to the severity of disease it causes in humans, and presently there is no standard test assay for the Ebola virus under representative conditions even in a high-level containment facility. Bacterial spores, therefore, provide the standard assay for assuring sterility with deployed FDKs and other decontamination technologies, because of the extreme resistance of bacterial spores to chemical (and other) decontamination methods. Consonant thereto, we review bacterial spore properties, structures, and resistance mechanisms and focus on the mechanisms through which

ClO_2 inactivates bacterial spores as the indicators of efficient bio-decontamination.

4.1. Bacterial Spores – Background

Members of bacteria of *Bacillus* and *Clostridium* species and their close relatives can form metabolically dormant spores, generally when the environment no longer can support growth (Setlow, 2006; Leggett et al., 2012; Setlow and Johnson, 2012). Spores are extremely resistant to all manner of harmful treatments, and can remain dormant for years. However, if given the proper stimuli, generally nutrients such as amino acids or sugars, spores can return to life in minutes and outgrow into vegetative cells (Setlow, 2013). During subsequent growth, the growing or stationary phase cells of some spore-formers can secrete toxins or deleterious enzymes, and these agents can cause food spoilage or food poisoning and other human or animal diseases. As a consequence, there is much interest in the mechanisms of spore resistance to and the killing of spores by different treatments including high pressure, radiation, chemicals, plasma, or heat. One factor that is often overlooked in thinking about these mechanisms is the significant heterogeneity in properties of individual spores in genetically homogeneous spore populations, and this is seen in levels of spore resistance and in rates of spore germination (Setlow et al., 2012). The reasons for some of this heterogeneity are known, in particular some of the causes of the variability in rates of germination between individual spores. However, the reasons for the variability in resistance properties between individual spores in populations are not clear.

4.2. Spore Structure, Components, Properties, and their Role(s) in Spore Resistance

A number of novel spore structures and many spore-specific components and features are crucial for imparting various spore resistance properties (**Table 7**). Spores of all species appear to have a generally similar basic structure (**Figure 16**), with the exception of the outermost exosporium layer that is present in spores of some species, but not all (Gerhardt, 1967; Henriques and Moran, 2007; Driks, 2009; McKenney et al., 2013). The large balloon-like exosporium is composed of proteins, sugars, and lipid, and it is present in many spore formers that can cause disease, including the members of the *Bacillus cereus* group and *Clostridium difficile*, but is absent from the model spore former *B. subtilis*. This structure appears to be important in spore adhesion properties, and in restricting access of antibodies to the spore coat layer below the exosporium, but plays no major role in most spore resistance properties (Setlow, 2006; Henriques and Moran, 2007). The exosporium may also play some role in virulence of spores, but this role is not yet clear.

Moving from the exterior inward, the spore coat is found under the exosporium, and is composed of many spore-specific proteins assembled in a number of layers (Henriques and Moran, 2007; McKenney et al., 2013). The coat can contain up to 50% of total spore proteins, and protects more inner spore layers such as peptidoglycan from attack by lytic enzymes such as lysozyme, and also by lytic enzymes of predatory eukaryotes. The coat is also probably crucial in the resistance of spores to a variety of chemicals, including chlorine dioxide

TABLE 7 | Mechanisms of spore killing by and resistance to various agents*.

Type of agent	Mechanisms of killing	Mechanisms of resistance
HPP	Probably protein damage	Low core water content
UV, γ-radiation, nitrite, formaldehyde	DNA damage	α/β-type SASP, DNA repair, and perhaps IM impermeability
Oxidizing agents (OCl^-, ClO_2, O_3)	IM damage	Spore coat/outer membrane
H_2O_2	Probably core protein damage	α/β-type SASP, low core water
OH^-, wet heat, some oxidizing agents	Inability to germinate	Spore coat/outer membrane
Strong mineral acid	Explosive rupture of IM	Not studied
Plasma	Protein or DNA damage[1]	α/β-type SASP, spore coat, DNA repair
Wet heat	Protein damage	α/β-type SASP, low core water content
Dry heat	DNA damage	α/β-type SASP, DPA, DNA repair

*Data for these conclusions are from Gerhardt and Marquis (1989), Setlow (1994,2001,2006; 2007a), Coleman et al. (2007, 2010), Leggett et al. (2012), Setlow and Johnson (2012), Reineke et al. (2013a,b), Moeller et al. (2014), Setlow et al. (2014).
[1]Plasma with UV-C kills spores by DNA damage, while UV-C free plasma likely kills by protein damage.

(ClO_2) and sodium hypochlorite (bleach, OCl^-), that are widely used for spore decontamination. As a consequence, spores that have defective coats because of mutation(s) or chemical de-coating are sensitized to these oxyhalogens and to many other types of chemicals, although not to all. Indeed, the coat plays only a minor role in spore resistance to chemicals such as H_2O_2, nitrous acid, acid and alkali, and DNA alkylating agents (**Table 7**). In a few cases, enzymes present in spore coats, superoxide dismutases (SODs catalyze the dismutation of superoxide, O_2^-, to O_2 or to H_2O_2) and catalases (catalyze

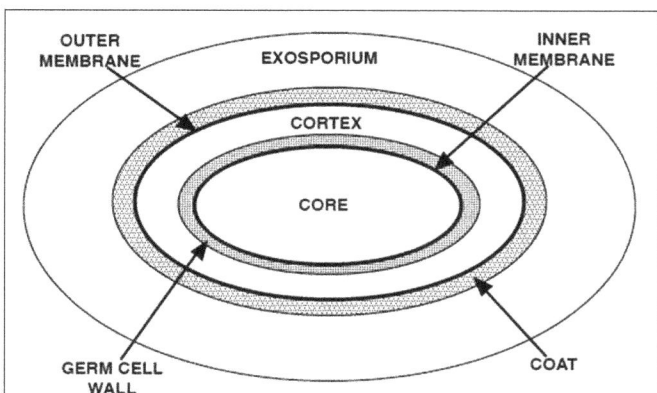

FIGURE 16 | Spore structure. The various layers of a typical spore are shown in schematic form, and the various spore layers are not drawn to scale. Note that the exosporium is not present in spores of all species.

the disproportionation of H_2O_2 to O_2 and H_2O), have been shown to play roles in spore resistance to some oxidizing agents, both *in vitro* and in *in vivo*-like environments (Henriques and Moran, 2007; Cybulski et al., 2009; McKenney et al., 2013).

Beneath the spore coat is found the outer membrane, which is a critical element in spore formation. While it is possible that the outer membrane is a permeability barrier in dormant spores (Gerhardt and Black, 1961; Rode et al., 1962), this role is not clear, and there are no mutants that specifically affect the spore's outer membrane. The outer membrane is also lost either partially or completely in spores with defective coats, and the outer membrane is also removed by chemical de-coating treatments (Buchanan and Neyman, 1986). In general, the outer membrane specifically is thought not to play a major role in spore resistance properties, but it is actually difficult to separate the roles of the outer membrane and coat in determining spore resistance properties.

Under the outer membrane are two layers of peptidoglycan (PG); first, the large spore cortex, then the thinner germ cell wall that comprises the minority of total spore PG (Popham, 2002). The germ cell wall PG has a structure that appears identical to that of growing cell wall PG structure, and cortex PG has a structure similar to that of the germ cell wall PG, but with several cortex-specific modifications. As a consequence, in the process of spore germination, cortex-lytic enzymes (CLEs) degrade cortex PG, while leaving the germ cell wall intact and allowing this structure to form the cell wall of the outgrowing spore (Setlow, 2013). The cortex's main role in spore resistance appears to be to maintain and possibly help establish the extremely low water content in the central spore core (25-50% of wet wt) that is crucial in spore resistance to wet heat (**Table 7**) and probably in spore dormancy (see below). There are a number of *B. subtilis* mutants that affect cortex PG structure. Some changes in cortex PG are associated with changes in spore wet heat resistance, and, in some cases, changes in cortex PG are associated with changes in spore core water content (Zhang et al., 2012).

Under the germ cell wall is found the second spore membrane called the inner membrane (IM). This IM has a phospholipid and fatty acid composition similar to the plasma membrane of the growing or sporulating cell (Griffiths and Setlow, 2009). However, lipid probes in the IM are immobile, suggesting the IM is not as fluid as membranes in growing cells, although IM fluidity is restored when spores complete germination and the core expands (Cowan et al., 2004). This expansion takes place in the absence of any ATP synthesis as the volume enclosed by the IM increases more than twofold. As with IM lipid mobility, IM permeability also rises dramatically, when the IM enclosed volume increases during spore germination. The low permeability of the dormant spore IM appears to contribute significantly to spore resistance to hydrophilic chemicals, as changes in dormant spore IM permeability due to changes in sporulation conditions, primarily temperature, parallel changes in IM permeability (Cortezzo and Setlow, 2004). It has been suggested that the IM is held in some sort of compressed state in the dormant spore, and this suggested structure is consistent with the extremely low permeability of the IM observed in dormant spores, even to a hydrophobic molecule such as methylamine, and perhaps even to water (Setlow and Setlow, 1980; Swerdlow et al., 1981; Cortezzo and Setlow, 2004; Sunde et al., 2009; Ghosal et al., 2010; Kaieda et al., 2013; Kong et al., 2013). Unfortunately, the precise structure of dormant spores' IM and how this structure influences IM permeability are still not clear. This is important information, because a number of key proteins in spore germination are in the IM (Setlow and Johnson, 2012; Setlow, 2013), and IM properties will most likely influence the function of these germination proteins.

The final spore layer is the central core, the site of spore DNA, RNA, and most spore enzymes. As noted above, the core has low water content, and this undoubtedly is the reason that soluble, mobile and active proteins in growing cells are immobile and inactive in the spore core (Gerhardt and Marquis, 1989; Setlow, 1994; Cowan et al., 2003). Indeed, while the dormant spore core has enzymes such as catalases and superoxide dismutases that play major roles in the resistance of growing cells to oxidizing agents, these enzymes play no role in dormant spore resistance, presumptively because they are inactive in the environment of the dormant spore core (Setlow, 2006; Leggett et al., 2012). The low core water content is most likely the major factor in spore resistance to wet heat (Gerhardt and Marquis, 1989). There are also at least three other unique spore core features to consider. First, the core pH is ~1.5 units lower than the pH in a growing cell or in the mother cell compartment of the sporulating cell (Setlow and Setlow, 1980). The importance of this low core pH in spore resistance is not clear, since the core pH can be elevated 1.5 units with no effects on spore dormancy or resistance (Swerdlow et al., 1981). However, the decrease in spore core pH during sporulation appears to be important in the modulation of enzyme activity in the developing spore late in sporulation (Setlow, 1994). Enzymes that are modulated in this way include the zymogen of the GPR protease that auto-activates at a low pH, and phosphoglycerate mutase [catalyzes the interconversion of 3-phosphoglycerate (3-PGA) to 2-phosphoglycerate] that becomes much less active at pH ~6.3, thus causing the accumulation of a large amount of 3-PGA in the dormant spore (Illades-Aguiar and Setlow, 1994; Setlow, 1994). This 3-PGA depot is rapidly catabolized to generate ATP in the first minutes of spore germination, during which core pH rises to ~7.8 and core water content rises to ~80% of wet wt (Setlow and Johnson, 2012; Setlow, 2013).

A second unique feature of the spore core is the high level (~20% of core dry wt) of pyridine-2,6-dicarboxylic acid [dipicolinic acid (DPA)], which exists as a 1:1 chelate with divalent cations, predominantly Ca^{2+} (CaDPA; Gerhardt and Marquis, 1989; Setlow, 2006; Setlow and Johnson, 2012). DPA is made in the mother cell compartment of the sporulating cell, and the accumulation of CaDPA in the core late in sporulation plays a significant role in reducing spore core water content (Paidhungat et al., 2000). CaDPA in the core also has significant affects on spore DNA photochemistry, and thus CaDPA accumulation in the core alters spore resistance to UV radiation, but actually sensitizes DNA to UV radiation at some wavelengths (Setlow, 2001, 2007b). While the spores' huge CaDPA depot is stable for long periods in spores suspended in water even at relatively high temperatures (75–80 °C for B. subtilis spores), the spores' entire CaDPA depot is released rapidly in the first minutes of spore germination (Setlow, 2013). In addition, treatment with a number of oxidizing agents alters spores in some fashion such that their CaDPA depot is released completely even at 75°C (Cortezzo et al., 2004).

Another notable unique feature of the spore core is the saturation of spore DNA with a group of novel small DNA-binding proteins, termed Small, Acid-Soluble Proteins (SASP) of the α/β-type, so named because the two major proteins of this type in B. subtilis were initially termed SASP-α and SASP-β (Setlow, 2001, 2007b). The amino acid sequences of these ~60–75 residue proteins are unique and extremely well-conserved both within and across spore forming species, including members of both Clostridiales and Bacillales. The α/β-type SASP play major roles in protecting spore DNA against UV radiation (~260 nm) by changing DNA structure from the B-form to an A-like form in which the UV photoproducts formed in growing cell DNA, including cyclobutane pyrimidine dimers and 6–4 photoproducts between adjacent pyrimidines, are not generated, and some of the latter UV lesions are extremely mutagenic and thus potentially lethal. In contrast, with α/β-type SASP saturated DNA the major UV photoproduct generated by ~ 260 nm radiation is a novel thymine-thymine adduct initially called the spore photoproduct (SP), and spores have a number of enzymes that repair SP in a relatively error-free manner. Thus α/β-type SASP are a major factor in spore resistance to UV radiation and also to γ-radiation, although repair of DNA damage in spore outgrowth is also important (Setlow, 2001, 2007b; Moeller et al., 2014). The α/β-type SASP also protect spore DNA against other types of damage, particularly depurination by wet and dry heat, and also against a number of genotoxic chemicals, including nitrous acid and formaldehyde (Setlow, 2006). The structure of a complex of an α/β-type SASP bound to a short DNA fragment in conjunction with modeling studies based on this structure have elucidated the causes of these effects of α/β-type SASP on DNA properties at the atomic level (Lee et al., 2008).

The α/β-type SASP are degraded in the first minutes of spore germination and outgrowth (Setlow, 1994, 2013) in a process initiated by an endoprotease termed GPR that is specific for a conserved sequence found in all α/β-type SASP. Oligopeptides generated by GPR cleavage are then degraded rapidly to free amino acids that are re-utilized by the outgrowing spore for metabolism and protein synthesis. Spores in which α/β-type SASP are not degraded rapidly following spore germination, either because GPR is inactive or because an α/β-type SASP binds too tightly to DNA, exhibit slow outgrowth and in some cases decreased viability (Setlow, 1994; Hayes and Setlow, 2001).

4.3. Mechanisms of Spore Killing

Work primarily with B. subtilis spores, and some with spores of other species (Setlow, 2006; Setlow et al., 2014), has identified five different mechanisms for killing of spores by various agents (Table 7). These are: (1) DNA damage; (2) damage to the spores' IM; (3) damage to some essential spore core protein; (4) explosive rupture of the dormant spores' IM; and (5) destruction of one or more spore components that are essential for spore germination. It is important to note that spores apparently inactivated by the last mechanism may not actually be dead, but just cannot return to active growth, since spore germination is blocked. There are a number of examples of spores that appear dead because they can't germinate, but return to life, when assisted in spore germination, generally by provision with a lytic enzyme that can degrade spores' PG cortex (Setlow et al., 2002; Paredes-Sabja et al., 2009a,b; Burns et al., 2010). This has been observed with spores that are apparently inactivated by treatment with chemical agents that either remove and/or inactivate enzymes in spores' outer layers such as CLEs and/or proteases that can activate CLE zymogens in spores of *Clostridium* species. Thus, it is crucial to ensure that a spore killing regimen has not simply rendered spores incapable of germinating, such that the treated spores cannot be revived by artificial germination treatments; a number of tests have been used to demonstrate this (Setlow et al., 2014).

Among the other four methods of spore killing, the rarest is probably explosive rupture of the dormant spores' outer layers. This phenomenon has only been seen with spores incubated in rather high concentrations of strong mineral acids such as HCl or HNO_3 (Setlow et al., 2002; Setlow, 2006). As would be expected, UV and γ-radiation kill spores largely by DNA damage, since loss of DNA repair capacity greatly decreases spore resistance to these treatments, and survivors of these radiation treatments also accumulate high levels of mutations. A number of genotoxic chemicals, including nitrous acid and formaldehyde, also kill spores by DNA damage, as does dry heat, which generates significant levels of abasic sites in DNA. However, the potentially genotoxic agent, H_2O_2, does not kill spores by DNA damage, because of the strong protection of DNA against H_2O_2 damage by the α/β-type SASP.

While the mechanisms of spore killing and of spore resistance are known for many of the agents used to kill spores in applied settings – wet heat, radiation, chemicals such as hypochlorite or chlorine dioxide - there is a lack of definitive studies on the exact mechanisms of spore killing by and resistance to high hydrostatic pressure (HP) and various types of plasmas. In the case of HP, treatments at HP $\approx$400–700MPa (levels proposed for food processing) do not efficiently kill spores, unless accompanied by elevated temperature ($T = 90-121°C$). Spore killing by HP treatments takes place in (at least) two steps; first, spores are germinated by the HP, then the much less resistant germinated spores are killed at the elevated temperatures at which HP treatments are carried out.

Generally, moderate HPs (100–350 MPa, $T = 37°C$) trigger germination by activating the same GRs that trigger spore germination with nutrients, while HPs of 400–900 MPa ($T \geq 50°C$) appear to directly open the channel through which CaDPA is normally released during germination (Setlow, 2007a; Reineke et al., 2013a,b). Recent studies have indicated that HP treatments in the regime $P = 550-692$ MPa and $T = 75-112°C$ induce spore activation (potentiate spores for germination), and this can even increase apparent spore titers (Luu et al., 2015).

This latter effect has been seen with spores of the model spore former, *B. subtilis*, and also with spores of *B. amyloliquefaciens*, a suggested surrogate for *Clostridium botulinum* spores in HP studies (Margosch et al., 2004; Sevenich et al., 2014). Other recent work (Kong et al., 2014) with *B. subtilis* spores has shown that: (i) 140 MPa HP treatments as short as 10 s can commit spores to germinate for up to 10 min during holding at atmospheric pressure; and (ii) almost all spores go on to germinate over >45 min after a 10 s 140 MPa HP treatment during holding at 1 MPa, a pressure that alone does not induce germination. These latter findings indicate that at least some of the effects of 140 MPa of HP that can lead to spore germination are reversible, and thus the reversible events take place before events that irreversibly commit spores to germinate. It will be of interest to examine these same phenomena in spores exposed to HPs $\geq$500 MPa (the HP that opens channels for CaDPA release in *B. subtilis* spores).

In the case of cold plasma, when significant levels of UV photons are associated with a particular type of plasma, there is evidence that these plasmas kill spores by DNA damage and α/β-type SASP and DNA repair are important in spore resistance to such plasmas (Roth et al., 2010; Yardimci and Setlow, 2010; Klämpfl et al., 2012; van Bokhorst-van de Veen et al., 2015). For plasmas that have minimal associated UV photons, mechanisms of spore killing and resistance are less clear. There is certainly evidence that plasmas with oxygen in the feed gas can cause significant physical damage to spores by etching the spores' outer layers. However, it is likely that this physical damage is not what actually kills spores, and research is needed to establish definitively mechanisms of spore killing by and resistance to various plasmas, including those that contain minimal levels of UV photons.

Surprisingly, wet heat treatment does not kill spores by DNA damage, even though the temperatures used for spore killing, $\geq$90°C, might be expected to generate many abasic sites in DNA due to depurination. However, α/β-type SASP protect DNA so well-against damage by wet heat that wet heat almost certainly kills spores by damage to one or more essential core proteins (Setlow, 2006; Coleman et al., 2007, 2010; Leggett et al., 2012). As might be expected, spores lacking the majority of their α/β-type SASP, termed $\alpha^-\beta^-$ spores, are more sensitive to wet heat, which kills $\alpha^-\beta^-$ spores by DNA damage including depurination. Some peroxides (e.g., H_2O_2), also kill $\alpha^-\beta^-$ spores by DNA damage, although they probably kill wild-type spores by damage to some core protein(s).

While some toxic chemicals kill spores either by DNA damage or damage to some spore core protein(s), many toxic chemicals used in the killing of spores, including chlorine dioxide (ClO_2), hypochlorite (OCl^-), and iodine, kill spores by damaging the spores' IM, such that the IM ruptures when the treated spores germinate (Cortezzo et al., 2004; Setlow, 2006; Leggett et al., 2012). Treatment with these agent(s) also seems to somehow alter the IM in dormant spores, such that: (i) the IM permeability increases dramatically; and (ii) the IM is less able to withstand a thermal stress, and DPA is released at much lower temperatures than corresponding untreated spores. Further indications of damage to spores by a number of oxidizing agents are that: (i) treated spores germinate quite slowly, even if some of the germinated spores do ultimately give rise to colonies upon plating; and (ii) recovery of colonies from such treated spores is extremely sensitive to increased salt concentrations in the recovery media. However, the nature of the damage affecting the spores' IM leading to these effects is not known.

4.4. Summary

The AFM methods discussed previously provide information complementary to the spore inactivation mechanism discussion above by visualizing and characterizing spore morphological, dimensional, and coat structural attributes, particularly in response to sterilization and/or decontamination treatments by gamma-irradiation or ClO_2. For example, irradiation treatments induced pronounced morphological and structural differences compared with native spores, including loss of refractility, decreased size, and damage to spore internal structural integrity, which is consistent with, but not necessary revealing of, DNA damage, the primary mechanism of bacterial spore inactivation by γ-irradiation. It is also possible that the sterilization treatment by γ-irradiation is extensive and induces further structural damage to already-killed spores as additional post-mortem events. Interestingly, ClO_2-treated spores showed similar height, morphology, and high-resolution spore coat architecture and topology compared to native, untreated spores. Such observations with AFM are consistent with the known mechanisms of spore inactivation by ClO_2 by acting on the IM as mentioned above, such that ClO_2-killed spores treatments retain refractility (phase-bright) and size, but falter during germination and are unable to grow out (Young and Setlow, 2003).

Atomic force microscopy measurements also showed the reversibility of a size transition in fully hydrated, dehydrated, and re-hydrated environments. AFM characterization also showed that spore size decreases in dehydrated environments (air versus water), most likely due to contraction of the spore core and/or cortex. In contradistinction, the spore coat is more intractable, not shrinking or expanding with dehydration or hydration, respectively, but providing some flexibility to accommodate internal volume changes of spores by surface folding and the formation of ridges in air-dried spores that are absent from the surface of fully hydrated spores. Such results are complementary to and consistent with the development of spore BIs mentioned above. The spore BIs tested a number of strains, doses (ClO_2 concentration × time profiles) and humid environments and found that %RH $\geq$ 70 was needed presumptively to promote spore swelling and facilitate internalization of ClO_2 molecules to effectuate spore inactivation and oxidizing proteins associated with the spore IM. It is absolutely essential for the sterilization of Ebola contaminated devices and equipment in that the spore BIs provide a full and reliable assay of sterilization, to prevent the spread of disease, protect international healthcare workers on the front lines of global public health crises, and to care for the afflicted and save lives. Thus, the complementary information provided through bacterial spore mechanistic studies, AFM

morphological and structural characterization, and spore BI are needed to provide the full assurance that NSRDEC's novel decontamination technologies, to include the FDKs that operate at 70–90 %RH and > 7000 ppm ClO_2, are achieving sterilization when deployed in global crises, and the results reported above provide science-based solutions that satisfy this mission with confidence.

4.5. Conclusion

Moving exciting research findings from the benchtop to the hands of the end-user as novel technologies is a paramount goal of Science and Technology (S&T) for many government research and development (R&D) facilities. The novel ClO_2 decontamination technologies invented at NSRDEC serve as a model for successful federal research and development work, Technology Transfer, commercialization, and deployment for actual use in the field. In fact, this work began with basic research in the mechanisms of bacterial spore inactivation, then transitioned to early exploratory research and development work focused on the chemical reaction kinetics and mechanisms of ClO_2 formation occurring through a unique chemical effector. Later development work consisted of inventions, validating efficacy, obtaining patents, and eventually transitioning to private industry through Technology Transfer for commercialization in the marketplace. This technology has become a COTS item that enables its recent culmination as FDKs deployed at clinical sites in West Africa for the sterilization of Ebola-contaminated medical devices at a time when concern of the Ebola crisis in those regions and the fears of the virus spreading internationally were at their highest.

Scientists at NIH/NIAID and USAMRIID (The U.S. Army Medical Research Institute of Infectious Diseases) adapted the COTS version of NSRDECs novel decontamination technologies and developed compact, lightweight, easy-to-carry FDKs for use by global public health organizations such as MSF, the WHO, Public Health Canada, NIH, and the U.S. government to improve hygienic conditions in remote clinics where little infrastructure. existed by sterilizing medical equipment and help fight the spread of a deadly disease that cost thousands of lives and untold costs in medical expenses caring for the afflicted.

Natick Soldier RD&E Center's novel decontamination technologies were readily available to help others meet and overcome the challenges and concerns of this international public health through R&D that strives for such preparedness. These novel technologies decontaminate microbes in myriad applications, such as rinses, sprays, and gas for fresh produce; food contact and handling surfaces; PPE; textiles used in clothing, uniforms, tents, and shelters; graywater; airplanes; surgical instruments; and hard surfaces in latrines, laundries, and deployable medical facilities.

One strength of NSRDECs novel decontamination technologies is that the mechanisms of bacterial spore inactivation by ClO_2 and the use of spore bio-indicators for ClO_2 in decontamination procedures have been investigated thoroughly at University of Connecticut Health Center, Lawrence

Livermore National Laboratories, Children's Hospital of Oakland Research Institute, Stonehill College, and Brandeis University, making the fundamental science of ClO_2 generation and decontamination well-understood, and NSRDECs technologies more assured and reliable. We have also reviewed mechanisms of bacterial spore inactivation by novel, emerging, and established non-thermal technologies for food preservation, such as HPP, irradiation, cold plasma, and chemical sanitizers, using an array of *B. subtilis* mutants to probe mechanisms of spore germination and inactivation, demonstrating that understanding the basic science allows the development of innovative technologies that can find solutions in diverse applications outside of the pasteurization and sterilization of foodstuffs.

Acknowledgments

We express our gratitude and appreciation to the US Army Natick Soldier RD&E Center, to the Joint Science and Technology Office for Chemical and Biological Defense – Integrated Protective Fabric System program (managed at NSRDEC), and to Mr. James M. Doona (Mathematician) for helpful discussions and Mr. Murphy C. Doona for counting spores.

Work carried out in the Setlow laboratory on spore resistance and killing has received generous support over many years from the Army Research Office, the Defense Threat Reduction Agency, and the NIH.

The AFM work in the Malkin laboratory was performed under the auspices of the U.S. Department of Energy by Lawrence Livermore National Laboratory under Contract DE-AC52-07NA27344 and supported by a grant from the Federal Bureau of Investigation and by Lawrence Livermore National Laboratory through Laboratory Directed Research and Development Grant 04-ERD-002. The authors are grateful to Marco Plomp for his critical contributions in the AFM characterization and data analysis.

Work in the Leighton laboratory was funded by The Defense Advanced Research Projects Agency (DARPA). We acknowledge with pleasure and gratitude the contributions of Katie Wheeler and Gordon Eggum to the study of ClO_2 bio-indicators.

The Natick Soldier Research, Development and Engineering Center is part of the U.S. Army Research, Development and Engineering Command, which has the mission to develop technology and engineering solutions for America's Soldiers.

RDECOM is a major subordinate command of the U.S. Army Materiel Command. AMC is the Army's premier provider of materiel readiness – technology, acquisition support, materiel development, logistics power projection, and sustainment – to the total force, across the spectrum of joint military operations. If a Soldier shoots it, drives it, flies it, wears it, eats it or communicates with it, AMC provides it.

References

Aieta, E. M., and Berg, J. D. (1986). A review of chlorine dioxide in drinking water treatment. *J. Am. Water Works Assoc.* 78, 62–72.

Anelich, L., and Moy, G. G. (2014). *Ebola Virus Disease (EVD): Important Aspects for the Food Science and Technology Community. IUFoST Scientific Information Bulletin.* Available at: http://www.iufost.org/iufostftp/IUFoST%20SIB%20%20-%20Ebola%20Virus%20Disease%20update%201.pdf [accessed March 12, 2015].

Bruch, C. W. (1973). "Sterilization of plastics: toxicity of ethylene oxide residues," in *Industrial Sterilization: International Symposium*, eds G. B. Phillips and W. S. Miller (Amsterdam: Duke University Press), 49–77.

Buchanan, C. E., and Neyman, S. L. (1986). Correlation of penicillin-binding protein composition with different functions of two membranes in *Bacillus subtilis* forespores. *J. Bacteriol.* 165, 498–503.

Burns, D. A., Heap, J. T., and Minton, N. P. (2010). SleC is essential for germination of *Clostridium difficile* spores in nutrient-rich medium supplemented with the bile salt taurocholate. *J. Bacteriol.* 192, 657–664. doi: 10.1128/JB.01209-09

Carroll, A. M., Plomp, M., Malkin, A. J., and Setlow, P. (2008). Protozoal digestion of coat-defective *Bacillus subtilis* spores produces "rinds" composed of insoluble coat protein. *Appl. Environ. Microbiol.* 74, 5875–5881. doi: 10.1128/AEM.01228-08

Chapman, J., Lee, W., Youkilis, E., and Martis, L. (1986). Animal model for ethylene oxide (EtO) associated hypersensitivity reactions. *ASAIO Trans.* 32, 482–485. doi: 10.1097/00002480-198609000-00019

Cléry-Barraud, C., Gaubert, A., Masson, P., and Vidal, D. (2004). Combined effects of high hydrostatic pressure and temperature for inactivation of *Bacillus anthracis* spores. *Appl. Environ. Microbiol.* 70, 635–637. doi: 10.1128/AEM.70.1.635-637.2004

Coleman, W. H., Chen, D., Li, Y.-Q., Cowan, A. E., and Setlow, P. (2007). How moist heat kills spores of *Bacillus subtilis. J. Bacteriol.* 189, 8458–8466. doi: 10.1128/JB.01242-07

Coleman, W. H., Zhang, P., Li, Y.-Q., and Setlow, P. (2010). Mechanism of killing of spores of *Bacillus cereus* and *Bacillus megaterium* by wet heat. *Lett. Appl. Microbiol.* 50, 507–514. doi: 10.1111/j.1472-765X.2010.02827.x

Cortezzo, D. E., Koziol-Dube, K., Setlow, B., and Setlow, P. (2004). Treatment with oxidizing agents damages the inner membrane of spores of *Bacillus subtilis* and sensitizes spores to subsequent stress. *J. Appl. Microbiol.* 97, 838–852. doi: 10.1111/j.1365-2672.2004.02370.x

Cortezzo, D. E., and Setlow, P. (2004). Analysis of factors that influence the sensitivity of spores of *Bacillus subtilis* to DNA damaging chemicals. *J. Appl. Microbiol.* 98, 606–617. doi: 10.1111/j.1365-2672.2004.02495.x

Cowan, A. E., Koppel, D. E., Setlow, B., and Setlow, P. (2003). A soluble protein is immobile in dormant spores of *Bacillus subtilis* but is mobile in germinated spores: implications for spore dormancy. *Proc. Natl. Acad. Sci. U.S.A.* 100, 4209–4214. doi: 10.1073/pnas.0636762100

Cowan, A. E., Olivastro, E. M., Koppel, D. E., Loshon, C. A., Setlow, B., and Setlow, P. (2004). Lipids in the inner membrane of dormant spores of *Bacillus* species are largely immobile. *Proc. Natl. Acad. Sci. U.S.A.* 101, 7733–7738. doi: 10.1073/pnas.0306859101

Curtin, M. A., Dwyer, S., Bukvic, D., Doona, C. J., and Kustin, K. (2014). Kinetics and mechanism of the reduction of sodium chlorite by sodium hydrogen ascorbate in aqueous solution at near neutral pH. *Int. J. Chem. Kinet.* 46, 216–219. doi: 10.1002/kin.20847

Curtin, M. A., Taub, I. A., Kustin, K., Sao, N., Duvall, J. R., Davies, K. I., et al. (2004). Ascorbate-induced oxidation of formate by peroxodisulfate: product yields, kinetics, and mechanism. *Res. Chem. Intermed.* 30, 647–661. doi: 10.1163/1568567041570384

Cybulski, R. J., Sanz, P., Alem, F., Stibitz, S., Bull, R. L., and O'Brien, A. D. (2009). Four superoxide dismutases contribute to *Bacillus anthracis* virulence and provide spores with redundant protection against oxidative stress. *Infect. Immun.* 77, 271–285. doi: 10.1128/IAI.00515-08

Dauphin, L. A., Newton, B. R., Rasmussen, M. V., Meyer, R. F., and Bowen, M. D. (2008). Gamma irradiation can be used to inactivate *Bacillus anthracis* spores without compromising the sensitivity of diagnostic assays. *Appl. Environ. Microbiol.* 74, 4427–4433. doi: 10.1128/AEM.00557-08

Dolovich, J., Marshall, C. P., Smith, E. K. M., Shimizu, A., Pearson, F. C., Sugona, M. A., et al. (1984). Allergy to ethylene oxide chronic hemodyalysis patients. *Artif. Organs* 8, 334–337. doi: 10.1111/j.1525-1594.1984.tb04301.x

Doona, C. J., Feeherry, F. E., Setlow, P., Malkin, A. J., and Leighton, T. J. (2014). The portable chemical sterilizer (PCS), D-FENS, and D-FEND ALL: novel chlorine dioxide decontamination technologies for the military. *J. Vis. Exp.* 88, e4354. doi: 10.3791/4354

Driks, A. (2003). The dynamic spore. *Proc. Natl. Acad. Sci. U.S.A.* 100, 3007–3009. doi: 10.1073/pnas.0730807100

Driks, A. (2009). The *Bacillus anthracis* spore. *Mol. Aspects Med.* 30, 368–373. doi: 10.1016/j.mam.2009.08.001

Feeherry, F. E., Munsey, D. T., and Rowley, D. B. (1987). Thermal inactivation and injury of *Bacillus stearothermophilus* spores. *Appl. Environ. Microbiol.* 53, 365–370.

Foran, A. (2013). *NSRDEC Patents Help Army into 'Top 100 Global Innovators.'* Available at: http://www.army.mil/article/99816/ [accessed March 12, 2015].

Gerhardt, P. (1967). Cytology of *Bacillus anthracis. Fed. Proc.* 26, 1504–1517.

Gerhardt, P., and Black, S. H. (1961). Permeability of bacterial spores. II. Molecular variables affecting solute permeation. *J. Bacteriol.* 82, 750–760.

Gerhardt, P., and Marquis, R. E. (1989). "Spore thermoresistance mechanisms," in *Regulation of Prokaryotic Development*, eds I. Smith, R. A. Slepecky, and P. Setlow (Washington, DC: American Society for Microbiology), 43–63.

Ghosal, S., Leighton, T. J., Wheeler, K. E., Hutcheon, I. D., and Weber, P. K. (2010). Spatially-resolved characterization of water and ion incorporation in *Bacillus* spores. *Appl. Environ. Microbiol.* 76, 3275–3282. doi: 10.1128/AEM. 02485-09

Ghosh, S., Setlow, B., Wahome, P. G., Cowan, A. E., and Plomp, M. (2008). Characterization of spores of *Bacillus subtilis* that lack most coat layers. *J. Bacteriol.* 190, 6741–6748. doi: 10.1128/JB.00896-08

Griffiths, K., and Setlow, P. (2009). Effects of modification of membrane lipid composition on *Bacillus subtilis* sporulation and spore properties. *J. Appl. Microbiol.* 106, 2064–2078. doi: 10.1111/j.1365-2672.2009.04176.x

Han, Y., Floros, J. D., Linton, R. H., Nielsen, S. S., and Nelson, P. E. (2001). Response surface modeling for the inactivation of *Escherichia coli* O157:H7 on green peppers (*Capsicum annuum* L.) by chlorine dioxide gas treatments. *J. Food Prot.* 64, 1128–1131.

Han, Y., Guentert, A. M., Smith, R. S., Linton, R. H., and Nelson, P. E. (1999). Efficacy of chlorine dioxide gas as a sanitizer for tanks used for aseptic juice storage. *Food Microbiol.* 16, 53–61. doi: 10.1006/fmic.1998.0211

Hayes, C. S., and Setlow, P. (2001). An α/β-type small, acid-soluble spore protein which has a very high affinity for DNA prevents outgrowth of *Bacillus subtilis* spores. *J. Bacteriol.* 183, 2662–2666. doi: 10.1128/JB.183.8.2662-2666.2001

Henriques, A. O., and Moran, C. P. (2007). Structure, assembly, and function of the spore surface layers. *Ann. Rev. Microbiol.* 61, 555–588. doi: 10.1146/annurev.micro.61.080706.093224

Horváth, A. K., Nagypál, I., Peintler, G., Epstein, I. R., and Kustin, K. (2003). Kinetics and mechanism of the decomposition of chlorous acid. *J. Phys. Chem. A* 107, 6966–6973. doi: 10.1021/jp027411h

Illades-Aguiar, B., and Setlow, P. (1994). Autoprocessing of the protease that degrades small, acid-soluble proteins of spores of *Bacillus* species is triggered by low pH, dehydration and dipicolinic acid. *J. Bacteriol.* 176, 7032–7037.

Kaieda, S., Setlow, B., Setlow, P., and Halle, B. (2013). Mobility of core water in *Bacillus subtilis* spores by 2H NMR. *Biophys. J.* 105, 2016–2023. doi: 10.1016/j.bpj.2013.09.022

Kingsley, D. H., Vincent, E. M., Meade, G. K., Watson, C. L., and Fan, X. (2014). Inactivation of human norovirus using chemical sanitizers. *Int. J. Food Microbiol.* 171, 94–99. doi: 10.1016/j.ijfoodmicro.2013.11.018

Klämpfl, T. G., Isbary, G., Shimizu, T., Li, Y.-F., Zimmerman, J. L., Stolz, W., et al. (2012). Cold atmospheric plasma and plasma sterilization against spores and other microorganisms of interest. *Appl. Environ. Microbiol.* 78, 5077–5082. doi: 10.1128/AEM.00583-12

Kong, L., Doona, C. J., Setlow, P., and Li, Y.-Q. (2014). Monitoring rates and heterogeneity of high pressure germination of *Bacillus* spores using phase contrast microscopy of individual spores. *Appl. Environ. Microbiol.* 80, 345–353. doi: 10.1128/AEM.03043-13

Kong, L., Setlow, P., and Li, Y.-Q. (2013). Direct analysis of water content and movement in single dormant bacterial spores using confocal Raman microspectroscopy. *Anal. Chem.* 85, 7094–7101. doi: 10.1021/ac400516p

Lee, K. S., Bumbaca, D., Kosman, J., Setlow, P., and Jedrzejas, M. J. (2008). Structure of a protein-DNA complex essential for DNA protection in spores of *Bacillus* species. *Proc. Natl. Acad. Sci. U.S.A.* 105, 2806–2811. doi: 10.1073/pnas.0708244105

Leggett, M., McDonnell, G., Denyer, S., Setlow, P., and Maillard, J.-Y. (2012). Bacterial spore structures and their protective role in biocide resistance. *J. Appl. Microbiol.* 113, 485–499. doi: 10.1111/j.1365-2672.2012.05336.x

Lettman, S. F., Boltansky, H., Alter, H. J., Pearson, F. C., and Kaliner, M. A. (1986). Allergic reaction in healthy plateletpheresis donors caused by sensitization to ethylene oxide gas. *N. Engl. J. Med.* 315, 1192–1196. doi: 10.1056/NEJM198611063151904

Luu, S., Cruz-Mora, J., Setlow, B., Feeherry, F. E., Doona, C. J., and Setlow, P. (2015). The effects of heat activation on *Bacillus* spore germination, with nutrients or under high-pressure, with and without germination proteins. *Appl. Environ. Microbiol.* 81, 2927–2938. doi: 10.1128/AEM.00193-15

Lykins, B. W., Koffskey, W. E., and Patterson, K. S. (1994). Alternative disinfectants for drinking water treatment. *J. Environ. Eng.* 120, 745–758. doi: 10.1061/(ASCE)0733-9372(1994)120:4(745)

Malkin, A. J. (2011). "Resolving the high-resolution architecture, assembly and functional repertoire of bacterial systems by *in vitro* atomic force microscopy," in *Life at the Nanoscale: Atomic Force Microscopy of Live Cells*, ed. Y. Dufresne (Singapore: Pan Stanford Publishing), 71–99. doi: 10.1201/b11404-5

Malkin, A. J., and Plomp, M. (2010). "High-resolution architecture and structural dynamics of microbial and cellular system: insights from high-resolution in vitro atomic force microscopy," in *Scanning Probe Microscopy of Functional Materials: Nanoscale Imaging and Spectroscopy*, eds S. V. Kalinin and A. Gruverman (New York, NY: Springer), 39–68.

Margosch, D., Ehrmann, M. A., Gänzle, M. G., and Vogel, R. F. (2004). Comparison of pressure and heat resistance of *Clostridium botulinum* and other endospores in mashed carrots. *J. Food Prot.* 67, 2530–2537.

McKenney, P. T., Driks, A., and Eichenberger, P. (2013). The *Bacillus subtilis* endoscpore and functions of the multilayered coat. *Nat. Rev. Microbiol.* 11, 33–44. doi: 10.1038/nrmicro2921

Moeller, R., Raguse, M., Reitz, G., Okayasu, R., Li, Z., Klein, S., et al. (2014). Resistance of *Bacillus subtilis* spore DNA to lethal ionizing radiation damage relies primarily on spore core components and DNA repair, with minor effects of oxygen radical detoxification. *Appl. Environ. Microbiol.* 80, 104–109. doi: 10.1128/AEM.03136-13

Muttamara, S., Sales, C. I., and Gazali, Z. (1995). The formation of trihalomethane from chemical disinfectants and humic substances in drinking water. *Water Supply* 13, 105–117.

Paidhungat, M., Setlow, B., Driks, A., and Setlow, P. (2000). Characterization of spores of *Bacillus subtilis* which lack dipicolinic acid. *J. Bacteriol.* 182, 5505–5512. doi: 10.1128/JB.182.19.5505-5512.2000

Paredes-Sabja, D., Setlow, P., and Sarker, M. R. (2009a). The protease CspB is essential for initiation of cortex hydrolysis and DPA release during germination of spores of *Clostridium perfringens. Microbiology* 155, 3464–3472. doi: 10.1099/mic.0.030965-0

Paredes-Sabja, D., Setlow, P., and Sarker, M. R. (2009b). SleC is essential for cortex peptidoglycan hydrolysis during germination of spores of the pathogenic bacterium *Clostridium perfringens. J. Bacteriol.* 191, 2711–2720. doi: 10.1128/JB.01832-08

Plomp, M., Leighton, T. J., Wheeler, K. E., Hill, H. D., and Malkin, A. J. (2007a). In vitro high-resolution structural dynamics of single germinating bacterial spores. *Proc. Natl. Acad. Sci. U.S.A.* 104, 9644–9649. doi: 10.1073/pnas.0610626104

Plomp, M., McCaffery, J. M., Cheong, I., Huang, X., and Bettegowda, C. (2007b). Spore coat architecture of *Clostridium novyi* NT spores. *J. Bacteriol.* 189, 6457–6468. doi: 10.1128/JB.00757-07

Plomp, M., Leighton, T. J., Wheeler, K. E., and Malkin, A. J. (2005a). The high-resolution architecture and structural dynamics of *Bacillus* spores. *Biophys. J.* 88, 603–608. doi: 10.1529/biophysj.104.049312

Plomp, M., Leighton, T. J., Wheeler, K. E., and Malkin, A. J. (2005b). Architecture and high-resolution structure of *Bacillus thuringiensis* and *Bacillus cereus* spore coat surfaces. *Langmuir* 21, 7892–7898. doi: 10.1021/la050412r

Plomp, M., Leighton, T. J., Wheeler, K. E., Pitesky, M. E., and Malkin, A. J. (2005c). *Bacillus atrophaeus* outer spore coat assembly and ultrastructure. *Langmuir* 21, 10710–10716. doi: 10.1021/la0517437

Plomp, M., and Malkin, A. J. (2009). Mapping of proteomic composition on the surfaces of *Bacillus* spores by atomic force microscopy. *Langmuir* 25, 403–409. doi: 10.1021/la803129r

Plomp, M., Monroe, C., Setlow, P., and Malkin, A. J. (2014). Architecture and assembly of the *Bacillus subtilis* spore coat. *PLoS ONE* 9:e108560. doi: 10.1371/journal.pone.0108560

Popham, D. L. (2002). Specialized peptidoglycan of the bacterial endospore: the inner wall of the lockbox. *Cell. Mol. Life Sci.* 59, 426–433. doi: 10.1007/s00018-002-8435-5

Reineke, K., Mathys, A., Heinz, V., and Knorr, D. (2013a). Mechanisms of endospore inactivation under high pressure. *Trends Microbiol.* 21, 296–304. doi: 10.1016/j.tim.2013.03.001

Reineke, K., Schlumbach, K., Baier, D., Mathys, A., and Knorr, D. (2013b). The release of dipicolinic acid – the rate-limiting step of *Bacillus* endospore inactivation during the high pressure thermal sterilization process. *Int. J. Food Microbiol.* 162, 55–63. doi: 10.1016/j.ijfoodmicro.2012.12.010

Rode, L. J., Lewis, C. W., and Foster, J. W. (1962). Electron microscopy of spores of *Bacillus megaterium* with special emphasis to the effects of fixation and thin sectioning. *J. Cell. Biol.* 13, 423–435. doi: 10.1083/jcb.13.3.423

Rosenblatt, D. H., Rosenblatt, A. A., and Knapp, J. A. (1987). *Use of Chlorine Dioxide Gas as a Chemosterilizing Agent.* US patent# 4681739.

Roth, S., Feichtinger, J., and Hertel, C. (2010). Characterization of *Bacillus subtilis* spore inactivation in low-pressure, low temperature gas plasma sterilization processes. *J. Appl. Microbiol.* 108, 521–531. doi: 10.1111/j.1365-2672.2009.04453.x

Setlow, B., Loshon, C. A., Genest, P. C., Cowan, A. E., Setlow, C., and Setlow, P. (2002). Mechanisms of killing of spores of *Bacillus subtilis* by acid, alkali and ethanol. *J. Appl. Microbiol.* 92, 362–375. doi: 10.1046/j.1365-2672.2002.01540.x

Setlow, B., Parish, S., Zhang, P., Li, Y.-Q., Neely, W. C., and Setlow, P. (2014). Mechanism of killing of spores of *Bacillus anthracis* in a high temperature gas environment, and analysis of DNA damage generated by various decontamination treatments of spores of *Bacillus anthracis*, *Bacillus subtilis* and *Bacillus thuringiensis*. *J. Appl. Microbiol.* 116, 805–814. doi: 10.1111/jam.12421

Setlow, B., and Setlow, P. (1980). Measurements of the pH within dormant and germinated spores of *Bacillus megaterium*. *Proc. Natl. Acad. Sci. U.S.A.* 77, 2744–2746. doi: 10.1073/pnas.77.5.2474

Setlow, P. (1994). Mechanisms which contribute to the long-term survival of spores of *Bacillus* species. *J. Appl. Bacteriol.* 76, 49S–60S. doi: 10.1111/j.1365-2672.1994.tb04357.x

Setlow, P. (2001). Resistance of spores of *Bacillus* species to ultraviolet light. *Environ. Mol. Mutagen.* 38, 97–104. doi: 10.1002/em.1058

Setlow, P. (2006). Spores of *Bacillus subtilis*: their resistance to radiation, heat and chemicals. *J. Appl. Microbiol.* 101, 514–525. doi: 10.1111/j.1365-2672.2005.02736.x

Setlow, P. (2007a). "Germination of spores of *Bacillus subtilis* by high pressure," in *High Pressure Processing of Foods*, eds C. J. Doona and F. E. Feeherry (London: Blackwell Publishing), 15–40.

Setlow, P. (2007b). I will survive: DNA protection in bacterial spores. *Trends Microbiol.* 15, 172–180. doi: 10.1016/j.tim.2007.02.004

Setlow, P. (2013). When the sleepers wake: the germination of bacterial spores. *J. Appl. Microbiol.* 115, 1251–1268. doi: 10.1111/jam.12343

Setlow, P., Doona, C. J., Feeherry, F. E., Kustin, K., Sisson, D., and Chandra, S. (2009). "Enhanced safety and extended shelf-life of fresh produce for the military," in *Microbial Safety of Fresh Produce*, eds X. Fan, B. A. Niemira, C. J. Doona, F. E. Feeherry, and R. B. Gravani (Ames, IA: IFT Press/Wiley-Blackwell), 263–289.

Setlow, P., and Johnson, E. A. (2012). "Spores and their significance," in *Food Microbiology: Fundamentals and Frontiers*, 4th Edn, eds M. P. Doyle and R. Buchanan (Washington, DC: ASM Press).

Setlow, P., Liu, J., and Faeder, J. R. (2012). "Heterogeneity in bacterial spore populations," in *Bacterial Spores: Current Research and Applications*, ed. E. Abel-Santos (Norwich: Horizon Scientific Press).

Sevenich, R., Kleinsteuck, E., Crews, C., Anderson, W., Pye, C., Riddellova, K., et al. (2014). High-pressure thermal sterilization: food safety and food quality of baby food purée. *J. Food Sci.* 79, M230–M237. doi: 10.1111/1750-3841.12345

Simpson, G. D., Miller, R. F., Laxton, G. D., and Clements, W. R. (1993). "A focus on chlorine dioxide: the "ideal" biocide," in *Proceedings of the Volume, Annual NACE Corrosion Conference 93, Paper No. 472*, New Orleans, LA. Available at: http://www.clo2.gr/en/pdf/secure/chlorinedioxideidealbiocide.pdf [accessed March 18, 2015].

Stubblefield, J. M., and Newsome, A. L. (2015). Potential biodefense model applications for portable chlorine dioxide gas production. *Health Secur.* 13, 20–28. doi: 10.1089/hs.2014.0017

Sunde, E. P., Setlow, P., Hederstedt, L., and Halle, B. (2009). The physical state of water in bacterial spores. *Proc. Natl. Acad. Sci. U.S.A.* 106, 19334–19339. doi: 10.1073/pnas.0908712106

Swerdlow, B. M., Setlow, B., and Setlow, P. (1981). Levels of H^+ and other monovalent cations in dormant and germinating spores of *Bacillus megaterium*. *J. Bacteriol.* 148, 20–29.

van Bokhorst-van de Veen, H., Xie, H., Esveld, E., Abee, T., Mastwijk, H., and Groot, M. N. (2015). Inactivation of chemical and heat-resistant spores of *Bacillus* and *Geobacillus* by nitrogen cold atmospheric plasma evokes distinct changes in morphology and integrity of spores. *Food Microbiol.* 45, 26–33. doi: 10.1016/j.fm.2014.03.018

Westphal, A. J., Price, P. B., Leighton, T. J., and Wheeler, K. E. (2003). Kinetics of size changes of individual *Bacillus thuringiensis* spores in response to changes in relative humidity. *Proc. Natl. Acad. Sci. U.S.A.* 100, 3461–3466. doi: 10.1073/pnas.232710999

World Health Organization [WHO]. (2014). *Ebola Virus Disease.* Available at: http://www.who.int/mediacentre/factsheets/fs103/en/ [accessed March 12, 2015].

World Health Organization [WHO]. (2015a). *2014 Ebola Outbreak in West Africa - Case Counts.* Available at: http://www.cdc.gov/vhf/ebola/outbreaks/2014-west-africa/case-counts.html [accessed March 23, 2015].

World Health Organization [WHO]. (2015b). *The Five Keys to Safer Food Programme.* Available at: http://www.who.int/mediacentre/factsheets/fs103/en/ [accessed March 12, 2015].

Yardimci, O., and Setlow, P. (2010). Plasma sterilization: opportunities and microbial assessment strategies in medical device manufacturing. *IEEE Trans. Plasma Sci.* 38, 973–981. doi: 10.1109/TPS.2010.2041674

Young, S. B., and Setlow, P. (2003). Mechanisms of killing of *Bacillus subtilis* spores by hypochlorite and chlorine dioxide. *J. Appl. Microbiol.* 95, 54–67. doi: 10.1046/j.1365-2672.2003.01960.x

Zhang, P., Thomas, S., Li, Y.-Q., and Setlow, P. (2012). Effects of cortex peptidoglycan structure and cortex hydrolysis on the kinetics of Ca^{2+}-dipicolinic acid release during *Bacillus subtilis* spore germination. *J. Bacteriol.* 194, 646–652. doi: 10.1128/JB.06452-11

Conflict of Interest Statement: Christopher J. Doona, Florence E. Feeherry and Kenneth Kustin, as inventors from the U.S. Army Natick Soldier RD&E Center, receive royalties for patented technologies licensed to ClorDiSys Solutions, Inc. and used in the Field Decontamination Kits. The other authors declare that the research was conducted in the absence of any commercial or financial relationships that could be construed as a potential conflict of interest.

Permissions

The contributors of this book come from diverse backgrounds, making this book a truly international effort. This book will bring forth new frontiers with its revolutionizing research information and detailed analysis of the nascent developments around the world.

We would like to thank all the contributing authors for lending their expertise to make the book truly unique. They have played a crucial role in the development of this book. Without their invaluable contributions this book wouldn't have been possible. They have made vital efforts to compile up to date information on the varied aspects of this subject to make this book a valuable addition to the collection of many professionals and students.

This book was conceptualized with the vision of imparting up-to-date information and advanced data in this field. To ensure the same, a matchless editorial board was set up. Every individual on the board went through rigorous rounds of assessment to prove their worth. After which they invested a large part of their time researching and compiling the most relevant data for our readers.

The editorial board has been involved in producing this book since its inception. They have spent rigorous hours researching and exploring the diverse topics which have resulted in the successful publishing of this book. They have passed on their knowledge of decades through this book. To expedite this challenging task, the publisher supported the team at every step. A small team of assistant editors was also appointed to further simplify the editing procedure and attain best results for the readers.

Apart from the editorial board, the designing team has also invested a significant amount of their time in understanding the subject and creating the most relevant covers. They scrutinized every image to scout for the most suitable representation of the subject and create an appropriate cover for the book.

The publishing team has been an ardent support to the editorial, designing and production team. Their endless efforts to recruit the best for this project, has resulted in the accomplishment of this book. They are a veteran in the field of academics and their pool of knowledge is as vast as their experience in printing. Their expertise and guidance has proved useful at every step. Their uncompromising quality standards have made this book an exceptional effort. Their encouragement from time to time has been an inspiration for everyone.

The publisher and the editorial board hope that this book will prove to be a valuable piece of knowledge for researchers, students, practitioners and scholars across the globe.

List of Contributors

Carla Solórzano , Sonia Paytubi and Cristina Madrid
Departament de Microbiologia, Universitat de Barcelona, Barcelona, Spain

Shabarinath Srikumar and Rocío Canals
Institute of Integrative Biology, University of Liverpool, Liverpool, UK

Antonio Juárez
Departament de Microbiologia, Universitat de Barcelona, Barcelona, Spain
Institut de Bioenginyeria de Catalunya, Parc Científic de Barcelona, Barcelona, Spain

Christian Hertwig, Kai Reineke, Antje Rademacher, Michael Klocke and Oliver Schlüter
Leibniz Institute for Agricultural Engineering, Potsdam-Bornim, Germany

Veronika Steins and Cornelia Rauh
Department of Food Biotechnology and Food Process Engineering, Berlin University of Technology, Berlin, Germany

Lucía Guadamuro, Susana Delgado, Ana B. Flórez and Baltasar Mayo
Departamento de Microbiologíay Bioquímica, Instituto de Productos Lácteosde Asturias–Consejo Superior de Investigaciones Científicas, Villaviciosa, Spain

Begoña Redruello
Servicios Científico-Técnicos, Instituto de Productos Lácteos de Asturias– Consejo Superior de Investigaciones Científicas, Villaviciosa, Spain

Adolfo Suárez
Servicio de Digestivo, Hospital de Cabueñes, Gijón, Spain

Pablo Martínez-Camblor
Consorcio de Apoyoala Investigación Biomédicaen Red, Hospital Universitario Central de Asturias, Oviedo, Spain
Facultad de Ciencias de la Educación, Universidad Autónoma de Chile, Santiago, Chile

Jia Hu
School of Food Science, Washington State University, Pullman, WA, USA
Department of Animal Science, University of Wyoming, Laramie, WY, USA

Mei-Jun Zhu
School of Food Science, Washington State University, Pullman, WA, USA

Miguel A. Dela Cruz
Unidad de Investigación Médicaen Enfermedades Infecciosasy Parasitarias, Centro Médico Nacional Siglo XX1-IMSS, México DF, Mexico

Deyanira Pérez-Morales, Irene J. Palacios, Marcos Fernández-Mora, Edmundo Calva and Víctor H. Bustamante
Departamento de Microbiología Molecular, Instituto de Biotecnología, Universidad Nacional Autónoma de México, Cuernavaca, Morelos, Mexico

Cátia Pinto, Diogo Pinho, Remy Cardoso, Valéria Custódio, Joana Fernandes, Susana Sousa and Ana C. Gomes
Genomics Unit, Biocant–Biotechnology Innovation Center, Cantanhede, Portugal

Miguel Pinheiro and Conceição Egas
GenoInSeq Unit, Biocant– Biotechnology Innovation Center, Cantanhede, Portugal

Sophie Roussel, Benjamin Felix, Noémie Vingadassalon, Joël Grout, Jacques-Antoine Hennekinne, Laurent Guillier, Anne Brisabois and Fréderic Auvray
Université Paris-Est, ANSES, Food Safety Laboratory, European Union Reference Laboratory for Coagulase Positive Staphylococci, Maisons-Alfort, France

Zizhong Liu, Haili Wang, Hongduo Wang, Yujing Bi, Xiaoyi Wang, Ruifu Yang and Yanping Han
State Key Laboratory of Pathogen and Biosecurity, Beijing Institute of Microbiology and Epidemiology, Beijing, China

Jing Wang
State Key Laboratory of Pathogen and Biosecurity, Beijing Institute of Microbiology and Epidemiology, Beijing, China
Animal Husbandry Base Teaching and Research Section, College of Animal Science and Technology, Hebei North University, Zhangjiakou, China

Simone Pelliciari, Andrea Vannini, Davide Roncarati and Alberto Danielli
Department of Pharmacy and Biotechnology (FaBiT), University of Bologna, Bologna, Italy

Hailan Piao and Thomas Henick-Kling
Department of Viticulture and Enology, Washington State University, Richland, WA, USA

Erik Hawley
Zea Chem Inc., Boardman, OR, USA

Scott Kopf and Steven Sealock
Pacific Rim Winemakers, West Richland, WA, USA

Richard De Scenzo
ETS Laboratories, Saint Helena, CA, USA

Matthias Hess
Functional Systems Microbiology Laboratory, University of California, Davis, Davis, CA, USA
Department of Energy Joint Genome Institute, Walnut Creek, CA, USA

Lovisa Eliasson, Sven Isaksson, Maria Lövenklev and Lilia Ahrné
Food and Bioscience, SP Technical Research Institute of Sweden, Gothenburg, Sweden

Genia Lücking and Andrea Rütschle
Department of Microbiology, Central Institute for Food and Nutrition Research (Zentralinstitut für Ernährungs- und Lebensmittel for schung), Technische Universität München, Freising, Germany

Elrike Frenzel and Monika Ehling-Schulz
Functional Microbiology, Institute of Microbiology, Department of Pathobiology, University of Veterinary Medicine Vienna, Vienna, Austria

Sandra Marxen, Timo D. Stark and Thomas Hofmann
Chair of Food Chemistry and Molecular Sensory Science, Technische Universität München, Freising, Germany

Siegfried Scherer
Lehrstuhl für Mikrobielle Ökologie, Wissenschaftszentrum Weihenstephan, Technische Universität München, Freising, Germany

Alessandra Pezzuto, Alessia Piovesana, Renzo Mioni and Michela Favretti
Optimization and Control of Food Production Laboratory, Istituto Zooprofilattico Sperimentale delle Venezie, San Donàdi Piave, Italy

Carmen Losasso, Marzia Mancin, Federica Gallocchio and Antonia Ricci
Department of Food Safety, Istituto Zooprofilattico Sperimentale delle Venezie, Legnaro, Italy

Giovanni Binato and Albino Gallina
Laboratory of Chemistry, Istituto Zooprofilattico Sperimentale delle Venezie, Legnaro, Italy

Alberto Marangon
Sensory Analysis Laboratory, Veneto Agricoltura, Istitutoper la Qualitàele Tecnologie Agroalimentari, Thiene, Italy

Nadja Bier, Keike Schwartz, Beatriz Guerra and Eckhard Strauch
Department of Biological Safety, Federal Institute for Risk Assessment, Berlin, Germany

Hilla Oknin
Department of Food Quality and Safety, Institute for Postharvest Technology and Food Sciences, Agricultural Research Organization, The Volcani Center, Bet-Dagan, Israel
Biofilm Research Laboratory, Faculty of Dental Medicine, Institute of Dental Sciences, Hebrew University-Hadassah, Jerusalem, Israel

Doron Steinberg
Biofilm Research Laboratory, Faculty of Dental Medicine, Institute of Dental Sciences, Hebrew University-Hadassah, Jerusalem, Israel

Moshe Shemesh
Department of Food Quality and Safety, Institute for Postharvest Technology and Food Sciences, Agricultural Research Organization, The Volcani Center, Bet-Dagan, Israel

Ryan G. Mercer, Rigoberto Garcia-Hernandez, Michael G. Gänzle and Lynn M. Mc Mullen
Department of Agricultural, Food and Nutritional Science, University of Alberta, Edmonton, AB, Canada

Jinshui Zheng and Lifang Ruan
State Key Laboratory of Agricultural Microbiology, Huazhong Agricultural University, Wuhan, China

Suja Senan and Hasmukh A. Patel
Department of Dairy Science, South Dakota State University, Brookings, SD, USA

Jashbhai B. Prajapati and V. Sreeja
Department of Dairy Microbiology, Anand Agricultural University, Anand, India

Chaitanya G. Joshi
Department of Animal Biotechnology, Anand Agricultural University, Anand, India

Manisha K. Gohel and Uday Shankar Singh
Department of Community Medicine, H. M Patel Center for Medical Care and Education, Karamsad, India

Sunil Trivedi5 and Rupal M. Patel
Department of Microbiology, H. M Patel Center for Medical Care and Education, Karamsad, India,

Himanshu Pandya
Department of Medicine, H. M Patel Center for Medical Care and Education, Karamsad, India,

Ajay Phatak
Central Research Services, Charutar Arogya Mandal, Karamsad, India

Amarela Terzić-Vidojević, Katarina Veljović, Jelena Begović, Dušanka Popović, Maja Tolinački, Marija Miljković, Milan Kojić and Nataša Golić
Laboratory for Molecular Microbiology, Institute of Molecular Genetics and Genetic Engineering, University of Belgrade, Belgrade, Serbia

Brankica Filipić
Laboratory for Molecular Microbiology, Institute of Molecular Genetics and Genetic Engineering, University of Belgrade, Belgrade, Serbia
Faculty of Pharmacy, University of Belgrade, Belgrade, Serbia

Hikmate Abriouel, Leyre Lavilla Lerma, María del Carmen Casado Muñoz, Beatriz Pérez Montoro, Antonio Gálvez and Nabil Benomar
Área de Microbiología, Departamento de Ciencias de la Salud, Facultad de Ciencias Experimentales, Universidad de Jaén, Jaén, Spain

Jan Kabisch, Rohtraud Pichner, Gyu-Sung Cho, Horst Neve and Charles M. A. P. Franz
Department of Microbiology and Biotechnology, Federal Research Institute of Nutrition and Food, Max Rubner-Institut, Kiel, Germany

Vincenzina Fusco
Institute of Sciences of Food Production, National Research Council of Italy, Bari, Italy

Giulia Tabanelli
Centro Interdipartimentale di Ricerca Industriale Agroalimentare, Università degli Studidi Bologna, Cesena, Italy

Pamela Vernocchi
Centro Interdipartimentale di Ricerca Industriale Agroalimentare, Università degli Studidi Bologna, Cesena, Italy
Unit of Metagenomics, Bambino Gesù Children's Hospital, IRCCS, Rome, Italy

Francesca Patrignani
Dipartimento di Scienzee Tecnologie Agro-alimentari, Università degli Studi di Bologna–Sededi Cesena, Cesena, Italy

Federica Del Chierico
Unit of Metagenomics, Bambino Gesù Children's Hospital, IRCCS, Rome, Italy

Lorenza Putignani
Unit of Metagenomics, Bambino Gesù Children's Hospital, IRCCS, Rome, Italy
Unit of Parasitology, Bambino Gesù Children's Hospital, IRCCS, Rome, Italy

GabrielVinderola and JorgeA.Reinheimer
Facultad deIngeniería Química, Instituto de Lactología Industrial (INLAIN, UNL-CONICET), Universidad Nacionaldel Litoral, Santa Fe, Argentina

FaustoGardini and Rosalba Lanciotti
Centro Interdipartimentale di Ricerca Industriale Agroalimentare, Università degli Studi di Bologna, Cesena, Italy
Dipartimento di Scienzee Tecnologie Agro-alimentari, Università degli Studi di Bologna–Sededi Cesena, Cesena, Italy

Christopher J. Doona and Florence E. Feeherry
U. S. Army Natick–Soldier RD&E Center, Warfighter Directorate, Natick, MA, USA

Kenneth Kustin
Department of Chemistry, Emeritus, Brandeis University, Waltham, MA, USA

Gene G. Olinger
National Institute of Allergy and Infectious Diseases, Integrated Research Facility– Division of Clinical Research, Fort Detrick, MD, USA

Peter Setlow
Department of Molecular Biology and Biophysics, University of Connecticut Health Center, Farmington, CT, USA

Alexander J. Malkin
Biosciences and Biotechnology Division, Physical and Life Sciences Directorate, Lawrence Livermore National Laboratory, Livermore, CA, USA

Terrance Leighton
Children's Hospital–Oakland Research Institute, University of California San Francisco-Benioff, Oakland, CA, USA

www.ingramcontent.com/pod-product-compliance
Lightning Source LLC
Chambersburg PA
CBHW070151240326
41458CB00126B/4079